MW00835268

Continuous Multivariate
Distributions

Continuous Multivariate Distributions

Volume 1: Models and Applications

SECOND EDITION

SAMUEL KOTZ
N. BALAKRISHNAN
NORMAN L. JOHNSON

A Wiley-Interscience Publication
JOHN WILEY & SONS, INC.
New York • Chichester • Weinheim • Brisbane • Singapore • Toronto

Copyright © 2000 by John Wiley & Sons, Inc. All rights reserved.

Published simultaneously in Canada.

For ordering or customer information, please call 1-800-CALL-WILEY.

Library of Congress Cataloging in Publication Data:

Kotz, Samuel.
    Continuous multivariate distributions : Volume 1: models and applications / Samuel Kotz, N. Balakrishnan, Norman L. Johnson — 2nd ed.
        p. cm — (Wiley series in probability and statistics)
    Rev. ed. of: Continuous multivariate distributions / Norman L. Johnson, Samuel Kotz. 1972.
    "A Wiley-Interscience publication."
    Includes bibliographical references and index.
    ISBN 0-471-18387-3 (cloth : alk. paper)
        1. Distribution (Probability theory)   2. Multivariate analysis.   I. Johnson, Norman Lloyd.   II. Balakrishnan, N., 1956–   III. Johnson, Norman Lloyd. Continuous multivariate distributions.   IV. Title.   V. Series.

QA273.6   K68 2000
519.5'35—dc21   99-089636

Printed in the United States of America

10 9 8 7 6 5 4 3 2

To the loving memory of my father,

Sri R. Narayanaswami Iyer,

who is responsible for who I am and
what I am today.

N.B.

Contents

Preface   xv

List of Tables   xix

List of Figures   xxi

44 Systems of Continuous Multivariate Distributions   1

  1 Introduction, 1

  2 Historical Remarks, 4

  3 Multivariate Generalization of Pearson System, 6

  4 Series Expansions and Multivariate Central Limit
     Theorems, 10

  5 Translation Systems, 20

  6 Multivariate Linear Exponential-Type Distributions, 28

  7 Sarmanov's Distributions, 30

  8 Multivariate Linnik's Distributions, 33

  9 Multivariate Kagan's Distributions, 34

 10 Generation of Multivariate Nonnormal Random
     Variables, 36

 11 Fréchet Bounds, 44

 12 Fréchet, Plackett and Mardia's Systems, 47

 13 Farlie-Gumbel-Morgenstern Distributions, 51

 14 Multivariate Phase-Type Distributions, 62

 15 Chebyshev-Type and Bonferroni Inequalities, 63

 16 Singular Distributions, 67

 17 Distributions with Almost-Lack of Memory, 69

 18 Distributions with Specified Conditionals, 70

 19 Distributions with Given Marginals, 73

 20 Measures of Multivariate Skewness and Kurtosis, 77

Bibliography, 85

45 Multivariate Normal Distributions   105

1 Introduction and Genesis, 105

2 Definition and Moments, 107

3 Other Properties, 111

4 Order Statistics, 116

5 Evaluation of Multivariate Normal Probabilities, 121
5.1 Reduction Formulas, 125
5.2 Orthant Probabilities, 127
5.3 Some Special Cases, 133
5.4 Approximations, 139

6 Quadrivariate Normal Orthant Probabilities, 145

7 Characterizations, 151

8 Estimation, 161
8.1 Estimation of $\boldsymbol{\xi}$, 161
8.2 Estimation of $\boldsymbol{V}$, 178
8.3 Estimation of Correlations, 182
8.4 Estimation Under Missing Data, 183
8.5 Estimation Under Special Structures, 187
8.6 Estimation of Functions of $\boldsymbol{\xi}$ and $\boldsymbol{V}$, 196

9 Tolerance Regions, 200

10 Truncated Multivariate Normal Distributions, 204

11 Related Distributions, 211

12 Mixtures of Multivariate Normal Distributions, 220

13 Complex Multivariate Normal Distributions, 222
Bibliography, 223

46 Bivariate and Trivariate Normal Distributions   251

1 Definition and Applications, 251

2 Historical Remarks, 252

3 Properties and Moments, 253

4 Bivariate Normal Integral–Tables and
Approximations, 264

5 Characterizations, 277

6 Order Statistics, 281

7 Trivariate Normal Integral, 287

8 Estimation, 293
   8.1 Bivariate Normal Distribution, 293
   8.2 Trivariate Normal Distributions, 310

9 Truncated Bivariate and Trivariate Normal
     Distributions, 311

10 Dichotomized Variables, 320
   10.1 Tetrachoric Correlation, 320
   10.2 Biserial Correlation, 324

11 Related Distributions, 326
   11.1 Mixtures of Bivariate Normal Distributions, 326
   11.2 Bivariate Half Normal Distribution, 326
   11.3 Distribution of Ratios, 327
   11.4 "Bivariate Normal" Distribution with
         Centered Normal Conditionals, 328
   11.5 Bivariate Skew-Normal Distribution, 331

   Bibliography, 333

47 Multivariate Exponential Distributions   349

1 Introduction, 349

2 Bivariate Exponential Distributions, 350
   2.1 Introduction, 350
   2.2 Gumbel's Bivariate Exponentials, 350
   2.3 Freund's Bivariate Exponential (Bivariate
         Exponential Mixture Distributions), 355
   2.4 Marshall and Olkin's Bivariate Exponential, 362
   2.5 Friday and Patil's Bivariate Exponential, 369
   2.6 Arnold and Strauss' Bivariate Exponential, 370
   2.7 Moran and Downton's Bivariate Exponential, 371
   2.8 Singpurwalla and Youngren's Bivariate
         Exponential, 377
   2.9 Raftery's Bivariate Exponential, 377
   2.10 Hayakawa's Bivariate Exponential, 379
   2.11 Lindley and Singpurwalla's Bivariate
         Exponential Mixture, 380
   2.12 Ghurye's Extended Bivariate Lack
         of Memory Distribution, 383
   2.13 Cowan's Bivariate Exponential, 385
   2.14 Infinite Divisibility, 385
   2.15 Characterizations, 386

3 Multivariate Exponential Distributions, 387
   3.1 Freund's Multivariate Exponential, 388
   3.2 Marshall and Olkin's Multivariate Exponential, 391
   3.3 Block and Basu's Multivariate Exponential, 396
   3.4 Olkin and Tong's Multivariate Exponential, 398
   3.5 Marshall and Olkin's Multivariate Exponential
      with Limited Memory, 399
   3.6 Moran and Downton's Multivariate Exponential, 400
   3.7 Raftery's Multivariate Exponential, 401
   3.8 Krishnamoorthy and Parthasarathy's
      Multivariate Exponential, 403
   3.9 Characterizations, 403
4 Multivariate Weibull Distributions, 407
5 Bivariate Distributions Induced by Frailties, 412
Bibliography, 416

48 Multivariate Gamma Distributions   **431**

1 Introduction, 431
2 Bivariate Gamma Distributions, 432
   2.1 McKay's Bivariate Gamma, 432
   2.2 Cheriyan and Ramabhadran's Bivariate Gamma, 432
   2.3 Kibble and Moran's Bivariate Gamma, 436
   2.4 Sarmanov's Bivariate Gamma, 437
   2.5 Jensen's Bivariate Gamma, 438
   2.6 Royen's Bivariate Gamma, 440
   2.7 Farlie–Gumbel–Morgenstern-Type Bivariate
      Gamma, 441
   2.8 Prékopa and Szántai's Bivariate Gamma, 442
   2.9 Smith, Adelfang and Tubb's Bivariate
      Gamma, 443
   2.10 Dussauchoy and Berland's Bivariate Gamma, 445
   2.11 Becker and Roux and Steel and le Roux's
      Bivariate Gamma, 446
   2.12 Schmeiser and Lal's Bivariate Gamma, 450
   2.13 Bivariate Chi-Square Distributions, 451
3 Multivariate Gamma Distributions, 454
   3.1 Cheriyan and Ramabhadran's Multivariate Gamma, 454
   3.2 Gaver's Multivariate Gamma, 456
   3.3 Krishnamoorthy and Parthasarathy's Multivariate
      Gamma and Its Extension, 457

3.4 Prékopa and Szántai's Multivariate Gamma, 458

3.5 Kowalczyk and Tyrcha's Multivariate Gamma, 459

3.6 Royen's Multivariate Gamma, 461

3.7 Mathai and Moschopoulos's Multivariate Gamma, 465

3.8 Dussauchoy and Berland's Multivariate Gamma, 470

4 Multivariate (Jensen-Type) Chi-Square
    Distributions, 471

5 Noncentral Multivariate Chi-Square (Gamma)
    Distributions, 475

6 Infinite Divisibility of Multivariate Gamma, 477

Bibliography, 478

49 Dirichlet and Inverted Dirichlet Distributions   **485**

1 Dirichlet Distributions, 485

2 Inverted Dirichlet Distribution, 491

3 Characteristic Functions, 495

4 Evaluation of Probability Integrals of Dirichlet
    Distributions, 497

5 Characterizations, 500

6 Estimation, 504

7 Applications, 507

8 Generalizations, 512
    8.1 Generalized Dirichlet Distribution as a Prior for
        Multinomial Parameters, 512
    8.2 Antelman's Generalization, 514
    8.3 Johnson and Kotz's Generalization, 515
    8.4 Delta-Dirichlet Distributions, 515
    8.5 Connor and Mosimann's Generalization, 519
    Bibliography, 523

50 Multivariate Liouville Distributions   **529**

1 Introduction, 529

2 Definitions, 530

3 Properties, 531

4 Moments and Covariance Structure, 536

5 Characterizations, 538

6 Estimation and Applications, 541

7 Sign-Symmetric Dirichlet and Liouville
     Distributions, 544

8 Multivariate p-order Liouville Distributions, 547
   Bibliography, 548

51 Multivariate Logistic Distributions   **551**

1 Introduction, 551

2 Gumbel–Malik–Abraham Distribution, 552

3 Frailty and Archimedean Distributions, 559

4 Farlie–Gumbel–Morgenstern Distributions, 561

5 Differences of Extreme Value Variables, 562

6 Mixture Forms, 564

7 Geometric Minima and Maxima, 565

8 A General Flexible Model, 568

9 Conditionally Specified Logistic Model, 569

10 Some Other Generalizations, 570
    Bibliography, 573

52 Multivariate Pareto Distributions   **577**

1 Introduction, 577

2 Forms of Univariate Pareto Distributions, 577

3 Bivariate Pareto Distributions, 579
   3.1 Bivariate Pareto of the First Kind, 579
   3.2 Mardia's Bivariate Pareto of the Second Kind, 584
   3.3 Bivariate Pareto of the Fourth Kind, 585
   3.4 Conditionally Specified Bivariate Pareto, 587
   3.5 Muliere and Scarsini's Bivariate Pareto, 595
   3.6 Bilateral Bivariate Pareto, 596
   3.7 Bivariate Semi-Pareto, 598

4 Multivariate Pareto Distributions, 599
   4.1 Multivariate Pareto of the First Kind, 599
   4.2 Multivariate Pareto of the Second Kind, 602
   4.3 Multivariate Pareto of the Third Kind, 605
   4.4 Multivariate Pareto of the Fourth Kind, 606
   4.5 Conditionally Specified Multivariate Pareto, 611
   4.6 Marshall–Olkin-Type Multivariate Pareto, 612
   4.7 Multivariate Semi-Pareto, 614

Bibliography, 615

53 Bivariate and Multivariate Extreme Value Distributions   621

1 General Bivariate Extreme Value Distributions, 622
1.1 Definition, 622
1.2 Properties, 622
2 Special Bivariate Extreme Value Distributions, 624
2.1 Type A Distributions, 626
2.2 Type B Distributions, 628
2.3 Type C Distributions, 629
2.4 Normal-Like Bivariate Extreme Value
    Distributions, 632
2.5 Estimation, 637
3 Multivariate Extreme Value Distributions, 640
Bibliography, 652

54 Natural Exponential Families   659

1 Introduction, 659
2 Multivariate Natural Exponential Families, 660
3 Multivariate General Exponential Families, 665
4 Parameterization by the Mean and Steepness, 667
5 Variance Function, 669
6 Affine Transformations and Convolution:
    The Exponential Dispersion Model, 670
7 Some Statistical Results for NEFs, 673
7.1 Sufficiency, MLE, and UMVUE, 673
7.2 Cuts in NEFs, 675
7.3 Bayesian Theory, 676
8 NEFs with Quadratic Variance Function, 678
8.1 Morris Class, 678
8.2 Multivariate Case, 681
8.3 Characterizations, 689
8.4 Extensions, 691
Bibliography, 691

Author Index   697

Subject Index   713

Preface

This is the fifth volume in the second edition of the collection of four books on *Distributions in Statistics* coauthored by Norman L. Johnson and Samuel Kotz and published in 1969–1972. The first four volumes in the second edition are:

Univariate Discrete Distributions
Continuous Univariate Distributions, Volume 1
Continuous Univariate Distributions, Volume 2
Discrete Multivariate Distributions

The present volume is a thorough and comprehensive revision of the last book in the original series, on *Continuous Multivariate Distributions*. Professor N. Balakrishnan, who joined the original authors for all but the first volume of the new series (which was coauthored with Professor Adrienne W. Kemp), has played a prominent role, once again, in this revision.

This volume contains eleven chapters, numbered from 44 to 54 in continuation of the chapter numbers in the four previous volumes of the new series. This compares with nine chapters (numbered 34 to 42) in the original *Continuous Multivariate Distributions* book. However, this does not provide an adequate representation of the scope of the revisions that have been carried out. Three chapters (numbered 37 to 39) in the original book, on multivariate t, Wishart, and sampling distributions associated with multivariate normal distributions, have been omitted from this volume. On the other hand, there are now separate chapters on multivariate exponential and multivariate extreme value distributions (originally combined into Chapter 41), and likewise on multivariate beta (Dirichlet) and multivariate gamma distributions (originally combined into Chapter 40). Furthermore, there are now separate chapters on multivariate logistic and multivariate Pareto distributions, which constituted only relatively short sections in the first edition of the book. The final chapter (numbered 54) on multivariate natural exponential families is new. It reflects on the remarkable growth of this topic, which came to general notice only in the 1970s.

We sincerely hope that the drastic changes outlined above reflect, adequately, changes in the direction and the scope of research over the last quarter of a century. A particular feature of change in the field of multivariate distributions has been the availability of electronic computing aids of increasing power. It is remarked, in the Preface to the first edition, that "Much of multivariate theory has become practically useful only with the advent of electronic computers." This comment, made in 1972, is remarkably apt in 2000. The unprecedented advances in this field posed a serious dilemma to us in regard to the scope of reasonable and appropriate revisions. It has become less necessary to provide extensive references to published tables but also, perhaps, more desirable to describe more ambitious practical applications. We have tried to avoid superficial historical evocations and "minor updating," while providing more attention to radical and conceptual changes in the ways in which multivariate distributions have been investigated and employed during the last quarter of the twentieth century.

As in the first edition of this book as well as in our last volume on *Discrete Multivariate Distributions*, matrices (including vectors) are denoted by boldface type. Random variables are usually assigned capital letters.

In a volume of this size and nature, there will inevitably be omission of some papers containing important results that could have been included in this volume. These should be considered as consequences of our ignorance and not of personal nonscientific antipathy.

Our special sincere thanks go to Professor Muriel Casalis for providing a write-up of Chapter 54 and to Professor Gerard Letac for his valuable comments and suggestions. We are also thankful to anonymous reviewers who provided valuable suggestions that led to a considerable improvement in the organization and presentation of the material. We express our sincere gratitude to the authors from all over the world, too numerous to be cited individually, who were kind enough to provide us with copies of their published papers and technical reports. We are also indebted to the librarians at McMaster University, Hamilton, Ontario; George Washington University, Washington, DC; and University of North Carolina, Chapel Hill, who assisted us in our extensive literature search. Special thanks are due to Mrs. Debbie Iscoe (Mississauga, Ontario, Canada) for typesetting the entire volume, and to Dr. Khalaf Sultan for assisting us in the literature survey.

Thanks are due to Gordon and Breach Science Publishers, Elsevier Science Publishers, Marcel Dekker, Inc., and the Editors of *Statistica Sinica* for giving permission to reproduce previously published tables and figures.

We are happy to acknowledge the support and encouragement of Mr. Steve Quigley of John Wiley & Sons, Inc., throughout the course of this project. The managerial and editorial help provided by Ms. Lisa Van Horn and Ms. Heather Haselkorn of John Wiley & Sons, Inc., are also gratefully acknowledged.

As always, we welcome readers to comment on the contents of this volume and are grateful in advance for informing us of any errors, misrepresentations, or omissions that they may notice.

SAMUEL KOTZ
N. BALAKRISHNAN
NORMAN L. JOHNSON

Washington, DC
Hamilton, Ontario
Chapel Hill, NC

February 2000

List of Tables

TABLE 44.1	Bivariate Pearson Surfaces	9
TABLE 44.2	Median Regressions for S_{IJ} Distributions	23
TABLE 44.3	Multivariate Chebyshev-Type Inequalities	67
TABLE 45.1	Estimates $\hat{\Phi}_k(x; \mathbf{R})$, where $\rho_{ij} = \rho$ $\forall\, i \neq j = 1, \ldots, k$, for Selected Values of k, x and ρ and the Corresponding True Values of $\Phi_k(x; \mathbf{R})$ Taken from Gupta (1963)	132
TABLE 45.2	Values of the Integrals $I_i(\rho)$, $i = 2, 3, 4$, Used in Evaluation of Multivariate Normal Probabilities	137
TABLE 45.3	Some Orthant Probabilities	150
TABLE 45.4	Quality of Drezner's Quadrivariate Approximations	151
TABLE 45.5	Corrective Multipliers for $\hat{\rho}_{ij}$	183
TABLE 45.6	Minimum Variance Estimators of Multivariate Normal Density Functions	197
TABLE 45.7	Approximate Values of K	201
TABLE 46.1	Approximation of Bivariate Normal Probabilities	276
TABLE 46.2	Smith's Suggestion for Estimators in the Bivariate Case	301
TABLE 46.3	Optimal Values of h	303
TABLE 46.4	Correlation in Truncated Bivariate Normal Population	313
TABLE 46.5	Constants for Elliptical Truncation of Bivariate Normal	320
TABLE 47.1	Simulated Bias of the Estimators $\tilde{\rho}_1, \tilde{\rho}_2, \tilde{\rho}_3, \tilde{\rho}_4, \tilde{\rho}_5, \tilde{\rho}_6, \tilde{\rho}_{5,J}$, and $\tilde{\rho}_{6,J}$	376
TABLE 47.2	Simulated Mean Square Error of the Estimators $\tilde{\rho}_1, \tilde{\rho}_2, \tilde{\rho}_3, \tilde{\rho}_4, \tilde{\rho}_5, \tilde{\rho}_6, \tilde{\rho}_{5,J}$, and $\tilde{\rho}_{6,J}$	376

TABLE 53.1   Modes of the Bivariate Extreme Value
                Distributions   627
TABLE 53.2   Variances of Estimators of the Parameters   639
TABLE 53.3   Trivariate Estimation Results   646

List of Figures

FIGURE 44.1 Median Regression Functions for Some S_{IJ} Distributions 24

FIGURE 44.2 Bivariate Edgeworth Series Density Function when $\rho = 0.1$ 41

FIGURE 44.3 Bivariate Edgeworth Series Density Function when $\rho = 0.8$ 42

FIGURE 44.4 Bivariate Normal Density Function when $\rho = 0.1$ 43

FIGURE 44.5 Bivariate Normal Density Function when $\rho = 0.8$ 43

FIGURE 44.6 49

FIGURE 44.7 The Local Dependence Function of the Standard Bivariate Cauchy Density Function 75

FIGURE 46.1 Contours of Equal Density of Bivariate Normal Distributions 256

FIGURE 46.2 Plots of Standardized Bivariate Normal Density Function 257

FIGURE 46.3 Shaded Region for the Integral of the Circular Normal Probability Density Function for $L(h, k; \rho)$ when $h > 0$, $k > 0$ and $\rho > 0$ 266

FIGURE 46.4 The Triangle for the Integral of the Standardized Circular Normal Distributions of $V(h, k)$ when $h > 0$ and $k > 0$ 268

FIGURE 46.5 271

FIGURE 46.6 271

FIGURE 46.7 Equidistributional Contours $(L(x, y; \rho) = \alpha)$ for the Sandard Bvariate Normal Distribution with $\alpha = 0.25$ 272

FIGURE 46.8   Scheme of Grouped Data   302
FIGURE 46.9   321

FIGURE 51.5   Contours of Bivariate Logistic Distribution
                with $p_{X,Y}(x,y) = c$ 553
FIGURE 51.2   Regression Curves for the Bivariate Logistic
                and Bivariate Normal Distributions   555
FIGURE 51.3   Contours of Constant Density Corresponding
                to (51.57) for Selected Values of λ 572

FIGURE 53.1   Bivariate Density along the Diagonal
                $(x = y)$—Type A. Bivariate Density along the
                Diagonal $(x = y)$—Type B   631

CHAPTER 44

Systems of Continuous Multivariate Distributions

1 INTRODUCTION

The multivariate normal distribution (which will be the subject of Chapter 45) has been studied far more extensively than any other continuous multivariate distribution. Indeed, its position of preeminence among continuous multivariate distributions was until at least the early 1990s more marked than that of the normal among continuous univariate distributions. However, the need for usable alternatives to the multivariate normal distribution has been recognized since the publication of the first edition of this volume in 1972. This has resulted in a significant growth in work relating to multivariate nonnormal distributions as will be clearly evident from materials presented after Chapter 46 of this volume. In the present chapter, we describe some systems of multivariate distributions that may provide acceptable models for practical use.

There is some parallelism between this chapter and Chapter 12. In particular, we shall describe systems of distributions based on

(a) generalizations of Pearson's differential equation (Section 4.1, Chapter 12)

(b) series expansions, especially Gram–Charlier and Edgeworth expansions (Section 4.2, Chapter 12), and

(c) transformation to multivariate normal joint distributions (Section 4.3, Chapter 12).

Basic concepts relevant to multivariate distributions have been introduced in Chapters 1, 12, and 34. Some essential features distinguishing

1

multivariate from univariate studies have also been mentioned in Chapter 34. Correlation and regression, which are among these features, appear throughout this volume with some regularity. Here we introduce three new functions—the *scedastic*, *clisy* (or *clitic*), and *kurtic* functions of a random variable Y given the values $x_1, x_2, \ldots, x_s$ of random variables $X_1, X_2, \ldots, X_s$, which are

$$\mathrm{var}(Y \mid x_1, x_2, \ldots, x_s),$$

$$\sqrt{\beta_1}(Y \mid x_1, x_2, \ldots, x_s),$$

and

$$\beta_2(Y \mid x_1, x_2, \ldots, x_s),$$

respectively. We recall [Eq. (34.7)] the definition of central product moments. The $m \times m$ matrix with (i, j)th element equal to the covariance of X_i and X_j [and (i, i)th element equal to the variance of X_i] is called the *variance–covariance* matrix of $\boldsymbol{X}' = (X_1, X_2, \ldots, X_m)$—sometimes written $\mathbf{Var}(\boldsymbol{X})$.

Mixtures of multivariate distributions are formed as for univariate distributions; see Section 4.1 of Chapter 34. If $X_1, \ldots, X_k$ have a joint distribution that is a mixture of m distributions with cumulative distribution functions $\{F_j(x_1, \ldots, x_k)\}$, with weights $\{a_j\}$

$$\left(j = 1, \ldots, m; \ a_j > 0; \ \sum_{j=1}^{m} a_j = 1 \right),$$

then

$$F_{X_1, \ldots, X_k}(x_1, \ldots, x_k) = \sum_{j=1}^{m} a_j F_j(x_1, \ldots, x_k). \tag{44.1}$$

The joint distribution of any subset of the X's is also a mixture with m components and the same weights $\{a_j\}$. In particular,

$$F_{X_1, \ldots, X_s}(x_1, \ldots, x_s) = \sum_{j=1}^{m} a_j F_j(x_1, \ldots, x_s) \tag{44.2}$$

and

$$F_{X_1}(x_1) = \sum_{j=1}^{m} a_j F_j(x_1), \tag{44.3}$$

where, for $1 \leq s < k$,

$$F_j(x_1, \ldots, x_s) = \lim_{\substack{x_i \to \infty \\ i=s+1,\ldots,k}} F_j(x_1, \ldots, x_k).$$

All the multivariate distributions encountered in this volume are purely continuous. However, it should be noted that it is possible for multivariate distributions to be of mixed type, even when each marginal distribution is continuous. Some interesting early examples are given by Koopmans (1969).

The concept of exchangeability is of substantial importance. Variables $X_1, X_2, \ldots, X_k$ are said to be *exchangeable* if

$$\Pr\left[\bigcap_{j=1}^{k}(X_j \leq x_j)\right] = \Pr\left[\bigcap_{j=1}^{k}(X_j \leq x_{a_j})\right], \qquad (44.4)$$

where $(a_1, \ldots, a_k)$ is any permutation of the integers $(1, \ldots, k)$.

The joint distribution is also, rather inappropriately, sometimes called *exchangeable* in these circumstances. A better term, used by Lancaster (1965), is *symmetrical*. Necessary and sufficient conditions are given for a distribution to be symmetrical in Section 4 of this Chapter.

The *characteristic coefficients* of distributions, constructed by Sarmanov (1965), have been extended to bivariate distributions by Abazaliev (1968). The characteristic coefficients of the joint distribution of X_1 and X_2 are

$$\lambda_{g_1,g_2}(r_1, r_2; X_1, X_2) = E\left[\exp\left\{2\pi i \sum_{j=1}^{2} r_j[g_j(X_j) - \frac{1}{2}]\right\}\right], \qquad (44.5)$$

where $g_j(y)$ is an increasing function of y with $\lim_{y \to -\infty} g_j(y) = 0$, $\lim_{y \to \infty} g_j(y) = 1$. If X_1 and X_2 are mutually independent,

$$\lambda_{g_1,g_2}(r_1, r_2; X_1, X_2) = \prod_{j=1}^{2} \lambda_{g_j}(r_j; X_j),$$

where $\lambda_{g_j}(r; X_j)$ is a characteristic coefficient for the distribution of X_j, as defined by Sarmanov (1965).

In particular, $g_j(x_j)$ may be taken as the cumulative distribution function of X_j $(j = 1, 2,)$. Then

$$
\begin{aligned}
F_{X_1,X_2}(x_1, x_2) = {} & F_{X_1}(x_1)F_{X_2}(x_2) \\
& + \pi^{-1} \sum_{r_1=1}^{\infty} \sum_{r_2=1}^{\infty} (r_1 r_2)^{-1}\lambda(r_1, r_2) \sin r_1 z_1 \sin r_2 z_2,
\end{aligned}
$$

where $\lambda(r_1, r_2)$ is an abbreviation for (44.5) and $z_j = 2F_{X_j}(x_j) - 1$; note the connection of this to Farlie–Gumbel–Morgenstern distributions discussed in Section 44.12.

Abazaliev (1968) has shown that any bivariate distribution is determined by the coefficients (44.5), if the functions $g_1(\cdot)$ and $g_2(\cdot)$ are known.

2   HISTORICAL REMARKS

Although the bivariate normal distribution (see Chapters 45 and 46) had been studied at the beginning of the nineteenth century, interest in multivariate distributions remained at a low level until it was stimulated by the work of Galton (1877) in the last quarter of the century. He did not, himself, introduce new forms of joint distribution, but he developed the ideas of correlation and regression and focused attention on the need for greater knowledge of possible forms of multivariate distribution.

Investigation of nonnormal joint distributions, or *skew frequency surfaces* (as nonsymmetrical forms have been termed), has generally followed lines suggested by previous work on univariate distributions. Early work in this field followed rather different lines, but was not very successful. Karl Pearson (1905), whose first investigations appear to have been prompted by noting distinctly nonnormal properties of some observed joint distributions, initially tried to proceed by an analogy with the bivariate normal surface. For this distribution (see Section 1 of Chapter 46), it is possible to replace a pair of correlated variables by a pair of independent ones, using a transformation corresponding to a rotation of axes. Pearson attempted to construct general systems for which this property holds. He found, however, that this method was unpromising. In fact, in general, the property cannot hold since, for independence, the rotation must produce uncorrelated variables, but this is not sufficient to ensure independence.

Pearson (1923a,b) and, later, Neyman (1926) also considered methods of construction of joint distributions, starting from certain requirements on the regression and scedastic functions. This was an extension of work initiated by Yule (1897), who showed that assuming multiple linear regression (i.e., $E[Y \mid x_1, x_2, \ldots, x_s]$ to be a linear function of $x_1, x_2, \ldots, x_s$), the multiple regression function obtained by the method of least squares is identical to that of a multivariate normal distribution. Although some useful results, not requiring detailed knowledge of the actual form of distribution, were obtained by this method, calculation of derived probabilities was not usually sufficiently precise for practical purposes. Narumi (1923) used the stronger requirement that the shape of each conditional (*array*) distribution of one variable, given the others, should be the same for all values of the conditioning variables. He also placed requirements on the *median regression* function μ (the median of Y, given $X_1 = x_1, \ldots, X_s = x_s$). In this way, he did construct some definite distributions. However, the requirement of unchanging shape for the conditional distribution of Y given $x_1, \ldots, x_s$ was clearly not in accord with features of many observed distributions. One distribution that can be constructed in this way (although this was not the way in which the author, in fact, approached it) is the *Rhodes' distribution*; see Rhodes (1923). Let X_1, X_2 be independent gamma variables with pdf's (see Chapter 17)

$$p_{X_j}(x_j) = \frac{1}{\delta_j^{\alpha_j} \Gamma(\alpha_j)} \, x_j^{\alpha_j - 1} \, e^{-x_j/\delta_j}, \ 0 < x_j < \infty, \ \alpha_j > 0, \ \delta_j > 0, j = 1, 2.$$

Then the joint density function of Y_1 and Y_2, where

$$X_1 = 1 - a_1^{-1} Y_1 + a_2^{-1} Y_2 \text{ and } X_2 = 1 - a_2'^{-1} Y_2 + a_1'^{-1} Y_1, \qquad (44.6)$$

is given by

$$
\begin{aligned}
p_{Y_1, Y_2}(y_1, y_2) \ = \ & \frac{e^{-(\delta_1 - 1 + \delta_2 - 1)}}{\delta_1^{\alpha_1} \delta_2^{\alpha_2} \Gamma(\alpha_1) \Gamma(\alpha_2)} \left| \frac{a_1' a_2 - a_1 a_2'}{a_1 a_2 a_1' a_2'} \right| \\
& \times (1 - a_1^{-1} y_1 + a_2^{-1} y_2)^{\alpha_1 - 1} (1 - a_2'^{-1} y_2 + a_1'^{-1} y_1)^{\alpha_2 - 1} \\
& \times e^{-\lambda_1 y_1 - \lambda_2 y_2}
\end{aligned}
\qquad (44.7)
$$

with

$$1 - a_1^{-1} y_1 + a_2^{-1} y_2 > 0 \quad \text{and} \quad 1 - a_2'^{-1} y_2 + a_1'^{-1} y_1 > 0.$$

Here, $\lambda_1 = (a_1 \delta_1)^{-1} - (a_1' \delta_2)^{-1}$ and $\lambda_2 = (a_2' \delta_2)^{-1} - (a_2 \delta_1)^{-1}$. The special case when $\delta_j = (\alpha_j - 1)^{-1}$ is of particular interest being the form originally proposed by Rhodes; see also Mardia (1970a–c).

Inverting the (linear) transformation in (44.6), we obtain

$$Y_1 = \{a_2(X_1 - 1) + a_2'(X_2 - 1)\}/(a_1'^{-1} a_2'^{-1} - a_1^{-1} a_2^{-1}) \qquad (44.8)$$

and

$$Y_2 = \{a_1(X_1 - 1) + a_1'(X_2 - 1)\}/(a_1^{-1} a_2^{-1} - a_1'^{-1} a_2'^{-1}). \qquad (44.9)$$

Hence,

$$
\begin{aligned}
\operatorname{var}(Y_1) \ &= \ \frac{1}{K^2} \, [a_2^2 \delta_1^2 \alpha_1 + a_2'^2 \delta_2^2 \alpha_2], \\
\operatorname{var}(Y_2) \ &= \ \frac{1}{L^2} \, [a_1^2 \delta_1^2 \alpha_1 + a_1' \delta_2^2 \alpha_2], \\
\operatorname{cov}(Y_1, Y_2) \ &= \ \frac{1}{KL} \, [a_1 a_2 \delta_1^2 \alpha_1 + a_1' a_2' \delta_2^2 \alpha_2],
\end{aligned}
\qquad (44.10)
$$

and

$$\text{corr}(Y_1, Y_2) = \frac{a_1 a_2 \delta_1^2 \alpha_1 + a_1' a_2' \delta_2^2 \alpha_2}{\sqrt{(a_2^2 \delta_1^2 \alpha_1 + a_2'^2 \delta_2^2 \alpha_2)(a_1^2 \delta_1^2 \alpha_1 + a_1'^2 \delta_2^2 \alpha_2)}},$$

where K and L are the denominators in (44.8) and (44.9), respectively.

Multivariate extension of Gram–Charlier and Edgeworth series expansions is the subject of Section 4 of this chapter. Work on these forms of distribution appears to have commenced rather suddenly about 1910 and continued, with slowly decreasing intensity after 1920, until Pretorius (1930) gave a comprehensive survey of results available in 1930. Since that time, interest has remained steady, but at a rather low level, with a continued interest exemplified by papers of some generality by Chambers (1967) and more recently Skovgaard (1986).

In Chapter 12 (Section 4.3) we have already discussed systems of distributions constructed by supposing certain (fairly simple) functions of variables to be normally distributed. It is natural to consider what forms of joint distribution one can construct by supposing certain functions of the original variables to have a joint multivariate normal distribution. Although, in the general case, we should consider situations where $Z_i = g_i(X_1, X_2, \ldots, X_s)$ $(i = 1, \ldots, s)$ have a joint multivariate normal distribution, we shall, in Section 5, consider only those cases in which $Z_i = g_i(X_i)$—that is, when each of the original variables $(X_1, \ldots, X_s)$ is transformed separately to a normal variable.

Edgeworth (1896, 1917) used cubic polynomial transformations for each of two variables separately. He also considered composite polynomial transformations. Wicksell (1917, 1923) supposed $\log X_1$ and $\log X_2$ to have a bivariate normal distribution (the *logarithmic surface*). This distribution is discussed in Section 5 as is the *semilogarithmic surface* in which X_1 and $\log X_2$ have a bivariate normal distribution; see Jørgensen (1916). More recent work has tended to aim at building up multivariate distributions having specified structures. These are also discussed in this chapter.

3 MULTIVARIATE GENERALIZATION OF PEARSON SYSTEM

The univariate Pearson system of distributions has been discussed in Chapter 12 (Section 4.1). Successes achieved, using these distributions, led to attempts to extend them to multivariate (in particular, bivariate)

distributions. Clearly, there can be considerable variety in possible bivariate distributions with each of the marginal distributions being one or
the other of the Pearson types. However, it is reasonable to restrict ourselves to consideration of systems derived from differential equations that
are natural generalizations of (12.33); after all, the differential equation
(12.33) was used to generate the univariate Pearson system of distributions.

We describe here some investigations reported by van Uven (1925–
1926, 1929, 1947–1948). These are not the only studies of this kind; see,
for example, Risser (1945, 1947, 1950) and Risser and Traynard (1957).
But they seem to be the most exhaustive and systematic studies known
to us. We start from the pair of differential equations

$$\frac{\partial \log p}{\partial x_j} = \frac{L_j(x_1, x_2)}{Q_j(x_1, x_2)}, \qquad j = 1, 2, \tag{44.11}$$

where $p \equiv p_{X_1, X_2}(x_1, x_2)$ is the joint probability density function of X_1
and X_2, and L_j and Q_j are linear and quadratic functions, respectively, of
their arguments. On fixing either x_1 or x_2, it is clear that the conditional
(*array*) distributions of either variable, given the other, satisfy differential equations of the form (12.33), hence belong to the Pearson system.
However, because the values of the constants depend on the value of the
conditioning variable, the array distributions do not, in general, all have
the same shape.

From (44.11), we see that

$$\frac{\partial^2 \log p}{\partial x_1 \partial x_2} = \frac{\partial}{\partial x_1}\left(\frac{L_2(x_1, x_2)}{Q_2(x_1, x_2)}\right) = Q_2^{-2}\left(Q_2 \frac{\partial L_2}{\partial x_1} - L_2 \frac{\partial Q_2}{\partial x_1}\right).$$

Note that arguments (x_1, x_2) have been omitted for convenience. Also

$$\frac{\partial^2 \log p}{\partial x_1 \partial x_2} = \frac{\partial}{\partial x_2}\left(\frac{L_1}{Q_1}\right) = Q_1^{-2}\left(Q_1 \frac{\partial L_1}{\partial x_2} - L_1 \frac{\partial Q_1}{\partial x_2}\right).$$

Hence,

$$Q_1^{-2}\left(Q_1 \frac{\partial L_1}{\partial x_2} - L_1 \frac{\partial Q_1}{\partial x_2}\right) = Q_2^{-2}\left(Q_2 \frac{\partial L_2}{\partial x_1} - L_2 \frac{\partial Q_2}{\partial x_1}\right),$$

showing that the L_j's and Q_j's cannot be chosen in a completely arbitrary
manner.

From (44.11), with $j = 1$, integrating over the range of variation of x_1,
we find

$$\int_{-\infty}^{\infty} L_1 p \, dx_1 = \int_{-\infty}^{\infty} Q_1 \frac{\partial p}{\partial x_1} \, dx_1 = Q_1 p \big|_{-\infty}^{\infty} - \int_{-\infty}^{\infty} \frac{\partial Q_1}{\partial x_1} p \, dx_1,$$

so that

$$\int_{-\infty}^{\infty} \left(L_1 + \frac{\partial Q_1}{\partial x_1} \right) p \, dx_1 = Q_1 p|_{-\infty}^{\infty}. \tag{44.12}$$

The quantity on the right-hand side of (44.12) is to be calculated as

$$\lim_{x_1 \to \infty} Q_1 p - \lim_{x_1 \to -\infty} Q_1 p.$$

Very often, each of these limits is zero (whenever $\lim_{x_1 \to \infty} x_1^2 p = 0 = \lim_{x_1 \to -\infty} x_1^2 p$, in fact). Because L_1 and $\partial Q_1/\partial x_1$ are each linear functions of x_1 and x_2, it follows that (44.12) can be written in the form

$$E[X_1 \mid x_2] = \alpha_1 + \beta_1 x_2$$

if $Q_1 p|_{-\infty}^{\infty} = 0$. This means that the regression of X_1 on X_2 is linear. Similarly, the regression of X_2 on X_1 is linear if $Q_2 p|_{-\infty}^{\infty} = 0$.

Note that if, for given $X_2 = x_2$, the range of X_1 is finite, $a_2(x_2) < X_1 < a_1(x_2)$, then

$$Q_1 p|_{-\infty}^{\infty} = \lim_{x_1 \to a_1(x_2)} Q_1 p - \lim_{x_1 \to a_2(x_2)} Q_1 p.$$

The condition $Q_1 p|_{-\infty}^{\infty} = 0$ is satisfied if $\lim x^2 p$ is zero at each end of the range of variation.

Table 44.1, reproduced from Elderton and Johnson (1969), gives the more important members of this system of bivariate Pearson distributions. They all have linear regression of either variable on the other. Most of these distributions are treated in detail in later chapters. Two-dimensional extensions of the Pearson system have also been discussed by Sagrista (1952). Kotz (1975) discussed the moments and shape properties of the multivariate forms of Pearson distributions. Johnson (1987) presented graphs of bivariate surfaces and also simulational algorithms for some bivariate Pearson distributions.

Steyn (1960) has extended this kind of analysis to joint distributions of more than two variables. Starting with a set of k equations obtained by letting j, in (44.11), run from 1 to k, he showed that if p vanishes at the extremes of the range of variation of x_i, the regression of X_i on the other $(k-1)$ variables is linear.

TABLE 44.1

Bivariate Pearson Surfaces

Type	Equation $y =$	Conditions	Marginal Types x_1	Marginal Types x_2
I	$f(x_1)f(x_2)$	(Independent variables with frequencies $f(x_1)$, $f(x_2)$)		
IIaα	$\frac{\Gamma(m_1+m_2+m_3)}{\Gamma(m_1)\Gamma(m_2)\Gamma(m_3)}$ $\times\, x_1^{m_1-1}x_2^{m_2-1}(1-x_1-x_2)^{m_3-1}$	$m_1,m_2,m_3 > 0$ $x_1,x_2 > 0;$ $x_1+x_2 \le 1$	I or II	I or II
IIaβ	$\frac{\Gamma(-m_3+1)x_1^{m_1-1}x_2^{m_2-1}(1+x_1+x_2)^{m_3-1}}{\Gamma(m_1)\Gamma(m_2)\Gamma(-m_1-m_2-m_3+1)}$	$m_1,m_2 > 0;$ $m_1+m_2+m_3 < 1$ $x_1,x_2 > 0$	VI	VI
IIaγ	$\frac{\Gamma(-m_2+1)x_1^{m_1-1}x_2^{m_2-1}(-1-x_1+x_2)^{m_3-1}}{\Gamma(m_1)\Gamma(m_3)\Gamma(-m_1-m_2-m_3+1)}$	$m_1,m_3 > 0;$ $m_1+m_2+m_3 < 1$ $x_2-1 > x_1 > 0$	VI	VI
IIaδ	$\frac{\Gamma(-m_1+1)x_1^{m_1-1}x_2^{m_2-1}(-1+x_1-x_2)^{m_3-1}}{\Gamma(m_2)\Gamma(m_3)\Gamma(-m_1-m_2-m_3+1)}$	$m_1,m_3 > 0;$ $m_1+m_2+m_3 < 1$ $x_1-1 > x_2 > 0$	VI	VI
IIb	$\frac{x_1^{m_1-1}x_2^{m_2-1}\exp[-(x_1+1)/x_2]}{\Gamma(m_1)\Gamma(-m_1-m_2)}$	$m_1 > 0;\ m_1+m_2 < 0$ $x_1,x_2 > 0$	VI	V
IIIaα	$\frac{-m\sqrt{(1-\rho^2)}}{\pi k^m}\,(k+x_1^2+2\rho x_1 x_2+x_2^2)^{m-1}$	$m < 0;\ |\rho| < 1;\ k > 0$	VII	VII
IIIaβ	$\frac{m\sqrt{(1-\rho^2)}}{\pi k^m}\,(k-x_1^2+2\rho x_1 x_2-x_2^2)^{m-1}$	$m > 0;\ |\rho| < 1;\ k > 0$ $x_1^2-2\rho x_1 x_2+x_2^2 < k$	II	II
IVa	$\frac{x_1^{m_1-1}(x_2-x_1)^{m_2-1}e^{-x_2}}{\Gamma(m_1)\Gamma(m_2)}$	$m_1,m_2 > 0$ $0 < x_1 < x_2$	III	III
VI	$\frac{1}{2\pi\sqrt{(1-\rho^2)}}$ $\times \exp\left[-\frac{1}{2(1-\rho^2)}(x_1^2-2\rho x_1 x_2+x_2^2)\right]$	$|\rho| < 1$	Normal	Normal

Source: Elderton and Johnson (1969), with permission.

By considering the multivariate Pearson distribution (of Type IIIaβ above) with joint density function

$$p_{\boldsymbol{X}}(\boldsymbol{x}) = C(\alpha,\beta)|\boldsymbol{\Omega}|^{-1/2}\{1-(\boldsymbol{x}-\boldsymbol{\theta})^T\boldsymbol{\Omega}^{-1}(\boldsymbol{x}-\boldsymbol{\theta})\}^{(\alpha-k)/2},$$

where $\boldsymbol{x}$ is a $(k\times 1)$ vector, $\boldsymbol{\theta}$ is a $(k\times 1)$ vector of location parameters, $\boldsymbol{\Omega}$ is a $(k\times k)$ positive definite matrix of scale parameters, $(\boldsymbol{x}-\boldsymbol{\theta})^T\boldsymbol{\Omega}^{-1}(\boldsymbol{x}-\boldsymbol{\theta}) \le 1$, and $\alpha \ge k$, Joarder (1997) has established the following results:

(i) The characteristic function of $\boldsymbol{X}$ is

$$\phi_{\boldsymbol{X}}(\boldsymbol{t}) = e^{i\boldsymbol{t}^T\boldsymbol{\theta}}\ {}_0F_1\left(\frac{\alpha}{2}+1; -\frac{\|\boldsymbol{\Omega}^{1/2}\boldsymbol{t}\|^2}{4}\right),$$

where $_pF_q(a_1, \ldots, a_p; b_1, \ldots, b_q; z)$ is the generalized hypergeometric function defined by

$$_pF_q(a_1, \ldots, a_p; b_1, \ldots, b_q; z) = \sum_{\ell=0}^{\infty} \frac{a_1^{[\ell]} \, a_2^{[\ell]} \, \cdots \, a_p^{[\ell]}}{b_1^{[\ell]} \, b_2^{[\ell]} \, \cdots \, b_q^{[\ell]}} \frac{z^{\ell}}{\ell!}$$

with $a^{[\ell]} = a(a+1) \, \cdots \, (a + \ell - 1)$;

(ii) If we partition

$$X = \begin{pmatrix} X_1 \\ X_2 \end{pmatrix}, \quad \theta = \begin{pmatrix} \theta_1 \\ \theta_2 \end{pmatrix}, \quad t = \begin{pmatrix} t_1 \\ t_2 \end{pmatrix}, \quad \text{and} \quad \Omega = \begin{pmatrix} \Omega_{11} & \Omega_{12} \\ \Omega_{21} & \Omega_{22} \end{pmatrix},$$

then X_2 has the same multivariate Pearson distribution with $\theta = \theta_2$ and $\Omega = \Omega_{22}$, and X_1, given $X_2 = x_2$, has the same multivariate Pearson distribution with θ replaced by $\theta_1 + \Omega_{12}\Omega_{22}^{-1}(x_2 - \theta_2)$ (say, θ^*), Ω replaced by $\{1 - (x_2 - \theta_2)^T \Omega_{22}^{-1}(x_2 - \theta_2)\}(\Omega_{11} - \Omega_{12}\Omega_{22}^{-1}\Omega_{21})$ (say, Ω^*), and α replaced by $\alpha - h$, where h is the dimension of the random vector X_2;

(iii) The conditional mean is $E[X_1|X_2 = x_2] = \theta^*$ and the conditional variance–covariance matrix is

$$\text{Var}(X_1|X_2 = x_2) = \frac{1}{2k}B\left(2 + \frac{\alpha - h}{2}, \, 1 + \frac{\alpha - k}{2}\right)\Omega^*.$$

4   SERIES EXPANSIONS AND MULTIVARIATE CENTRAL LIMIT THEOREMS

Some interesting general results on expansions of multivariate density functions have been obtained by Lancaster (1963); see also Lancaster (1969). For continuous distributions, some of these results can be expressed in relatively simple form. We first need to introduce the concept of an *orthonormal set of functions* on the distribution of a random variable X. These are simply an infinite sequence of functions $\{X_{(j)}\}$ of X such that $E[X_{(j)}^2] = 1$, $E[X_{(i)}X_{(j)}] = 0$ if $i \neq j$. If the density function of X is differentiable, the functions can be defined by

$$x_{(j)} = \frac{1}{p_X(x)} \frac{d^j p_X(x)}{dx^j} \, .$$

Then, the bivariate joint density function of X_1 and X_2 can be written as [Lancaster (1963)]

$$p_{X_1,X_2}(x_1, x_2)$$
$$= p_{X_1}(x_1)p_{X_2}(x_2)\left[\sum_{j_1=0}^{\infty}\sum_{j_2=0}^{\infty}\rho_{(j_1,j_2)}x_{1(j_1)}x_{2(j_2)}\right], \qquad (44.13)$$

where $x_{t(0)} = 1$, $\rho_{(00)} = 1$, and

$$\rho_{(j_1,j_2)} = E[X_{1(j_1)}X_{2(j_2)}]$$

is called the *generalized correlation coefficient* of order (j_1, j_2) between X_1 and X_2, provided that

$$\phi^2 = E\left[\frac{p_{X_1,X_2}(X_1, X_2)}{p_{X_1}(X_1)p_{X_2}(X_2)}\right] - 1 = \sum\sum_{j_1+j_2>0}\rho_{(j_1,j_2)}^2 \qquad (44.14)$$

is finite. The coefficient ϕ^2 was originally introduced by Pearson as a *contingency coefficient*; see also Hirschfeld (1935).

It follows directly that a necessary and sufficient set of conditions for such a continuous bivariate distribution to be symmetrical (in the sense defined at the end of Section 1 of this chapter) is that

(a) the marginal distributions be identical

and

(b) $\rho_{(j_1,j_2)} = \rho_{(j_2,j_1)}$ for all j_1, j_2.

Equation (44.13) can be extended in a rational fashion to the joint distribution of k random variables $X_1, \ldots, X_k$. The conditions for symmetry of a k-variate distribution with finite

$$\phi^2 = E\left[p_{\boldsymbol{X}}(\boldsymbol{X})\left\{\prod_{j=1}^{m}p_{X_j}(X_j)\right\}^{-1}\right] - 1 \qquad (44.15)$$

are that $\rho_{(j_1,\ldots,j_k)}$ shall be unchanged for every permutation of $j_1, \ldots, j_k$, for any given set of values $(j_1, \ldots, j_k)$; see Lancaster (1965) and Eagleson (1964).

Several of the expansions encountered in this volume will be recognized as being of the kind just described.

Jensen (1971) has shown that if two random variables X_1 and X_2 have a joint distribution that can be expanded in an orthonormal series with all the (generalized correlation) coefficients positive *and with identical marginal distributions*, then

$$\Pr[(X_1 \in A) \cap (X_2 \in A)] \le \Pr[X_1 \in A]\Pr[X_2 \in A]$$

for all sets A for which the probabilities exist. The condition of identical marginals can probably be relaxed in some cases.

Griffiths (1970) has shown that under fairly broad conditions any sequence of positive numbers $\rho_1, \rho_2, \ldots$, with $\sum_{j=1}^{\infty} \rho_j^2$ finite can be canonical correlations of a symmetric bivariate distribution with ϕ^2 finite.

Mihaĭla (1968) has given explicit formulas for Gram–Charlier expansions of trivariate density functions. In terms of standardized variables, we can write

$$p(x_1, x_2, x_3)$$
$$= \left[\sum_{j_1=0}^{\infty} \sum_{j_2=0}^{\infty} \sum_{j_3=0}^{\infty} C_{j_1, j_2, j_3} \frac{\partial^{j_1+j_2+j_3}}{\partial x_1^{j_1} \partial x_2^{j_2} \partial x_3^{j_3}} \right] Z_3(\boldsymbol{x}; \mathbf{O}; \mathbf{R}), \quad (44.16)$$

where $Z_3(\boldsymbol{x}; \mathbf{O}; \mathbf{R})$ is a standardized trivariate normal density function with correlation matrix $\mathbf{R}$; see Chapter 46. The expansion can be expressed in terms of trivariate Hermite polynomials as

$$\frac{\partial^{j_1+j_2+j_3}}{\partial x_1^{j_1} \partial_2^{j_2} \partial_3^{j_3}} Z_3(\boldsymbol{x}; \mathbf{O}; \mathbf{R})$$
$$= (-1)^{j_1+j_2+j_3} H_{j_1, j_2, j_3}(\boldsymbol{x}) Z_3(\boldsymbol{x}; \mathbf{O}; \mathbf{R}). \quad (44.17)$$

These polynomials have coefficients that depend on the correlation matrix $\mathbf{R}$. They are most conveniently expressed in terms of the elements of the inverse matrix $\boldsymbol{A} = \mathbf{R}^{-1}$. We have (up to the fourth order)

$$
\begin{aligned}
H_{000} &= 1, \qquad H_{100} = x_1, \qquad H_{200} = x_1^2 - a_{11}, \\
H_{110} &= x_1 x_2 - a_{12}, \qquad H_{300} = x_1^3 - 3a_{11}x_1, \\
H_{210} &= x_1^2 x_2 - 2a_{12}x_1 - a_{11}x_2, \\
H_{111} &= x_1 x_2 x_3 - a_{23}x_1 - a_{13}x_2 - a_{12}x_3, \\
H_{400} &= x_1^4 - 6a_{11}x_1^2 + 3a_{11}^2, \\
H_{310} &= x_1^3 x_2 - 3a_{12}x_1^2 - 3a_{11}x_1 x_2 + 3a_{11}a_{12}, \\
H_{200} &= x_1^2 x_2^2 - a_{22}x_1^2 - a_{11}x_2^2 - 4a_{12}x_1 x_2 + a_{11}a_{22} + 2a_{12}^2, \\
H_{211} &= x_1^2 x_2 x_3 - a_{23}x_1^2 - 2a_{13}x_1 x_2 - 2a_{12}x_1 x_3 - a_{11}x_2 x_3 \\
&\quad + a_{11}a_{23} + 2a_{12}a_{13}.
\end{aligned}
$$

Other expressions can be obtained by permutation of subscripts; for example,

$$H_{201} = x_1^2 x_3 - 2a_{13}x_1 - a_{11}x_3.$$

The coefficients $C_{j_1 j_2 j_3}$ are given by the following formulas, in which $\mu_{r_1 r_2 r_3}$ denotes

$$E\left[\prod_{j=1}^{3}(X_j - E[X_j])^{r_j}\right].$$

(If a nonstandardized distribution is being fitted, then $\mu_{r_1 r_2 r_3}$ should be replaced by $\beta_{r_1 r_2 r_3} = \mu_{r_1 r_2 r_3}/\sigma_1^{r_1}\sigma_2^{r_2}\sigma_3^{r_3}$ in an obvious notation.) $C_{000} = 1$, $C_{100} = 0$, $C_{200} = 0$, $C_{110} = \mu_{110} - \rho_{12}$. (Note that $C_{110} = 0$ if we choose ρ_{12} as the actual correlation between X_1 and X_2.)

$$C_{300} = -\frac{1}{6}\mu_{300}, \qquad C_{210} = -\frac{1}{2}\mu_{210}, \qquad C_{111} = \mu_{111},$$

$$C_{400} = \frac{1}{24}(\mu_{400} - 3), \qquad C_{310} = \frac{1}{6}(\mu_{310} - 3\mu_{110}),$$

$$C_{220} = \frac{1}{4}(\mu_{220} - \mu_{200} - \mu_{020} - 4\rho_{12}\mu_{110} - 1 + 2\rho_{12}^2),$$

$$C_{211} = \frac{1}{2}(\mu_{211} - \rho_{23}\mu_{200} - 2\rho_{13}\mu_{110} - 2\rho_{12}\mu_{101} - \mu_{011} + \rho_{23} + \rho_{12}\rho_{13}).$$

As in the case of the H's, further values can be obtained by permutation of the subscripts.

Formulas for the bivariate case can be obtained from those for trivariate distributions in the following simple way. To obtain H_{r_1,r_2}, C_{r_1,r_2}, take the formula for $H_{r_1,r_2,0}$, $C_{r_1,r_2,0}$, respectively, and replace $\mu_{s_1,s_2,0}$ by μ_{s_1,s_2}. [Of course, a_{ij} are now elements of a 2×2 matrix and, in fact, $a_{11} = a_{22} = (1 - \rho_{12}^2)^{-1}$; $a_{12} = -\rho_{12}(1 - \rho_{12}^2)^{-1}$.]

As k increases, the algebra rapidly becomes more complex, but the formulas are similar and the method of fitting remains the same. For details, see Guldberg (1920) and Meixner (1934).

Sarmanov and Bratoeva (1967) presented conditions for

$$\frac{1}{2\pi} e^{-\frac{1}{2}(x^2+y^2)}\left[1 + \sum_{j=1}^{\infty} C_j H_j(x) H_j(y)\right]$$

to be nonnegative for all x and y.

They showed that a necessary and sufficient condition is that $\{C_j\}$ be the moment sequence of some distribution with range contained in the interval $[-1, 1]$.

Chambers (1967) presented an algorithm for the construction of Edgeworth-type expansions (see Chapter 12, Section 4) for a general k-variate distribution with joint characteristic function $\varphi(t)$ and cumulant generating function $K(t) = \log \varphi(t)$; see Section 44.9 for further discussion on this topic as applied to simulation of data.

Generalizing (44.17), we define the k-variate Hermite polynomial $H(x; r; A)$ by the equation

$$H(x; r; A) \exp\left(-\frac{1}{2} x^T A x\right)$$
$$= (-1)^{\Sigma r_i} \frac{\partial^{\Sigma r_i}}{\partial x_1^{r_1} \cdots \partial x_k^{r_k}} \exp\left(-\frac{1}{2} x^T A x\right). \tag{44.18}$$

The rth Edgeworth approximation to the joint density function $p_X(x)$ is constructed as follows:

(i) Calculate the polynomial in $t = (t_1, \ldots, t_k)$; $Q^{(r)}(t) =$ terms of $K(t)$ from order 3 to order $(r+2)$ inclusive.

(ii) Expand $[\exp Q^{(r)}(t)]$ to terms of order $n^{-(1/2)r}$ assuming that cumulants of order s are of order $n^{1-(1/2)s}$ for $s \geq 2$, and first-order cumulants are of order $n^{-1/2}$. We denote the resulting expansion by

$$P^{(r)}(t) = 1 + \sum_{j=1}^{r} P_j(t) n^{-(1/2)j}.$$

(iii) Replace the product $\prod_{j=1}^{k} t_j^{r_j}$ in $P^{(r)}(t)$ by $(-1)^{\Sigma r_j} H(x; r; V^{-1})$ for all r.

Denote the result by $R_{(r)}(x)$. Then the required approximation is $R_{(r)}(x) \times Z_k(x; O, V)$.

Chambers (1967) also gave some account of formal convergence of the series expansion (though for practical purposes, when the expansion is fitted, rather few terms are used, and this question is of little importance). Bikelis (1968a,b) studied the problem in some detail and has obtained the following result. If $X_1, \ldots, X_k$ are standardized variables and $X_j = (X_{1j}, \ldots, X_{kj})$ $(j = 1, 2, \ldots, n)$ are independent vectors each having the same distribution as $X = (X_1, \ldots, X_k)$, then the joint characteristic function of $S_n = n^{-1/2} \sum_{j=1}^{n} X_j$, which is $\{E[\exp(it^T X/\sqrt{n})]\}^n$, can be expressed in the form

$$e^{-(1/2)Q(t)} \left[1 + \sum_{j=1}^{s-3} P_j(it) n^{-(1/2)j} \right] + R_s, \tag{44.19}$$

where the P_j's are certain polynomials, $Q(t)$ is a positive definite quadratic form in t, and

$$|R_s| \leq (2/0.99)^{s-1} n^{-\frac{1}{2}(s+1)} E[|t^T X|^s] e^{-(1/4)Q(t)}$$

provided that

$$\frac{E[|t^T X|^s]}{Q(t)} \leq \left(\frac{\sqrt{n}}{8}\right)^{s-2}. \tag{44.20}$$

Here, s can be chosen arbitrarily, but X must possess finite moments of order s. The condition in (44.20) is a limitation on values of t.

Bikelis (1968a,b, 1970a,b) used (44.19) to obtain expansions for the difference between the densities, and between the cumulative distribution functions, of S_n and a multivariate normal distribution. Bikelis (1970a,b) showed that if the joint density function of X has an upper bound C, and the expected value vector is 0, then the modulus of the characteristic function $|E[\exp(it^T X)]|$ cannot exceed

$$\exp\left[-\frac{\pi^2}{27C^2} \frac{\{2^{k-1}(k-1)!\}^2 t^T V t}{(8\pi)^k k^{k-1} |V| \{2\pi + \sqrt{k} \sqrt{t^T V t}\}^2}\right]$$

and used this expression to obtain another upper bound for $|R_s|$. Bikelis (1970b) showed that for sums of n independent and identically distributed random vectors, with finite third moments, the error of approximation to the probability integral, using a transformation truncated at $s = 4$ is $o(n^{-1/2})$ uniformly in all k variables; see also Bikelis and Mogyoródi (1970). This result is evidently a multivariate relationship analogous to the univariate central limit theorems. Note that the characteristic function of a standardized multivariate normal distribution is of the form $e^{-(1/2)Q(t)}$, as will be seen in Chapter 45.

Among further work on multivariate central limit theorems, we take note of a paper by Sazonov (1967). He showed that if all moments of the third order of X exist, the difference between the probability that S falls in a region E, and the integral over E of the standardized multivariate normal density function with the same correlation matrix as each X_j, is less than

$$n^{-1/2} C(k,r) \sup_{\ell \neq 0} \sqrt{\beta_1}(|\ell^T X|),$$

where $C(k,r)$ is a constant that may depend on k and r, and E is the intersection of r sets defined by $a_h^T X \gtrless \alpha_j$ ($h = 1, 2, \ldots, r$). Sazonov suggested that $C(k, r)$ may be replaced by a constant depending only on k. Paulauskas (1970) generalized these results to a wider class of sets E.

Zolotarev (1966) showed that, if $\boldsymbol{X}_1, \boldsymbol{X}_2, \ldots$ are independent and identically distributed k-dimensional vectors, composed of correlated elements each having zero mean and unit variance, then provided the vector lengths $|\boldsymbol{X}_j|$ have finite fourth moments,

$$\lim_{n \to \infty} \sqrt{n} \sup_A \left[\Pr[n^{-1/2} \Sigma \boldsymbol{X}_j \in A] - (2\pi)^{-(1/2)k} \right.$$

$$\left. \times \int_A \cdots \int \exp\left(-\frac{1}{2} \sum_{j=1}^k x_j^2 \right) dx_1 \ldots dx_k \right]$$

$$\leq \frac{1}{6\sqrt{2\pi}} [1 + 2(1+\theta)e^{-3/2}] \nu_3, \qquad (44.21)$$

where $\nu_3 = E[|\boldsymbol{X}_j|^3]$ and $\theta = 0$ or 1 according as the regions A are restricted to being simply connected or not. Note that the upper bound in (44.21) does not depend on k. Of course, the region A is supposed to be such that the integral in (44.21) exists.

Dunnage (1970a) showed that if (X_{1j}, X_{2j}) each have expected value vector $(0,0)$ and are mutually independent $(j = 1, 2, \ldots, n)$ and $S_{in} = \sum_{j=1}^n X_{ij}$ $(i = 1, 2)$ then, for all $s_1, s_2,$

$$\left| F_{S_{1n}, S_{2n}}(s_1, s_2) - \Phi\left(\frac{s_1}{\sigma_1}, \frac{s_2}{\sigma_2}; \rho_n \right) \right|$$

$$\leq K \frac{\nu_3^{1/3}}{\min(\sigma_1, \sigma_2)} + \frac{n\nu^{1/2}}{k^{3/2}} + \frac{n\nu \log n}{k^{3/2}}, \qquad (44.22)$$

where K is an absolute constant and

$$\sigma_i^2 = \sum_{j=1}^n \text{var}(X_{ij}) = \text{var}(S_{in}),$$

$$\rho_n = \text{correlation between } S_{1n} \text{ and } S_{2n},$$

$$\nu_3 = \max_{i,j} E[|X_{ij}|^3],$$

$$\nu = n^{-1} \sum_{j=1}^n \max\{E[|X_{1j}|^3], E[|X_{2j}|^3]\},$$

and

$$k = \frac{1}{\sqrt{2}} [\sigma_1^2 + \sigma_2^2 - \{(\sigma_1^2 - \sigma_2^2)^2 + 4\rho_n^2\sigma_1^2\sigma_2^2\}^{1/2}].$$

Note that this result still holds even if some of the correlations between X_{1j}, X_{2j} are numerically equal to 1 in absolute value. Note also that k lies between $\frac{1}{2}(1 - \rho_n^2)\min(\sigma_1^2, \sigma_2^2)$ and $(1 - \rho_n^2)\min(\sigma_1^2, \sigma_2^2)$.

Dunnage (1970b) showed that the right-hand side of (44.22) may be replaced by

$$\frac{n\nu}{\{\min(\sigma_1,\sigma_2)\}^3(1-\rho_n^2)^{3/2}}\left[24+\frac{4}{5}\log\left\{\frac{(\min(\sigma_1,\sigma_2))^3(1-\rho_n^2)^{1/2}}{n\nu}\right\}\right]$$

$$+\max\left[\frac{2\nu_3^{1/2}}{\min(\sigma_1,\sigma_2)}\,,\,\frac{48n\nu}{\{\min(\sigma_1,\sigma_2)\}^3}\,(1-\rho_n^2)\right].\qquad(44.23)$$

The concept of a *stable* distribution (Section 4.5 of Chapter 12) can be directly generalized to sets of k variables. If $\boldsymbol{X}_1, \boldsymbol{X}_2$, and $\boldsymbol{X}$ have the same joint distribution, with $\boldsymbol{X}_1, \boldsymbol{X}_2$ independent, and for any nonsingular $\boldsymbol{A}_1, \boldsymbol{A}_2, \boldsymbol{B}_1(k \times k)$ and $\boldsymbol{B}_2(k \times k)$ it is possible to find $\boldsymbol{A}, \boldsymbol{B}$ such that $\boldsymbol{B}(\boldsymbol{X} - \boldsymbol{A})$ has the same distribution as $\boldsymbol{B}_1(\boldsymbol{X}_1 - \boldsymbol{A}_1) + \boldsymbol{B}_2(\boldsymbol{X}_2 - \boldsymbol{A}_2)$, then the common joint distribution is said to be *stable*. The general forms of such distributions are discussed by Kalinauskaité (1970a,b). Kalinauskaité (1970b) also discussed the *symmetrical stable distributions*, which have characteristic functions $\exp\left[-\left(\sum_{j=1}^n t_j^2\right)^{\alpha/2}\right]$ $(0 < \alpha \le 2)$. The case $\alpha = 2$ gives a multivariate normal distribution; for $0 < \alpha < 2$, the joint density function is

$$\alpha^{-1}(2\pi)^{-\frac{1}{2}k}\sum_{j=0}^{\infty}\left(-\frac{1}{4}\right)^j\frac{\Gamma((2j+s)\alpha^{-1})}{\Gamma(j+1)\Gamma(j+\frac{1}{2}s)}\left(\sum_{j=1}^k x_j\right)^j.$$

Lévy (1937) and Feldheim (1937) presented a general form for the characteristic function of a multivariate stable distribution under an integral form. The results of Press (1972) and Paulauskas (1976) seem to indicate that closed-form expressions for the characteristic function of a multivariate stable distribution analogous to the univariate case are still unknown; see also De Silva (1978). An especially valuable source on these distributions is the Japanese monograph by Sato (1981). In particular, Sato (1981) has shown that a necessary and sufficient condition for a characteristic function $\phi(\boldsymbol{t}) = e^{h(\boldsymbol{t})}$ of a k-variate distribution with zero mean to be the characteristic function of k-variate stable distribution (with characteristic exponent v) is that $h(\boldsymbol{t})$ satisfies

$$h(c\boldsymbol{t}) = |c|^v h(\boldsymbol{t}) \qquad \text{for all } c \in \mathbb{R}.$$

Chikuse (1990) has shown that the characteristic exponent v satisfies $1 < v \le 2$.

Nolan (1996) has pointed out that a k-dimensional α-stable random vector is determined by a spectral measure Γ (a finite Borel measure on the unit sphere in $\mathbb{R}^k$) and a shift vector $\boldsymbol{\mu}^0 \in \mathbb{R}^k$. He has used the notation S_k for the unit sphere in $\mathbb{R}^k$ and $\boldsymbol{X} \stackrel{d}{=} S_{\alpha,k}(\Gamma, \boldsymbol{\mu}^0)$ to denote a stable random vector.

When $\boldsymbol{X} \in \mathbb{R}^k$ has a multivariate α-stable distribution $(0 < \alpha < 2)$, the characteristic function of $\boldsymbol{X}$ is [see Samorodnitsky and Taqqu (1994) and Gupta, Nguyen, and Zeng (1995)]

$$\phi_{\boldsymbol{X}}(\boldsymbol{t}) \ E\left[e^{i\langle \boldsymbol{X}, \boldsymbol{t}\rangle}\right] = \exp\{-I_{\boldsymbol{X}}(\boldsymbol{t}) + i\langle \boldsymbol{\mu}, \boldsymbol{t}\rangle\},$$

where

$$I_{\boldsymbol{X}}(\boldsymbol{t}) = \int_{S^k} \psi_\alpha(\langle \boldsymbol{t}, \boldsymbol{s}\rangle)\Gamma(d\boldsymbol{s}),$$

S^k is the unit sphere in $\mathbb{R}^k$, Γ_k is the spectral measure of $\boldsymbol{X}$, $\boldsymbol{\mu}$ is a vector in $\mathbb{R}^k$, $\langle \boldsymbol{t}, \boldsymbol{s}\rangle = \sum_{i=1}^k t_i s_i$ is the inner product, and

$$\psi_\alpha(u) = \begin{cases} |u|^\alpha \left(1 - i \ \mathrm{sgn}(u) \tan \frac{\pi\alpha}{2}\right), & \alpha \neq 1 \\[2mm] |u| \left(1 + i \ \frac{2}{\pi} \ \mathrm{sgn}(u) \ln |u|\right), & \alpha = 1. \end{cases}$$

Furthermore, for any $\boldsymbol{t} \in S^k$, the projection $\langle \boldsymbol{t}, \boldsymbol{X}\rangle$ of the random vector $\boldsymbol{X}$ on $\boldsymbol{t}$ is an univariate stable random variable with characteristic function $E[e^{iu\langle \boldsymbol{t}, \boldsymbol{X}\rangle}] = \exp\{-I_{\boldsymbol{X}}(u\boldsymbol{t})\}$.

A stable distribution is symmetric if and only if $I_{\boldsymbol{X}}(\boldsymbol{t})$ is real. Discrete spectral measures Γ with a finite number of point masses is represented as

$$\Gamma(\cdot) = \sum_{i=1}^\ell \gamma_i \ \delta_{\boldsymbol{s}_i}(\cdot),$$

where γ_i's are the weights and $\delta_{\boldsymbol{s}_i}$'s are point masses at points $\boldsymbol{s}_i \in S_k$, $j = 1, \ldots, \ell$. Such spectral measures arise naturally in particular when the components of $\boldsymbol{X}$ are independent.

Byczkowski, Nolan, and Rajput (1993) have presented an approximation for enabling the numerical computation of multivariate stable densities. Employing the inversion formula for characteristic functions, they obtained

$$\begin{aligned} p_{\boldsymbol{X}}(\boldsymbol{x}) &= \frac{1}{(2\pi)^k} \int_{\mathbb{R}^k} e^{-i\langle \boldsymbol{x}, \boldsymbol{t}\rangle} \ \exp\{-I_{\boldsymbol{X}}(u\boldsymbol{t})\} \ dt \\[2mm] &= \frac{1}{(2\pi)^k} \int_{\mathbb{R}^k} \cos\left[\langle \boldsymbol{x}, \boldsymbol{t}\rangle + I \ I_{\boldsymbol{X}}(\boldsymbol{t})\right] \exp\{-R I_{\boldsymbol{X}}(\boldsymbol{t})\} \ dt. \end{aligned}$$

Modarres and Nolan (1994) have given an algorithm for simulating a class of multivariate α-stable random vectors $(0 < \alpha < 2)$ with dependent components using a method that is based on a representation for this class as a linear combination of vector multiples of independent univariate stable terms. Nolan and Rajput (1995) have described the calculation of multivariate stable densities by numerically inverting the above given characteristic function. Nolan and Panorska (1997) have proposed methods of exploratory data analysis for testing the suitability of a joint stable distribution for a multivariate data set. Abdul-Hamid and Nolan (1999) have expressed the density function of a general k-dimensional stable random vector $\boldsymbol{X}$ as an integral over the sphere in $\mathbb{R}^k$ of a function of the parameters of the univariate projections of $\boldsymbol{X}$, which is useful for numerical calculations. A lucid review of all these works on multivariate stable distributions has been prepared recently by Nolan (1998).

Ghosh (1990) and Zeng (1995) have presented some characterization results for the multivariate stable distributions. Specifically, Ghosh (1990) has established the characterization result that the random vector $\boldsymbol{X}$ has a k-dimensional stable distribution if and only if the distribution of $Y = \boldsymbol{\ell}^T \boldsymbol{X}$ is univariate stable for all nonzero vectors $\boldsymbol{\ell} \in \mathbb{R}^k$. Zeng (1995) has characterized the multivariate stable distributions through the independence of the linear statistic $\boldsymbol{U} = \sum_{i=1}^{n} Y_i \boldsymbol{X}_i$ and the random coefficient vector $\boldsymbol{Y} = (Y_1, \ldots, Y_n)^T \in \mathbb{R}^n$, where $\boldsymbol{X}_1, \boldsymbol{X}_2, \ldots, \boldsymbol{X}_n$ are independent and identically distributed random vectors in $\mathbb{R}^k$, independently of the coefficient vector $\boldsymbol{Y}$.

Nguyen (1995) has proved a conditional characterization of multivariate stable distributions similar to that of multivariate normal (see Chapter 45). Let $\boldsymbol{X} = (X_1, \ldots, X_k)^T$ and $\boldsymbol{Y} = (Y_1, \ldots, Y_\ell)^T$ be two random vectors. Suppose $\boldsymbol{Y}$ has a multivariate α-stable distribution, and that $\boldsymbol{X}|\boldsymbol{Y} = \boldsymbol{y}$ also has a multivariate α-stable distribution depending on $\boldsymbol{y}$ only through a location vector under the form $\boldsymbol{AY} + \boldsymbol{a}$, where $\boldsymbol{A}$ is a $k \times \ell$ matrix of constants and $\boldsymbol{a} \in \mathbb{R}^k$ is a constant vector. Then, $\boldsymbol{X}$ and $\boldsymbol{Y}$ have a joint multivariate α-stable distribution.

A generalization of this result, also due to Nguyen (1995), is as follows. Let $\boldsymbol{X}$ and $\boldsymbol{Y}$ be two $k \times 1$ identically distributed random vectors. Suppose that $\boldsymbol{X}|\boldsymbol{Y} = \boldsymbol{y}$ has a multivariate α-stable distribution depending on $\boldsymbol{y}$ only through a location vector under the form $\boldsymbol{Ay} + \boldsymbol{a}$, where $\boldsymbol{A}$ is a $k \times k$ matrix of constants and $\boldsymbol{a} \in \mathbb{R}^k$ is a constant vector. Then, $\boldsymbol{X}$ and $\boldsymbol{Y}$ have a joint multivariate α-stable distribution.

The importance of stable random vectors is mainly to achieve a source of multivariate noise with heavy tails in order to use it in evaluating the robustness of multivariate statistical methods.

Just as the multivariate stable distributions provide heavy-tailed alter-

natives to the multivariate normal, some other models have been proposed
in the Bayesian literature as heavy-tailed distributions for developing inferential procedures which are robust to outliers. For example, Fernandez,
Osiewalski, and Steel (1995) have proposed the class of *v-spherical distributions* with joint densities of the form

$$p(\boldsymbol{x}) = g\left(v(\boldsymbol{x} - \boldsymbol{\xi})\right),$$

where the function $v(\cdot)$ operates like a metric with the property $v(a\boldsymbol{z}) = av(\boldsymbol{z})$, but is otherwise arbitrary and $g(\cdot)$ is an arbitrary univariate function (provided it results in a proper density function) that can be chosen
so that $p(\boldsymbol{x})$ is heavy-tailed. For the same purpose, O'Hagan and Le (1994,
1999) and Le and O'Hagan (1998) have studied bivariate density functions
of the form

$$(1 + x_1^2)^{-c_1/2}(1 + x_2^2)^{-c_2/2}(1 + x_1^2 + x_2^2)^{-c/2}$$

for some c_1, c_2, and c. The first two components corresponds to independent univariate t distributions, while the last component corresponds to
a bivariate t distribution. Le and O'Hagan (1998) have discussed various
properties of this family of distributions.

5 TRANSLATION SYSTEMS

We have already described (in Chapter 12, Section 4.3) systems of distributions constructed by supposing certain (fairly simple) functions of
variables to be normally distributed. It is natural to consider what forms
of joint distribution one can construct in this way. First we suppose that
it is possible to normalize the marginal distributions by simple univariate
transformations, and then we consider how to construct joint distributions
with such marginal distributions.

If each of a set of k variables is normally distributed, it is not necessary
that their joint distribution should be multivariate normal. However, it
is possible for this to be so, and some systems of distributions have been
constructed in this way.

Johnson (1949) studied bivariate distributions, S_{IJ}, in which one variable X_1 has an S_I distribution and the other, X_2, an S_J distribution,
where I, J can take the values B, U, L, and N. S_B, S_U have been defined
in Section 4.3 of Chapter 12; S_L means *log-normal* and S_N means *normal*.
Thus, the variables

$$
\begin{aligned}
Z_1 &= \gamma_1 + \delta_1 f_I((X_1 - \xi_1)/\lambda_1), \\
Z_2 &= \gamma_2 + \delta_2 f_J((X_2 - \xi_2)/\lambda_2),
\end{aligned}
$$

where $f_B(y) = \log\{y/(1-y)\}$, $f_U(y) = \sinh^{-1} y$, $f_L(y) = \log y$, and $f_N(y) = y$, are standardized (unit normal) variables, with a joint bivariate normal distribution with correlation coefficient ρ. We also take $\delta_1 > 0$, $\delta_2 > 0$ by convention.

The joint distribution of X_1 and X_2, so defined, has nine parameters γ_1, γ_2, δ_1, δ_2, ξ_1, ξ_2, λ_1, λ_2 and ρ. The *shape* of the distribution depends only on the five parameters γ_1, γ_2, δ_1, δ_2, and ρ. The *standard form* of the distribution is obtained by taking $\xi_1 = \xi_2 = 0$; $\lambda_1 = \lambda_2 = 1$. This form, which is convenient for algebraic treatment, will be used in the discussion that follows.

The random variable

$$T = (1 - \rho^2)^{-1}(Z_1^2 - 2\rho Z_1 Z_2 + Z_2^2)$$

has a χ^2 distribution with 2 degrees of freedom. It is thus quite easy to construct regions in the X_1, X_2 plane containing specified proportions (α) of the distribution, which have boundaries with equations

$$(1 - \rho^2)^{-1}[\{\gamma_1 + \delta_1 f_I(X_1)\}^2 - 2\rho\{\gamma_1 + \delta_1 f_J(X_1)\}\{\gamma_2 + \delta_2 f_J(X_2)\}$$
$$+ \{\gamma_2 + \delta_2 f_J(X_2)\}^2] = \chi_{2,\alpha}^2 = -2\log(1 - \alpha). \tag{44.24}$$

It should be realized, however, that these boundaries are not (except in the case of the bivariate normal distribution) contours on which the probability density function is constant.

The conditional distribution of Z_2, given $Z_1 = z_1$, is normal with expected value ρz_1 and standard deviation $\sqrt{1 - \rho^2}$ (see Chapter 46). This is thus the conditional distribution of $\gamma_2 + \delta_2 f_J(X_2)$ given $X_1 = x_1$. The conditional distribution of

$$(1 - \rho^2)^{-1/2}[\gamma_2 - \rho(\gamma_1 + \delta_1 f_I(x_1)) + \delta_2 f_J(X_2)]$$

is therefore standard normal. This means that the condition (*array*) distribution of X_2, given $X_1 = x_1$, is of the same system (S_J) as X_2, but with γ_2, δ_2 replaced by $(1 - \rho^2)^{-1/2}[\gamma_2 - \rho\{\gamma_1 + \delta_1 f_I(x_1)\}]$ and $(1 - \rho^2)^{-1/2}\delta_2$, respectively. All these array distributions have the same δ-parameter, but (if $\rho \neq 0$) the γ parameter varies from $-\infty$ to $+\infty$. The shape (and variance) of the array distributions of X_2 therefore change with x_1. In particular, when the sign of the skewness depends on the γ-parameter (for example, when $I \equiv U$ or $I \equiv B$), there will be a change in sign of skewness of the array distributions. This feature is, in fact, observed in empirical joint distributions.

A further consequence is that when $J \equiv B$ there may be a range of values of X_1 for which the array distribution of X_2 is bimodal but outside of which it is unimodal.

It is an easy matter to calculate any required percentage points of the array distributions from the formula

$$f_J(X_{2,\alpha}) = \left[U_\alpha \sqrt{1 - \rho^2} - \gamma_2 + \rho\{\gamma_1 + \delta_1 f_I(x_1)\} \right] \delta_2^{-1}. \qquad (44.25)$$

In particular, we have the median regression

$$M(X_2 \mid X_1) = f_J^{-1}[(\rho\gamma_1 - \gamma_2)\delta_2^{-1} + (\rho\delta_1/\delta_2) f_I(x_1)]. \qquad (44.26)$$

This depends on I, J and the two parameters $\gamma = (\rho\gamma_1 - \gamma_2)\delta_2^{-1}$ and $\phi = \rho\delta_1/\delta_2$.

Taking all possible combinations $(I, J \equiv N, L, B, U)$, we have 16 possible median regression functions that are set out in Table 44.2, taken from Johnson (1949).

The graphical forms of these regressions, also taken from Johnson (1949), are shown in Figure 44.1. The examples in these diagrams should suffice to show the effect of reversing the sign of ϕ.

The bivariate S_{BB} distribution (with marginal distributions of X_1 and X_2 being both S_B) have been applied successfully to describe stand structure of tree heights and diameters by Schreuder and Hafley (1977) and Knoebel and Burkhart (1991).

The system S_{NL} (*normal lognormal*) has been discussed in some detail by Crofts (1969), who has presented formulas for the maximum likelihood estimators of the parameters. Crofts has also considered the more general estimation when X_1 is normal, the conditional distribution of X_2 given $X_1 = x_1$ is (three-parameter) lognormal, and the regression of X_2 on X_1 has a specified form.

In the bivariate lognormal distribution (S_{LL}), it is possible to obtain reasonably simple expressions for the ordinary regression function. Assuming that $\log X_1$, $\log X_2$ have a joint bivariate normal distribution with expected values (ζ_1, ζ_2), variances (σ_1^2, σ_2^2), and correlation ρ, the conditional distribution of $\log X_2$, given $X_1 = x_1$, is normal with expected value

$$\zeta_2 + (\rho\sigma_2/\sigma_1)(\log X_1 - \zeta_1) = \zeta_2(X_1)$$

and variance $(1 - \rho^2)\sigma_2^2 = \sigma_2'^2$. The conditional distribution of X_2, given $X_1 = x_1$, is therefore lognormal with parameters $\zeta_2(x_1), \sigma_2'$. It follows that the regression of X_2 on X_1 is (from Chapter 14)

$$\begin{aligned}
E[X_2 \mid X_1 = x_1] &= e^{\zeta_2(x_1) + \frac{1}{2}\sigma_2'^2} \\
&= x_1^{\rho\sigma_2/\sigma_1} e^{\frac{1}{2}(1-\rho^2)\sigma_2^2 + \zeta_2 - \rho\sigma_2\zeta_1/\sigma_1}. \qquad (44.27)
\end{aligned}$$

TABLE 44.2

Median Regressions for S_{IJ} Distributions

Distribution of		Median of X_2
X_2	X_1	when $X_1 = x_1$
S_N	S_N	$\log\theta + \phi x_1$
S_N	S_L	$\log\theta + \phi\log x_1$
S_L	S_N	$\theta\,e^{\phi x_1}$
S_N	S_B	$\log\theta + \phi\log\{x_1/(1-x_1)\}$
S_B	S_N	$[1 + \theta^{-1}\,e^{-\phi x_1}]^{-1}$
S_N	S_U	$\log\theta + \phi\log[x_1 + \sqrt{(x_1^2+1)}]$
S_U	S_N	$\frac{1}{2}[\theta\,e^{\phi x_1} - \theta^{-1}\,e^{-\phi x_1}]$
S_L	S_L	$\theta\,x_1^{\phi}$
S_L	S_B	$\theta[x_1/(1-x_1)]^{\phi}$
S_B	S_L	$[1 + \theta^{-1}x_1^{-\phi}]^{-1}$
S_L	S_U	$\theta[x_1 + \sqrt{(x_1^2+1)}]^{\phi}$
S_U	S_L	$\frac{1}{2}[\theta x_1^{\phi} - \theta^{-1}x_1^{-\phi}]$
S_B	S_B	$\theta x_1^{\phi}[(1-x_1)^{\phi} + \theta x_1^{\phi}]^{-1}$
S_B	S_U	$\left[1 + \theta^{-1}\left\{\sqrt{(x_1^2+1)} - x_1\right\}^{\phi}\right]^{-1}$
S_U	S_B	$\frac{1}{2}[\theta x_1^{2\phi} - \theta^{-1}(1-x_1)^{2\phi}]x_1^{-\phi}(1-x_1)^{-\phi}$
S_U	S_U	$\frac{1}{2}\left[\theta\left\{x_1 + \sqrt{(x_1^2+1)}\right\}^{\phi} - \theta^{-1}\left\{\sqrt{(x_1^2+1)} - x_1\right\}^{\phi}\right]$

(*Note:* $\theta = \exp[(\rho\gamma_1 - \gamma_2)/\delta_2]$; $\phi = \rho\delta_1/\delta_2$.)

Source: Johnson (1949), with permission.

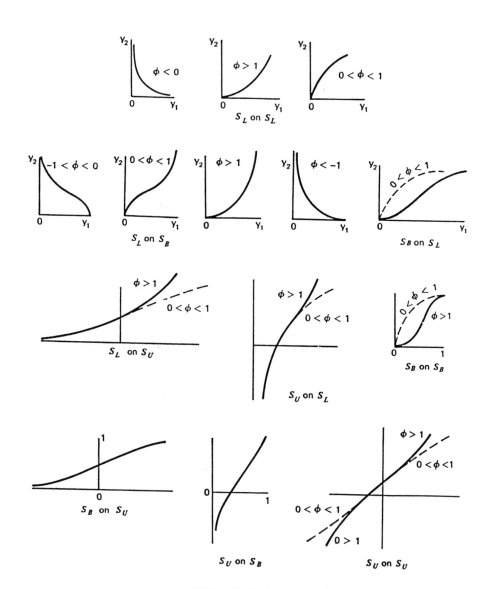

FIGURE 44.1

Median Regression Functions for Some S_{IJ} Distributions. [From Johnson (1949), with permission.]

The array variance is

$$\operatorname{var}(X_2 \mid X_1 = x_1) = \omega'(\omega' - 1)x_1^{2\rho\sigma_2/\sigma_1} \, e^{2(\zeta_2 - \rho\sigma_2\zeta_1/\sigma_1)}, \qquad (44.28)$$

where $\omega' = e^{(1-\rho^2)\sigma_2^2}$.

Good (1983b), correcting an earlier result reported in Good (1983a), presented an asymptotic expression for the (r, s)-th cumulant of the bivariate lognormal distribution as

$$\kappa_{rs} \sim \sigma_{12}(s\sigma_{12} + r\sigma_1^2)^{r-1}(r\sigma_{12} + s\sigma_2^2)^{s-1}e^{r\zeta_1 + s\zeta_2}, \quad r \neq 0, \ s \neq 0,$$

where $\sigma_{12} = \rho\sigma_1\sigma_2$. He has also pointed out an interesting connection of this formula with the combinatorial problem of enumerating colored trees, along the lines of the cumulants of an univariate lognormal distribution being connected with the enumeration of labelled trees; see Mallows and Riordan (1968).

Mostafa and Mahmoud (1964) constructed an unbiased estimator of the function (44.27), based on a random sample giving n pairs of observed values (X_{1j}, X_{2j}) $(j = 1, 2, \ldots, n)$ of X_1, X_2. Since $(Z_{1j}, Z_{2j}) = (\log X_{1j}, \log X_{2j})$ have a joint bivariate normal distribution, we can obtain the maximum likelihood estimators of $\zeta_1, \zeta_2, \sigma_1, \sigma_2$, and ρ using the standard formulas given in Chapter 46. Mostafa and Mahmoud (1964) showed that

$$E\left[e^{\hat{\zeta}_2 + (\hat{\rho}\hat{\sigma}_2/\hat{\sigma}_1)(\log x_1 - \hat{\zeta}_1)}\right] = e^{\zeta_2(x_1) + \frac{1}{2}\sigma_2'^2 n^{-1}(1 + nK)},$$

where

$$K = (\log x_1 - \zeta_1)^2 \left[\sum_{j=1}^{n}(\log X_{1j} - \zeta_1)^2\right]^{-1}.$$

They then sought to find a function $g(\hat{\sigma}_2'^2)$ of the residual mean square

$$\hat{\sigma}_2'^2 = (n - 2)^{-1} \sum_{j=1}^{n} \left\{\log X_{2j} - \hat{\zeta}_2 - (\hat{\rho}\hat{\sigma}_2/\hat{\sigma}_1)(\log X_{1j} - \hat{\zeta}_1)\right\}^2,$$

which shall have expected value $e^{\frac{1}{2}\{1 - n^{-1}(1 + nK)\}\sigma_2'^2}$. Then, since $\hat{\sigma}_2'^2$ and $[\hat{\zeta}_2 + (\hat{\rho}\hat{\sigma}_2/\hat{\sigma}_1)(\log x_1 - \hat{\zeta}_1)]$ are independent, the product

$$g(\hat{\sigma}_2'^2) \, e^{\hat{\zeta}_2 + (\hat{\rho}\hat{\sigma}_2/\hat{\sigma}_1)(\log x_1 - \hat{\zeta}_1)} \tag{44.29}$$

will be an unbiased estimator of $E[X_2 \mid X_1 = x_1]$. They found

$$g(\hat{\sigma}_2'^2) = \sum_{j=0}^{\infty}(\lambda^j/j!)\{(n-2)\hat{\sigma}_2'^2\}^j\Gamma\left(\frac{1}{2}n - 1\right)\left\{\Gamma\left(\frac{1}{2}n + j - 1\right)\right\}^{-1}, \tag{44.30}$$

where $\lambda = \frac{1}{2}[1 - n^{-1}(1 + nK)]$. For practical calculations, they suggested the approximation

$$g(\hat{\sigma}_2'^2) = [1 - n^{-1}\{\hat{\sigma}_2'^2 + (1 - K)^2\hat{\sigma}_2'^4\}]\, e^{(1-K)\hat{\sigma}_2'^2}. \tag{44.31}$$

The variance of the estimator is approximately

$$\{E[X_2 \mid X_1 = x_1]\}^2 \left[\left\{ 1 + n^{-1}\sigma_2'^2 + \frac{1}{2}n^{-1}(1-K)^2\sigma_2'^4 \right\} e^{K\sigma_2'^2 - 1} \right]. \quad (44.32)$$

Mostafa and Mahmoud (1964) also gave formulas for estimators of the median regression and of the *modal regression*:

$$\text{Mode}[X_2 \mid X_1 = x_1] = e^{\zeta_2 + (\rho\sigma_2/\sigma_1)(\log x_1 - \zeta_1) - \sigma_2^2(1-\rho^2)}. \quad (44.33)$$

By considering $X_{1:2} = \min(X_1, X_2)$ when $(X_1, X_2)^T$ is jointly distributed as bivariate lognormal, Lien (1986) has derived explicit formulas for the moments $E[X_{1:2}^r]$, $E[X_{1:2}^r X_1^s]$, and $E[X_{1:2}^r X_2^s]$. For example, he has shown that

$$E[X_{1:2}] = e^{\zeta_1 + \sigma_1^2/2} \, \Phi\left(- \frac{\zeta_1 - \zeta_2 + \sigma_1^2 - \sigma_{12}}{\sqrt{\sigma_1^2 + \sigma_2^2 - 2\sigma_{12}}} \right)$$

$$+ e^{\zeta_2 + \sigma_2^2/2} \, \Phi\left(- \frac{\zeta_2 - \zeta_1 + \sigma_2^2 - \sigma_{12}}{\sqrt{\sigma_1^2 + \sigma_2^2 - 2\sigma_{12}}} \right)$$

and

$$E[X_{1:2}^2] = e^{\zeta_1 + 2\sigma_1^2} \, \Phi\left(- \frac{\zeta_1 - \zeta_2 + 2\sigma_1^2 - 2\sigma_{12}}{\sqrt{\sigma_1^2 + \sigma_2^2 - 2\sigma_{12}}} \right)$$

$$+ e^{\zeta_2 + 2\sigma_2^2} \, \Phi\left(- \frac{\zeta_2 - \zeta_1 + 2\sigma^2 - 2\sigma_{12}}{\sqrt{\sigma_1^2 + \sigma_2^2 - 2\sigma_{12}}} \right),$$

where $\sigma_{12} = \rho\sigma_1\sigma_2$. Similar formulas can be presented for $X_{2:2} = \max(X_1, X_2)$ as well. Lien and Rearden (1996) have discussed the coefficients of variation of $X_{1:2}$ and $X_{2:2}$ for the case when $E[X_1] = E[X_2]$. Specifically, they have shown that both $X_{1:2}$ and $X_{2:2}$ have a smaller coefficient of variation than X_1 and X_2 in the case when $\text{var}(X_1) = \text{var}(X_2)$. In the case when $\text{var}(X_1) \neq \text{var}(X_2)$, as the variance ratio increases, $X_{1:2}(X_{2:2})$ has a smaller coefficient of variation if $\text{corr}(X_1, X_2)$ is small (small) and the variances are large (small). Lien and Rearden (1998) have also derived moment formulas for the extremes in the case of trivariate lognormal distributions. They have illustrated the usefulness of these results in evaluating and comparing the hedging effectiveness of futures markets.

Lien (1985) has derived explicit expressions for the moments of truncated bivariate lognormal distributions by exploiting the joint moment-generating function of truncated bivariate normal distributions. He has also applied these results to test the Houthakker effect in futures markets.

To form the *k-variate lognormal distribution*, we suppose the variables $Z_j = \log X_j$ $(j = 1, \ldots, k)$ to have a joint multivariate normal distribution with expected value vector $\boldsymbol{\zeta} = (\zeta_1, \ldots, \zeta_k)$ and variance–covariance matrix $\boldsymbol{V}$.

By an analysis similar to that used in the bivariate case, it can be seen that the conditional distribution of X_1, given $(X_2, \ldots, X_k) = (x_2, \ldots, x_k)$, is lognormal. The moments and product moments of the X's are derived straightforwardly from the moment-generating function of the Z's, because

$$
\begin{aligned}
\mu_{r_1, r_2, \ldots, r_k}(\boldsymbol{X}) &= E\left[\prod_{j=1}^{k} X_j^{r_j}\right] \\
&= E[\exp(\boldsymbol{r}^T \boldsymbol{Z})] \\
&= \exp\left(\boldsymbol{r}^T \boldsymbol{\zeta} + \frac{1}{2} \boldsymbol{r}^T \boldsymbol{V} \boldsymbol{r}\right). \quad (44.34)
\end{aligned}
$$

Putting $r_j = r_k = 1$ and all the other r's equal to zero, we find

$$
\text{cov}(X_j, X_k) = \{e^{\rho_{jk}\sigma_j\sigma_k} - 1\} e^{\zeta_j + \zeta_k + \frac{1}{2}(\sigma_j^2 + \sigma_k^2)}, \quad (44.35)
$$

where $\rho_{jk} = \text{corr}(Z_j, Z_k)$, from which

$$
\text{corr}(X_j, X_k) = \{e^{\rho_{jk}\sigma_j\sigma_k} - 1\} \left[\{e^{\sigma_j^2} - 1\}\{e^{\sigma_k^2} - 1\}\right]^{-1/2}; \quad (44.36)
$$

see Jones and Miller (1966).

Let us consider the case when the bivariate random vector $(X_1, X_2)^T$ is such that

$$
\begin{aligned}
\Pr[X_1 = X_2 = 0] &= \delta_0, \\
\Pr[0 < X_1 \leq x_1, \ X_2 = 0] &= \delta_1 \, F_1(x_1), \quad x_1 > 0, \\
\Pr[X_1 = 0, \ 0 < X_2 \leq x_2] &= \delta_2 \, F_2(x_2), \quad x_2 > 0,
\end{aligned}
$$

and

$$
\Pr[0 < X_1 \leq x_1, \ 0 < X_2 \leq x_2] = \delta_3 \, F(x_1, x_2), \quad x_1, x_2 > 0,
$$

where $0 \leq \delta_i < 1$ $(i = 0, 1, 2)$, $\delta_3 = 1 - \delta_0 - \delta_1 - \delta_2 > 0$, F_1 and F_2 are univariate lognormal distributions, and F is a bivariate lognormal distribution. The above joint distribution of $(X_1, X_2)^T$, first considered by Shimizu and Sagae (1990), is called a *bivariate mixed lognormal distribution* because it is a mixture of discrete and continuous distributions. If $\delta_0 = \delta_1 = \delta_2 = 0$ (so that $\delta_3 = 1$), the distribution simply reduces to a bivariate lognormal distribution. If $\delta_1 = \delta_2 = 0$, the distribution becomes a *bivariate delta distribution* introduced by Iwase, Shimizu, and Suzuki (1982) and is called as a *bivariate delta-lognormal distribution* by Crow and Shimizu (1988). The general bivariate mixed lognormal distribution presented above has been utilized by Shimizu (1993) as a probability model for representing rainfalls, containing zeros, measured at two monitoring sites. Shimizu has also discussed the maximum likelihood estimation of all the parameters of this distribution.

R. L. Obenchain (personal communication) suggested another multivariate extension of the S_B distributions, which might be appropriate when the range of variation is restricted to a simplex (for example, $0 \leq \sum_{j=1}^{k} X_j \leq 1; X_j > 0, j = 1, \ldots, k$). He considered the joint distribution of

$$X_j = e^{Y_j} \left[\sum_{i=1}^{k+1} e^{Y_i} \right]^{-1}, \qquad j = 1, \ldots, k,$$

when $Y_1, Y_2, \ldots, Y_{k+1}$ have a multivariate normal distribution. By putting $Y_i^* = Y_i - Y_{k+1}$, we have

$$X_j = e^{Y_j^*} \left[1 + \sum_{i=1}^{k} e^{Y_i^*} \right]^{-1}, \qquad j = 1, \ldots, k, \qquad (44.37)$$

with $Y_1^*, \ldots, Y_k^*$ jointly distributed as multivariate normal. In the case $k = 1$, (44.37) gives an S_B distribution. For $k \geq 2$, the marginal distributions are not, in general, S_B.

Obenchain made some detailed investigations of the bivariate ($k = 2$) case and developed methods of fitting the distributions to data (i.e., estimating the underlying population parameters).

6 MULTIVARIATE LINEAR EXPONENTIAL-TYPE DISTRIBUTIONS

Bildikar and Patil (1968) defined a k-variate *exponential-type* distribution, in a general way, as a distribution with joint likelihood function of the form

$$L_{\boldsymbol{X}}(\boldsymbol{x}) = h(\boldsymbol{x}) \exp[\boldsymbol{x}^T \boldsymbol{t} - q(\boldsymbol{\theta})], \qquad (44.38)$$

where $\boldsymbol{X}^T = (X_1, \ldots, X_k)$ represents random variables and $\boldsymbol{\theta}^T = (\theta_1, \ldots, \theta_k)$ represents parameters. We are concerned here with continuous multivariate distributions, and so we regard $L_{\boldsymbol{X}}(\boldsymbol{x})$ as a density function.

The following results have been obtained by Bildikar and Patil (1968). The moment-generating function of $X_1, \ldots, X_k$ is

$$
\begin{aligned}
E[\exp(\boldsymbol{t}^T \boldsymbol{X})] &= e^{-q(\boldsymbol{\theta})} \int_{-\infty}^{+\infty} \cdots \int_{-\infty}^{\infty} h(\boldsymbol{x})\, e^{\boldsymbol{x}^T(\boldsymbol{\theta}+\boldsymbol{t})}\, d\boldsymbol{x} \\
&= e^{q(\boldsymbol{\theta}+\boldsymbol{t})-q(\boldsymbol{\theta})}
\end{aligned}
\tag{44.39}
$$

since $\int_{-\infty}^{\infty} \cdots \int_{-\infty}^{\infty} L_{\boldsymbol{X}}(\boldsymbol{x}, \boldsymbol{\theta}) d\boldsymbol{x} = 1$ for all $\boldsymbol{\theta}$. The cumulant generating function is, therefore,

$$
\Psi(\boldsymbol{t}) = q(\boldsymbol{\theta} + \boldsymbol{t}) - q(\boldsymbol{\theta}).
\tag{44.40}
$$

From (44.40), one can deduce the recurrence relation

$$
\kappa_{r_1,\ldots,r_j-1,r_j+1,r_{j+1},\ldots,r_k} = \partial \kappa_{r_1,\ldots,r_k}/\partial \theta_j.
\tag{44.41}
$$

Taking $k = 2$, we see that if κ_{11} is zero (i.e., X_1 and X_2 are uncorrelated), then so are κ_{21} and κ_{12}, hence also κ_{r_1,r_2} for any $r_1 \geq 1, r_2 \geq 1$. In fact, X_1 and X_2 are independent. This property can be extended: if $X_1, \ldots, X_k$ have a joint exponential-type distribution, they form a mutually independent set if and only if they are pairwise independent.

If

$$
p(x_1, \ldots, x_k) = h(x_1, \ldots, x_k)\, e^{\sum_{j=1}^{k} \theta_j x_j - q(\theta_1,\ldots,\theta_k)},
\tag{44.42}
$$

then (for $s < k$) the joint density function of $X_1, \ldots, X_s$ has the form

$$
p(x_1, \ldots, x_s) = h(x_1, \ldots, x_s)\, e^{\sum_{j=1}^{s} \theta_j x_j - q_1(\theta_1,\ldots,\theta_s)},
\tag{44.43}
$$

where

$$
h_1(x_1, \ldots, x_s) = \int_{-\infty}^{\infty} \cdots \int_{-\infty}^{\infty} h(x_1, \ldots, x_k) e^{\sum_{j=s+1}^{k} \theta_j x_j}\, dx_{s+1} \ldots dx_k
$$

and $q_1(\theta_1, \ldots, \theta_s)$ is simply $q(\theta_1, \ldots, \theta_k)$ regarded as a function of $\theta_1, \ldots, \theta_s$ (i.e., $\theta_{s+1}, \ldots, \theta_k$ regarded as pure constants, rather than parameters).

Comparison of (44.42) with (44.43) shows that the joint distribution of $X_1, \ldots, X_s$ (hence of any subset of $X_1, \ldots, X_k$) is of the exponential type.

The conditional joint density of $X_{s+1}, \ldots, X_k$, given $X_1 = x_1, X_2 = x_2, \ldots, X_s = x_s$, is of the form

$$
h_s(x_{s+1}, \ldots, x_k) e^{\sum_{j=s+1}^{k} \theta_j x_j},
\tag{44.44}
$$

since $q_1(\boldsymbol{\theta}) \equiv q(\boldsymbol{\theta})$. (The function $h_s(\cdot)$ depends also on $x_1, \ldots, x_s$, but the θ's do not.) This is also of exponential type, but is restricted by the requirement that $q(\theta) \equiv 0$.

Seshadri and Patil (1964) showed that if $p_{X_1}(x_1)$ and $p_{X_1|X_2}(x_1 \mid x_2)$ are given, a sufficient condition for $p_{X_2}(x_2)$ to be unique is that the conditional density function $p_{X_1|X_2}(x_1 \mid x_2)$ is of (univariate) exponential form. Roux (1971) extended this result to sets of k (≥ 2) variables. He showed that, given $p_{X_1,\ldots,X_{k-1}}(x_1, \ldots, x_{k-1})$ and $p_{X_1,\ldots,X_{k-1}|X_k}(x_1, \ldots, x_{k-1} \mid x_k)$, a sufficient condition for $p_{X_k}(x_k)$ [hence $p_{X_1,\ldots,X_k}(x_1, \ldots, x_k)$] to be unique is that the conditional density function $p_{X_1,\ldots,X_{k-1}|X_k}(x_1, \ldots, x_{k-1} \mid x_k)$ is of exponential form.

In view of the occurrence of a linear function of $\boldsymbol{x}$ in the exponent of (44.44), such distributions may be called *linear exponential-type* distributions [see Wani (1968)] to distinguish them from the more general *quadratic exponential-type* distributions [Day (1969)] for which

$$L_{\boldsymbol{X}}(\boldsymbol{x}) = h(\boldsymbol{x}) \, e^{-(\boldsymbol{x}-\boldsymbol{\xi})^T \boldsymbol{A}(\boldsymbol{x}-\boldsymbol{\xi}) - q(\boldsymbol{\theta})}, \tag{44.45}$$

where $\boldsymbol{A}$ is positive definite and $\boldsymbol{A}$, $\boldsymbol{\xi}$, and $\boldsymbol{\theta}$ are sets of parameters.

More detailed discussion on *natural* exponential families is presented in Chapter 54.

7 SARMANOV'S DISTRIBUTIONS

Let $f_1(x_1)$ and $f_2(x_2)$ be univariate probability density functions with supports $A_1 \subseteq \mathbb{R}$ and $A_2 \subseteq \mathbb{R}$, respectively. Let $\phi_1(t)$ and $\phi_2(t)$ be bounded nonconstant functions such that

$$\int_{-\infty}^{\infty} \phi_i(t) \, f_i(t) \, dt = 0 \qquad \text{for } i = 1, 2.$$

Then, Sarmanov (1966) introduced the bivariate distribution with joint density function

$$f(x_1, x_2) = f_1(x_1) f_2(x_2) \{1 + \omega \, \phi_1(x_1) \phi_2(x_2)\}, \tag{44.46}$$

where ω is a real number such that $1 + \omega \phi_1(x_1)\phi_2(x_2) \geq 0 \; \forall \; (x_1, x_2)$. This distribution, under a more general construction, was proposed independently in the physics literature by Cohen (1984). Lee (1996) examined some properties of this distribution and also suggested a k-variate version of it.

Let us denote, for $i = 1, 2$,

$$\mu_i = \int_{-\infty}^{\infty} t\, f_i(t)\, dt,$$

$$\sigma_i^2 = \int_{-\infty}^{\infty} (t - \mu_i)^2 f_i(t)\, dt,$$

$$\nu_i = \int_{-\infty}^{\infty} t\, \phi_i(t)\, f_i(t)\, dt$$

and

$$\eta_i = \int_{-\infty}^{\infty} t^2 \phi_i(t)\, f_i(t)\, dt.$$

Then, for the distribution in (44.46), it can be shown that

$$E[X_1 X_2] = \mu_1 \mu_2 + \omega \nu_1 \nu_2,$$
$$\text{cov}(X_1, X_2) = \omega \nu_1 \nu_2,$$
$$\text{corr}(X_1, X_2) = \rho = \frac{\omega \nu_1 \nu_2}{\sigma_1 \sigma_2},$$
$$|\rho| \leq |\omega| \sqrt{E\{\phi_1^2(X_1)\} E\{\phi_2^2(X_2)\}},$$
$$\Pr[X_2 \leq x_2 \mid X_1 = x_1] = F_2(x_2) - \omega \phi_1(x_1) \int_{x_2}^{\infty} f_2(t)\, \phi_2(t)\, dt$$

where $F_2(x_2) = \Pr[X_2 \leq x_2]$, and

$$E[X_2 \mid X_1 = x_1] = \mu_2 + \omega \nu_2 \phi_1(x_1).$$

Note that X_1 and X_2 are independent if $\omega = 0$. Also, observe the similarity of this distribution with the FGM distributions (see Section 12) for which, however, $|\rho| \leq \frac{1}{3}$.

A special case of interest involves the beta marginal distribution with density functions (see Chapter 25)

$$f_i(x_i) = \frac{1}{B(a_i, b_i)} x_i^{a_i - 1} (1 - x_i)^{b_i - 1}, \qquad 0 < x_i < 1,\ a_i > 0,\ b_i > 0,$$

and linear mixing functions $\phi_i(x_i) = x_i - \mu_i$, where $\mu_i = \frac{a_i}{a_i + b_i}$ $(i = 1, 2)$ are the means of the above beta distributions. The corresponding $f(x_1, x_2)$ derived from (44.46), a bivariate beta density function, can be expressed as a linear combination of products of univariate beta density functions; see Lee (1996). Hence, if this bivariate density function is used as a prior, the posterior will be pseudoconjugate to the prior; the same property also holds for bivariate distribution constructed with gamma marginal distributions. For an application of this bivariate beta distribution in analyzing incompletely observed longitudinal binary store display data, one may refer to Cole *et al.* (1995).

Lee (1996) proposed a k-variate Sarmanov's distribution as one with the joint density function

$$f(x_1, \ldots, x_k) = \left\{ \prod_{i=1}^k f_i(x_i) \right\} \{1 + R_{\phi_1, \ldots, \phi_k, \Omega_k}(x_1, \ldots, x_k)\}, \qquad (44.47)$$

where

$$\begin{aligned} R_{\phi_1, \ldots, \phi_k, \Omega_k}(x_1, \ldots, x_k) \\ = \sum_{1 \le i_1 < i_2 \le k} \sum w_{i_1, i_2} \phi_{i_1}(x_{i_1}) \phi_{i_2}(x_{i_2}) \\ + \sum_{1 \le i_1 < i_2 < i_3 \le k} \sum \sum w_{i_1, i_2, i_3} \phi_{i_1}(x_{i_1}) \phi_{i_2}(x_{i_2}) \phi_{i_3}(x_{i_3}) \\ + \cdots \\ + w_{1, 2, \ldots, k} \prod_{i=1}^k \phi_i(x_i), \end{aligned}$$

and $\Omega_k = \{w_{i_1, i_2}, w_{i_1, i_2, i_3}, \ldots, w_{1, 2, \ldots, k}\}$. The set of real numbers Ω_k is such that $1 + R_{\phi_1, \ldots, \phi_k, \Omega_k}(x_1, \ldots, x_k) \ge 0 \ \forall \ (x_1, \ldots, x_k) \in \mathbb{R}^k$. Compare the form of the joint density function in (44.47) with that of the extended FGM distribution, due to Johnson and Kotz (1977), presented in Section 12 of this chapter.

Lee (1996) has suggested the use of $f_i(x_i \mid \theta_i)$ belonging to the ℓ-parameter exponential family of distributions. If the joint prior distribution of $(\theta_1, \theta_2, \ldots, \theta_k)$ is with the density function

$$\begin{aligned} \Pi(\theta_1, \ldots, \theta_k \mid t_{i,1}, \ldots, t_{i,\ell+1} \text{ for } 1 \le i \le k) \\ = \prod_{i=1}^k \Pi_i(\theta_i \mid t_{i,1}, \ldots, t_{i,\ell+1}) \{1 + R_{\phi_1, \ldots, \phi_k, \Omega_k}(\theta_1, \ldots, \theta_k)\}, \end{aligned}$$

the posterior density function is then the pseudoconjugate to the above prior, which turns out to be a linear combination of products of univariate density functions from the univariate natural exponential family.

Bairamov and Kotz (1999) studied a subclass of the Sarmanov family of the form

$$f_\alpha(x, y) = 1 + \alpha\, A(x) A(y), \qquad 0 < x, y \le 1,$$

where $A(x) = \phi(x) - \phi(1 - x)$, and $\phi(x) \ (0 < x \le 1)$ is a continuous function.

8   MULTIVARIATE LINNIK'S DISTRIBUTIONS

Univariate Linnik's distribution [see Linnik (1963)] has been discussed by Johnson, Kotz, and Balakrishnan (1994). Its characteristic function is given by

$$\phi_X(t) = \frac{1}{1 + |t|^\alpha}, \qquad 0 < \alpha \le 2.$$

This distribution is known to be closed under geometric compounding.

Anderson (1992) defined *multivariate Linnik's distribution* through the joint characteristic function

$$\phi_{\boldsymbol{X}}(\boldsymbol{t}) = \frac{1}{1 + \left(\sum_{i=1}^m \boldsymbol{t}^T \boldsymbol{\Omega}_i \boldsymbol{t}\right)^{\alpha/2}},$$

where $0 < \alpha \le 2$, $\boldsymbol{\Omega}_i$'s are $k \times k$ positive semi-definite matrices and no two of $\boldsymbol{\Omega}_i$'s are proportional. This distribution is also closed under geometric compounding. Anderson (1992) has also shown that

$$\phi_{\boldsymbol{X}}(\boldsymbol{t}) = \frac{1}{1 + |\boldsymbol{t}^T \boldsymbol{\Sigma} \boldsymbol{t}|^{\alpha/2}},$$

where $0 < \alpha \le 2$, $\boldsymbol{\Sigma}$ is a positive $k \times k$ matrix and $\boldsymbol{t} \in \mathbb{R}^k$, is a characteristic function of a k-dimensional random variable.

The special case of $\alpha = 2$ yields a *multivariate Laplace distribution* with joint density function

$$p_{\boldsymbol{X}}(\boldsymbol{x}) = (2\pi)^{-k/2} |\boldsymbol{\Omega}|^{-1/2} \int_0^\infty \exp\left(-\frac{Q}{2u} - u\right) u^{\frac{2-k}{2} - 1}\, du,$$

where $\boldsymbol{\Omega}$ is a positive definite matrix and $Q = \boldsymbol{x}^T \boldsymbol{\Omega}^{-1} \boldsymbol{x}$. This density function can alternatively be written as

$$p_{\boldsymbol{X}}(\boldsymbol{x}) = (2\pi)^{-k/2} |\boldsymbol{\Omega}|^{-1/2}\, 2 \left(\sqrt{\frac{Q}{2}}\right)^{(2-k)/2} K_{\frac{2-k}{2}}\left(\sqrt{2Q}\right) \sqrt{2Q},$$

where $K_\nu(z)$ is the modified Bessel function of the third kind. This density, however, does not include all multivariate models with Laplace marginals. Anderson (1992) has provided an example of bivariate Gumbel type Laplace model with density function

$$
\begin{aligned}
p_{X_1, X_2}&(x_1, x_2) \\
&= \frac{1}{4} \left\{ (1 + \theta|x_1|)(1 + \theta|x_2|) - \theta \right\} \\
&\quad \times \exp\left\{ -(1 + |x_1| + |x_2| + \theta|x_1||x_2|) \right\}, \qquad (x_1, x_2)^T \in \mathbb{R}^2.
\end{aligned}
$$

This bivariate distribution has Laplace marginals but is different than the bivariate case of the multivariate Laplace distribution given above.

Ostrovskii (1995) has noted that the study of multivariate Linnik's distributions can be restricted to the case when $\mathbf{\Sigma} = \mathbf{I}$, yielding a joint characteristic function

$$\phi_{\mathbf{X}}(\mathbf{t}) = \frac{1}{1 + |\mathbf{t}|^{\alpha}}, \qquad 0 < \alpha \leq 2, \quad \mathbf{t} \in \mathbb{R}^{k},$$

where $|\mathbf{t}|$ denotes the Euclidean norm of the vector $\mathbf{t}$. He has shown that the distribution defined by this characteristic function is absolutely continuous with respect to the k-dimensional Lebesgue measure and that the corresponding density function possesses spherical symmetry (since the function $\{1 + |\mathbf{t}|^{\alpha}\}^{-1}$ has this property).

9   MULTIVARIATE KAGAN'S DISTRIBUTIONS

The multivariate distribution of a random vector $\mathbf{X}$ of dimension m is said to belong to the class $\mathcal{D}_{m,k}$ ($k = 1, 2, \ldots, m$, $m = 1, 2, \ldots,$) if its characteristic function $\phi_{\mathbf{X}}(\mathbf{t})$ allows the factorization

$$\phi_{\mathbf{X}}(\mathbf{t}) = \phi_{\mathbf{X}}(t_1, \ldots, t_m) = \prod_{1 \leq i_1 < \cdots < i_k \leq m} \phi_{i_1, \ldots, i_k}(t_{i_1}, \ldots, t_{i_k}), \quad (44.48)$$

where $(t_1, \ldots, t_m) \in \mathbb{R}^m$ and $\phi_{i_1, \ldots, i_k}$ are continuous complex-valued functions with $\phi_{i_1, \ldots, i_k}(0, \ldots, 0) = 1$ for any $1 \leq i_1 < \cdots < i_k \leq m$. In this case, we denote $\mathbf{X} \in \mathcal{D}_{m,k}$. If the factorization in (44.48) holds in a neighborhood of the origin, then $\mathbf{X}$ is said to belong to the class $\mathcal{D}_{m,k}(\text{loc})$ and is denoted by $\mathbf{X} \in \mathcal{D}_{m,k}(\text{loc})$.

These two families of distributions were introduced by Kagan (1988) and they generalize the concept of the distribution of a random vector $\mathbf{X}$ with independent components. Clearly, for any $k = 1, 2, \ldots, m$, we have $\mathcal{D}_{m,k} \subset \mathcal{D}_{m,k}(\text{loc})$.

Wesolowski (1991a) has shown that if the characteristic function of $\mathbf{X}$ does not vanish and $\mathbf{X} \in \mathcal{D}_{m,k}$, then its distribution is uniquely determined by all its k-dimensional distributions. If $\mathbf{X} \in \mathcal{D}_{m,k}(\text{loc})$ and all its k-dimensional marginal distributions are Gaussian, then $\mathbf{X}$ is a Gaussian random vector.

With ϕ denoting the characteristic function of $\mathbf{X} = (X_1, \ldots, X_m)^T$ and $\phi_{i_1, \ldots, i_k}$ denoting similarly the characteristic function of $(X_{i_1}, \ldots, X_{i_k})^T$ for any $1 \leq i_1 < \cdots < i_k \leq m$, for $\boldsymbol{X} \in \mathcal{D}_{m,k}(\text{loc})$ in a neighborhood of the origin $V \subset \mathbb{R}^m$ it follows that

$$\phi_{\boldsymbol{X}}(\boldsymbol{t}) = \prod_{r=1}^{k} \left\{ \prod_{1 \leq i_1 < \cdots < i_r \leq m} \phi_{i_1,\ldots,i_r}(t_{i_1},\ldots,t_{i_r}) \right\}^{a_{m,k,r}},$$

where

$$a_{m,k,r} = - \sum_{j=1}^{m-k} (-1)^{k-r+j} \binom{m-r}{k-r+j}, \qquad r = 1, 2, \ldots, k;$$

for $\boldsymbol{X} \in \mathcal{D}_{m,k}(\text{loc})$, we have (in V) as expected

$$\phi_{\boldsymbol{X}}(\boldsymbol{t}) = \prod_{i=1}^{m} \phi_i(t_i)$$

while for $\boldsymbol{X} \in \mathcal{D}_{3,2}(\text{loc})$ we have

$$\phi_{\boldsymbol{X}}(t_1, t_2, t_3) = \frac{\phi(t_1, t_2, 0)\, \phi(t_1, 0, t_3)\, \phi(0, t_2, t_3)}{\phi(t_1, 0, 0)\, \phi(0, t_2, 0)\, \phi(0, 0, t_3)};$$

more generally, for $\boldsymbol{X} \in \mathcal{D}_{m,m-1}(\text{loc})$, we have $a_{m,m-1,r} = (-1)^{m-r+1}$ for $r = 1, 2, \ldots, m-1$ and

$$\begin{aligned}
\phi_{\boldsymbol{X}}(t_1, \ldots, t_m) \;=\;\; & \prod_{1 \leq i_1 < \cdots < i_{m-1} \leq m} \phi_{i_1,\ldots,i_{m-1}}(t_{i_1},\ldots,t_{i_{m-1}}) \\
& \times \left\{ \prod_{1 \leq i_1 < \cdots < i_{m-2} \leq m} \phi_{i_1,\ldots,i_{m-2}}(t_{i_1},\ldots,t_{i_{m-2}}) \right\}^{-1} \\
& \times \cdots \\
& \times \left\{ \prod_{1 \leq i_1 \leq m} \phi_{i_1}(t_{i_1}) \right\}^{(-1)^m}.
\end{aligned}$$

A related concept is the *Gaussian conditional structure of the second order* wherein a random element $\boldsymbol{X} = (X_\alpha)_{\alpha \in A}$ is said to have a Gaussian conditional structure of the second order if for any $\alpha_1, \alpha_2, \ldots, \alpha_n \in A$, $n = 2, 3, \ldots,$

  (i) the random variables $X_{\alpha_1}, \ldots, X_{\alpha_n}$ are linearly independent and pairwise correlated (a technical condition to avoid the case of independence);

  (ii) $E[X_{\alpha_1} \mid X_{\alpha_2}, \ldots, X_{\alpha_n}]$ is a linear function of $X_{\alpha_2}, \ldots, X_{\alpha_n}$;

and

(iii) $\text{var}(X_{\alpha_1} \mid X_{\alpha_2}, \ldots, X_{\alpha_n})$ is nonrandom.

In this case, we denote $\boldsymbol{X} \in GCS_2(A)$. If A is a finite subset of N (the set of all integers), then we will use the notation $GCS_2(n)$ where n is the number of elements in the set A. We will also denote a n-dimensional random vector $GCS_2(n)$ with mean vector $\boldsymbol{\mu}$ and variance–covariance matrix $\boldsymbol{\Sigma}$ by $GCS_2(n; \boldsymbol{\mu}, \boldsymbol{\Sigma})$. Then, Bryc (1985) and Bryc and Plucińska (1985) proved that

(i) $GCS_2(N) \equiv \text{Gauss}(N)$

and

(ii) If $\boldsymbol{X}, \boldsymbol{Y} \in GCS_2(n; \boldsymbol{\mu}, \boldsymbol{\Sigma})$, then for every c_1 and c_2 such that $c_1 + c_2 = 1$ we have $c_1 \boldsymbol{X} + c_2 \boldsymbol{Y} \in GCS_2(n; \boldsymbol{\mu}, \boldsymbol{\Sigma})$.

Kagan (1988) has shown that $\text{Gauss}(n) \subset \mathcal{D}_{n,2}(\text{loc}) \subset \mathcal{D}_{n,k}(\text{loc})$ for any $k \geq 2$. The conjecture that $\text{Gauss}(n) \not\equiv GCS_2(n)$ seems to be still open, except for the case $n = 2$ when it was proved to be true by Bryc and Plucińska (1985). For $n \geq 3$, Wesolowski (1991b) has established that

$$GCS_2(n) \cap \mathcal{D}_{n,2}(\text{loc}) \equiv \text{Gauss}(n)$$

based on some properties of

$$H(t_1, t_2, t_3) = \psi(t_1, t_2, t_3) - \psi(t_1, t_2, 0) - \psi(t_1, 0, t_3) - \psi(0, t_2, t_3),$$

where $\boldsymbol{X}$ is a three-dimensional random vector and $\psi(\boldsymbol{t}) = \log E(e^{i\boldsymbol{t}^T \boldsymbol{X}})$ in some neighborhood of the origin $V \subset \mathbb{R}^3$.

10 GENERATION OF MULTIVARIATE NONNORMAL RANDOM VARIABLES

Vale and Maurelli (1983) suggested generating multivariate nonnormal random variables with a specified correlation structure by combining the matrix decomposition procedure and a method devised by Fleishman (1978).

Fleishman's (1978) method of generating univariate nonnormal random variables is based on the variable Y defined as

$$Y = a + bX + cX^2 + dX^3, \qquad (44.49)$$

where X is a standard normal random variable, and a, b, c and d are constants chosen in such a way that Y has the desired coefficients of skewness and kurtosis (β_1 and β_2). For a standard distribution (with mean 0 and variance 1), after using the first fourteen moments of the standard normal variable (see Chapter 13) and doing considerable algebraic manipulation, Fleishman (1978) showed that $a = -c$ and the constants b, c, and d need to be determined by simultaneously solving the following three nonlinear equations:

$$b^2 + 6bd + 2c^2 + 15d^2 - 1 = 0,$$
$$2c(b^2 + 24bd + 105d^2 + 2) - \beta_1 = 0,$$
$$24\{bd + c^2(1 + b^2 + 28bd) + d^2(12 + 48bd + 141c^2 + 225d^2)\} - \beta_2 = 0.$$

Then, Fleishman's method of generating univariate nonnormal random variables is to generate a standard normal variable X and to transform it to Y through (44.49) by using the constants a, b, c and d determined from the above equations. This procedure can be combined with the matrix decomposition method in order to generate a multivariate nonnormal random variable as follows: Let X_1, X_2 be two standard normal variables, and Y_1, Y_2 be the two nonnormal variables determined from them as

$$Y_1 = a_1 + b_1 X_1 + c_1 X_1^2 + d_1 X_1^3, \quad Y_2 = a_2 + b_2 X_2 + c_2 X_2^2 + d_2 X_2^3.$$

Then, it is easy to show that the correlation coefficient between Y_1 and Y_2 is

$$\begin{aligned}
\rho_{Y_1,Y_2} &= \rho_{X_1,X_2}(b_1 b_2 + 3b_1 d_2 + 3b_2 d_1 + 9d_1 d_2) \\
&+ \rho_{X_1,X_2}^2 2c_1 c_2 + \rho_{X_1,X_2}^3 6d_1 d_2.
\end{aligned}$$

Solving the above cubic equation for ρ_{X_1,X_2}, one obtains the required correlation coefficient between the two standard normal variables (X_1 and X_2) in order to achieve the specified correlation coefficient between the two nonnormal variables Y_1 and Y_2.

This method seems to produce bivariate random numbers with univariate moments and intercorrelation near the specified values. Although the shortcomings of Fleishman's (1978) method, pointed out by Tadikamalla (1980), also apply here, this method does provide a way of generating bivariate nonnormal random variables. Simple extensions of other univariate methods are not available yet. Steyn (1993) has used this method in his construction of multivariate distributions with coefficient of kurtosis greater than one.

Bélisle, Romeijn, and Smith (1990) proposed a general class of "hit-and-run algorithms," for generating absolutely continuous distributions on $\mathbb{R}^k$. Given a bounded open set S in $\mathbb{R}^k$. Given a bounded open set S in $\mathbb{R}^k$, an absolutely continuous probability distribution on p on S (the target distribution), and an arbitrary probability distribution f on the boundary of the k-dimensional unit sphere centered at the origin (the direction distribution), the (f, p)-hit-and-run algorithm produces a sequence of iteration points as follows. Given the nth iteration point $\boldsymbol{x}$, choose a direction $\boldsymbol{\theta}$ according to the distribution f and then choose the $(n + 1)$th iteration point according to the conditionalization of the distribution p along the line $\{\boldsymbol{x} + \lambda\boldsymbol{\theta}; \ \lambda \in \mathbb{R}\}$.

Another method of simulating bivariate nonnormal data discussed by Kocherlakota, Kocherlakota, and Balakrishnan (1986) is through bivariate Edgeworth series distribution. The joint density function of the bivariate Edgeworth series distribution is [Gayen (1951)]

$$g(x, y) = \left\{ 1 + \sum_{\substack{j,k=0 \\ j+k=3,4,6}}^{3} \frac{(-1)^{j+k} A_{j,k}}{j!k!} \, D_x^j \, D_y^k \right\} f(x, y), \qquad (44.50)$$

where $f(x, y)$ is the standard bivariate normal density function given by (see Chapters 45 and 46)

$$f(x, y) = \frac{1}{2\pi\sqrt{1 - \rho^2}} \, e^{-\frac{1}{2(1-\rho^2)} (x^2 - 2\rho xy + y^2)},$$

A_{jk}'s are functions of the population cumulants, and D_x, D_y are partial derivative operators. The joint characteristic function of (44.50) can be shown to be

$$\phi(t_1, t_2) = \left\{ 1 + \sum_{\substack{j,k=0 \\ j+k=3,4,6}}^{3} \frac{i^{j+k} A_{jk}}{j!k!} \, t_1^j t_2^k \right\} e^{-\frac{1}{2}(t_1^2 + t_2^2 + 2\rho t_1 t_2)}. \qquad (44.51)$$

Although it is difficult to generate samples from (44.50) with specified parameters A_{jk}, it is possible to consider the distribution with prescribed marginals as shown by Kocherlakota, Kocherlakota, and Balakrishnan (1986).

Starting now with the probability density function of the univariate Edgeworth series distribution given by (see Chapter 12)

$$g(u) = \left\{ 1 - \frac{\lambda_3}{6} \, D^3 + \frac{\lambda_4}{24} \, D^4 + \frac{\lambda_3^2}{72} \, D^6 \right\} \frac{1}{\sqrt{2\pi}} \, e^{-u^2/2},$$

where $\lambda_3(=\sqrt{\beta_1})$ is the coefficient of skewness and $\lambda_4(=\beta_2-3)$ is the coefficient of kurtosis, the characteristic function of the linear function $X = aU + bV$, with U and V being independently distributed as univariate Edgeworth series, is

$$\phi_X(t) = \left\{1 + \frac{(it)^3}{6}(a^3\lambda_{3U} + b^3\lambda_{3V}) + \frac{(it)^4}{24}(a^4\lambda_{4U} + b^4\lambda_{4V}) \right.$$
$$\left. + \frac{(it)^6}{72}(a^3\lambda_{3U} + b^3\lambda_{3V})^2\right\}e^{-\frac{1}{2}(a^2+b^2)t^2}.$$

Clearly, with $X = a_1U + b_1V$ and $Y = a_2U + b_2V$, we have the joint characteristic function of $(X,Y)^T$ as

$$\phi_{X,Y}(t_1, t_2) = \left[1 + \frac{i^3}{6}\left\{t_1^3(a_1^3\lambda_{3U} + b_1^3\lambda_{3V}) + t_2^3(a_2^3\lambda_{3U} + b_2^3\lambda_{3V})\right.\right.$$
$$\left. + 3t_1^2t_2(a_1^2a_2\lambda_{3U} + b_1^2b_2\lambda_{3V}) + 3t_1t_2^2(a_1a_2^2\lambda_{3U} + b_1b_2^2\lambda_{3V})\right\}$$
$$+ \frac{i^4}{24}\left\{t_1^4(a_1^4\lambda_{4U} + b_1^4\lambda_{4V}) + t_2^4(a_2^4\lambda_{4U} + b_2^4\lambda_{4V})\right.$$
$$+ 4t_1^3t_2(a_1^3a_2\lambda_{4U} + b_1^3b_2\lambda_{4V}) + 4t_1t_2^3(a_1a_2^3\lambda_{4U} + b_1b_2^3\lambda_{4V})$$
$$\left. + 6t_1^2t_2^2(a_1^2a_2^2\lambda_{4U} + b_1^2b_2^2\lambda_{4V})\right\}$$
$$+ \frac{i^6}{72}\left\{t_1^6(a_1^3\lambda_{3U} + b_1^3\lambda_{3V})^2 + t_2^6(a_2^3\lambda_{3U} + b_2^3\lambda_{3V})^2\right.$$
$$+ 6t_1^5t_2(a_1^3\lambda_{3U} + b_1^3\lambda_{3V})(a_1^2a_2\lambda_{3U} + b_1^2b_2\lambda_{3V})$$
$$+ 6t_1t_2^5(a_2^3\lambda_{3U} + b_2^3\lambda_{3V})(a_1a_2^2\lambda_{3U} + b_1b_2^2\lambda_{3V})$$
$$+ t_1^4t_2^2\{6(a_1^3\lambda_{3U} + b_1^3\lambda_{3V})(a_1a_2^2\lambda_{3U} + b_1b_2^2\lambda_{3V})$$
$$+ 9(a_1^2a_2\lambda_{3U} + b_1^2b_2\lambda_{3V})^2\}$$
$$+ t_1^2t_2^4\{6(a_2^3\lambda_{3U} + b_2^3\lambda_{3V})(a_1^2a_2\lambda_{3U} + b_1^2b_2\lambda_{3V})$$
$$+ 9(a_1a_2^2\lambda_{3U} + b_1b_2^2\lambda_{3V})^2\}$$
$$+ t_1^3t_2^3\{2(a_1^3\lambda_{3U} + b_1^3\lambda_{3V})(a_2^3\lambda_{3U} + b_2^3\lambda_{3V})$$
$$\left.\left.\left. + 18(a_1^2a_2\lambda_{3U} + b_1^2b_2\lambda_{3V})(a_1a_2^2\lambda_{3U} + b_1b_2^2\lambda_{3V})\right\}\right\}\right]$$
$$\times\, e^{-\frac{1}{2}\{(a_1^2+b_1^2)t_1^2+(a_2^2+b_2^2)t_2^2+2(a_1a_2+b_1b_2)t_1t_2\}}. \tag{44.52}$$

A comparison of (44.52) with the characteristic function (44.51) shows that the coefficients A_{ij} can be expressed as follows:

$$
\begin{aligned}
A_{30} &= a_1^3 \lambda_{3U} + b_1^3 \lambda_{3V}, & A_{03} &= a_2^3 \lambda_{3U} + b_2^3 \lambda_{3V}, \\
A_{21} &= a_1^2 a_2 \lambda_{3U} + b_1^2 b_2 \lambda_{3V}, & A_{12} &= a_1 a_2^2 \lambda_{3U} + b_1 b_2^2 \lambda_{3V}, \\
A_{40} &= a_1^4 \lambda_{4U} + b_1^4 \lambda_{4V}, & A_{04} &= a_2^4 \lambda_{4U} + b_2^4 \lambda_{4V}, \\
A_{31} &= a_1^3 a_2 \lambda_{4U} + b_1^3 b_2 \lambda_{4V}, & A_{13} &= a_1 a_2^3 \lambda_{4U} + b_1 b_2^3 \lambda_{4V}, \\
A_{22} &= a_1^2 a_2^2 \lambda_{4U} + b_1^2 b_2^2 \lambda_{4V},
\end{aligned}
$$

while

$$
\begin{aligned}
A_{60} &= 10 A_{30}^2, & A_{06} &= 10 A_{03}^2, \\
A_{51} &= 10 A_{30} A_{21}, & A_{15} &= 10 A_{03} A_{12}, \\
A_{42} &= 6 A_{21}^2 + 4 A_{30} A_{12}, & A_{24} &= 6 A_{12}^2 + 4 A_{03} A_{21}, \\
A_{33} &= A_{30} A_{03} + 9 A_{21} A_{12}.
\end{aligned}
$$

Note that the parameters of the marginal distributions of X and Y are

$$
\lambda_{30} = A_{30}, \quad \lambda_{03} = A_{03}, \quad \lambda_{40} = A_{40}, \quad \lambda_{04} = A_{04},
$$

and the standardized form of the bivariate Edgeworth series distribution corresponds to the choice

$$
a_1^2 + b_1^2 = 1, \quad a_2^2 + b_2^2 = 1, \quad a_1 a_2 + b_1 b_2 = \rho. \tag{44.53}
$$

The bivariate Edgeworth series random variable $(X, Y)^T$ is then generated by taking independent univariate Edgeworth series random variables U and V with parameters $(\lambda_{3U}, \lambda_{4U})$ and $(\lambda_{3V}, \lambda_{4V})$, respectively, and then taking $b_2 = 0$, $a_2 = \pm 1$, $a_1 \pm \rho$, $b_1 = \pm \sqrt{1 - \rho^2}$. Under the conditions in (44.53), there are only four possible choices of the coefficients. Upon basing the required generation of univariate Edgeworth series random variables on the inverse cdf method, Kocherlakota, Kocherlakota, and Balakrishnan (1986) have given a Fortran source code for the generation of bivariate Edgeworth series random variables. In Figures 44.2 and 44.3, taken from Kocherlakota, Kocherlakota, and Balakrishnan (1986), bivariate Edgeworth series densities with $\lambda_{3U} = \lambda_{3V} = \lambda_3$ and $\lambda_{4U} = \lambda_{4V} = \lambda_4$ are presented for $\rho = 0.1$ and $\rho = 0.8$, respectively, each for four cases: (i) $\lambda_3 = -0.4$, $\lambda_4 = 0.8$, (ii) $\lambda_3 = 0.4$, $\lambda_4 = 0.8$, (iii) $\lambda_3 = -0.4$, $\lambda_4 = 1.6$, and (iv) $\lambda_3 = 0.4$, $\lambda_4 = 1.6$. The corresponding bivariate normal density functions are presented in Figures 44.4 and 44.5; see also Chapter 46 for some more figures.

A look at these figures readily reveals that the bivariate Edgeworth series distribution is unimodal in all the cases considered and also remains quite similar to the corresponding bivariate normal distribution. For this reason, the bivariate Edgeworth series distribution has been used in robustness studies; see, for example, Kocherlakota, Kocherlakota, and Balakrishnan (1985).

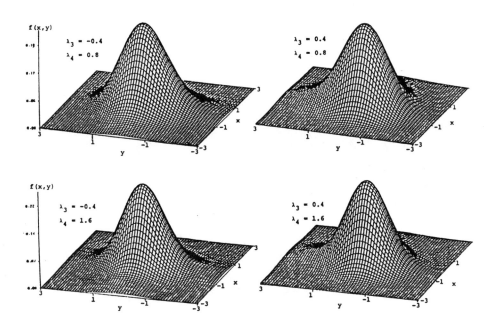

FIGURE 44.2
Bivariate Edgeworth Series Density Function when $\rho = 0.1$.

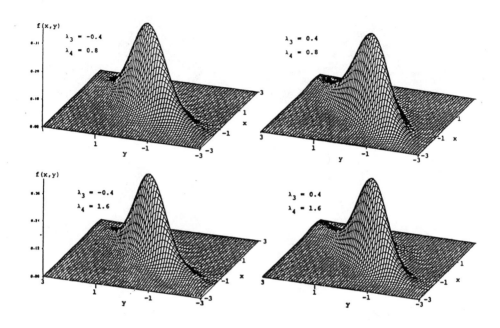

FIGURE 44.3
Bivariate Edgeworth Series Density Function when $\rho = 0.8$.

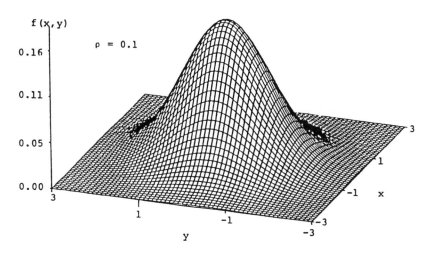

FIGURE 44.4
Bivariate Normal Density Function when $\rho = 0.1$.

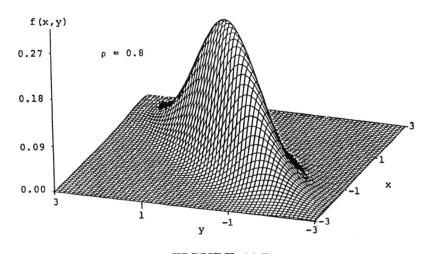

FIGURE 44.5
Bivariate Normal Density Function when $\rho = 0.8$.

Recall that Box and Muller's (1958) method [see Chapter 13 of Johnson, Kotz, and Balakrishnan (1994)] of generating standard normal variables is based on the relations

$$X_1 = \sqrt{-2\log U_1}\cos(2\pi U_2) \quad \text{and} \quad X_2 = \sqrt{-2\log U_1}\sin(2\pi U_2),$$

where U_1 and U_2 are independent Uniform$(0,1)$ random variables. Of course, X_1 and X_2 are independent standard normal variables. Thus, to generate a pair of independent standard normal variables, one needs to (i) randomly generate the ordinate of the density and solve for equidensity contour associated with that ordinate and (ii) randomly generate a point on that contour.

Troutt (1993) generalized this approach by showing that if $\boldsymbol{X} = (X_1, \ldots, X_k)^T$ has density function $p(\boldsymbol{x})$ and the univariate random variable $V = p(\boldsymbol{X})$ possesses density $g(v)$ with $A(v)$ being the Lebesgue measure of the set

$$S(v) = \{\boldsymbol{x} \in \mathbb{R}^k : p(\boldsymbol{x}) \geq v\}$$

and $a'(v)$ existing, then $g(v) = -vA'(v)$. Kotz and Troutt (1996) applied this result to obtain the so-called *vertical density representation* $g(v)$ of several univariate distributions. Kotz, Fang and Liang (1997) generalized this result for spherically symmetric multivariate distributions with density $p(\boldsymbol{x}) = h(\boldsymbol{x}^T\boldsymbol{x})$, $\boldsymbol{x} \in \mathbb{R}^k$, when $h(\cdot)$ is strictly decreasing and differentiable and $V = p(\boldsymbol{x})$ possesses density function $g(v)$. With

$$S(v) = \{\boldsymbol{x} \in \mathbb{R}^k : \boldsymbol{x}^T\boldsymbol{x} \leq h^{-1}(v)\},$$

where $h^{-1}(\cdot)$ is the inverse function of $h(\cdot)$, and $A(v)$ being the Lebesgue measure of $S(v)$, the modified vertical density representation involves $W = V/\max V$. For the multivariate normal distribution, for example, we have

$$p(w) = \frac{2}{(2\pi)^{k/2}\Gamma(k/2)}(-2\log w)^{\frac{k}{2}-1}, \qquad 0 < w < 1,$$

so that $Z = -2\log W$ has a chi-square distribution with k degrees of freedom.

11   FRÉCHET BOUNDS

Fréchet (1951) noted that since

$$\Pr[(X_1 \leq x_1) \cap (X_2 \leq x_2)] \leq \min\{\Pr[X_1 \leq x_1], \ \Pr[X_2 \leq x_2]\},$$

the relationship

$$F_{X_1,X_2}(x_1, x_2) \leq \min[F_{X_1}(x_1), F_{X_2}(x_2)] \tag{44.54}$$

must hold for all pairs of random variables X_1 and X_2, and for all x_1, x_2.

In a similar way, since

$$\Pr[(X_1 > x_1) \cup (X_2 > x_2)] \le \Pr[X_1 > x_1] + \Pr[X_2 > x_2],$$

it follows that

$$1 - F_{X_1,X_2}(x_1, x_2) \le \{1 - F_{X_1}(x_1)\} + \{1 - F_{X_2}(x_2)\},$$

that is

$$F_{X_1,X_2}(x_1, x_2) \ge F_{X_1}(x_1) + F_{X_2}(x_2) - 1. \tag{44.55}$$

Warmuth (1988) extended these bounds for k-dimensional distributions as follows. Let $m < k$ be an integer, and let

$$
\begin{aligned}
I &= (i_1, i_2, \ldots, i_m) \text{ with } i_1 < i_2 < \cdots < i_m, \\
\boldsymbol{x}_I &= (x_{i_1}, x_{i_2}, \ldots, x_{i_m})^T, \\
\boldsymbol{X}_I &= (X_{i_1}, X_{i_2}, \ldots, X_{i_m})^T,
\end{aligned}
$$

and J_m^k be the set of all ordered m-tuples I with $i_j \in \{1, 2, \ldots, k\}$ for $j = 1, 2, \ldots, m$. Then, since

$$F_{X_1,X_2,\ldots,X_k}(x_1, x_2, \ldots, x_k) \le \min\{F_{X_1}(x_1), F_{X_2}(x_2), \ldots, F_{X_k}(x_k)\}, \tag{44.56}$$

it immediately follows that

$$
\begin{aligned}
&F_{X_1,X_2,\ldots,X_k}(x_1, x_2, \ldots, x_k) \\
&\le \min_{I \in J_m^k} F_{\boldsymbol{X}_I}(\boldsymbol{x}_I) \equiv F_{X_1,X_2,\ldots,X_k}^{U(m)}(x_1, x_2, \ldots, x_k). \tag{44.57}
\end{aligned}
$$

Warmuth (1988) has pointed out that $F_{X_1,X_2,\ldots,X_k}^{U(m)}$ is a k-dimensional distribution function with m-dimensional marginal distribution $F_{\boldsymbol{X}_I}$, $I \in J_m^k$, and that it serves as an upper bound on $F_{X_1,X_2,\ldots,X_k}(x_1, x_2, \ldots, x_k)$. Moreover, the m-dimensional marginal distributions are the same as those of $F_{X_1,X_2,\ldots,X_k}$.

As to the lower bound, Warmuth (1988) has shown that

$$
\begin{aligned}
&F_{X_1,X_2,\ldots,X_k}(x_1, x_2, \ldots, x_k) \\
&\ge \max\{F_{X_1}(x_1) + \cdots + F_{X_k}(x_k) - k + 1, 0\}, \tag{44.58}
\end{aligned}
$$

but did not mention that the right-hand side of (44.58) is not necessarily a cumulative distribution function for $k > 2$; see Kemp (1973) and also the discussion below. Also, Warmuth (1988) has observed that for $k = 2r$

$$F_{X_1,\ldots,X_k}(x_1,\ldots,x_k)$$
$$\geq \max\left\{\sum_{i_1=1}^{2r} F_{X_{i_1}}(x_{i_1}) - \sum_{1\leq i_1<i_2\leq 2r} F_{X_{i_1},X_{i_2}}(x_{i_1},x_{i_2}) + \cdots - 1, 0\right\},$$

(44.59)

and for $k = 2r+1$

$$F_{X_1,\ldots,X_k}(x_1,\ldots,x_k)$$
$$\geq \max\left\{1 - \sum_{i_1=1}^{2r+1} F_{X_{i_1}}(x_{i_1}) + \sum_{1\leq i_1<i_2\leq 2r+1} F_{X_{i_1},X_{i_2}}(x_{i_1},x_{i_2})\right.$$
$$- \cdots + \sum_{1\leq i_1<i_2<\cdots<i_{2r}\leq 2r+1} F_{X_{i_1},\ldots,X_{i_{2r}}}(x_{i_1},\ldots,x_{i_{2r}})$$
$$\left. - \min_{i=1,\ldots,2r+1} \bar{F}_{X_1,\ldots,X_{i-1},X_{i+1},\ldots,X_{2r+1}}(x_1,\ldots,x_{i-1},x_{i+1},\ldots,x_{2r+1}), 0\right\}.$$

(44.60)

Denoting the bounds in (44.59) and (44.60) by

$$F^{L(2r-1)}_{X_1,\ldots,X_{2r}}(x_1,\ldots,x_{2r}) \text{ and } F^{L(2r)}_{X_1,\ldots,X_{2r+1}}(x_1,\ldots,x_{2r+1}),$$

respectively, and iterating these bounds, one gets the lower bound $F^{L(m)}_{X_1,\ldots,X_k}$ for all $m < n$. Warmuth (1988) has claimed that this $F^{L(m)}_{X_1,\ldots,X_k}$ is a k-dimensional distribution function with m-dimensional marginal distributions $F_{\mathbf{X}_I}$, $I \in J_m^k$, and has referred to $F^{U(m)}_{X_1,\ldots,X_k}$ and $F^{L(m)}_{X_1,\ldots,X_k}$ as *marginal Fréchet bounds*. There is evidently a lacuna in Warmuth's proof. Note that these bounds involve singular distribution functions and the surfaces described by them contain hyperplanes parallel to marginal spaces.

The upper Fréchet bound in (44.56) for the class of all k-dimensional distributions, denoted by $F^+(x_1, x_2, \ldots, x_k)$, is a k-variate maximal distribution that always exists; see Dall'Aglio (1960) and Kemp (1973); also, all correlations of this maximal distribution are maximal.

The lower Fréchet bound in (44.58), denoted by $F^-(x_1, x_2, \ldots, x_k)$, does not always define a distribution function; see, for example, Cuadras (1981) and Tiit (1984). Dall'Aglio (1960, 1991) presented necessary and sufficient conditions on the marginal distributions $F_{X_i}(x_i)$ so that

$F^-(x_1, x_2, \ldots, x_k)$ would be a distribution function. These are given by

$$F_{X_1}(a_1+) + \cdots + F_{X_k}(a_k+) \geq k - 1 \text{ or } F_{X_1}(b_1) + \cdots + F_{X_k}(b_k) \leq 1,$$

$$(44.61)$$

where

$$a_i = \inf\{x : F_{X_i}(x) > 0\} \text{ and } b_i = \sup\{x : F_{X_i}(x) < 1\}.$$

Also, if the minimal distribution exists, then it is unique and all the correlations are minimal; see Rüschendorf (1983). Helemäe and Tiit (1996) have shown that the multivariate minimal distribution having equal marginals is exchangeable, and the minimal possible value of the correlation coefficient is $r^- = -1/(k-1)$; see also Kotz and Tiit (1992) and Shaked and Tong (1991). Moreover, the minimal distribution is degenerate in the $(k-1)$-variate space $\mathbb{R}^{k-1}$; additionally, for equal and symmetric marginals only a bivariate minimal distribution exists. Hence, it follows that for the majority of commonly used univariate marginal distributions the multivariate minimal distribution does not exist.

12 FRÉCHET, PLACKETT, AND MARDIA'S SYSTEMS

Fréchet (1951) suggested that any system of bivariate distributions with specified marginal distributions $F_{X_1}(x_1), F_{X_2}(x_2)$ should include the limits in (44.54) and (44.55) as limiting cases. In particular, he suggested the system

$$\begin{aligned}
F_{X_1,X_2}(x_1, x_2) &= \theta \max\{F_{X_1}(x_1) + F_{X_2}(x_2) - 1, 0\} \\
&\quad + (1-\theta) \min\{F_{X_1}(x_1), F_{X_2}(x_2)\}, \quad 0 \leq \theta \leq 1.
\end{aligned}$$

$$(44.62)$$

This system does not, however, include the case when X_1 and X_2 are independent. A system that does include this case, but not the limits in (44.54) and (44.55), is the Farlie–Gumbel–Morgenstern system of distributions that is described in the next section.

Mardia (1970a) [see also Nataf (1962)] pointed out that there is a simple way of constructing systems that include the limits in (44.54) and (44.55) and also the case of independence. This is done by finding the transformations $Y_j = g_j(X_j)$, $j = 1, 2$, which make Y_1, Y_2 standard normal variables (as in Section 5), and then ascribing a joint bivariate normal distribution to Y_1 and Y_2. [If X_1 and X_2 are each continuous random variables, there is always such a pair of transformations, defined by $F_{X_j}(x_j) = \Phi(g(x_j))$, $j = 1, 2$.] It is not necessary that the transformation be to bivariate normality. Many other standard joint distributions may be used for this purpose, each one will give rise to a different system of distributions.

Plackett (1965) constructed another such system that has some intrinsic interest, though it is more complicated than Mardia's system. The joint cumulative distribution function $F_{X_1,X_2}(x_1, x_2)$ is required to satisfy the equation

$$\psi = \frac{F_{X_1,X_2}(x_1, x_2)\{1 - F_{X_1}(x_1) - F_{X_2}(x_2) + F_{X_1,X_2}(x_1, x_2)\}}{\{F_{X_1}(x_1) - F_{X_1,X_2}(x_1, x_2)\}\{F_{X_2}(x_2) - F_{X_1,X_2}(x_1, x_2)\}} \quad (44.63)$$

with $\psi(x_1, x_2) > 0$. For different values of ψ, different members of Plackett's system are obtained. For example, if $\psi = 1$ in (44.63), then $F_{X_1,X_2}(x_1, x_2) = F_{X_1}(x_1)F_{X_2}(x_2)$ and so X_1 and X_2 are independent. If $\psi = 0$ in (44.63), then $F_{X_1,X_2}(x_1, x_2)$ equals the lower limit in (44.55); if $\psi = \infty$ in (44.63), then $F_{X_1,X_2}(x_1, x_2)$ equals the upper limit in (44.54). In general, there is just one value of $F_{X_1,X_2}(x_1, x_2)$, between the lower and upper limits, which satisfies (44.63). This may be seen by noting that, for fixed $F_{X_1}(x_1)$ and $F_{X_2}(x_2)$, the right-hand side of (44.63) is an increasing function of $F_{X_1,X_2}(x_1, x_2)$, increasing from 0 to ∞ as $F_{X_1,X_2}(x_1, x_2)$ increases from the lower limit in (44.55) to the upper limit in (44.54). Furthermore, as $F_{X_1}(x_1)$ increases (with $F_{X_2}(x_2)$ remaining fixed), $F_{X_1,X_2}(x_1, x_2)$ increases; similarly, as $F_{X_2}(x_2)$ increases (with $F_{X_1}(x_1)$ remaining fixed), $F_{X_1,X_2}(x_1, x_2)$ increases. Note also that when x_1, x_2 take on the corresponding median values, so that $F_{X_1}(x_1) = F_{X_2}(x_2) = \frac{1}{2}$, then $F_{X_1,X_2}(x_1, x_2) = \frac{\sqrt{\psi}}{2(1+\sqrt{\psi})}$.

The conditional cumulative distribution function of X_1, given $X_2 = x_2$, is

$$\Pr[X_1 \leq x_1 \mid X_2 = x_2]$$
$$= \frac{\psi F_{X_1}(x_1) - (\psi - 1)F_{X_1,X_2}(x_1, x_2)}{1 + (\psi - 1)[F_{X_1}(x_1) + F_{X_2}(x_2) - 2F_{X_1,X_2}(x_1, x_2)]} . \quad (44.64)$$

The median regression $X_{1,0.5}(x_2)$ of X_1 on X_2 is obtained by equating this to $\frac{1}{2}$, leading to

$$(\psi + 1)F_{X_1}(X_{1,0.5}(x_2)) = 1 + (\psi - 1)F_{X_2}(x_2).$$

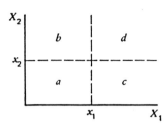

FIGURE 44.6

Note that as $F_{X_2}(x_2) \to 0$ [so that $F_{X_1,X_2}(x_1, x_2) \to 0$ also],

$$\Pr[X_1 \le x_1 \mid X_2 = x_2] \to \frac{\psi F_{X_1}(x_1)}{1 + (\psi - 1) F_{X_1}(x_1)} , \qquad (44.65)$$

that is, there is a nondegenerate limiting conditional distribution. Similarly, as $F_{X_2}(x_2) \to 1$ [and so $F_{X_1,X_2}(x_1, x_2) \to F_{X_1}(x_1)$], we obtain

$$\Pr[X_1 \le x_1 \mid X_2 = x_2] \to \frac{F_{X_1}(x_1)}{1 + (\psi - 1)\{1 - F_{X_1}(x_1)\}} . \qquad (44.66)$$

If $F_{X_j}(x_j)$ is taken to be equal to $\Phi(x_j)$ for $j = 1, 2$, we have a bivariate distribution with standard normal marginal distributions which differs from the standardized bivariate normal distribution. Plackett (1965) has provided numerical comparisons of the two distributions. He has also discussed estimation of ψ. It is clear from (44.63) that, for any double dichotomy as in Figure 44.6, $\tilde{\psi} = ad/bc$ (where $a, b, c,$ and d are the observed frequencies in the cells indicated) should give a good estimator of ψ. The variance of $\tilde{\psi}$ may be estimated as

$$\tilde{\psi}^2(a^{-1} + b^{-1} + c^{-1} + d^{-1}).$$

Note that these formulas do not require a knowledge of $F_{X_1}(x_1)$ and $F_{X_2}(x_2)$. The functions can be estimated separately from the observed marginal distributions.

By writing the joint cumulative distribution function $F_{X_1,X_2}(x_1, x_2)$ from (44.63) as

$$
\begin{aligned}
&F_{X_1,X_2}(x_1, x_2) \\
&= \frac{S_{12}(x_1, x_2) - \{S_{12}^2(x_1, x_2) - 4\psi(\psi - 1)F_{X_1}(x_1)F_{X_2}(x_2)\}^{1/2}}{2(\psi - 1)} , \quad \psi > 0,
\end{aligned}
$$

where $S_{12}(x_1, x_2) = 1 + (\psi - 1)(F_{X_1}(x_1) + F_{X_2}(x_2))$, we have the associated copula as

$$C_{12}(u, v) = \frac{S(u, v) - \{S^2(u, v) - 4\psi(\psi - 1)uv\}^{1/2}}{2(\psi - 1)},$$

where $S(u, v) = 1 + (\psi - 1)(u + v)$, $0 \le u \le 1$, $0 \le v \le 1$. According to Sklar (1959), there then exists a copula C_{12,x_3} such that the trivariate cumulative distribution function is

$$F_{X_1,X_2,X_3}(x_1, x_2, x_3)$$
$$= C_{12,x_3}\left(\frac{C_{13}(F_{X_1}(x_1), F_{X_3}(x_3))}{F_{X_3}(x_3)}, \frac{C_{23}(F_{X_2}(x_2), F_{X_3}(x_3))}{F_{X_3}(x_3)}\right) F_{X_3}(x_3);$$

using this form, Chakak and Koehler (1995) have presented the trivariate Plackett distribution as

$$F_{X_1,X_2,X_3}(x_1, x_2, x_3)$$
$$= \frac{S_{12,x_3}(x_1, x_2, x_3)}{2(\psi_{12,x_3} - 1)}$$
$$- \frac{\{S_{12,x_3}^2(x_1, x_2, x_3) - 4\psi_{12,x_3}(\psi_{12,x_3} - 1)F_{X_1,X_3}(x_1, x_3)F_{X_2,X_3}(x_2, x_3)\}^{1/2}}{2(\psi_{12,x_3} - 1)},$$

where $X_{12,x_3} = F_{X_3}(x_3) + (\psi_{12,x_3} - 1)(F_{X_1,X_3}(x_1, x_3) + F_{X_2,X_3}(x_2, x_3))$. This trivariate distribution has all its three bivariate marginal distributions to be of Plackett type. In particular, when the association parameter ψ_{12,x_3} is a constant (equaling its limit $\psi = \psi_{12} = \lim_{x_3 \to \infty} \psi_{12,x_3}$), then the above trivariate Plackett distribution reduces to

$$F_{X_1,X_2,X_3}(x_1, x_2, x_3)$$
$$= \frac{S_{12.3}(x_1, x_2, x_3)}{2(\psi - 1)}$$
$$- \frac{\{S_{12.3}^2(x_1, x_2, x_3) - 4\psi(\psi - 1)F_{X_1,X_3}(x_1, x_3)F_{X_2,X_3}(x_2, x_3)\}^{1/2}}{2(\psi - 1)},$$

where $S_{12.3}(x_1, x_2, x_3) = F_{X_3}(x_3) + (\psi - 1)(F_{X_1,X_3}(x_1, x_3) + F_{X_2,X_3}(x_2, x_3))$. This trivariate distribution with bivariate Plackett marginal distributions $F_{X_1,X_2}(x_1, x_2)$, $F_{X_1,X_3}(x_1, x_3)$ and $F_{X_2,X_3}(x_2, x_3)$ can also be obtained from conditional odds ratios. As suggested by Conway (1979), the trivariate Plackett distribution should be based on odds ratios for a $2 \times 2 \times 2$ contingency table. For any fixed values of X_3, constraining the conditional odds ratio ψ_{12,x_3} to be constant for any choice of (x_1, x_2) and then expressing the various probabilities as functions of cumulative distribution functions, we obtain

$$
\begin{aligned}
\psi_{12,x_3} &= [F_{X_1,X_2,X_3}(x_1,x_2,x_3)\{F_{X_3}(x_3) - F_{X_1,X_3}(x_1,x_3) \\
&\quad - F_{X_2,X_3}(x_2,x_3) + F_{X_1,X_2,X_3}(x_1,x_2,x_3)\}] \\
&\quad \times [\{F_{X_1,X_3}(x_1,x_3) - F_{X_1,X_2,X_3}(x_1,x_2,x_3)\} \\
&\qquad \times \{F_{X_2,X_3}(x_2,x_3) - F_{X_1,X_2,X_3}(x_1,x_2,x_3)\}]^{-1}.
\end{aligned}
$$

Setting now $\psi_{12,x_3} \equiv \psi$ and solving for $F_{X_1,X_2,X_3}(x_1,x_2,x_3)$, we obtain the above given trivariate Plackett distribution. From it, we readily obtain the following special cases:

(i) If $\psi_{12,x_3} = \psi_{13} = 1$, then $F_{X_1,X_2,X_3}(x_1,x_2,x_3) = F_{X_1}(x_1)F_{X_2,X_3}(x_2,x_3)$;

(ii) If $\psi_{12,x_3} = \psi_{23} = 1$, then $F_{X_1,X_2,X_3}(x_1,x_2,x_3) = F_{X_2}(x_2)F_{X_1,X_3}(x_1,x_3)$;

(iii) If $\psi_{13} = \psi_{23} = 1$, then $F_{X_1,X_2,X_3}(x_1,x_2,x_3) = F_{X_3}(x_3)F_{X_1,X_2}(x_1,x_2)$;

(iv) If $\psi_{12,x_3} = \psi_{13} = \psi_{23} = 1$, then $F_{X_1,X_2,X_3}(x_1,x_2,x_3) = F_{X_1}(x_1)F_{X_2}(x_2)$ $F_{X_3}(x_3)$ in which case the univariate marginal distribution functions are all independent;

(v) If $\psi_{12,x_3} = 1$, then $F_{X_1,X_2,X_3}(x_1,x_2,x_3) = \frac{F_{X_1,X_3}(x_1,x_3)F_{X_2,X_3}(x_2,x_3)}{F_{X_3}(x_3)}$ in which case F_{X_1} and F_{X_2} are conditionally independent for any event of the form $(X_3 \leq x_3)$.

Molenberghs and Lesaffre (1994) have presented another construction of the multivariate Plackett distribution. Their distribution is defined using the set of $2^k - 1$ generalized cross-ratios with values in $[0, \infty]$: $\psi_i(1 \leq i \leq k), \psi_{ij}(1 \leq i < j \leq k), \ldots, \psi_{i_1 i_2 \ldots i_\ell}(1 \leq i_1 < i_2 < \cdots < i_\ell \leq k), \ldots, \psi_{12\ldots k}$. The k-dimensional probabilities can be computed if all lower-dimensional probabilities together with the global cross-ratio of dimension k are known. Though the set of $2^k - 1$ generalized cross-ratios fully specifies the k-dimensional Plackett distribution, the existence and uniqueness of such a distribution are not guaranteed. Even if they were, the calculation of the distribution is not clear in higher dimensions ($k \geq 4$) as they are specified only implicitly.

13 FARLIE–GUMBEL–MORGENSTERN DISTRIBUTIONS

As noted in the last section, although the Fréchet class of distributions in (44.62) includes the lower and upper limits in (44.55) and (44.54), it does not include the case when X_1 and X_2 are independent. Morgenstern (1956) defined the class of distributions

$$
\begin{aligned}
F_{X_1,X_2}(x_1, x_2) \\
= \; F_{X_1}(x_1)F_{X_2}(x_2)[1 + \alpha\{1 - F_{X_1}(x_1)\}\{1 - F_{X_2}(x_2)\}], \; |\alpha| < 1,
\end{aligned}
$$

$$(44.67)$$

which does include the case of independence, but does not include the lower and upper limits in (44.55) and (44.54). This class of distributions was extended by Farlie (1960) to the general form

$$
F_{X_1,X_2}(x_1, x_2) = F_{X_1}(x_1)F_{X_2}(x_2)\{1 + \alpha g_1(x_1)g_2(x_2)\}, \qquad (44.68)
$$

where $g_1(x_1)$ and $g_2(x_2)$ are more general functions than $1 - F_{X_1}(x_1)$ and $1 - F_{X_2}(x_2)$.

The class of distributions in (44.67), proposed by Morgenstern (1956) and extended by Farlie (1960) to the form in (44.68), is nowadays known as the *Farlie–Gumbel–Morgenstern* (FGM) class of distributions. This class of distributions, having a simple natural form with given univariate marginals, was further generalized to include distributions with a stronger correlation structure; see, for example, Johnson and Kotz (1975, 1977). More recent discussions on this family of distributions are due to Lin (1987), Kotz and Seeger (1993), Cambanis (1993), and Huang and Kotz (1999).

A k-dimensional FGM distribution can be defined in a manner analogous to (44.67) as

$$
\begin{aligned}
F_{X_1,\ldots,X_k}(x_1, \ldots, x_k) \; = \; \prod_{i_1=1}^{k} F_{X_{i_1}}(x_i)\Bigg[1 + \sum_{1 \le i_1 < i_2 \le k} a_{i_1 i_2} \\
\times \left\{1 - F_{X_{i_1}}(x_{i_1})\right\}\left\{1 - F_{X_{i_2}}(x_{i_2})\right\}\Bigg]
\end{aligned}
$$

$$(44.69)$$

for all vectors $\boldsymbol{x} = (x_1, \ldots, x_k)^T \in \mathbb{R}^k$, where the $\binom{n}{2}$ coefficients $a_{i_1 i_2}$ are suitable constants so that $F_{X_1,\ldots,X_k}(x_1, \ldots, x_k)$ in (44.69) is a distribution function. The univariate marginals of F are the given F_{X_i}. The constants $a_{i_1 i_2}$ are admissible if the following 2^k inequalities hold:

$$
1 + \sum_{1 \le i_1 < i_2 \le k} \varepsilon_{i_1}\varepsilon_{i_2}a_{i_1 i_2} \ge 0 \qquad (44.70)
$$

for all $\varepsilon_i = -M_i$ or $1 - m_i$, where M_i and m_i are the supremum and the infimum of the set

$$
\{F_{X_i}(x), \; -\infty < x < \infty\} \setminus \{0, 1\}.
$$

If F_{X_i} is absolutely continuous, we have $M_i = 1$ and $m_i = 0$, and hence $\varepsilon = \pm 1$. Then the inequalities in (44.70) imply that the coefficients are all bounded; for example,

$$|a_{i_1, i_2}| \leq \frac{1}{[\min\{M_{i_1}, M_{i_2}, 1 - m_{i_1}, 1 - m_{i_2}\}]^2} , \tag{44.71}$$

which follows immediately from the bivariate distributions. We assume that the marginal distributions F_{X_i} are nondegenerate with $\inf_{i \geq 1} M_i > 0$. Observe that the multivariate distributions are determined by the bivariate marginals (by the coefficients $a_{i_1 i_2}$ and the univariate distributions F_{X_i}) and that their ℓ-dimensional marginals are also of the same type.

Hüsler (1996) investigated the extreme values from the multivariate FGM class and showed that they behave as if no dependence exists between its components.

Various other forms of multivariate FGM distributions are also available in the literature. For example, Cambanis (1977) defined a *general system structure* as

$$
\begin{aligned}
F_{X_1, \ldots, X_k}(x_1, \ldots, x_k) = {} & \prod_{i_1=1}^{k} F(x_{i_1}) \Big[1 + \sum_{i_1=1}^{k} a_{i_1} \{1 - F(x_{i_1})\} \\
& + \sum_{1 \leq i_1 < i_2 \leq k} a_{i_1 i_2} \{1 - F(x_{i_1})\}\{1 - F(x_{i_2})\} \\
& + \cdots \\
& + a_{1\, 2 \cdots k} \prod_{i=1}^{k} \{1 - F(x_i)\} \Big].
\end{aligned}
\tag{44.72}
$$

see Peristiani (1991). Of course, a more general form of (44.72) can be obtained by using F_i ($i = 1, 2, \ldots, k$) instead of a common F. Johnson and Kotz (1975, 1977) considered such a general system, but with the special choice of $a_i = 0$ (for $i = 1, 2, \ldots, k$) because in this case the univariate marginal distributions are equal to the given distributions F_i. So, their k-dimensional FGM distribution has the form

$$
\begin{aligned}
& F_{X_1, \ldots, X_k}(x_1, \ldots, x_k) \\
& = \prod_{i_1=1}^{k} F_{X_{i_1}}(x_{i_1}) \Big[1 + \sum_{1 \leq i_1 < i_2 \leq k} a_{i_1 i_2} \{1 - F_{X_{i_1}}(x_{i_1})\}\{1 - F_{X_{i_2}}(x_{i_2})\} \\
& \quad + \cdots + a_{1\, 2 \cdots k} \prod_{i=1}^{k} \{1 - F_{X_i}(x_i)\} \Big].
\end{aligned}
\tag{44.73}
$$

The coefficients $a_{i_1 \ldots i_\ell}$ are all real numbers with constraints on them in order to ensure that $F_{X_1, \ldots, X_k}(x_1, \ldots, x_k)$ in (44.73) is a nondecreasing function in each of $x_1, \ldots, x_k$. These constraints are

$$1 + \sum \sum_{1 \leq i_1 < i_2 \leq k} \varepsilon_{i_1} \varepsilon_{i_2} a_{i_1 i_2} + \cdots + \varepsilon_1 \varepsilon_2 \cdots \varepsilon_k \, a_{1\,2\ldots k} \geq 0, \qquad (44.74)$$

where $\varepsilon_i = \pm 1$. For this form of the multivariate FGM distribution, Hashorva and Hüsler (1999) have shown that the extreme values behave as if no dependence exists between its components.

If the marginal distributions are not absolutely continuous, the constraints in (44.74) are not a necessary set of conditions for $F_{X_1, \ldots, X_k}(x_1, \ldots, x_k)$ in (44.73) to be a distribution function. We shall illustrate this with the following well-known example. Let

$$F_{X_1}(x) = F_{X_2}(x) = \begin{cases} 0 & \text{if } x < 0 \\ p & \text{if } 0 \leq x < 1 \\ 1 & \text{if } x \geq 1 \end{cases}$$

with $0 < p < 1$. Note that, for $i = 1, 2$,

$$F_{X_i}(x)\{1 - F_{X_i}(x)\} = \begin{cases} p(1-p) & \text{if } 0 \leq x < 1 \\ 0 & \text{otherwise .} \end{cases}$$

Then, with $a_{12} = \frac{1}{p(1-p)}$, the distribution in (44.73) corresponds to the bivariate distribution

$$\Pr[X_1 = X_2 = 0] = p \quad \text{and} \quad \Pr[X_1 = X_2 = 1] = 1 - p$$

even though $a_{12} > 1$.

Observe that if $a_{i_1 i_2} = a_{(2)}$, $a_{i_1 i_2 i_3} = a_{(3)}, \ldots, a_{1\,2\ldots k} = a_{(k)}$ (i.e., all coefficients with ℓ subscripts have a common value $a_{(\ell)}$ for $\ell = 2, 3, \ldots, k$), then the constraints in (44.74) become

$$\prod_{i=1}^{k}(1 + \varepsilon_i a) \geq 0$$

where a^ℓ is to be interpreted as $a_{(\ell)}$ and $a_{(1)} = 0$.

If in the definition of $F_{X_1, \ldots, X_k}(x_1, \ldots, x_k)$ the term $\sum_{i=1}^{k} a_i\{1 - F_{X_i}(x_i)\}$ had been retained, then the univariate marginal distribution of X_i would not have been $F_{X_i}(x_i)$, but rather

$$F_{X_i}(x_i)[1 + a_i\{1 - F_{X_i}(x_i)\}]. \qquad (44.75)$$

Note, however, that any distribution $G(x)$ can always be written as $G(x) = F(x)[1 + b\{1 - F(x)\}]$ for some distribution $F(x)$ and some real number b.

In fact, given any distribution $G(x)$ and any real b, there is a distribution $F(x)$ satisfying the above relation, given by

$$F(x) = \frac{1 + b - \sqrt{(1+b)^2 - 4bG(x)}}{2b}.$$

If the relevant densities exist, we have the joint density function of the k-dimensional FGM distribution in (44.73) to be

$$
\begin{aligned}
& f_{X_1,\ldots,X_k}(x_1,\ldots,x_k) \\
&= \prod_{i_1=1}^{k} f_{X_{i_1}}(x_{i_1}) \Bigg[1 + \sum_{1 \le i_1 < i_2 \le k} a_{i_1 i_2} \{1 - 2F_{X_{i_1}}(x_{i_1})\}\{1 - 2F_{X_{i_2}}(x_{i_2})\} \\
&\quad + \cdots + a_{1\,2\,\cdots\,k} \prod_{i=1}^{k} \{1 - 2F_{X_i}(x_i)\} \Bigg].
\end{aligned}
\tag{44.76}
$$

Note from (44.76) that if $x_i = \mathrm{Median}(X_i) = x_i^*$ (say) for all $i = 1, 2, \ldots, k$, then

$$f_{X_1,\ldots,X_k}(x_1^*,\ldots,x_k^*) = \prod_{i=1}^{k} f_{X_i}(x_i^*) \tag{44.77}$$

for all values of the coefficients a's.

Multivariate FGM distributions can be defined in terms of survival functions [instead of distribution functions as in (44.73)] as

$$
\begin{aligned}
& \bar{F}_{X_1,\ldots,X_k}(x_1,\ldots,x_k) \\
&= \Pr[X_1 > x_1,\ldots,X_k > x_k] \\
&= \prod_{i=1}^{k} \bar{F}_{X_i}(x_i) \Bigg[1 + \sum_{1 \le i_1 < i_2 \le k} a_{i_1 i_2} \{1 - \bar{F}_{X_{i_1}}(x_{i_1})\}\{1 - \bar{F}_{X_{i_2}}(x_{i_2})\} \\
&\quad - \sum_{1 \le i_1 < i_2 < i_3 \le k} a_{i_1 i_2 i_3} \{1 - \bar{F}_{X_{i_1}}(x_{i_1})\}\{1 - \bar{F}_{X_{i_2}}(x_{i_2})\}\{1 - \bar{F}_{X_{i_3}}(x_{i_3})\} \\
&\quad + \cdots + (-1)^k a_{1\,2\,\cdots\,k} \prod_{i=1}^{k} \{1 - \bar{F}_{X_i}(x_i)\} \Bigg].
\end{aligned}
\tag{44.78}
$$

Note that the signs of successive orders of a-terms alternate. Then, the following two conditional distributions ought to be distinguished:

(i) $\displaystyle \bar{F}_{X_1}\left(x_1 \mid \bigcap_{i=2}^{k}(X_i > x_i) \right)$

$$= \Pr\left[X_1 > x_1 \mid \bigcap_{i=2}^{k}(X_i > x_i) \right]$$

$$= \bar{F}_{X_1}(x_1)\Bigg[1 + \sum_{1 \leq i_1 < i_2 \leq k} a_{i_1 i_2} F_{X_{i_1}}(x_{i_1}) F_{X_{i_2}}(x_{i_2})$$

$$- \sum_{1 \leq i_1 < i_2 < i_3 \leq k} a_{i_1 i_2 i_3} F_{X_{i_1}}(x_{i_1}) F_{X_{i_2}}(x_{i_2}) F_{X_{i_3}}(x_{i_3})$$

$$+ \cdots + (-1)^k a_{12\cdots k} \prod_{i=1}^{k} F_{X_i}(x_i)\Bigg] \Big/ B^{(1)}, \qquad (44.79)$$

where $B^{(1)}$ is obtained from the expression in square brackets on the right-hand side of (44.79) by setting each a, which has 1 among its subscripts, equal to 0. Of course, in the case $k = 2$, we have

$$\bar{F}_{X_1}(x_1 \mid X_2 > x_2) = \Pr[X_1 > x_1 \mid X_2 > x_2]$$
$$= \bar{F}_{X_1}(x_1)[1 + a_{12} F_{X_1}(x_1) F_{X_2}(x_2)].$$

(ii) $F_{X_1}\left(x_1 \mid \bigcap_{i=2}^{k}(X_i \leq x_i)\right)$

$$= \Pr\left[X_1 \leq x_1 \mid \bigcap_{i=2}^{k}(X_i \leq x_i)\right]$$

$$= F_{X_1}(x_1)\Bigg[1 + \sum_{1 \leq i_1 < i_2 \leq k} a_{i_1 i_2} \bar{F}_{X_{i_1}}(x_{i_1}) \bar{F}_{X_{i_2}}(x_{i_2})$$

$$+ \cdots + a_{12\cdots k} \prod_{i=1}^{k} \bar{F}_{X_i}(x_i)\Bigg] \Big/ C^{(1)}, \qquad (44.80)$$

where $C^{(1)}$ is obtained from the expression in square brackets on the right hand side of (44.80) by setting each a, which has 1 among its subscripts, equal to 0. Of course, in the case $k = 2$, we have

$$F_{X_1}(x_1 \mid X_2 \leq x_2) = \Pr[X_1 \leq x_1 \mid X_2 \leq x_2]$$
$$= F_{X_1}(x_1)[1 + a_{12} \bar{F}_{X_1}(x_1) \bar{F}_{X_2}(x_2)].$$

Note that, in general, (44.79) and (44.80) are different distributions.

For the case when the marginal F_{X_i}'s are all the same but the coefficients a's are all different, a mixture of any number of k-variate FGM distributions has once again a k-variate FGM distribution with the same F and with a as the average of the corresponding a's.

For the bivariate FGM distribution, we have

$$E[X_1 \mid X_2 = x_2]$$

$$= E[X_1] + a\{1 - 2F_{X_2}(x_2)\} \int x_1\{1 - 2F_{X_1}(x_1)\} f_{X_1}(x_1) \, dx_1$$

which is linear in $F_{X_2}(x_2)$. More general formulas of this nature are given by Johnson and Kotz (1977). The ratio $\mathrm{var}(E[X_1 \mid X_2 = x_2])/\mathrm{var}(X_1) \leq a^2/3$ so that at most one-third of the variance of X_1 can be explained by X_2; see Schucany, Parr, and Boyer (1978). Thus, the correlation is given by $\mathrm{corr}(X_1, X_2) \leq 1/3$, with the maximum being attained for the uniform marginals. The correlation value is $1/\pi$ for normal marginals, $1/4$ for exponential marginals, and 0.281 for Laplace and gamma, with 2 as shape parameter marginals; see, for example, Schucany, Parr, and Boyer (1978).

Nelsen (1994) has shown that among all absolutely continuous bivariate distributions with fixed marginals and a given value, ρ_0, of Spearman's correlation coefficient

$$
\begin{aligned}
&\rho(X_1, X_2) \\
&= 12 \iint\limits_{\mathbb{R}^2} \left\{ F_{X_1}(x_1) - \frac{1}{2} \right\} \left\{ F_{X_2}(x_2) - \frac{1}{2} \right\} f_{X_1,X_2}(x_1, x_2) \, dx_1 \, dx_2,
\end{aligned}
$$

where $|\rho_0| \leq 1/3$, the one whose joint density is closest (in the sense of χ^2-divergence) to the density of independent random variables is the bivariate FGM distribution with coefficient $3\rho_0$. More specifically, the distance between the bivariate probability density function $f_{X_1,X_2}(x_1, x_2)$ (with marginal density functions $f_{X_1}(x_1)$ and $f_{X_2}(x_2)$) and the density function $f_{X_1}(x_1)f_{X_2}(x_2)$ can be measured by χ^2-divergence measure defined as

$$
\chi^2(f; f_1, f_2) = \iint\limits_{\mathbb{R}^2} \left\{ \frac{f_{X_1,X_2}(x_1, x_2)}{f_{X_1}(x_1)f_{X_2}(x_2)} - 1 \right\}^2 f_{X_1}(x_1)f_{X_2}(x_2) \, dx_1 \, dx_2.
$$

$$(44.81)$$

Note that the unconstrained minimum of $\chi^2(f; f_1, f_2)$ is 0, corresponding to $f_{X_1,X_2}(x_1, x_2) = f_{X_1}(x_1)f_{X_2}(x_2)$. Also if $|\rho_0| > 1/3$, the function

$$
f(x_1, x_2) = f_{X_1}(x_1)f_{X_2}(x_2) \left[1 + 3\rho_0 \{1 - 2F_{X_1}(x_1)\}\{1 - 2F_{X_2}(x_2)\} \right]
$$

still minimizes (44.81) subject to $\rho(X_1, X_2) = \rho_0$, but this function fails to be a joint probability density function.

Huang and Kotz (1984) studied an *iterated FGM distribution* with

$$
\begin{aligned}
F_{X_1,X_2}(x_1, x_2) =\ & F_{X_1}(x_1)F_{X_2}(x_2) + a_1 F_{X_1}(x_1)F_{X_2}(x_2)\bar{F}_{X_1}(x_1)\bar{F}_{X_2}(x_2) \\
& + a_2 \{F_{X_1}(x_1)F_{X_2}(x_2)\}^2 \bar{F}_{X_1}(x_1)\bar{F}_{X_2}(x_2),
\end{aligned}
\qquad (44.82)
$$

which is obtained by replacing $\bar{F}_{X_1}(x_1)\bar{F}_{X_2}(x_2)$ in the FGM distribution by $\bar{F}_{X_1}(x_1)\bar{F}_{X_2}(x_2)\{1 + b F_{X_1}(x_1) F_{X_2}(x_2)\}$. For (44.82) to be a proper distribution function, we must have

$$|a_1| \le 1, \ a_1 + a_2 \le -1 \ \text{ and } \ a_2 \le \frac{1}{2}\{3 - a_1(9 - 6a_1 - 3a_1^2)^{1/2}\}.$$

In this case, $\text{corr}(X_1, X_2) = \frac{a_1}{3} + \frac{a_2}{12}$ for uniform marginals, yielding $\text{corr}(X_1, X_2) \le 0.5072$, and $\text{corr}(X_1, X_2) = \frac{a_1}{\pi} + \frac{a_2}{4\pi}$ for normal marginals, yielding $\text{corr}(X_1, X_2) \le \frac{\sqrt{13}-1}{2\pi} = 0.4147$.

Elandt-Johnson (1976) studied in detail the trivariate FGM distribution with exponential marginals and paid special attention to various hazard rates in connection with competing risks with dependent failure times. From (44.78), we have the joint trivariate survival function as

$$\begin{aligned}
\bar{F}_{X_1,X_2,X_3}(x_1, x_2, x_3) &= \bar{F}_{X_1}(x_1)\bar{F}_{X_2}(x_2)\bar{F}_{X_3}(x_3)[1 + a_{12}F_{X_1}(x_1)F_{X_2}(x_2) \\
&\quad + a_{13}F_{X_1}(x_1)F_{X_3}(x_3) + a_{23}F_{X_2}(x_2)F_{X_3}(x_3) \\
&\quad + a_{123}F_{X_1}(x_1)F_{X_2}(x_2)F_{X_3}(x_3)] \qquad (44.83)
\end{aligned}$$

with the restriction

$$|a_{13} + a_{23} \pm a_{123}| \le 1 + a_{12}$$

and two similar restrictions, and

$$|a_{123}| \le 1 + a_{12} + a_{13} + a_{23}.$$

Now taking $\bar{F}_{X_i}(x_i) = e^{-x_i}$, $x_i \ge 0$ (standard exponential), and all a's to be equal—namely, $a_{12} = a_{13} = a_{23} = a_{123} = a$—we get from (44.83)

$$\begin{aligned}
\bar{F}_{X_1,X_2,X_3}(x_1, x_2, x_3) &= e^{-(x_1+x_2+x_3)}[1 + a\{(1 - e^{-x_1})(1 - e^{-x_2}) \\
&\quad + (1 - e^{-x_1})(1 - e^{-x_3}) + (1 - e^{-x_2})(1 - e^{-x_3}) \\
&\quad - (1 - e^{-x_1})(1 - e^{-x_2})(1 - e^{-x_3})\}] \qquad (44.84)
\end{aligned}$$

with $0 < a < 1/2$. Then, the survival function from all causes at observed time t is obtained from (44.84) as

$$\bar{F}_{X_1,X_2,X_3}(t, t, t) = e^{-3t}\{1 + a(2 + e^{-t})(1 - e^{-t})^2\}.$$

The hazard rate due to all causes is

$$h_A(t) = -\frac{d\log \bar{F}_{X_1,X_2,X_3}(t, t, t)}{dt} = 3\left\{1 - \frac{ae^{-t}(1 - e^{-2t})}{1 + a(2 + e^{-t})(1 - e^{-t})^2}\right\},$$

while the hazard rate for cause C_1 (associated with failure time X_1) is

$$
\begin{aligned}
h_{C_1}(t) &= \left. -\frac{\partial \log \bar{F}_{X_1, X_2, X_3}(x_1, x_2, x_3)}{\partial x_1} \right|_{x_1 = x_2 = x_3 = t} \\
&= 1 - \frac{ae^{-t}(1 - e^{-2t})}{1 + a(2 + e^{-t})(1 - e^{-t})^2} \\
&= \frac{1}{3} h_A(t).
\end{aligned}
$$

We thus have

$$
h_{C_1}(t) = h_{C_2}(t) = h_{C_3}(t) = \frac{1}{3} h_A(t),
$$

that is, the hazard rates are proportional over the whole range $(0, \infty)$ with equal proportionality coefficients of $1/3$.

Shaked (1975) investigated the relationship between the k-dimensional FGM distribution and the family of k-dimensional distributions of the form

$$
F_{X_1, \ldots, X_k}(x_1, \ldots, x_k) = \int_\Omega \left\{ \prod_{i=1}^{k} F^{(\omega)}(x_i) \right\} dG(\omega), \tag{44.85}
$$

where $\{F^{(\omega)}; \omega \in \Omega\}$ is a family of univariate distributions and G is a probability measure on Ω which is assumed to be a subset of a finite-dimensional Euclidean space. Distributions of the form are said to be *positively dependent by mixture* (PDM); see Shaked (1975). These distributions arise in reliability theory as joint distributions of life-lengths of identical components operating in a random environment. Then, *exchangeable FGM distributions* can be introduced since only these distributions can be PDM. They are of the form

$$
\begin{aligned}
&F_{X_1, \ldots, X_k}(x_1, \ldots, x_k) \\
&= \prod_{i_1=1}^{k} F(x_{i_1}) \Bigg[1 + a_2 \sum_{1 \le i_1 < i_2 \le k} \{1 - F(x_{i_1})\}\{1 - F(x_{i_2})\} \\
&\quad + a_3 \sum_{1 \le i_1 < i_2 < i_3 \le k} \{1 - F(x_{i_1})\}\{1 - F(x_{i_2})\}\{1 - F(x_{i_3})\} \\
&\quad + \cdots + a_k \prod_{i=1}^{k} \{1 - F(x_i)\} \Bigg] \\
&= \prod_{i=1}^{k} F(x_i) \left[1 + \sum_{i=2}^{k} a_i \phi_{i,k} \Big(1 - F(x_1), \ldots, 1 - F(x_k) \Big) \right], \tag{44.86}
\end{aligned}
$$

where

$$\phi_{i,k}(z_1, \ldots, z_k) = \sum_{1 \le j_1 < \cdots < j_i \le k} z_{j_1} z_{j_2} \cdots z_{j_i}$$

is the ith elementary symmetric function of $z_1, \ldots, z_k$; in other words, $a_{12} = \cdots = a_{k-1,k} = a_2$, $a_{123} = \cdots = a_{k-2,k-1,k} = a_3, \ldots, a_{12\cdots k} = a_k$, and all the univariate marginal distributions are equal to F.

Alternatively, the k-dimensional survival function is of the form

$$\bar{F}_{X_1, \ldots, X_k}(x_1, \ldots, x_k) = \prod_{i=1}^{k} \bar{F}(x_i) \left[1 + \sum_{i=2}^{k} b_i \phi_{i,k} \Big(F(x_1), \ldots, F(x_k) \Big) \right].$$

$$(44.87)$$

Now, if $F_{X_1, \ldots, X_k}(x_1, \ldots, x_k)$ is an exchangeable FGM distribution and if $(1, 0, a_2, \ldots, a_k) = (\mu_0, \mu_1, \mu_2, \ldots, \mu_k)$ where μ_i is the ith moment of a probability measure Ψ on $[-1, 1]$, then the distribution $F_{X_1, \ldots, X_k}(x_1, \ldots, x_k)$ is PDM. Conversely, if $F_{X_1, \ldots, X_k}(x_1, \ldots, x_k)$ is an exchangeable FGM distribution and is also PDM, then $(1, 0, a_2, \ldots, a_k) = (\mu_0, \mu_1, \mu_2, \ldots, \mu_k)$, where μ_i is the ith moment of some probability measure Ψ on $\mathbb{R}$. Consequently, the k-dimensional exchangeable FGM distributions of the form

$$F_{X_1, \ldots, X_k}(x_1, \ldots, x_k) = \prod_{i=1}^{k} F(x_i) \left[1 + a \prod_{i=1}^{k} \{1 - F(x_i)\} \right], \qquad (44.88)$$

where F is some univariate distribution, cannot be PDM. Essentially, this model is "too simple" to be used as the joint distribution of life-lengths of identical components operating in a random environment.

Lee (1994) defined a general family of multivariate density functions with preassigned marginals as follows:

(i) Let $f_{X_1}(x_1), \ldots, f_{X_k}(x_k)$ be univariate probability density functions with supports defined on $A_1, \ldots, A_k \subseteq \mathbb{R}$, respectively. Let $\phi_i(x)$, $i = 1, \ldots, k$, be a set of bounded non-constant functions such that

$$\int_{-\infty}^{\infty} \phi_i(x) f_{X_i}(x) dx = 0 \quad \text{for all } i = 1, \ldots, k \text{ and } x \in \mathbb{R}. \quad (44.89)$$

Then, the function

$$f(x_1, \ldots, x_k) = \prod_{i=1}^{k} f_{X_i}(x_i) \left\{ 1 + \frac{1}{\alpha_k} R_{\phi_1, \ldots, \phi_k, \Omega_k}(x_1, \ldots, x_k) \right\}$$

$$(44.90)$$

is a multivariate density function, where

$$
\begin{aligned}
R_{\phi_1,\ldots,\phi_k,\Omega_k}&(x_1,\ldots,x_k)\\
&= \sum_{1\leq i_1<i_2\leq k} \omega_{i_1 i_2}\phi_{i_1}(x_{i_1})\phi_{i_2}(x_{i_2})\\
&\quad + \sum_{1\leq i_1<i_2<i_3\leq k} \omega_{i_1 i_2 i_3}\phi_{i_1}(x_{i_1})\phi_{i_2}(x_{i_2})\phi_{i_3}(x_{i_3})\\
&\quad + \cdots + \omega_{12\cdots k}\prod_{i=1}^{k}\phi_i(x_i)
\end{aligned}
\tag{44.91}
$$

and $\Omega_k = \{\omega_{i_1 i_2}, \omega_{i_1 i_2 i_3}, \ldots, \omega_{12\cdots k}\}$. The set of real numbers Ω_k and α_k are chosen such that $|R_{\phi_1,\ldots,\phi_k,\Omega_k}(x_1,\ldots,x_k)| \leq \alpha_k$ holds for all $x_i \in \mathbb{R}$, $i = 1,\ldots,k$. Furthermore, $f(x_1,\ldots,x_k)$ in (44.90) has specified marginal densities $f_{X_1}(x_1),\ldots,f_{X_k}(x_k)$,

(ii) If $|\phi_i(x)| \leq C_i$ for all $x \in \mathbb{R}$, $i = 1,\ldots,k$, then Ω_k and α_k can be chosen such that

$$
|\omega_{i_1 i_2}| \leq \frac{1}{C_{i_1}C_{i_2}}\,,\, |\omega_{i_1 i_2 i_3}| \leq \frac{1}{C_{i_1}C_{i_2}C_{i_3}}\,,
$$
$$
\ldots, |\omega_{12\cdots k}| \leq \frac{1}{C_1 C_2 \cdots C_k}\,,
$$

and α_k can be chosen as the number of nonzero ω's in the set Ω_k, with $1 \leq \alpha_k \leq 2^k - k - 1$.

(iii) If all of the ω's are taken to be 0, the density function in (44.90) reduces to the case of independence.

Restricting to the case $|\phi_i(x)| \leq 1$, $i = 1, 2, \ldots, k$, without loss of generality, it can be shown that for any subset of $(X_1, \ldots, X_k)$, say $(X_{i_1}, X_{i_2}, \ldots, X_{i_\ell})$ where $1 \leq i_1 < i_2 < \cdots < i_\ell \leq k$, the corresponding joint density function is

$$
f(x_{i_1},\ldots,x_{i_\ell}) = \prod_{j=1}^{\ell} f_{i_j}(x_{i_j})\left\{1 + \frac{1}{\alpha_k}R_{\phi_{i_1},\ldots,\phi_{i_\ell},\Omega_\ell}(x_{i_1},\ldots,x_{i_\ell})\right\},
$$

where $R_{\phi_1,\Omega_1} = 0$, and Ω_ℓ is a subset of Ω_k such that subscripts of ω's involve combinations of integers $i_1,\ldots,i_\ell$ only.

From (44.90), we note that in the case when $\phi_i(x_i) = 1 - 2F_{X_i}(x_i)$, where $F_{X_i}(x_i)$ is the distribution function of X_i, we obtain the FGM distribution. Lee (1994) has provided numerous examples, but unfortunately the Weibull case is restricted only to FGM family. This method can be used to construct pseudoconjugate distributions for multivariate distributions with natural exponential family marginals.

Huang and Kotz (1999) studied the distributions with uniform marginals and the cumulative distribution functions

$$F_\alpha(x, y) = xy\{1 + \alpha(1 - \boldsymbol{x}^p)(1 - y^p)\} \quad \text{and}$$
$$F_\alpha(x, y) = xy\{1 + \alpha(1 - x)^p(1 - y)^p\}p > 0, \quad 0 < x, y < 1.$$

14   MULTIVARIATE PHASE-TYPE DISTRIBUTIONS

Mutivariate phase-type (MPH) distributions were introduced by Assaf *et al.* (1984) as a natural extension of the univariate phase-type (PH) distributions of Neuts (1975) in the following way. Suppose $\{V(t) : t > 0\}$ is a regular Markov chain with finite state-space E. Let $\Gamma_1, \ldots, \Gamma_k$ be k nonempty subsets of E such that once V enters Γ_i it never leaves. Suppose that $\bigcup_{i=1}^k \Gamma_i$ consists of one state Δ, into which absorption is certain. Let β be an initial probability vector on E, which puts all its mass on states in $E \backslash \{\Delta\}$. The generator of V is of the form

$$\begin{bmatrix} \boldsymbol{A} & \boldsymbol{A}\boldsymbol{1} \\ \boldsymbol{0}^T & 0 \end{bmatrix},$$

where $\boldsymbol{A}$ is a square matrix, $\boldsymbol{1}$ is a column vector of ones, and $\boldsymbol{0}$ is a column vector of zeros. Let $T_i = \inf\{t : V(t) \in \Gamma_i\}$ for $i = 1, \ldots, k$. Then the distribution of the vector $(T_1, \ldots, T_k)^T$ is MPH.

The MPH family is closed under many operations of interest in reliability. Perhaps the most striking property is that the joint distribution of the life functions of a finite number of coherent structure functions on an MPH random vector is again MPH. In particular, the minimal component of an MPH random vector has a PH distribution.

Generally, if one seeks flexible models for nonnegative multivariate data, then it is quite reasonable to require that the family of distributions considered be closed under scaling of the data; in other words, one should be able to change the units of the components independently of one another without having to consider a different family of distributions. Unforunately, the MPH family it not closed under such scaling; O'Cinneide (1987) has attributed this to a feature of MPH distributions termed "simultaneosity."

15 CHEBYSHEV-TYPE AND BONFERRONI INEQUALITIES

Inequalities satisfied by univariate distribution functions under fairly general conditions have been described in Chapter 33. We now describe some extensions of these inequalities to multivariate distributions. These are naturally more complicated, and often present more difficulties to intuitive comprehension than do the univariate inequalities. While the univariate formulas are usually expressed in terms of moments of a single variable, it is only to be expected that to obtain good inequalities in the multivariate case, not only moments of single variables, but also product moments will be needed in the formulas.

We first note useful multivariate forms of Bonferroni's inequalities, given by Meyer (1969). For the bivariate case, we consider two classes of events:

$$\{E_{11}, \ldots, E_{1N_1}\} \quad \text{and} \quad \{E_{21}, \ldots, E_{2N_2}\}.$$

Then the probability $P[n_1, n_2]$ that exactly n_1 of the first class and n_2 of the second class of events occur lies between

$$\sum_{t=n_1+n_2}^{n_1+n_2+2m+1} \sum_{i+j=t} f(i,j;t) \quad \text{and} \quad \sum_{t=n_1+n_2}^{n_1+n_2+2m} f(i,j;t) \quad \text{for any } m > 0,$$

(44.92)

where $f(i,j;t) = (-1)^{t-(n_1+n_2)} \binom{i}{n_1} \binom{j}{n_2} S_{i,j}$ with

$$S_{i,j} = \sum_{t=i+j}^{N_1+N_2} \sum_{g+h=t} \binom{g}{i} \binom{h}{j} P[g,h].$$

The simplest way of deriving multivariate inequalities is to combine univariate inequalities, using the formula

$$\Pr\left[\bigcap_{j=1}^{k} E_j\right] = 1 - \Pr\left[\bigcup_{j=1}^{k} \bar{E}_j\right] \geq 1 - \sum_{j=1}^{k} \Pr[\bar{E}_j]. \qquad (44.93)$$

Taking $E_j \equiv (|X_j - E[X_j]| \leq t_j \sqrt{\text{var}(X_j)})$ and using Chebyshev's inequality, we obtain

$$\Pr\left[\bigcap_{j=1}^{k} \left(|X_j - E[X_j]| \leq t_j \sqrt{\text{var}(X_j)}\right)\right] \geq 1 - \sum_{j=1}^{k} t_j^{-2}. \qquad (44.94)$$

Of course, for this formula to be of any use we must have $\sum_{j=1}^{k} t_j^{-2} < 1$, and preferably the sum should be rather small. As is to be expected, this becomes more difficult to ensure as k increases [if $t_1 = t_2 = \cdots = t_k = t$, for instance, the right-hand side of (44.94) is kt^{-2}]. The inequality (44.94) (and similar ones that may be obtained from other formulas) is of rather general applicability. No assumption of independence among the variables is made, nor, indeed, is any specific form of dependence assumed. If independence can be assumed, then

$$\Pr\left[\bigcap_{j=1}^{k}\left(|X_j - E[X_j]| \leq t_j\sqrt{\operatorname{var}(X_j)}\right)\right] \geq \prod_{j=1}^{k}(1 - t_j^{-2}). \qquad (44.95)$$

Provided that $t_j > 1$ for all j, this inequality gives a nontrivial lower bound. Note that $\prod_{j=1}^{k}(1 - t_j^{-2}) \geq 1 - \sum_{j=1}^{k} t_j^{-2}$, so that (44.95) gives a larger lower bound than (44.94), as is to be expected, because the former inequality requires a restriction (namely, independence) on the joint distribution of $X_1, X_2, \ldots, X_k$.

The simplest inequality introducing correlation was obtained by Berge (1938). This is

$$\Pr\left[\bigcap_{j=1}^{2}\left(\left|\frac{X_j - E[X_j]}{\sigma_j}\right| < t\right)\right] \geq 1 - t^{-2}(1 + \sqrt{1 - \rho^2}), \qquad (44.96)$$

where $\sigma_j = \sqrt{\operatorname{var}(X_j)}$ and $\rho = \operatorname{corr}(X_1, X_2)$. This is an improvement over (44.94) with $k = 2$, $t_1 = t_2 = t$, in that the lower bound $1 - 2t^{-2}$ is replaced by $1 - (1 + \sqrt{1 - \rho^2})t^{-2}$. For $\rho = 1$, when X_1 and X_2 may be regarded as linear functions of each other, we obtain the univariate Chebyshev formula. On the other hand, when $\rho = 0$ we obtain (44.94) and not the better lower bound $(1 - t^{-2})^2$ corresponding to independence of X_1 and X_2. However, it should be remembered that a zero value for ρ need not imply independence between X_1 and X_2.

Berge (1938) obtained the inequality (44.96) in the following way. We will use the notation $Y_j = (X_j - E[X_j])/\sqrt{\operatorname{var}(X_j)}$ for the standardized X_j variate. Consider the statistic $H = Y_1^2 + Y_2^2 + 2aY_1Y_2$ with $|a| < 1$. If either $|Y_1| \geq t$ or $|Y_2| \geq t$ then $H \geq (1-a^2)t^2$; and also $H \geq 0$ for all Y_1, Y_2. Hence, $E[H] = 2(1 + a\rho) \geq (1 - a^2)t^2 \Pr\left[\bigcup_{j=1}^{2}(|Y_j| \geq t)\right]$. Remembering that $\Pr\left[\bigcap_{j=1}^{2}(|Y_j| < t)\right] = 1 - \Pr\left[\bigcup_{j=1}^{2}(|Y_j| \geq t)\right]$, we have

$$\Pr\left[\bigcap_{j=1}^{2}(|Y_j| < t)\right] \geq 1 - 2(1 + a\rho)(1 - a^2)^{-1}t^{-2}.$$

This formula is valid for any $|a| < 1$. The best value to take for a will minimize the multiplier $2(1 + a\rho)(1 - a^2)^{-1}$. This is effected by taking $a = -\rho^{-1}(1 - \sqrt{1 - \rho^2})$ leading to (44.96).

By an extension of this argument, Olkin and Pratt (1958) obtained the inequality

$$\Pr\left[\bigcap_{j=1}^{k}(|Y_j| < t_j)\right] \geq 1 - \frac{1}{k^2}\left\{\sqrt{u} + \sqrt{(k-1)}\sqrt{\left(k\sum_{j=1}^{k}t_j^{-2} - u\right)}\right\}^2,$$
(44.97)

where $u = \sum_{j=1}^{k}t_j^{-2} + 2\sum\sum_{i<j}\rho_{ij}t_i^{-1}t_j^{-1}$ with $\rho_{ij} = \mathrm{corr}(X_i, X_j)$. For $k = 2$, this gives

$$\Pr\left[\bigcap_{j=1}^{2}(|Y_j| < t_j)\right]$$
$$\geq 1 - \frac{1}{2}(t_1 t_2)^{-2}\left\{t_1^2 + t_2^2 + \sqrt{(t_1^2 + t_2^2)^2 - 4\rho_{12}^2 t_1^2 t_2^2}\right\}, \quad (44.98)$$

a result obtained earlier by Lal (1955). Olkin and Pratt (1958) pointed out that it is possible to improve (44.97), but the necessary calculations will usually be heavy. Godwin (1964) generalized these results to sets of more than two variables.

A further generalization, due to Isii (1959), gives bounds for

$$P = \Pr\left[\bigcap_{j=1}^{2}(-k_1 < X_j < k_2)\right]$$

with $0 < k_1 \leq k_2$.

(a) If $2k_1^2 > 1 - \rho$ and $\frac{1}{2}(k_2 - k_1) \geq \lambda$ with

$$\lambda = \frac{k_1(1 + \rho) + [(1 - \rho^2)(k_1^2 + \rho)]^{1/2}}{2k_1^2 - 1 + \rho},$$

then

$$P \leq 2\lambda^2(2\lambda^2 + 1 + \rho)^{-1}. \tag{44.99}$$

(b) If conditions in (a) are not satisfied, and also $k_1 k_2 \geq 1$ and

$$2(k_1 k_2 - 1)^2 \geq 2(1 - \rho^2) + (1 - \rho)(k_2 - k_1)^2, \tag{44.100}$$

then

$$\begin{aligned}P \leq {} &(k_1 + k_2)^{-2}[(k_2 - k_1)^2 + 4 + \{16(1 - \rho^2)\\ &+ 8(1 - \rho)(k_2 - k_1)\}^{1/2}].\end{aligned} \tag{44.101}$$

In all other cases, there is no universal upper bound for P (other than 1).

An extension of the Gauss–Camp type of inequality, due to Leser (1942), is of some interest. He obtained bounds for

$$P = \Pr\left[\sum_{j=1}^{k} \lambda_j^{-2} Y_j^2 \le k\right].$$

In the univariate case, the Gauss–Camp inequalities are derived on the assumption that the density function of the standardized variable Y is in some sense decreasing as $|Y|$ increases. Leser generalized this by requiring the conditional average of the joint density function $f(y_1, \ldots, y_k)$, given the value of

$$R^2 = \lambda_0^2 \sum_{j=1}^{k} \lambda_j^{-2} y_j^2,$$

(where λ_0^2 is the harmonic mean of $\lambda_1^2, \ldots, \lambda_k^2$) to be a nondecreasing function of R^2 for R^2 less than $k\kappa^2$ (for some $\kappa > 0$). The inequalities are summarized in Table 44.3. Note that as k increases, the range $1 \le \kappa \le \sqrt{1 + 2/k}$ becomes narrower. Also, for "really unimodal" distributions, κ can be quite large.

A very useful discussion of multivariate Chebyshev-type inequalities for quite general regions is given by Karlin and Studden (1966), and a valuable discussion of Bonferroni-type inequalities is presented by Galambos and Simonelli (1996). Numerous results on multivariate Bonferroni inequalities are due to Lee (1992, 1996) and Galambos and Lee (1992, 1994). A typical result is

$$\begin{aligned}
q(1,1) \le\ & S_{1,1} - \sum_{i=1}^{n_1} \sum_{j=1}^{n_2-1} \Pr\left[A_{i1} A_{j2} A_{(j+1)2}\right] \\
& - \sum_{i=1}^{n_1-1} \sum_{j=1}^{n_2} \Pr\left[A_{i1} A_{(i+1)1} A_{j2}\right] \\
& + \sum_{i=1}^{n_1-1} \sum_{j=1}^{n_2-1} \Pr\left[A_{i1} A_{(i+1)1} A_{j2} A_{(j+1)2}\right];
\end{aligned}$$

TABLE 44.3
Multivariate Chebyshev-Type Inequalities

κ	λ_0	P
$\kappa \leq 1$	$\lambda_0 \leq 1$	≥ 0
	$\lambda_0 \geq 1$	$\geq 1 - \lambda_0^{-2}$
$1 \leq \kappa \leq \sqrt{1 + 2/k}$	$\lambda_0 \leq (1 + \frac{1}{2}k)^{-1/k}\kappa$	$\geq (\frac{1}{2}k + 1)(1 - \kappa^{-2})(\lambda_0/\kappa)^k$
	$(1 + \frac{1}{2}k)^{-1/k}\kappa \leq \lambda_0 \leq \kappa$	$\geq 1 - \kappa^{-2}$
	$\lambda_0 \geq \kappa$	$\geq 1 - \lambda_0^{-2}$
$\kappa \geq \sqrt{1 + 2/k}$	$\lambda_0 \leq (1 + \frac{1}{2}k)^{-1/k}\sqrt{1 + 2/k}$	$\geq (1 + 2/k)^{-k/2}\lambda_0^k$
	$(1 + \frac{1}{2}k)^{-1/k}\sqrt{1 + 2/k} \leq \lambda_0$ $\leq (1 + \frac{1}{2}k)^{-1/k}\kappa$	$\geq 1 - (1 + \frac{1}{2}k)^{-2/k}\lambda_0^{-2}$
	$(1 + \frac{1}{2}k)^{-1/k}\kappa \leq \lambda_0 \leq \kappa$	$\geq 1 - \kappa^{-2}$
	$\lambda_0 \geq \kappa$	$> 1 - \lambda_0^{-2}$

here we have two sequences A_{ij}, $1 \leq i \leq n_j$, $j = 1, 2$, of events on a given probability space, $\nu_{n_j}(A; j)$ (for $j = 1, 2$) denotes the number of those A_{ij}'s $(1 \leq i \leq n_j)$ that occur, and $q(1, 1) = \Pr[\nu_{n_1} \geq 1, \nu_{n_2} \geq 1]$. By defining A_{ij}'s suitably, this inequality can be easily transformed to one involving distribution functions.

16 SINGULAR DISTRIBUTIONS

It sometimes happens that one or more mathematical relations hold precisely among k random variables $X_1, X_2, \ldots, X_k$. In such cases, the joint distribution is said to be *singular*. We shall give no direct discussion of singular distributions, though they will be referred to in Section 1 of Chapter 45, for example.

Consider the case, for example, when there are just r distinct linear relations among k random variables $X_1, X_2, \ldots, X_k$. These relations can then be derived from the variance–covariance matrix of the X's, by replacing an arbitrary row in the left-hand side determinant of each of the r equations

$$\begin{vmatrix} \mu_{r,i} & \mu_{r,r+1} & \cdots & \mu_{r,k} \\ \mu_{r+1,i} & \mu_{r+1,r+1} & \cdots & \mu_{r+1,k} \\ \vdots & \vdots & & \vdots \\ \mu_{k,i} & \mu_{k,r+1} & \cdots & \mu_{k,k} \end{vmatrix} = 0, \qquad i = 1, 2, \ldots, r,$$

by $X_i, X_{r+1}, \ldots, X_k$ (the X's being so ordered that

$$\begin{vmatrix} \mu_{r+1,r+1} & \mu_{r+1,r+2} & \cdots & \mu_{r+1,k} \\ \mu_{r+2,r+1} & \mu_{r+2,r+2} & \cdots & \mu_{r+2,k} \\ \vdots & \vdots & & \vdots \\ \mu_{k,r+1} & \mu_{k,r+2} & \cdots & \mu_{k,k} \end{vmatrix} > 0).$$

In the formulas above, $\mu_{ij} = E[(X_i - E[X_i])(X_j - E[X_j])]$ is the covariance of X_i and X_j.

Harris and Helvig (1966) derived marginal and conditional distributions and their means and variance–covariance matrices when the joint distribution is possbily singular. Though their results are stated in terms of multivariate normal distributions, they hold for any multivariate distribution for which zero correlation and independence are equivalent. Let X be a k-dimensional random vector with mean ξ and variance–covariance matrix V (possibly singular). Let X, ξ, and V have the following partitionings:

$$X = \begin{pmatrix} X_1 \\ X_2 \end{pmatrix}, \quad \xi = \begin{pmatrix} \xi_1 \\ \xi_2 \end{pmatrix}, \quad \text{and} \quad V = \begin{pmatrix} V_{11} & V_{12} \\ V_{12} & V_{22} \end{pmatrix}.$$

Then, the equation $V_{11}M + V_{12} = 0$ has at least one solution and the random vectors X_1 and $X_2 + M^T X_1$ are independent; also

$$E[X_2 | X_1 = x_1] = \xi_2 - M^T(x_1 - \xi_1)$$

and

$$\text{Var}(X_2 | X_1 = x_1) = V_{22} + M^T V_{12}.$$

Marsaglia (1964) had earlier shown that with A^+ denoting the pseudo-inverse of A (in the sense of Penrose) we obtain $V_{11}(-V_{11}^+ V_{12}) + V - 12 = 0$ (i.e., that one choice for M is $-V_{11}^+ V_{12}$), and hence that X_1 and $X_2 - V_{21}V_{11}^+ X_1$ are independent, and he concluded from this that

$$E[X_2 | X_1 = x_1] = \xi_2 + V_{21}V_{11}^+(x_1 - \xi_1)$$

and

$$\text{Var}(X_2 | X_1 = x_1) = V_{22} + V_{21}V_{11}^+ V_{12}.$$

If V_{11} is nonsingular, then $V_{11}^+ = V_{11}^{-1}$ and M is unique, and in this case we get the standard formulas. However, if V_{11} is singular, then if the equation $V_{11}M + V_{12} = 0$ has one solution, it has many solutions.

More details on singular multivariate normal distributions are presented in Chapter 45.

17 DISTRIBUTIONS WITH ALMOST-LACK OF MEMORY

First, we say that the distribution of a nonnegative two-dimensional random vector $(X_1, X_2)^T$ has *lack of memory* property iff

$$\Pr[X_1 \geq a_1 + x_1, \ X_2 \geq a_2 + x_2 \mid X_1 \geq a_1, X_2 \geq a_2] = \Pr[X_1 \geq x_1, \ X_2 \geq x_2]$$

for any $x_1 \geq 0$ and $x_2 \geq 0$ and any a_1, a_2. This bivariate distribution is closely associated with bivariate exponential distributions discussed in Chapter 47.

Generalizing the above defined concept, Chukova and Dimitrov (1992), Chukova, Dimitrov, and Khalil (1993), and Dimitrov, Chukova, and Khalil (1994) defined the distribution of a nonnegative two-dimensional random vector $(X_1, X_2)^T$ to have *almost-lack of memory* (ALM) property iff

$$\begin{aligned}
\Pr[X_1 \geq a_1 + x_1, \ &X_2 \geq a_2 + x_2 \mid X_1 \geq a_1, \ X_2 \geq a_2] \\
&= \Pr[X_1 \geq x_1, \ X_2 \geq x_2]
\end{aligned} \tag{44.102}$$

for any $x_1 \geq 0$ and $x_2 \geq 0$ and for infinitely many nonnegative choices of a_1 and a_2. This class of distributions is denoted by $K_0(a_1, a_2)$. The random vector $(X_1, X_2)^T$ belongs to the subclass $K(a_1, a_2)$ iff, for any x_1, x_2 $(0 \leq x_1 \leq a_1, 0 \leq x_2 \leq a_2)$,

$$\Pr[X_1 \geq n_1 a_1 + x_1, \ X_2 \geq x_2] = \Pr[X_1 \geq n_1 a_1] \Pr[X_1 \geq x_1, \ X_2 \geq x_2] \tag{44.103}$$

and

$$\Pr[X_1 \geq x_1, \ X_2 \geq n_2 a_2 + x_2] = \Pr[X_1 \geq x_1, \ X_2 \geq x_2] \Pr[X_2 \geq n_2 a_2] \tag{44.104}$$

are both satisfied for all nonnegative integers n_1, n_2.

Dimitrov, Chukova, and Khalil (1994) have shown that if $(X_1, X_2)^T$, belonging to the class $K(a_1, a_2)$, has a continuous distribution, then its probability density function is of the form

$$f_{X_1, X_2}(x_1, x_2)$$
$$= \begin{cases} \alpha_1^{n_1} \alpha_2^{n_2}(1 - \alpha_1)(1 - \alpha_2) f^*(x_1 - n_1 a_1, x_2 - n_2 a_2) \\ \quad \text{for } n_1 a_1 \leq x_1 \leq (n_1 + 1)a_1, n_2 a_2 \leq x_2 \leq (n_2 + 1)b_2 \\ \quad\quad\quad 0 \quad \text{otherwise} \end{cases}$$

for real α_1, α_2 with $0 < \alpha_1, \alpha_2 < 1$, where $f^*(x_1, x_2)$ is a probability density function with support $[0, a_1) \times [0, a_2)$. Furthermore, in this case, $(X_1, X_2)^T$ can be represented as

$$\begin{pmatrix} X_1 \\ X_2 \end{pmatrix} \stackrel{d}{=} \begin{pmatrix} Y_1 \\ Y_2 \end{pmatrix} + \begin{pmatrix} Z_1 \\ Z_2 \end{pmatrix},$$

where $(Y_1, Y_2)^T$ has the joint survival function $\Pr[Y_1 > y_1, Y_2 > y_2]$ satisfying $\Pr[0 \le Y_1 < a_1, 0 \le Y_2 < a_2] = 1$ and $(Z_1, Z_2)^T$ are two independent geometric random variables over $\{0, a_1, 2a_1, \ldots\}$ and $\{0, a_2, 2a_2, \ldots, \}$ with parameters α_1 and α_2.

Examples of this family of distributions include:

(i) *Truncated bivariate exponential distribution* with joint survival function

$$\frac{(e^{-\lambda_1 x_1} - e^{-\lambda_1 a_1})(e^{-\lambda_2 x_2} - e^{-\lambda_2 a_2})}{(1 - e^{-\lambda_1 a_1})(1 - e^{-\lambda_2 a_2})} \quad \text{for } x_1 \in [0, a_1), \; x_2 \in [0, a_2),$$

where $\alpha_1 = e^{-\lambda_1 a_1}$ and $\alpha_2 = e^{-\lambda_2 a_2}$.

Any other choice of α_1 and α_2 gives rise to continuous bivariate distributions from the class $K(a_1, a_2)$ which are not exponential.

(ii) *Generalized uniform distribution* with joint density function

$$f_{X_1, X_2}(x_1, x_2) = \frac{\alpha_1^{n_1} \alpha_2^{n_2}(1 - \alpha_1)(1 - \alpha_2)}{a_1 a_2}$$
$$\text{for } x_1 \in [n_1 a_1, (n_1 + 1)a_1),$$
$$x_2 \in [n_2 a_2, (n_2 + 1)a_2),$$

for $\alpha_1 \in (0, 1)$, $\alpha_2 \in (0, 1)$ and $n_1, n_2 = 0, 1, 2, \ldots$. Note that this continuous bivariate distribution, which has the almost-lack of memory property, is not exponential.

18   DISTRIBUTIONS WITH SPECIFIED CONDITIONALS

Knowledge of the marginal distributions $F_X(x)$ and $F_Y(y)$ has long been known to be inadequate to determine the joint distribution function $F_{X,Y}(x, y)$. As already seen in some of the earlier sections, a variety of joint distribution functions with given marginal distribution functions have been developed over the years; see, for example, the surveys of Mardia (1970), Ord (1972a–c), and Hutchinson and Lai (1990). However, if we specify marginals and also incorporate some conditional specification, then it will be possible to determine the joint distributions.

First of all, it is clear that knowledge of the marginal distribution $F_X(x)$ and the conditional distribution $F_{X|Y}(x|y) = \Pr[X \leq x | Y = y]$ (for every y) will completely determine the joint distribution $F_{X,Y}(x, y)$. In some situations, for example, the specification of the above conditional distribution and the fact that $X \overset{d}{=} Y$ may characterize the joint distribution $F_{X,Y}(x, y)$, as shown by Arnold and Pourahmadi (1988).

The following joint density function, given by Castillo and Galambos (1989),

$$f_{X,Y}(x, y) = C\, e^{-\{x^2 + y^2 + 2xy(x + y + xy)\}} \quad \forall\, x, y,$$

where $C > 0$ is the normalizing constant, possesses conditional density functions $f_{X|Y}(x|y)$ and $f_{Y|X}(y|x)$ that are each normal. It provides a simple example for the case when the two conditionals are normal and yet the joint distribution is not bivariate normal. In this case, it may be verified that $E[Y|X = x] = \frac{-x^2}{1 + 2x + 2x^2}$ and $\mathrm{var}(Y|X = x) = \frac{1}{2 + 4x + 4x^2}$. It is possible to construct many such joint distributions with bivariate marginals; see also Castillo and Galambos (1987).

Castillo and Galambos (1989) have established the following conditional characterization of the bivariate normal distribution (see Chapter 46). $f_{X,Y}(x, y)$ is a classical bivariate normal density function if and only if the conditional densities of X given Y and Y given X are both normal, and any one of the following four conditions hold:

(i) $\mathrm{var}(Y|X = x)$ or $\mathrm{var}(X|Y = y)$ is constant;

(ii) $\lim_{y \to \infty} y^2 \mathrm{var}(X|Y = y) = \infty$ or $\lim_{x \to \infty} x^2 \mathrm{var}(Y|X = x) = \infty$;

(iii) $\liminf_{y \to \infty} \mathrm{var}(X|Y = y) \neq 0$ or $\liminf_{x \to \infty} \mathrm{var}(Y|X = x) \neq 0$;

and

(iv) $E[Y|X = x]$ or $E[X|Y = y]$ is linear and nonconstant.

Multivariate distributions with specified conditional distributions have been discussed quite extensively in the literature. The monograph by Arnold, Castillo, and Sarabia (1992) provides an excellent discussion on this topic. (Availability of this monograph has prompted us not to discuss this topic in detail so as to avoid unnecessary repetition of all the results from an easily accessible source.) If both families of conditional distributions, $F_{X|Y}(x|y)$ for every possible value y of Y and $F_{Y|X}(y|x)$ for every possible value x of X, are given and we assume that these two families of conditional distributions are compatible [see Chapter 2 of Arnold, Castillo, and Sarabia (1992)] and that a related Markov process is indecomposable, then it is known that those two families of conditional distributions will uniquely determine the joint distribution function $F_{X,Y}(x, y)$; see Arnold and Press (1989) for a review of results on such characterizations.

The value of conditionally specified models is due to the fact that it is often easier to envision the characteristics of the conditional distributions when modeling a bivariate experiment. For example, in the conditional construction of *bivariate Cauchy distributions*, Anderson and Arnold (1991) specified that

$$p_{X|Y}(x|y) = \frac{1}{\pi} \frac{\tau(y)}{(x^2 + \tau^2(y))} \quad \text{and} \quad p_{Y|X}(y|x) = \frac{1}{\pi} \frac{\gamma(x)}{(\gamma^2(x) + y^2)} ,$$

where $\gamma(x)$ and $\tau(y)$ are parametric functions. These are centered Cauchy densities in the sense that their location parameters are constrained to be zero. In this case, we are led to the functional equation

$$\phi(y)\gamma^2(x) + \phi(y)y^2 = \psi(x)\tau^2(y) + \psi(x)x^2,$$

where $\phi(y) = p_Y(y)\tau(y)$ and $\psi(x) = p_X(x)\gamma(x)$. In the independent case when $\tau(y) = \tau$ (implying $\gamma(x) = \gamma$), we obtain

$$p_{X,Y}(x, y) = \frac{1}{\pi^2} \frac{\tau\gamma}{(\tau^2 + x^2)(\gamma^2 + y^2)} .$$

Otherwise, as Anderson and Arnold (1991) have shown, we obtain

$$p_{X,Y}(x, y) = \frac{C(\lambda_1, \lambda_2, \lambda_{12})}{1 + \lambda_1 x^2 + \lambda_2 y^2 - \lambda_{12} x^2 y^2} ,$$

where $\lambda_1, \lambda_2,$ and λ_{12} are positive. The normalizing constant is expressed in terms of a *complete elliptical integral of the first kind* given by

$$K(\beta) = \int_0^{\pi/2} \frac{1}{\sqrt{1 - \beta^2 \sin^2 u}} \, du.$$

In fact,

$$C(\lambda_1, \lambda_2, \lambda_{12}) = \begin{cases} \dfrac{\sqrt{\lambda_1 \lambda_2}}{2\pi K\left(\sqrt{\frac{\lambda_1 \lambda_2 - \lambda_{12}}{\lambda_1 \lambda_2}}\right)} & \text{if } \lambda_1 \lambda_2 \geq \lambda_{12} \\[4mm] \dfrac{\sqrt{\lambda_{12}}}{2\pi K\left(\sqrt{\frac{\lambda_{12} - \lambda_1 \lambda_2}{\lambda_{12}}}\right)} & \text{if } \lambda_1 \lambda_2 < \lambda_{12} \end{cases}$$

and the marginal densities of X and Y turn out to be

$$p_X(x) = \frac{\pi C(\lambda_1, \lambda_2, \lambda_{12})}{\sqrt{(\lambda_2 + \lambda_{12}x^2)(1 + \lambda_1 x^2)}}$$

and

$$p_Y(y) = \frac{\pi C(\lambda_1, \lambda_2, \lambda_{12})}{\sqrt{(\lambda_1 + \lambda_{12}y^2)(1 + \lambda_2 y^2)}}.$$

Anderson and Arnold (1991) applied htis bivariate Cauchy distribution to model body mass index and cholesterol ratios for the data obtained from 43 male and 42 femail subjects at the Riverside Student Health Center, University of California, in 1990.

We close this section by noting that in all the subsequent chapters, we will provide discussions on relevant multivariate distributions constructed through conditional specification. In this connection, Arnold and Strauss (1988, 1991) would be of special importance.

19 DISTRIBUTIONS WITH GIVEN MARGINALS

In Sections 11 and 12, we discussed some methods of constructing bivariate distributions with specified marginal distributions; see also Johnson and Tenenbein (1981), Johnson (1987), and Marshall and Olkin (1988). The association structure of these distributions, however, is either indirectly modeled or difficult to interpret.

Defining the *local dependence function* of a bivariate distribution as the rate of change of the local cross-ratio, Holland and Wang (1987) showed that, under some mild regularity conditions, any bivariate distribution may be specified by marginal distributions and the local dependence function. Wang (1993) proposed a theoretical method, called *iterative marginal replacement* algorithm, in order to determine a bivariate density function given its marginal density functions and its local dependence function.

To fix ideas, let us consider an $r \times c$ table with cell probabilities p_{ij} for $1 \leq i \leq r$, $1 \leq j \leq c$. For any two pairs of indices (i, j) and (k, ℓ), the cross-product ratio is defined as

$$\alpha_{ij,k\ell} = \frac{p_{ij}p_{k\ell}}{p_{i\ell}p_{kj}}.$$

The *local cross-product ratios* are defined as

$$\alpha_{ij} = \frac{p_{ij}p_{i+1,j+1}}{p_{i,j+1}p_{i+1,j}} \quad \text{for } 1 \leq i \leq r-1 \text{ and } 1 \leq j \leq c-1.$$

The local log cross-ratios $\gamma_{ij} = \log \alpha_{ij}$ can be considered instead, for the added advantage of linearity on the log scale. The set $(\alpha_{ij})_{ij}$ or $(\gamma_{ij})_{ij}$ together with the marginal probabilities $(p_{i\cdot})_i$ and $(p_{\cdot j})_j$ completely determine the table of cell probabilities. Holland and Wang (1987) and Wang (1993) extended this construction to continuous bivariate density functions as follows.

Let us consider a bivariate density function $p(x, y)$ with support $S = \{(x, y) : p(x, y) > 0\}$. Let us imagine partitioning the support S by an infinitesimal rectangular grid. The probability of the rectangle $[x, x+dx] \times [y, y + dy]$ is $p(x, y)dx\,dy$. The cross-ratio of two pairs of points (x_1, y_1) and (x_2, y_2) is defined as

$$\alpha(x_1, y_1; x_2, y_2) = \frac{p(x_1, y_1)p(x_2, y_2)}{p(x_1, y_2)p(x_2, y_1)}$$

and the local cross-ratio at (x, y) is then defined as

$$\gamma_p(x, y) = \lim_{dx \to 0,\, dy \to 0} \frac{\log \alpha(x, y; x + dx, y + dy)}{dx\,dy} .$$

The function $\gamma_p(x, y)$ is called the *local dependence function*. It is easy to note that

$$\gamma_p(x, y) = \frac{\partial^2 \log p(x, y)}{\partial x \partial y} .$$

The local dependence function $\gamma_p(x, y)$, defined whenever $\log p(x, y)$ is a mixed differentiable function, has the following properties:

(i) The random variables X and Y with joint density function $p(x, y)$ are independent iff $\gamma_p(x, y) \equiv 0$;

(ii) $\gamma_p(x, y)$ is margin free in the sense that $\gamma_p = \gamma_q$ if $q(x, y) = p(x, y) \phi_1(x) \cdot \phi_2(y)$.

(iii) If $p_{1|2}$ and $p_{2|1}$ are the conditional density functions, then $\gamma_p = \gamma_{p_{1|2}} = \gamma_{p_{2|1}}$.

For the standard bivariate normal density function (see Chapter 46)

$$p(x, y) = \frac{1}{2\pi\sqrt{1 - \rho^2}} \exp\left\{ -\frac{1}{2(1 - \rho^2)} (x^2 - 2\rho xy + y^2) \right\},$$

the local dependence function is simply $\gamma_p(x, y) = \frac{\rho}{1 - \rho^2}$. For the standard bivariate Cauchy density function

$$p(x, y) = \frac{c}{\pi (x^2 + y^2 + c^2)^{3/2}}, \qquad c > 0,$$

the local dependence function is $\gamma_p(x, y) = \frac{6xy}{(x^2+y^2+c^2)^2}$. Figure 44.7, taken from Molenberghs and Lesaffre (1997), gives the local dependence function of the bivariate Cauchy density function.

Holland and Wang (1987) then proved the following result: For any integrable local dependence function $\gamma(x, y)$ defined over $S =]a, b[\times]c, d[$, and any given continuous marginal density functions $p_X(x)$ and $p_Y(y)$ defined over $]a, b[$ and $]c, d[$, respectively, there exists a unique bivariate density function $p(x, y)$ defined over S such that

(i) $\gamma(x, y) = \dfrac{\partial^2 \log p(x, y)}{\partial x \partial y} \qquad \forall\, (x, y) \in S$

and

(ii) $p_X(x)$ and $p_Y(y)$ are the marginal density functions of $p(x, y)$.

Wang (1987, 1993) and Molenberghs and Lesaffre (1997) have all discussed numerical methods to approximate the bivariate density function $p(x, y)$, given the marginal density functions $p_X(x)$ and $p_Y(y)$ and the local dependence function $\gamma(x, y)$.

Jones (1996) motivated $\gamma(x, y)$ from the point of view of localizing the Pearson correlation coefficient p. In a recently published paper, Jones (1999) has shown that all distributions with constant local dependence (which includes the normal distribution) involve a linear exponential family conditional distribution with its canonical parameter being a linear function.

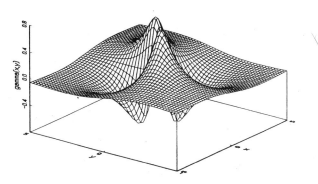

FIGURE 44.7
The Local Dependence Function of the Standard Bivariate Cauchy Density Function. [From Molenberghs and Lasaffre (1997), with permission.]

Koehler and Symanowski (1995) presented a method for constructing multivariate distributions with specified univariate marginal distributions. Suppose k univariate distribution functions $F_1(\cdot), \ldots, F_k(\cdot)$ are
given. Then, a k-variate distribution is constructed by taking the joint
distribution of

$$X_i = F_i^{-1}\left(\left(1 + \frac{Y_i}{G_{i+}}\right)^{-\alpha_{i+}}\right), \quad 1 = 1, 2, \ldots, k,$$

where $\alpha_{i+} = \sum_{j=1}^k \alpha_{ij}$, α_{ij} are chosen constants with $\alpha_{ij} = \alpha_{ji} \geq 0$; furthermore, $Y_1, \ldots, Y_k$ are independent standard exponential random variables,
$G_{i+} = \sum_{j=1}^k G_{ij}$, $G_{ij} = G_{ji}$ (for $i \geq j$), G_{ij} are independent gamma random variables with scale parameter 1 and shape parameter α_{ij}, and $\{Y_i\}$
and $\{G_{ij}\}$ are independent. Then, the marginal distribution function of
X_i is $F_i(\cdot)$, and the constants α_{ij} determine the correlation structure of
$X_1, \ldots, X_p$.

Shaked (1979) presented a method of constructing bivariate distributions, with given marginals, possessing positive dependence. Let $\Psi(u)$ be
a probability generating function of a nonnegative integer-valued random
variable, and let $F(x)$ be an univariate distribution function. Then,

$$G(x_1, x_2) = \Psi\left(F(x_1)F(x_2)\right)$$

is an exchangeable bivariate distribution with marginals $\Psi(F(x))$. The
bivariate distribution G remains well-defined if $\Psi(u)$ is of the form

$$\Psi(u) \doteq \int_0^\infty u^x d\psi(x),$$

where $\psi(x)$ is a probability measure on $[0, \infty)$. Thus, if $\tilde{F}$ is a given
univariate distribution that can be expressed as $\tilde{F}(x) = \Psi(F(x))$ for some
nontrivial F and Ψ of the form $\Psi(u) = \int_0^\infty u^x d\psi(x)$, then

$$G(x_1, x_2) = \Psi(F(x_1)F(x_2))$$

defines a bivariate distribution, with $\tilde{F}$ as its marginals, which is positively dependent in mixture. It should be mentioned that Gumbel's bivariate logistic distribution in Chapter 51 and the bivariate extreme value
distributions in Chapter 53 are special cases of the above bivariate form.

Another method suggested by Shaked (1979) for generating bivariate
distributions with specified marginals is based on the bivariate function

$$\Psi(u_1, u_2) = \int_0^\infty \int_0^\infty u_1^x u_2^y \, d\psi(x, y),$$

where ψ is a probability measure on $[0, \infty) \times [0, \infty)$. If $F(x)$ is an univariate distribution function, then

$$G(x_1, x_2) = \Psi\left(F(x_1), F(x_2)\right)$$

is also a bivariate distribution function with marginals $F(x)$ which is exchangeable if $\psi(x, y)$ is exchangeable, and it is positively dependent in mixture if $\psi(x, y)$ possesses this property.

20 MEASURES OF MULTIVARIATE SKEWNESS AND KURTOSIS[1]

Let X be a k-dimensional random variable with mean vector μ and covariance matrix $\Sigma = E[(X - \mu)(X - \mu)^T]$. Let us also use $\tilde{X}$ to denote the centered and reduced random vector $\Sigma^{-1/2}(X - \mu)$. In a series of papers, Mardia (1970, 1974, 1975) defined and discussed the properties of two nonregular affine invariant measures given by

$$\beta_{1,k} = E[\{(X - \mu)^T \Sigma^{-1}(Y - \mu)\}^3] \quad \text{(for skewness)}$$

where Y is independent and identically distributed as X, and

$$\beta_{2,k} = E[\{(X - \mu)^T \Sigma^{-1}(X - \mu)\}^2] \quad \text{(for kurtosis)}.$$

For the univariate case, we readily have $\beta_{1,1}$ and $\beta_{2,1}$ as the coefficients of skewness and kurtosis, β_1 and β_2, introduced in Chapter 12. For multivariate normal distributions (discussed in Chapter 45), we have $\beta_{2,k} = k(k + 2)$. Furthermore, for any centrally symmetric random variable X, we have $\beta_{1,k} = 0$. Note that the above measures of skewness and kurtosis can be written equivalently, as

$$\beta_{1,k} = E[(\tilde{X}\,\tilde{Y})^3] \quad \text{and} \quad \beta_{2,k} = E[(\|\tilde{X}\|)^4],$$

where $\tilde{X}$ and $\tilde{Y}$ are the centered and reduced random vectors $\Sigma^{-1/2}(X - \mu)$ and $\Sigma^{-1/2}(Y - \mu)$, respectively, and $\|\cdot\|$ is the Euclidean norm. The above expression of $\beta_{2,k}$ readily reveals that it only depends on the radial part of the distribution.

---

[1]Our sincere thanks go to Dr. Jean Avérous (Laboratoire de Statistique et Probabilités, Université Paul Sabatier, 31062 Toulouse Cedex, France) for providing us an original write-up of this section.

The sample estimates (based on a sample $X_1, X_2, \ldots, X_n$) of these measures are

$$b_{1,k} = \frac{1}{n^2} \sum_{i=1}^{n} \sum_{j=1}^{n} \left\{ (X_i - \overline{X})^T S^{-1} (X_j - \overline{X}) \right\}^3$$

and

$$b_{2,k} = \frac{1}{n} \sum_{i=1}^{n} \left\{ (X_i - \overline{X})^T S^{-1} (X_i - \overline{X}) \right\}^2,$$

where $\overline{X}$ and S are the sample mean vector and covariance matrix given by

$$\overline{X} = \frac{1}{n} \sum_{i=1}^{n} X_i \quad \text{and} \quad S = \frac{1}{n-1} \sum_{i=1}^{n} (X_i - \overline{X})(X_i - \overline{X})^T.$$

If $n \geq k+1$, then S is almost surely nonsingular. The above expressions for $\beta_{1,k}$ and $b_{1,k}$ were derived by Mardia (1974) from more complex ones given earlier by Mardia (1970c) using moments and cumulants of X and of the empirical distribution. Mardia and Zemroch (1975) presented algorithms for computing $b_{1,k}$ and $b_{2,k}$ from a given sample data.

Under the assumption that X is distributed as multinormal (see Chapter 45), Mardia (1970c) established the following asymptotic distributional results for $b_{1,k}$ and $b_{2,k}$:

(i) $\frac{n}{6} b_{1,k}$, asymptotically (as $n \to \infty$), has a central chi-square distribution with $k(k+1)(k+2)/6$ degrees of freedom.

(ii) $\sqrt{n} \left\{ \dfrac{b_{2,k} - \frac{n-1}{n+1} k(k+2)}{\sqrt{8k(k+2)}} \right\}$, asymptotically, has a standard normal distribution.

Mardia (1974) and Mardia and Kanazawa (1983) presented improved approximations for these limit distributions.

Based on $b_{1,k}$ and $b_{2,k}$, some tests for multinormality have been proposed in the literature. It has been shown in particular that, as in the case of t-test in the univariate situation, the Hotelling's T^2-test is more sensitive to $\beta_{1,k}$ than to $\beta_{2,k}$. Mardia and Foster (1983), in addition to deriving the covariance between $b_{1,k}$ and $b_{2,k}$, also presented some other tests for multinormality based on both $b_{1,k}$ and $b_{2,k}$.

The asymptotic properties of $b_{1,k}$ and $b_{2,k}$ have also been studied without the assumption of multinormality for X; for example, Baringhaus and Henze (1992) and Henze (1994a) assumed an elliptical distribution for X

and then established some asymptotic results for $b_{1,k}$ and $b_{2,k}$, respectively, which are as follows. If X has an elliptical distribution such that $E[\{(X-\mu)^T \Sigma^{-1}(X-\mu)\}^3] < \infty$, then $nb_{1,k}$ is asymptotically distributed as a linear combination of two χ^2 variables, one with k and the other with $k(k-1)(k+4)/6$ degrees of freedom. Baringhaus and Henze (1992) also presented closed-form expressions for the two coefficients involved in this asymptotic result. For more general distributions with $\beta_{1,k} = 0$, the asymptotic distribution of $nb_{1,k}$ is also a linear combination of χ^2 distributions but with more than two terms; closed-form expressions are not yet available in such cases.

If $\beta_{1,k} > 0$, under the conditions that $E(X) = 0$, $E(XX^T) = I_k$, $E[\|X\|^6] < \infty$, and $\mathrm{var}(h_1(X)) > 0$, where h_1 is the function defined on $\mathbf{R}^k$ as $h_1(x) = E[(x^T X)^3]$, then $\sqrt{n}(b_{1,k} - \beta_{1,k})$ is asymptotically distributed as $N(0, \sigma^2)$, where σ^2 takes on a rather complicated form. This, incidentally, is a generalization of the corresponding univariate result of Gastwirth and Owens (1977) in which σ^2 takes the form

$$\sigma^2 = \mu_6 - 6\,\mu_4 + 11\,\mu_3^2 - 3\,\mu_3\mu_5 + \frac{9}{4}\,\mu_3^2(\mu_4 - 1) + 9,$$

where μ_i denotes the ith central moment of X.

Similar results for $b_{2,k}$ have been established by Henze (1994a), who has shown that if $E[X] = 0$, $E[XX^T] = I_k$, and $E[\|X^8\|] < \infty$, then $\sqrt{n}(b_{2,k} - \beta_{2,k})$ is asymptotically distributed as $N(0, \sigma^2)$, where σ^2 again takes on quite a complicated form and can be expressed using $(k+2)$-dimensional vectors. This is also a generalization of the corresponding univariate result of Gastwirth and Owens (1977) in which σ^2 takes the form

$$\sigma^2 = \mu_8 - \mu_4^2 + 2\,\mu_4(\mu_4^2 - \mu_6) - 8\,\mu_3\mu_5 + 16\,\mu_3^2(\mu_4 + 1).$$

It needs to be mentioned that Baringhaus and Henze (1992) and Henze (1994a) also studied the power performances of Mardia's tests based on $b_{1,k}$ and $b_{2,k}$ and presented the conditions for consistency against specific alternatives.

Ardanuy and Sánchez (1993) showed that the distributions of $b_{1,k}$ and $b_{2,k}$ for singular multinormal distributions (i.e., multinormal distributions with singular covariance matrix Σ; see Chapter 45) are exactly the same as those obtained from a random sample from a multinormal distribution with covariance matrix as an identity matrix of dimension $\mathrm{rank}(\Sigma)$.

Another interesting multivariate generalization of β_1 and β_2, in the spirit of projection pursuit, was given by Malkovich and Afifi (1973) as follows: For any vector c of the unit k-sphere S^{k-1}, the measures of skewness and kurtosis of the univariate projection $c^T X$ are

$$\beta_1(c) = \frac{\{E[(c^T X - c^T \mu)^3]\}^2}{\{\text{var}(c^T X)\}^3} \quad \text{and} \quad \beta_2(c) = \left\{ \frac{E[(c^T X - c^T \mu)^4]}{\{\text{var}(c^T X)\}^2} \right\}^2.$$

Malkovich and Afifi (1973) used these measures and defined

$$\beta_1^* = \max_{c \in S^{k-1}} \beta_1(c) \quad \text{and} \quad \beta_2^{*2} = \max_{c \in S^{k-1}} \{\beta_2(c) - 3\}^2$$

as the measures of multivariate skewness and kurtosis, respectively. Baringhaus and Henze (1991) studied the asymptotic distributions of the sample counterparts of these measures for the case of elliptically symmetric distributions which involve supremum of Gaussian processes. Cox and Small (1978) discussed a method similar to the one proposed by Malkovich and Afifi (1973).

Srivastava (1984), using principal components of the covariance matrix Σ, defined measures of multivariate skewness and kurtosis as follows. Let $\lambda_1, \ldots, \lambda_k$ be the eigenvalues of Σ and let $\gamma_1, \ldots, \gamma_k$ be the columns of a matrix Γ such that $\Gamma^T \Sigma \Gamma = \text{diag}(\lambda_1, \ldots, \lambda_k)$. Let $Y_i = \gamma_i^T X$ and $\theta_i = \gamma_i^T \mu$ for $i = 1, 2, \ldots, k$. Then, Srivastava (1984) proposed

$$\bar{\beta}_{1,k}^2 = \frac{1}{k} \sum_{i=1}^{k} \left\{ \frac{E[(Y_i - \theta_i)^3]}{\lambda_i^{3/2}} \right\}^2 \quad \text{and} \quad \bar{\beta}_{2,k} = \frac{1}{k} \sum_{i=1}^{k} \frac{E[(Y_i - \theta_i)^4]}{\lambda_i^2}$$

as the measures of skewness and kurtosis of X, respectively. Let $\bar{b}_{1,k}^2$ and $\bar{b}_{2,k}$ be the corresponding sample measures of multivariate skewness and kurtosis. Then, it has been shown that, under multivariate normality, $\frac{nk}{6} \bar{b}_{1,k}^2$ is asymptotically distributed as χ_k^2 and that $\sqrt{\frac{nk}{24}} (\bar{b}_{2,k} - 3)$ is asymptotically distributed as standard normal.

Lütkepohl and Theilen (1991) used the Choleski decomposition $\Sigma = PP^T$ to define $\nu_j = (\nu_{1j}, \ldots, \nu_{kj})^T = P^{-1}(X_j - \bar{X})$, and $b_{i1} = \frac{1}{n} \sum_{j=1}^{n} \nu_{ij}^3$ and $b_{i2} = \frac{1}{n} \sum_{j=1}^{n} \nu_{ij}^4$. Then, by denoting $b_1 = (b_{11}, \ldots, b_{k1})^T$ and $b_2 = (b_{12}, \ldots, b_{k2})^T$, Lütkepohl and Theilen (1991) observed that Mardia's measures can be expressed as

$$b_{1,k} = \frac{1}{n^2} \sum_{i=1}^{n} \sum_{j=1}^{n} (\nu_i^T \nu_j)^3 \quad \text{and} \quad b_{2,k} = \frac{1}{n} \sum_{i=1}^{n} (\nu_i^T \nu_i)^2,$$

and they also proposed the following alternate measures of multivariate skewness and kurtosis:

$$b_{1,k}^* = \frac{n}{6} b_1^T b_1 \quad \text{and} \quad b_{2,k}^* = \frac{n}{24} (b_2 - 3 \mathbf{1}_k)^T (b_2 - 3 \mathbf{1}_k),$$

where $\mathbf{1}_k$ is a column vector of k 1's. Under the assumption of multivariate normality, it has been shown that the asymptotic joint distribution of $\sqrt{n}(\boldsymbol{b}_1, \boldsymbol{b}_2 - 3\,\mathbf{1}_k)^T$ is multivariate normal with mean $\mathbf{0}$ and covariance matrix $\begin{pmatrix} 6\,\boldsymbol{I}_{k \times k} & \mathbf{0} \\ \mathbf{0} & 24\,\boldsymbol{I}_{k \times k} \end{pmatrix}$, so that $b_{1,k}^*$ and $b_{2,k}^*$ are both asymptotically distributed as χ_k^2.

Koziol (1986, 1987) considered sample measures of multivariate skewness and kurtosis, $\hat{U}_3^2$ and $\hat{U}_4^2$, based on the notion of Neyman's smooth test. In this case, $\hat{U}_3^2$ is equal to $\frac{n}{6}\,b_{1,k}$ and $\hat{U}_4^2$ is related to $b_{2,k}$ as

$$\frac{n}{24}\,\hat{b}_{2,k} = \hat{U}_4^2 + \frac{n}{4}\,b_{2,k} - \frac{nk(k+2)}{8}.$$

The variant $\hat{b}_{2,k} = \frac{1}{n^2}\sum_{i=1}^{n}\sum_{j=1}^{n}\{(\boldsymbol{X}_i - \overline{\boldsymbol{X}})^T \boldsymbol{S}^{-1}(\boldsymbol{X}_j - \overline{\boldsymbol{X}})\}^4$ of $b_{2,k}$, as a higher-degree analog of $b_{1,k}$, was introduced by Koziol (1989), who also established that, under multivariate normality, the asymptotic distribution of $\frac{n}{24}\,\hat{b}_{2,k}$ is noncentral chi-square [see Chapter 29 of Johnson, Kotz and Balakrishnan (1995)] with $\frac{k(k+1)(k+2)(k+3)}{24}$ degrees of freedom and non-centrality parameter $\frac{nk(k+2)}{8}$. In fact, under quite general conditions, Henze (1994b) established that $\sqrt{n}(\hat{b}_{2,k} - \hat{\beta}_{2,k})$ is asymptotically distributed as $N(0, \sigma^2)$, where $\hat{\beta}_{2,k}$ is the population counterpart of $\hat{b}_{2,k}$ defined as $E[\{(\boldsymbol{X} - \boldsymbol{\mu})^T \boldsymbol{\Sigma}^{-1}(\boldsymbol{X} - \boldsymbol{\mu})\}^4]$ and $\sigma^2 = \boldsymbol{c}^T E[\boldsymbol{Z}_3 \boldsymbol{Z}_3^T]\boldsymbol{c}$ with $\boldsymbol{Z}_3$ being a rather complicated $(1 + k + k^2)$-dimensional random vector involving $\boldsymbol{X}_i$'s. In the univariate case, σ^2 takes on the form

$$\sigma^2 = 4\,\mu_4^2\{\mu_8 - \mu_4^2 + 4\,\mu_4(\mu_4^2 - \mu_6) - 8\,\mu_3\mu_5 + 16\,\mu_3^2(1 + \mu_4)\}.$$

Isogai (1983) discussed an elaborate moment evaluation procedure from which measures of skewness and kurtosis follow as special cases. Specifically, measures of multivariate skewness and kurtosis are obtained from eigenvalues of matrices involving third- and fourth-order cumulants. The measure of skewness $b_{1,k}$ and the measure of kurtosis proposed by Box and Watson (1962) are special cases. Under multivariate normality, many of these measures have been shown to be asymptotically distributed as χ^2. A Monte Carlo study comparing the powers of fourteen measures has also been carried out. By making use of dependence measures between influence functions IF$(\boldsymbol{X}; \boldsymbol{\mu})$ and IF$(\boldsymbol{X}; \boldsymbol{\Sigma})$, Isogai (1989) also proposed a general approach in which several previously defined measures are incorporated.

Móri, Rohatgi, and Székely (1994) considered tensors of orders 3 and 4 as multivariate analogs of univariate measures of skewness and kurtosis. From these tensors, they defined the vector $\boldsymbol{A} = E[\|\tilde{\boldsymbol{X}}\|^2 \tilde{\boldsymbol{X}}]$ and the matrix $\boldsymbol{B} = E[\tilde{\boldsymbol{X}}\tilde{\boldsymbol{X}}^T\tilde{\boldsymbol{X}}\tilde{\boldsymbol{X}}^T] - (k+2)\boldsymbol{I}_{k\times k}$, with $\alpha = \|\boldsymbol{A}\|$ and $\beta = $ trace($\boldsymbol{B}$) as real measures of multivariate skewness and kurtosis; here, $\tilde{\boldsymbol{X}} = \boldsymbol{\Sigma}^{-1}(\boldsymbol{X} - \boldsymbol{\mu})$ as defined earlier. Móri, Rohatgi, and Székely (1993) also established some inequalities between α^2 and β, in particular, for central convex unimodal distributions and for infinitely divisible distributions.

A supplementary measure of multivariate skewness was defined by Davis (1980) in studies of the effects of moderate nonnormality on the MANOVA test criterion.[2] In the expansion of the joint density of the latent roots, it was observed that the skewness terms involved not only Mardia's skewness measure $\beta_{1,k}$, but also the quantity

$$
\begin{aligned}
G_k \;=\;\; & E[(\boldsymbol{X} - \boldsymbol{\mu})^T\boldsymbol{\Sigma}^{-1}(\boldsymbol{Y} - \boldsymbol{\mu})(\boldsymbol{X} - \boldsymbol{\mu})^T\boldsymbol{\Sigma}^{-1}(\boldsymbol{X} - \boldsymbol{\mu}) \\
& \times (\boldsymbol{Y} - \boldsymbol{\mu})^T\boldsymbol{\Sigma}^{-1}(\boldsymbol{Y} - \boldsymbol{\mu})],
\end{aligned}
$$

where $\boldsymbol{X}$ and $\boldsymbol{Y}$ are i.i.d. random vectors with mean vector $\boldsymbol{\mu}$ and variance-covariance matrix $\boldsymbol{\Sigma}$. Noting that $G_k = \beta_{1,k}$ when the components of the standardized vector $\boldsymbol{\Sigma}^{-1/2}\boldsymbol{X}$ are independently distributed, the difference

$$
\psi_k = G_k - \beta_{1,k}
$$

was introduced as a supplementary skewness measure. Like $\beta_{1,k}$, ψ_k is invariant under a linear transformation of $\boldsymbol{X}$. The influence of ψ_k is apparently negligible, in general, which means that Mardia's measure $\beta_{1,k}$ expresses the basic effects of skewness on the sampling distribution of the test criteria. As an example, let us consider Marshall–Olkin's bivariate exponential distribution (see Chapter 47)

$$
\Pr[X_1 \geq x_1, X_2 \geq x_2] = e^{-x_1 - x_2 - \lambda \max(x_1, x_2)}, \quad 0 < x_1, x_2 < \infty;
$$

it can be shown in this case that

$$
\beta_{1,2} = \frac{2(3\rho^4 + 9\rho^3 + 15\rho^2 + 12\rho + 4)}{(1+\rho)^3} \quad \text{and} \quad \psi_2 = \frac{2\rho^2(2+\rho)}{(1+\rho)^2},
$$

where the correlation coefficient is $\rho = \frac{\lambda}{\lambda+2}$. When $\lambda = \rho = 0$, X_1 and X_2 are independent, in which case $\beta_{1,2} = 8$ and $\psi_2 = 0$. When $\lambda = 2$ so that $\rho = \frac{1}{2}$, we have $\beta_{1,2} = 8.926$ and $\psi_2 = 0.556$.

Another class of distributions for which the supplementary measure of multivariate skewness vanishes has been given by Davis (1998) as follows. Consider the k-dimensional random vector $\boldsymbol{X} = \boldsymbol{U} + \boldsymbol{V}\boldsymbol{a}$, where $\boldsymbol{U}$

---

[2]Thanks to Dr. A. W. Davis for bringing this development to our attention.

is distributed as k-dimensional normal with mean vector $\mathbf{0}$ and variance–covariance matrix $\mathbf{\Sigma}^*$ (see Chapter 45), V is an univariate random variable distributed independently of $\mathbf{U}$, and $\mathbf{a}$ is a k-dimensional vector of constants. Suppose the moment-generating function of V is $M_V(t)$. Then, the joint moment-generating function of $\mathbf{X}$ is

$$
\begin{aligned}
M_{\mathbf{X}}(\mathbf{t}) = E[e^{\mathbf{t}^T \mathbf{X}}] &= E[e^{\mathbf{t}^T \mathbf{U}}]E[e^{\mathbf{t}^T \mathbf{a}V}] \\
&= e^{\frac{1}{2}\mathbf{t}^T \mathbf{\Sigma}^* \mathbf{t}}\, M_V(\mathbf{t}^T \mathbf{a});
\end{aligned}
\tag{44.105}
$$

see Chapter 45. If the random variable V has cumulants $m, \sigma^2, \kappa_3, \kappa_4, \ldots$, then we obtain from (44.105) the joint cumulant-generating function of $\mathbf{X}$:

$$
\begin{aligned}
K_{\mathbf{X}}(\mathbf{t}) &= \log M_{\mathbf{X}}(\mathbf{t}) \\
&= \mathbf{t}^T \boldsymbol{\mu} + \frac{1}{2}\mathbf{t}^T \mathbf{\Sigma}\mathbf{t} + \sum_{i=3}^{m} \kappa_i(\mathbf{t}^T \mathbf{a})^i/i!\,,
\end{aligned}
\tag{44.106}
$$

where

$$
\boldsymbol{\mu} = E[\mathbf{X}] = m\mathbf{a} \quad \text{and} \quad \mathbf{\Sigma} = \mathrm{Var}(\mathbf{X}) = \mathbf{\Sigma}^* + \sigma^2 \mathbf{a}\mathbf{a}^T.
$$

The cumulants of order 3 or more are thus proportional to products of the components of $\mathbf{a}$ (or $\boldsymbol{\mu}$). Hence, if $\mathbf{X}$ and $\mathbf{Y}$ are independent and identically distributed, then Mardia's measure becomes

$$
\begin{aligned}
E\left[\left\{(\mathbf{X} - \boldsymbol{\mu})^T \mathbf{\Sigma}^{-1}(\mathbf{Y} - \boldsymbol{\mu})\right\}^3\right] &= \kappa_3^2(\mathbf{a}^T \mathbf{\Sigma}^{-1}\mathbf{a})^3 \\
&= \kappa_3^2\left\{\frac{\mathbf{a}^T \mathbf{\Sigma}^{*-1}\mathbf{a}}{1 + \sigma^2(\mathbf{a}^T \mathbf{\Sigma}^{*-1}\mathbf{a})}\right\}^3.
\end{aligned}
\tag{44.107}
$$

Since $E[(\mathbf{X} - \boldsymbol{\mu})^T \mathbf{\Sigma}^{-1}(\mathbf{Y} - \boldsymbol{\mu})(\mathbf{X} - \boldsymbol{\mu})^T \mathbf{\Sigma}^{-1}(\mathbf{X} - \boldsymbol{\mu})(\mathbf{Y} - \boldsymbol{\mu})^T \mathbf{\Sigma}^{-1}(\mathbf{Y} - \boldsymbol{\mu})]$ has the same value, the supplementary measure vanishes.

All the measures described above generalize the univariate measures of skewness and kurtosis, and most of them have been utilized to develop tests for multivariate normality. An empirical comparison of some of these tests has been made by Horswell and Looney (1992, 1993), for example.

In spite of all these developments, not much attention has been paid to generalization of Pearson's measures of skewness and kurtosis in the literature; for example, in the review article of Schwager (1985), only the papers of Isogai (1982) and Oja (1983) are mentioned as ones dealing with such generalizations. Oja (1983) proposed a direction for multivariate skewness as $\boldsymbol{\mu}_2 - \boldsymbol{\mu}_1$, where $\boldsymbol{\mu}_2$ and $\boldsymbol{\mu}_1$ are a generalized mean vector and a generalized median vector, respectively, which are defined by minimization of mean volume of simplexes.

A natural extension of kurtosis and right (left) skewness to multivariate distributions is through a quantile-based approach. MacGillivray (1986) and Balanda and MacGillivray (1990) provided comprehensive discussions on this topic for the univariate case. Extensions to the multivariate case have been discussed by Avérous and Meste (1994, 1997a). The univariate spread function $|F^{-1}(\alpha) - F^{-1}(1 - \alpha)|$, $\alpha \in]0, \frac{1}{2}]$, which is essential for the study of univariate kurtosis (see Chapter 12), has been generalized by Avérous and Meste (1994) using multivariate interquartile regions introduced earlier by the authors and called *median balls*; see Avérous and Meste (1997b). Avérous and Meste (1994) also extended the kurtosis measures and kurtosis orders along the lines of those for the univariate case discussed by Balanda and MacGillivray (1990).

Avérous and Meste (1997a) discussed two ways of extending the comparative, qualitative, and quantitative concepts of weak skewness to the right to the corresponding concepts of weak multivariate skewness in a given direction $h \in S^{k-1}(\| \cdot \|)$, where $\| \cdot \|$ is a given norm; one uses the location of the median balls, while the other introduces *weight of a tail in the direction h* and then uses the difference between the weights of tails in opposite directions. These two extensions are not equivalent even though they extend the two equivalent definitions of univariate weak skewness to the right:

(i) $\forall \alpha \in]0, 1[$, $\frac{F^{-1}(\alpha) + F^{-1}(1-\alpha)}{2} - \text{Median}(F) \geq 0$;

(ii) $\forall x \in \mathbb{R}$, $1 - F(\text{Median}(F) + x) - F(\text{Median}(F) - x) \geq 0$.

These measures and orders are related to the spatial median m associated with an arbitrary norm $\| \cdot \|$ and defined as

$$m = \text{argmin}_{\boldsymbol{c} \in \mathbb{R}^k} E[\|\boldsymbol{X} - \boldsymbol{c}\| - \|\boldsymbol{X}\|].$$

For kurtosis measures, the invariance properties depend on the scaling technique; the skewness measures are translation and homothety invariant as well as rotationally equivariant. Based on depths or on penalized optimization, other multivariate quantiles similar to multivariate extensions of the univariate median have been developed, but their application to multivariate skewness and kurtosis has not been explored very much. One exception is the work of Chaudhuri (1996) in which a measure based on another family of interquantile regions for the spatial median has been introduced.

BIBLIOGRAPHY

(Some bibliographical items not mentioned in the text are included here for completeness.)

Abazaliev, A. K. J. (1968). Characteristic coefficients of two-dimensional distributions and their applications, *Soviet Mathematics-Doklady*, **9**, 52–56 [English translation; Russian original in *Doklady Akademii Nauk SSSR*, **178** (1968)].

Abdul-Hamid, H., and Nolan, J. P. (1999). Multivariate stable densities as functions of one dimensional projections, *Journal of Multivariate Analysis* (to appear).

Anderson, D. N. (1992). A multivariate Linnik distribution, *Statistics & Probability Letters*, **14**, 333–336.

Anderson, D. N., and Arnold, B. C. (1991). Centered distributions with Cauchy conditionals, *Communications in Statistics—Theory and Methods*, **20**, 2881–2889.

Ardanuy, R., and Sánchez, J. M. (1993). Multivariate skewness and kurtosis for singular distributions, *Extracta Mathematica*, **8**, 98–101.

Arnold, B. C., Castillo, E., and Sarabia, J.-M. (1992). *Conditionally Specified Distributions*, Lecture Notes in Statistics—**73**, New York: Springer-Verlag.

Arnold, B. C., and Pourahmadi, M. (1988). Conditional characterizations of multivariate distributions, *Metrika*, **35**, 99–108.

Arnold, B. C., and Press, S. J. (1989). Compatible conditional distributions, *Journal of the American Statistical Association*, **84**, 152–156.

Arnold, B. C., and Strauss, D. J. (1988). Bivariate distributions with conditional exponentials, *Journal of the American Statistical Association*, **83**, 522–527.

Arnold, B. C., and Strauss, D. J. (1991). Bivariate distributions with conditionals in prescribed exponential families, *Journal of the Royal Statistical Society, Series B*, **53**, 365–375.

Assaf, D., Langberg, N. A., Savits, T. H., and Shaked M. (1984). Multivariate phase type distribution, *Operations Research*, **32**, 688–702.

Avérous, J., and Meste, M. (1994). Multivariate kurtosis in L_1 sense, *Statistics & Probability Letters*, **19**, 281–284.

Avérous, J., and Meste, M. (1997a). Skewness for multivariate distributions: Two approaches, *Annals of Statistics*, **25**,

Avérous, J., and Meste, M. (1997b). Median balls: An extension of the interquantile intervals to multivariate distributions, *Journal of Multivariate Analysis*, **63**, 222–241.

Azzalini, A., and Dalla Valle, A. (1983). The multivariate skewnormal distribution, *Biometrika*, **4**, 715–726.

Bairamov, I. G., and Kotz, S. (1999). On a generalization of FGM and Sarmanov-Lee class of distributions, *Life-time Data Analysis* (to appear).

Balanda, K. P., and MacGillivray, H. L. (1990). Kurtosis and spread, *Canadian Journal of Statistics*, **18**, 17–30.

Banys, M. I. (1970). The integral multivariate limit theorem for convergence to a stable limiting distribution, *Lietuvos Matematikos Rinkinys*, **10**, 665–672 (in Russian).

Baringhaus, L., and Henze, N. (1991). Limit distributions for measures of multivariate skewness and kurtosis based on projections, *Journal of Multivariate Analysis*, **38**, 51–69.

Baringhaus, L., and Henze, N. (1992). Limit distributions for Mardia's measure of multivariate skewness, *Annals of Statistics*, **20**, 1889–1902.

Bélisle, C. J. P., Romeijn, H. E., and Smith, R. L. (1990). Hit-and-run algorithms for generating multivariate distributions, *Technical Report No. 90-18*, Department of Industrial and Operations Engineering, University of Michigan, Ann Arbor, MI.

Berge, P. O. (1938). A note on a form of Tchebycheff's theorem for two variables, *Biometrika*, **29**, 405–406.

Bikelis, A. (1968a). On the multivariate characteristic functions, *Lietuvos Matematikos Rinkinys*, **8**, 21–39 (in Russian).

Bikelis, A. (1968b). The asymptotic expansions of the distribution function and of density functions for the sums of independent identically distributed random variables, *Lietuvos Matematikos Rinkinys*, **8**, 405–422.

Bikelis, A. (1970a). Two inequalities for the multivariate characteristic function, *Lietuvos Matematikos Rinkinys*, **10**, 1–12 (in Russian).

Bikelis, A. (1970b). Asymptotic expansions of the distribution function of the sum of independent identically distributed non-lattice random vectors, *Lietuvos Matematikos Rinkinys*, **10**, 673–679 (in Russian).

Bikelis, A., and Mogyoródi, J. (1970). On asymptotic expansion of the convolution of n multidimensional distribution functions, *Lietuvos Matematikos Rinkinys*, **10**, 433–443.

Bildikar, S., and Patil, G. P. (1968). Multivariate exponential-type distributions, *Annals of Mathematical Statistics*, **9**, 1316–1326.

Box, G. E. P., and Muller, M. E. (1958). A note on the generation of random normal deviates, *Annals of Mathematical Statistics*, **29**, 610–611.

Box, G. E. P., and Watson, G. S. (1962). Robustness to non-normality of regression tests, *Biometrika*, **49**, 93–106.

Browne, M. W., and Shapiro, A. (1987). Adjustments for kurtosis in factor analysis with elliptically distributed errors, *Journal of the Royal Statistical Society, Series B*, **49**, 346–352.

Bryc, W. (1985). Some remarks on random vectors with nice enough behaviour of conditional moments, *Bulletin of the Polish Academy of Sciences, Mathematics*, **33**, 677–684.

Bryc, W., and Plucińska, A. (1985). A characterization of infinite Gaussian sequences by conditional moments, *Sankhyā, Series A*, **47**, 166–173.

Byczkowski, T., Nolan, J. P., and Rajput, B. (1993). Approximation of multidimensional stable densities, *Journal of Multivariate Analysis*, **46**, 13–31.

Cambanis, S. (1977). Some properties and generalizations of multivariate Eyraud–Gumbel–Morgenstern distributions, *Journal of Multivariate Analysis*, **7**, 551–559.

Cambanis, S. (1993). On Eyraud–Gumbel–Morgenstern random processes, in *Stochastic Inequalities*, IMS Lecture Notes—**22**, Hayward, California.

Castillo, E., and Galambos, J. (1987). Bivariate distributions with normal conditionals, in *Proceedings of the International Association of Science and Technology for Development*, pp. 59–62, Anaheim, California: Acta Press.

Castillo, E., and Galambos, J. (1989). Conditional distributions and the bivariate normal distribution, *Metrika*, **36**, 209–214.

Chakak, A., and Koehler, K. J. (1995). A strategy for constructing multivariate distributions, *Communications in Statistics—Simulation and Computation*, **24**, 537–550.

Chambers, J. M. (1967). On methods of asymptotic approximation for multivariate distributions, *Biometrika*, **54**, 367–383.

Chaudhuri, P. (1996). On a geometric notion of quantiles for multivariate data, *Journal of the American Statistical Association*, **91**, 862–872.

Chikuse, Y. (1990). Functional forms of characteristic functions and characterizations of multivariate distributions, *Econometric Theory*, **6**, 445–458.

Chukova, S., and Dimitrov, B. (1992). On the distributions having almost lack-of-memory property, *Journal of Applied Probability*, **29**, 691–698.

Chukova, S., Dimitrov, B., and Khalil, Z. (1993). Probability distributions similar to the exponential, *Canadian Journal of Statistics*, **21**, 269–276.

Cohen, L. (1984). Probability distributions with given multivariate marginals, *Journal of Mathematical Physics*, **25**, 2402–2403.

Cole, B. F., Lee, M-L. T., Whitmore, G. A., and Zaslavsky, A. M. (1995). An empirical Bayes model for Markov-dependent binary sequences with randomly missing observations, *Journal of the American Statistical Association*, **90**, 1364–1372.

Conway, D. (1979). Multivariate distributions with specified marginals, Ph.D. Dissertation, Stanford University, Stanford, California.

Conway, D. (1983). Farlie–Gumbel–Morgenstern distributions, In *Encyclopedia of Statistical Sciences—3* (S. Kotz and N. L. Johnson, eds.), pp. 28–31, New York: John Wiley & Sons.

Cooper, P. W. (1963). Multivariate extensions of univariate distributions, *IEEE Transactions on Electronic Computers*, **12**, 572–573.

Cox, D. R., and Small, N. J. H. (1978). Testing multivariate normality, *Biometrika*, **65**, 263–272.

Crofts, A. E. (1969). *An Investigation of Normal Lognormal Distributions*, Technical Report No. 32, Themis Contract, Department of Statistics, Southern Methodist University, Dallas, Texas.

Crow, E. L., and Shimizu, K. (eds.) (1988). *Lognormal Distributions: Theory and Applications*, New York: Marcel Dekker.

Cuadras, C. M. (1981). *Methods de analisis multivariante*, Barcelona, Spain: University of Barcelona.

Dall'Aglio, G. (1960). Les fonctions extremes de la classe de Fréchet à trois dimensions, *Inst. Stat. Univ. Paris*, **9**, 175–188.

Dall'Aglio, G. (1991). Fréchet classes. The beginnings, in *Advances in Probability Distributions with Given Marginals* (G. Dall'Aglio et al., eds.), pp. 1–12, Dordrecht, The Netherlands: Kluwer Academic Publishers.

Davis, A. W. (1980). On the effects of moderate multivariate nonnormality on Wilks's likelihood ratio criterion, *Biometrika*, **67**, 419–427.

Davis, A. W. (1998). Private communication.

Day, N. E. (1969). Linear and quadratic discrimination in pattern recognition, *IEEE Transactions on Information Theory*, **15**, 419–421.

De Silva, B. M. (1978). A class of multivariate symmetric stable distributions, *Journal of Multivariate Analysis*, **8**, 335–345.

Dimitrov, B., Chukova, S., and Khalil, Z. (1994). Bivariate probability distributions similar to exponential, in *Approximation, Probability and Related Fields*, pp. 167-178, New York: Plenum.

Dunnage, J. E. A. (1970a). The speed of convergence of the distribution functions in the two-dimensional central limit theorem, *Proceedings of the London Mathematical Society, Series 3*, **20**, 33–59.

Dunnage, J. E. A. (1970b). On the remainder in the two-dimensional central limit theorem, *Proceedings of the Cambridge Philosophical Society*, **68**, 455–458.

Eagleson, G. K. (1964). Polynomial expansions of bivariate distributions, *Annals of Mathematical Statistics*, **35**, 1208–1215.

Edgeworth, F. Y. (1896). The compound law of error, *Philosophical Magazine, 5th Series*, **41**, 207–215.

Edgeworth, F. Y. (1917). On the mathematical representation of statistical data, *Journal of the Royal Statistical Society, Series A*, **80**, 266–288.

Elandt-Johnson, R. C. (1976). Conditional failure time distributions under competing risk theory with dependent failure times and proportional hazard rates, *Scandinavian Actuarial Journal*, 37–51.

Elderton, W. P., and Johnson, N. L. (1969). *Systems of Frequency Curves*, London: Cambridge University Press.

Farlie, D. J. G. (1960). The performance of some correlation coefficients for a general bivariate distribution, *Biometrika*, **47**, 307–323.

Feldheim, M. E. (1937). Étude de la stabilité des lois de probabilités, *Thèse de la Faculté des Sciences de Paris*, Paris, France.

Fernandez, C., Osiewalski, J., and Steel, M. F. J. (1995). Modeling and inference with v-spherical distributions, *Journal of the American Statistical Association*, **90**, 1331–1340.

Fleishman, A. I. (1978). A method for simulating nonnormal distributions, *Psychometrika*, **43**, 521–532.

Fréchet, M. (1951). Sur les tableaux de corrélation dont les marges sont données, *Annales de l'Université de Lyon, Section A, Series 3*, **14**, 53–77.

Galambos, J., and Lee, M.-Y. (1992). Extensions of some univariate Bonferroni-type inequalities to multivariate setting, in *Probability Theory and Applications* (J. Galambos and I. Kátai, eds.), pp. 143–154, Dordrecht, The Netherlands: Kluwer Academic Publishers.

Galambos, J., and Lee, M.-Y. (1994). Further studies of bivariate Bonferroni-type inequalities, *Journal of Applied Probability*, **31**, 63–69.

Galambos, J., and Simonelli, I. (1996). *Bonferroni-type Inequalities with Applications*, New York: Springer-Verlag.

Galton, F. (1877). Typical laws of heredity in man, *Proceedings of the Royal Institution of Great Britain*, **8**, 282–301.

Gastwirth, J. L., and Owens, M. E. B. (1977). On classical tests of normality, *Biometrika*, **64**, 135–139.

Gayen, A. K. (1951). The frequency distribution of the product-moment correlation coefficient in random samples of any size drawn from non-normal universes, *Biometrika*, **38**, 219-247.

Ghosh, P. (1990). Multivariate stable distributions, a natural generalization of Cauchy law and some related questions, *Calcutta Statistical Association Bulletin*, **40**, 153–161.

Godwin, H. J. (1964). *Inequalities on Distribution Functions*, London: Griffin.

Good, I. J. (1983a). Conjectures concerning the cumulants of the multivariate lognormal distribution (C180), *Journal of Statistical Computation and Simulation*, **18**, 239–241.

Good, I. J. (1983b). The multivariate lognormal distribution and colored trees (C182), *Journal of Statistical Computation and Simulation*, **18**, 316–322.

Goodman, I. (1972). Ph.D. dissertation, Department of Mathematics, Temple University, Philadelphia.

Griffiths, R. C. (1970). Positive definite sequences and canonical correlation coefficients, *Australian Journal of Statistics*, **12**, 162–165.

Guldberg, S. (1920). Application des polynômes d'Hermite à un problème de statistique, *Proceedings of the International Congress of Mathematicians*, Strasbourg, pp. 552–560.

Gupta, A. K., Nguyen, T. T., and Zeng, W. B. (1995). Conditions for stability of laws with all projections stable, *Sankhyā, Series A*, **57**,

Harris, R. (1970). A multivariate definition for increasing hazard rate distribution functions, *Annals of Mathematical Statistics*, **41**, 713–717.

Harris, W. A., and Helvig, T. N. (1966). Marginal and conditional distributions of singular distributions, *Publications of the Research Institute for Mathematical Sciences, Series A*, **1**, 199–204.

Hashorva, E., and Hüsler, J. (1999). Extreme values in FGM random sequences, *Journal of Multivariate Analysis*, **68**, 212–225.

Helemäe, H.-L., and Tiit, E.-M. (1996). Multivariate minimal distributions, *Proc. Estonian Acad. Sci. Phys. Math.*, **45**, 317–322.

Henze, N. (1994a). On Mardia's kurtosis test for multivariate normality, *Communications in Statistics—Theory and Methods*, **23**, 1031–1045.

Henze, N. (1994b). The asymptotic behavior of a variant of multivariate kurtosis, *Communications in Statistics—Theory and Methods*, **23**, 1047–1061.

Hirschfeld, H. O. (Hartley, H. O.) (1935). A connection between correlation and contingency, *Proceedings of the Cambridge Philosophical Society*, **31**, 520–524.

Holland, P. W., and Wang, Y. J. (1987). Dependence function for continuous bivariate densities, *Communications in Statistics—Theory and Methods*, **16**, 863–876.

Horswell, R. L., and Looney, S. W. (1992). A comparison of tests for multivariate normality that are based on measures of multivariate skewness and kurtosis, *Journal of Statistical Computation and Simulation*, **42**, 21–38.

Horswell, R. L., and Looney, S. W. (1993). Diagnostic limitation of skewness coefficients in assessing departures from univariate and multivariate normality, *Communications in Statistics—Simulation and Computation*, **22**, 435–459.

Huang, J. S., and Kotz, S. (1984). Correlation structure in the iterated Farlie–Gumbel–Morgenstern distributions, *Biometrika*, **71**, 633–636.

Huang, J. S., and Kotz, S. (1999). On modifications of the Farlie–Gumbel–Morgenstern distributions. A tough hill to climb, *Metrika*, **49**, 135–145.

Hüsler, J. (1996). Multivariate option price models and extremes, *Communications in Statistics—Theory and Methods*, **25**, 853–869.

Hutchinson, T. P., and Lai, C. D. (1990). *Continuous Bivariate Distributions, Emphasizing Applications*, Adelaide, Australia: Rumsby Scientific Publishing.

Isii, K. (1959). On a method for generalizations of Tchebycheff's inequality, *Annals of the Institute of Statistical Mathematics*, **10**, 65–88.

Isogai, T. (1982). On a measure of multivariate skewness and a test for multivariate normality, *Annals of the Institute of Statistical Mathematics*, **34**, 531–541.

Isogai, T. (1983). On measures of multivariate skewness and kurtosis, *Mathematica Japonica*, **28**, 251–261.

Isogai, T. (1989). On using influence functions for testing multivariate normality, *Annals of the Institute of Statistical Mathematics*, **41**, 169–186.

Iwase, K., Shimizu, K., and Suzuki, M. (1982). On UMVU estimators for the multivariate lognormal distribution and their variances, *Communications in Statistics—Theory and Methods*, **11**, 687–697.

Jensen, D. R. (1971). A note on positive dependence and the structure of bivariate distributions, *SIAM Journal of Applied Mathematics*, **20**, 749–752.

Joarder, A. H. (1997). On the characteristic function of the multivariate Pearson type II distribution, *Journal of Information and Optimization Sciences*, **18**, 177-182.

Johnson, M. E. (1987). *Multivariate Statistical Simulation*, New York: John Wiley & Sons.

Johnson, M. E., and Tenenbein, A. (1981). A bivariate distribution family with specified marginals, *Journal of the American Statistical Association*, **76**, 198–201.

Johnson, N. L. (1949). Bivariate distributions based on simple translation systems, *Biometrika*, **36**, 297–304.

Johnson, N. L., and Kotz, S. (1975). On some generalized Farlie-Gumbel-Morgenstern distributions, *Communications in Statistics*, **4**, 415–427.

Johnson, N. L., and Kotz, S. (1977). On some generalized Farlie-Gumbel-Morgenstern distributions II. Regressions, correlation and further generalizations, *Communications in Statistics*, **6**, 485–496.

Johnson, N. L., Kotz, S., and Balakrishnan, N. (1994). *Continuous Univariate Distributions*, Vol. 1, second edition, New York: John Wiley & Sons.

Johnson, N. L., Kotz, S., and Balakrishnan, N. (1995). *Continuous Univariate Distributions*, Vol. 2, second edition, New York: John Wiley & Sons.

Jones, M. C. (1996). The local dependence function, *Biometrika*, **83**, 899–904.

Jones, M. C. (1999). Personal communcation.

Jones, R. M., and Miller, K. S. (1966). On the multivariate lognormal distribution, *Journal of Industrial Mathematics*, **16**, 63–76.

Jørgensen, N. R. (1916). *Undersøgelser over Frequensflader of Correlation*, Copenhagen: Busch.

Kagan, A. M. (1988). New classes of dependent random variables and generalization of the Darmois-Skitovitch theorem to the case of several forms, *Teor. Veroyatnost. i Primenen.*, **33**, 305–315 (in Russian).

Kalinauskaité, N. (1970a). On some expansions for the multidimensional stable densities with parameter $\alpha > 1$, *Lietuvos Matematikos Rinkinys*, **10**, 490–495 (in Russian).

Kalinauskaité, N. (1970b). On some expansions for the multidimensional symmetric stable densities, *Lietuvos Matematikos Rinkinys*, **10**, 726–731 (in Russian).

Karlin, S., and Studden, W. J. (1966). *Tchebycheff Systems: With Applications in Analysis and Statistics*, New York: John Wiley & Sons.

Kemp, J. F. (1973). Advanced Problem 5894, *American Mathematical Monthly*, **80**.

Knoebel, B. R., and Burkhart, H. E. (1991). A bivariate distribution approach to modeling forest diameter distributions at two points in time, *Biometrics*, **47**, 241–253.

Kocherlakota, S., Kocherlakota, K. and Balakrishnan, N. (1985). Effects of nonnormality on the SPRT for the correlation coefficient: Bivariate Edgeworth series distribution, *Journal of Statistical Computation and Simulation*, **23**, 41–51.

Kocherlakota, K., Kocherlakota, S., and Balakrishnan, N. (1986). Random number generation from a bivariate Edgeworth series distribution, *Computational Statistics Quarterly*, **2**, 97–105.

Koehler, K. J., and Symanowski, J. T. (1995). Constructing multivariate distributions with specific marginal distributions, *Journal of Multivariate Analysis*, **55**, 261–282.

Koopmans, L. H. (1969). Some simple singular and mixed probability distributions, *American Mathematical Monthly*, **76**, 297–299.

Kotz, S. (1975). Multivariate distributions at a cross roads, n *Statistical Distributions in Scientific Work*, Vol. 1 (G. P. Patil, C. Taillie, and B. Baldessari, eds.), pp. 247–270, Dordrecht, The Netherlands: D. Reidel.

Kotz, S., Fang, K. T., and Liang, J.-J. (1997). On multivariate vertical density representation and its application to random number generation, *Statistics*, **30**, 163–180.

Kotz, S., and Seeger, J. P. (1991). A new approach to dependence in multivariate distributions, n *Advances in Probability Distributions with Given Marginals* (G. Dall'Aglio *et al.*, eds.), pp. 113–127, Dordrecht, The Netherlands: Kluwer Academic Publishers.

Kotz, S., and Tiit, E.-M. (1992). Bounds in multivariate dependence, *Acta et Comment. Univ. Tartuensis*, **942**, 35–45.

Kotz, S., and Troutt, M. D. (1996). Vertical density representation, *Statistics*, **28**, 241–247.

Koziol, J. A. (1986). A note on the asymptotic distribution of Mardia's measure of multivariate kurtosis, *Communications in Statistics—Theory and Methods*, **15**, 1507–1513.

Koziol, J. A. (1987). An alternative formulation of Neyman's smooth goodness of fit test under composite alternatives, *Metrika*, **34**, 17–24.

Koziol, J. A. (1989). A note on measures of multivariate kurtosis, *Biometrical Journal*, **31**, 619–624.

Lal, D. N. (1955). A note on a form of Tchebycheff's inequality for two or more variables, *Sankhyā*, **15**, 317–320.

Lancaster, H. O. (1957). Some properties of the bivariate normal distribution considered in the form of a contingency table, *Biometrika*, **44**, 289–292.

Lancaster, H. O. (1958). The structure of bivariate distributions, *Annals of Mathematical Statistics*, **29**, 719–736.

Lancaster, H. O. (1963). Correlations and canonical forms of bivariate distributions, *Annals of Mathematical Statistics*, **34**, 532–538.

Lancaster, H. O. (1965). Symmetry in multivariate distributions, *Australian Journal of Statistics*, **7**, 115–126.

Lancaster, H. O. (1969). *The Chi-Squared Distribution*, New York: John Wiley & Sons.

Le, H., and O'Hagan, A. (1998). A class of bivariate heavy-tailed distributions, *Sankhyā, Series B*, Special Issue on Bayesian Analysis, **60**, 82–100.

Lee, M-L. T. (1994). A family of multivariate density functions with given marginals, Preprint.

Lee, M-L. T. (1996). Properties and applications of the Sarmanov family of bivariate distributions, *Communications in Statistics—Theory and Methods*, **25**, 1207-1222.

Lee, M.-Y. (1992). Bivariate Bonferroni inequalities, *Aequationes Mathematicae*, **44**, 220–225.

Lee, M.-Y. (1996). Improved bivariate Bonferroni-type inequalities, *Statistics & Probability Letters*, **31**, 359–364.

Leser, C. E. V. (1942). Inequalities for multivariate frequency distributions, *Biometrika*, **32**, 284-293.

Lévy, P. (1937). *Théorie de l'Addition des Variables Aléatoires*, second edition, Paris: Gauthier Villars.

Lien, D.-H. D. (1985). Moments of truncated bivariate log-normal distributions, *Economics Letters*, **19**, 243–247.

Lien, D.-H. D. (1986). Moments of ordered bivariate log-normal distributions, *Economics Letters*, **20**, 45–47.

Lien, D.-H. D., and Rearden, D. (1996). A property of coefficients of variation for ordered bivariate log-normal variables, *Communications in Statistics—Theory and Methods*, **25**, 1903–1916.

Lien, D.-H. D., and Rearden, D. (1998). Cross moments of extreme observations from a multivariate lognormal distribution, *Communications in Statistics—Theory and Methods*, **27**, 601–607.

Lin, G. D. (1987). Relationship between two extensions of Farlie–Gumbel–Morgenstern distributions, *Annals of the Institute of Statistical Mathematics*, **39**, 129–140.

Linnik, Yu. V. (1963). Linear forms and statistical criteria: I and II, *Selected Translations in Mathematical Statistics and Probability*, **3**, 1–40, 41–90.

Lukomski, J. (1939). On some properties of multidimensional distributions, *Annals of Mathematical Statistics*, **10**, 236–246.

Lütkepohl, H., and Theilen, B. (1991). Measures of multivariate skewness and kurtosis, *Statistical Papers*, **32**, 179–193.

MacGillivray, H. L. (1986). Skewness and asymmetry: Measures and orderings, *Annals of Statistics*, **14**, 994–1011.

Malkovich, J. F., and Afifi, A. A. (1973). On tests for multivariate normality, *Journal of the American Statistical Association*, **68**, 176–179.

Mallows, C. L., and Riordan, J. (1968). The inversion enumerator for labelled trees, *Bulletin of the American Mathematical Society*, **74**, 92–94.

Mamatkulov, K. K. (1970). Quadratic estimation of the density function of bivariate lognormal density function from a sample, *Izvestia Akademii Nauk Uzbek SSR, Seria Fiziko-Matematicheskikh Nauk*, **14**, No. 1, 17–20 (in Russian).

Mardia, K. V. (1970a). A translation family of bivariate distributions and Fréchet's bounds, *Sankhyā, Series A*, **32**, 119–121.

Mardia, K. V. (1970b). *Families of Bivariate Distributions*, London: Griffin.

Mardia, K. V. (1970c). Measures of multivariate skewness and kurtosis with applications, *Biometrika*, **57**, 519–530.

Mardia, K. V. (1974). Applications of some measures of multivariate skewness and kurtosis for testing normality and robustness studies, *Sankhyā, Series B*, **36**, 115–128.

Mardia, K. V. (1975). Assessment of multinormality and the robustness of the Hotelling's T^2 test, *Applied Statistics*, **24**, 163–171.

Mardia, K. V., and Foster, K. (1983). Omnibus tests of multinormality based on skewness and kurtosis, *Communications in Statistics—Theory and Methods*, **12**, 207–221.

Mardia, K. V., and Kanazawa, M. (1983). The null distribution of multivariate kurtosis, *Communications in Statistics—Simulation and Computation*, **12**, 569–576.

Mardia, K. V., and Zemroch, P. J. (1975). Measures of multivariate skewness and kurtosis, *Applied Statistics*, **24**, 262–265.

Marsaglia, G. (1964). Conditional means and covariances of normal variables with singular covariance matrix, *Journal of the American Statistical Association*, **59**, 1203–1204.

Marshall, A. W. and Olkin, I. (1988). Families of multivariate distributions, *Journal of the American Statistical Association*, **83**, 834–841.

Maung, K. (1941). Measurement of association in a contingency table with special reference to the pigmentation of hair and eye colours of Scottish school children, *Annals of Eugenics*, **11**, 189–223.

Meixner, J. (1934). Orthogonale Polynomsyteme mit einer besonderen Gestalt des erzeugenden Funktion, *Journal of the London Mathematical Society*, **9**, 6–13.

Meyer, R. M. (1969). Note on a "multivariate" form of Bonferroni's inequalities, *Annals of Mathematical Statistics*, **40**, 692–693.

Mihaĭla, I. M. (1968). Development of the trivariate frequency function in Gram–Charlier series, *Revue Roumaine de Mathématiques Pures et Appliquées*, **13**, 803–813.

Modarres, R., and Nolan, J. P. (1994). A method for simulating stable random vectors, *Computational Statistics*, **9**, 11–19.

Molenberghs, G., and Lesaffre, E. (1994). Marginal modeling of correlated ordinal data using a multivariate Plackett distribution, *Journal of the American Statistical Association*, **89**, 633–644.

Molenberghs, G., and Lesaffre, E. (1997). Non-linear integral equations to approximate bivariate densities with given marginals and dependence function, *Statistica Sinica*, **7**, 713–738.

Morgenstern, D. (1956). Einfache Beispiele zweidimensionaler Verteilungen, *Mitteilungsblatt für Mathematische Statistik*, **8**, 234–235.

Móri, T. F., Rohatgi, V., and Székely, G. J. (1994). On multivariate skewness and kurtosis, *Theory of Probability and Its Applications*, **38**, 547–551.

Mostafa, M. D., and Mahmoud, M. W. (1964). On the problem of estimation for the bivariate lognormal distribution, *Biometrika*, **51**, 522–527.

Narumi, S. (1923). On the general forms of bivariate frequency distributions which are mathematically possible when regression and variation are subjected to limiting conditions, *Biometrika*, **15**, 77–88, 209–221.

Nataf, A. (1962). Détermination des distributions dont les marges sont données, *Comptes Rendus de l'Académie des Sciences, Paris*, **225**, 42–43.

Nelsen, R. G. (1994). A characterization of Farlie-Gumbel-Morgenstern distributions via Spearman rho and chi-square divergence, *Sankhyā, Series A*, **56**, 476–479.

Neuts, M. F. (1975). Probability distributions of phase type, in *Liber Amicorum Prof. H. Florin*, pp. 173–206, Belgium, University of Louvain.

Neyman, J. (1926). Further notes on non-linear regression, *Biometrika*, **18**, 257–262.

Nguyen, T. T. (1995). Conditional distributions and characterizations of multivariate stable distributions, *Journal of Multivariate Analysis*, **53**, 181–193.

Nguyen, T. T., and Sampson, A. R. (1991). A note on characterizations of multivariate stable distributions, *Annals of the Institute of Statistical Mathematics*, **43**, 793–801.

Nolan, J. P. (1996). An overview of multivariate stable distributions, Preprint, American University, Washington, D.C.

Nolan, J. P. (1998). Multivariate stable distributions: approximation, estimation, simulation and identification, in *A Practical Guide to Heavy Tails*, (R. Adler, R. E. Feldman, and M. S. Taqqu, eds.), pp. 509–525, Boston: Birkhaüser.

Nolan, J. P., and Panorska, A. K. (1997). Data analysis for heavy tailed multivariate samples, *Communications in Statistics—Stochastic Models*, **13**, 687–702.

Nolan, J. P., and Rajput, B. (1995). Calculation of multidimensional stable densities, *Communications in Statistics—Simulation and Computation*, **24**, 551–566.

O'Cinneide, C. A. (1987). On the limitations of the multivariate phase-type family, Technical Report, University of Arkansas.

O'Hagan, A., and Le, H. (1994). Conflicting information and a class of bivariate heavy-tailed distributions, in *Aspects of Uncertainty: A Tribute to D. V. Lindley* (P. R. Freeman and A. F. M. Smith, eds.), pp. 311–327, Chichester, England: John Wiley & Sons.

O'Hagan, A., and Le, H. (1999). Conflicting information and a class of bivariate heavy-tailed distributions, preprint.

Oja, H. (1983). Descriptive statistics for multivariate distributions, *Statistics & Probability Letters*, **1**, 327–332.

Olkin, I., and Pratt, J. W. (1958). A multivariate Tchebycheff inequality, *Annals of Mathematical Statistics*, **29**, 226–234.

Ord, J. K. (1972). *Families of Frequency Distributions*, London: Griffin.

Ostrovskii, I. V. (1995). Analytic and asymptotic properties of multivariate Linnik's distribution, *Mathematical Physics, Analysis, and Geometry*, **2**, 436–455 (in English).

Paulauskas, V. J. (1970). On the multidimensional central limit theorem, *Lietuvos Matematikos Rinkinys*, **10**, 783–789 (in Russian).

Paulauskas, V. J. (1976). Some remarks on multivariate stable distributions, *Journal of Multivariate Analysis*, **6**, 356–368.

Pearson, K. (1905). On the general theory of skew correlation and nonlinear regression, *Drapers' Company Research Memoirs, Biometric Series*, **2**.

Pearson, K. (1923a). Notes on skew frequency surfaces, *Biometrika*, **15**, 222–230.

Pearson, K. (1923b). On non-skew frequency surfaces, *Biometrika*, **15**, 231–244.

Peristiani, S. (1991). The F-system distribution as an alternative multivariate normality: An application in multivariate models with qualitative dependent variables, *Communications in Statistics—Theory and Methods*, **20**, 147–163.

Plackett, R. L. (1965). A class of bivariate distributions, *Journal of the American Statistical Association*, **60**, 516–522.

Press, S. J. (1972). Multivariate stable distributions, *Journal of Multivariate Analysis*, **2**, 444–462.

Pretorius, S. J. (1930). Skew bivariate frequency surfaces, examined in the light of numerical illustrations, *Biometrika*, **22**, 109-223.

Rhodes, E. C. (1923). On a certain skew correlation surface, *Biometrika*, **14**, 355–377.

Risser, R. (1945). Sur l'équation caractéristique des surfaces de probabilité, *Comptes Rendus, Académie des Sciences, Paris*, **220**, 31–32.

Risser, R. (1947). D'un certain mode de recherche des surfaces de probabilité, *Comptes Rendus, Académie des Sciences, Paris*, **225**, 1266–1268.

Risser, R. (1950). Calcul des constantes de certaines surfaces de distribution, *Bulletin des Actuaires Français*, **191**, 141–232.

Risser, R., and Traynard, C. E. (1957). *Les Principes de la Statistique Mathématique*, **2** (Part 2), Paris: Gauthier Villars.

Roux, J. J. J. (1971). A characterization of a multivariate distribution, *South African Statistical Journal*, **5**, 27–36.

Rüschendorf, L. (1983). Construction of multivariate distribution with given marginals, *Annals of the Institute of Statistical Mathematics*, **37**, 225–233.

Sagrista, S. N. (1952). On a generalization of Pearson's curves to the two-dimensional case, *Trabajos de Estadistica*, **3**, 273–314 (in Spanish).

Samorodnitsky, G., and Taqqu, M. S. (1994). *Stable Non-Gaussian Random Processes*, New York: Chapman and Hall.

Sarmanov, O. V. (1958). Maximal correlation coefficient. Symmetrical case, *Doklady Akademii Nauk SSSR*, **120**, 715–718 (in Russian).

Sarmanov, O. V. (1965). On the method of characteristic coefficients, *Soviet Mathematics-Doklady*, **6**, 1083–1091 [English translation, Russian original in *Doklady Akademii Nauk SSSR*, **163** (1956)].

Sarmanov, O. V. (1966). Generalized normal correlation and two-dimensional Fréchet classes, *Soviet Mathematics—Doklady*, **7**, 596–599 [English translation; Russian original in *Doklady Akademii Nauk SSSR*, **108** (1966)].

Sarmanov, O. V., and Bratoeva, Z. N. (1967). Probabilistic properties of bilinear expansions in Hermite polynomials, *Teoriya Veroyatnostei i ee Primeneniya*, **12**, 520–531 (in Russian; English translation, pp. 470–481).

Sato, K. (1981). Infinitely divisible distributions, *Seminar on Probability*, **52** (in Japanese).

Sazonov, V. V. (1967). On the rate of convergence in the multidimensional central limit theorem, *Teoriya Veroyatnostei i ee Primeneniya*, **12**, 82–95 (in Russian; English translation, pp. 77–89).

Schreuder, H. T., and Hafley, W. L. (1977). A useful bivariate distribution for describing stand structure of tree heights and diameters, *Biometrics*, **33**, 471–478.

Schucany, W. R., Parr, W. C., and Boyer, J. E. (1978). Correlation structure in Farlie–Gumbel–Morgenstern distributions, *Biometrika*, **65**, 650–653.

Schwager, S. J. (1985). Multivariate skewness and kurtosis, in *Encyclopedia of Statistical Sciences*, Vol. 6 (S. Kotz, N. L. Johnson, and C. B. Read, eds.), pp. 122–125, New York: John Wiley & Sons.

Seshadri, V., and Patil, G. P. (1964). A characterization of a bivariate distribution by the marginal and the conditional distributions of the same component, *Annals of the Institute of Statistical Mathematics*, **15**, 215–221.

Shaked, M. (1975). A note on the exchangeable generalized Farlie–Gumbel–Morgenstern distributions, *Communications in Statistics*, **4**, 711–721.

Shaked, M. (1979). Some concepts of positive dependence for bivariate interchangeable distributions, *Annals of the Institute of Statistical Mathematics*, **31**, 67–84.

Shaked, M., and Tong, Y. L. (1991). Positive dependence of random variables with a common marginal distribution, *Communications in Statistics—Theory and Methods*, **20**, 4299–4313.

Shimizu, K. (1993). A bivariate mixed lognormal distribution with an analysis of rainfall data, *Journal of Applied Meteorology*, **32**, 161–171.

Shimizu, K., and Sagae, M. (1990). Modelling bivariate data containing zeroes, with an analysis of daily rainfall data, *Japanese Journal of Applied Statistics*, **19**, 19–31.

Sklar, A. (1959). Fonctions de repartition et leurs marges, *Publications of the Institute of Statistics, Université de Paris*, **8**, 229–231.

Skovgaard, Ib. M. (1986). On multivariate Edgeworth expansions, *International Statistical Review*, **54**, 169–186.

Srivastava, M. S. (1984). A measure of skewness and kurtosis and a graphical method for assessing multivariate normality, *Statistics & Probability Letters*, **2**, 263–267.

Steyn, H. S. (1960). On regression properties of multivariate probability functions of Pearson's types, *Proceedings of the Royal Academy of Sciences, Amsterdam*, **63**, 302–311.

Steyn, H. S. (1993). On the problem of more than one kurtosis parameter in multivariate analysis, *Journal of Multivariate Analysis*, **44**, 1–22.

Tadikamalla, P. R. (1980). On simulating nonnormal distributions, *Psychometrika*, **45**, 273–279.

Tiit, E.-M. (1984). Definition of random vectors with given marginal distributions and given correlation matrix, *Acta et Comment. Univ. Tartuensis*, **685**, 21–36.

Troutt, M. D. (1993). Vertical density representation and a further remark on the Box–Muller method, *Statistics*, **24**, 81–83.

Vale, C. D., and Maurelli, V. A. (1983). Simulating multivariate nonnormal distributions, *Psychometrika*, **48**, 465–471.

van Uven, M. J. (1925–1926). On treating skew correlation, I–III, *Proceedings of the Royal Academy of Sciences, Amsterdam*, **28**, 797–811, 919–935; **29**, 580–590.

van Uven, M. J. (1929). Skew correlation between three and more variables, I–III, *Proceedings of the Royal Academy of Sciences, Amsterdam*, **32**, 793–807, 995–1007, 1085–1103.

van Uven, M. J. (1947–1948). Extension of Pearson's probability distributions of two variables, I–IV, *Proceedings of the Royal Academy of Sciences, Amsterdam*, **50**, 1063–1070, 1252–1264; **51**, 41–52, 191–196.

Wang, Y. J. (1987). The probability integrals of bivariate normal distribution: a contingency table approach, *Biometrika*, **74**, 185–190.

Wang, Y. J. (1993). Construction of continuous bivariate density functions, *Statistica Sinica*, **3**, 173–187.

Wani, J. K. (1968). On the linear exponential family, *Proceedings of the Cambridge Philosophical Society*, **64**, 481–485.

Warmuth, W. (1988). Marginal-Fréchet-bounds for multidimensional distribution functions, *Statistics*, **19**, 283–294.

Weiss, L., and Wolfowitz, J. (1966). Generalized maximum likelihood estimators, *Theory of Probability and Its Applications*, **11**, 58–81 (reprinted from *Teoriya Veroyatnostei i ee Primeneniya*, **11**, 68–93).

Weiss, L., and Wolfowitz, J. (1967). Maximum probability estimators, *Annals of the Institute of Statistical Mathematics*, **19**, 193–206.

Wesolowski, J. (1991a). On determination of multivariate distribution by its marginals for the Kagan class, *Journal of Multivariate Analysis*, **36**, 314–319.

Wesolowski, J. (1991b). Gaussian conditional structure of the second order and the Kagan classification of multivariate distributions, *Journal of Multivariate Analysis*, **39**, 79-86.

Wicksell, S. D. (1917). On the genetic theory of frequency, *Arkiv for Matematik Astronomi och Fysik*, **12**, No. 20.

Wicksell, S. D. (1923). Contributions to the analytical theory of sampling, *Arkiv for Mathematik Astronomi och Fysik*, **17**, No. 19.

Yule, G. U. (1897). On the significance of Bravais' formulae for regression in the case of skew correlation, *Proceedings of the Royal Society of London*, **60**, 477-489.

Zeng, W.-B. (1995). On characterization of multivariate stable distributions via random linear statistics, *Journal of Theoretical Probability*, **8**, 1–15.

Zolotarev, V. M. (1966). A multidimensional analogue of the Berry–Esseen inequality for sets with bounded diameter, *Theory of Probability and Its Applications*, **11**, 447–454.

CHAPTER 45

Multivariate Normal Distributions

1 INTRODUCTION AND GENESIS

Perhaps a word of explanation is needed about the reason for discussing general multivariate normal distributions here and the simpler cases of bivariate and trivariate normal distributions in the next chapter. The principal reason is that there is still a greater volume of results on latter special cases than on general multivariate distributions. Their review therefore must be more detailed. By examining the general distribution first, we are able to concentrate on details specific to the bivariate and trivariate cases in Chapter 46. Historical remarks will be found in Section 2 of Chapters 44 and 46.

If $U_1, U_2, \ldots, U_k$ are independent standard normal variables, their joint density is

$$p_{\boldsymbol{U}}(\boldsymbol{u}) = (2\pi)^{-k/2} \exp\left(-\frac{1}{2}\sum_{j=1}^{k} u_j^2\right) = (2\pi)^{-k/2}\exp\left(-\frac{1}{2}\,\boldsymbol{u}^T\boldsymbol{u}\right).$$

Applying the nonsingular linear transformation

$$\boldsymbol{U}^T = \boldsymbol{Y}^T\boldsymbol{H}^T \qquad \text{with } |\boldsymbol{H}| \neq 0$$

to $\boldsymbol{Y}^T = (Y_1, \ldots, Y_k)$, we find that $\boldsymbol{Y}$ has joint density function

$$
\begin{aligned}
p_{\boldsymbol{Y}}(\boldsymbol{y}) &= (2\pi)^{-k/2}|\boldsymbol{H}|\exp\left(-\frac{1}{2}\,\boldsymbol{y}^T\boldsymbol{H}^T\boldsymbol{H}\boldsymbol{Y}\right) \\
&= (2\pi)^{-k/2}|\boldsymbol{A}|^{1/2}\exp\left(-\frac{1}{2}\,\boldsymbol{y}^T\boldsymbol{A}\boldsymbol{Y}\right) \qquad \text{with } \boldsymbol{A} = \boldsymbol{H}^T\boldsymbol{H},
\end{aligned}
$$

so that $\boldsymbol{A}$ is positive definite. This is a special case of multivariate normal distribution. The variance–covariance matrix of $\boldsymbol{Y}$ is (since $E[\boldsymbol{Y}] = \boldsymbol{0}$)

$$
\begin{aligned}
\text{Var}(\boldsymbol{Y}) = E[\boldsymbol{Y}\boldsymbol{Y}^T] &= E\left[\boldsymbol{H}^{-1}\boldsymbol{U}\boldsymbol{U}^T(\boldsymbol{H}^T)^{-1}\right] \\
&= \boldsymbol{H}^{-1}E[\boldsymbol{U}\boldsymbol{U}^T](\boldsymbol{H}^T)^{-1} \\
&= \boldsymbol{H}^{-1}(\boldsymbol{H}^T)^{-1} \\
&= \boldsymbol{A}^{-1}.
\end{aligned}
$$

If we now consider the joint distribution of $\boldsymbol{Z}^T$, where $\boldsymbol{U}^T + \boldsymbol{\zeta}^T = \boldsymbol{Z}^T\boldsymbol{H}^T$, we obtain

$$
p_{\boldsymbol{Z}}(\boldsymbol{z}) = (2\pi)^{-k/2}|\boldsymbol{A}|^{1/2}\exp\left\{-\frac{1}{2}(\boldsymbol{z}-\boldsymbol{\zeta})^T\boldsymbol{A}(\boldsymbol{z}-\boldsymbol{\zeta})\right\}
$$

with a more general form [in fact, the most general form, as will be seen shortly] of multivariate normal distribution. Here, $E[\boldsymbol{Z}] = \boldsymbol{\zeta}$ and $\text{Var}(\boldsymbol{Z}) = \boldsymbol{A}^{-1}\;(=\text{Var}(\boldsymbol{Y}))$.

The multivariate normal distribution is a limiting form of the multinomial distribution; see Chapter 35 of Johnson, Kotz and Balakrishnan (1997). If $X_1, X_2, \ldots, X_{k+1}$ have a joint multinomial distribution with parameters $n, p_1, p_2, \ldots, p_{k+1}$ $\left(\sum_{j=1}^{k+1} p_j = 1\right)$, then the limiting joint distribution, as $n \to \infty$, of the standardized variables

$$
Y_j = \frac{X_j - np_j}{\sqrt{np_j(1-p_j)}} \qquad (j = 1, \ldots, k)
$$

is multivariate normal. Note that only the k variables $Y_1, Y_2, \ldots, Y_k$ are included here. The joint distribution of $Y_1, \ldots, Y_k, Y_{k+1}$ is a *singular* multivariate normal distribution (see Section 2) because there is a fixed linear relation among the $(k+1)$ variables $(\sum_{j=1}^{k+1} Y_j\{p_j(1-p_j)\}^{1/2} = 0)$.

The multivariate normal distribution is also the limiting joint distribution (as $n \to \infty$) of standardized variables corresponding to $S_1, S_2, \ldots, S_k$, where

$$
S_j = \sum_{i=1}^{n} X_{ji}
$$

and $(X_{1i}, \ldots, X_{ki})$ have the same joint distribution with finite means and variances for all $i = 1, 2, \ldots, k$; and $(X_{1i}, \ldots, X_{ki})$ and $(X_{1i'}, \ldots, X_{ki'})$ are mutually independent if $i \neq i'$. (See also Chapter 44.)

2   DEFINITION AND MOMENTS

The random variables $X_1, X_2, \ldots, X_k$ have a *joint multivariate normal distribution* if their joint probability density function can be written in the form

$$p_{X_1, \ldots, X_k}(x_1, \ldots, x_k)$$
$$= C \exp\{-(\text{positive definite quadratic form in } x_1, \ldots, x_k)\},$$
$$(45.1)$$

where C is an appropriate constant. Writing now the exponent as $-\frac{1}{2}(\boldsymbol{x} - \boldsymbol{\xi})^T \boldsymbol{A}(\boldsymbol{x} - \boldsymbol{\xi})$ where $\boldsymbol{A}$ is a real symmetric positive definite matrix, we see that C must be a function of $\boldsymbol{\xi}$ and $\boldsymbol{A}$. In order to find the value of C, we evaluate the joint moment generating function of $\boldsymbol{X} = (X_1, \ldots, X_k)^T$, given by

$$M_{\boldsymbol{X}}(t_1, \ldots, t_k) = E\left[e^{\boldsymbol{t}^T \boldsymbol{X}}\right].$$

We have

$$M_{\boldsymbol{X}}(\boldsymbol{t}) = C \int_{-\infty}^{\infty} \cdots \int_{-\infty}^{\infty} \exp\left\{-\frac{1}{2}(\boldsymbol{x} - \boldsymbol{\xi})^T \boldsymbol{A}(\boldsymbol{x} - \boldsymbol{\xi}) + \boldsymbol{t}^T \boldsymbol{x}\right\} d\boldsymbol{x}.$$

Making the transformation $\boldsymbol{y} = \boldsymbol{x} - \boldsymbol{\xi}$, we obtain

$$M_{\boldsymbol{X}}(\boldsymbol{t}) = C e^{\boldsymbol{t}^T \boldsymbol{\xi}} \int_{-\infty}^{\infty} \cdots \int_{-\infty}^{\infty} \exp\left\{-\frac{1}{2}\boldsymbol{y}^T \boldsymbol{A}\boldsymbol{y} + \boldsymbol{t}^T \boldsymbol{y}\right\} d\boldsymbol{y}.$$

Since $\boldsymbol{A}$ is positive definite, $\boldsymbol{A} = \boldsymbol{H}^T \boldsymbol{H}$ with, of course, $|\boldsymbol{A}| = |\boldsymbol{H}|^2$. Making the transformation $\boldsymbol{z}^T = \boldsymbol{y}^T \boldsymbol{H}^T$ (with Jacobian $\partial(\boldsymbol{z}^T)/\partial(\boldsymbol{y}^T) = |\boldsymbol{H}|$), we have

$$
\begin{aligned}
M_{\boldsymbol{X}}(\boldsymbol{t}) &= C|\boldsymbol{H}|^{-1} e^{\boldsymbol{t}^T \boldsymbol{\xi}} \int_{-\infty}^{\infty} \cdots \int_{-\infty}^{\infty} \exp\left\{-\frac{1}{2}\boldsymbol{z}^T \boldsymbol{z} + \boldsymbol{t}^T (\boldsymbol{H}^T)^{-1} \boldsymbol{z}\right\} d\boldsymbol{z} \\
&= C|\boldsymbol{H}|^{-1} e^{\boldsymbol{t}^T \boldsymbol{\xi}} \int_{-\infty}^{\infty} \cdots \int_{-\infty}^{\infty} \exp\left\{-\frac{1}{2}\sum_{j=1}^{k}(z_j^2 + 2b_j z_j)\right\} dz_1 \cdots dz_k
\end{aligned}
$$
$$(45.2)$$

with $\boldsymbol{b}^T = (b_1, \ldots, b_k) = \boldsymbol{t}^T (\boldsymbol{H}^T)^{-1}$. Since $z_j^2 + 2b_j z_j = (z_j + b_j)^2 - b_j^2$ and $\boldsymbol{b}^T \boldsymbol{b} = \sum_{j=1}^{k} b_j^2$, (45.2) can be written as

$$
\begin{aligned}
M_{\boldsymbol{X}}(\boldsymbol{t}) &= C|\boldsymbol{H}|^{-1} \exp\left\{\boldsymbol{t}^T \boldsymbol{\xi} + \frac{1}{2}\boldsymbol{b}^T \boldsymbol{b}\right\} \prod_{j=1}^{k} \int_{-\infty}^{\infty} \exp\left\{-\frac{1}{2}(z_j + b_j)^2\right\} dz_j \\
&= C|\boldsymbol{H}|^{-1}(2\pi)^{k/2} \exp\left\{\boldsymbol{t}^T \boldsymbol{\xi} + \frac{1}{2}\boldsymbol{b}^T \boldsymbol{b}\right\}.
\end{aligned}
$$

Finally, since $|\boldsymbol{H}| = |\boldsymbol{A}|^{1/2}$ and $\boldsymbol{b}^T\boldsymbol{b} = \boldsymbol{t}^T(\boldsymbol{H}^T)^{-1}\boldsymbol{H}^{-1}\boldsymbol{t} = \boldsymbol{t}^T\boldsymbol{A}^{-1}\boldsymbol{t}$, we have

$$M_{\boldsymbol{X}}(\boldsymbol{t}) = C|\boldsymbol{A}|^{-1/2}(2\pi)^{k/2} \exp\left(\boldsymbol{t}^T\boldsymbol{\xi} + \frac{1}{2}\,\boldsymbol{t}^T\boldsymbol{A}^{-1}\boldsymbol{t}\right).$$

Since $M_{\boldsymbol{X}}(0) = 1$, if follows that $C = |\boldsymbol{A}|^{1/2}(2\pi)^{-k/2}$, so that the joint density function is

$$p_{\boldsymbol{X}}(\boldsymbol{x}) = (2\pi)^{-k/2}|\boldsymbol{A}|^{1/2} \exp\left\{-\frac{1}{2}(\boldsymbol{x} - \boldsymbol{\xi})^T\boldsymbol{A}(\boldsymbol{x} - \boldsymbol{\xi})\right\} \qquad (45.3)$$

and

$$M_{\boldsymbol{X}}(\boldsymbol{t}) = \exp\left(\boldsymbol{t}^T\boldsymbol{\xi} + \frac{1}{2}\,\boldsymbol{t}^T\boldsymbol{A}^{-1}\boldsymbol{t}\right). \qquad (45.4)$$

From (45.4), we find

$$E[\boldsymbol{X}] = \boldsymbol{\xi}; \qquad (45.5)$$

and the variance–covariance matrix of $\boldsymbol{X}$ is $\boldsymbol{A}^{-1}$, or symbolically

$$V(\boldsymbol{X}) = \boldsymbol{A}^{-1}. \qquad (45.6)$$

[Sometimes the notation $\mathbf{Var}(\boldsymbol{X})$ is used; sometimes $(\boldsymbol{X})$ is omitted.] In terms of $\boldsymbol{V}$, (45.3) becomes

$$p_{\boldsymbol{X}}(\boldsymbol{x}) = (2\pi)^{-k/2}|\boldsymbol{V}|^{-1/2} \exp\left\{-\frac{1}{2}(\boldsymbol{x} - \boldsymbol{\xi})^T\boldsymbol{V}^{-1}(\boldsymbol{x} - \boldsymbol{\xi})\right\}. \qquad (45.7)$$

Furthermore, all cumulants and cross-cumulants of order higher than 2 are zero. We will use the notation $\boldsymbol{X} \overset{d}{=} N_k(\boldsymbol{\mu}, \boldsymbol{V})$ for this variable.

Holmquist (1988) has provided compact expressions in vector notation for central as well as noncentral moments of the multivariate normal distribution in (45.7). Let $\boldsymbol{\xi}$ and $\boldsymbol{v}$ denote the vector (column) arrangements of means and variances of the components of $\boldsymbol{X}$. Let $\boldsymbol{A} \otimes \boldsymbol{B}$ denote the Kronecker product of matrices $\boldsymbol{A} \otimes \boldsymbol{B} = (a_{ij}\boldsymbol{B})$, and let us use the notation

$$\bigotimes_{i=1}^{r} \boldsymbol{A}_i = \boldsymbol{A}_1 \otimes \boldsymbol{A}_2 \otimes \cdots \otimes \boldsymbol{A}_r \text{ and } \boldsymbol{A}^{(r)} = \bigotimes_{i=1}^{r} \boldsymbol{A}.$$

Let $\boldsymbol{E}_{i,j}$ be a $k^r \times k^r$ matrix with the (i, j)th element equal to 1, being the only nonzero element. Define

$$q\left(\boldsymbol{i}_r; k\,\boldsymbol{1}_r\right) = 1 + \sum_{j=1}^{r}(i_j - 1)k^{j-1},$$

where $\boldsymbol{i}_r = (i_1, \ldots, i_r)$ and $\boldsymbol{1}_r = (1, \ldots, 1)$ (r times). The direct product permuting matrix with r degrees k is defined by

$$Q_{k\boldsymbol{1}_r}(\pi) = \sum_{\boldsymbol{i}_r \in \{1, \ldots, k\}^r} E_{q(\pi\boldsymbol{i}_r;k\boldsymbol{1}_r), q(\boldsymbol{i}_r;k\boldsymbol{1}_r)}$$

with $\pi\boldsymbol{i}_r = (i_{\pi^{-1}(1)}, i_{\pi^{-1}(2)}, \ldots, i_{\pi^{-1}(r)})$ (π is the argument in the symmetric group of order r), and the symmetrizer $S_{k\boldsymbol{1}_r}$ is defined by

$$S_{k\boldsymbol{1}_r} = \sum_{\pi} Q_{k\boldsymbol{1}_r}(\pi)/r!,$$

where the summation extends over all $r!$ permutations π in the symmetric group of order r. Then, Holmquist (1988) has given an expression for the raw moment $E[\boldsymbol{X}^{\langle r \rangle}]$ (for $r \geq 1$) as

$$
\begin{aligned}
E[\boldsymbol{X}^{\langle r \rangle}] &= \sum_{j=1}^{r} \sum_{\substack{j_1, j_2 \\ j_1 + j_2 = j \\ j_1 + 2j_2 = r}} \frac{r!}{j_1! j_2! 2^{j_2}} S_{k\boldsymbol{1}_r} \left(\boldsymbol{\xi}^{\langle j_1 \rangle} \otimes \boldsymbol{v}^{\langle j_2 \rangle} \right) \\
&= r! \sum_{i=0}^{[r/2]} S_{k\boldsymbol{1}_r} \frac{\left(\boldsymbol{\xi}^{\langle r-2i \rangle} \otimes \boldsymbol{v}^{\langle i \rangle} \right)}{i!(r-2i)! 2^i} .
\end{aligned}
\tag{45.8}
$$

In particular, by denoting $\boldsymbol{v} = \boldsymbol{\sigma}_2$, we have

$$E[\boldsymbol{X}^{\langle 2 \rangle}] = S_{k\boldsymbol{1}_2}(\boldsymbol{\xi}^{\langle 2 \rangle} + \boldsymbol{\sigma}_2) = \boldsymbol{\xi}^{\langle 2 \rangle} + \boldsymbol{\sigma}_2$$

and

$$E[\boldsymbol{X}^{\langle 4 \rangle}] = S_{k\boldsymbol{1}_4} \left(\boldsymbol{\xi}^{\langle 4 \rangle} + 6\boldsymbol{\xi}^{\langle 2 \rangle} \otimes \boldsymbol{\sigma}_2 + 3\boldsymbol{\sigma}_2^{\langle 2 \rangle} \right).$$

A similar but simpler expression for the central moment $E\left[(\boldsymbol{X} - \boldsymbol{\xi})^{\langle r \rangle} \right]$ is

$$
\begin{aligned}
E\left[(\boldsymbol{X} - \boldsymbol{\xi})^{\langle r \rangle} \right] &= 0 && \text{if } r \text{ is odd} \\
&= (r-1)!! \, S_{k\boldsymbol{1}_r} \boldsymbol{v}^{\langle r/2 \rangle} && \text{if } r \text{ is even,}
\end{aligned}
\tag{45.9}
$$

where $(r-1)!! = (r-1)(r-3)\cdots 1$. An alternate expression for the central moment is

$$E\left[(\boldsymbol{X} - \boldsymbol{\xi})^{\langle 2r \rangle} \right] = \frac{(2r)!}{r! 2^r} S_{k\boldsymbol{1}_{2r}} \boldsymbol{v}^{\langle r \rangle} = (2r-1)!! S_{k\boldsymbol{1}_{2r}} \boldsymbol{v}^{\langle r \rangle}. \tag{45.10}$$

Note that since we can *always* find $\boldsymbol{H}$ such that $\boldsymbol{A} = \boldsymbol{H}^T \boldsymbol{H}$, *any* multivariate normal distribution can be constructed as the joint distribution of linear functions of independent normal variables, as described in Section 1.

A derivation of the value of C, by Todhunter (1869), is of some historical interest.

If $\boldsymbol{A}$ is only positive semidefinite (i.e., $|\boldsymbol{A}| = 0$), the joint distribution of $X_1, X_2, \ldots, X_k$ is called *singular multivariate normal*.

Note that since we have $(\boldsymbol{X} - \boldsymbol{\xi})^T \boldsymbol{A}(\boldsymbol{X} - \boldsymbol{\xi}) = \boldsymbol{Z}^T \boldsymbol{Z}$ with $\boldsymbol{Z}^T = (Z_1, \ldots, Z_k)$ comprised of independent unit normal variables, this quadratic form is distributed as χ^2 with k degrees of freedom [see Chapter 17 of Johnson, Kotz, and Balakrishnan (1994)].

The entropy of the distribution in (45.3) (with $\boldsymbol{V} = \boldsymbol{A}^{-1}$) is

$$-E[\log p_{\boldsymbol{X}}(\boldsymbol{X})] = \frac{1}{2} \, k \log 2\pi + \frac{1}{2} \, \log |\boldsymbol{V}| + \frac{1}{2} \, k.$$

Rao (1965) has shown that this is the maximum entropy possible for any random vector of k dimensions with specified variance–covariance matrix $\boldsymbol{V}$. No other distribution attains this maximum.

Dowson and Landau (1982) calculated the Fréchet distance, $d(F, G)$, between two normal distributions. This distance is defined by

$$d^2(F, G) = \min_{X,Y} E|X - Y|^2,$$

where the minimization is taken over all random variables X and Y having the distributions F and G, respectively. The bivariate distribution that minimizes the Fréchet distance is the Fréchet lower bound

$$H(x, y) = \min\{F(x), G(y)\}$$

as seen earlier in Chapter 44. In the case when F and G belong to a family of distributions closed with respect to changes of location and scale, we obtain

$$d^2(F, G) = (\xi_X - \xi_Y)^2 + (\sigma_X - \sigma_Y)^2, \qquad (45.11)$$

where (ξ_X, ξ_Y) and (σ_X, σ_Y) are the respective means and standard deviations of F and G, respectively. Multivariate generalization (when F and G belong to a family of k-dimensional distributions) is straightforward if the family to which F and G belong is closed with respect to linear transformations of the random vector. It is given by [see Dowson and Landau (1982)]

$$d^2(F, G) = |\boldsymbol{\xi}_{\boldsymbol{X}} - \boldsymbol{\xi}_{\boldsymbol{Y}}|^2 + \text{tr}\left\{\boldsymbol{V}_{\boldsymbol{X}} + \boldsymbol{V}_{\boldsymbol{Y}} - 2(\boldsymbol{V}_{\boldsymbol{X}}\boldsymbol{V}_{\boldsymbol{Y}})^{1/2}\right\}, \; (45.12)$$

where $(\boldsymbol{\xi}_{\boldsymbol{X}}, \boldsymbol{\xi}_{\boldsymbol{Y}})$ and $(\boldsymbol{V}_{\boldsymbol{X}}, \boldsymbol{V}_{\boldsymbol{Y}})$ are the respective mean vectors and variance–covariance matrices of F and G, respectively. The result holds, under certain conditions, more generally, for any two distributions from a family of real elliptically contoured distributions. As a consequence, we find

$$\rho = \mathrm{tr}(\boldsymbol{V_X V_Y})^{1/2} \left\{ \mathrm{tr}\, \boldsymbol{V_X} \cdot \mathrm{tr}\, \boldsymbol{V_Y} \right\}^{-1/2} \tag{45.13}$$

is the largest correlation coefficient possible between two random vectors $\boldsymbol{X}$ and $\boldsymbol{Y}$ having prescribed covariance matrices $\boldsymbol{V_X}$ and $\boldsymbol{V_Y}$, respectively.

3   OTHER PROPERTIES

From the form of the density function in (45.1), it is clear that if any subset—$X_1, \ldots, X_s$, say—of variables is eliminated by "integrating out", the remaining variables $X_{s+1}, X_{s+2}, \ldots, X_k$ have a joint density function of the same form. This means that $X_{s+1}, X_{s+2}, \ldots, X_k$ also have a joint multivariate normal distribution. In particular, each variable has a normal distribution. The parameters of each distribution are given by (45.5) and (45.6) with appropriate modifications.

The other parameters (correlations) of the joint distribution of X_{s+1}, $X_{s+2}, \ldots, X_k$ could also be found from (45.5) and (45.6). The following argument, however, yields concise formulas for the parameters, and also derives the conditional joint distribution of $X_1, \ldots, X_s$, given $X_{s+1}, X_{s+2}, \ldots,$ X_k.

We partition the matrix $\boldsymbol{A}$ at the sth row and column to give

$$\boldsymbol{A} = \begin{pmatrix} \boldsymbol{A}_{11} & \boldsymbol{A}_{12} \\ \boldsymbol{A}_{21} & \boldsymbol{A}_{22} \end{pmatrix}. \tag{45.14}$$

(Note that $\boldsymbol{A}_{21} = \boldsymbol{A}_{12}^T$.) The similarly partitioned original $k \times k$ matrix

$$\boldsymbol{C} = \begin{pmatrix} \boldsymbol{I} & \boldsymbol{0} \\ -\boldsymbol{A}_{21}\boldsymbol{A}_{11}^{-1} & \boldsymbol{I} \end{pmatrix}$$

satisfies the equation

$$\boldsymbol{CAC}^T = \begin{pmatrix} \boldsymbol{A}_{11} & \boldsymbol{0} \\ \boldsymbol{0} & \boldsymbol{A}_{22} - \boldsymbol{A}_{21}\boldsymbol{A}_{11}^{-1}\boldsymbol{A}_{12} \end{pmatrix}. \tag{45.15}$$

Hence, making the transformation

$$(\boldsymbol{x} - \boldsymbol{\xi})^T = \boldsymbol{y}^T \boldsymbol{C},$$

we have

$$(\boldsymbol{x} - \boldsymbol{\xi})^T \boldsymbol{A}(\boldsymbol{x} - \boldsymbol{\xi}) = \boldsymbol{y}^T \boldsymbol{C} \boldsymbol{A} \boldsymbol{C}^T \boldsymbol{y} = \boldsymbol{y}_{(1)}^T \boldsymbol{A}_{11} \boldsymbol{y}_{(1)} + \boldsymbol{y}_{(2)}^T \boldsymbol{D} \boldsymbol{y}_{(2)},$$

where $\boldsymbol{y}_{(1)}^T = (y_1, y_2, \ldots, y_s)$, $\boldsymbol{y}_{(2)}^T = (y_{s+1}, \ldots, y_k)$ and

$$\boldsymbol{D} = \boldsymbol{A}_{22} - \boldsymbol{A}_{21} \boldsymbol{A}_{11}^{-1} \boldsymbol{A}_{12}.$$

The joint density function of the variate $\boldsymbol{Y}^T = (Y_1, Y_2, \ldots, Y_k)$ defined by $(\boldsymbol{X} - \boldsymbol{\xi})^T = \boldsymbol{Y}^T \boldsymbol{C}$ is

$$p_{\boldsymbol{Y}}(\boldsymbol{y}_{(1)}, \boldsymbol{y}_{(2)}) = \frac{|\boldsymbol{A}_{11}|^{1/2}}{(2\pi)^{s/2}} \exp\left(-\frac{1}{2} \boldsymbol{y}_{(1)}^T \boldsymbol{A}_{11} \boldsymbol{y}_{(1)}\right)$$

$$\times \frac{|\boldsymbol{D}|^{1/2}}{(2\pi)^{(k-s)/2}} \exp\left(-\frac{1}{2} \boldsymbol{y}_{(2)}^T \boldsymbol{D} \boldsymbol{y}_{(2)}\right), \quad (45.16)$$

since $|\boldsymbol{A}| = |\boldsymbol{A}_{11}| |\boldsymbol{D}|$ [from (45.15), noting that $|\boldsymbol{C}| = 1$].

It follows from (45.16) that the sets $\boldsymbol{Y}_{(1)}^T = (Y_1, \ldots, Y_s)$ and $\boldsymbol{Y}_{(2)}^T = (Y_{s+1}, \ldots, Y_k)$ are independent of each other, and each has a joint multivariate normal distribution. Examining $\boldsymbol{C}$ more closely, we see that $\boldsymbol{Y}_{(2)}^T = \boldsymbol{X}_{(2)}^T$ (i.e., $Y_j = X_j - \xi_j$ for $j = s+1, s+2, \ldots, k$), while $\boldsymbol{Y}_{(1)}^T = (\boldsymbol{X}_{(1)} - \boldsymbol{\xi}_{(1)})^T + (\boldsymbol{X}_{(2)} - \boldsymbol{\xi}_{(2)})^T \boldsymbol{A}_{21} \boldsymbol{A}_{11}^{-1}$. Thus, we can restate our results as follows:

(i) X_{s+1}, X_{s+2}, X_k have a joint multivariate normal distribution with expected values $\xi_{s+1}, \ldots, \xi_k$ and variance–covariance matrix $(\boldsymbol{A}_{22} - \boldsymbol{A}_{21} \boldsymbol{A}_{11}^{-1} \boldsymbol{A}_{12})^{-1}$.

(ii) The conditional joint distribution of $\boldsymbol{X}_{(1)}^T = (X_1, X_2, \ldots, X_s)$, given $X_{s+1}, \ldots, X_k$, is multivariate normal with expected value

$$\boldsymbol{\xi}_{(1)}^T - (\boldsymbol{X}_{(2)} - \boldsymbol{\xi}_{(2)})^T \boldsymbol{A}_{21} \boldsymbol{A}_{11}^{-1} \quad (45.17)$$

and variance–covariance matrix $\boldsymbol{A}_{11}^{-1}$. Expression (45.17) shows that the regression of each of $X_1, X_2, \ldots, X_s$ on the set $\boldsymbol{X}_{(2)}^T = (X_{s+1}, \ldots, X_k)$ is *linear* and *homoscedastic* (since $\boldsymbol{A}_{11}^{-1}$ does not depend on $\boldsymbol{X}_{(2)}^T$).

From Eq. (45.16), it can be seen that the joint distribution of any linear functions of the X's will be a (singular or nonsingular) multivariate normal distribution.

Many simple and interesting examples have been presented in the literature in order to illustrate that (i) nonnormal multivariate distributions can have normal marginals and (ii) uncorrelated normal random variables need not be independent; see, for example, Lancaster (1959), Pierce and Dykstra (1969), Kowalski (1973), Melnick and Tenenbein (1982), Anderson (1984), and Broffitt (1986). With regard to (ii), let X_1 be a standard normal random variable and $X_2 = ZX_1$, where Z (independently of X_1) has a probability mass function $\Pr[Z = 1] = \Pr[Z = -1] = 1/2$. Then, since $\Pr[X_2 \leq x | Z = 1] = \Pr[X_2 \leq x | Z = -1]$, X_2 and Z are independent. Also, since $\Pr[X_2 \leq x] = \Pr[X_2 \leq x | Z = 1] = \Pr[X_1 \leq x]$, X_2 is distributed as standard normal. Clearly, the distribution of $X_2 | X_1 = x$ depends on x and also $E[X_1 X_2] = 0$. Thus, X_1 and X_2 are uncorrelated normal random variables, but are dependent. More recent examples provided by Johnson and Kotz (1999) can easily be adjusted to normal marginals.

Šidák (1967) [see also Dunn (1958)] has shown that if $X_1, X_2, \ldots, X_k$ have a joint multivariate normal distribution, then

$$\Pr\left[\bigcap_{j=1}^{k}(|X_j - \xi_j| \leq c_j)\right] \geq \prod_{j=1}^{k} \Pr[|X_j - \xi_j| \leq c_j] \qquad (45.18)$$

for any set of positive constants $c_1, c_2, \ldots, c_k$.

Scott (1967) [see also Khatri (1967) for a particular case] has proved the inequality obtained by replacing $\leq c_j$ in (45.18) by $\geq c_j$ (twice).

Gupta (1969) has proved the more general results that if $\mathcal{C}_1$ and $\mathcal{C}_2$ are convex sets, symmetrical about $(\xi_1, \ldots, \xi_{k_1})$, $(\xi_{k_1+1}, \ldots, \xi_k)$ in the spaces of $\boldsymbol{X}_{(1)} = (X_1, \ldots, X_{k_1})$, $\boldsymbol{X}_{(2)} = (X_{k_1+1}, \ldots, X_k)$, respectively, then

$$\Pr\left[(\boldsymbol{X}_{(1)} \in \mathcal{C}_1) \cap (\boldsymbol{X}_{(2)} \in \mathcal{C}_2)\right] \geq \Pr\left[\boldsymbol{X}_{(1)} \in \mathcal{C}_1\right] \Pr\left[\boldsymbol{X}_{(2)} \in \mathcal{C}_2\right] \qquad (45.19)$$

and

$$\Pr\left[(\boldsymbol{X}_{(1)} \in \bar{\mathcal{C}}_1) \cap (\boldsymbol{X}_{(2)} \in \bar{\mathcal{C}}_2)\right] \geq \Pr\left[\boldsymbol{X}_{(1)} \in \bar{\mathcal{C}}_1\right] \Pr\left[\boldsymbol{X}_{(2)} \in \bar{\mathcal{C}}_2\right], \qquad (45.20)$$

where $\bar{\mathcal{C}}_j$ denotes the complement of $\mathcal{C}_j$ $(j = 1, 2)$.

Slepian (1962) showed that (for any $c_1, c_2, \ldots, c_k$) the derivative of $\Pr\left[\bigcap_{j=1}^{k}(X_j - \xi_j \leq c_j)\right]$ with respect to $\rho_{ii'}$ is nonnegative for all i, i'. [He used (45.45) below to establish this result.] Jogdeo (1970) showed that if ρ_{1i} $(= \rho_{i1})$ is increased by a multiplier λ (other ρ's remaining the same), then

$$\frac{d}{d\lambda} \Pr\left[\bigcap_{j=1}^{k}(|X_j - \xi_j| \leq c_j)\right] \geq 0.$$

For the particular (symmetric) case, when all correlation coefficients are equal and positive, Tong (1970) has obtained inequalities between certain probabilities relating to different numbers of variables. In particular, for $k \geq m \geq 2$ we have

$$
\Pr\left[\bigcap_{j=1}^{k}(X_j - \xi_j \leq d\sigma_j)\right]
$$

$$
\geq \left\{\Pr\left[\bigcap_{j=1}^{m}(X_j - \xi_j \leq d\sigma_j)\right]\right\}^{k/m}
$$

$$
\geq \left\{\Phi(d) + \left\{\Pr\left[\bigcap_{j=1}^{2}(X_j - \xi_j \leq d\sigma_j)\right] - [\Phi(d)]^2\right\}\right\}^{k/2}, \quad (45.21)
$$

where $\Phi(\cdot)$ is the cumulative distribution function of the univariate standard normal distribution. The same inequalities hold (for $d > 0$) with $X_j - \xi_j$ replaced by $|X_j - \xi_j|$.

Anderson (1955) has shown that if $\boldsymbol{X} \overset{d}{=} N_k(\boldsymbol{0}, \boldsymbol{V})$, then $\boldsymbol{V}_1 \leq \boldsymbol{V}_2$ (i.e. $\boldsymbol{V}_2 - \boldsymbol{V}_1$ is positive definite) implies that

$$
\Pr_{\boldsymbol{V}_1}(C) \geq \Pr_{\boldsymbol{V}_2}(C)
$$

for every centrally symmetric (i.e., $-C = C$) convex set $C \subseteq \mathbb{R}^k$. Here, $\Pr_{\boldsymbol{V}}(C) = \Pr[\boldsymbol{X} \in C]$. In other words, $\Pr_{\boldsymbol{V}_1}$ is *more concentrated about* $\boldsymbol{0}$ than $\Pr_{\boldsymbol{V}_2}$. The above inequality of Anderson (1955) for centrally symmetric convex set C has been extended by Eaton and Perlman (1991) to the case when the condition $-C = C$ is replaced by the invariance of C under a group G of $k \times k$ matrices for $\boldsymbol{V}_1$.

Multivariate *Mills' ratio* is defined as

$$
M(\boldsymbol{x}, \boldsymbol{V}) = \int_{x_k}^{\infty} \cdots \int_{x_1}^{\infty} \frac{p(\boldsymbol{y})}{p(\boldsymbol{x})} \, dy_1 \cdots dy_k, \quad (45.22)
$$

where $p(\boldsymbol{x})$ is the density function of the multivariate normal distribution with mean $\boldsymbol{0}$ and variance–covariance matrix $\boldsymbol{V}$. Fang and Xu (1987) have shown that $M(\boldsymbol{x}, \boldsymbol{V})$ in (45.22) can be written as

$$
M(\boldsymbol{x}, \boldsymbol{V}) = \int_{\mathbb{R}_+^k} \exp\left\{-\boldsymbol{x}^T \boldsymbol{V}^{-1} \boldsymbol{z} - \frac{1}{2} \boldsymbol{z}^T \boldsymbol{V}^{-1} \boldsymbol{z}\right\} d\boldsymbol{z}, \quad (45.23)
$$

where $\mathbb{R}_+^k = \left\{\boldsymbol{z} : \boldsymbol{z} = (z_1, \ldots, z_k)^T, \ z_i \geq 0 \text{ for } i = 1, 2, \ldots, k\right\}$, and that it is a convex function of $\boldsymbol{x}$. With $\boldsymbol{e}_i$ denoting a column vector of dimension

k with 1 in the ith place and 0 in all other places, Fang and Xu (1987) have obtained the formula

$$\frac{\partial M(\boldsymbol{x}, \boldsymbol{V})}{\partial v_{ij}} = \left(1 - \frac{\delta_{ij}}{2}\right) \left\{ \frac{\partial^2 M(\boldsymbol{x}, \boldsymbol{V})}{\partial x_i \partial x_j} - 2\boldsymbol{x}^T \boldsymbol{V}^{-1} \boldsymbol{e}_i \frac{\partial M(\boldsymbol{x}, \boldsymbol{V})}{\partial x_j} \right\},$$

(45.24)

where $\delta_{ij} = 1$ if $i = j$ and 0 if $i \neq j$.

In the bivariate case, taking $\boldsymbol{V} = \begin{pmatrix} 1 & \rho \\ \rho & 1 \end{pmatrix}$ and using $M(x)$ to denote the Mills' ratio of the univariate standard normal distribution [see Chapter 13 of Johnson, Kotz and Balakrishnan (1994)], Fang and Xu (1987) have used (45.23) to obtain the following bounds for $M(\boldsymbol{x}, \boldsymbol{V})$ when $\rho \leq 0$:

(i) For the case when $\boldsymbol{x} = (x, x)^T$, we have

$$M(\boldsymbol{x}, \boldsymbol{V}) \leq \frac{(1 - \rho^2)\sqrt{1 + \rho}}{x} M\left(\frac{x}{\sqrt{1 + \rho}}\right), \quad x > 0. \quad (45.25)$$

(ii) For the case when $\boldsymbol{x} = (x, 0)^T$, we obtain

$$M(\boldsymbol{x}, \boldsymbol{V})$$
$$\geq \frac{1 - \rho^2}{x^2 + 1 - \rho^2} \left\{ \sqrt{1 - \rho^2}(1 + \rho^2)x M\left(\frac{-x\rho}{\sqrt{1 - \rho^2}}\right) + C\rho \right\}.$$

(45.26)

(iii) For $\boldsymbol{x} = (x_1, x_2)^T$, we have

$$\frac{M(\boldsymbol{x}, \boldsymbol{V})}{1 - \rho^2}$$
$$\geq \frac{e^{-x_1 x_2/(1 - \rho)}}{4\bar{x}^2 + 1 - \rho^2} \left\{ 2\sqrt{1 - \rho^2}(1 + \rho^2)\bar{x} M\left(\frac{-2\bar{x}\rho}{\sqrt{1 + \rho}}\right) + (1 - \rho^2)\rho \right\}$$

(45.27)

and

$$\frac{M(\boldsymbol{x}, \boldsymbol{V})}{1 - \rho^2}$$
$$\leq \frac{\sqrt{1 + \rho}}{\bar{x}} \exp\left\{ \frac{-(x_1 - x_2)^2}{4(1 - \rho)} \right\} M\left(\frac{\bar{x}}{\sqrt{1 + \rho}}\right), \quad \bar{x} > 0,$$

(45.28)

where $\bar{x} = (x_1 + x_2)/2$; see also Steck (1979).

Let $\boldsymbol{X}$ and $\boldsymbol{Y}$ be jointly distributed as multivariate normal with mean vector $\boldsymbol{0}$ and covariance matrix $\boldsymbol{V} = \begin{pmatrix} \boldsymbol{V}_{11} & \boldsymbol{V}_{12} \\ \boldsymbol{V}_{21} & \boldsymbol{V}_{22} \end{pmatrix}$, let k be the dimension of $\boldsymbol{X}$, and let ℓ be the dimension of $\boldsymbol{Y}$. Let $F_{\boldsymbol{V}}(\cdot)$ and $p_{\boldsymbol{V}}(\cdot)$ denote the cumulative distribution function and density function of a multivariate normal distribution with mean vector $\boldsymbol{0}$ and covariance matrix $\boldsymbol{V}$. Then, Joe (1994) has noted that

$$\frac{\partial^\ell F_{\boldsymbol{V}}(\boldsymbol{x}, \boldsymbol{y})}{\partial y_1 \cdots \partial y_\ell} = F_{\boldsymbol{V}_{11} - \boldsymbol{V}_{12} \boldsymbol{V}_{22}^{-1} \boldsymbol{V}_{21}}(\boldsymbol{x} - \boldsymbol{V}_{12} \boldsymbol{V}_{22}^{-1} \boldsymbol{y}) \, p_{\boldsymbol{V}_{22}}(\boldsymbol{y}).$$

As defined earlier in Chapter 44, the joint multivariate hazard rate of k jointly absolutely continuous random variables $X_1, \ldots, X_k$ is the vector

$$h(\boldsymbol{x}) = \left(-\frac{\partial}{\partial x_1}, \cdots, -\frac{\partial}{\partial x_k} \right) \ln S(\boldsymbol{x}), \qquad (45.29)$$

where $S(\boldsymbol{x}) = \Pr[X_1 > x_1, \ldots, X_k > x_k]$ is the joint survival function. Johnson and Kotz (1975) established that for bivariate normal distributions, the joint hazard rate is an increasing function of $\boldsymbol{x}$ when the correlation coefficient is positive. Extending this result, Gupta and Gupta (1997) proved that the joint hazard rate is increasing without any condition on the correlation coefficient. They have also generalized this result to multivariate normal distributions. Ma (1997) has obtained this result as a consequence of a more general result that the hazard gradient of multivariate log-concave distributions (which includes multivariate normal) is increasing.

4   ORDER STATISTICS

Let $\boldsymbol{X} \overset{d}{=} N_k(\boldsymbol{\mu}, \boldsymbol{V})$. Assuming that X_i $(i = 1, 2, \ldots, k)$ are distinct (otherwise, $\Pr[\max \boldsymbol{X} = X_i]$ should be divided by the number of elements in the various distinct classes), Houdré (1995) has shown that

$$\mathrm{var}\left(\max_{1 \le i \le k} X_i \right) \le \sum_{i=1}^k \mathrm{var}(X_i) \Pr[\max \boldsymbol{X} = X_i]$$

$$\le \max_{1 \le i \le k} \mathrm{var}(X_i). \qquad (45.30)$$

In fact, the inequality $\mathrm{var}(\max_{1 \le i \le k} X_i) \le \max_{1 \le i \le k} \mathrm{var}(X_i)$ is originally due to Cirel'son, Ibragimov, and Sudakov (1976). Also,

$$\mathrm{var}\left(\max_{1 \le i \le k} |X_i| \right) \le \sum_{i=1}^k \mathrm{var}(X_i) \Pr[\max |\boldsymbol{X}| = |X_i|]$$

$$\leq \max_{1 \leq i \leq k} \text{var}(X_i), \tag{45.31}$$

and in the case when $\boldsymbol{X} \overset{d}{=} N_k(\boldsymbol{0}, \boldsymbol{V})$

$$\text{var}\left(\sum_{i=1}^{k} |X_i|\right) \leq \frac{2}{\pi} \sum_{i=1}^{k} \sum_{j=1}^{k} v_{ij} \sin^{-1}\left(\frac{v_{ij}}{\sqrt{v_{ii}v_{jj}}}\right), \tag{45.32}$$

where v_{ij} are the elements of the variance–covariance matrix $\boldsymbol{V}$. Next, when $\boldsymbol{X} \overset{d}{=} N_k(\boldsymbol{\mu}, \boldsymbol{V})$ and k is odd we obtain

$$\text{cov}\left(\sum_{i=1}^{k} X_i, \text{ med } \boldsymbol{X}\right) = \sum_{i=1}^{k} \sum_{j=1}^{k} v_{ij} \Pr[\text{med} \boldsymbol{X} = X_j], \tag{45.33}$$

where med $\boldsymbol{X}$ denotes the median of $\boldsymbol{X}$, that is, the $(k+1)/2$th order statistic among $X_1, X_2, \ldots, X_k$.

For a multivariate normal random variable with arbitrary mean and covariance structure, Siegel (1993) established that

$$\text{cov}(X_1, \min_{1 \leq i \leq k} X_i) = \sum_{j=1}^{k} \text{cov}(X_1, X_j) \Pr\left[X_j = \min_{1 \leq i \leq k} X_i\right]. \tag{45.34}$$

This means that the covariance between the first and the smallest elements of a multivariate normal vector is the weighted average of covariances between the first and the jth element of the vector weighted according to the probability that the jth element is the minimum. In particular, if X_1 is independent of $(X_2, \ldots, X_k)^T$, then (45.34) yields

$$\text{cov}(X_1, \min_{1 \leq i \leq k} X_i) = \text{var}(X_1) \Pr\left[X_1 = \min_{1 \leq i \leq k} X_i\right]. \tag{45.35}$$

Furthermore, upon replacing $(X_1, X_2, \ldots, X_k)^T$ by $(-X_1, -X_2, \ldots, -X_k)^T$ in (45.34), we get the identity

$$\text{cov}(X_1, \min_{1 \leq i \leq k} X_i) = \sum_{j=1}^{k} \text{cov}(X_1, X_j) \Pr\left[X_j = \max_{1 \leq i \leq k} X_i\right]. \tag{45.36}$$

Extending Siegel's identity, Rinott and Samuel-Cahn (1994) proved that

$$\text{cov}(X_1, X_{(r)}) = \sum_{j=1}^{k} \text{cov}(X_1, X_j) \Pr\left[X_j = X_{(r)}\right], \tag{45.37}$$

where $X_{(r)}$ is the rth-order statistic in $\boldsymbol{X}$. The main step involved is in showing that, for X_i's independently distributed as $N(\mu_i, \sigma_i^2)$, we have

$$\text{cov}(X_1, X_{(r)}) = \sigma_1^2 \Pr[X_1 = X_{(r)}].$$

Clearly, the index 1 in all the above formulas can be replaced by any other fixed index.

Olkin and Viana (1995) established the following results along these lines:

(i) If the joint distribution of $X_1, \ldots, X_k$, Y is multivariate normal where $X_1, \ldots, X_k$ are exchangeable, with (X_i, Y) and (X_j, Y) having the same distribution for all $i \neq j$, we obtain

$$\text{cov}(\boldsymbol{X}, Y) = \text{cov}(\boldsymbol{X}_{(\ )}, Y),$$

where $\boldsymbol{X}_{(\ )} = (X_{(1)}, \ldots, X_{(k)})^T$ is the vector of order statistics associated with the random vector $\boldsymbol{X} = (X_1, \ldots, X_k)^T$.

(ii) If $\boldsymbol{X} = (X_1, \ldots, X_k)^T$ has a multivariate normal distribution with common mean ξ, common variance σ^2, and common correlation ρ, then the variance–covariance matrix of the vector of order statistics $\boldsymbol{X}_{(\ )}$ is

$$\text{var}(\boldsymbol{X}_{(\ )}) = \sigma^2 \{\rho \boldsymbol{e}\boldsymbol{e}^T + (1 - \rho)\boldsymbol{\Sigma}\},$$

where $\boldsymbol{\Sigma}$ is the variance–covariance matrix of order statistics from k independent standard normal variables, and $\boldsymbol{e} = (1, 1, \ldots, 1)_{1 \times k}$.

Olkin and Viana (1995) have also extended Siegel's result to elliptically contoured distributions.

Vitale (1996) has used the theory of Steiner points of convex bodies (i.e., compact convex subsets of $\mathbb{R}^k$) to provide an insight into Siegel's identity in (45.34) and also a new proof for this result. Vitale (1996) also utilized a vector analogue of the Euler–Schläfli identity for polytopes to derive a generalization of the result in (45.34).

Gupta and Gupta (1998) have proved that the distributions of the minimum and the maximum of a multivariate normal distribution [which possesses the IFR (increasing failure rate) property, as mentioned earlier at the end of Section 3] retain the IFR property.

Let $\boldsymbol{X}_j = (X_{1j}, X_{2j}, \ldots, X_{kj})^T$, $j = 1, 2, \ldots, n$, be independent observations from a multivariate normal, $N_k(\boldsymbol{\xi}, \boldsymbol{V})$, population. Let $P = \{i_1, i_2, \ldots, i_m\}$ $(m \geq 1)$ be a partition of $\{1, 2, \ldots, k\}$ and let Q be its complementary partition. Let $\boldsymbol{C} = (c_1, c_2, \ldots, c_k)^T$ be a vector of nonzero constants. Furthermore, let us define

$$X_j = \sum_{i \in P} c_i X_{ij} \quad \text{and} \quad Y_j = \sum_{i \in Q} c_i X_{ij}$$

for $j = 1, 2, \ldots, n$. Equivalently, if we define

$$\boldsymbol{C}_P^T = (0, \ldots, 0, c_{i_1}, 0, \ldots, 0, c_{i_2}, \ldots, 0, \ldots, 0, c_{i_m}, 0, \ldots, 0)_{1 \times k}$$

and

$$\boldsymbol{C}_Q^T = (c_1, \ldots, c_{i_1-1}, 0, c_{i_1+1}, \ldots, c_{i_2-1}, 0, \ldots, c_{i_m-1}, 0, c_{i_m+1}, \ldots, c_k)_{1 \times k},$$

we can write

$$X_j = \boldsymbol{C}_P^T \boldsymbol{X}_j \quad \text{and} \quad Y_j = \boldsymbol{C}_Q^T \boldsymbol{X}_j \quad \text{for} \quad j = 1, \ldots, n. \tag{45.38}$$

From (45.38), we get

$$
\begin{aligned}
\mu_X &= E[X_j] = \boldsymbol{C}_P^T \boldsymbol{\xi}, \quad \mu_Y = E[Y_j] = \boldsymbol{C}_Q^T \boldsymbol{\xi}, \\
\sigma_X^2 &= \text{var}(X_j) = \boldsymbol{C}_P^T \boldsymbol{V} \boldsymbol{C}_P, \quad \sigma_Y^2 = \text{var}(Y_j) = \boldsymbol{C}_Q^T \boldsymbol{V} \boldsymbol{C}_Q, \\
\sigma_{X,Y} &= \text{cov}(X_j, Y_j) = \boldsymbol{C}_P^T \boldsymbol{V} \boldsymbol{C}_Q,
\end{aligned}
$$

and

$$\rho = \frac{\sigma_{X,Y}}{\sigma_X \sigma_Y} = \frac{\boldsymbol{C}_P^T \boldsymbol{V} \boldsymbol{C}_Q}{\{(\boldsymbol{C}_P^T \boldsymbol{V} \boldsymbol{C}_P)(\boldsymbol{C}_Q^T \boldsymbol{V} \boldsymbol{C}_Q)\}^{1/2}} \cdot$$

Let

$$S_j = X_j + Y_j = \boldsymbol{C}_P^T \boldsymbol{X}_j + \boldsymbol{C}_Q^T \boldsymbol{X}_j = \boldsymbol{C}^T \boldsymbol{X}_j, \quad 1 \leq j \leq n \tag{45.39}$$

and let $S_{1:n} \leq S_{2:n} \leq \cdots \leq S_{n:n}$ be the order statistics of S_j's in (45.39). Let $\boldsymbol{X}_{[\ell:n]}$ be the induced ℓth multivariate order statistic; that is, $\boldsymbol{X}_{[\ell:n]} = \boldsymbol{X}_j$ whenever $S_{\ell:n} = S_j$. In this setup, Balakrishnan (1993) has studied these multivariate order statistics, $\boldsymbol{X}_{[\ell:n]}$, induced by ordering the linear combinations S_j's.

Evidently, from (45.38) and (45.39), we have

$$X_{[\ell:n]} = \boldsymbol{C}_P^T \boldsymbol{X}_{[\ell:n]} \quad \text{and} \quad Y_{[\ell:n]} = \boldsymbol{C}_Q^T \boldsymbol{X}_{[\ell:n]}$$

and, hence,

$$
\begin{aligned}
E[X_{[\ell:n]}] &= E\left[\boldsymbol{C}_P^T \boldsymbol{X}_{[\ell:n]}\right] \\
&= \boldsymbol{C}_P^T \boldsymbol{\xi} + \left\{ \frac{\boldsymbol{C}_P^T \boldsymbol{V} \boldsymbol{C}_P + \boldsymbol{C}_P^T \boldsymbol{V} \boldsymbol{C}_Q}{(\boldsymbol{C}_P^T \boldsymbol{V} \boldsymbol{C}_P + \boldsymbol{C}_Q^T \boldsymbol{V} \boldsymbol{C}_Q + 2\boldsymbol{C}_P^T \boldsymbol{V} \boldsymbol{C}_Q)^{1/2}} \right\} \alpha_{\ell:n}
\end{aligned}
$$

$$(45.40)$$

and

$$
\begin{aligned}
& \operatorname{var}(X_{[\ell:n]}) \\
&= \operatorname{var}\left(\boldsymbol{C}_P^T \boldsymbol{X}_{[\ell:n]}\right) \\
&= \frac{\beta_{\ell,\ell:n}(\boldsymbol{C}_P^T \boldsymbol{V} \boldsymbol{C}_P + \boldsymbol{C}_P^T \boldsymbol{V} \boldsymbol{C}_Q)^2 + (\boldsymbol{C}_P^T \boldsymbol{V} \boldsymbol{C}_P)(\boldsymbol{C}_Q^T \boldsymbol{V} \boldsymbol{C}_Q) - (\boldsymbol{C}_P^T \boldsymbol{V} \boldsymbol{C}_Q)^2}{\boldsymbol{C}_P^T \boldsymbol{V} \boldsymbol{C}_P + \boldsymbol{C}_Q^T \boldsymbol{V} \boldsymbol{C}_Q + 2\boldsymbol{C}_P^T \boldsymbol{V} \boldsymbol{C}_Q},
\end{aligned}
$$

$$(45.41)$$

where $\alpha_{\ell:n}$ and $\beta_{\ell,\ell:n}$ ($\ell = 1, 2, \ldots, n$) denote the mean and variance of the ℓth order statistic among n independent standard normal variables.

Balakrishnan (1993) has further derived the following formulas, where v_{rs} are the elements of the variance-covariance matrix $\boldsymbol{V}$:

$$
E[X_{i[\ell:n]}] = \xi_i + \left\{ \frac{\sum_{r=1}^k c_r v_{ir}}{\left(\sum_{s=1}^k \sum_{r=1}^k c_r c_s v_{rs}\right)^{1/2}} \right\} \alpha_{\ell:n},
$$

$$i = 1, \ldots, k, \ \ell = 1, \ldots, n,$$

$$
\operatorname{var}(X_{i[\ell:n]}) = v_{ii} - \left\{ \frac{\left(\sum_{r=1}^k c_r v_{ir}\right)^2}{\sum_{s=1}^k \sum_{r=1}^k c_r c_s v_{rs}} \right\} (1 - \beta_{\ell,\ell:n}),
$$

$$i = 1, \ldots, k, \ \ell = 1, \ldots, n, \quad (45.42)$$

$$
\operatorname{cov}(X_{i[\ell:n]}, X_{j[\ell:n]}) = v_{ij} - \left\{ \frac{\sum_{s=1}^k \sum_{r=1}^k c_r c_s v_{ir} v_{js}}{\sum_{s=1}^k \sum_{r=1}^k c_r c_s v_{rs}} \right\} (1 - \beta_{\ell,\ell:n}),
$$

$$1 \le i < j \le k, \ \ell = 1, \ldots, n,$$

$$
\operatorname{cov}(X_{i[\ell:n]}, X_{i[\ell':n]}) = \frac{\left(\sum_{r=1}^k c_r v_{ir}\right)^2}{\sum_{s=1}^k \sum_{r=1}^k c_r c_s v_{rs}} \beta_{\ell,\ell':n},
$$

$$i = 1, \ldots, k, \ 1 \le \ell < \ell' \le n,$$

and

$$\text{cov}(X_{i[\ell:n]}, X_{j[\ell':n]}) = \frac{\sum_{s=1}^{k} \sum_{r=1}^{k} c_r c_s v_{ir} v_{js}}{\sum_{s=1}^{k} \sum_{r=1}^{k} c_r c_s v_{rs}} \beta_{\ell,\ell':n},$$
$$1 \leq i < j \leq k, \ 1 \leq \ell < \ell' \leq n.$$

Using the results that $\beta_{j,\ell:n} > 0$ [see Bickel (1967)] and $\sum_{j=1}^{n} \beta_{j,\ell:n} = 1$ for $1 \leq \ell \leq n$ [see David (1981) and Balakrishnan and Cohen (1991)], we readily have $0 < \beta_{\ell,\ell:n} < 1$, and therefore it follows from (45.42) that $\text{var}(X_{i[\ell:n]}) < v_{ii}$ for $i = 1, 2, \ldots, k$ and all $\ell = 1, 2, \ldots, n$, as noted by Balakrishnan (1993). This shows that the variability of the ith component of ℓth induced order statistic is less than the variability of the ith component of $\boldsymbol{X}$ for any $\ell = 1, 2, \ldots, n$.

Along these lines, Bairamov and Gebizlioglu (1997) have discussed norm order statistics when the multivariate i.i.d. random variables are ordered by the magnitudes of their norm in a normed space. For general continuous distributions, they have investigated some distributional properties of these norm order statistics (under the Euclidean norm) and also discussed their applications.

5 EVALUATION OF MULTIVARIATE NORMAL PROBABILITIES

In this section we consider, for the most part, *standardized* multivariate normal distributions, that is, density functions of the form

$$p_{\boldsymbol{X}}(\boldsymbol{x}) = \frac{|\boldsymbol{R}|^{1/2}}{(2\pi)^{k/2}} \exp\left(-\frac{1}{2} \boldsymbol{x}^T \boldsymbol{R}^{-1} \boldsymbol{x}\right), \tag{45.43}$$

where $\boldsymbol{X}^T \equiv (X_1, X_2, \ldots, X_k)$, $\boldsymbol{x}^T = (x_1, x_2, \ldots, x_k)$ and $\boldsymbol{R}$ is the *correlation* matrix of $\boldsymbol{X}$. The notation $Z_k(\boldsymbol{x}; \boldsymbol{R})$ is often used for this function. We shall do so.

Generalizing the univariate probability integral $\Phi(\cdot)$, we define

$$\Phi_k(h_1, \ldots, h_k; \boldsymbol{R})$$
$$= \text{Pr}\left[\bigcap_{j=1}^{k}(X_j \leq h_j)\right]$$
$$= \frac{|\boldsymbol{R}|^{1/2}}{(2\pi)^{k/2}} \int_{-\infty}^{h_k} \cdots \int_{-\infty}^{h_1} \exp\left(-\frac{1}{2} \boldsymbol{x}^T \boldsymbol{R}^{-1} \boldsymbol{x}\right) dx_1 \cdots dx_k. \tag{45.44}$$

(Later, as in Section 2, we shall use $d\boldsymbol{x}$ as an abbreviation for $dx_1 \cdots dx_k$). This definition applies also when $\boldsymbol{R}$ is a variance–covariance matrix. There does not appear to be any simple procedure for evaluating (45.44) in the general case. Reduction formulas developed by Plackett (1954), Steck (1958), and John (1959), as well as others that are described below, are somewhat laborious in practice when ρ is greater than $\frac{1}{2}$. There are, however, simplifications in certain special cases, which will be described later in this section. Also, some calculations are practicable for smaller values of k. Those for $k = 4$ are described in Section 5 of this chapter; detailed discussion of the cases $k = 2$ and $k = 3$ appears in Chapter 46.

We shall not discuss calculation of multivariate normal probabilities other than those of form (45.44). Evaluation of multivariate normal probabilities over convex polyhedra has been described by John (1966); see also van der Vaart (1953, 1955). The special case of convex polygons will be discussed in Chapter 46.

Evaluation of multivariate normal integrals by some form of Monte Carlo technique has been studied by Escoufier (1967) and Abbe (1964), among others. Escoufier (1967) obtained some simplification in evaluation of integrals over regions bounded by planes $(\boldsymbol{a}^T \boldsymbol{X} = \boldsymbol{c})$ by transforming to independent variables $\boldsymbol{Z}$ as in Section 2. Abbe (1964) discussed the use of varying sampling rates in different parts of the region of integration.

Over the years, several methods have been proposed for the computation of multivariate normal probabilities; see, for example, the survey paper by Martynov (1981) and the book by Tong (1990). Many of these methods were mentioned in the first edition of this volume. They are presented here for historical as well as mathematical interest, even though the practical relevance of some of them has certainly diminished, due to the advances in numerical analysis facilitated by modern computer technology. We now note several of the key developments and results in this direction.

Firstly, the tetrachoric series and Kibble's series (described later in this section) for orthant probabilities converge very slowly unless all ρ_{ij}'s are small; see also Stuart and Ord (1994, p. 513) for similar comments. Harris and Soms (1980) reexamined the convergence of the tetrachoric series and concluded that, unless certain severe limitations on the correlation matrix are satisfied, the tetrachoric series will, in fact, diverge. Specifically, for the case of orthant probabilities, the tetrachoric series will converge if $|\rho_{ij}| < \frac{1}{k-1}$ for $1 \le i < j \le k$, but it will diverge whenever k is even $(k \ge 4)$ and $|\rho_{ij}| > \frac{1}{k-1}$ or when k is odd $(k \ge 5)$ and $|\rho_{ij}| > \frac{1}{k-2}$ for $1 \le i < j \le k$. One of the corollaries of Harris and Soms's (1980) results is that if $\rho_{ij} = \rho$, the tetrachoric series for orthant probabilities converges absolutely whenever $\frac{1}{k-1} < \rho < \frac{1}{k-2}$ for any odd $k \geq 5$; also, when $k = 3$, the series converges absolutely. It should be mentioned that even though these results are comprehensive, they do not cover all possibilities.

Moran (1985) dealt with the special case

$$
\begin{aligned}
\rho_{ij} &= \rho \quad \text{if } |i - j| = 1 \\
&= 0 \quad \text{if } |i - j| > 1,
\end{aligned}
$$

and noted that his method is applicable more generally. He considered the case $|\rho| \leq \frac{1}{2}$, which assures positive-definiteness of the correlation matrix. Let A be a positive number, and let $Y_0, Y_1, \ldots, Y_k$ be independent standard normal random variables. Let $X_i = AY_{i-1} + Y_i$ for $i = 1, 2, \ldots, k$. The variables X_i's then possess the required correlation structure with $\rho = A/(1+A^2)$. Note that for any given ρ, there are two possible reciprocal values of A. We then have

$$
\Pr[a_1 \leq X_1 \leq b_1, \ldots, a_k \leq X_k \leq b_k] = \int_{-\infty}^{\infty} p_k(y) \, dy,
$$

where

$$
p_k(y) = \Pr[y < Y_k \leq y + dy, a_1 \leq X_1 \leq b_1, \ldots, a_k \leq X_k \leq b_k].
$$

In particular,

$$
\begin{aligned}
p_1(y) &= \varphi(y) \Pr[a_1 \leq AY_0 + y \leq b_1] \\
&= \varphi(y) \left\{ \Phi\left(\frac{b_1 - y}{A}\right) - \Phi\left(\frac{a_1 - y}{A}\right) \right\}
\end{aligned}
$$

and

$$
p_k(y) = \varphi(y) \int_{(a_k-y)/A}^{(b_k-y)/A} p_{k-1}(u) \, du,
$$

where $\varphi(\cdot)$ is the univariate standard normal density function. If $a_i = 0$ and $b_i = \infty$ for all i, then we get

$$
p_k(y) = \varphi(y) \int_{-y/A}^{\infty} p_{k-1}(u) \, du.
$$

Moran (1985) has mentioned that this method runs into difficulties (resulting in inaccurate numerical integration) when ρ is small, due to discontinuity of the probability density function of Y_k conditional on $X_i \geq 0$ for $i = 1, \ldots, k$.

Dunnett (1989) considered the case when $\rho_{ij} = \rho_i \rho_j$ (for $i \neq j$ and $-1 < \rho_\ell < 1$) and expressed the variables X_i's in terms of $k+1$ independent standard normal variables $Y_1, \ldots, Y_k, Z$ as

$$X_i = \sqrt{1 - \rho_i^2} Y_i + \rho_i Z \qquad \text{for } i = 1, 2, \ldots, k.$$

Using this form, Dunnett (1989) presented a formula for the multivariate normal probability as a single integral as

$$\Pr[a_1 \leq X_1 \leq b_1, \ldots, a_k \leq X_k \leq b_k]$$
$$= \frac{1}{\sqrt{\pi}} \int_0^\infty \left[\prod_{i=1}^k \left\{ \Phi\left(\frac{b_i - \sqrt{2}\rho_i z}{\sqrt{1 - \rho_i^2}} \right) - \Phi\left(\frac{a_i - \sqrt{2}\rho_i z}{\sqrt{1 - \rho_i^2}} \right) \right\} \right.$$
$$\left. + \prod_{i=1}^k \left\{ \Phi\left(\frac{b_i + \sqrt{2}\rho_i z}{\sqrt{1 - \rho_i^2}} \right) - \Phi\left(\frac{a_i + \sqrt{2}\rho_i z}{\sqrt{1 - \rho_i^2}} \right) \right\} \right] e^{-z^2} dz.$$

Dunnett's program ($MVNPRD$) makes use of Simpson's rule for the required single integration in such a way that the prescribed accuracy is achieved. The program uses numerical integration only for the variables that have nonzero values of ρ_i, while the variables that have zero correlation between themselves are factored out and their contribution computed as univariate normal integrals.

Another algorithm due to Schervish (1984), known as $MULNOR$, does not require the special correlation structure in Dunnett's (1989) algorithm. However, computational times for Schervish's algorithm increase rapidly with k, making it impractical to use for "dimensions much higher than 5 or 6." In comparison, Dunnett's $MVNPRD$ algorithm has no restriction on the value of k.

It has to be mentioned that Milton (1972) indicated a direct method for computing

$$\Pr[X_1 \leq x_1, \ldots, X_k \leq x_k]$$
$$= \frac{1}{(2\pi)^{k/2} |\boldsymbol{R}|^{1/2}} \int_{-\infty}^{x_1} \cdots \int_{-\infty}^{x_k} e^{-\boldsymbol{u}^T \boldsymbol{R}^{-1} \boldsymbol{u}/2} \, du_k \cdots du_1,$$

where $\boldsymbol{R}$ is the correlation matrix of $\boldsymbol{X}$. His method is based on the observation that the distribution of $X_1, \ldots X_{k-1}$, conditional on X_k, is multivariate normal; and hence, effectively, his method is iterative, computing upwards through $2, 3, \ldots$, dimensions to k. In order to achieve the desired accuracy, a multidimensional adaptive quadrature procedure based on Simpson's rule is suggested.

Along the lines of Dunnett (1989), Soong and Hsu (1997) considered the case when the correlation matrix is singular and is of the form $R = D - \eta\eta^T$ and discussed the evaluation of the multivariate normal rectangular probability. In this case also, the multivariate normal rectangular probability can be expressed as a single integral, but with complex variables in the integrand. Soong and Hsu (1997) have demonstated how this complex integral can be computed using Romberg integration of complex variables when the dimension is low.

5.1 Reduction Formulas

Plackett (1954) based his reduction formula on the differential equation

$$\frac{\partial Z_k(x; R)}{\partial \rho_{ij}} = \frac{\partial^2 Z_k(x; R)}{\partial x_i \partial x_j}. \tag{45.45}$$

Suppose that $\Phi_k(h; R)$ can be evaluated for $R = R_0$. Then it follows from (45.45) that

$$\Phi_k(h; R) = \Phi(h, R_0) + \sum_{i<j}\sum \int_{\rho_{ij0}}^{\rho_{ij}} \frac{\partial \Phi_k(h; tR + (1-t)R_0)}{\partial \lambda_{ij}} d\lambda_{ij} \tag{45.46}$$

with

$$\lambda_{ij} = t\rho_{ij} + (1-t)\rho_{ij0},$$

where ρ_{ij0} denotes the (i,j)th element of R_0.

Plackett also derived the formula

$$\frac{\partial \Phi_k(h; R)}{\partial \rho_{12}} = Z_2(h_1, h_2; \rho_{12})\Phi_{k-2}(h_3 - \bar{h}_3, \ldots, h_k - \bar{h}_k; V_{(2)}), \tag{45.47}$$

where

$$\bar{h}_j = \frac{(\rho_{1j} - \rho_{2j}\rho_{12})h_1 + (\rho_{2j} - \rho_{1j}\rho_{12})h_2}{\sqrt{1 - \rho_{12}^2}} \qquad (j \neq 1, 2)$$

and $V_{(2)}$ is the conditional variance–covariance matrix of $X_3, \ldots, X_k$ given X_1 and X_2; see also Poznyakov (1971).

Steck (1958) noted that if $h_i h_j$ is not negative for any pair (i, j), then

$$\Phi_k(h; R) = \sum_{j=1}^{k} \Pr\left[(X_j \leq h_j) \bigcap_{\substack{i=1 \\ i \neq j}}^{k}(X_i < X_j h_i/h_j)\right] \tag{45.48}$$

(where $0/0$ is interpreted as 1). Each term on the right-hand side of (45.48) can be expressed in terms of a multivariate normal integral involving only $(k-1)$ variables; for example,

$$\Pr\left[(X_k \leq h_k) \bigcap_{i=1}^{k-1}(X_i \leq X_k h_i/h_k)\right]$$

$$= \int_{-\infty}^{h_k} \Pr\left[\bigcap_{i=1}^{k-1}(X_i \leq x h_i/h_k)\right] Z(x)\, dx. \qquad (45.49)$$

By repeated application of (45.49), evaluation of $\Phi_k(\boldsymbol{h}; \boldsymbol{R})$ can be made to depend on quadratures involving only univariate normal integrals. The process will become somewhat cumbersome if k is not rather small ($k > 5$, say); the limitation on values of the h's should also be noted.

John (1959) used a probabilistic argument to express integrals of the multivariate normal density in terms of integrals of multivariate normal densities with fewer variables. The event $\bigcap_{j=1}^{k}(X_j \leq h_j)$ is equivalent to the event $\max_{1 \leq j \leq k}(X_j - h_j) \leq 0$. Hence, if $L \equiv L(h_1, \dots, h_k) = \max_{1 \leq j \leq k}(X_j - h_j)$,

$$\begin{aligned}
\Phi_k(\boldsymbol{h}; \boldsymbol{R}) &= \Pr\left[\bigcap_{j=1}^{k}(X_j \leq h_j)\right] \\
&= \Pr[L \leq 0] \\
&= \sum_{j=1}^{k} \int_{-\infty}^{0} \Pr\left[\bigcap_{\substack{i=1 \\ i \neq j}}^{k}(X_i < h_i) \mid X_j = h_j + t\right] Z(t + h_j)\, dt.
\end{aligned}$$

$$(45.50)$$

These methods, as applied in the special cases $k = 2$, $k = 3$, are discussed further in Chapter 46.

In order to render calculations simpler, Marsaglia (1963) has utilized the relationship

$$\Phi_k(\boldsymbol{h}; \boldsymbol{A} + \boldsymbol{R}) = E[\Phi(\boldsymbol{h} - \boldsymbol{Y}; \boldsymbol{R})],$$

where $\boldsymbol{Y}$ has a joint multivariate normal distribution with variance–covariance matrix $\boldsymbol{A}$ and expected value vector $\boldsymbol{0}$. [We have already noted in Eq. (45.44) that the definition of $\Phi_k(\cdot)$ also applies when $\boldsymbol{R}$ is a variance-covariance matrix.] Choice of $\boldsymbol{A}$ and $\boldsymbol{R}$ has been discussed by Anderson (1970).

5.2 Orthant Probabilities

The problem of evaluating $\Phi_k(h; R)$ may be specialized with respect to either h or R, or both. If we take $h = 0$, we have the problem of evaluating orthant probabilities. Since R is unchanged if each X_j is replaced by $-X_j$, we have

$$\Phi_k(0; R) = \frac{|R|^{1/2}}{(2\pi)^{k/2}} \int_0^\infty \int_0^\infty \cdots \int_0^\infty \exp\left(-\frac{1}{2} x^T R^{-1} x\right) dx. \quad (45.51)$$

The integral in (45.51) was studied as early as 1858, when Schläfli (1858) obtained a differential equation, corresponding to (45.46) with $h = 0$.

A direct method of calculation can be based on an expansion of the ratio of the density function (45.43) to the density function with $R = I$. This was obtained by Mehler (1866) for bivariate distributions and was generalized by Kibble (1945) to the multivariate normal case. The expansion is

$$Z_k(x; 0; R) = Z_k(x; 0; I) \sum_{j=0}^\infty (j!)^{-1} \sum_i {}^* C_i \prod_{\alpha\beta} \rho_{\alpha\beta} \prod_{t=1}^k H_{i_t}(x_t), \quad (45.52)$$

where

$\sum^*$ is a sum over all possible sets of j $\rho_{\alpha\beta}$'s (including repeated values),

C_i is the number of different permutations of the $\rho_{\alpha\beta}$'s—that is, $j! \left(\prod_m j_m!\right)^{-1}$ where the same ρ is repeated $j_1, j_2, \ldots, j_m$ times $(\sum_m j_m = j)$ $(\alpha < \beta)$,

i_t is the number of times t occurs among the suffices α, β in the ith term of $\sum^*$,

and

$H_r(x) =$ the rth Hermite polynomial (see Chapter 1).

A relatively simple derivation of (45.52) for the case $k = 2$ is given by Brown (1968).

Since each term of (45.52) is a product of functions of the form

$$\text{constant} \times H_\ell(x_\ell) e^{-(1/2)x_\ell^2},$$

it is possible to integrate term-by-term and thus obtain a series expansion for $\Phi_k(0; R)$.

Kendall (1941) has obtained an equivalent series expansion by working with the inversion formula for the density in terms of the characteristic function. From (45.4), we have

$$Z_k(\boldsymbol{x}; \boldsymbol{R}) = (2\pi)^{-k/2} \int_{-\infty}^{\infty} \cdots \int_{-\infty}^{\infty} \exp\left(-i\boldsymbol{t}^T\boldsymbol{x} - \frac{1}{2}\,\boldsymbol{t}^T\boldsymbol{R}^{-1}\boldsymbol{t}\right) dt.$$

(45.53)

Kendall used the formula $(\alpha < \beta)$

$$\exp\left(-\frac{1}{2}\,\boldsymbol{t}^T\boldsymbol{R}^{-1}\boldsymbol{t}\right) = \exp\left(-\frac{1}{2}\,\boldsymbol{t}^T\boldsymbol{t}\right) \sum_{j=0}^{\infty}(-1)^j \sum{}^* \prod_{m=1}^{k} t_m^{jm} \prod_{\alpha,\beta}(\rho_{\alpha\beta}^{j_{\alpha\beta}}/j_{\alpha\beta}!),$$

(45.54)

where $\sum^*$ now denotes summation over all $\{j_{\alpha,\beta}\}$ for which

$$\sum_{\alpha=1}^{k} j_\alpha = 1 \text{ and } \sum_{\beta}(j_{\alpha\beta} + j_{\beta\alpha}) = j_\alpha.$$

This gives, after some reduction,

$$\begin{aligned}
\Phi_k(\boldsymbol{0}; \boldsymbol{R}) &= \int_{-\infty}^{\infty} \cdots \int_{-\infty}^{\infty} Z_k(\boldsymbol{x}; \boldsymbol{R})\, d\boldsymbol{x} \\
&= \sum_{j=0}^{\infty}(-1)^j \sum{}^* \left(\prod_{m=1}^{k} A_{jm}\right) \prod_{\alpha,\beta}(\rho_{\alpha\beta}^{j_{\alpha\beta}}/j_{\alpha\beta}!), \quad (45.55)
\end{aligned}$$

where

$$A_t = \begin{cases} \frac{1}{2} & \text{if } t = 0, \\ 0 & \text{if } t \text{ is even,} \\ \frac{1}{i\sqrt{2\pi}} \, \frac{(t-1)!}{2^{(t-1)/2}\left[\frac{1}{2}(t-1)\right]!} & \text{if } t \text{ is odd.} \end{cases}$$

Note that since each $\rho_{\alpha\beta}$ is counted twice in j, j must be even. This ensures that $\prod_{m=1}^{k} A_{jm}$ must be real. It *also* means that $(-1)^j$ can be omitted from (45.55).

Unfortunately, these series converge very slowly unless all the ρ_{ij}'s are small. An approximate formula is presented in Section 5.4.

We now consider results that can be obtained by giving $\boldsymbol{R}$ special forms.

Sun (1988a) has presented a Fortran subroutine for computing normal orthant probabilities for dimensions up to 9 based on the formulas

$$\Phi_{2k}(\boldsymbol{0}; \boldsymbol{R}_{2k}) = \frac{1}{2^{2k}} + \frac{1}{2^{2k-1}\pi} \sum_{1 \leq i < j \leq 2k} \sin^{-1}(\rho_{ij})$$

$$+\sum_{j=2}^{k}\frac{1}{2^{2k-j}\pi^j}\sum_{1\le i_1<\cdots<i_{2j}\le 2k}I_{2j}\left(\boldsymbol{R}^{(i_1,\ldots,i_{2j})}\right)$$

$$(45.56)$$

and

$$\Phi_{2k+1}(0;\boldsymbol{R}_{2k+1})=\frac{1}{2^{2k+1}}+\frac{1}{2^{2k}\pi}\sum_{1\le i<j\le 2k+1}\sin^{-1}(\rho_{ij})$$

$$+\sum_{j=2}^{k}\frac{1}{2^{2k+1-j}\pi^j}\sum_{1\le i_1<\cdots<i_{2j}\le 2k+1}I_{2j}\left(\boldsymbol{R}^{(i_1,\ldots,i_{2j})}\right),$$

where $\boldsymbol{R}_{2j}^{(i_1,\ldots,i_{2j})}$ denotes the submatrix consisting of $(i_1,\ldots,i_{2j})$th rows and columns of the correlation matrix $\boldsymbol{R}$, and

$$I_{2\ell}(\boldsymbol{\Lambda}_{2\ell})=(-2\pi)^{-\ell}\int_{-\infty}^{\infty}\prod_{i=1}^{2\ell}\frac{1}{w_i}\exp\left\{-\frac{1}{2}\ \boldsymbol{w}^T\boldsymbol{\Lambda}_{2\ell}\boldsymbol{w}\right\}dw_1\ldots dw_{2\ell},$$

with $\boldsymbol{\Lambda}_{2\ell}$ being a covariance matrix of 2ℓ variates. For $\ell=1$, we of course have

$$I_2(\boldsymbol{\Lambda}_2)=\sin^{-1}\left(\frac{\lambda_{12}}{\sqrt{\lambda_{11}\lambda_{22}}}\right);$$

see Chapter 46. These formulas are generalizations of those given by Childs (1967). Sun (1988b) has also given a recursive formula

$$I_{2\ell}(\boldsymbol{\Lambda}_{2\ell})=\int_0^1\sum_{i=2}^{2\ell}\frac{\lambda_{1i}}{\sqrt{\lambda_{11}\lambda_{ii}-\lambda_{1i}^2t^2}}\ I_{2\ell-2}\left(\boldsymbol{\Lambda}_{2\ell-2}^{1,i}\right)dt\quad(\ell>1).$$

It can be easily seen that $I_{2\ell}$ can be reduced to a total number M of $(\ell-1)$th-order multivariate integrals, where $M=(2\ell-1)(2\ell-3)\cdots 1=(2\ell-1)!!$.

Evans and Swartz (1988) reduced (45.51) to the form

$$\frac{\Gamma(k/2)}{(2\pi)^{k/2}}|\boldsymbol{W}|\int_{S_R}\cdots\int\|\boldsymbol{W}\boldsymbol{b}\|^{-k}J(\boldsymbol{b})\ d\boldsymbol{b},\qquad(45.57)$$

where $\boldsymbol{W}=\boldsymbol{R}^{-1/2}\mathrm{diag}(\|\boldsymbol{a}_1\|^{-1},\ldots,\|\boldsymbol{a}_k\|^{-1})$, $\boldsymbol{R}^{-1/2}=(\boldsymbol{a}_1,\ldots,\boldsymbol{a}_k)$, $\boldsymbol{R}^{1/2}$ is any matrix such that $(\boldsymbol{R}^{1/2})^T(\boldsymbol{R}^{1/2})=\boldsymbol{R}$, $\|\cdot\|$ is the Euclidean norm on $\mathbb{R}^k$, and

$$S_R=\left\{\boldsymbol{b}\in\mathbb{R}^k|\left(\sum_{i\in R_1}b_i\right)^2+\cdots+\left(\sum_{i\in R_{k-1}}b_i\right)^2+\sum_{i\in R_k}b_i^2=1,\ b_i\ge 0\right\}$$

(some of the sums may be empty), with $R_1, \ldots, R_k$ being such that $\bigcup_{i=1}^{k} R_i = \{1, 2, \ldots, k\}$ and $R_i \cap R_j = \emptyset$ $(i \neq j)$. To see this, we first need to transform $\boldsymbol{x}$ to $\boldsymbol{y} = \text{diag}(\|\boldsymbol{a}_1\|, \ldots, \|\boldsymbol{a}_k\|)\boldsymbol{x}$ and then transform $\boldsymbol{y}$ to $(r, b_1, \ldots, b_{k-1})^T$, where

$$r = \left\{ \left(\sum_{i \in R_1} y_i \right)^2 + \cdots + \left(\sum_{i \in R_{k-1}} y_i \right)^2 + \sum_{i \in R_k} y_i^2 \right\}^{1/2}$$

and $b_i = y_i/r$ $(i = 1, \ldots, k)$, with the Jacobian of the second transformation being $r^{k-1} J(\boldsymbol{b})$. Making use of the representation in (45.57), Evans and Swartz (1988) have developed an efficient unbiased Monte Carlo estimator of (45.51) via the Dirichlet distribution (see Chapter 49).

Solow (1990) has provided a simple method for approximating multivariate normal orthant probabilities which is based on decomposing the orthant probability into a product of conditional probabilities and approximating the terms in the product using a linear model. (In fact, the method may be applied to approximate distribution functions for arbitrary arguments in cases other than the normal.) For $\boldsymbol{X} = (X_1, \ldots, X_k)^T$ distributed as $N_k(\boldsymbol{0}, \boldsymbol{R})$, let us denote as usual

$$\Phi_k(x; \boldsymbol{R}) = \Pr[X_i \leq x \text{ for } i = 1, 2, \ldots, k]$$

and

$$P_i(x) = \Pr[X_i \leq x \mid X_j \leq x \text{ for } j = i + 1, \ldots, k].$$

If we define

$$I_i(x) = \begin{cases} 1 & \text{if } X_i \leq x, \\ 0 & \text{otherwise} \end{cases}$$

for $i = 1, 2, \ldots, k$, we have

$$E[I_i(x)] = \Pr[X_i \leq x] = \Phi_1(x)$$

and

$$\begin{aligned} \text{cov}(I_i(x), I_j(x)) &= \Pr[X_i \leq x, X_j \leq x] - \Pr[X_i \leq x]\Pr[X_j \leq x] \\ &= \Phi_2(x; \rho_{ij}) - \Phi_1^2(x) \\ &= C_{ij}(x), \end{aligned}$$

where $\Phi_1(x)$ and $\Phi_2(x; \rho)$ denote the cumulative distribution function of the standard univariate and bivariate normal distributions. Evidently,

$$P_i(x) = E[I_i(x)|I_j(x) = 1 \text{ for } j = i + 1, \ldots, k].$$

Now a linear regression approximation to $P_i(x)$ is

$$\hat{P}_i(x) = \Phi_1(x) + \sum_{j=i+1}^{k} b_j^{(i)}(x)\{1 - \Phi_1(x)\}, \qquad (45.58)$$

where

$$b^{(i)}(x) = \left(C^{(i)}(x)\right)^{-1} k^{(i)}(x)$$

with $C_{j\ell}^{(i)}(x) = C_{j\ell}(x)$ for $j, \ell = i+1, \ldots, k$, and $k_j^{(i)}(x) = C_{ij}(x)$ for $j = i+1, \ldots, k$. The approximation to $\Phi_k(x; R)$ is then

$$\hat{\Phi}_k(x; R) = \prod_{i=1}^{k} \hat{P}_i(x). \qquad (45.59)$$

In the case of equicorrelation (namely, $\rho_{ij} = \rho \; \forall \; i, j$), (45.59) gives

$$\hat{\Phi}_k(x; R) = \prod_{i=1}^{k} \left[\Phi_1(x) + (k-i)w_i(x)\{1 - \Phi_1(x)\}\right], \qquad (45.60)$$

where $w_i(x) = \frac{C(x)}{V(x)+(k-i-1)C(x)}$ with $C(x) = \Phi_2(x; \rho) - \Phi_1^2(x)$ and $V(x) = \Phi_1(x) - \Phi_1^2(x)$. Evidently, $\hat{\Phi}_k(0; R) = \Phi_k(0; R)$ when $\rho = 0.5$.

Table 45.1, taken from Solow (1990), provides some comparisons between the exact and approximate values. There is slight degradation for moderate x and large k.

Next, for $X = (X_1, \ldots, X_k)^T$ distributed as $N_k(0, R)$, let $A_k(a) = \{X : \cap_{j=1}^{k}(|X_j| \le a)\}$ for $a > 0$, and that $\rho_{ij} = -b_i b_j$, where $b_i > 0$ for $i = 1, \ldots, k$ and $\sum_{j=1}^{k} \frac{b_j^2}{1+b_j^2} = 1$. For any given α, let $h_\alpha > 0$ be such that

$$\Pr[X \in A_k(h_\alpha)] = 1 - \alpha.$$

This correlation matrix can be seen to be positive semidefinite of rank $k - 1$. Also, Kwong (1995) and Kwong and Iglewicz (1996) have shown that:

(i)

$$\Pr\left[\bigcap_{j=1}^{2}(-c_j \le X_j \le a_j); \rho_{12} = -1\right] = \Phi_1(c_1) + \Phi_1(c_2) - 1, \qquad (45.61)$$

where $a_j, c_j \ge 0$ (for $j = 1, 2$) and $\min(c_1, c_2) \ge \max(a_1, a_2)$;

TABLE 45.1

Estimates $\hat{\Phi}_k(x; \boldsymbol{R})$, where $\rho_{ij} = \rho \; \forall \; i \neq j = 1, \ldots, k$, for Selected Values of k, x and ρ and the Corresponding True Values of $\Phi_k(x; \boldsymbol{R})$ Taken from Gupta (1963)

ρ	x	$k = 4$ $\hat{\Phi}_k(x)$	$\Phi_k(x)$	$k = 8$ $\hat{\Phi}_k(x)$	$\Phi_k(x)$	$k = 12$ $\hat{\Phi}_k(x)$	$\Phi_k(x)$
0.2	−1.0	0.004	0.004	−	−	−	−
	0.0	0.113	0.113	0.031	0.030	0.012	0.012
	1.0	0.551	0.551	0.361	0.357	0.260	0.253
	2.0	0.917	0.917	0.850	0.849	0.795	0.793
0.4	−1.0	0.013	0.014	0.002	0.003	−	−
	0.0	0.169	0.169	0.079	0.079	0.049	0.048
	1.0	0.603	0.601	0.461	0.453	0.384	0.371
	2.0	0.924	0.924	0.874	0.872	0.838	0.832
0.6	−2.0	0.001	0.001	−	−	−	−
	−1.0	0.030	0.031	0.011	0.013	0.006	0.008
	0.0	0.233	0.233	0.149	0.149	0.113	0.113
	1.0	0.656	0.654	0.557	0.548	0.501	0.488
	2.0	0.935	0.934	0.902	0.898	0.880	0.872
0.8	−2.0	0.004	0.005	0.002	0.002	−	−
	−1.0	0.060	0.061	0.037	0.040	0.027	0.031
	0.0	0.314	0.314	0.247	0.248	0.215	0.217
	1.0	0.718	0.716	0.657	0.652	0.624	0.615
	2.0	0.948	0.948	0.931	0.927	0.920	0.914

Source: Solow (1990), with permission.

(ii)

$$\Pr[\boldsymbol{X} \in A_3(h_\alpha); (\rho_{ij} = -b_i b_j)]$$

$$= 2 \int_0^{h_\alpha} \left\{ 2 \sum_{1 \leq j < k \leq 3} \Phi_1 \left(s \sqrt{\frac{1 - b_j b_k}{1 + b_j b_k}} \right) - 3 \right\} \varphi(s) \, ds,$$

$$(45.62)$$

where, as above, $\varphi(\cdot)$ is the univariate standard normal density function.

If, for example, $b_1 = b_2 = b_3 = \frac{1}{\sqrt{2}}$, (45.62) reduces to

$$\Pr\left[\mathbf{X} \in A_3(z_3(\alpha)); \left(\rho_{ij} = -\frac{1}{2}\right)\right]$$

$$= 6\int_0^{z_3(\alpha)} \left\{2\Phi_1\left(\frac{s}{\sqrt{3}}\right) - 1\right\} \varphi(s)\, ds, \qquad (45.63)$$

where $z_k(\alpha)$ denotes the two-sided $100(1-\alpha)\%$ point of the standardized k-variate normal distribution with the singular negative equi-correlated structure.

Kwong (1995) and Kwong and Iglewicz (1996) have presented results for the case $k = 4$.

5.3   Some Special Cases

The matrix $\mathbf{R}$ may be specialized in a number of ways. Ihm (1959) has obtained a general formula for $\Pr[(X_1, X_2, \ldots, X_k) \in \Omega]$ which applies when the *variance–covariance* matrix is of form $\Delta + c^2 \mathbf{1}\mathbf{1}^T$, where Δ is a positive definite diagonal matrix and $\mathbf{1}^T = (1, 1, \ldots, 1)$—that is,

$$\begin{aligned}
\text{var}(X_j) &= \delta_{jj} + c^2, \\
\text{cov}(X_i, X_j) &= c^2.
\end{aligned} \qquad (45.64)$$

Ihm showed that if $E[\mathbf{X}^T] = \mathbf{0}$, then

$$\Pr[(X_1, X_2, \ldots, X_k) \in \Omega]$$
$$= \frac{c}{\sqrt{2\pi}} \int_{-\infty}^{\infty} e^{-(1/2)c^2 t^2} \underbrace{\int \cdots \int}_{\Omega} \left[(2\pi)^{k/2} |\Delta|^{1/2}\right]^{-1}$$

$$\times \exp\left\{-\frac{1}{2}\sum_{j=1}^{k} \delta_{jj}^{-1}(y_j - t)^2\right\} d\mathbf{y}\, dt. \qquad (45.65)$$

Although this is a multiple integral of $(k+1)$th order, which is greater than the order (k) of the original multivariate normal integral, the integral is, in general, of simpler form. If the correlations can be expressed in the form $\rho_{ij} = \lambda_i \lambda_j$, for all i and j, then $X_1, X_2, \ldots, X_k$ can be represented as

$$X_j = \lambda_j U_0 + \sqrt{1 - \lambda_j}\, U_j \qquad (j = 1, 2, \ldots, k),$$

where $U_0, U_1, \ldots, U_k$ are independent unit normal variables. This representation greatly facilitates calculation of probabilities. The inequality $(X_j \le h_j)$ is equivalent to

$$U_j \le (h_j - \lambda_j U_0)/\sqrt{1 - \lambda_j} ,$$

and hence

$$\Pr\left[\bigcap_{j=1}^{k}(X_j \le h_j)\right] = \int_{-\infty}^{\infty} Z(u_0) \prod_{j=1}^{k} \Phi\left(\frac{h_j - \lambda_j u_0}{\sqrt{1 - \lambda_j}}\right) du_0; \qquad (45.66)$$

see Dunnett and Sobel (1955).

For the special case $\rho_{ij} = \lambda_i/\lambda_j$ for all $i \le j$, Curnow and Dunnett (1962) have found reduction formulas for $k = 3, 4, 5$.

If all the correlations are equal and positive ($\rho_{ij} = \rho > 0$ for all i, j), then we have the representation

$$X_j = \sqrt{\rho}\, U_0 + \sqrt{1 - \rho}\, U_j \qquad (j = 1, 2, \ldots, k) \qquad (45.67)$$

obtained by putting $\lambda_j = \sqrt{\rho}$. The inequality $X_j \le h_j$ is equivalent to $U_j \le (h_j - \sqrt{\rho}\, U_0)/\sqrt{1 - \rho}$, and

$$\Pr\left[\bigcap_{j=1}^{k}(X_j \le h_j)\right] = \int_{-\infty}^{\infty} Z(u_0) \prod_{j=1}^{k} \Phi\left(\frac{h_j - \sqrt{\rho}\, u_0}{\sqrt{1 - \rho}}\right) du_0. \qquad (45.68)$$

In the general case, this must still be evaluated by numerical quadrature, but the reduction to a single integral makes the calculation much simpler. If also $h_1 = h_2 = \cdots = h_k = h$, we have

$$\Pr[\max(X_1, \ldots, X_k) \le h] = \int_{-\infty}^{\infty} Z(u_0)\left[\Phi\left(\frac{h - \sqrt{\rho}\, u_0}{\sqrt{1 - \rho}}\right)\right]^{k} du_0.$$
$$(45.69)$$

This formula has been obtained in a number of equivalent forms by Das (1956), Dunnett (1955), Dunnett and Sobel (1955), Gupta (1963), Ihm (1959), Moran (1956), Ruben (1961), and Stuart (1958), among others. Steck and Owen (1962) have shown that this formula is valid for negative ρ as well, even though the integrand on the right-hand side is complex. These authors also obtain a useful recurrence relation (valid for ρ positive or negative). Denoting the probability in (45.65) by $F(h \mid \rho, k)$, they show that

$$F(h \mid \rho, k) = \sum_{j=1}^{k} (-1)^{j+1} \binom{k}{j} F(\alpha h \mid \rho', j) F(h \mid \rho, k - j), \qquad (45.70)$$

where

$$\alpha = \left[\frac{1 - \rho}{\{1 + (k-1)\rho\}\{1 + (k-2)\rho\}}\right]^{1/2}$$

and
$$\rho' = -\rho\{1 + (k-2)\rho\}^{-1}.$$

When $\rho_{ij} = \rho$ for all i, j, and $h = 0$, a number of simplifications are possible. We denote $\Phi_k(0; \mathbf{R})$ in this case by $L_k(\rho)$, for convenience.

Kwong and Iglewicz (1996) have pointed out that Bland and Owen's (1966) claim that Steck and Owen's (1962) result for nonsingular negative equicorrelated case can be extended to the singular case is in doubt. Kwong (1995) has also corrected Nelson's (1991) numerical evaluation of multivariate normal integrals with correlation matrix $\rho_{ij} = -b_i b_j$.

Sampford, quoted by Moran (1956), showed that if $\rho > 0$, then

$$L_k(\rho) = \frac{1}{\sqrt{\pi}} \int_{-\infty}^{\infty} e^{-t^2} [1 - \Phi(at)]^k dt, \qquad (45.71)$$

where $a = 2\rho/(1-\rho)$. Although the integral must be evaluated by quadrature, accurate values are easily obtained from simple summation formulas.

From Plackett's formula (45.46), putting $\rho_{ij} = \rho$ for all i, j and adding, we obtain

$$\frac{\partial L_k(\rho)}{\partial \rho} = \frac{k(k-1)}{4\pi(1-\rho^2)^{1/2}} L_{k-2}\left(\frac{\rho}{1+2\rho}\right), \qquad (45.72)$$

from which [noting that $L_k(0) = 2^{-k}$] we have

$$L_k(\rho) = \left(\frac{1}{2}\right)^k + \frac{k(k-1)}{4\pi} \int_0^r L_{k-2}\left(\frac{r}{1+2r}\right)(1-r^2)^{-1/2}\, dr; \quad (45.73)$$

see Ruben (1961). Using the known values of $L_2(\rho)$ and $L_3(\rho)$ (see Chapter 46), we find

$$\begin{aligned}
L_k(\rho) &= 2^{-k}\left[1 + \frac{k^{(2)}}{\pi}\sin^{-1}\rho + \frac{k^{(4)}}{\pi^2}\int_0^\rho \frac{\sin^{-1}[r_1/(1+2r_1)]}{(1-r_1^2)^{1/2}}\, dr_1 \right. \\
&\quad \left. + \frac{k^{(6)}}{\pi^3}\int_0^\rho \int_0^{r_2/(1+2r_2)} \frac{\sin^{-1}[r_1/(1+2r_1)]}{(1-r_1^2)^{1/2}} \frac{dr_1\, dr_2}{(1-r_2^2)^{1/2}} + \cdots \right].
\end{aligned}$$
$$(45.74)$$

The $(j+1)$th term in the series on the right-hand side of (45.74) is

$$\frac{k^{(2j)}}{\pi^j} I_j(\rho),$$

where

$$I_j(\rho) = \int_0^\rho \int_0^{r_j/(1+2r_j)} \cdots \int_0^{r_2/(1+2r_2)} \frac{\sin^{-1}[r_1/(1+2r_1)]}{\prod_{i=1}^j (1-r_i^2)^{1/2}}\, dr_1\, dr_2 \cdots dr_j.$$

Bacon (1963) gives a table of values of $I_2(\rho)$, $I_3(\rho)$, and $I_4(\rho)$ (Table 45.2). David and Six (1971) have shown that when $\rho_{ij} = \frac{1}{2}$ for all i, j, then

$$\Pr\left[\bigcap_{t=1}^{u}(X_t \le 0) \bigcap_{t=u+1}^{k} (X_t > 0)\right] = (k+1)^{-1} \quad \text{for } u = 0, 1, \ldots, k.$$

(45.75)

They also give tables of the probability that u or fewer of $X_1, X_2, \ldots, X_k$ are positive when $\rho_{ij} = \rho$, to three decimal places for $\rho = 0.4(0.025)0.5$; $k = 12, 14, 16, 20, 24, 36, 48, 96$ for various values of u.

Das (1956) reduced the evaluation of L to an integral of the density function of $k + m$ *independent* normal variables, where m need not exceed k (multiplicity of smallest eigenvalue of $\boldsymbol{R}$); see Marsaglia (1963).

To perform the reduction, it is necessary to express $\boldsymbol{R}$ in the form

$$\boldsymbol{R} = c^2\boldsymbol{I}_m + \boldsymbol{B}\boldsymbol{B}^T,$$

(45.76)

where $c > 0$ and $\boldsymbol{B}$ is a real $k \times m$ matrix. If (45.76) holds, then $X_1, X_2, \ldots, X_k$ can be represented by

$$c(Y_1, \ldots, Y_k) - (Z_1, \ldots, Z_m)\boldsymbol{B}^T$$

with $Y_1, \ldots, Y_k, Z_1, \ldots, Z_m$ independent unit normal variables. Hence

$$\Pr\left[\bigcap_{j=1}^{k}(X_j < h_j)\right]$$

$$= \Pr\left[\bigcap_{j=1}^{k}\left(Y_j \le \left\{h_j + \sum_{i=1}^{m} b_{ji}Z_i\right\}c^{-1}\right)\right]$$

$$= (2\pi)^{-m/2}\int_{-\infty}^{\infty}\cdots\int_{-\infty}^{\infty} e^{-\frac{1}{2}\sum_1^m z_i^2}\prod_{j=1}^{k}\Phi\left(c^{-1}\left\{h_j + \sum_{i=1}^{m} b_{ji}z_i\right\}\right) d\boldsymbol{z}.$$

(45.77)

This is an m-fold integral, so that it is desirable to make m as small as possible.

Webster (1970) has extended Das' method, replacing (45.76) by

$$\boldsymbol{R} = \boldsymbol{D}_{c^2} + \boldsymbol{B}\boldsymbol{B}^T$$

with $\boldsymbol{D}_{c^2}$ a diagonal matrix with diagonal elements $c_1^2, c_2^2, \ldots, c_m^2$. By appropriate choice of values of the c_j's, it may be possible to obtain a smaller value for m.

For the particular case when all correlations are equal to the least possible common value $[-(k-1)^{-1}]$, so that the distribution is singular, Bland and Owen (1966) have utilized the recurrence relation

$$\Pr\left[\bigcap_{j=1}^{k}(X_j \leq x)\right] = \sum_{m=0}^{k-1}(-1)^{m+1}\binom{k}{m}\Pr\left[\bigcap_{j=1}^{k-m}(X_j \leq x)\right].$$

TABLE 45.2

Values of the Integrals $I_i(\rho)$, $i = 2, 3, 4$, Used in Evaluation of Multivariate Normal Probabilities

ρ	$I_2(\rho)$	$I_3(\rho)$	$I_4(\rho)$
.00	.000000	.000000	.000000
.05	.001172	.000017	.000002
.10	.004404	.000117	.000022
.15	.009477	.000343	.000083
.20	.016067	.000712	.000203
.25	.024057	.001232	.000397
.30	.033375	.001907	.000673
.35	.043812	.002727	.001037
.40	.055459	.003706	.001495
.45	.068254	.004843	.002052
.50	.082247	.006152	.002714
.55	.097454	.007628	.003486
.60	.114012	.009291	.004379
.65	.132053	.011154	.005404
.70	.151813	.013243	.006580
.75	.173640	.015607	.007934
.80	.198120	.018306	.009506
.85	.226180	.021464	.011370
.90	.259820	.025310	.013668
.95	.303950	.030429	.016770
1.00	.411234	.043064	.024159

In the special case when all the off-diagonal elements of $\boldsymbol{R}$ are equal [Butler and Moffit (1982)] or, more generally, $\rho_{ij} = \delta_i\delta_j$, $|\delta_i| \leq 1$, there exists a simple decomposition of the multivariate normal integral. This form has been called an ℓ- *structure* by Peristiani (1991). In this case, we have the decomposition

$$\Pr[X_1 \leq x_1, \ldots, X_k \leq x_k] = \int_{-\infty}^{\infty}\prod_{i=1}^{k}\Phi_1\left(\frac{x_i - \delta_i y}{\sqrt{1-\delta_i^2}}\right)\varphi(y)\,dy;$$

$$(45.78)$$

see also (45.66). If, in addition, $\rho_{ij} = \rho$ and $x_i = 0$ for all i and j, then (45.78) simplifies to [see also (45.68)]

$$\Pr\left[\bigcap_{j=1}^{k}(X_j \leq 0)\right]$$

$$= \Pr[X_1 \leq 0, \ldots, X_k \leq 0] = \int_{-\infty}^{\infty}\left\{\Phi_1\left(-\sqrt{\frac{\rho}{1-\rho}}\, y\right)\right\}^k \varphi(y)\, dy.$$

$$(45.79)$$

When $\rho = 0.5$, the right-hand side of (45.79) becomes

$$\int_{-\infty}^{\infty}\{\Phi_1(-y)\}^k \varphi(y)dy = \frac{1}{k+1}\, ;$$

also see (45.75). Evaluation of (45.79) may be carried out by means of the iterative trapezoidal rule, as suggested by Peristiani (1991).

Lohr (1993) presented an algorithm for computing $\Pr[\boldsymbol{X} \in A]$ when $\boldsymbol{X} \stackrel{d}{=} N_k(\boldsymbol{0}, \boldsymbol{V})$, where $\boldsymbol{V}$ is a positive semidefinite covariance matrix of rank q and A is a compact region which is *star-shaped* with respect to $\boldsymbol{0}$ (i.e., if $\boldsymbol{x} \in A$, then any point between $\boldsymbol{0}$ and $\boldsymbol{x}$ is also in A). Then, by making use of the Cholesky decomposition of $\boldsymbol{V} = \boldsymbol{T}^T\boldsymbol{T}$, where $\boldsymbol{T}$ is a triangular matrix, we can write

$$\Pr[\boldsymbol{X} \in A] = \Pr[\boldsymbol{T}^T\boldsymbol{J}\boldsymbol{Z} \in A],$$

where $\boldsymbol{Z} \stackrel{d}{=} N_q(\boldsymbol{0}, \boldsymbol{I})$, and $J_{ij} = 1$ if the ith diagonal entry of $\boldsymbol{T}$ is the jth nonzero diagonal entry of $\boldsymbol{T}$ and $J_{ij} = 0$ otherwise. Consider a new random variable $\boldsymbol{Y}$ that is uniformly distributed on the surface of the unit q-dimensional sphere S. $\boldsymbol{T}^T\boldsymbol{J}S$ is the unit contour of constant density for the $N_k(\boldsymbol{0}, \boldsymbol{V})$ distribution and $\boldsymbol{T}^T\boldsymbol{J}\boldsymbol{Y}$ is distributed on $\boldsymbol{T}^T\boldsymbol{J}S$ in proportion to the volume enclosed in $\boldsymbol{T}^T\boldsymbol{J}S$. Let $\boldsymbol{y}$ be a realization of $\boldsymbol{Y}$ and $s(\boldsymbol{y})$ be the distance from the origin to the boundary of A in the direction specified by the k-dimensional vector $\boldsymbol{T}^T\boldsymbol{J}\boldsymbol{y}$. Finally, let

$$r(\boldsymbol{y}) = \frac{s(\boldsymbol{y})}{\text{length of } \boldsymbol{T}^T\boldsymbol{J}\boldsymbol{Y}}\, .$$

Evidently,

$$\Pr[\boldsymbol{X} \in A \mid \boldsymbol{Y} = \boldsymbol{y}] = \Pr[\chi_q^2 \leq r^2(\boldsymbol{y})].$$

Hence,

$$\Pr[\boldsymbol{X} \in A] = E_{\boldsymbol{Y}}\left[\Pr[\chi_q^2 \leq r^2(\boldsymbol{y})]\right],$$

which may be estimated by

$$\Pr[\widehat{\boldsymbol{X} \in A}] = \frac{1}{\ell} \sum_{i=1}^{\ell} \Pr[\chi_q^2 \le r^2(\boldsymbol{y}_i)], \tag{45.80}$$

where $\boldsymbol{y}_i$'s are randomly oriented vectors of length 1. Lohr (1993) has, in fact, recommended using antithetic variates instead of the crude Monte Carlo estimate in (45.80) [see, for example, Deàk (1980a,b)] for calculating the multivariate normal probability. A Fortran 77 program (Subroutine *MULNOR*) has been presented by Lohr (1993). This algorithm is most useful for calculating multivariate normal probabilities of oddly shaped regions or in high dimensions, as well as for calculating probabilities for indefinite covariance matrices. Related, but less general in form, are algorithms due to Donnelly (1973), DiDonato, Jarnagin, and Hageman (1980) and DiDonato and Hageman (1982) for calculating probability of a bivariate normal random variable falling in an arbitrary polygon. As already mentioned in the beginning of Section 5, Schervish (1984) and Dunnett (1989) have provided algorithms for calculating multivariate normal probabilities for rectangular regions.

5.4 Approximations

Because of the difficulties of exact evaluation of multivariate normal probabilities, reasonably accurate approximate formulas would be valuable. The only general formulas available are due to Bacon (1963). He obtained, on empirical grounds, the formula

$$\begin{aligned}
\Pr &\left[\bigcap_{j=1}^{k} (X_j > 0) \right] \\
&= \left(\frac{1}{2}\right)^k \left[1 + 2\sum \theta_{ij} + 4\sum \theta_{ij}\theta_{m\ell} \left(1 + \sum{}^*\theta_{uv}\right)^{-1} \right. \\
&\quad + 8\sum \theta_{ij}\theta_{m\ell}\theta_{rs} \left(1 + \frac{1}{3}\sum{}^*\theta_{uv}\right)^{-1} \left(1 + \frac{2}{3}\sum{}^*\theta_{u'v'}\right)^{-1} + \cdots \\
&\quad + 2^k \sum \theta_{ij} \cdots \theta_{rs} \left(1 + \frac{2}{k(k-1)}\sum{}^*\theta_{uv}\right)^{-1} \cdots \\
&\quad \left. \times \left(1 + \frac{2(k-1)}{k(k-1)}\sum{}^*\theta_{u'v'}\right)^{-1} \right], \tag{45.81}
\end{aligned}$$

where $\theta_{ij} = \pi^{-1}\sin^{-1}\rho_{ij}$; the summations $\sum$ are over $i \ne j$, $m \ne \ell$, $(i,j) \ne (m,\ell)$, and so on, and the summations $\sum{}^*$ are over subscript pairs that do not appear in the corresponding numerator. Error bounds are not available, but some numerical examples, given by Bacon (1963), indicate an accuracy of about 0.002 for values of the probability in the range 0.1 to 0.2.

(For the special cases $k = 2, 3, 4$, there are better approximate formulas. The case $k = 4$ will be discussed in Section 6; the cases $k = 2, 3$ are the subjects of Chapter 46.)

Bacon also obtained the following formula for the equally correlated case:

$$\Pr\left[\bigcap_{j=1}^{k}(X_j \le 0)\right]$$

$$\doteq 1 + \sum_{j=1}^{[(1/2)k]} \frac{k^{(2j)}}{j!} \frac{\theta^j}{(1+4\theta)(1+8\theta)\cdots(1+4[j-1]\theta)} \left(\frac{1}{2}\right)^k,$$

(45.82)

where $\theta = \pi^{-1}\sin^{-1}\rho$. This approximation is exact for $\rho = 0, \frac{1}{2}$, and 1.

The orthant probability $\Phi_k(\mathbf{0}; \mathbf{R})$ can be expressed in terms of multivariate normal integrals with $(k-1)$ variables, provided that k is odd. The relationship is obtained by an ingenious use of Boole's formula, due to David (1953) [see also Schläfli (1858)]. We have

$$\Phi_k(\mathbf{0}; \mathbf{R}) = \Pr\left[\bigcap_{j=1}^{k}(X_j \le 0)\right] = 1 - \Pr\left[\bigcup_{j=1}^{k}(X_j > 0)\right]$$

$$= 1 - \Pr\left[\bigcup_{j=1}^{k}(X_j \le 0)\right]$$

(noting that probabilities and correlations are unchanged by replacing each X_j by $-X_j$).

Using Boole's formula [see Chapter 1 of Johnson, Kotz, and Kemp (1992)], we obtain

$$\Phi_k(\mathbf{0}; \mathbf{R}) = 1 - \sum_{j=1}^{k}\Pr[X_j \le 0] + \sum_{j<j'}^{k}\sum^{k}\Pr[(X_j \le 0) \cap (X_{j'} \le 0)]$$

$$- \cdots + (-1)^k \Pr\left[\bigcap_{j=1}^{k}(X_j \le 0)\right].$$

(45.83)

If k is odd, then (45.83) is equivalent to

$$\Phi_k(\mathbf{0}; \mathbf{R}) = \frac{1}{2}\left\{-\frac{k}{2} + 1 + \sum_{j<j'}\sum\Pr[(X_j \le 0) \cap (X_{j'} \le 0)]\right.$$

$$- \cdots + \sum_{j_1 < j_2 < \cdots < j_{k-1}} \Pr \left[\bigcap_{i=1}^{k-1} (X_{j_i} \le 0) \right] \bigg\} \qquad (45.84)$$

upon noting that $\Pr[X_j \le 0] = \frac{1}{2}$ and $\Pr \left[\bigcap_{j=1}^{k} (X_j \le 0) \right] = \Phi_k(\mathbf{0}; \mathbf{R})$. In the special case when $\rho_{ij} = \rho$ for all i, j, $\Pr[(X_j \le 0) \cap (X_{j'} \le 0)]$ has the same value, $L_2(\rho)$ for all j, j' and so on, and (45.84) becomes

$$L_k(\rho) = \frac{1}{2} \left\{ -\frac{k}{2} + 1 + \binom{k}{2} L_2(\rho) - \binom{k}{3} L_3(\rho) + \cdots \right.$$
$$\left. + \binom{k}{k-1} L_{k-1}(\rho) \right\}. \qquad (45.85)$$

Unfortunately, there is no such simple formula when k is even.

When the common value, ρ, of the ρ_{ij}'s is equal to $\frac{1}{2}$, we have the simple result [a special case of (45.75)]

$$L_k \left(\frac{1}{2} \right) = (k+1)^{-1}; \qquad (45.86)$$

see Moran (1948). The orthant probability has the same value when

$$\mathbf{R}^{-1} = \begin{pmatrix} 1 & \frac{1}{2} & 0 & \cdot & \cdot & \cdots & 0 & 0 \\ \frac{1}{2} & 1 & \frac{1}{2} & \cdot & \cdot & \cdots & 0 & 0 \\ 0 & \frac{1}{2} & 1 & \cdot & \cdot & \cdots & 0 & 0 \\ \cdot & & & \cdot & \cdot & \cdots & \cdot & \cdot \\ 0 & 0 & 0 & \cdot & \cdot & \cdots & 1 & \frac{1}{2} \\ 0 & 0 & 0 & \cdot & \cdot & \cdots & \frac{1}{2} & 1 \end{pmatrix};$$

see Anis and Lloyd (1953).

Moran (1983) provided an expansion for the multivariate normal integral

$$\bar{\Phi}_k(\mathbf{x}_i, \ldots, \mathbf{x}_k; \mathbf{V}) = \frac{1}{(2\pi)^{k/2} |\mathbf{V}|^{1/2}} \int_{x_1}^{\infty} \cdots \int_{x_k}^{\infty} e^{-\mathbf{x}^T \mathbf{V}^{-1} \mathbf{x}} dx_k \cdots dx_1,$$
$$(45.87)$$

which has some similarities to the tetrachoric series but is based on a different idea and has a different region of convergence. Without loss of any generality, by means of transformation from $\mathbf{V}^{-1} = (v^{ij})$ to $\left(w_{ij} = \frac{v^{ij}}{\sqrt{v^{ii} v^{jj}}} \right)$, it is sufficient to deal with the integral

$$J = \frac{1}{(2\pi)^{k/2}} \int_{u_1}^{\infty} \cdots \int_{u_k}^{\infty} \exp \left(-\frac{1}{2} \sum w_{ij} x_i x_j \right) dx_k \cdots dx_1. \qquad (45.88)$$

Define

$$B_n(u) = \frac{1}{\sqrt{2\pi}} \int_u^\infty x^n e^{-x^2/2} \, dx$$

and

$$B_n(0) = \frac{(2m)!}{2^{m+1}m!} \quad \text{if } n = 2m,$$

$$= \frac{2^m m!}{\sqrt{2\pi}} \quad \text{if } n = 2m+1.$$

The integral in (45.88) can then be written as

$$J = \sum_{n_{12}=0}^{\infty} \cdots \sum_{n_{k-1,k}=0}^{\infty} \frac{(-1)^n w_{12}^{n_{12}} w_{13}^{n_{13}} \cdots w_{k-1,k}^{n_{k-1,k}}}{n_{12}! n_{13}! \cdots n_{k-1,k}!} B_{n_1}(u_1) \cdots B_{n_k}(u_k),$$

$$(45.89)$$

where $n = \sum n_{ij}$, $n_i = \sum_{j(<i)} n_{ij} + \sum_{j(>i)} n_{ij}$ for $i = 1, \ldots, k$, and $w_{ij}^{n_{ij}}$ is taken as 1 if $w_{ij} = n_{ij} = 0$. This series has been shown by Moran (1983) to be convergent if $\sum_{j=1}^{k} |w_{ij}| < 2$ for all $i = 1, \ldots, k$. In the equicorrelated case (i.e. $\rho_{ij} = \rho$ for $i \neq j$), the series in (45.89) is convergent for $0 \leq \rho < 1$ and $-\frac{1}{2k-3} < \rho < 0$, but is divergent for $-\frac{1}{k-1} < \rho < -\frac{1}{2k-3}$.

Seneta (1987) presented an approximate expression for

$$\bar{\Phi}_k(a, \ldots, a; \rho) = \Pr[X_1 \geq a, \ldots, X_k \geq a] \qquad (45.90)$$

when $\boldsymbol{X} = (X_1, \ldots, X_k)^T$ has mean $\boldsymbol{0}$, variances 1, and positive equal correlation ρ. For the case $a = 0$ and $k = 2, 3$, we have (see Chapter 46)

$$\Pr[X_1 \geq 0, \ X_2 \geq 0] = \frac{1}{4} + \frac{1}{2\pi} \sin^{-1} \rho,$$

$$\Pr[X_1 \geq 0, \ X_2 \geq 0, \ X_3 \geq 0] = \frac{1}{8} + \frac{3}{4\pi} \sin^{-1} \rho. \qquad (45.91)$$

Denote $C_i = \{X_i \in A\}$ for $i = 1, 2, \ldots, k$, where A is a Borel set in $\mathbb{R}$. Let $\Pr[X_1 \in A] = \Pr[C_1] = \gamma_1$ and $\gamma_i = \Pr[C_i \mid C_{i-1}, \ldots, C_1]$ for $i = 2, 3, \ldots, k$. Then, the following recurrence relation holds:

$$\gamma_{i+1} = \alpha_i(1 - \gamma_i) + \gamma_i, \quad i = 1, \ldots, k-1, \qquad (45.92)$$

where

$$\alpha_1 = \text{corr}\left(I(C_2), I(C_1)\right) = \frac{\Pr(C_1 C_2) - \gamma_1^2}{\gamma_1(1 - \gamma_1)}$$

and

$$\alpha_i = \text{corr}\left(I(C_{i+1}), I(C_i) \mid I(C_{i-1}) = 1, \ldots, I(C_1) = 1\right).$$

Seneta (1987) has suggested approximating α_i by

$$\alpha_i = \rho/\{1 + (i-1)\rho\}, \qquad i \geq 1.$$

Numerical calculations show that for fixed k and $a \geq 0$, the approximation becomes worse as ρ increases from 0 to 0.8 and then improves, which is in agreement with (45.91).

For $\boldsymbol{X} \stackrel{d}{=} N_k(\boldsymbol{0}, \boldsymbol{R})$, Joe (1995) observed that (for $k \geq 3$)

$$\Pr\left[\bigcap_{j=1}^{k}(a_j < X_j \leq b_j)\right]$$

$$= \Pr\left[\bigcap_{\ell=1}^{2}(a_{j_\ell} < X_{j_\ell} \leq b_{j_\ell})\right]$$

$$\times \prod_{\ell=3}^{k} \Pr\left[a_{j_\ell} < X_{j_\ell} \leq b_{j_\ell} \,\Big|\, \bigcap_{m=1}^{\ell-1}(a_{j_m} < X_{j_m} \leq b_{j_m})\right], \quad (45.93)$$

where $(j_1, \ldots, j_k)$ is a permutation of $(1, 2, \ldots, k)$ with $j_1 < j_2$. There are $k!/2$ permutations that could be considered. We define $I_i = I(a_i < X_i \leq b_i)$, $i = 1, \ldots, k$, where $I(A)$ is the indicator of the event A. Clearly, $E[I_i] = \Phi(b_i) - \Phi(a_i)$, where $\Phi(\cdot)$ denotes the cumulative distribution function of an univariate standard normal variable. Now, using the well-known formula (in an obvious notation)

$$E[Y_2 | \boldsymbol{Y}_1 = \boldsymbol{y}_1] = \mu_2 + \boldsymbol{\Sigma}_{21}\boldsymbol{\Sigma}_{11}^{-1}(\boldsymbol{y}_1 - \boldsymbol{\mu}_1),$$

we have

$$\Pr[a_\ell < X_\ell \leq b_\ell | a_1 < X_1 \leq b_1, \ldots, a_{\ell-1} < X_{\ell-1} \leq b_{\ell-1}]$$

$$= E[I_\ell | I_1 = 1, \ldots, I_{\ell-1} = 1] \qquad (45.94)$$

$$= E[I_\ell] + \boldsymbol{\Sigma}_{21}\boldsymbol{\Sigma}_{11}^{-1}[1 - E[I_1], \ldots, 1 - E[I_{\ell-1}]]^T, \qquad (45.95)$$

where $\boldsymbol{\Sigma}_{21}$ is a row vector consisting of $\mathrm{cov}(I_\ell, I_i) = E[I_\ell I_i] - E[I_\ell]E[I_i]$ for $i = 1, \ldots, \ell-1$, and $\boldsymbol{\Sigma}_{11}$ is a $(\ell-1) \times (\ell-1)$ matrix with its (i, j)th element as $\mathrm{cov}(I_i, I_j) = E[I_i I_j] - E[I_i]E[I_j]$ for $1 \leq i, j \leq \ell - 1$. Joe (1995) has suggested averaging all the $k!/2$ values of (45.94) in order to arrive at an overall value for (45.93). Solow (1990) has used this approach described above (with mixed results) without the averaging.

The second-order improvement is obtained by using

$$E[I_\ell | I_1 = 1, \ldots, I_{\ell-1} = 1]$$

$$= E[I_\ell | I_i \text{ for } i = 1, \ldots, \ell-1; I_{ij} = 1 \text{ for } 1 \leq i < j \leq \ell - 1]$$

$$= E[I_\ell] + \boldsymbol{\Sigma}_{21}^*(\boldsymbol{\Sigma}_{11}^*)^{-1}[1 - E[I_1], \ldots, 1 - E[I_{\ell-1}],$$

$$1 - E[I_{12}], \ldots, 1 - E[I_{\ell-2, \ell-1}]]^T, \qquad (45.96)$$

where

$$\boldsymbol{\Sigma}_{21}^* = (\boldsymbol{\Sigma}_{21}, \boldsymbol{A}),$$

$$\boldsymbol{\Sigma}_{11}^* = \begin{pmatrix} \boldsymbol{\Sigma}_{11} & \boldsymbol{B} \\ \boldsymbol{B}^T & \boldsymbol{C} \end{pmatrix},$$

$\boldsymbol{A}$ is a row vector consisting of $\mathrm{cov}(I_\ell, I_{ij}) = E[I_\ell I_{ij}] - E[I_\ell]E[I_{ij}]$ for $1 \le i < j \le \ell - 1$, $\boldsymbol{B}$ is a $(\ell - 1) \times \binom{\ell-1}{2}$ dimensional matrix with entries $\mathrm{cov}(I_m, I_{ij}) = E[I_m I_{ij}] - E[I_m]E[I_{ij}]$ for $1 \le m \le \ell - 1$ and $1 \le i < j \le \ell - 1$, and $\boldsymbol{C}$ is a $\binom{\ell-1}{2} \times \binom{\ell-1}{2}$-dimensional matrix with entries $\mathrm{cov}(I_{ij}, I_{i'j'}) = E[I_{ij}I_{i'j'}] - E[I_{ij}]E[I_{i'j'}]$ for $1 \le i < j \le \ell - 1$ and $1 \le i' < j' \le \ell - 1$. In this case, the decomposition used is

$$\Pr\left[\prod_{j=1}^k (a_j < X_j \le b_j) \right]$$

$$= \Pr\left[\bigcap_{\ell=1}^n (a_{j_\ell} < X_{j_\ell} \le b_{j_\ell}) \right]$$

$$\times \prod_{\ell=n+1}^k \Pr\left[a_{j_\ell} < X_{j_\ell} \le b_{j_\ell} \middle| \bigcap_{m=1}^{\ell-1} (a_{j_m} < X_{j_m} \le b_{j_m}) \right], \quad (45.97)$$

where $(j_1, \ldots, j_k)$ is a permutation of $(1, \ldots, k)$ with $j_1 < j_2 < \cdots < j_n$. For each of the $k^{(n)} = k!/n!$ permutations, each conditional probability can be determined by (45.96) and then the average of these $k!/n!$ values can be used to determine (45.97). This way, all three- and four-dimensional marginal probabilities may be used when $k > 4$, and all three-dimensional marginal probabilities may be used when $k = 4$. Joe (1995) has also suggested to use a random set of permutations (instead of using all permutations) when k is greater than 7 or 8. He has recommended that the second-order approximation should be used for dimensions of $k = 12$ or more; in this case, it provides close to four-decimal accuracy and is also faster than a numerical quadrature and the Monte Carlo simulation procedure. The first-order approximation is suitable if correlations are not large, but has the advantage in its speed even for k more than 20. However, both approximations become inaccurate as the correlations increase.

Genz (1992) and Hickernell and Hong (1997) have presented methods of computing multivariate normal probabilities; the latter authors suggested a rank-1 lattice quadrature rule. These authors have shown that this method is particularly useful when either high accuracy is required or the dimension k is large. Genz (1993) has compared the performance of several methods for computing multivariate normal probabilities.

Vijverberg (1997) has discussed the simulation of multivariate normal probabilities of high-order dimensions by developing a family of simulators which are derived from a Cholesky decomposition of the covariance matrix and combined with a suitable choice of an importance sampling distribution.

6   QUADRIVARIATE NORMAL ORTHANT PROBABILITIES

While it is possible to compute integrals of bivariate and trivariate normal density functions with some facility, using auxiliary tables to be described in Chapter 46, this is not the case for quadrivariate normal densities. If such integrals can be calculated, calculation of integrals of five-variate normal integrals is straightforward, using (45.84). We present here several methods of calculation and approximation that can be helpful.

Moran (1956) evaluated the first few terms in Kendall's (1941) series in (45.55) for the case $k = 4$, obtaining

$$\Phi(0; \boldsymbol{R}) = \frac{1}{16} + \frac{1}{8\pi} \sum \rho_{ij} + \frac{1}{4\pi^2} \sum \rho_{ij}\rho_{ij'} - \cdots . \tag{45.98}$$

Cheng (1969) has obtained the following formulas for orthant probabilities when the correlation matrix has certain specific forms. If $\rho_{12} = \rho_{34} = \alpha$ and all four other correlations are equal to β, with $-\frac{1}{3} < \alpha < 1$ and $|\beta| \leq 1$, then, denoting the orthant probability by $L(\alpha, \beta)$ we have

$$L(\alpha, \beta) = L(\alpha, 0) + \int_0^\beta \frac{\partial L(\alpha, b)}{\partial b}\, db \tag{45.99}$$

and

$$\frac{\partial L(\alpha, b)}{\partial b} = \alpha \left[\frac{\partial}{\partial \rho_{13}} + \frac{\partial}{\partial \rho_{14}} + \frac{\partial}{\partial \rho_{23}} + \frac{\partial}{\partial \rho_{24}} \right] L(\alpha, b). \tag{45.100}$$

Using Eq. (45.46), we find

$$
\begin{aligned}
L(\alpha, \beta) &= \frac{1}{16} + \frac{1}{4\pi}[\sin^{-1}\alpha + 2\sin(\alpha\beta)] + \frac{1}{4\pi^2}[\sin^{-1}\alpha]^2 \\
&\quad - \frac{1}{\pi^2}\int_0^{\alpha\beta}(1 - t^2)^{-1/2}\sin^{-1}\{g(t)\}\, dt,
\end{aligned}
\tag{45.101}
$$

where $g(t) = t[1 - 2(1 + \alpha)^{-1}][1 - 2t^2(1 + \alpha)^{-1}]^{-1}$. Cheng shows how to evaluate the final term in (45.101) in terms of the *dilogarithmic function*

$Li_2(z)$ defined by

$$Li_2(z) = -\int_0^z v^{-1} \log(1-v) \, dv. \tag{45.102}$$

The z may be real or complex. If $z = re^{i\theta}$ with r, θ real, then the real part of $Li_2(z)$ is

$$-\frac{1}{2}\int_0^z v^{-1} \log(1 - 2v\cos\theta + v^2) \, dv, \tag{45.103}$$

which is denoted by $Li_2(r, \theta)$.

Lewin (1958) gives tables of $Li_2(z)$ to five decimal places for $z = 0.00(0.01)1.00$, and of $Li_2(r, \theta)$ to six decimal places for $r = 0.00(0.01)1.00$; $\theta = 0^0(5^0)180^0$. [Of course, $Li_2(r, 0) = Li_2(r)$, so that the latter table includes the former.] In order to evaluate $Li_2(z)$ for values of z between -1 and 0, the equation

$$Li_2(-z) = \frac{1}{2} Li_2(z^2) - Li_2(z) \qquad (z > 0) \tag{45.104}$$

can be used. Note that for $|z| \leq 1$, we have

$$Li_2(z) = \sum_{j=1}^{\infty} j^{-2} z^j. \tag{45.105}$$

In particular, $Li_2(1) = \frac{1}{6}\pi^2$, $Li_2(\frac{1}{2}) = \frac{1}{12}\pi^2 - \frac{1}{2}(\log 2)^2$.

In terms of this function, we have

$$
\begin{aligned}
L(\alpha, \beta) &= \frac{1}{16} + \frac{1}{4\pi}\left[\sin^{-1}\alpha + 2\sin^{-1}(\alpha\beta)\right] \\
&\quad + \frac{1}{4\pi^2}\left[\{\sin^{-1}\alpha\}^2 - 2\{\sin^{-1}(\alpha\beta)\}^2\right] \\
&\quad + \frac{1}{\pi^2}\left[2Li_2(f, \cos^{-1}(\alpha\beta)) - Li_2(f^2, \cos^{-1}\alpha) + \frac{1}{2}Li_2(-f^2)\right],
\end{aligned}
\tag{45.106}
$$

where

$$f = (2\alpha\beta)^{-1}\left[1 + \alpha - \sqrt{(1+\alpha)^2 - 4\alpha^2\beta^2}\right]$$

with $f = 0$ when $\alpha = 0$ and $\beta = 0$.

For the equally correlated case, $\beta = 1$ and $\alpha = \rho$; hence,

$$
\begin{aligned}
L(\rho, 1) &= \frac{1}{16} + \frac{3}{4\pi}\sin^{-1}\rho - \frac{1}{4\pi^2}(\sin^{-1}\rho)^2 \\
&\quad + \frac{1}{\pi^2}\left[2Li_2(f, \cos^{-1}\rho) - Li_2(f^2, \cos^{-1}\rho) + \frac{1}{2}Li_2(-f^2)\right],
\end{aligned}
\tag{45.107}
$$

where

$$f = (2\rho)^{-1}\left[1 + \rho - \sqrt{(1-\rho)(1+3\rho)}\right] \qquad \text{for } -\frac{1}{3} \le \rho \le 1.$$

The corresponding value for the five-variate orthant probability, with all correlations equal to ρ, is easily obtained from the recurrence relation (45.85). It is

$$\frac{1}{32} + \frac{5}{8\pi}(\sin^{-1}\rho)\left\{1 - \frac{1}{\pi}\sin^{-1}\rho\right\}$$
$$+ \frac{5}{2\pi^2}\left\{2Li_2(f, \cos^{-1}\rho) - Li_2(f^2, \cos^{-1}\rho) + \frac{1}{2}Li_2(-f^2)\right\},$$
$$\tag{45.108}$$

with f as in (45.107).

For the case $\rho_{12} = \rho_{34} = \alpha$, $\rho_{13} = \rho_{24} = \beta$, $\rho_{14} = \rho_{23} = \alpha\beta$ (with $|\alpha| \le 1$, $|\beta| \le 1$), the orthant probability is

$$\frac{1}{16} + \frac{1}{4\pi}\left\{\sin^{-1}\alpha + \sin^{-1}\beta + \sin^{-1}(\alpha\beta)\right\}$$
$$+ \frac{1}{4\pi^2}\left\{(\sin^{-1}\alpha)^2 + (\sin^{-1}\beta)^2 - \{\sin^{-1}(\alpha\beta)\}^2\right\};$$
$$\tag{45.109}$$

see Cheng (1969). Note that this expression does not include dilogarithmic functions but only easily computable inverse sine functions.

For the case $\rho_{12} = \alpha$, $\rho_{13} = \rho_{24} = \beta$, $\rho_{14} = \rho_{23} = \alpha\beta$, $\rho_{34} = \alpha\beta^2$ (with $|\alpha| \le 1$, $|\beta| \le 1$), the orthant probability has the more complicated expression

$$\frac{1}{16} + \frac{1}{4\pi}\left\{\frac{1}{2}\sin^{-1}\alpha + \sin^{-1}\beta + \sin^{-1}(\alpha\beta) + \frac{1}{2}\sin^{-1}(\alpha\beta^2)\right\}$$
$$+ \frac{1}{4\pi^2}\left\{(\sin^{-1}\beta)^2 - \frac{1}{2}(\sin^{-1}(\alpha\beta^2))^2 + 2Li_2(f, \cos^{-1}(\alpha\beta^2))\right.$$
$$\left. - Li_2(f^2, \cos^{-1}(2\beta^2 - 1)) + \frac{1}{2}Li_2(-f^2)\right\} \tag{45.110}$$

with $f = \alpha^{-1}\left(1 - \sqrt{1 - \alpha^2}\right)$ ($= 0$ if $\alpha = 0$).

For the case $\rho_{12} = \rho_{13} = \rho_{24} = \rho_{34} = \rho$, $\rho_{14} = \rho_{23} = 0$ (with $|\rho| < \frac{1}{2}$), the orthant probability is

$$\frac{1}{16} + \frac{\sin^{-1}\rho}{2\pi}\left\{1 - \frac{\sin^{-1}\rho}{\pi}\right\}$$
$$+ \frac{1}{\pi^2}\left\{2Li_2(f, \cos^{-1}\rho) + \frac{1}{2}Li_2(-f^2) - \frac{1}{4}Li_2(-f^4)\right\}$$
$$\tag{45.111}$$

with $f = (2\rho)^{-1}[1 - \sqrt{1 - 4\rho^2}]$.

David and Mallows (1961) give series expansions for certain of these expressions. Abrahamson (1964) has shown that the orthant probability for the general quadrivariate normal probability can be expressed as a linear function of six orthant probabilities of the kind obtained with $\rho_{13} = \rho_{14} = \rho_{24} = 0$ (i.e., only ρ_{12}, ρ_{23}, and ρ_{34} nonzero). See Drezner's (1990) approximation discussed below.

Unfortunately, closed-form expressions for such orthant probabilities are not available, though Cheng has shown that for the special case when $\rho_{12} = \rho_{34}$ the orthant probability is

$$
\begin{aligned}
&\frac{1}{16} + \frac{1}{4\pi}\left\{ \sin^{-1}\rho_{12} + \frac{1}{2}\sin^{-1}\rho_{23} \right\} \\
&+ \frac{1}{4\pi^2}\left\{ (\sin^{-1}\rho_{13})^2 - \frac{1}{2}(\sin^{-1}\rho_{23})^2 \right\} \\
&+ \frac{1}{4\pi^2}\left\{ 2Li_2(f, \cos^{-1}\rho_{23}) - Li_2(f^2, \cos^{-1}(1 - 2\rho_{12}^2)) + \frac{1}{2}Li_2(-f^2) \right\}
\end{aligned}
$$

$$(45.112)$$

with $f = \beta^{-1}[1 - \sqrt{1 - \beta^2}]$ $(= 0$ if $\beta = 0)$, where $\beta = \rho_{23}/(1 - \rho_{12}^2)$.

Approximate expressions derived by McFadden (1956, 1960) and Sondhi (1961) give five-decimal accuracy for the orthant probability when all six simple correlations are equal to ρ. For $0 \leq \rho \leq \frac{1}{2}$, the formula to use is

$$\frac{1}{16} + \frac{1}{4\pi}\phi + \frac{1}{4\pi^2}\frac{\phi^2(3 + 5\phi)}{(1 + \phi)(1 + 2\phi)} \qquad (45.113)$$

where $\phi = \sin^{-1}\rho$; see McFadden (1956). For $\frac{1}{2} < \rho < 1$, the formula is

$$\frac{1}{2} - \frac{3\phi'}{2\pi^2}\left(\frac{1}{2}\pi + \sin^{-1}\frac{1}{3}\right) + \frac{3\phi'^3}{\pi^2\sqrt{8}}\left(\frac{1}{36} + \phi'^2\frac{(ac - b^2)\phi'^2 - ab}{c\phi'^2 - b}\right),$$

$$(45.114)$$

where $\phi' = \cos^{-1}\rho$,

$$
\begin{aligned}
a &= (1/5!)\left(\tfrac{23}{48}\right) = 0.00399306, \\
b &= (1/7!)\left(\tfrac{3727}{1152}\right) = 0.00064191, \\
c &= (1/9!)\left(\tfrac{3320309}{82944}\right) = 0.00011031,
\end{aligned}
$$

so that (45.114) can be calculated as

$$\frac{1}{2} - 0.275\phi' + 0.003\phi'^3 + 0.000028\phi'^5(\phi'^2 - 90.20)(\phi'^2 - 5.82)^{-1};$$

see Sondhi (1961).

The range of values $-\frac{1}{3} \le \rho < 0$ can be covered by using the relation

$$L_4\left(-\frac{\rho}{1+2\rho}\right) + L_4(\rho) = \frac{1}{8} + \frac{3}{4\pi}\left\{\sin^{-1}\rho - \sin^{-1}\left(\frac{\rho}{1+2\rho}\right)\right\}$$

$$+ \frac{3}{2\pi^2}\sin^{-1}\rho \,\sin^{-1}\left(\frac{\rho}{1+2\rho}\right).$$

These expressions are considerably better approximations than the general formulas of Section 5.4 (though the latter can be employed for $k > 4$). They can be used in conjunction with (45.85) to give approximations for $L_5(\rho)$.

Poznyakov (1971) obtained the exact formula

$$L_4(\rho) = \frac{1}{16} + \frac{3}{4\pi}\sin^{-1}\rho + \frac{3}{2\pi^2}\int_0^\rho (1-u^2)^{-1/2}\sin^{-1}\left(\frac{u}{1+2u}\right)du.$$

David and Mallows (1961) have given formulas for quadrivariate orthant probabilities in a number of special cases, each involving, at most, univariate integrals needing evaluation by quadrature. We give a few examples in Table 45.3.

Here,

$$I_j = \int_0^{\sin^{-1}\rho} \sin^{-1}(g_j(t))\,dt$$

with

$$g_1(t) = \frac{\sin 2t}{\sqrt{1+2\cos 2t}}, \qquad g_2(t) = \frac{\sin 2t}{2\sqrt{\cos 2t}},$$

$$g_3(t) = \frac{3\sin t - \sin 3t}{4\cos 2t}, \qquad g_4(t) = \frac{\sin t}{\cos 2t},$$

and

$$g_5(t) = \frac{3+2\cos 2t}{1+2\cos 2t}.$$

Drezner (1990) provided approximations to

$$L_k(\boldsymbol{h};\boldsymbol{R}) = \frac{1}{(2\pi)^{k/2}|\boldsymbol{R}|^{1/2}}\int_{h_1}^\infty \cdots \int_{h_k}^\infty \exp\left\{-\frac{1}{2}\,\boldsymbol{x}^T\boldsymbol{R}^{-1}\boldsymbol{x}\right\}d\boldsymbol{x}$$

for the quadrivariate case ($k = 4$) with

$$\boldsymbol{R} = \begin{pmatrix} 1 & \rho_{12} & \rho_{13} & \rho_{14} \\ \rho_{12} & 1 & \rho_{23} & \rho_{24} \\ \rho_{13} & \rho_{23} & 1 & \rho_{34} \\ \rho_{14} & \rho_{24} & \rho_{34} & 1 \end{pmatrix}.$$

TABLE 45.3
Some Orthant Probabilities

ρ_{12}	ρ_{13}	ρ_{14}	ρ_{23}	ρ_{24}	ρ_{34}	$\Pr\left[\bigcap_{j=1}^{4}(X_j \leq 0)\right]$
ρ	0	0	0	0	ρ	$\left(\frac{1}{4} + \frac{1}{2\pi}\sin^{-1}\rho\right)^2$
ρ	0	0	0	0	$\frac{1}{2}\rho$	$\left(\frac{1}{4} + \frac{1}{2\pi}\sin^{-1}\frac{1}{2}\rho\right)\left(\frac{1}{4} + \frac{1}{2\pi}\sin^{-1}\rho\right)$
ρ	$\frac{1}{2}$	0	0	0	ρ	$\frac{1}{12} + \frac{1}{4\pi}\sin^{-1}\rho + \frac{1}{2\pi^2}I_1$
ρ	0	0	ρ	0	ρ	$\frac{1}{16} + \frac{3}{8\pi}\sin^{-1}\rho + \frac{1}{4\pi^2}(2I_2 + I_3)$
ρ	0	ρ	ρ	0	ρ	$\frac{1}{16} + \frac{1}{2\pi}\sin^{-1}\rho + \frac{1}{\pi^2}I_4$
ρ	$\frac{1}{2}$	0	0	$\frac{1}{2}$	ρ	$\frac{1}{9} + \frac{1}{4\pi}\sin^{-1}\rho + \frac{1}{2\pi^2}I_5$
ρ	$-\frac{1}{2}$	$-\rho$	0	0	ρ	$\frac{1}{24} + \frac{1}{8\pi}\sin^{-1}\rho + \frac{1}{4\pi^2}I_1$
ρ	$-\frac{1}{2}$	0	$-\rho$	$\frac{1}{2}$	ρ	$\frac{1}{18} + \frac{1}{8\pi}\sin^{-1}\rho + \frac{1}{4\pi^2}I_5$
ρ	0	ρ	$-\rho$	$-\frac{1}{2}$	ρ	$\frac{1}{24} + \frac{1}{4\pi}\sin^{-1}\rho$
ρ	$\frac{1}{2}$	$\frac{1}{2}\rho$	$\frac{1}{2}\rho$	$\frac{1}{2}$	ρ	$\frac{1}{9} + \frac{1}{4\pi}\left(\sin^{-1}\rho + \sin^{-1}\frac{1}{2}\rho\right)$ $+ \frac{1}{4\pi^2}\left\{(\sin^{-1}\rho)^2 - \left(\sin^{-1}\frac{1}{2}\rho\right)^2\right\}$
$\frac{1}{2}\rho$	$-\frac{1}{2}$	$\frac{1}{2}\rho$	$-\rho$	$-\frac{1}{2}$	$\frac{1}{2}\rho$	$\frac{1}{36} + \frac{1}{8\pi}\left(3\sin^{-1}\frac{1}{2}\rho - \sin^{-1}\rho\right)$ $+ \frac{1}{8\pi^2}\left\{\left(\sin^{-1}\frac{1}{2}\rho\right)^2 - (\sin^{-1}\rho)^2\right\}$
ρ	$-\frac{1}{2}$	$\frac{1}{2}\rho$	$-\frac{1}{2}\rho$	$\frac{1}{2}$	$\frac{1}{2}\rho$	$\frac{1}{18} + \frac{1}{8\pi}\left(\sin^{-1}\rho + \sin^{-1}\frac{1}{2}\rho\right)$ $+ \frac{1}{8\pi^2}\left\{(\sin^{-1}\rho)^2 - \left(\sin^{-1}\frac{1}{2}\rho\right)^2\right\}$
ρ	$-\frac{1}{2}$	$-\frac{1}{2}\rho$	$-\frac{1}{2}\rho$	$-\frac{1}{2}$	ρ	$\frac{1}{36} + \frac{1}{4\pi}\left(\sin^{-1}\rho - \sin^{-1}\frac{1}{2}\rho\right)$ $+ \frac{1}{4\pi^2}\left\{(\sin^{-1}\rho)^2 - \left(\sin^{-1}\frac{1}{2}\rho\right)^2\right\}$

These approximations, in obvious notations, are

$$
\begin{aligned}
L_4(\boldsymbol{h}; \boldsymbol{R}) \;\doteq\;& L_2(h_1, h_2; \rho_{12}) L_2(h_3, h_4; \rho_{34}) + L_2(h_1, h_3; \rho_{13}) L_2(h_2, h_4; \rho_{24}) \\
& + L_2(h_1, h_4; \rho_{14}) L_2(h_2, h_3; \rho_{23}) \\
& - 2 L_1(h_1) L_1(h_2) L_1(h_3) L_1(h_4)
\end{aligned}
\tag{45.115}
$$

and

$$
\begin{aligned}
L_4(\boldsymbol{h}; \boldsymbol{R}) \;\doteq\;& \sum_{1 \le i_1 < i_2 \le 4} L_2(h_{i_1}, h_{i_2}; \rho_{i_1 i_2}) L_1(h_{i_3}) L_1(h_{i_4}) \\
& - 5\, L_1(h_1) L_1(h_2) L_1(h_3) L_1(h_4).
\end{aligned}
\tag{45.116}
$$

The accuracy of (45.115) is slightly better than that of (45.116), but the difference is insignificant; see Table 45.4, taken from Drezner (1990).

TABLE 45.4
Quality of Drezner's Quadrivariate Approximations

ρ_{max}	Appr. in (45.115)		Appr. in (45.116)	
	Ave. error	Max. error	Ave. error	Max. error
0.1	0.00003	0.00031	0.00003	0.00030
0.2	0.00012	0.00101	0.00013	0.00133
0.3	0.00021	0.00339	0.00022	0.00257
0.4	0.00050	0.00453	0.00053	0.00579
0.5	0.00075	0.00806	0.00078	0.00792

Source: Drezner (1990), with permission.

7   CHARACTERIZATIONS

The literature on characterizations of multivariate normal distribution is rather extensive. We provide here a brief survey in a rough chronological order.

Fréchet (1951) showed that if $X_1, X_2, \ldots, X_k$ are random variables, and the distribution of $\sum_{j=1}^{k} a_j X_j$ is normal for *any* set of real numbers $a_1, a_2, \ldots, a_k$ (not all zero), then the joint distribution of $X_1, X_2, \ldots, X_k$ must be multivariate normal. This property has been used by Rao (1965) and other authors as a *definition* of multivariate normal distributions.

Basu (1956) showed that if $\boldsymbol{X}_1^T, \ldots, \boldsymbol{X}_n^T$ are independent $1 \times k$ vectors and there are two sets of n constants $(a_1, \ldots, a_n)$, $(b_1, \ldots, b_n)$ such that the vectors $\sum a_j \boldsymbol{X}_j$ and $\sum b_j \boldsymbol{X}_j$ are mutually independent, then the distribution of all $\boldsymbol{X}_i$'s for which $a_i b_i \neq 0$ must be multivariate normal.

This is a generalization of the univariate Darmois–Skitovitch theorem [see Chapter 13 of Johnson, Kotz, and Balakrishnan (1994)]. Ghurye and Olkin (1962) have demonstrated another generalization of this theorem. They show that if there exist two sets of nonsingular $k \times k$ matrices $(\boldsymbol{A}_1, \boldsymbol{A}_2, \ldots, \boldsymbol{A}_n)$, $(\boldsymbol{B}_1, \boldsymbol{B}_2, \ldots, \boldsymbol{B}_n)$ such that

$$\sum_{j=1}^{n} \boldsymbol{A}_j \boldsymbol{X}_j \quad \text{and} \quad \sum_{j=1}^{n} \boldsymbol{B}_j \boldsymbol{X}_j$$

are mutually independent, then each $\boldsymbol{X}_j$ has a multivariate normal distribution. Additional generalizations of Darmois–Skitovitch theorem are mentioned below.

Lukacs (1956) and Laha (1955) have shown that if $X_1, X_2, \ldots, X_k$ have a finite variance–covariance matrix $\boldsymbol{V}$ and $\boldsymbol{X}_j^T \equiv (X_{1j}, \ldots, X_{kj})$ $(j = 1, \ldots, n)$ are independent vectors each having the same distribution as $X_1, X_2, \ldots, X_k)$, then if there is a fixed $k \times n$ matrix $\boldsymbol{A}$ such that

$$E[(X_1, X_2, \ldots, X_k)^T \boldsymbol{A} (X_1, X_2, \ldots, X_k) \mid \boldsymbol{T}] = \boldsymbol{V},$$

where

$$\boldsymbol{T} = \left(\sum_{j=1}^{n} X_{1j}, \sum_{j=1}^{n} X_{2j}, \ldots, \sum_{j=1}^{n} X_{kj} \right),$$

the common distribution must be multivariate normal. This is a generalization of a univariate result due to Geary (1933) and Laha (1953). A simplified proof has been given by Basu (1956).

Invariance under linear combination can be used in characterizing multivariate normal distributions. As before, let $\boldsymbol{X}_1^T, \ldots, \boldsymbol{X}_n^T$ be n independent $1 \times k$ random vectors, with a common distribution. Furthermore, suppose that $\boldsymbol{B}_1, \boldsymbol{B}_2, \ldots, \boldsymbol{B}_n$ are symmetric nonsingular $k \times k$ matrices and $\boldsymbol{b}$ a vector such that $\left(\sum_{j=1}^{n} \boldsymbol{X}_j^T \boldsymbol{B}_j + \boldsymbol{b} \right)$ has the same distribution as each of the $\boldsymbol{X}$'s. Then

(i) if $\sum_{j=1}^{n} \boldsymbol{B}_j^2 - \boldsymbol{I}$ is positive definite, each $\boldsymbol{X}_j$ is equal to a constant vector with probability 1;

(ii) if $\sum_{j=1}^{n} \boldsymbol{B}_j^2 - \boldsymbol{I}$ is positive semidefinite and $\left| \sum_{j=1}^{n} \boldsymbol{B}_j^2 - \boldsymbol{I} \right| = 0$, each $\boldsymbol{X}_j$ has a multivariate normal distribution, with variance-covariance matrix $\boldsymbol{V}$ satisfying the equation

$$\boldsymbol{V} = \sum_{j=1}^{n} \boldsymbol{B}_j \boldsymbol{V} \boldsymbol{B}_j;$$

see Eaton (1966). Shimizu (1962) established a similar result, assuming $\boldsymbol{X}_j^T$ to have finite first and second moments. In the univariate case, if the conditional distribution of W, given $(W + Z)$, is normal, then both W and Z are normally distributed [e.g., Patil and Seshadri (1964)]. A multivariate extension, due to Seshadri (1966), is natural, but requires more complicated conditions. Suppose that W and Z are independent $k \times 1$ random vectors, each with a continuous density that is not zero when $W = 0$ (or $Z = 0$). The C and V are nonsingular $k \times k$ matrices, V is symmetrical and positive definite, and $V^{-1}C$ is symmetrical, satisfying either (i) $V^{-1}(I - C)$ is positive definite or (ii) the eigenvalues of C lie in the open interval 0 to 1. Then, if the conditional distribution of W given $W + Z = K$ is multivariate normal with expected value vector CK^T and variance–covariance matrix V, both W and Z have multivariate normal distributions.

Fisk (1970) has shown that multivariate normal distributions can be characterized by linear regression and by homogeneity of conditional *distribution* (not just homoscedasticity) subject only to the requirement of finiteness of each absolute first moment. More precisely, if X_1, X_2 are nondegenerate random vectors with all absolute first moments finite, and the conditional distribution of X_j, given X_ℓ $(j, \ell = 1, 2; j \neq \ell)$, depends on X_ℓ only through the conditional expected value

$$E[X_j \mid X_\ell] = A_j + B_j X_\ell,$$

where each row and column of B_j contains at least one nonzero element, and $B_1 B_2 \neq I$, $B_2 B_1 \neq I$, then the joint distribution of X_1, X_2 is multivariate normal. Fisk (1970) also gives a generalization of this result to m sets of variables. Kagan, Linnik and Rao (1965) have shown that the condition of

$$E[\bar{X} \mid X_2 - X_1, \ldots, X_k - X_1] = \text{constant}$$

suffices to ensure multivariate normality.

Khatri (1971) has shown that the conditioning set $(X_2 - X_1, \ldots, X_k - X_1)$ may be replaced by two or more linear sets of functions of the X's, subject to certain conditions on the coefficients.

Bildikar and Patil (1968) have obtained the following characterizations. A k-variate exponential-type distribution (see Chapter 44) is multivariate normal if and only if (i) all cumulants of order 3 are zero or (ii) the regression of one variable on the remaining $(k - 1)$ variables is linear, *and* every pair of these $(k - 1)$ variables has a bivariate normal joint distribution.

Anderson (1971) has shown that if (i) the joint density function

$p_{\boldsymbol{X}}(\boldsymbol{x} \mid \boldsymbol{\theta})$, with $\boldsymbol{\theta}^T = (\theta_1, \ldots, \theta_k)$, is such that

$$\int_{-\infty}^{\infty} \frac{\partial p_{\boldsymbol{X}}(\boldsymbol{x} \mid \boldsymbol{\theta})}{\partial \theta_j} \, d\boldsymbol{x} = \frac{\partial}{\partial \theta_j} \int_{-\infty}^{\infty} p_{\boldsymbol{X}}(\boldsymbol{x} \mid \boldsymbol{\theta}) \, d\boldsymbol{x} = 0 \qquad \text{for all } j$$

and (ii) $E[\boldsymbol{X}] = \boldsymbol{M}(\theta)$ with the Jacobian $\partial \boldsymbol{M}/\partial \boldsymbol{\theta}$ nonsingular, then $\boldsymbol{X}$ has a multivariate normal distribution if and only if

$$p_{\boldsymbol{X}}(\boldsymbol{x} \mid \boldsymbol{\theta}) = \exp[\boldsymbol{x}^T \boldsymbol{B} \boldsymbol{M}(\theta) + S(\boldsymbol{x}) + Q(\boldsymbol{\theta})],$$

where $\boldsymbol{B}$ is a $k \times k$ matrix not depending on $\boldsymbol{\theta}$ or $\boldsymbol{x}$.

The relation (45.45) used by Plackett in obtaining a computational formula for the integral of a standardized multivariate normal density function can be extended to general multivariate normal density functions. For such functions, we have

$$\frac{\partial p}{\partial v_{rs}} = \frac{\partial^2 p}{\partial x_r \partial x_s} \qquad (r \neq s), \tag{45.117}$$

$$\frac{\partial p}{\partial v_{rr}} = \frac{1}{2} \frac{\partial^2 p}{\partial x_r^2}, \tag{45.118}$$

where v_{rs}, the covariance between X_r and X_s, is the (r, s)th element of $\boldsymbol{V}$. Patil and Boswell (1970) showed that the relations (45.117) and (45.118) suffice to ensure that $X_1, X_2, \ldots, X_n$ have a joint multivariate normal distribution.

The multivariate normal distribution can also be characterized by the property of "radial symmetry" of the joint distribution of $\boldsymbol{X}_1, \ldots, \boldsymbol{X}_n$, where each $\boldsymbol{X}_j$ (with $m \leq n$ elements) has the same distribution, and the $\boldsymbol{X}_j$'s are mutually independent. If the joint distribution is a function only of elements of the matrix $\boldsymbol{X} \boldsymbol{X}^T$ where $\boldsymbol{X} = (\boldsymbol{X}_1, \boldsymbol{X}_2, \ldots, \boldsymbol{X}_n)$, then the common distribution of the $\boldsymbol{X}_j$'s is multivariate normal. A proof, on the assumption that the distribution has a continuous density function, was originally given by Kendall and Stuart (1963) and reproduced by Stuart and Ord (1994). Proofs requiring only the existence of a density have been given by James (1954) and Thomas (1970).

Zinger and Linnik (1964) have shown that a limited kind of symmetry suffices to characterize multivariate normality. In fact, if $X_1, \ldots, X_k$ have a continuous density function and identical marginal distributions, then equality of density at (a) three points $(x_1, x_2, \ldots, x_k)$, (b) three points obtained from (a) by replacing (x_4, x_5, x_6) with (x_1, x_2, x_3), or (c) three points obtained from (a) by replacing (x_1, x_2, x_3) with (x_4, x_5, x_6) with

$x_1^r + x_2^r + x_3^r = x_4^r + x_5^r + x_6^r$ ($r = 1, 2$) ensures multivariate normality. Clearly, this applies only when $k \geq 6$.

The property of maximum entropy for specified variance–covariance matrix, mentioned at the end of Section 2, also characterizes multivariate normal distributions.

Fairweather (1973) generalized Csörgő and Seshadri's (1971) characterization of the univariate normal distribution [see Chapter 13 of Johnson, Kotz and Balakrishnan (1994)] as follows. Let $\boldsymbol{Z}_i$ ($1 \leq i \leq 2m\ell$) be independent and identically distributed as $N_k(\boldsymbol{\xi}, \boldsymbol{V})$. Let

$$\boldsymbol{X}_i = (\boldsymbol{Z}_{2i-1} - \boldsymbol{Z}_{2i})/\sqrt{2}, \qquad i = 1, 2, \ldots, m\ell,$$

and let the matrix $\boldsymbol{W}_j$ be expressed as

$$\boldsymbol{W}_j = \sum_{i=(j-1)m+1}^{jm} \boldsymbol{X}_i \boldsymbol{X}_i^T, \qquad j = 1, 2, \ldots, \ell.$$

In addition to $\boldsymbol{X}_i$'s and $\boldsymbol{W}_j$'s being independent and identically distributed by construction, we also have $\boldsymbol{X}_i \overset{d}{=} N_k(\boldsymbol{0}, \boldsymbol{V})$ and $\boldsymbol{W}_j \overset{d}{=} Wishart(\boldsymbol{V}, m, k)$. Fairweather (1973) has then proved that $\boldsymbol{X}_i \overset{d}{=} N_k(\boldsymbol{0}, \boldsymbol{V})$ and $\boldsymbol{W}_j \overset{d}{=} Wishart(\boldsymbol{V}, m, k)$ if and only if $\boldsymbol{Z}_i \overset{d}{=} N_k(\boldsymbol{\xi}, \boldsymbol{V})$. This result has been used by Fairweather (1973) for testing multivariate normality of $\boldsymbol{Z}_i$'s.

Let X_i ($i = 1, \ldots, k$) be independent and identically distributed random variables, let $\boldsymbol{Y} = (Y_1, \ldots, Y_m)^T$ and $\boldsymbol{X} = (X_1, \ldots, X_k)^T$ be independently distributed, and let $\boldsymbol{A} = (a_{ij})$ be a $k \times k$ random coefficient matrix with $a_{ij} = a_{ij}(\boldsymbol{Y})$ for $1 \leq i, j \leq k$. Let $\boldsymbol{U} = \boldsymbol{AX}$. Then, Kingman and Graybill (1970) have shown that $\boldsymbol{U} \overset{d}{=} N_k(\boldsymbol{0}, \boldsymbol{I})$ if and only if $\boldsymbol{X} \overset{d}{=} N_k(\boldsymbol{0}, \boldsymbol{I})$, provided that certain conditions defined in terms of a_{ij}'s are satisfied.

Li (1978) relaxed the conditions in Kingman and Graybill's result and also generalized it to the vector case as follows. Let $\boldsymbol{X} = (\boldsymbol{X}_1^T, \ldots, \boldsymbol{X}_n^T)^T$, where $\boldsymbol{X}_i = (X_{1i}, \ldots, X_{k_i i})^T$ ($i = 1, \ldots, n$) are independent with $k = \sum_{i=1}^n k_i$. The elements within each $\boldsymbol{X}_i$ need not be independent and identically distributed, but for each i, X_{ai} ($a = 1, \ldots, k_i$) satisfy

$$E\left[\prod_{a \neq b} X_{ai}^{t_a} X_{bi}^{t_b}\right]$$

$$= E[X_{bi}] E\left[\prod_{a \neq b} X_{ai}^{t_a} X_{bi}^{t_b - 1}\right] \quad \text{if } t_b \text{ is odd}$$

$$= E[X_{bi}^{t_b}]E\left[\prod_{a\neq b} X_{ai}^{t_a}\right] \text{ if } t_b \text{ is even,}$$

where a may take on some or all values of $1, 2, \ldots, k_i$, and $t_a = 0, 1, \ldots$. Let $\mathbf{A} = (a_{ij})$ be an orthogonal matrix with $a_{ij} = a_{ij}(\mathbf{Y})$ for $1 \leq i, j \leq k$, $\mathbf{Y}$ and $\mathbf{X}_i$ ($i = 1, \ldots, n$) be independently distributed, and $\mathbf{U} = \mathbf{AX}$. Then, Li (1978) has shown that $\mathbf{U} \stackrel{d}{=} N_k(\mathbf{0}, \mathbf{I})$ if and only if $\mathbf{X} \stackrel{d}{=} N_k(\mathbf{0}, \mathbf{I})$.

Note that these results can be viewed as generalizations of Skitovič's (1954) theorem [see Section 6 of Chapter 13 of Johnson, Kotz, and Balakrishnan (1994)].

Let $\mathbf{X} = (X_1, \ldots, X_k)^T$ be a random vector whose arbitrarily dependent components have finite second moments. Then, Kagan (1998) has recently proved that all uncorrelated pairs of linear forms $\sum_{i=1}^{k} a_i X_i$ and $\sum_{i=1}^{k} b_i X_i$ are independent iff $\mathbf{X}$ is distributed as multivariate normal. This result should be compared to the classical Darmois–Skitovič theorem in which the components of $\mathbf{X}$ are required to be independent.

Rao (1969a,b) has given an example to show that if $\mathbf{X}$ is a multivariate normal random vector and $\mathbf{X} = \mathbf{AY}$, where $\mathbf{A}$ is a matrix and $\mathbf{Y}$ is a vector of standard normal components, then neither the matrix $\mathbf{A}$ nor the number of components of $\mathbf{Y}$ is unique.

Let $\mathbf{X}_1, \ldots, \mathbf{X}_n$ be independent random variables in finite-dimensional Euclidean spaces $E_1, \ldots, E_n$, respectively, with scalar product $\langle \cdot, \cdot \rangle$ and norm $\| \cdot \|$. Let $R = \{\sum_{i=1}^{n} \|\mathbf{X}_i\|^2\}^{\frac{1}{2}}$ and let $(R^{-1}\mathbf{X}_1, \ldots, R^{-1}\mathbf{X}_n)^T$ be uniformly distributed on a sphere of $\bigoplus_{i=1}^{n} E_i$, the direct orthogonal sum of spaces E_i's. Letac (1981) has shown that $\mathbf{X}_i$'s are normal if $n \geq 3$. This characterization involves the concepts of isotropy and sphericity, which have been discussed in detail by Fang, Kotz, and Ng (1989).

Nguyen and Sampson (1991) have established a characterization of multivariate normal distribution based on distributions of linear combinations of two multivariate random vectors as follows. Let $\mathbf{X}$ and $\mathbf{Y}$ be two independent nondegenerate k-dimensional random vectors with finite covariance matrices, and let $f(\cdot)$ be a nonnegative function with domain an interval of $\mathbb{R}$. If the distribution of $\lambda\mathbf{X} + f(\lambda)\mathbf{Y}$ does not depend on λ for all λ in the domain of $f(\cdot)$, then

(i) $f(\lambda) = \sqrt{a - b\lambda^2}$ for some $a, b > 0$ and

(ii) $\mathbf{X}$ and $\mathbf{Y}$ have multivariate normal distributions with mean vectors $\mathbf{0}$ and covariance matrices $\mathbf{V_X}$ and $\mathbf{V_Y}$, respectively, where $\mathbf{V_X} = b\mathbf{V_Y}$.

Another characterization of multivariate normal distribution (with independent components) due to Nguyen and Sampson (1991) is along the lines of Eaton's (1966) characterization and is as follows. Let $\boldsymbol{X}$ and $\boldsymbol{Y}$ be independent k-dimensional random vectors with finite covariance matrices. If the distributions of $\boldsymbol{AX} + (a\boldsymbol{I} - b\boldsymbol{AA}^T)^{1/2}\boldsymbol{Y}$ do not depend on $\boldsymbol{A}$, for some $a, b > 0$ and for all $k \times k$ matrices $\boldsymbol{A}$ such that $a\boldsymbol{I} - b\boldsymbol{AA}^T$ is non-negative definite, then $\boldsymbol{X} \overset{d}{=} N_k(\boldsymbol{0}, \sigma_{\boldsymbol{X}}^2\boldsymbol{I})$ and $\boldsymbol{Y} \overset{d}{=} N(\boldsymbol{0}, \sigma_{\boldsymbol{Y}}^2\boldsymbol{I})$, where $\sigma_{\boldsymbol{X}}^2 = b\sigma_{\boldsymbol{Y}}^2$.

These characterizations deal with the following type of question. Let $g_\lambda(u, v) = \lambda u + f(\lambda)v$ be viewed as a parametric family of functions from $\mathbb{R}^k \times \mathbb{R}^k$ to $\mathbb{R}^k$. Nguyen and Sampson's (1991) characterization concern the nondependence of the distribution of $g_\lambda(\boldsymbol{X}, \boldsymbol{Y})$ upon λ.

Ahsanullah's (1985) conjecture, as modified by Arnold and Pourahmadi (1988), is as follows [see also Ahsanullah and Sinha (1986), and Hamedani (1984, 1992)]. If $X_1, \ldots, X_k$ are jointly distributed random variables with $(X_1, \ldots, X_{k-1})^T \overset{d}{=} (X_2, \ldots, X_k)^T$ and if

$$X_k|(X_1 = x_1, \ldots, X_{k-1} = x_{k-1}) \overset{d}{=} N\left(\alpha + \sum_{i=1}^{k-1} \beta_i x_i, \sigma^2\right),$$

then $(X_1, \ldots, X_k)^T$ are jointly multivariate normal. This conjecture was proved rigorously by Arnold and Pourahmadi (1988) who also established the following characterization result. Given $\boldsymbol{X} = (X_1, \ldots, X_k)^T$ is a k-dimensional random vector, if the conditions

(a) for $i = 2, 3, \ldots, k$,

$$X_i|(X_1 = x_1, \ldots, X_{i-1} = x_{i-1})$$
$$\overset{d}{=} N\left(\beta_i + \sum_{j=1}^{i-1} \alpha_{j,i-1} x_j, \sigma_i^2\right)$$

for all real numbers $x_1, \ldots, x_{i-1}$, where $\beta_i, \alpha_{j,i-1}$ and σ_i^2 are some real numbers,

and

(b) X_1, X_2 belong to a location-scale family (i.e., $X_2 = aX_1 + b$ for some a and b)

are satisfied, then $\boldsymbol{X}$ has a k-dimensional normal distribution, and conversely.

Castillo and Galambos (1989) provided an example of a two-dimensional random vector $\boldsymbol{X} = (X_1, X_2)^T$ with normal conditional densities $p_{X_1|X_2}(x_1|x_2)$ and $p_{X_2|X_1}(x_2|x_1)$, but $\boldsymbol{X}$ is not bivariate normal. They provided necessary and sufficient conditions for $\boldsymbol{X}$ to have bivariate normal. Bischoff and Fieger (1991) generalized Castillo and Galambos' (1989) result to multivariate random vectors as follows. Let $\boldsymbol{X} = (X_1, \ldots, X_k)^T$ and $\boldsymbol{Y} = (Y_1, \ldots, Y_\ell)^T$. Suppose the conditional densities $p_{\boldsymbol{X}|\boldsymbol{Y}}(\boldsymbol{x}|\boldsymbol{y})$ and $p_{\boldsymbol{Y}|\boldsymbol{X}}(\boldsymbol{y}|\boldsymbol{x})$ are both multivariate normal, and $\boldsymbol{V}(\boldsymbol{x})$ is the covariance matrix of $p_{\boldsymbol{Y}|\boldsymbol{X}}(\boldsymbol{y}|\boldsymbol{x})$. Then, the following three statements are equivalent:

(i) The joint density function $p_{\boldsymbol{X},\boldsymbol{Y}}(\boldsymbol{x}, \boldsymbol{y})$ is multivariate normal.

(ii) $\boldsymbol{V}(\boldsymbol{x})$ is constant on $\mathbb{R}^k$.

(iii) For the minimal eigenvalue $\lambda(\boldsymbol{x})$ of the positive definite matrix $\boldsymbol{V}(\boldsymbol{x})$, we have

$$\alpha^2 \lambda(\alpha \cdot b_j) \to \infty \quad \text{as } \alpha \to \infty \qquad \text{for } j = 1, \ldots, k,$$

where $b_1, \ldots, b_k$ is an arbitrary (but fixed) basis of $\mathbb{R}^k$.

Another generalization of Castillo and Galambos' (1989) result, due to Arnold, Castillo, and Sarabia (1994a), is as follows. Suppose that for each i and for each $\boldsymbol{x}_{(i)} \in \mathbb{R}^{k-1}$, the conditional distribution of X_i given $\boldsymbol{X}_{(i)} = \boldsymbol{x}_{(i)}$, is univariate normal and, in addition, that the regression of each X_i on $\boldsymbol{X}_{(i)}$ is linear (or the conditional variance of X_i given $\boldsymbol{X}_{(i)} = \boldsymbol{x}_{(i)}$ does not depend on $\boldsymbol{x}_{(i)}$), then $\boldsymbol{X}$ has a k-variate normal distribution.

Arnold, Castillo, and Sarabia (1994a,b) also provided conditional characterizations of the multivariate normal distribution which are somewhat similar to (but more general than) the earlier-stated Arnold and Pourahmadi's (1988) characterization. Let $\boldsymbol{X}$ denote a k-dimensional random vector and let $\boldsymbol{X}^{(i,j)}$ denote the vector $\boldsymbol{X}$ with ith and jth components deleted for each i and j. Let a similar definition hold for $\boldsymbol{x}^{(i,j)}$ obtained from $\boldsymbol{x}$, a generic point in $\mathbb{R}^k$. If, for each $\boldsymbol{x}^{(i,j)} \in \mathbb{R}^{k-2}$, the conditional distribution of $(X_i, X_j)^T$, given $\boldsymbol{X}^{(i,j)} = \boldsymbol{x}^{(i,j)}$, is bivariate normal with mean vector $(\xi_i(\boldsymbol{x}^{(i,j)}), \xi_j(\boldsymbol{x}^{(i,j)}))^T$ and variance–covariance matrix

$$\begin{pmatrix} v_{11}(\boldsymbol{x}^{(i,j)}) & v_{12}(\boldsymbol{x}^{(i,j)}) \\ v_{21}(\boldsymbol{x}^{(i,j)}) & v_{22}(\boldsymbol{x}^{(i,j)}) \end{pmatrix},$$

then Arnold, Castillo, and Sarabia (1994a) have proved that $\boldsymbol{X}$ has a k-variate normal distribution. In other words, bivariate conditionals seem to provide a "pleasant surprise."

A modification of this result, due to Arnold, Castillo, and Sarabia (1994a), is as follows. Suppose that for each $i = 1, \ldots, k$ and for each $\boldsymbol{x}^{(i)} \in \mathbb{R}^{k-1}$, the conditional distribution of X_i, given $\boldsymbol{X}^{(i)} = \boldsymbol{x}^{(i)}$, is $N(\xi_i(\boldsymbol{x}^{(i)}), v_i^2(\boldsymbol{x}^{(i)}))$. In addition, suppose that for each $i, j = 1, 2, \ldots, k$ and for each $\boldsymbol{x}^{(i,j)} \in \mathbb{R}^{k-2}$, the conditional distributions of X_i, given $\boldsymbol{X}^{(i,j)} = \boldsymbol{x}^{(i,j)}$, is $N(\xi_{ij}(\boldsymbol{x}^{(i,j)}), v_{ij}^2(\boldsymbol{x}^{(i,j)}))$. Then, $\boldsymbol{X} = (X_1, \ldots, X_k)^T$ (with $k \geq 3$) has a k-variate normal distribution.

Note that the joint density

$$p_{\boldsymbol{X}}(\boldsymbol{x}) = \frac{1}{(2\pi)^{k/2}} \, e^{-\sum_{i=1}^{k} x_i^2/2} \left\{ 1 + \left(\prod_{i=1}^{k} x_i \right) I(|x_i| < 1 \, \forall \, i) \right\}$$

serves as an example of a k-dimensional *nonnormal* distribution whose marginals of all orders less than k are normal and also the conditional distribution of X_i, given $\boldsymbol{X}^{(i)} = \boldsymbol{x}^{(i)}$, is normal; see Arnold, Castillo, and Sarabia (1992).

A variant on this theme is a characterization result of Ahsanullah and Wesolowski (1994), who considered the following conditions:

(i) The conditional distribution of X_k, given $(X_1 = x_1, \ldots, X_{k-1} = x_{k-1})$, is $N\left(\alpha_0 + \sum_{i=1}^{k-1} \alpha_i x_i, \beta^2 \right)$, where $\alpha_0, \alpha_1, \ldots, \alpha_{k-1}$ and $\beta^2 (> 0)$ are some real constants.

(ii) The random variables $X_1, \ldots, X_k$ are identically distributed.

these authors showed that these conditions do not characterize the joint normality of X's; see, for example, Ahsanullah and Sinha (1986) and Arnold and Pourahmadi (1988) discussed above. However, if condition (ii) is replaced by

(ii)' $(X_0, X_1, \ldots, X_i)^T \overset{d}{=} (X_0, X_1, \ldots, X_{i-1}, X_{i+1})^T$ for $i = 1, \ldots, k-1$

with $X_0 = 0$ a.s., then as has been shown by Ahsanullah and Wesolowski (1994), (i) and (ii)' do characterize multivariate normality. These authors have also noted that the Markovian type property $(X_1, \ldots, X_{k-1})^T \overset{d}{=} (X_2, \ldots, X_k)$ would not on its own result in characterization unless conditions of the type $\rho_{12} = \rho_{23} = \rho_{34} = \cdots$ are imposed, where ρ_{ij} is the correlation coefficient between X_i and X_j; for details and counterexamples, see Ahsanullah and Wesolowski (1994). The proofs are based on manipulations of characteristic functions.

It seems likely that an abbreviated list of conditional normals should characterize a k-variate normal distribution; however, as Arnold (1997) has pointed out, the nature of that minimal sufficient list is not known as yet.

Wang (1997) obtained a characterization of a multivariate normal density function $p_{\boldsymbol{X}}(x_1, \ldots, x_k)$ $(k \geq 3)$ by the following two conditions:

(i) The third-order partial derivatives of $\log p_{\boldsymbol{X}}(\boldsymbol{x})$ are null functions.

(ii) Each of the $\binom{k}{2}$ two-dimensional marginal densities of $p_{\boldsymbol{X}}(\boldsymbol{x})$ is a bivariate normal density function.

Note that the normality of bivariate marginals is a strong basis for the normality of k-dimensional distributions. Wang (1997) has also presented another similar characterization of the multivariate normal density function $p_{\boldsymbol{X}}(\boldsymbol{x})$ in which condition (ii) above is replaced by the following two conditions:

(ii) The second-order partial derivative of the logarithm of each of the $\binom{k}{2}$ two-dimensional marginal densities of $p_{\boldsymbol{X}}(\boldsymbol{x})$ is a constant function.

(iii) The k univariate marginal densities of $p_{\boldsymbol{X}}(\boldsymbol{x})$ are normal.

These characterizations are in the same vein as the earlier characterization for the bivariate normal distribution also due to Wang (1987).

Stadje (1993) generalized to the multivariate case the well-known "maximum likelihood" characterization of the univariate normal distribution due to Teicher (1961); see Chapter 13 of Johnson, Kotz, and Balakrishnan (1994). [It should be mentioned here that even though Teicher (1961) required lower semicontinuity of the density $p(x)$ at 0, Findeisen (1982) has shown that measurability of $p(\cdot)$ is sufficient for the characterization result.] Stadje's (1993) result is as follows. Let $\boldsymbol{X}_1, \ldots, \boldsymbol{X}_n$ be a sample from a population with density $p(\cdot)$ in $\mathbb{R}^k$, and let the mean $\bar{\boldsymbol{X}}$ be the maximum likelihood estimator of parameter $\boldsymbol{\theta} \in \mathbb{R}^k$ of the translation family $p\{\boldsymbol{x} - \boldsymbol{\theta})\}$, that is,

$$\prod_{i=1}^{n} p(\boldsymbol{x}_i - \bar{\boldsymbol{x}}) \geq \prod_{i=1}^{k} p(\boldsymbol{x}_i - \boldsymbol{\theta}) \quad \text{for all } \boldsymbol{\theta} \in \mathbb{R}^k.$$

Then, $p(\boldsymbol{x}) = c \exp(-\boldsymbol{x}^T \boldsymbol{A} \boldsymbol{x})$, $\boldsymbol{x} \in \mathbb{R}^k$, for some $c > 0$ and a nonnegative definite $k \times k$ matrix $\boldsymbol{A}$. The assumption on $p(\cdot)$ is that it is a Borel-measurable nonnegative function on $\mathbb{R}^k$ and that $\mu^k(p(\boldsymbol{x}) > 0) > 0$, where μ^k is the k-dimensional Lebesgue measure. An earlier proof by Campbell (1970) required $p(\cdot)$ to be positive and also twice differentiable.

Hamedani (1992) provided a list of eighteen characterizations of the bivariate normal distribution (see Chapter 46) supplemented by an extensive bibliography and an equal number of characterizations of the multivariate normal distribution. Many of these characterizations have been discussed above. An important negative result is that the multivariate normality of all subsets $(r < k)$ of the normal variables $X_1, \ldots, X_k$ together with the normality of an infinite number of linear combinations of them do not guarantee the joint normality of these variables; see also Hamedani (1984).

8 ESTIMATION

8.1 Estimation of $\boldsymbol{\xi}$

The parameters of the marginal (normal) distributions of each X_j may be estimated, using the observed values of X_j alone, by any of the methods described earlier in Chapter 13 of Johnson, Kotz, and Balakrishnan (1994).

Given a random sample, representable as observed values of n independent random vectors $\boldsymbol{X}_t^T = (X_{1t}, X_{2t}, \ldots, X_{kt})$ $(t = 1, \ldots, n)$ with

$$p_{\boldsymbol{X}_t}(\boldsymbol{x}_t) = \frac{|\boldsymbol{V}|^{-1/2}}{(2\pi)^{m/2}} \exp\left\{-\frac{1}{2}(\boldsymbol{x}_t - \boldsymbol{\xi})^T \boldsymbol{V}^{-1}(\boldsymbol{x}_t - \boldsymbol{\xi})\right\}, \qquad (45.119)$$

the maximum likelihood estimators are

$$\hat{\boldsymbol{\xi}} = \bar{\boldsymbol{X}} = \frac{1}{n}\sum_{t=1}^{n} \boldsymbol{X}_t, \qquad \hat{\boldsymbol{V}} = \frac{1}{n}\boldsymbol{S} \qquad (45.120)$$

with

$$S_{ij} = \sum_{t=1}^{n}(X_{ti} - \bar{X}_i)(X_{tj} - \bar{X}_j). \qquad (45.121)$$

Many, more or less arbitrary, criteria have been set up to measure the overall inaccuracy of sets of estimators $(\tilde{\boldsymbol{\xi}})$ of the expected value vector $(\boldsymbol{\xi})$. Among these we may note:

(a) The determinant of the variance–covariance matrix of $\tilde{\boldsymbol{\xi}}$.

(b) The expected value of $(\tilde{\boldsymbol{\xi}} - \boldsymbol{\xi})^T \boldsymbol{V}^{-1}(\tilde{\boldsymbol{\xi}} - \boldsymbol{\xi})$.

James and Stein (1955, 1961) have shown that if criterion (b) is used and there is available a matrix $\boldsymbol{S}$ independent of $\bar{\boldsymbol{X}}$ and having a Wishart distribution with parameters ν $(> k - 3)$ and $\boldsymbol{V}$, then the vector

$$\left(1 - \frac{k-2}{\nu - k + 3} \times \frac{1}{n\bar{\boldsymbol{X}}^T \boldsymbol{S}^{-1}\bar{\boldsymbol{X}}}\right)\bar{\boldsymbol{X}}$$

has a smaller value of (b) than $\bar{X}$.

In fact, for this vector, the value of (b) is

$$\left\{ k - \frac{\nu - k + 1}{\nu - k + 3} (k - 2)^2 E[(k - 2 + 2\phi)^{-1}] \right\} \bigg/ n, \qquad (45.122)$$

where ϕ has a Poisson distribution with expected value $\frac{n}{2} \boldsymbol{\xi}^T \boldsymbol{V}^{-1} \boldsymbol{\xi}$, while the value for $\bar{X}$ is just k/n.

In many cases, $\boldsymbol{S}$ may be taken as the matrix of sums of squares and products for sample means. Then, $\nu = n - 1$.

If it is known that $\boldsymbol{V}$ is of form $\boldsymbol{I}\sigma^2$—that is, the variates are independent and all have the same variance—then in place of $\boldsymbol{S}$, we may use a statistic T distributed as $\chi^2_\nu \sigma^2 / n$. In this case, Baranchik (1970) has shown that any estimator of the form

$$\left(1 - g \left(\frac{\bar{X}^T \bar{X}}{T} \right) \times \frac{T}{\bar{X}^T \bar{X}} \right) \bar{X} \qquad (45.123)$$

has a smaller value of (b) than $\bar{X}$, provided that $g(\cdot)$ is a positive monotonic nondecreasing function that is less than $2(k-2)/(\nu+2)$.

A substantial amount of research in the last 25 years has been devoted to the estimation of multivariate normal mean based on Bayesian and decision-theoretic frameworks. Let $\boldsymbol{X}_i$, $i = 1, 2, \ldots, n$, be a random sample from $N_k(\boldsymbol{\xi}, \boldsymbol{V})$, where $\boldsymbol{V}$ is known. Let the prior of $\boldsymbol{\xi}$ be $N_k(\boldsymbol{\alpha}, \boldsymbol{M})$ (the conjugate prior). Then, DeGroot (1970) has shown that the posterior distribution of $\boldsymbol{\xi}$, given $\boldsymbol{X}_1, \ldots, \boldsymbol{X}_n$, is $N_k(\boldsymbol{\mu}, \boldsymbol{\Sigma})$, where

$$\boldsymbol{\mu} = (\boldsymbol{M} + n\boldsymbol{V})^{-1}(\boldsymbol{M}\boldsymbol{\alpha} + n\boldsymbol{V}\bar{X}) \quad \text{and} \quad \boldsymbol{\Sigma} = \boldsymbol{M} + n\boldsymbol{V}.$$

$\boldsymbol{\mu}$ is a function of $\bar{X}$ while $\boldsymbol{\Sigma}$ is a function of n only; see also Kunte and Rattihalli (1989). Now, consider the case when $\boldsymbol{V}$ is also unknown. Let the prior joint distribution of $\boldsymbol{\xi}$ and $\boldsymbol{V}$ be as follows: The conditional distribution of $\boldsymbol{\xi}$, given $\boldsymbol{V}$, is $N_k(\boldsymbol{\beta}, \nu\boldsymbol{V})$ ($\nu > 0$), and the marginal distribution of $\boldsymbol{V}$ is Wishart with α degrees of freedom and the precision matrix $\boldsymbol{T}$ ($\alpha > k - 1$ and $\boldsymbol{T}$ is symmetric positive definite). Then, the joint posterior distribution of $\boldsymbol{\xi}$ and $\boldsymbol{V}$, given $\boldsymbol{X} = (X_1, \ldots, X_n)$, is such that the conditional distribution of $\boldsymbol{\xi}$, given $\boldsymbol{V}$, is $N_k(\boldsymbol{\mu}, (n+\nu)\boldsymbol{V})$; furthermore, the marginal distribution of $\boldsymbol{V}$ is Wishart with $n + \alpha$ degrees of freedom and precision matrix $\boldsymbol{\Sigma}$, where $\boldsymbol{\mu} = (\nu\boldsymbol{\beta} + n\bar{X})/(\nu + n)$ and $\boldsymbol{\Sigma} = \boldsymbol{T} + \boldsymbol{S} + \frac{\nu n}{\nu + n} (\boldsymbol{\beta} - \bar{X})(\boldsymbol{\beta} - \bar{X})^T$; here, $\bar{X} = \frac{1}{n} \sum_{i=1}^{n} X_i$ and $\boldsymbol{S} = \sum_{i=1}^{n} (X_i - \bar{X})(X_i - \bar{X})^T$. The posterior marginal density of $\boldsymbol{\xi}$, given

$\boldsymbol{X}$, is

$$p(\boldsymbol{\xi}|\boldsymbol{X}) = C_N \left\{ 1 + \frac{1}{N}(\boldsymbol{\xi} - \boldsymbol{\mu})^T \boldsymbol{D}(\boldsymbol{\xi} - \boldsymbol{\mu}) \right\}^{-(N+k)/2},$$

where $N = \alpha + n - k + 1$, $\boldsymbol{D} = (n + \nu)N\boldsymbol{\Sigma}^{-1}$, and

$$\frac{1}{C_N} = \frac{B\left(\frac{N}{2}, \frac{k}{2}\right) \pi^{k/2}}{\Gamma(k/2)} \left| \frac{1}{N} \boldsymbol{D} \right|^{-1/2},$$

which is incidentally a multivariate t-distribution with N degrees of freedom; see also Rattihalli (1994). Kunte and Rattihalli (1989) derived fixed sample size Bayes estimators for $\boldsymbol{\xi}$ when $\boldsymbol{V}$ is known by using the conjugate prior distribution of DeGroot (1970), while Rattihalli (1994) derived Bayes estimators for $\boldsymbol{\xi}$ (once again by using the conjugate prior) when $\boldsymbol{V}$ is unknown.

Let $\boldsymbol{X} \overset{d}{=} N_k(\boldsymbol{\xi}, \boldsymbol{V})$, where $\boldsymbol{V}$ is known. Assuming a quadratic loss

$$L(\boldsymbol{\xi}, \boldsymbol{\delta}) = (\boldsymbol{\xi} - \boldsymbol{\delta})^T \boldsymbol{Q}(\boldsymbol{\xi} - \boldsymbol{\delta}),$$

where $\boldsymbol{\delta}(\boldsymbol{X}) = (\delta_1(\boldsymbol{X}), \dots, \delta_k(\boldsymbol{X}))^T$ is an estimator of $\boldsymbol{\xi}$ and $\boldsymbol{Q}$ is a known positive definite matrix, we evaluate an estimator by its risk function (or the expected loss)

$$R(\boldsymbol{\xi}, \boldsymbol{\delta}) = E_{\boldsymbol{\xi}}[L(\boldsymbol{\xi}, \boldsymbol{\delta}(\boldsymbol{X}))],$$

which becomes mean square error when $\boldsymbol{Q} = \boldsymbol{I}_{k \times k}$. As mentioned earlier, the estimator $\boldsymbol{\delta}_0(\boldsymbol{X}) = \boldsymbol{X}$ is inadmissible for $k \geq 3$ whenever $\boldsymbol{Q} = \boldsymbol{V} = \boldsymbol{I}_{k \times k}$ [Stein (1955)]. Estimators having uniformly smaller risk than $\boldsymbol{\delta}_0$ have been developed by many authors including Hudson (1974), Shinozaki (1974), Berger (1976, 1979, 1980, 1982), and Casella (1977).

Let $\boldsymbol{\mu}$ and $\boldsymbol{A}$ be the prior mean and covariance matrix of $\boldsymbol{\xi}$. Hudson (1974) and Berger (1976) independently derived the minimax estimator

$$\boldsymbol{\delta}_{\mathrm{HB}}(\boldsymbol{X}) = \left\{ \boldsymbol{I}_{k \times k} - \frac{r(\|\boldsymbol{X} - \boldsymbol{\mu}\|^2)}{\|\boldsymbol{X} - \boldsymbol{\mu}\|^2} \boldsymbol{Q}^{-1}\boldsymbol{V}^{-1} \right\}(\boldsymbol{X} - \boldsymbol{\mu}) + \boldsymbol{\mu},$$

$$\text{(45.124)}$$

where

$$\|\boldsymbol{X} - \boldsymbol{\mu}\|^2 = (\boldsymbol{X} - \boldsymbol{\mu})^T \boldsymbol{V}^{-1} \boldsymbol{Q}^{-1} \boldsymbol{V}^{-1} (\boldsymbol{X} - \boldsymbol{\mu})$$

and $r(\cdot)$ is any positive nondecreasing function less than or equal to $2k - 4$. The estimator $\boldsymbol{\delta}_{\mathrm{HB}}(\boldsymbol{X})$ has some pitfalls, the main one being that usually the improvement over $\boldsymbol{\delta}_0$ is quite minor in the "desired region."

Berger (1980) proposed a robust generalized Bayes estimator

$$\boldsymbol{\delta}_{\text{RB}} = \left\{ \boldsymbol{I}_{k \times k} - \frac{r((\boldsymbol{X} - \boldsymbol{\mu})^T (\boldsymbol{V} + \boldsymbol{A})^{-1} (\boldsymbol{X} - \boldsymbol{\mu}))}{(\boldsymbol{X} - \boldsymbol{\mu})^T (\boldsymbol{V} + \boldsymbol{A})^{-1} (\boldsymbol{X} - \boldsymbol{\mu})} \, \boldsymbol{V} (\boldsymbol{V} + \boldsymbol{A})^{-1} \right\}$$
$$\times \, (\boldsymbol{X} - \boldsymbol{\mu}) + \boldsymbol{\mu}, \qquad (45.125)$$

where $r(\cdot)$ is an increasing function. The estimator provides a significant improvement in risk over $\boldsymbol{\delta}_0$ in the region $\{\boldsymbol{\xi} : (\boldsymbol{\xi} - \boldsymbol{\mu})^T \boldsymbol{A}^{-1} (\boldsymbol{\xi} - \boldsymbol{\mu}) \leq k\}$ specified by $\boldsymbol{\mu}$ and $\boldsymbol{A}$, but it is not a minimax estimator. In addition, Berger (1982) proposed an estimator that is a combination of $\boldsymbol{\delta}_{\text{HB}}(\boldsymbol{X})$ and the minimax estimator of Bhattacharya (1966). Suppose it is desired to estimate $\boldsymbol{\xi}$ under the loss

$$L(\boldsymbol{\xi}, \boldsymbol{\delta}) = \sum_{i=1}^{k} q_i^* (\xi_i - \delta_i)^2, \quad \text{where } q_1^* \geq q_2^* \geq \cdots \geq q_k^*,$$

and it is known that, in estimating $\boldsymbol{\xi}_j = (\xi_1, \ldots, \xi_j)^T$ under sum of squares error loss, $\boldsymbol{\delta}^{(j)}(\boldsymbol{X}) = (\delta_1^{(j)}(\boldsymbol{X}), \ldots, \delta_j^{(j)}(\boldsymbol{X}))^T$ is minimax. Define componentwise

$$\delta_{i,\text{MB}}(\boldsymbol{X}) = q_i^{*-1} \sum_{j=1}^{k} (q_j^* - q_{j+1}^*) \delta_i^{(j)}(\boldsymbol{X}) \qquad (q_{k+1}^* = 0).$$

The estimator $\boldsymbol{\delta}_{\text{MB}}$ is also minimax. This fact yields the following specific minimax estimator $(\boldsymbol{\delta}_{\text{MB}})$, under the loss given above:

$$\left\{ \boldsymbol{B}^{-1} \boldsymbol{\delta}_{\text{MB}}(\boldsymbol{X}) \right\}_i = q_i^{*-1} \sum_{j=1}^{k} (q_j^* - q_{j+1}^*) \delta_i^{(j)} \left((\boldsymbol{B}\boldsymbol{X})^j \right)$$

(defined componentwise), where $\boldsymbol{\Lambda}$ is the $k \times k$ orthogonal matrix such that

$$\boldsymbol{Q}^* = \text{diag}(q_1^*, \ldots, q_k^*) = \boldsymbol{\Lambda} (\boldsymbol{V} + \boldsymbol{A})^{-1/2} \boldsymbol{V} \boldsymbol{Q} \boldsymbol{V} (\boldsymbol{V} + \boldsymbol{A})^{-1/2} \boldsymbol{\Lambda}$$

with $q_1^* \geq q_2^* \geq \cdots \geq q_k^*$,

$$\boldsymbol{B} = \boldsymbol{\Lambda} (\boldsymbol{V} + \boldsymbol{A})^{1/2} \boldsymbol{V}^{-1}, \quad \boldsymbol{X}^* = \boldsymbol{B}\boldsymbol{X}, \quad \boldsymbol{\xi}^* = \boldsymbol{B}\boldsymbol{\xi},$$
$$\boldsymbol{V}^* = \boldsymbol{B}\boldsymbol{V}\boldsymbol{B}^T, \quad \boldsymbol{\mu}^* = \boldsymbol{B}\boldsymbol{\mu}, \quad \boldsymbol{A}^* = \boldsymbol{B}\boldsymbol{A}\boldsymbol{B}^T,$$

and

$$\boldsymbol{\delta}^{(j)}(\boldsymbol{X}^{*j}) = \left\{ \boldsymbol{I}_{j \times j} - \frac{\min\{2(j-2)^+, \|\boldsymbol{X}^{*j} - \boldsymbol{\mu}^{*j}\|^2\}}{\|\boldsymbol{X}^{*j} - \boldsymbol{\mu}^{*j}\|^2} \, \boldsymbol{V}_j^{*-1} \right\}$$
$$\times (\boldsymbol{X}^{*j} - \boldsymbol{\mu}^{*j}) + \boldsymbol{\mu}^{*j};$$

here, $(j-2)^+ = j-2$ if $j \geq 2$ and 0 if $j = 1$.

Berger (1982) has noted that minimaxity of δ_{MB} assures certain "safety" that can be interpreted as safety with respect to misspecification of the prior. However, if q_1^* and q_2^* are large compared to the remaining q_i^*, then δ_{MB} is worse than δ_{RB} from the Bayesian point of view. (Insisting on minimaxity seems to eliminate most of the potential gains available from prior information when the coordinates are disparate in terms of q_i^*.)

Karunamuni and Schmuland (1995) considered the estimation of $\boldsymbol{\xi}$ under quadratic loss assuming that $\boldsymbol{V}$ is known. Let $\boldsymbol{X}$ be an observation from $N_k(\boldsymbol{\xi}, \boldsymbol{V})$, where $\boldsymbol{V}$ is a known positive definite variance-covariance matrix. $\boldsymbol{\xi}$ needs to be estimated using an estimator $\boldsymbol{\delta}(\boldsymbol{X}) = (\delta_1(\boldsymbol{X}), \ldots, \delta_k(\boldsymbol{X}))^T$ under the quadratic loss $(\boldsymbol{\delta} - \boldsymbol{\xi})^T \boldsymbol{Q}(\boldsymbol{\delta} - \boldsymbol{\xi})$, where $\boldsymbol{Q}$ is a positive definite matrix of dimension $k \times k$. These authors considered the generalized prior density

$$\pi_\alpha(\boldsymbol{\xi}) = \int_{\mathbb{R}^k} |\boldsymbol{\lambda}|^{-(k-\alpha)} \exp\left\{-\frac{1}{2}(\boldsymbol{\xi} - \boldsymbol{\lambda})^T(\boldsymbol{\xi} - \boldsymbol{\lambda})\right\} d\boldsymbol{\lambda},$$

where $|\boldsymbol{\lambda}|$ is the Euclidean norm of $\boldsymbol{\lambda}$ in $\mathbb{R}^k$ and $0 < \alpha < k$. Note that $\pi_\alpha(\boldsymbol{\xi})$ does not depend on $\boldsymbol{V}$. In contrast, the generalized prior density used by Berger (1980) discussed above,

$$g_n(\boldsymbol{\xi}) = \int_0^1 [\det\{\boldsymbol{B}(\lambda)\}]^{-1/2} \exp\left\{-\frac{1}{2}\boldsymbol{\xi}^T(\boldsymbol{B}(\lambda))^{-1}\boldsymbol{\xi}\right\} \lambda^{n-1-k/2} d\lambda,$$

where $\boldsymbol{B}(\lambda) = \frac{1}{\lambda}\boldsymbol{C} - \boldsymbol{V}$ for $0 < \lambda < 1$ and $n > 0$ with $\boldsymbol{C}$ being a $k \times k$ symmetric matrix such that $\boldsymbol{C} - \boldsymbol{V}$ is positive semidefinite, depends on the variance-covariance matrix $\boldsymbol{V}$. In addition, like $g_n(\boldsymbol{\xi})$, $\pi_\alpha(\boldsymbol{\xi})$ is a hierarchical prior density. Note that the hierarchical variable $\boldsymbol{\lambda}$ in $\pi_\alpha(\boldsymbol{\xi})$ is the location parameter of the variable $\boldsymbol{\xi}$, whereas λ in $g_n(\boldsymbol{\xi})$ is the scale parameter of $\boldsymbol{\xi}$. The prior $\pi_\alpha(\boldsymbol{\xi})$ is constructed by convoluting a prior of the form $\frac{1}{|\boldsymbol{\xi}|^{k-\alpha}}$ with a normal kernel in that it produces a smooth bounded prior with the same tail behavior $\frac{1}{|\boldsymbol{\xi}|^{k-\alpha}}$.

The Bayes estimator $\boldsymbol{\delta}_\pi(\boldsymbol{X})$ with respect to the prior π_α is

$$\boldsymbol{\delta}_\pi(\boldsymbol{X}) = \frac{\int \boldsymbol{\xi} \exp\left\{-\frac{1}{2}(\boldsymbol{X} - \boldsymbol{\xi})^T \boldsymbol{V}^{-1}(\boldsymbol{X} - \boldsymbol{\xi})\right\} \pi_\alpha(\boldsymbol{\xi})\, d\boldsymbol{\xi}}{\int \exp\left\{-\frac{1}{2}(\boldsymbol{X} - \boldsymbol{\xi})^T \boldsymbol{V}^{-1}(\boldsymbol{X} - \boldsymbol{\xi})\right\} \pi_\alpha(\boldsymbol{\xi})\, d\boldsymbol{\xi}}. \qquad (45.126)$$

Karunamuni and Schmuland (1995) have shown that

$$\boldsymbol{\delta}_\pi(\boldsymbol{X}) = \left(\boldsymbol{I} - \frac{T_\alpha(\boldsymbol{X})}{U_\alpha(\boldsymbol{X})}\right) \boldsymbol{X},$$

where

$$T_\alpha(\boldsymbol{X}) = \int_0^\infty \frac{\delta^{(-k+\alpha-2)/2}}{\left\{\det\left(\boldsymbol{A}^{-1} + \frac{1}{\delta}\boldsymbol{I}\right)\right\}^{1/2}} \boldsymbol{V}(\boldsymbol{A} + \delta\boldsymbol{I})^{-1}$$
$$\times \exp\left\{-\frac{1}{2}\boldsymbol{X}^T(\boldsymbol{A} + \delta\boldsymbol{I})^{-1}\boldsymbol{X}\right\} d\delta$$

and

$$U_\alpha(\boldsymbol{X}) = \int_0^\infty \frac{\delta^{(-k+\alpha-2)/2}}{\left\{\det\left(\boldsymbol{A}^{-1} + \frac{1}{\delta}\boldsymbol{I}\right)\right\}^{1/2}} \exp\left\{-\frac{1}{2}\boldsymbol{X}^T(\boldsymbol{A} + \delta\boldsymbol{I})^{-1}\boldsymbol{X}\right\} d\delta.$$

When $\boldsymbol{V} = \boldsymbol{I}$, $T_\alpha(\boldsymbol{X})$ and $U_\alpha(\boldsymbol{X})$ become

$$T_\alpha(\boldsymbol{X}) = 2^{(\alpha-2)/2)} \int_0^1 y^{(k-\alpha)/2}(1-y)^{(\alpha-2)/2} \exp\left\{-\frac{1}{4}|\boldsymbol{X}|^2 y\right\} dy$$

and

$$U_\alpha(\boldsymbol{X}) = 2^{\alpha/2} \int_0^1 y^{(k-\alpha-2)/2}(1-y)^{(\alpha-2)/2} \exp\left\{-\frac{1}{4}|\boldsymbol{X}|^2 y\right\} dy.$$

These authors have shown that $\boldsymbol{\delta}_\pi(\boldsymbol{X})$ is minimax for $\alpha = 2$ and admissible for $0 \le \alpha \le 2$ and $k \ge 3$. Denoting

$$W_\alpha(v) = \frac{v \int_0^1 y^{(k-\alpha)/2}(1-y)^{(\alpha-2)/2} \exp\left\{-\frac{1}{4}vy\right\} dy}{2 \int_0^1 y^{(k-\alpha-2)/2}(1-y)^{(\alpha-2)/2} \exp\left\{-\frac{1}{4}vy\right\} dy},$$

the estimator $\boldsymbol{\delta}_\pi(\boldsymbol{X})$ can be rewritten as

$$\boldsymbol{\delta}_\pi(\boldsymbol{X}) = \left(\boldsymbol{I} - \frac{W_\alpha(|\boldsymbol{X}|^2)}{|\boldsymbol{X}|^2}\right)\boldsymbol{X}. \qquad (45.127)$$

In this form, for the case $\alpha = 2$, it is similar to the Stein estimator $(\boldsymbol{I} - (k-2)/|\boldsymbol{X}|^2)\boldsymbol{X}$ since $\lim_{v\to\infty} W_2(v) = k - 2$. However, no comparison of Berger's and Karunamuni and Schmuland's estimators has been made yet.

Chen (1983, 1988) proposed several "compromise" estimators for $\boldsymbol{\xi}$, when the variance-covariance matrix $\boldsymbol{V}$ is known, under the quadratic loss function $L(\boldsymbol{\xi}, \boldsymbol{\delta}) = (\boldsymbol{\xi} - \boldsymbol{\delta})^T\boldsymbol{Q}(\boldsymbol{\xi} - \boldsymbol{\delta})$ when prior beliefs concerning $\boldsymbol{\xi}$ are approximately modeled by a conjugate prior distribution π which is $N_k(\boldsymbol{\theta}, \boldsymbol{A})$ with known $\boldsymbol{\theta}$ and $\boldsymbol{A}$. The compromise is between a strict Bayes estimator $\boldsymbol{\delta}_\pi$ minimizing the Bayes risk—that is, $r(\pi, \boldsymbol{\delta}_\pi) = \min_{\boldsymbol{\delta}} r(\pi, \boldsymbol{\delta})$, where $r(\pi, \boldsymbol{\delta}) = E_\pi[R(\boldsymbol{\xi}, \boldsymbol{\delta})]$ and $R(\boldsymbol{\xi}, \boldsymbol{\delta}) = E_{\boldsymbol{\xi}}[L(\boldsymbol{\xi}, \boldsymbol{\delta}(\boldsymbol{X}))]$—and a min-imax estimator that protects against the worst possible state of nature that may, however, be inadmissible. First, restrict the risk $R(\boldsymbol{\xi}, \boldsymbol{\delta})$ so that $R(\boldsymbol{\xi}, \boldsymbol{\delta}) - R(\boldsymbol{\xi}, \boldsymbol{\delta}_0) \leq c$ for all $\boldsymbol{\xi} \in \mathbb{R}^k$, where $\boldsymbol{\delta}_0(\boldsymbol{X}) = \boldsymbol{X}$ is the maximum likelihood estimator of $\boldsymbol{\xi}$ and c is a given non-negative constant. Evidently, $R(\boldsymbol{\xi}, \boldsymbol{\delta}_0) = \text{trace}(\boldsymbol{QV})$. Let $\boldsymbol{Q}, \boldsymbol{V}$, and $\boldsymbol{A}$ be all diagonal matrices with diagonal elements q_i, σ_i^2 and a_i, respectively. Arrange the components X_i of $\boldsymbol{X}$ so that $d_1 \geq d_2 \geq \cdots \geq d_k$, where $d_i = q_i \sigma_i^4 (\sigma_i^2 + a_i)$, $i = 1, \ldots, k$, and $d_{k+1} = 0$. Let us now define the ith component of $\boldsymbol{\delta}_{\text{MB},c}$ as

$$X_i - \frac{\sigma_i^2}{\sigma_i^2 + a_i} (X_i - \xi_i) \sum_{j=i}^k \left(\frac{d_j - d_{j+1}}{d_i} \right) \min \left(1, \rho_c^{(j)}(r_j) \right),$$

where

$$\rho_c^{(j)}(r_j) = \begin{cases} \dfrac{2(j-2)^+}{r_j} & \text{if} \quad c = 0, \\[2ex] \dfrac{c}{2d_1 t_j} \cdot \dfrac{K_{v_j+1}(t_j)}{K_{v_j}(t_j)} & \text{if} \quad c > 0 \text{ and } j = 1, \\[2ex] \dfrac{c}{2d_1 t_j} \cdot \dfrac{K_{v_j-1}(t_j)}{K_{v_j}(t_j)} & \text{if} \quad c > 0 \text{ and } j \geq 2, \end{cases}$$

$$r_j = \sum_{i=1}^j \frac{(X_i - \xi_i)^2}{\sigma_i^2 + a_i}, \quad t_j = \frac{1}{2} \left(\frac{cr_j}{d_1} \right)^{1/2}, \quad v_j = \frac{1}{2} |j - 2|,$$

and $K_v(\cdot)$ is the modified Bessel function of the second kind of order v. If all $\rho_c^{(j)}(r_j) \geq 1$, then the estimator $\boldsymbol{\delta}_{\text{MB},c}$ is based on the conjugate prior Bayes rule. Berger (1982) provided motivation for estimators of this form; it can be shown that indeed $R(\boldsymbol{\xi}, \boldsymbol{\delta}_{\text{MB},c}) - R(\boldsymbol{\xi}, \boldsymbol{\delta}_0) \leq c$ for all $\boldsymbol{\xi}$.

Another estimator is a weighted minimax estimator used when d_1 (or possibly d_2) is much larger than all the other d_i's. Let $\boldsymbol{D} = (\boldsymbol{V} + \boldsymbol{A})^{-1/2} \boldsymbol{V} \boldsymbol{Q} \boldsymbol{V} (\boldsymbol{V} + \boldsymbol{A})^{-1/2}$ and let d_i's be the eigenvalues of $\boldsymbol{D}$. Let $\boldsymbol{W} = \frac{1}{k-2} \left\{ \left(\boldsymbol{I} + (k-2)y_0 \boldsymbol{D}^{-1} \right)^{1/2} - \boldsymbol{I} \right\}$, where y_0 is the positive solution of the equation

$$\frac{1}{k-2} \sum_{i=1}^k d_i \left[\left\{ 1 + (k-2) \frac{y_0}{d_i} \right\}^{1/2} - 1 \right] = y_0.$$

Define

$$\boldsymbol{\delta}_W(\boldsymbol{X}) = \boldsymbol{X} - \min \left(1, \frac{2(k-2)^+}{r} \right) \boldsymbol{W} \boldsymbol{V} (\boldsymbol{V} + \boldsymbol{A})^{-1} (\boldsymbol{X} - \boldsymbol{\xi}),$$

$$(45.128)$$

where $r = (\boldsymbol{X} - \boldsymbol{\xi})^T (\boldsymbol{V} + \boldsymbol{A})^{-1} (\boldsymbol{X} - \boldsymbol{\xi})$. Then, Chen (1988) has shown that $\boldsymbol{\delta}_W$ is a minimax estimator; also, numerical computations have revealed that when $\frac{d_k}{d_2}$ or $\frac{d_k}{d_3}$ is close to 1, $\boldsymbol{\delta}_W$ is better than $\boldsymbol{\delta}_{\text{MB},0}$, and when $k < 6$, $\boldsymbol{\delta}_W$ is better than $\boldsymbol{\delta}_{\text{MB},0}$ except when $\frac{d_{i+1}}{d_i}$ is very small (≤ 0.1) for some $3 \leq i \leq k - 1$. However, for $k \geq 6$ when four or more of $\frac{d_j}{d_1}$ are close to 1, $\boldsymbol{\delta}_{\text{MB},0}$ is better than $\boldsymbol{\delta}_W$. These comparisons were made based on the values of linearly transformed relative savings risk proposed by Efron and Morris (1971). Furthermore, using the fact that a convex combination of minimax estimators is also minimax, Chen (1983) provided some improvements on both $\boldsymbol{\delta}_{\text{MB},0}$ and $\boldsymbol{\delta}_W$; but the improvements are not substantial enough to justify the added complexity.

Lin and Tsai (1973) generalized the James–Stein estimator of $\boldsymbol{\xi}$ and obtained a class of minimax estimators of the form

$$\left(1 - \frac{r(T)}{T}\right) \bar{\boldsymbol{X}},$$

where $T = \bar{\boldsymbol{X}}^T \boldsymbol{S}^{-1} \bar{\boldsymbol{X}}$, $0 < r(\cdot) \leq \frac{(k-2)}{n(n-k+2)}$ (a constant) and $r(\cdot)$ is non-decreasing. Pal and Elfessi (1995) suggested an improvement over this estimator of the form

$$\left(1 - \frac{g}{T}\right) \bar{\boldsymbol{X}},$$

where $g = T + u(\boldsymbol{S}) \bar{\boldsymbol{X}}^T \bar{\boldsymbol{X}}$ and $u(\boldsymbol{X})$ is a scalar function. Several forms of $u(\boldsymbol{S})$ have been suggested by the authors including

(i) $u(\boldsymbol{S}) = v\, t(v)$, where $v = \text{trace}(\boldsymbol{S}^{-1})$, for a nondecreasing $t(v)$,

(ii) $u(\boldsymbol{S}) = v\, t(v)$, where $v = \{\text{trace}(\boldsymbol{S})\}^{-1}$, for a nondecreasing $t(v)$,

and

(iii) $u(\boldsymbol{S}) = v\, t(v)$, where $v = |\boldsymbol{S}|^{-1}$, for $t(v) = \left(\frac{1}{v}\right)^{1+(1/k)}$ for $k \geq 3$.

In addition, Pal and Elfessi (1995) proposed the estimator

$$\left\{1 - \frac{r(T)}{T + b\, \text{trace}(\boldsymbol{S}^{-1}) \bar{\boldsymbol{X}}^T \bar{\boldsymbol{X}}}\right\} \bar{\boldsymbol{X}},$$

where $r(\cdot)$ is as defined above and $b \geq 0$. For $b = 0$ and $r(\cdot) = \frac{k-2}{n(n-k+2)}$, this estimator becomes the James-Stein estimator. The authors have noted that "small" values of b give better result than "large" values, but the optimal selection of b seems to be an open problem. For $b = 0.01$ and

$r(\cdot) = \frac{k-2}{n(n-k+2)}$, the maximum relative risk improvement over the James–Stein estimator is about 39% when $k = 4$ and $n = 11$.

Perron and Giri (1990) derived the best equivariant estimator of the location parameter $\boldsymbol{\xi}$ when the covariance matrix is $\left(\frac{\boldsymbol{\xi}^T\boldsymbol{\xi}}{C^2}\right)\boldsymbol{I}_{k\times k}$, where C is a known coefficient of variation, under the loss function

$$L(\boldsymbol{\xi}, \boldsymbol{a}) = \frac{(\boldsymbol{\xi} - \boldsymbol{a})^T(\boldsymbol{\xi} - \boldsymbol{a})}{\boldsymbol{\xi}^T\boldsymbol{\xi}} .$$

Let $\boldsymbol{X}_1, \boldsymbol{X}_2, \ldots, \boldsymbol{X}_n$ be a random sample from this multivariate normal distribution. Then,

$$\boldsymbol{Y} = \sqrt{n}\bar{\boldsymbol{X}} \quad \text{and} \quad W = \text{trace}\left(\sum_{i=1}^{n}(\boldsymbol{X}_i - \bar{\boldsymbol{X}})(\boldsymbol{X}_i - \bar{\boldsymbol{X}})^T\right)$$

are sufficient statistics and

$$\boldsymbol{Y} \stackrel{d}{=} N_k\left(\sqrt{n}\boldsymbol{\xi}, \frac{\boldsymbol{\xi}^T\boldsymbol{\xi}}{C^2}\boldsymbol{I}\right) \quad \text{and} \quad \frac{C^2}{\boldsymbol{\xi}^T\boldsymbol{\xi}}W \stackrel{d}{=} \chi^2_{k(n-1)}$$

independently of each other. Under the loss function $L(\boldsymbol{\xi}, \boldsymbol{a})$ given above, the best equivariant estimator is $\delta_0(\boldsymbol{X}_1, \ldots, \boldsymbol{X}_n, C) = g_0(t)\bar{\boldsymbol{X}}$, provided that $m = \frac{k(n-1)}{2}$ is an integer, where

$$g_0(t) = \frac{\sum_{i=1}^{m+1} \frac{\left(\frac{i}{m+1}\right)\binom{m+1}{i}}{\Gamma(\frac{k}{2}+i)}\left(\frac{nC^2}{2}\right)^i t^i}{\left\{\sum_{j=0}^{m+1} \frac{\binom{m+1}{j}}{\Gamma(\frac{k}{2}+j)}\left(\frac{nC^2}{2}\right)^j t^{j+1}\right\}},$$

$v = \boldsymbol{Y}^T\boldsymbol{Y}/W$ and $t = \frac{v}{1+v}$. The function $g_0(t)$ is a strictly decreasing continuous function of t.

The maximum likelihood estimator of $\boldsymbol{\xi}$ in this case is

$$\delta_1(\boldsymbol{X}_1, \ldots, \boldsymbol{X}_n, C) = \left(\frac{\sqrt{1 + \frac{4k}{C^2 t}} - 1}{2k}\right)C^2\bar{\boldsymbol{X}};$$

it is equivariant and hence is inadmissible. The other three equivariant estimators are

$$\delta_2(\boldsymbol{X}_1, \ldots, \boldsymbol{X}_n, C) = \left\{1 - \frac{k-2}{(n-1)(k+2)v}\right\}\bar{\boldsymbol{X}}$$

$$\text{[James and Stein (1961) estimator]},$$

$$\delta_3(\boldsymbol{X}_1, \ldots, \boldsymbol{X}_n, C) = \max(\delta_2, 0),$$
$$\delta_4(\boldsymbol{X}_1, \ldots, \boldsymbol{X}_n, C) = \bar{\boldsymbol{X}}.$$

Perron and Giri (1990), who derived the estimators δ_0 and δ_1, observed that among all these estimators, δ_4 is the worst compared to δ_0. When C is small, δ_0 is markedly superior to all others; when C is large, all five estimators are more or less similar.

Efron and Morris (1973) showed that a necessary condition for an equivariant estimator of the form $g(t)\bar{X}$ to be minimax is that $g(t) \to 1$ as $t \to 1$. So, δ_0 is not minimax if we do not know the value of C.

An example of a model relating to this particular multivariate normal distribution (with mean $\boldsymbol{\xi}$ and variance–covariance matrix $\frac{\boldsymbol{\xi}^T\boldsymbol{\xi}}{C^2}\boldsymbol{I}_{k\times k}$) has been given by Kent, Briden, and Mardia (1983). It concerns the natural remanent magnetization in rocks.

Pal et $al.$ (1995) have extended the James–Stein estimator of $\boldsymbol{\xi}$ to the following situation. Let $\boldsymbol{X} \stackrel{d}{=} N_k(\boldsymbol{\xi}, \boldsymbol{I}_{k\times k})$. Assume that an independent nonnegative scalar observation U on $[0,\infty)$ is given with known density function $f(u)$. Then, Pal et $al.$ (1995) have proposed various estimators of the form $\hat{\boldsymbol{\xi}} = \hat{\boldsymbol{\xi}}(\boldsymbol{X}, U)$ and determined their risk functions under the square loss defined by

$$R(\hat{\boldsymbol{\xi}}, \boldsymbol{\xi}) = E\|\hat{\boldsymbol{\xi}} - \boldsymbol{\xi}\|^2.$$

The "simple" estimator is

$$\hat{\boldsymbol{\xi}}_{c,U} = \left(1 - \frac{c}{S+U}\right)\boldsymbol{X},$$

where $S = \|\boldsymbol{X}\|^2$ and c is a constant. This estimator is better than $\hat{\boldsymbol{\xi}}_0 = \boldsymbol{X}$ for any $0 < c \le 2(k-2)$, provided that $k \ge 3$. A generalization of $\hat{\boldsymbol{\xi}}_{C,U}$ is

$$\hat{\boldsymbol{\xi}}_{g,U} = \left(1 - \frac{g}{S+U}\right)\boldsymbol{X},$$

where $g \equiv g(S, U)$ is a nonnegative function of S and U. Pal et $al.$ (1995) have shown that the estimator $\hat{\boldsymbol{\xi}}_{g,U}$ dominates $\hat{\boldsymbol{\xi}}_0$ under the squared loss given above, provided that

(i) U is any nonnegative random variable independent of $\boldsymbol{X}$,

(ii) $g(S, U)$ is continuous, nondecreasing in S, and piecewise differentiable with respect to S, and

(iii) $0 < g(S, U) \le 2(k-2)$ when $k \ge 3$.

The case when $U \overset{d}{=} \text{Gamma}\left(\frac{r}{2}, \frac{1}{2}\right)$ (i.e., χ_r^2 where r is not necessarily an integer) has been treated separately. This gamma assumption simplifies the distribution of $S + U$. Using the optimal value of c, $c_{\text{opt}} = k + r - 2 = d$ (say), and denoting the estimator $\hat{\boldsymbol{\xi}}_{c,U}$ is the case by $\hat{\boldsymbol{\xi}}_{c,r,U}$, it turns out that, for large k, $\hat{\boldsymbol{\xi}}_{d,k-2,U}$ gives almost 50% risk reduction at $\boldsymbol{\xi} = 0$ compared to the James–Stein estimator

$$\hat{\boldsymbol{\xi}}_{\text{JS}} = \left(1 - \frac{k-2}{S}\right)\boldsymbol{X}.$$

However, the estimator $\hat{\boldsymbol{\xi}}_{d,k-2,U}$ is not uniformly better than $\hat{\boldsymbol{\xi}}_{\text{JS}}$.

Finally, using Rao–Blackwellization, Pal *et al.* (1995) proposed another estimator

$$\hat{\boldsymbol{\xi}}_{c,r} = E_U\left[\hat{\boldsymbol{\xi}}_{c,r,U}\right]$$

which is uniformly better than $\hat{\boldsymbol{\xi}}_{c,r,U}$. It should be noted that when we take the Rao–Blackwellized version, the value of c that minimizes the risk of $\hat{\boldsymbol{\xi}}_{c,r}$ at $\boldsymbol{\xi} = 0$ may not be equal to d. The use of $\hat{\boldsymbol{\xi}}_{d,k-2}$ in real applications can be justified further if we can show that the average (with respect to $\boldsymbol{\xi}$) risk improvement of $\hat{\boldsymbol{\xi}}_{d,k-2}$ is higher than the existing ones. Pal *et al.* (1995) investigated Bayes risks of estimators under a normal conjugate prior assuming $\boldsymbol{\xi} \overset{d}{=} \pi_0(\boldsymbol{\xi}) \equiv N_k(\boldsymbol{0}, \boldsymbol{I}_{k \times k})$. The Bayes risk of $\hat{\boldsymbol{\xi}}_{d,k-2,U}$ is

$$R(\hat{\boldsymbol{\xi}}_{d,k-2,U}, \pi_0) = k - k(k-2)\sum_{j=0}^{\infty}\{(k+j-1)(k+j-2)\}^{-1}c_j^*,$$

where

$$c_j^* = \left\{2^{\frac{k}{2}+j}B\left(\frac{k}{2}-1, j+1\right)\right\}^{-1},$$

and that of $\hat{\boldsymbol{\xi}}_{d,k-2}$ is

$$\begin{aligned}
R(\hat{\boldsymbol{\xi}}_{d,k-2}, \pi_0) &= k - 4(k-2)\sum_{j=0}^{\infty}(k+j)\{(k+j-1)(k+j-2)\}^{-1}c_j^* \\
&\quad + 8(k-2)\sum_{j=0}^{\infty}\{3(k-2)+2j\}^{-1}c_j^*\sum_{\ell=0}^{\infty}A_1/A_2,
\end{aligned}$$

where

$$A_1 = A_1(k, j, \ell) = B\left(\frac{k}{2}+j+1, \frac{k}{2}+\ell-1\right)B\left(k+j+\ell-1, \frac{k}{2}-1\right)$$

and

$$A_2 = A_2(k, j) = B\left(\frac{k}{2} + j,\ \frac{k}{2} - 1\right) B\left(k + j - 1,\ \frac{k}{2} - 1\right).$$

The Bayesian risk of $\hat{\boldsymbol{\xi}}_{JS}$ with respect to π_0 is $\frac{k}{2} + 1$; see Pal $et\ al.$ (1995). The estimators $\hat{\boldsymbol{\xi}}_{d,r,U}$ (for $0 \leq r \leq k - 2$) are all minimax and compete with $\hat{\boldsymbol{\xi}}_{JS}$.

Let $\boldsymbol{X} \overset{d}{=} N_k(\boldsymbol{\xi}, \boldsymbol{I})$. Conditional on $W = w$ ($0 < w \leq 1$), let the distribution of $\boldsymbol{\xi}$ be multivariate normal with mean $\boldsymbol{0}$ and variance–covariance matrix $\left(\frac{1-w}{w}\right)\boldsymbol{I}$, and let λ be the unconditional distribution of W. Then, $\hat{\boldsymbol{t}}(\boldsymbol{X}) = (1 - \hat{w})\boldsymbol{X}$ is the Bayes compound rule, where

$$\hat{w} = \frac{\int_0^1 w^{(p+2)/2} e^{-wS/2} d\lambda(w)}{\int_0^1 w^{p/2} e^{-wS/2} d\lambda(w)} \qquad (\text{with } S = \sum_{i=1}^k X_i^2);$$

see, for example, Li and Bhoj (1986). [In fact, taking λ as the Beta$(\alpha, 1)$ distribution, Strawderman (1971) was the first one to use this method to obtain a family of Bayes minimax estimators.] The risk of $\hat{\boldsymbol{t}}$ is then

$$R(\boldsymbol{\xi}, \hat{\boldsymbol{t}}) = k - 2E(\boldsymbol{X} - \boldsymbol{\xi})^T \boldsymbol{X}\hat{w} + E\hat{w}^2 S.$$

Now, let λ have a differentiable density $f(w)$ and $g(w) = wf'(w)/f(w)$. Then, Li and Bhoj (1991) have shown that if $g(w)$ is bounded and does not change sign, then $\hat{\boldsymbol{t}}(\boldsymbol{X})$ is minimax. In particular:

(i) If $f(w) = cw^\alpha e^{\beta w}$, $\alpha \geq 0$, $\beta \geq 0$, then $g(w) = \alpha + \beta w$ and $\hat{\boldsymbol{t}}$ is minimax and dominates $\boldsymbol{X}$ for $k - 6 - 4(\alpha + \beta) \geq 0$.

(ii) If $f(w) = ce^{\beta(1-w)^2}$, then $g(w) = -2\beta(1 - w)w$ and $\hat{\boldsymbol{t}}$ is minimax and dominates $\boldsymbol{X}$ for $k - 6 + 2\beta \geq 0$ if $\beta \leq 0$ and for $k - 6 - \beta \geq 0$ if $\beta \geq 0$.

(iii) If $f(w) = cw^{\alpha-1}(1 - w)^{\beta-1}$, $\alpha > 0$, $\beta \geq 1$, then $g(w) = (\alpha - 1) - (\beta - 1)\left(\frac{w}{1-w}\right)$ and $\hat{\boldsymbol{t}}$ is minimax for $k - 4 - 2\alpha \geq 0$.

(iv) If $f(w) = cw^{\alpha-1}e^{-w/\beta}$, $\alpha > 0$, $0 < \beta \leq \infty$, then $g(w) = \alpha - 1 - \frac{w}{\beta}$ and $\hat{\boldsymbol{t}}$ is minimax for $k - 4 - 2\alpha \geq 0$.

An adaptive (minimax or near-minimax) empirical Bayes estimator of $\boldsymbol{\xi}$, under quadratic loss, has been developed by Judge, Hill, and Bock (1990).

Let $X_1, \ldots, X_n$ be a random sample from $N_k(\xi, V)$, where V is a positive definite symmetric matrix. Also, let $(\xi | \tau, A)$ be distributed as $N_k(\tau \mathbf{1}, A)$, where $\mathbf{1}$ is a $k \times 1$ column vector of 1's and A is a positive definite symmetric matrix. If V is known, the posterior distribution of ξ is

$$N_k \left((I - B)\bar{X} + B\tau\mathbf{1}, (I - B)\frac{1}{n}V \right),$$

where $B = \frac{1}{n}V\left(A + \frac{1}{n}V\right)^{-1}$, and thus the empirical Bayes estimator of ξ is

$$\hat{\xi} = (I - B)\bar{X} + B\tau\mathbf{1}.$$

When V is known, and assuming that $V = \sigma^2 I$ and $A = aI$, we get (Lindley's modification to the James–Stein estimator)

$$\hat{B} = \frac{(k-3)\sigma^2}{n\sum_{i=1}^{k}(\bar{X}_i - \bar{\bar{X}})^2}I \quad \text{and} \quad \hat{\tau} = \bar{\bar{X}} = \frac{1}{k}\sum_{i=1}^{k}\bar{X}_i,$$

where $\bar{X}_i$ is the ith component of $\bar{X}$. When $V = \sigma^2 I$ is unknown, we need to use $\hat{\sigma}^2 = \frac{1}{np+2}\operatorname{trace}(V)$; see Efron and Morris (1973). Alternatively, assuming $V = \operatorname{diag}(\sigma_1^2, \ldots, \sigma_k^2)$ and $A \propto V$, we have (the proportional prior estimator)

$$\hat{B} = \frac{k-3}{n\sum_{i=1}^{k}\frac{1}{\sigma_i^2}(\bar{X}_i - \bar{\bar{X}})^2}I.$$

When σ_i^2's are unknown, we need to use $\hat{\sigma}_i^2 = \frac{1}{n+2}V_{ii}$. Some other situations that have been considered are: $V = \operatorname{diag}(\sigma_1^2, \ldots, \sigma_k^2)$ and a constant prior $A = aI$ [see Carter and Rolph (1974) and Efron and Morris (1973)] and the interclass prior $A = (a - b)I + b\mathbf{1}\mathbf{1}^T$ [see Haff (1978)].

Note that in the case when $V = \sigma^2 I$ and $A = aI$, the above given estimator $\hat{B}$ is similar to the full Bayesian estimator (with a vague prior)

$$\tilde{B} = \frac{1}{Y}\frac{\Gamma(q+1, Y)}{\Gamma(q, Y)}I,$$

where $q = \frac{k-3}{2}$, $Y = \frac{n}{2\sigma^2}\sum_{i=1}^{k}(\bar{X}_i - \bar{\bar{X}})^2$, and $\Gamma(\cdot)$ is the complete gamma function; see Leonard (1974, 1976).

Press and Rolph (1986) generalized the above results to the model $A = cV$ and proposed the method-of-moments empirical Bayes estimator of ξ of the form

$$\hat{\xi} = (I - \hat{B})\bar{X} + \hat{B}\hat{\tau}\mathbf{1}, \tag{45.129}$$

where $\hat{B}$ and $\hat{\tau}$ are estimated by moments as

$$\hat{B} = \frac{1}{n\hat{c}+1}I \text{ and } \hat{\tau} = \frac{1}{k}\mathbf{1}^T\bar{X},$$

where

$$\hat{c} = \frac{1}{\text{trace}\left(\frac{V}{n}\right)}(\bar{X} - \hat{\tau}\mathbf{1})^T(\bar{X} - \hat{\tau}\mathbf{1}) - \frac{1}{n}.$$

If this $\hat{c}$ turns out to be negative, we need to set it as 0. Press and Rolph (1986) have also presented expressions for the posterior risks.

Chen, Eichenauer-Hermann, and Lehn (1990) considered the Γ-minimax estimation of the mean $\boldsymbol{\xi}$. Let Π be the set of all priors π for which the vector $\boldsymbol{m}(\pi)$ of first moments and the symmetric positive semidefinite matrix $\boldsymbol{M}(\pi)$ of second moments exist. Consider a nonnull convex subset Γ of priors of the form

$$\Gamma = \{\pi \in \Pi | \boldsymbol{m}(\pi) \in \boldsymbol{C}, \ \boldsymbol{M}(\pi) \leq \boldsymbol{M}\}$$

for some closed convex set $\boldsymbol{C} \subset \mathbb{R}^k$ and some symmetric positive definite matrix; here, $\leq$ denotes the partial ordering on the set of symmetric $k \times k$ matrices defined as $\boldsymbol{A} \leq \boldsymbol{B}$ if $\boldsymbol{B} - \boldsymbol{A}$ is positive semidefinite. A Γ-minimax estimator $\boldsymbol{\delta}^*$ minimizes the maximum Bayes risk with respect to the elements of Γ; that is,

$$\sup_{\pi \in \Pi} r(\pi, \boldsymbol{\delta}^*) = \inf_{\boldsymbol{\delta} \in \Delta} \sup_{\pi \in \Pi} r(\pi, \boldsymbol{\delta}),$$

where Δ is the set of all estimators $\boldsymbol{\delta}$,

$$r(\pi, \boldsymbol{\delta}) = \int_{\mathbb{R}^k} R(\boldsymbol{\xi}, \boldsymbol{\delta})\pi(d\boldsymbol{\xi}).$$

Here, $R(\boldsymbol{\xi}, \boldsymbol{\delta})$ denotes the risk function of $\boldsymbol{\delta}$ given by

$$\begin{aligned}
R(\boldsymbol{\xi}, \boldsymbol{\delta}) &= \frac{1}{(2\pi)^{k/2}|V|^{1/2}} \int_{\mathbb{R}^k} (\boldsymbol{\xi} - \boldsymbol{\delta}(\boldsymbol{X}))^T Q(\boldsymbol{\xi} - \boldsymbol{\delta}(\boldsymbol{X})) \\
&\quad \times \exp\left\{-\frac{1}{2}(\boldsymbol{X} - \boldsymbol{\xi})^T V^{-1}(\boldsymbol{X} - \boldsymbol{\xi})\right\} d\boldsymbol{X}
\end{aligned}$$

when arbitrary squared error is assumed, with Q being a symmetric positive definite matrix.

Denoting $\boldsymbol{E}_{\boldsymbol{M}} = \{\boldsymbol{m} \in \mathbb{R}^k | \boldsymbol{m}\boldsymbol{m}^T \leq \boldsymbol{M}\}$ and $\boldsymbol{C}_{\boldsymbol{M}} = \boldsymbol{C} \cap \boldsymbol{E}_{\boldsymbol{M}}$, the set Γ can be written as

$$\Gamma = \{\pi \in \Pi | \boldsymbol{m}(\pi) \in \boldsymbol{C}_{\boldsymbol{M}}, \ \boldsymbol{M}(\pi) \leq \boldsymbol{M}\}.$$

For $m \in C_M$, the normal priors $\pi_m = N_k(m, M - mm^T) \in \Gamma$ are considered. Berger (1985) has shown that under the squared error loss the linear estimator

$$(I - V(V + M - mm^T)^{-1})X + V(V + M - mm^T)^{-1}m$$

(45.130)

is the Bayes estimator of ξ with respect to the normal prior π_m.

Chen, Eichenauer-Hermann, and Lehn (1990) have also provided a characterization for the Γ-minimax estimator. From a corollary of their result, it follows that if $V = \sigma^2 I$, $Q = qI$ and $M = mI$ for some $\sigma, q, m > 0$ and $C = \{m \in \mathbb{R}^k | (m - a)^T(m - a) \leq r^2\}$ is a k-dimensional sphere with center $a \in \mathbb{R}^k$, radius $r > 0$, $r^2 < a^T a \leq (r + \sqrt{m})^2$ (i.e., $0 \notin C$ and $C_M \neq \emptyset$), then $\tilde{m} = \left(1 - \frac{r}{\sqrt{a^T a}}\right) a$ is Γ-minimax; and, if $C = \{m \in \mathbb{R}^k | m^T a \geq 1\}$ with $\frac{1}{m} \leq a^T a$ (i.e., $C_M \neq \emptyset$), then $\tilde{m} = \left(\frac{1}{a^T a}\right) a$ is Γ-minimax – in other words, within the set of all estimators, a linear estimator is Γ-minimax.

With $X_1, \ldots, X_n$ being a random sample from $N_k(\xi, V)$, Becker and Roux (1995) assumed the prior knowledge about the parameter (ξ, V) to be given by the natural conjugate Bayesian densities

$$f(\xi|V) = \frac{1}{(2\pi)^{k/2}|a^{-1}V|^{1/2}} \exp\left\{-\frac{a}{2}(\xi - \theta)^T V^{-1}(\xi - \theta)\right\},$$

where $a > 0$, and

$$f(V) = \left|\frac{1}{2}U\right|^{m/2} |V|^{(m-k-1)/2} \text{etr}\left(\frac{1}{2}UV\right) \Big/ \Gamma_k\left(\frac{m}{2}\right),$$

where V and U are both positive definite and $m \geq k$. Note that the prior of ξ is $N_k(\theta, V/a)$ and the prior of V is Wishart rather than inverted Wishart. Becker and Roux (1995) then carried out the Bayesian analysis using a multivariate quadratic loss function. The marginal posterior density for ξ turns out to be

$$p(\xi|\bar{X}, S) = C\left\{1 + (\xi - b^*)^T W^{-1}(\xi - b^*)\right\}^{(m-n-1)/2}$$
$$\times B_{\frac{m-n-1}{2}}\left(\frac{1}{4}(n+a)U\left\{(\xi - b^*)(\xi - b^*)^T + W\right\}\right),$$

where

$$\bar{X} = \frac{1}{n}\sum_{i=1}^{n} X_i, \quad S = \sum_{i=1}^{n}(X_i - \bar{X})(X_i - \bar{X})^T, \quad b^* = \frac{1}{n+a}(n\bar{X} + a\theta),$$

$B_v(\boldsymbol{D})$ is the Bessel function of the second kind with matrix argument,

$$\boldsymbol{W} = \frac{1}{n+a}\boldsymbol{S} + \frac{na}{(n+a)^2}(\bar{\boldsymbol{X}} - \boldsymbol{\theta})(\bar{\boldsymbol{X}} - \boldsymbol{\theta})^T,$$

and C is the normalizing constant for which an explicit expression has been given by Becker and Roux (1995). This posterior density is in fact the density of a multivariate Bessel distribution with matrix parameters; in addition, it can be shown that $E[\boldsymbol{\xi}|\bar{\boldsymbol{X}},\boldsymbol{S}] = \boldsymbol{b}^*$ which is the Bayes estimator of $\boldsymbol{\xi}$ under the multivariate quadratic loss function, the same had we used the inverted Wishart prior distribution for $\boldsymbol{V}$.

Similarly, the marginal posterior density for $\boldsymbol{V}$ becomes

$$p(\boldsymbol{V}|\bar{\boldsymbol{X}},\boldsymbol{S}) = C_1|\boldsymbol{V}|^{(m-n-k-1)/2}\,\mathrm{etr}\left(-\frac{1}{2}(n+a)\boldsymbol{V}^{-1}\boldsymbol{W}\right)\,\mathrm{etr}\left(-\frac{1}{2}\boldsymbol{V}\boldsymbol{U}\right)$$

for $n > k$, where C_1 is the normalizing constant. From this, it can be shown that

$$E[|\boldsymbol{V}| \mid \bar{\boldsymbol{X}},\boldsymbol{S}] = \frac{B_{\frac{m-n+2}{2}}\left(\frac{1}{4}(n+a)\boldsymbol{U}\boldsymbol{W}\right)}{B_{\frac{m-n}{2}}\left(\frac{1}{4}(n+a)\boldsymbol{U}\boldsymbol{W}\right)}\,2^{-k}(n+a)^k|\boldsymbol{W}|.$$

For the bivariate case, Becker and Roux (1995) presented explicit but complicated expressions for $E[V_{ij} \mid \bar{\boldsymbol{X}},\boldsymbol{S}]$ $(i,j = 1,2)$ in terms of multiple infinite series in powers of the elements of $\boldsymbol{U}$ and of $\boldsymbol{W}$.

Instead of the above given Wishart prior for $\boldsymbol{V}$, if we use the more flexible prior motivated by the above marginal posterior density

$$f(\boldsymbol{V}) \propto |\boldsymbol{V}|^{(\ell-k-1)/2}\,\mathrm{etr}\left(-\frac{1}{2}\boldsymbol{V}^{-1}\boldsymbol{T}\right)\,\mathrm{etr}\left(-\frac{1}{2}\boldsymbol{V}\boldsymbol{U}\right)$$

for $\boldsymbol{V}$, where $\boldsymbol{V}$, $\boldsymbol{T}$ and $\boldsymbol{U}$ are all positive definite matrices, the marginal posterior density for $\boldsymbol{V}$ becomes

$$p(\boldsymbol{V}|\bar{\boldsymbol{X}},\boldsymbol{S}) \propto |\boldsymbol{V}|^{(\ell-n-k-1)/2}\,\mathrm{etr}\left(-\frac{1}{2}(n+a)\boldsymbol{V}^{-1}\boldsymbol{W}^*\right)\,\mathrm{etr}\left(-\frac{1}{2}\boldsymbol{V}\boldsymbol{U}\right),$$

where

$$\boldsymbol{W}^* = \frac{1}{n+a}\boldsymbol{S} + \frac{1}{n+a}\boldsymbol{T} + \frac{na}{(n+a)^2}(\bar{\boldsymbol{X}} - \boldsymbol{\theta})(\bar{\boldsymbol{X}} - \boldsymbol{\theta})^T.$$

Krishnamoorthy (1991) considered the estimation of the common mean $\boldsymbol{\xi}$ from two independent samples with $\boldsymbol{X}_1,\ldots,\boldsymbol{X}_n$ coming from $N_k(\boldsymbol{\xi},\boldsymbol{V}_1)$ and $\boldsymbol{Y}_1,\ldots,\boldsymbol{Y}_n$ coming from $N_k(\boldsymbol{\xi},\boldsymbol{V}_2)$. Chiou and Cohen (1985) had earlier shown that between two unbiased estimators $\hat{\boldsymbol{\xi}}_1$ and $\hat{\boldsymbol{\xi}}_2$ of $\boldsymbol{\xi}$, under the covariance criterion, $\hat{\boldsymbol{\xi}}_1$ is preferable to $\hat{\boldsymbol{\xi}}_2$ if $\mathbf{Var}(\hat{\boldsymbol{\xi}}_2) - \mathbf{Var}(\hat{\boldsymbol{\xi}}_1)$ is a positive definite matrix. Krishnamoorthy (1991) assumed the quadratic loss function

$$L(\boldsymbol{\xi}, \hat{\boldsymbol{\xi}}) = (\boldsymbol{\xi} - \hat{\boldsymbol{\xi}})^T \mathbf{V}_1^{-1} (\boldsymbol{\xi} - \hat{\boldsymbol{\xi}}).$$

Consider the transformation $\mathbf{U}_i = \mathbf{X}_i$ and $\mathbf{W}_i = \mathbf{X}_i - \mathbf{Y}_i$ for $i = 1, 2, \ldots, n$. Let

$$\bar{\mathbf{U}} = \frac{1}{n} \sum_{i=1}^{n} \mathbf{U}_i, \; \mathbf{S_U} = \sum_{i=1}^{n} (\mathbf{U}_i - \bar{\mathbf{U}})(\mathbf{U}_i - \bar{\mathbf{U}})^T,$$

$$\bar{\mathbf{W}} = \frac{1}{n} \sum_{i=1}^{n} \mathbf{W}_i, \; \mathbf{S_W} = \sum_{i=1}^{n} (\mathbf{W}_i - \bar{\mathbf{W}})(\mathbf{W}_i - \bar{\mathbf{W}})^T.$$

Noting that the estimator

$$\hat{\boldsymbol{\xi}} = \bar{\mathbf{U}} - \mathbf{V}_1 (\mathbf{V}_1 + \mathbf{V}_2)^{-1} \bar{\mathbf{W}}$$

is the best unbiased estimator under the above quadratic loss function, in the case when $\mathbf{V}_1$ and $\mathbf{V}_2$ are unknown, Krishnamoorthy (1991) suggested replacing $\mathbf{V}_1(\mathbf{V}_1 + \mathbf{V}_2)^{-1}$ by $a\mathbf{S_U}\mathbf{S_W}^{-1}$, where a is a positive constant, leading to the estimator

$$\hat{\boldsymbol{\xi}}_a = \bar{\mathbf{U}} - a\mathbf{S_U}\mathbf{S_W}^{-1}\bar{\mathbf{W}}, \tag{45.131}$$

where a is chosen by minimizing the risk. $\hat{\boldsymbol{\xi}}_a$ is an unbiased estimator of $\boldsymbol{\xi}$, and its risk under the above quadratic loss is

$$
\begin{aligned}
R(\boldsymbol{\xi}, &\hat{\boldsymbol{\xi}}_a) \\
= \; & R(\boldsymbol{\xi}, \bar{\mathbf{X}}) + \frac{ac_1}{n} [\{a(n-1)(n-2)(n+k) - 2(n-1)(n-k-1) \\
& \times (n-k-4)\} \mathrm{trace}(\mathbf{D}) \\
& + \{2(n-k-1)(n-k-4) + a(2n+k)(k+2-2n) \\
& \quad - 2a(2n-k-4)\} \mathrm{trace}(\mathbf{D}^2) \\
& + 4a(3n-2k-4)\mathrm{trace}(\mathbf{D}^3) \\
& + a(9n-6p-10)\mathrm{trace}(\mathbf{D})\mathrm{trace}(\mathbf{D}^2) \\
& + \{2(n-k-1)(n-k-4) - a(4n^2 - 8n - k^2 - 4k + 2)\} \\
& \quad \times \{\mathrm{trace}(\mathbf{D})\}^2 \\
& + a(3n-2k-6)\{\mathrm{trace}(\mathbf{D})\}^3],
\end{aligned}
$$

where

$$c_1 = \frac{1}{(n-k-1)(n-k-2)(n-k-4)}$$

and

$$D = (V_1 + V_2)^{-1/2} V_1 (V_1 + V_2)^{-1/2}.$$

Moreover, for $n \geq k+5$ and $a_0 = \frac{(n-k-1)(n-k-4)}{(n-2)(n+k)}$, $R(\boldsymbol{\xi}, \hat{\boldsymbol{\xi}}_{a_0}) < R(\boldsymbol{\xi}, \bar{\boldsymbol{X}})$ for all positive definite matrices V_1 and V_2. Also, $\bar{\boldsymbol{Y}}$ is inadmissible under the loss $(\boldsymbol{\xi} - \hat{\boldsymbol{\xi}})^T V_2^{-1} (\boldsymbol{\xi} - \hat{\boldsymbol{\xi}})$ for any $k \geq 1$ and $n \geq k+5$.

Let $\mathcal{P}$ denote the set of all permutations on the integers $1, \ldots, n-1$, and $P = \{i_1, \ldots, i_{n-1}\}$ be an element in $\mathcal{P}$. Furthermore, let

$$S_{\boldsymbol{W}(P)} = \sum_{j=1}^{n-1} (\boldsymbol{X}_j - \boldsymbol{Y}_{i_j} - \bar{\boldsymbol{W}})(\boldsymbol{X}_j - \boldsymbol{Y}_{i_j} - \bar{\boldsymbol{W}})^T$$

and

$$\hat{\boldsymbol{\xi}}_{a(P)} = \bar{\boldsymbol{U}} - a S_{\boldsymbol{U}} S_{\boldsymbol{W}(P)}^{-1} \bar{\boldsymbol{W}}.$$

Then, the estimator

$$\hat{\boldsymbol{\xi}}_a^* = \frac{1}{(n-1)!} \sum_{P \in \mathcal{P}} \hat{\boldsymbol{\xi}}_{a(P)}$$

is invariant under the permutations of the observations, and is also unbiased for $\boldsymbol{\xi}$. $\hat{\boldsymbol{\xi}}_a^*$ also dominates $\hat{\boldsymbol{\xi}}_a$. The estimator $\hat{\boldsymbol{\xi}}_a^*$ is itself inadmissible since it is not a function of the minimal sufficient statistics. However, some numerical computations have revealed that the percentage relative improvement of $\hat{\boldsymbol{\xi}}_{a_0}$ over $\bar{\boldsymbol{X}}$ is quite significant for moderately large values of trace($\boldsymbol{D}$), where $\boldsymbol{D} = V_1 (V_1 + V_2)^{-1}$.

8.2 Estimation of V

The maximum likelihood estimator of V is $\hat{V} = \frac{1}{n} S$, as mentioned earlier in (45.120). Under the loss functions

$$L_1(\hat{V}, V) = \text{trace}\left(\hat{V} V^{-1} - I_{k \times k}^2\right)$$

and

$$L_2(\hat{V}, V) = \text{trace}(\hat{V} V^{-1}) - \ln |\hat{V} V^{-1}| - k,$$

the best affine equivariant estimators of V are

$$\hat{V}_1 = \frac{1}{n+k} S \quad \text{and} \quad \hat{V}_2 = \frac{1}{n-1} S,$$

respectively. As far as V^{-1} is concerned, the affine equivariant estimator is of the form

$$\widehat{V^{-1}} = \text{const} \times S^{-1}.$$

For estimating V, a substantial amount of work has been done in the last three decades. If we consider the class of estimators depending only on S, then the estimators V_1 and V_2 are admissible only for $k = 1$. In this univariate case, Stein (1964) considered a larger class of estimators depending on both $\bar{X}$ and S and proved that the affine equivariant estimators are inadmissible in this class. [Recall that the group of affine transformations is $(\bar{X}, S) \rightarrow (A\bar{X} + b, ASA^T)$ for a nonsingular $A_{k \times k}$ and $b \in \mathbb{R}^k$.] When $k \geq 2$, the estimators V_1 and V_2 are inadmissible in the class of estimators depending on S alone, and often the improved estimators have simple structure that provides substantial risk improvements over the best affine equivariant estimators; for details, see Pal (1993). Motivated by Stein (1964), when $k \geq 2$, one can also use $\bar{X}$ to obtain improvements, but such improved estimators have one undesirable property in that they are nonanalytic and hence again inadmissible; see Sinha (1987) and Sinha and Ghosh (1987). One thing is clear, however: The use of $\bar{X}$ is always helpful in estimating V since it also contains some information about V.

Haff (1977, 1979a,b, 1980) derived a better estimator of V^{-1} of the form

$$\widehat{V^{-1}}_* = \widehat{V^{-1}} + u(S)I_{k \times k},$$

where $u(S)$ is a suitable scalar valued function of S.

If $\bar{X}$ in the definition of S is replaced by $(1 - \frac{c}{T})\bar{X}$, then we arrive at

$$S_* = S + \frac{\text{const}}{T^2}\, \bar{X}\bar{X}^T.$$

Based on this, Pal and Elfessi (1995) considered estimators of the form

$$\hat{V}_{i(c,\alpha)} = \hat{V}_i + \frac{c}{T^\alpha}\, \bar{X}\bar{X}^T, \qquad i = 1, 2, \tag{45.132}$$

where $\hat{V}_i$ are the best affine equivariant estimators of V under the loss functions L_1 and L_2 as presented above, and c and α are suitable real constants. The estimator $\hat{V}_{1(c,\alpha)}$ is uniformly better than V_1, provided that $1 \leq \alpha < 1 + \frac{k}{4}$ and c is such that

$$0 < c \leq 2^\alpha \left(\frac{d_1 - d_3/(n + k)}{d_2}\right) n^{1-\alpha}\varepsilon(k, \alpha),$$

where

$$d_1 = 2^\alpha \Gamma\left(\frac{n - k}{2} + \alpha\right) \Big/ \Gamma\left(\frac{n - k}{2}\right),$$

$$d_2 = 2^{2\alpha}\Gamma\left(\frac{n-k}{2}+2\alpha\right)\Big/\Gamma\left(\frac{n-k}{2}\right),$$

$$d_3 = (n-1+2\alpha)d_1,$$

and

$$\varepsilon(k,\alpha) = \frac{\Gamma\left(\frac{k}{2}+1-\alpha\right)}{\Gamma\left(\frac{k}{2}+2(1-\alpha)\right)}, \quad \alpha \geq 1.$$

The optimal value of c (for $\alpha = 1$) that gives the maximum risk improvement is $c = \frac{k-1}{(n+k)(n-k+2)}$. Although the relative risk improvement is small, it is more than what is obtained by the nonsmooth estimators (using both $\bar{X}$ and S) derived earlier by Kubokawa (1989) and Perron (1990).

Similar results obtained by Pal and Elfessi (1995) indicated that $\hat{V}_{2(c,\alpha)}$ is uniformly better than V_2 under the loss function L_2, provided that $1 \leq \alpha < 1+\frac{k}{2}$ and $0 < c < c_0$, where

$$c_0 = \frac{2^{\alpha-1}\Gamma\left(\frac{k}{2}\right)}{d_1 n^{\alpha-1}\Gamma\left(\frac{k}{2}-(\alpha-1)\right)}$$
$$\times\left\{1 - \frac{\frac{1}{n-1}d_1 d_4 \Gamma\left(\frac{k}{2}-(\alpha-1)\right)\Gamma\left(\frac{k}{2}+(\alpha-1)\right)}{\left\{\Gamma\left(\frac{k}{2}\right)\right\}^2}\right\}$$

and

$$d_4 = \frac{\Gamma\left(\frac{n-k}{2}\right)-(\alpha-1)}{2^{\alpha-1}\Gamma\left(\frac{n-k}{2}\right)}.$$

These improved estimators of V, however, are not location-invariant which is a resultant of using the James–Stein structure.

Based on a random sample $X_1,\ldots,X_n$ from $N_k(\boldsymbol{\xi},V)$, where $\boldsymbol{\xi}$ is known, the sample sum of squares and products matrix S is sufficient for V. Dickey, Lindley, and Press (1985) assumed an inverted-Wishart prior distribution on V [i.e., $V^{-1} \stackrel{d}{=}$ Wishart$(\frac{1}{\delta-2}\Omega^{-1}, k, v = \delta + k - 1)$], and the density of V is assumed to be

$$p(V|\Omega,\delta) \propto |\Omega|^{-(\delta+k-1)/2}|V|^{-(\delta+2k)/2}\exp\left\{-\frac{\delta-2}{2}\,\mathrm{trace}(\Omega V^{-1})\right\}$$

for $\delta > 2$, and $E[V] = \Omega$. These authors have provided a lengthy justification that the inverted-Wishart distribution for V has reasonable consistency properties in addition to being a natural conjugate. Assume additionally that all the variables are expected to have the same variance and that the off-diagonal elements of $E(V)$ are all the same; that is, $E(V) = \Omega$ has all its diagonal elements equal to σ^2 and all its off-diagonal elements equal to $\rho\sigma^2$, where $\rho > -1/(k-1)$. In other words, the matrix of the inverted-Wishart distribution is a scale matrix which has intraclass structure. For mathematical convenience, the precision matrix $\Lambda = V^{-1}$ is used rather than V. The posterior density is derived to be

$$p(V|S) \propto |V|^{-n_1/2} \left\{ a_0 + \frac{\delta-2}{k} \mathbf{1}^T V^{-1} \mathbf{1} \right\}^{-(n_1+\delta+k-1)/2}$$
$$\cdot \left\{ b_0 + (\delta-2)[\text{trace}(AV^{-1})] \right\}^{-n_2/2} \exp\left\{ -\frac{1}{2}\,\text{trace}(SV^{-1}) \right\},$$

where $n_1 = n + \delta + k - 1$, $n_2 = (\delta + k - 1)(k - 1) + b_1$, $A = I - \frac{1}{k}\mathbf{1}\mathbf{1}^T$, $\alpha = \sigma^2\{1 + (k-1)\rho\}$ and $\beta = \sigma^2(1-\rho)$, with α and β having joint density

$$p(\alpha, \beta) \propto \alpha^{(a_1/2)-1} \beta^{(b_1/2)-1} e^{-(a_0\alpha + b_0\beta)/2}, \qquad a_0, b_0, a_1, b_1 > 0.$$

Since the posterior density is quite complicated, Dickey, Lindley, and Press (1985) provided the following two forms of estimating equations for the posterior mode:

$$\hat{V} = \frac{S}{n_1} + \sigma_0^2\,\{(1-\rho_0)I + \rho_0\mathbf{1}\mathbf{1}^T\}, \qquad (45.133)$$

where

$$\sigma_0^2 = \frac{a(k-1)+bk}{k}, \qquad \rho_0 = \frac{bk-a}{a(k-1)+bk},$$

$$a = \frac{n_2(\delta-2)}{n_1\{b_0 + (\delta-2)\text{trace}(A\hat{\Lambda})\}} \quad \text{and} \quad b = \frac{(\delta+k+a_1-1)(\delta-2)}{n_1\{a_0k + (\delta-2)\mathbf{1}^T\hat{\Lambda}\mathbf{1}\}};$$

$$\hat{V} = \frac{S}{n_1} + aA + b\mathbf{1}\mathbf{1}^T. \qquad (45.134)$$

These equations need to be solved by iteration. The authors have also observed that

$$\hat{V} = \alpha_0\hat{V}_{\text{MLE}} + (1-\alpha_0)\Delta, \qquad 0 < \alpha_0 < 1, \qquad (45.135)$$

where $\hat{V}_{\text{MLE}} = \frac{1}{n}S$ and Δ denotes the positive semidefinite intraclass covariance matrix $\Delta = \sigma_1^2\{(1-\rho_0)I + \rho_0\mathbf{1}\mathbf{1}^T\}$, with $\sigma_1^2 = n_1\sigma_0^2/(n_1 - n)$.

The weight of the maximum likelihood estimator is proportional to the sample size and is given by $\alpha_0 = n/n_1$. It turns out that approximately

$$\hat{V} \doteq \alpha_0 \hat{V}_{\text{MLE}} + (1 - \alpha_0)\hat{V}'_{\text{Mode}},$$

where $\hat{V}'_{\text{Mode}}$ denotes the mode of the prior distribution.

This development is, of course, based on the assumption that $\boldsymbol{\xi}$ is known which, without loss of generality, is taken to be $\mathbf{0}$. The case in which $\boldsymbol{\xi}$ is unknown has been handled by Press (1975), who has derived point estimators of $\boldsymbol{\xi}$ and V as the joint mode of the posterior distribution.

8.3 Estimation of Correlations

The maximum likelihood estimator of the variance of X_{jt} is

$$\hat{\sigma}_{jj}^2 = \nu_{jj} = \frac{1}{n}\, S_{jj}$$

and the estimator of the correlation between X_{it} and X_{jt}:

$$\hat{\rho}_{ij} = \frac{\nu_{ij}}{\sqrt{\nu_{ii}\nu_{jj}}} = \frac{S_{ij}}{\sqrt{S_{ii}S_{jj}}}. \tag{45.136}$$

This is the ordinary sample product moment correlation. Here, we note a correction to reduce the bias in $\hat{\rho}_{ij}$ as an estimator of ρ_{ij}, suggested by Olkin and Pratt (1958). This consists of using the modified estimator

$$\hat{\rho}_{ij}\left\{1 + \frac{1 - \hat{\rho}_{ij}^2}{2(n-4)}\right\}. \tag{45.137}$$

Olkin and Pratt (1958) also give a table of corrective multipliers to apply to $\hat{\rho}_{ij}$. This is reproduced as our Table 45.5. The corrected estimator is the minimum variance unbiased estimator of ρ_{ij}.

Tallis (1967) has studied estimation of parameters of multivariate normal distributions from grouped data. He found that the univariate formulas

$$\text{sample variance} \; - \frac{1}{12}(\text{group width})^2$$

can be used for each variate, and the sample covariances do not need correction. He found that the variance of an estimated correlation (ρ) is increased by approximately

$$(12n)^{-1}(1 - \rho^4)$$

$\times$(sum of squares of standardized group widths of the two variates).

TABLE 45.5

Corrective Multipliers for $\hat{\rho}_{ij}$

n	0	.1	.2	.3	.4	.5	.6	.7	.8	.9	1.0
3	∞	10.000	5.000	3.333	2.500	2.000	1.667	1.429	1.250	1.111	1
5	1.571	1.478	1.398	1.327	1.265	1.209	1.159	1.114	1.073	1.035	1
7	1.178	1.173	1.161	1.144	1.125	1.105	1.083	1.062	1.041	1.020	1
9	1.104	1.103	1.098	1.090	1.080	1.068	1.056	1.042	1.028	1.014	1
11	1.074	1.073	1.070	1.065	1.058	1.050	1.042	1.032	1.022	1.011	1
13	1.057	1.056	1.054	1.050	1.046	1.040	1.033	1.026	1.018	1.009	1
15	1.046	1.046	1.044	1.041	1.038	1.033	1.027	1.021	1.015	1.008	1
17	1.039	1.039	1.037	1.035	1.032	1.028	1.023	1.018	1.013	1.006	1
19	1.034	1.033	1.032	1.030	1.028	1.024	1.020	1.016	1.011	1.006	1
21	1.030	1.029	1.028	1.027	1.024	1.022	1.018	1.014	1.010	1.005	1
23	1.027	1.026	1.025	1.024	1.022	1.019	1.016	1.013	1.009	1.005	1
25	1.024	1.024	1.023	1.022	1.020	1.018	1.015	1.012	1.008	1.004	1
27	1.022	1.022	1.021	1.020	1.018	1.016	1.014	1.011	1.007	1.004	1
29	1.020	1.020	1.019	1.018	1.017	1.017	1.012	1.010	1.007	1.004	1
31	1.019	1.018	1.018	1.017	1.015	1.014	1.012	1.009	1.006	1.003	1
∞	1	1	1	1	1	1	1	1	1	1	1

8.4 Estimation Under Missing Data

Problems of estimation peculiar to multivariate distributions arise when the sets of observations on some individuals are incomplete. We shall give a fairly detailed account of ways of dealing with this problem for the bivariate normal distribution in Chapter 46. In the general multivariate normal case, there is a wide variety of possible patterns and complete analysis would be lengthy; see, for example, Anderson (1957), Afifi and Elashoff (1966–1969), Lord (1955), Trawinski and Bargmann (1964), Bhargava (1975), Anderson and Olkin (1985), Haider (1991), Little and Rubin (1987), Jinadasa and Tracy (1992), and Fujisawa (1995).

Anderson and Olkin (1985) reviewed various methods of obtaining maximum likelihood estimators of the parameters of the multivariate normal distribution and pointed out that there is no single method that provides answer for all models. They also obtained the maximum likelihood estimators of the parameters with a two-step monotone missing data pattern using matrix derivatives. Additional discussion on this problem have been provided by Bhargava (1975), Jinadasa and Tracy (1992), and Fujisawa (1995). Extending the results of Anderson and Olkin (1985), Jinadasa and Tracy (1992) derived in explicit form the maximum likelihood estimator of $\boldsymbol{\xi}$ and $\boldsymbol{V}$ with an r-step monotone missing data pattern. Let $\boldsymbol{X} \overset{d}{=} N_k(\boldsymbol{\xi}, \boldsymbol{V})$, $\boldsymbol{X}_i = (\boldsymbol{X})_i$, the subvector of $\boldsymbol{X}$ containing the first k_i components of $\boldsymbol{X}$, and similarly $\boldsymbol{\xi}_i = (\boldsymbol{\xi})_i$ is the subvector of $\boldsymbol{\xi}$ containing the first k_i components of $\boldsymbol{\xi}$ (for $i = 1, \ldots, r$), where $k = k_1 > k_2 > \cdots > k_r > 0$. Let there be n_1 observations on $\boldsymbol{X}_1, n_2$ observations on $\boldsymbol{X}_2, \ldots, n_r$ observations on $\boldsymbol{X}_r$, with $n_1 > k$. If $\boldsymbol{X}_{ij}$ denotes the jth observation on $\boldsymbol{X}_i$, the sample of observations $\boldsymbol{X}_{ij}$ ($i = 1, \ldots, r$; $j = 1, \ldots, n_i$) is called a *monotone sample* [Srivastava and Carter (1983)], *monotone missing data pattern* [Little and Rubin (1987)], and an *r-step monotone missing data pattern* [Jinadasa and Tracy (1992)].

Let $V_1 = V$ and, for $i < j$, let $(V_i)_j$ be the principal submatrix of V_i of order $k_j \times k_j$ for $i = 1, \ldots, r$ and $j = i + 1, \ldots, r$. Then, in an obvious notation,

$$V_i = (V_1)_i, \quad V_1 = V = \begin{pmatrix} V_i & V_{i2} \\ V_{i2}^T & V_{i3} \end{pmatrix},$$

$$V_i = \begin{pmatrix} V_{i+1} & V_{(i,2)} \\ V_{(i,2)}^T & V_{(i,3)} \end{pmatrix}, \quad i = 1, 2, \ldots, r - 1.$$

Under this setup, Jinadasa and Tracy (1992) derived the maximum likelihood estimators of $\boldsymbol{\xi}$ and V as follows:

$$\hat{\boldsymbol{\xi}} = \sum_{i=1}^{r} \hat{\boldsymbol{f}}_i \quad \text{with } \hat{\boldsymbol{f}}_1 = \boldsymbol{d}_1, \ \hat{\boldsymbol{f}}_i = T_1 T_2 \cdots T_i \boldsymbol{d}_i \ (i = 2, \ldots, r),$$

where

$$\boldsymbol{d}_1 = \bar{\boldsymbol{X}}_1,$$

$$\boldsymbol{d}_i = \frac{n_i}{N_{i+1}} \left\{ \bar{\boldsymbol{X}}_i - \frac{1}{N_i} \sum_{j=1}^{i-1} n_j (\bar{\boldsymbol{X}}_j)_i \right\} \ (i = 2, \ldots, r),$$

$$\bar{\boldsymbol{X}}_i = \frac{1}{n_i} \sum_{j=1}^{n_i} \boldsymbol{X}_{ij}, \ T_1 = I_1,$$

$$T_{i+1} = \begin{pmatrix} I_{i+1} \\ \Sigma_{(i,2)}^T & \Sigma_{i+1}^{-1} \end{pmatrix} \ (i = 1, \ldots, r - 1),$$

and $N_\ell = \sum_{i=\ell}^{r} n_i$; $\hat{\boldsymbol{\xi}}$ is the solution of the equation

$$\sum_{i=1}^{r} n_i \begin{pmatrix} V_i^{-1}(\bar{\boldsymbol{X}}_i - \boldsymbol{\xi}_i) \\ \boldsymbol{0}_i \end{pmatrix} = \boldsymbol{0},$$

where $\boldsymbol{0}_i$ is the null vector of order $k - k_i$;

$$\hat{V} = \frac{1}{n_1} H_1 + \sum_{i=2}^{r} \frac{1}{N_{i+1}} F_i \left(H_i - \frac{n_i}{N_i} L_{i-1,1} \right) F_i^T,$$

where

$$H_1 = E_1, \ H_i = E_i + \frac{N_i N_{i+1}}{n_i} \boldsymbol{d}_i \boldsymbol{d}_i^T \quad (i = 2, 3, \ldots, r),$$

$$E_i = \sum_{j=1}^{n_i} (\boldsymbol{X}_{ij} - \bar{\boldsymbol{X}}_i)(\boldsymbol{X}_{ij} - \bar{\boldsymbol{X}}_i)^T,$$

$$\begin{aligned}
\boldsymbol{L}_1 &= \boldsymbol{H}_1, \ \boldsymbol{L}_i = (\boldsymbol{L}_{i-1})_i + \boldsymbol{H}_i & (i = 2, \ldots, r),
\end{aligned}$$

$$\begin{aligned}
\boldsymbol{L}_{i1} &= (\boldsymbol{L}_i)_{i+1}, \ \boldsymbol{L}_i = \begin{pmatrix} \boldsymbol{L}_{i1} & \boldsymbol{L}_{i2} \\ \boldsymbol{L}_{i2}^T & \boldsymbol{L}_{i3} \end{pmatrix} & (i = 1, \ldots, r-1),
\end{aligned}$$

$$\begin{aligned}
\boldsymbol{G}_1 &= \boldsymbol{I}_1, \ \boldsymbol{G}_{i+1} = \begin{pmatrix} \boldsymbol{I}_{i+1} & \\ \boldsymbol{L}_{i2}^T & \boldsymbol{L}_{i1}^{-1} \end{pmatrix} & (i = 1, \ldots, r-1),
\end{aligned}$$

and

$$\boldsymbol{F}_1 = \boldsymbol{G}_1, \ \boldsymbol{F}_i = \boldsymbol{F}_{i-1}\boldsymbol{G}_i \qquad (i = 2, \ldots, r).$$

$\hat{\boldsymbol{V}}$ is the solution of

$$\sum_{i=1}^r n_i \begin{pmatrix} \boldsymbol{V}_i^{-1} & 0 \\ 0 & 0 \end{pmatrix} = \sum_{i=1}^r \begin{pmatrix} \boldsymbol{V}_i^{-1} \boldsymbol{H}_i \boldsymbol{V}_i^{-1} & 0 \\ 0 & 0 \end{pmatrix},$$

where $\boldsymbol{0}$'s are null matrices of appropriate dimensions. Jinadasa and Tracy (1992) and Fujisawa (1995) have also shown that $\hat{\boldsymbol{\xi}}$ is an unbiased estimator of $\boldsymbol{\xi}$.

Dahel, Giri and Lapage (1985) and Dahel (1987) considered the maximum likelihood estimation of $\boldsymbol{\xi}$ with additional information. The estimator is computed on the basis of three independent samples: The first sample is drawn from all k variables while the other two samples are drawn on the first k_1 and the last $k_2 = k - k_1$ variables, respectively. Specifically, let $\boldsymbol{X} \stackrel{d}{=} N_k(\boldsymbol{\xi}, \boldsymbol{V})$, where $\boldsymbol{V}$ is positive definite. Let us partition $\boldsymbol{X}, \boldsymbol{\xi}$ and $\boldsymbol{V}$ as $\boldsymbol{X} = (\boldsymbol{X}_{(1)}^T, \boldsymbol{X}_{(2)}^T)^T$, $\boldsymbol{\xi} = (\boldsymbol{\xi}_{(1)}^T, \boldsymbol{\xi}_{(2)}^T)^T$ and $\boldsymbol{V} = \begin{pmatrix} \boldsymbol{V}_{11} & \boldsymbol{V}_{12} \\ \boldsymbol{V}_{21} & \boldsymbol{V}_{22} \end{pmatrix}$, where $\boldsymbol{X}_{(i)}$ and $\boldsymbol{\xi}_{(i)}$ are subvectors of dimension k_i and $\boldsymbol{V}_{ij}$ is a submatrix of dimension $k_i \times k_j$ for $i, j = 1, 2$. Now, let $\boldsymbol{X}_1, \ldots, \boldsymbol{X}_n$ be a random sample on $\boldsymbol{X}$, $\boldsymbol{Z}_{1(1)}, \ldots, \boldsymbol{Z}_{n_1(1)}$ be a random sample on $\boldsymbol{X}_{(1)}$, and $\boldsymbol{Z}_{1(2)}, \ldots, \boldsymbol{Z}_{n_2(2)}$ be a random sample on $\boldsymbol{X}_{(2)}$. Furthermore, let

$$\bar{\boldsymbol{X}} = \frac{1}{n}\sum_{i=1}^n \boldsymbol{X}_i, \quad \boldsymbol{S} = \sum_{i=1}^n (\boldsymbol{X}_i - \bar{\boldsymbol{X}})(\boldsymbol{X}_i - \bar{\boldsymbol{X}})^T,$$

$$\boldsymbol{Y} = \sqrt{n}\,\bar{\boldsymbol{X}}, \quad \bar{\boldsymbol{Z}}_{(j)} = \frac{1}{n_j}\sum_{i=1}^{n_j} \boldsymbol{Z}_{i(j)},$$

$$\boldsymbol{S}_{(j)} = \sum_{i=1}^{n_j} (\boldsymbol{Z}_{i(j)} - \bar{\boldsymbol{Z}}_{(j)})(\boldsymbol{Z}_{i(j)} - \bar{\boldsymbol{Z}}_{(j)})^T,$$

$$\boldsymbol{T}_{(j)} = \sqrt{n_j}\bar{\boldsymbol{Z}}_{(j)}, \ h_j = \sqrt{\frac{n_j}{n}} \quad \text{for } j = 1, 2,$$

and

$$\boldsymbol{\eta} = \sqrt{n}\boldsymbol{\xi}, \ \boldsymbol{B} = \boldsymbol{V}_{12}\boldsymbol{V}_{22}^{-1}, \text{ and } \boldsymbol{V}_{11\cdot2} = \boldsymbol{V}_{11} - \boldsymbol{B}\boldsymbol{V}_{22}\boldsymbol{B}^T.$$

Let us partition the vector $\boldsymbol{Y}$ similar to $\boldsymbol{X}$, the vector $\boldsymbol{\eta}$ as $\boldsymbol{\xi}$, and the matrix $\boldsymbol{S}$ similar to $\boldsymbol{V}$. Then, Dahel, Giri, and Lapage (1985) have shown that the maximum likelihood estimator $\hat{\boldsymbol{\eta}}$ of $\boldsymbol{\eta}$ is

$$
\begin{aligned}
\hat{\boldsymbol{\eta}}_{(1)} \;=\; & \boldsymbol{A}^{-1}\Big[\boldsymbol{V}_{11\cdot2}^{-1}\big\{(1+h_2^2)\boldsymbol{Y}_{(1)} + h_2\boldsymbol{B}(\boldsymbol{T}_{(2)} - h_2\boldsymbol{Y}_{(2)}) + h_1\boldsymbol{T}_{(1)}\big\} \\
& + h_1 h_2^2 \boldsymbol{V}_{11}^{-1}\boldsymbol{T}_{(1)}\Big]
\end{aligned}
$$

and

$$
\hat{\boldsymbol{\eta}}_{(2)} = \frac{h_1}{1+h_2^2}\, \boldsymbol{V}_{22}\boldsymbol{B}^T\boldsymbol{V}_{11}^{-1}\left(\boldsymbol{T}_{(1)} - h_1\hat{\boldsymbol{\eta}}_{(1)}\right) + \frac{1}{1+h_2^2}\left(\boldsymbol{Y}_{(2)} + h_2\boldsymbol{T}_{(2)}\right),
$$

where

$$
\boldsymbol{A} = (1+h_1^2+h_2^2)\boldsymbol{V}_{11\cdot2}^{-1} + h_1^2 h_2^2 \boldsymbol{V}_{11}^{-1}.
$$

Dahel (1987) has further shown that $\hat{\boldsymbol{\eta}}$ is an extended Bayes estimator and, hence, minimax with respect to the quadratic loss function $(\hat{\boldsymbol{\eta}}-\boldsymbol{\eta})^T(\hat{\boldsymbol{\eta}}-\boldsymbol{\eta})$.

By assuming that the data on first s and the last ℓ components of an $(s+r+k+\ell)$-dimensional normal random vector are missing for m observations while no components are missing for n other independent observations, Provost (1990) has derived explicit expressions for the maximum likelihood estimators of the mean vector $\boldsymbol{\xi}$ and the variance–covariance matrix $\boldsymbol{V}$. He has also proposed the likelihood ratio statistic to test the independence between the first $r+s$ and the last $k+\ell$ components.

Krishnamoorthy and Pannala (1999) have proposed a simple approximate confidence region for $\boldsymbol{\xi}$ when the available data has a monotone pattern. Specifically, they have assumed the data to consist of n independent observations and m additional observations on the first k_1 components; or, equivalently, m observations are missing at random on the last k_2 components. They have then assessed the validity of this approximation through Monte Carlo simulations. This approximate $100(1-\alpha)\%$ confidence region for $\boldsymbol{\xi}$ is essentially of the form

$$
(\hat{\boldsymbol{\xi}} - \boldsymbol{\xi})^T(\widehat{\mathbf{Var}(\hat{\boldsymbol{\xi}})})^{-1}(\hat{\boldsymbol{\xi}} - \boldsymbol{\xi}) \le dF_{k,c}(\alpha),
$$

where $0 < \alpha < 0.5$, $n > k+4$, $F_{k,c(\alpha)}$ denotes the $100(1-\alpha)$th percentage point of a central F-distribution with (k,c) degrees of freedom, and c and d are chosen by matching the first two moments. Here, $\widehat{\mathbf{Var}(\hat{\boldsymbol{\xi}})}$ is an estimator of $\mathbf{Var}(\hat{\boldsymbol{\xi}})$, $\hat{\boldsymbol{\xi}}$ being the MLE of $\boldsymbol{\xi}$, where the unknown parameters appearing in $\mathbf{Var}(\hat{\boldsymbol{\xi}})$ are replaced by their MLEs. An explicit expression has been provided by Krishnamoorthy and Pannala (1999). Evidently, the confidence region is an ellipsoid centered at $\hat{\boldsymbol{\xi}}$.

8.5 Estimation Under Special Structures

If it is known that

(i) all variances (σ^2) are the same and

(ii) all correlations (ρ) are the same,

then there is an orthogonal transformation

$$\boldsymbol{Y} = \boldsymbol{X}\boldsymbol{\Gamma} \qquad (\boldsymbol{\Gamma}\boldsymbol{\Gamma}^T = \boldsymbol{I})$$

such that $Y_1, \ldots, Y_k$ are mutually independent and

$$\text{var}(Y_1) = \{1 + (k-1)\rho\}\sigma^2; \quad \text{var}(Y_j) = (1-\rho)\sigma^2 \qquad (j \geq 2).$$

Applying this transformation to $\boldsymbol{X}_1, \ldots, \boldsymbol{X}_n$, we obtain $\boldsymbol{Y}_1, \ldots, \boldsymbol{Y}_n$. The ratio

$$\frac{1}{n-1}\sum_{j=1}^{n}(Y_{1j} - \bar{Y}_1)^2 \quad \text{to} \quad \frac{1}{(k-1)(n-1)}\sum_{i=2}^{k}\sum_{j=1}^{n}(Y_{ij} - \bar{Y}_i)^2$$

is distributed as (F with $(n-1), (k-1)(n-1)$ degrees of freedom) multiplied by $\{[1 + (k-1)\rho]/(1-\rho)\}$. Since

$$\frac{1}{n-1}\sum_{j=1}^{n}(Y_{1j} - \bar{Y}_1)^2 = S^2\{1 + (k-1)R\}$$

and

$$\frac{1}{(k-1)(n-1)}\sum_{i=2}^{k}\sum_{j=1}^{n}(Y_{ij} - \bar{Y}_i)^2 = S^2(1-R)$$

with

$$S^2 = \frac{1}{k(n-1)}\sum_{i=1}^{k}\sum_{j=1}^{n}(X_{ij} - \bar{X}_i)^2$$

and

$$R = \{k(k-1)(n-1)S^2\}^{-1}\sum_{i \neq \ell}\sum\sum_{j=1}^{n}(X_{ij} - \bar{X}_i)(X_{\ell j} - \bar{X}_\ell),$$

we have

$$\frac{(1-\rho)\{1 + (k-1)R\}}{(1-R)\{1 + (k-1)\rho\}}$$

distributed as $F_{n-1,(k-1)(n-1)}$. Confidence intervals for ρ with $100\alpha\%$ confidence coefficient are thus given by

$$1 - k(1 - R)F^*_{1-\alpha_1}\{1 + (k-1)[R + (1-R)F^*_{1-\alpha_1}]\}^{-1}$$

and

$$1 - k(1 - R)F^*_{\alpha_2}\{1 + (k-1)[R + (1-R)F^*_{\alpha_2}]\}^{-1}$$

with

$$F^*_\varepsilon = F_{n-1,(k-1)(n-1),\varepsilon} \quad \text{and} \quad \alpha_1 + \alpha_2 = \alpha;$$

see Geisser (1964). Confidence regions for $\boldsymbol{\xi}$ can be derived by noting that

$$\frac{n(\bar{Y}_1 - \eta_1)^2}{S^2\{1 + (k-1)R\}} \quad \text{and} \quad \frac{n\sum_{i=2}^{k}(\bar{Y}_i - \eta_i)^2}{(k-1)S^2(1-R)},$$

where $\boldsymbol{\eta} = \boldsymbol{\xi}\boldsymbol{\Gamma}^T$ are independently distributed as $F_{1,n-1}$ and $F_{k-1,(k-1)(n-1)}$, respectively.

We take note of formulas for maximum likelihood estimators in the highly symmetrical case when it is known that, in addition to (i) and (ii),

(iii) all expected values (ξ) are the same.

The formulas given by Kusunori (1967) are as follows:

$$\hat{\xi} = \bar{X} = \frac{1}{kn}\sum_{i=1}^{k}\sum_{j=1}^{n}X_{ij},$$

$$\hat{\sigma}^2 = S^2 = \frac{1}{kn}\sum_{i=1}^{k}\sum_{j=1}^{n}(X_{ij} - \bar{X})^2,$$

$$\hat{\rho} = [k(k-1)nS^2]^{-1}\sum_{i<i'}\sum\sum_{j=1}^{n}(X_{ij} - \bar{X})(X_{i'j} - \bar{X}).$$

Doktorov (1969) gave the following expression for the maximum likelihood estimator of σ_1, when the values of all the correlations and all other standard deviations $\sigma_2, \sigma_3, \ldots, \sigma_m$ are known:

$$\hat{\sigma}_1 = \frac{1}{2}\left(\sum_{j=2}^{k}\rho^{1j}\hat{v}_{ij}\sigma_j^{-1} + \left(\sum_{j=2}^{k}\rho^{1j}\hat{v}_{1j}\sigma_j^{-1}\right)^2 + 4\rho^{11}\hat{v}_{11}\right),$$

where

$$(\rho^{ij}) = \boldsymbol{R}^{-1} \quad \text{and} \quad \hat{v}_{ij} = \frac{1}{n}\sum_{\ell=1}^{n}(X_{i\ell} - \bar{X}_i)(X_{j\ell} - \bar{X}_j).$$

For n large,
$$n \operatorname{var}(\hat{\sigma}_1) \doteq \sigma_1^2 (1 + \rho^{11})^{-1}.$$

Further special cases are discussed by Styan (1968).

Krishnamoorthy and Rohatgi (1990) considered the unbiased estimation of the common mean. Let $(U_1, \ldots, U_{k+1})^T$ have a $(k+1)$-variate normal distribution with mean $(\xi, \ldots, \xi)^T$ and variance–covariance matrix $\boldsymbol{V}$. Then, the maximum likelihood estimation of ξ is equivalent to the estimation of the intercept in multiple regression with random regressors. Let $\boldsymbol{A} = (a_{ij})$ be a $(k+1) \times (k+1)$ matrix, where

$$a_{ij} = \begin{cases} 1 & \text{for } j = 1, \ i = 1, \ldots, k+1, \\ -1 & \text{for } j = i = 2, \ldots, k+1, \\ 0 & \text{otherwise.} \end{cases}$$

Consider the transformation

$$(Y, \boldsymbol{X}^T)^T = (Y, X_1, \ldots, X_k)^T = \boldsymbol{A}(U_1, \ldots, U_{k+1}).$$

Clearly,

$$(Y, \boldsymbol{X}^T)^T \stackrel{d}{=} N_{k+1}\left(\begin{pmatrix} \xi \\ 0 \\ \vdots \\ 0 \end{pmatrix}, \boldsymbol{W} \right), \qquad \text{where } \boldsymbol{W} = \boldsymbol{A}\boldsymbol{V}\boldsymbol{A}^T.$$

Let $\boldsymbol{W} = \begin{pmatrix} W_{YY} & \boldsymbol{W}_{XY}^T \\ \boldsymbol{W}_{XY} & \boldsymbol{W}_{XX} \end{pmatrix}$. Now, given n independent observations on $(Y, \boldsymbol{X}^T)$, let

$$(\bar{Y}, \bar{\boldsymbol{X}}^T) = \frac{1}{n} \sum_{i=1}^{n} (Y_i, \boldsymbol{X}_i^T) \quad \text{(the sample mean vector)}$$

and

$$\boldsymbol{S} = \begin{pmatrix} S_{YY} & \boldsymbol{S}_{XY}^T \\ \boldsymbol{S}_{XY} & \boldsymbol{S}_{XX} \end{pmatrix}$$
$$= \begin{pmatrix} \sum_{i=1}^{n}(Y_i - \bar{Y})^2 & \sum_{i=1}^{n}(Y_i - \bar{Y})(\boldsymbol{X}_i - \bar{\boldsymbol{X}})^T \\ \sum_{i=1}^{n}(Y_i - \bar{Y})(\boldsymbol{X}_i - \bar{\boldsymbol{X}}) & \sum_{i=1}^{n}(\boldsymbol{X}_i - \bar{\boldsymbol{X}})(\boldsymbol{X}_i - \bar{\boldsymbol{X}})^T \end{pmatrix}$$

(the sample of sum of squares and products matrix).

The maximum likelihood estimator of ξ is then $\bar{Y}(\boldsymbol{b}) = \bar{Y} - \boldsymbol{b}^T \bar{\boldsymbol{X}}$, where $\boldsymbol{b} = \boldsymbol{S}_{XX}^{-1} \boldsymbol{S}_{XY}$. This estimator is unbiased,

$$\operatorname{var}(\bar{Y}(\boldsymbol{b})) = \frac{1}{n}\left(1 + \frac{k}{n-k-2}\right) W_{YY \cdot X} \quad \text{(for } n > k+2),$$

where $W_{YY\cdot\mathbf{x}} = W_{YY}(1 - \rho_{Y\cdot\mathbf{x}}^2)$ and $\rho_{Y\cdot\mathbf{x}}^2 = \dfrac{\mathbf{W}_{\mathbf{X}Y}^T \mathbf{W}_{\mathbf{xx}}^{-1} \mathbf{W}_{\mathbf{X}Y}}{W_{YY}}$. It is easy to check that $\mathrm{var}(\bar{Y}(\mathbf{b})) < \mathrm{var}(\bar{Y})$ if and only if $n \geq k + 3$ and $\rho_{Y\cdot\mathbf{x}}^2 > \frac{k}{n-2}$. These results were also obtained by Baranchik (1973) and Gleser (1987).

Note that the estimator $\mathbf{b} = \mathbf{S}_{\mathbf{xx}}^{-1} \mathbf{S}_{\mathbf{X}Y}$ uses the cross-product matrix $\sum_{i=1}^n (\mathbf{X}_i - \bar{\mathbf{X}})(\mathbf{X}_i - \bar{\mathbf{X}})^T$ even though the mean vector of $\mathbf{X}$ is $\mathbf{0}$. So, Krishnamoorthy and Rohatgi (1990) suggested using

$$\mathbf{b}_0 = c \left(\sum_{i=1}^n \mathbf{X}_i \mathbf{X}_i^T \right)^{-1} \mathbf{S}_{\mathbf{X}Y} = c \left(\mathbf{S}_{\mathbf{xx}} + n\bar{\mathbf{X}}\bar{\mathbf{X}}^T \right)^{-1} \mathbf{S}_{\mathbf{X}Y},$$

where c is a constant, which yields the estimator

$$\hat{\xi}_c = \bar{Y} - \mathbf{b}_0^T \bar{\mathbf{X}} = \mathbf{Y} - c \left(\frac{1}{1 + T^2} \right) \mathbf{b}^T \bar{\mathbf{X}}, \tag{45.138}$$

where $T^2 = n\bar{\mathbf{X}}^T \mathbf{S}_{\mathbf{xx}}^{-1} \bar{\mathbf{X}}$. This estimators $\hat{\xi}_c$ is also unbiased, and

$$\begin{aligned}
\mathrm{var}(\hat{\xi}_c) =\ & \frac{1}{n} W_{YY} - \frac{2c(n-k)}{n(n+2)} W_{YY}\, \rho_{Y\cdot\mathbf{x}}^2 \\
& + \frac{c^2(n-k)}{n^2} \left\{ \frac{k}{n+2} + \frac{n-2k}{n+4} \rho_{Y\cdot\mathbf{x}}^2 \right\} W_{YY}.
\end{aligned}$$

For $c > 0$, $\mathrm{var}(\hat{\xi}_c) < \mathrm{var}(\bar{Y})$ if and only if $\rho_{Y\cdot\mathbf{x}}^2 > \frac{k}{n\left(\frac{k}{n} + \frac{2}{c} - \frac{n-k+2}{n+4}\right)}$. Choosing $c_0 = \frac{2n(n+4)}{9k(n+4)+n(n-k+2)}$, we have $\mathrm{var}(\hat{\xi}_{c_0}) < \mathrm{var}(\bar{Y})$ for $\rho_{Y\cdot\mathbf{x}}^2 > 0.1$, and $\hat{\xi}_{c_0}$ dominates both $\bar{Y}$ and $\bar{Y}(\mathbf{b})$ for $0.1 < \rho_{Y\cdot\mathbf{x}}^2 < 0.5$. Furthermore, the estimator $\hat{\xi}_1$ has a smaller variance than $\bar{Y}$ over a wide range of parameter values. The worst case is when $n = k + 3$ (the smallest possible value of n), in which case $\mathrm{var}(\hat{\xi}) < \mathrm{var}(\bar{Y})$ includes the set $|\rho_{Y\cdot\mathbf{x}}| > 0.58$.

Consider the estimation of $\boldsymbol{\xi}$ based on a random sample $\mathbf{X}_1, \ldots, \mathbf{X}_n$ from $N_k(\boldsymbol{\xi}, \mathbf{V})$ when it is suspected that $\xi_1 = \cdots = \xi_k = \xi$ (unknown). In this case, we have the unrestricted maximum likelihood estimator (UMLE) as

$$\tilde{\boldsymbol{\xi}} = (\tilde{\xi}_1, \ldots, \tilde{\xi}_k)^T, \qquad \text{where } \tilde{\xi}_i = \frac{1}{n} \sum_{j=1}^n X_{ij}, \ i = 1, \ldots, k,$$

and we obtain the restricted (or pooled) maximum likelihood estimator (RMLE) as

$$\hat{\boldsymbol{\xi}} = (\hat{\xi}_n, \ldots, \hat{\xi}_n)^T, \qquad \text{where } \hat{\xi}_n = \frac{1}{k} \sum_{i=1}^k \bar{X}_i = \frac{1}{k} \mathbf{1}^T \tilde{\boldsymbol{\xi}}.$$

The RMLE performs better than the UMLE when $\xi_1 = \cdots = \xi_k = \xi$, but if the components of $\boldsymbol{\xi}$ are different the RMLE becomes biased and inefficient.

Ahmed and Badahdah (1992) discussed a preliminary test for the null hypothesis $H_0 : \xi_1 = \cdots = \xi_k = \xi$ based on Hotelling's T_n^2 statistic, $T_n^2 = n\tilde{\boldsymbol{\xi}}^T \boldsymbol{C}^T \boldsymbol{S}^{-1} \boldsymbol{C}\tilde{\boldsymbol{\xi}}$, where

$$(n-1)\boldsymbol{S} = \boldsymbol{C}\left\{\sum_{i=1}^{n}(\boldsymbol{X}_i - \tilde{\boldsymbol{\xi}})(\boldsymbol{X}_i - \tilde{\boldsymbol{\xi}})^T\right\}\boldsymbol{C}^T,$$

where $\boldsymbol{C} = \boldsymbol{I}_{k \times k} - \frac{1}{k}\boldsymbol{1}_k\boldsymbol{1}_k^T$ is an idempotent matrix of rank $k-1$. Their preliminary test maximum likelihood estimator (PTMLE) is

$$\hat{\boldsymbol{\xi}}_P = \tilde{\boldsymbol{\xi}} - (\tilde{\boldsymbol{\xi}} - \hat{\xi}_n\boldsymbol{1}_k) \cdot I(T_n^2 \leq t_\alpha^2), \qquad (45.139)$$

where $I(A)$ is the indicator function of the set A, and t_α^2 is the critical value of the T_n^2 statistic (which has an F-distribution modified by a constant). This estimator was also studied by DaSilva and Han (1984) and by Ali and Saleh (1990).

Ahmed and Badahdah (1992), in addition, also suggested the estimator $\gamma\tilde{\boldsymbol{\xi}} + (1-\gamma)\hat{\xi}_n\boldsymbol{1}_k$ as a shrinkage restricted maximum likelihood estimator (SRMLE) of $\boldsymbol{\xi}$. The value of γ may be completely specified by the experimenter. The SRMLE yields smaller mean square error at and near $\xi_1 = \cdots = \xi_k = \xi$ at the cost of poor performance for the rest of the parameter space. However, the SRMLE provides a wider range than that of the RMLE in which it dominates the UMLE. These authors also proposed a shrinkage preliminary test estimator (SPTMLE) as

$$\hat{\boldsymbol{\xi}}_{\text{SP}} = \tilde{\boldsymbol{\xi}}I(T_n^2 \geq t_\alpha^2) + \{\gamma\tilde{\boldsymbol{\xi}} + (1-\gamma)\hat{\xi}_n\boldsymbol{1}_k\}I(T_n^2 < t_\alpha^2). \qquad (45.140)$$

The bias of this estimator is

$$E[\hat{\boldsymbol{\xi}}_{\text{SP}} - \hat{\boldsymbol{\xi}}] = -(1-\gamma)\boldsymbol{\delta}H_{q,m}(F^*; \Delta),$$

where $\Delta = n\boldsymbol{\delta}^T(\boldsymbol{CVC}^T)^{-1}\boldsymbol{\delta}$, $\boldsymbol{\delta} = \boldsymbol{C}\boldsymbol{\xi}$, $q = k+1$, $m = n - q = n - k - 1$, $H_{v_1,v_2}(\cdot;\Delta)$ is the cumulative distribution function of a noncentral F-distribution with degrees of freedom (v_1, v_2) and noncentrality parameter Δ, and $F^* = \frac{q-2}{q}F_{q-2,m,\alpha/2}$ with $F_{q-2,m,\alpha/2}$ being the upper $\alpha/2$ percentage point of a central F-distribution with degrees of freedom $(q-2, m)$. The mean square error matrix of $\hat{\boldsymbol{\xi}}_{\text{SP}}$ is

$$\begin{aligned}
\boldsymbol{\Gamma}^* &= E[n(\hat{\boldsymbol{\xi}}_{\text{SP}} - \boldsymbol{\xi})(\hat{\boldsymbol{\xi}}_{\text{SP}} - \boldsymbol{\xi})^T] \\
&= \boldsymbol{\Gamma}_1 - \sigma^2(1-\rho)(1-\gamma^2)\boldsymbol{C}H_{q,m}(F^*; \Delta) \\
&\quad + n\boldsymbol{\delta}\boldsymbol{\delta}^T\{2(1-\gamma)H_{q,m}(F^*; \Delta) - (1-\gamma^2)H_{q+2,m}(F_0; \Delta)\},
\end{aligned}$$

where

$$\Gamma_1 = E[n(\tilde{\xi} - \xi)(\tilde{\xi} - \xi)^T] = \sigma^2\{(1 - \rho)I_{k \times k} + \rho J\},$$

$$J = 1_k 1_k^T, \quad \text{and} \quad F_0 = \frac{q - 2}{q + 2} F_{q+2, m, \alpha/2}.$$

The optimal value of the shrinkage constant γ may be found, but it is not very useful since it depends on the unknown quantity Δ. Ahmed and Badahdah (1992) have recommended SRMLE over SPTMLE since it dominates over a wide range of the parameter space.

Wang (1991) examined admissibility of an estimator of the mean ξ when the variance–covariance matrix is of the form $\sigma^2 V$, where V is a known positive definite matrix while σ^2 is unknown. Let us take the quadratic loss function

$$L(\delta, \xi, \sigma^2) = (\delta - \xi)^T Q(\delta - \xi),$$

where Q is a positive definite matrix, and the risk

$$R(\delta, \xi, \sigma^2) = E_{\xi, \sigma^2}[L(\delta(X), \xi, \sigma^2)],$$

with $\delta(X)$ being an estimator of ξ. With X distributed as $N_k(\xi, \sigma^2 V)$, where V is a known positive definite matrix, necessary and sufficient conditions for $AX + a$ to be admissible for ξ are

$a \in \mathcal{C}(A - I),$ where $\mathcal{C}(A - I)$ denotes the column space of $A - I$,

$AV = VA^T$, $AVA^T \leq AV$, and the rank of $A - I \geq k - 2$.

A lot of discussion has focused on the estimation of σ^2 based on a random sample $X_1, \ldots, X_n$ from $N_k(\xi, \sigma^2 I)$, and also the estimation of $\theta = \sigma^{2\alpha}$, where $\alpha > 0$ is known. The equivariant estimator of θ is of the form $\hat{\theta}_c = cS^\alpha$, where

$$S = \sum_{i=1}^n \sum_{j=1}^k (X_{ij} - \bar{X}_j)^2,$$

$\bar{X}_j$ is the jth element of $\bar{X} = \frac{1}{n}\sum_{i=1}^n X_i$, and $c > 0$ is a real constant. Let $m - 1 = k(n - 1)$. Two popular choices for loss functions to estimate are

$$L_1(\hat{\theta}, \theta) = \left(\frac{\hat{\theta}}{\theta} - 1\right)^2 \quad \text{(quadratic loss function)}$$

and

$$L_2(\hat{\theta}, \theta) = \left(\frac{\hat{\theta}}{\theta}\right) - \ln\left(\frac{\hat{\theta}}{\theta}\right) - 1 \quad \text{(entropy loss function)}.$$

An unbiased estimator of σ^2 is $\hat{\sigma}_U^2 = S/(m-1)$, while the best affine equivariant estimator under L_1 is $\hat{\sigma}_1^2 = S/(m+1)$ which is inadmissible. Stein's (1964) improved estimator of σ^2 is

$$\hat{\sigma}_{1(S)}^2 = \min\left\{\frac{S}{m+1}, \frac{S + n\|\bar{X} - \xi_0\|^2}{m+k+1}\right\}, \quad (45.141)$$

where $\xi_0 \in \mathbb{R}^k$ is known (based on prior knowledge of ξ). Observe that $\left\{\frac{n\|\bar{X} - \xi_0\|^2}{S} > \frac{k}{m+1}\right\}$ can be treated as a rejection region for testing $H_0 : \xi = \xi_0$ vs. $H_1 : \xi \neq \xi_0$. If H_0 is rejected, then one uses the usual best affine equivariant estimator; otherwise, the variance is estimated by $\{S + n\|\bar{X} - \xi_0\|^2\}/(m+k+1)$, which is the best affine equivariant estimator when $\xi = \xi_0$. Unfortunately, $\hat{\sigma}_{1(S)}^2$ is also inadmissible since it is nonanalytic. In addition to Stein-type estimator under L_1, there are some other estimators available. Brewster and Zidek's (1974) analytic estimator [based on Brown's (1968) idea] of the form

$$\hat{\sigma}_{1(BZ)}^2 = \frac{S}{m+1}(1 - \phi_1(W)), \quad (45.142)$$

where

$$W = n\|\bar{X} - \xi_0\|^2/\{S + n\|\bar{X} - \xi_0\|^2\}$$

and

$$\phi_1(w) = \frac{\frac{2}{m+k+1} w^{k/2}(1-w)^{(m+1)/2}}{\int_0^w u^{(k/2)-1}(1-u)^{(m+1)/2} \, du},$$

is admissible and is a generalized Bayes estimator under a certain prior given by Rukhin and Ananda (1992).

Strawderman's (1974) minimax estimator is of the form

$$\hat{\sigma}_{1(ST)}^2 = \frac{S}{m+1}\{1 - U^\delta \varepsilon(U)\}, \quad (45.143)$$

where $U = 1 - W$, $\delta \geq 0$, $\varepsilon(\cdot)$ is nondecreasing with $0 \leq \varepsilon(U) \leq D(\delta)$, and

$$D(\delta) = \min\left\{\frac{1}{1+\delta}, \kappa\right\},$$

where $\kappa = \dfrac{2B\left(\frac{m-1}{2}+2+\delta,\frac{k}{2}\right)\left[B\left(\frac{m-1}{2}+1,1\right)-B\left(\frac{m-1}{2}+\delta+1,\frac{k}{2}\right)\right]B\left(\frac{m+k-1}{2}+1,1\right)}{B\left(\frac{m-1}{2}+1,1\right)B\left(\frac{m-1}{2}+2\delta+2,\frac{k}{2}\right)}$. Some of
the estimators in Strawderman's class are admissible, and Brewster and
Zidek's estimator is one of them.

Under the loss function L_2, the best affine equivariant estimator of σ^2 is
the unbiased estimator $\hat{\sigma}_U^2 = S/(m-1)$ and of $\theta = \sigma^{2\alpha}$ is $\hat{\theta}_2 = c_2 S^\alpha$, where
$c_2 = 2^{-\alpha}\Gamma\left(\frac{m-1}{2}\right)/\Gamma\left(\frac{m-1}{2}+\alpha\right)$. The Stein-type estimator (nonanalytic)
is [Sinha and Ghosh (1987)]

$$\hat{\sigma}_{2(S)}^2 = \min\left\{\frac{S}{m-1}, \frac{s+n\|\bar{X}-\xi_0\|^2}{m+k-1}\right\}. \tag{45.144}$$

The Brewster–Zidek-type estimator under the loss L_2 is

$$\hat{\sigma}_{2(\mathrm{BZ})}^2 = \frac{S}{m-1}(1-\phi_2(W)), \tag{45.145}$$

where

$$\phi_2(w) = \frac{\frac{2}{m+k-1}\,w^{k/2}(1-w)^{(m-3)/2}}{\int_0^w u^{(k/2)-1}(1-u)^{(m-3)/2}du}.$$

Pal and Ling (1995) have shown that the structure of Strawderman-
type improved minimax estimator of σ^2 (actually of $\sigma^{2\alpha}$) under the loss
function L_2 is the same as the corresponding estimator under the loss
function L_1.

Dey and Gelfand (1989) considered the estimation of the covariance
matrix of the form $\boldsymbol{V} = \sum_{i=1}^{\ell} \theta_i \boldsymbol{W}_i$, where $\boldsymbol{W}_i$'s form a known complete
orthogonal set of projection matrices and θ_i's are distinct eigenvalues of
$\boldsymbol{V}$. An example for this form is the equicorrelated case when

$$\boldsymbol{V} = \sigma^2\{(1-\rho)\boldsymbol{I}_{k\times k} + \rho\boldsymbol{J}\},$$

where $\boldsymbol{I}_{k\times k}$ is the identity matrix of dimension $k \times k$ and $\boldsymbol{J}$ is the $k \times k$
matrix of 1's. Dey and Gelfand (1989) took decision-theoretic approach
using the loss functions

$$\mathrm{trace}(\hat{\boldsymbol{V}} - \boldsymbol{V})^2 \qquad \text{(the squared error loss)}$$

and

$$\mathrm{trace}(\hat{\boldsymbol{V}}\boldsymbol{V}^{-1}) - \log|\hat{\boldsymbol{V}}\boldsymbol{V}^{-1}| - k \qquad \text{(entropy-like loss)},$$

which, for the estimators of the form $\sum_{i=1}^{\ell}\hat{\theta}_i \boldsymbol{W}_i$, become

$$L(\hat{\boldsymbol{\theta}},\boldsymbol{\theta}) = \sum_{i=1}^{\ell} k_i(\hat{\theta}_i - \theta_i)^2$$

and

$$L(\hat{\boldsymbol{\theta}}, \boldsymbol{\theta}) = \sum_{i=1}^{\ell} k_i \left\{ \frac{\hat{\theta}_i}{\theta_i} - \log \left(\frac{\hat{\theta}_i}{\theta_i} \right) - 1 \right\},$$

respectively, where $k_i = \text{rank}(\boldsymbol{W}_i)$, $i = 1, \ldots, \ell$, $\sum_{i=1}^{\ell} k_i = k$, and $\boldsymbol{\theta} = (\theta_1, \ldots, \theta_\ell)^T$. In the above-mentioned equicorrelated model, we have

$$\theta_1 = \sigma^2(1 - \rho), \qquad \theta_2 = \sigma^2\{1 + (k - 1)\rho\},$$

$$\boldsymbol{W}_1 = \boldsymbol{I}_{k \times k} - \frac{1}{k}\boldsymbol{J}, \qquad \boldsymbol{W}_2 = \frac{1}{k}\boldsymbol{J}, \qquad k_1 = k - 1 \text{ and } k_2 = 1.$$

Dey and Gelfand (1989) then showed that the estimation of the patterned covariance matrix $\boldsymbol{V}$ is dual to simultaneous estimation of scale parameters of independent χ^2 distributions.

The derivation of a confidence region of the mean vector $\boldsymbol{\xi}$ with fixed width d and confidence coefficient $1 - \alpha$ has been treated by many authors. Using Chow–Robbins (1965) theory, Srivastava (1967) derived an asymptotic confidence interval in the case when $\boldsymbol{V}$ is a general unknown variance–covariance matrix. Khan (1968) considered the case when $\boldsymbol{V} = \text{diag}(\sigma_1^2, \ldots, \sigma_k^2)$ where σ_i's are unknown. Mukhopadhyay and Al-Mousawi (1986) presented an asymptotic expansion for the coverage probability, in the case when $\boldsymbol{V} = \sigma^2\boldsymbol{W}$ with $\boldsymbol{W}$ being a known positive definite matrix, by applying Woodroofe's (1982) renewal theory. For a completely unknown matrix, Srivastava and Bhargava (1979) developed an asymptotic expansion for the coverage probability using martingale theory. Hyakutake, Takada, and Aoshima (1995) extended the result of Mukhopadhyay and Al-Mousawi (1986) to the case when the variance–covariance matrix $\boldsymbol{V}$ is of the so-called *intraclass correlation structure*, namely, $\boldsymbol{V} = \sigma^2\{(1 - \rho)\boldsymbol{I} + \rho\boldsymbol{J}\}$, where $\boldsymbol{I}$ is a $k \times k$ identity matrix and $\boldsymbol{J}$ is a $k \times k$ matrix with all its entries as 1. Recently, Nagao (1996) discussed the derivation of fixed width confidence region when $\boldsymbol{V}$ is a linear combination of known symmetric matrices with the combining coefficients being unknown, that is,

$$\boldsymbol{V} = \sigma_1\boldsymbol{A}_1 + \cdots + \sigma_\ell\boldsymbol{A}_\ell,$$

where $\boldsymbol{A}_i$'s are known symmetric matrices of rank k_i, $\sum_{i=1}^{\ell} \boldsymbol{A}_i = \boldsymbol{I}$, $\sum_{i=1}^{\ell} k_i = k$, and σ_i's are all unknown. Evidently, this includes many of the models mentioned above as special cases.

8.6   Estimation of Functions of $\boldsymbol{\xi}$ and $\boldsymbol{V}$

Sometimes, it is desired to estimate particular functions of the parameters. In particular, we may wish to examine

$$P(\Omega) = (2\pi)^{-k/2}|\boldsymbol{V}|^{-1/2} \int \cdots \int_{\Omega} \exp\left\{-\frac{1}{2}\,(\boldsymbol{x}-\boldsymbol{\xi})^T \boldsymbol{V}^{-1}(\boldsymbol{x}-\boldsymbol{\xi})\right\} d\boldsymbol{x}.$$

Lumel'skii (1968) has shown that the minimum variance unbiased estimator of $P(\Omega)$, based on a random sample of size n $(> k)$, is

$$\begin{aligned}
P(\Omega) &= [\pi(n-1)]^{-k/2}|\boldsymbol{V}|^{-1/2}\Gamma\left(\frac{1}{2}(n-1)\right)\left\{\Gamma\left(\frac{1}{2}(n-k-1)\right)\right\}^{-1} \\
&\quad \times \int \cdots \int_{\Omega} \{f(\boldsymbol{x})\}^{(1/2)(n-k-3)}\, d\boldsymbol{x},
\end{aligned}$$

where

$$f(\boldsymbol{x}) = \begin{cases}
1 - \frac{1}{n-1}\,(\boldsymbol{x}-\bar{\boldsymbol{X}})^T \boldsymbol{V}^{-1}(\boldsymbol{x}-\bar{\boldsymbol{X}}) \\
\quad \text{if } \boldsymbol{V} \text{ is positive definite and} \\
\quad (\boldsymbol{x}-\bar{\boldsymbol{X}})^T \boldsymbol{V}^{-1}(\boldsymbol{x}-\bar{\boldsymbol{X}}) < (n-1)^{-1}, \\
0 \quad \text{otherwise.}
\end{cases}$$

This may be regarded as a generalization of the corresponding result in Chapter 13 of Johnson, Kotz, and Balakrishnan (1994); see also Kabe (1968).

Ghurye and Olkin (1969) obtained formulas for minimum variance estimators of multivariate normal *density functions* [i.e., of (45.3)] under various conditions. Some of their results are summarized in Table 45.6.

Lumel'skii and Sapozhnikov (1969) also gave these formulas for the cases (i), (ii), and (iv).

Murray (1979) considered the problem of estimating the parametric density function $p(\boldsymbol{y}|\boldsymbol{\xi}, \boldsymbol{V})$ of a k-dimensional multivariate normal distribution with mean vector $\boldsymbol{\xi}$ and variance–covariance matrix $\boldsymbol{V}$, based on observed data $\boldsymbol{x}_1, \boldsymbol{x}_2, \ldots, \boldsymbol{x}_n$. Using the maximum likelihood estimates $\hat{\boldsymbol{\xi}}$ and $\hat{\boldsymbol{V}}$ in (45.120), the *estimative fit* is

$$\hat{p}(\boldsymbol{y}|\boldsymbol{x}_1, \ldots, \boldsymbol{x}_n) = p_{\boldsymbol{X}}\left(\hat{\boldsymbol{\xi}}, \frac{n}{n-1}\hat{\boldsymbol{V}}\right).$$

The *Bayesian predictive* method uses the estimate

$$\begin{aligned}
&\hat{p}_B(\boldsymbol{y}|\boldsymbol{x}_1, \ldots, \boldsymbol{x}_n) \\
&= \int_{\boldsymbol{\xi}, \boldsymbol{V}} p(\boldsymbol{y}|\boldsymbol{\xi}, \boldsymbol{V}) p(\boldsymbol{\xi}, \boldsymbol{V}|\boldsymbol{x}_1, \ldots, \boldsymbol{x}_n)\, d\boldsymbol{\xi}\, d\boldsymbol{V},
\end{aligned}$$

where $p(\boldsymbol{\xi}, \boldsymbol{V}|\boldsymbol{x}_1, \ldots, \boldsymbol{x}_n)$ is a Bayesian posterior density function of $(\boldsymbol{\xi}, \boldsymbol{V})$ based on a prior distribution $\pi(\boldsymbol{\xi}, \boldsymbol{V})$ and the data $\boldsymbol{x}_1, \ldots, \boldsymbol{x}_n$. If we use the vague prior for $(\boldsymbol{\xi}, \boldsymbol{V})$ that is proportional to $|\boldsymbol{V}|^{-(k+1)/2} d\boldsymbol{\xi}\, d\boldsymbol{V}$, then

$$\hat{p}_B(\boldsymbol{y}|\boldsymbol{x}_1, \ldots, \boldsymbol{x}_n) = f_{t_k}\left(n-1, \hat{\boldsymbol{\xi}}, \frac{n+1}{n-1}\,\hat{\boldsymbol{V}}\right),$$

where $f_{t_k}(a, \boldsymbol{b}, \boldsymbol{c})$ denotes the density function of a k-dimensional Student's t-distribution given by

$$\frac{\Gamma\left(\frac{a+1}{2}\right)}{\pi^{k/2}\Gamma\left(\frac{a-k+1}{2}\right)|a\boldsymbol{C}|^{1/2}\{1+(\boldsymbol{t}-\boldsymbol{b})^T(a\boldsymbol{C})^{-1}(\boldsymbol{t}-\boldsymbol{b})\}^{(a+1)/2}};$$

see Aitchison and Dunsmore (1975, p. 29).

TABLE 45.6
Minimum Variance Estimators of Multivariate Normal Density Functions

Known Parameters	Estimator
(i) $\boldsymbol{\xi} = \boldsymbol{\xi}_0$	$(2\pi)^{-k/2}[K_{n-1-k}/K_{n-1}]\|\boldsymbol{S}\|^{-\frac{1}{2}(n-k-2)}$ $\times \|\boldsymbol{S} - (\boldsymbol{x} - \boldsymbol{\xi}_0)(\boldsymbol{x} - \boldsymbol{\xi}_0)^T\|^{\frac{1}{2}(n-k-3)}$
(ii) $\boldsymbol{V} = \boldsymbol{V}_0$	$(2\pi)^{-k/2}\|\boldsymbol{V}_0\|^{-1/2}(1 - n^{-1})^{-k/2}$ $\times \exp\left\{-\frac{n}{2(n-1)}\,(\boldsymbol{x} - \bar{\boldsymbol{X}})^T \boldsymbol{V}_0^{-1}(\boldsymbol{x} - \bar{\boldsymbol{X}})\right\}$
(iii) $\boldsymbol{V} = \sigma^2\boldsymbol{V}_0$ (σ unknown)	$(2\pi)^{-k/2}2^{k/2}\Gamma\left(\frac{1}{2}(n-1)k\right)\left[\Gamma\left(\frac{1}{2}(n-2)k\right)\right]^{-1}$ $\times [\mathrm{tr}\,\boldsymbol{V}_0^{-1}\boldsymbol{S}]^{-\frac{1}{2}[(n-1)k-2]}$ $\times \|\boldsymbol{S} - n(\boldsymbol{x} - \bar{\boldsymbol{X}})\boldsymbol{V}_0^{-1}(\boldsymbol{x} - \bar{\boldsymbol{X}})^T\|^{\frac{1}{2}[(n-2)k-2]}$
(iv) None	$(2\pi)^{-k/2}[K_{n-1-k}/K_{n-1}](1 - n^{-1})^{-k/2}$ $\times \|\boldsymbol{S}\|^{-\frac{1}{2}(n-k-2)}$ $\times \left|\boldsymbol{S} - \frac{n-1}{n}\,(\boldsymbol{x} - \bar{\boldsymbol{X}})(\boldsymbol{x} - \bar{\boldsymbol{X}})^T\right|^{\frac{1}{2}(n-k-3)}$

Notes.

1. $K_\nu = \left[2^{(1/2)k\nu}\pi^{(1/4)k(k-1)}\prod_{j=1}^{k}\Gamma\left(\frac{1}{2}(\nu - j + 1)\right)\right]^{-1}$.

2. When a matrix of form $\boldsymbol{S} - \boldsymbol{A}$ is not positive definite, its determinant is to be replaced by zero.

Murray (1979) compared the estimative fit and the Bayesian predictive estimates for $k = 1$ and 8 when $n = 4, 6, 11, 14, 20$, and 50. For sample sizes slightly larger than $k + 3$, the superiority of the Bayesian predictive fit over the estimative fit has been clearly shown, with the effect being more marked for large k.

In the case of incomplete data, Murray (1979) first considered the nested case where we have n complete observations on all k variables and m extra observations on the first k_1 variables. It is possible in this case to evaluate $E[\log\{p(\boldsymbol{y}|\boldsymbol{\xi}, \boldsymbol{V})/\hat{p}_B(\boldsymbol{y}|\boldsymbol{x})\}]$, and it turns out that for the Bayesian predictive methods, the fit is very largely determined by the weakest link— the variables that are least observed. In the case of an arbitrary deletion pattern, it is not possible to evaluate the predictive density analytically. Among various alternatives, Murray (1979) has suggested to determine the maximum likelihood estimates of $\boldsymbol{\xi}$ and $\boldsymbol{V}$ based on the available data and then (in analogy to the case of complete data) to use the estimate

$$\hat{p}_B(\boldsymbol{y}|\text{data}) = f_{t_k}\left(\ell - 1, \hat{\boldsymbol{\xi}}, \frac{\ell + 1}{\ell - 1}\,\hat{\boldsymbol{V}}\right)$$

for some suitable choice of ℓ. No details have been given on the selection of ℓ. For certain data patterns in the three-dimensional case, this method turns out to be superior (in the sense of Kullback–Leibler information measure) than the nested pattern approach (after deleting a minimal amount of data to obtain a nested pattern).

Masuda (1980) discussed the estimation of the probability (p) that an observed value $\boldsymbol{x}$ is contained in domain A for $N_k(\boldsymbol{\xi}, \boldsymbol{V})$. The unique UMVUE, $\hat{p}(\bar{\boldsymbol{X}}, \boldsymbol{S})$, of this probability p is

$$\hat{p}(\bar{\boldsymbol{X}}, \boldsymbol{S}) = \int_{T_1 A} \frac{\Gamma\left(\frac{n-1}{2}\right)}{\Gamma\left(\frac{n-2-k+1}{2}\right)\pi^{k/2}}\,(1 - \boldsymbol{z}^T \boldsymbol{z})^{(n-2-k-1)/2}\,d\boldsymbol{z},$$

$$(45.146)$$

where

$$T_1 A = \left\{\boldsymbol{z} = \boldsymbol{D}\left(\sqrt{\lambda_i}\right)\boldsymbol{P}\boldsymbol{D}\left(\sqrt{\frac{n}{(n-1)S_{ii}}}\right)(\boldsymbol{X}_1 - \bar{\boldsymbol{X}})|\boldsymbol{X}_1 \in A\right\}.$$

Here, $\boldsymbol{D}\left(\frac{1}{\sqrt{S_{ii}}}\right)$ is the diagonal matrix with ith diagonal element as $\frac{1}{\sqrt{S_{ii}}}$, where S_{ii} is the ith diagonal element of $\boldsymbol{S} = \sum_{i=1}^n (\boldsymbol{X}_i - \bar{\boldsymbol{X}})(\boldsymbol{X}_i - \bar{\boldsymbol{X}})^T$, $\boldsymbol{R}$ is the sample correlation matrix, and $\boldsymbol{P}$ is an orthogonal matrix such that

$$\boldsymbol{P}\boldsymbol{R}^{-1}\boldsymbol{P}^T = \text{Diag}(\lambda_1, \ldots, \lambda_k),$$

where $\lambda_1, \ldots, \lambda_k$ are the eigenvalues of $\mathbf{R}^{-1}$. For the case when $k = 2$ and $\mathbf{V} = \begin{pmatrix} \frac{1}{\sqrt{2}} & \frac{1}{\sqrt{2}} \\ -\frac{1}{\sqrt{2}} & \frac{1}{\sqrt{2}} \end{pmatrix}$, Masuda (1980) has provided some numerical examples.

Ivshin and Lumel'skii (1995) considered the estimation of the probability $P = \Pr(\mathbf{a}^T \mathbf{X} + b > 0)$ when $\mathbf{X} \overset{d}{=} N_k(\boldsymbol{\xi}, \mathbf{V})$, where $\mathbf{a}$ is a given column vector and $b > 0$ is a scalar constant. Evidently, $P = \Phi\left(\frac{\mathbf{a}^T \boldsymbol{\xi} + b}{\sqrt{\mathbf{a}^T \mathbf{V} \mathbf{a}}} \right)$, where $\Phi(\cdot)$ denotes the cumulative distribution function of the univariate standard normal variable. Based on a random sample $\mathbf{X}_1, \ldots, \mathbf{X}_n$ from $N_k(\boldsymbol{\xi}, \mathbf{V})$, the maximum likelihood estimator of P is

$$\hat{P} = \begin{cases} \Phi\left(\dfrac{\mathbf{a}^T \bar{\mathbf{X}} + b}{\sqrt{\frac{1}{n} \mathbf{a}^T \mathbf{S} \mathbf{a}}} \right) & \text{if } \boldsymbol{\xi} \text{ and } \mathbf{V} \text{ are unknown,} \\[2ex] \Phi\left(\dfrac{\mathbf{a}^T \boldsymbol{\xi} + b}{\sqrt{\frac{1}{n} \mathbf{a}^T \mathbf{T} \mathbf{a}}} \right) & \text{if } \boldsymbol{\xi} \text{ is known and } \mathbf{V} \text{ is unknown,} \\[2ex] \Phi\left(\dfrac{\mathbf{a}^T \bar{\mathbf{X}} + b}{\sqrt{\mathbf{a}^T \mathbf{V} \mathbf{a}}} \right) & \text{if } \boldsymbol{\xi} \text{ is unknown and } \mathbf{V} \text{ is known,} \end{cases} \quad (45.147)$$

where $\mathbf{S} = \sum_{i=1}^n (\mathbf{X}_i - \bar{\mathbf{X}})(\mathbf{X}_i - \bar{\mathbf{X}})^T$ and $\mathbf{T} = \sum_{i=1}^n (\mathbf{X}_i - \boldsymbol{\xi})(\mathbf{X}_i - \boldsymbol{\xi})^T$. In the case when both $\boldsymbol{\xi}$ and $\mathbf{V}$ are unknown, an unbiased estimator of P is

$$\tilde{P} = \begin{cases} 0 & \text{if } R_1 \leq -1, \\[1ex] \frac{1}{2} + \frac{1}{\sqrt{\pi}} \frac{\Gamma\left(\frac{n-1}{2}\right)}{\Gamma\left(\frac{n-2}{2}\right)} R_1 F\left(\frac{1}{2}, \frac{4-n}{2}, \frac{3}{2}; R_1^2\right) & \text{if } |R_1| < 1, \\[1ex] 1 & \text{if } R_1 \geq 1, \end{cases} \quad (45.148)$$

where $R_1 = (\mathbf{a}^T \bar{\mathbf{X}} + b)/\sqrt{\frac{n-1}{n} \mathbf{a}^T \mathbf{S} \mathbf{a}}$, and $F(\alpha, \beta, \gamma; z)$ is the Gaussian hypergeometric function; see Chapter 1 of Johnson, Kotz, and Kemp (1992). For even $n > 4$, we have

$$F\left(\frac{1}{2}, \frac{4-n}{2}, \frac{3}{2}; R_1^2\right) = \sum_{j=0}^{(n-4)/2} \binom{\frac{n-4}{2}}{j} \frac{(-1)^j}{2j+1} R_1^{2j+1}.$$

In the case when $\boldsymbol{\xi}$ is known and $\mathbf{V}$ is unknown, an unbiased estimator of P is

$$\tilde{P} = \begin{cases} 0 & \text{if } R_2 \leq -1, \\[1ex] \frac{1}{2} + \frac{1}{\sqrt{\pi}} \frac{\Gamma\left(\frac{n}{2}\right)}{\Gamma\left(\frac{n-1}{2}\right)} R_2 F\left(\frac{1}{2}, \frac{3-n}{2}, \frac{3}{2}; R_2^2\right) & \text{if } |R_2| < 1, \\[1ex] 1 & \text{if } R_2 \geq 1, \end{cases} \quad (45.149)$$

where $R_2 = (a^T \xi + b)/\sqrt{a^T T a}$. Finally, for the case when ξ is unknown and V is known, an unbiased estimator of P is

$$\tilde{P} = \frac{1}{2} + \frac{1}{\sqrt{\pi}} \sum_{j=0}^{\infty} (-1)^j \frac{\alpha^{2j+1}}{j!(2j+1)} \,, \tag{45.150}$$

where $\alpha = (a^T \bar{X} + b)/\sqrt{\frac{2(n-1)}{n} a^T V a}$. More general results have also been provided by Ivshin and Lumel'skii (1995).

9 TOLERANCE REGIONS

The normal density function in (45.1) is clearly a decreasing function of $(x - \xi)^T V^{-1}(x - \xi)$. The region R_β with smallest "volume" that contains a specified proportion, say β, of the distribution is therefore

$$(X - \xi)^T V^{-1}(X - \xi) \leq \chi^2_{k,\beta}.$$

It is sometimes desired to "estimate" R_β, in some sense.

It is natural to construct a "tolerance region" from a given set of r ($> k$) random sample values $X_1, \ldots, X_n$ of the form

$$(X - \bar{X})^T \hat{V}^{-1}(X - \bar{X}) \leq K, \tag{45.151}$$

where $\bar{X} = \frac{1}{n} \sum_{j=1}^{n} X_n$ and $\hat{V}$ is an independent Wishart matrix with $(n - 1)$ degrees of freedom and the same variancecovariance matrix as each X_j. The constant K is to be chosen in (45.151) so that

$$\Pr[\Pr[(X - \bar{X})^T \hat{V}^{-1}(X - \bar{X}) \leq K] \geq \beta] = \gamma. \tag{45.152}$$

The value of K depends on n, β, and γ (and, of course, on k). Although we can quite simply make the *expected value* of the inner probability in (45.152) equal to specified value, say α, by taking

$$K = k(n^2 - 1)n^{-1}(n - k)^{01} F_{k,n-k,\alpha} \tag{45.153}$$

[see Fraser and Guttman (1956)], the solution of (45.152) for K is difficult. Guttman (1970a) has approximated K by finding approximations to the mean and variance of

$$P = \Pr[(X - \bar{X})^T \hat{V}^{-1}(X - \bar{X}) \leq K]$$

and then fitting a beta distribution [see Chapter 25 of Johnson, Kotz, and Balakrishnan (1995)] with the same mean, variance, and range of variation (0 to 1). He has given tables of approximate values of K to four decimal places for $k = 2(1)4$; $n = 100(20)1000, \infty$; $\beta, \gamma = 0.75, 0.90,$ 0.95, and 0.99. (The $n = \infty$ value is $K = \chi^2_{k,\beta}$.) The accuracy of the values should increase with n. Comparison with Wald and Wolfowitz's approximation for the univariate ($k = 1$) case [see Chapter 13 of Johnson, Kotz and Balakrishnan (1994)] confirmed this and appeared to indicate a satisfactory absolute level of accuracy. A few values from the tables Guttman (1970a) are reproduced in Table 45.7.

TABLE 45.7
Approximate Values of K

β		0.95				
γ		0.75			0.90	
n/k	2	3	4	2	3	4
100	6.3737	8.2290	9.9296	6.7582	8.6300	10.3460
400	6.1889	8.0267	9.7122	6.3775	8.2253	9.9196
∞	5.9915	7.8147	9.4877	5.9915	7.8147	9.4877
β		0.99				
100	9.7329	11.8881	13.8403	10.2901	12.4404	14.3949
400	9.4977	11.6381	13.5771	9.7798	11.9195	13.8608
∞	9.2103	11.3449	13.2767	9.2103	11.3449	13.2767

Note. Interpolation with respect to $n^{-1/2}$ gives very useful results.

It is of interest to note the formulas for mean and variance of $\hat{P}$ used by Guttman (1970a):

$$E[\hat{P}] = \Pr(\chi^2_k \le K) - K^{k/2} \left[2^{k/2} \Gamma \left\{ \frac{1}{2} k \right\} \right]^{-1} e^{-(1/2)K} n^{-1} + o(n^{-1}),$$

$$\mathrm{var}(\hat{P}) = K^k \left[2^{k-1} k \left\{ \Gamma \left(\frac{1}{2} k \right) \right\}^2 \right]^{-1} e^{-K} n^{-1} + o(n^{-1}).$$

John (1968) has considered construction of a region of form (45.151) which shall include *all* of R_β with specified probability δ. He presented two approximate formulas for K:

$$\left[\sqrt{\frac{(n-1)\chi^2_{k,\beta}}{n-k-2}} + \sqrt{\frac{(n-1)kF_{k,n-k,\delta}}{n(n-k)F_{k,n-k,\delta}}} \right]^2, \tag{45.154}$$

$$\left[\sqrt{\frac{(n-1)\chi^2_{k,\beta}}{\lambda_{k,0.5}}} + \sqrt{\frac{(n-1)kF_{k,n-k,\delta}}{n(n-k)}} \right]^2, \tag{45.155}$$

where, in (45.155), $\lambda_{k,0.5}$ is the median of the distribution of the smallest canonical root of a Wishart matrix with variance–covariance matrix I_k and $(n-1)$ degrees of freedom.

Of these two formulas, (45.155) is the more accurate, but (45.154) does not require knowledge of $\lambda_{k,0.5}$.

In addition to these results, some other works dealing with the computation or application of tolerance factors have appeared in the literature. Following the work of John (1963) who not only developed the theoretical framework for this problem but also provided simple and easy-to-use approximations for computing the tolerance factors, some other approximations were suggested by Siotani (1964), Chew (1966), Guttman (1970a,b), and Krishnamoorthy and Mathew (1999); see also the early work of Wald (1942) in this direction. For the bivariate case, Hall and Sheldon (1979) used Monte Carlo methods to estimate the tolerance factors. Now, let $\boldsymbol{\Sigma} = \boldsymbol{V}^{-1/2}\boldsymbol{S}\boldsymbol{V}^{-1/2}$, where $\boldsymbol{S} = \sum_{i=1}^{n}(\boldsymbol{X}_i - \bar{\boldsymbol{X}})(\boldsymbol{X}_i - \bar{\boldsymbol{X}})^T$, and let $\ell_1 > \cdots > \ell_k > 0$ be the ordered eigenvalues of $\boldsymbol{\Sigma}$. Furthermore, let $\boldsymbol{\ell} = (\ell_1, \ldots, \ell_k)^T$. John's (1963) main result is as follows. Let $\xi(\boldsymbol{\ell})$ be a real-valued function of $\boldsymbol{\ell}$ such that $\ell_k < \xi(\boldsymbol{\ell}) < \ell_1$. Then, an approximate expression for K that satisfies (45.152) is

$$K = (n-1)\chi_{k,\beta}'^2(k/n)\Big/v,$$

where $\chi_{k,\beta}'^2(\lambda)$ denotes the 100βth percentile of a noncentral chi-square distribution with k degrees of freedom and the noncentrality parameter λ [see Chapter 29 of Johnson, Kotz, and Balakrishnan (1995)], and v is the $100(1-\gamma)$th percentile of $\xi(\boldsymbol{\ell})$. Since John (1963) has used $\lambda/2$ to denote the noncentrality parameter of a noncentral chi-square distribution instead of λ, some confusion seems to have occurred in the literature [for example, in the work of Fuchs and Kenett (1987, 1988)], as pointed out by Krishnamoorthy and Mathew (1999).

Through different choices of $\xi(\boldsymbol{\ell})$ or by some appropriate compromise, Krishnamoorthy and Mathew (1999) have considered the following approximations:

$$K_{\text{am}} = \frac{k(n-1)\chi_{k,\beta}'^2\left(\frac{k}{n}\right)}{\chi_{(n-1)k,1-\gamma}^2} \quad \text{(Arithmetic-Mean Approximation)},$$

where $\chi_{v,1-\gamma}^2$ is the $100(1-\gamma)$th percentile of central chi-square distribution with v degrees of freedom:

$$K_{\mathrm{gm}} = \frac{\frac{k}{2}\left\{1 - \frac{(k-1)(k-2)}{2n}\right\}^{1/k}(n-1){\chi'}^2_{k,\beta}\left(\frac{k}{n}\right)}{\Gamma_{\frac{k(n-k)}{2},1-\gamma}}$$

(Geometric-Mean approximation),

where $\Gamma_{k,1-\gamma}$ denotes the $100(1-\gamma)$th percentile of the gamma distribution with shape parameter k;

$$K_{\mathrm{HM}} = \frac{k(n-1){\chi'}^2_{k,\beta}\left(\frac{k}{n}\right)}{\chi^2_{(n-1)k-k(k+1)+2,1-\gamma}}$$

(Harmonic-Mean approximation),

$$K_{\mathrm{MHM}} = \frac{a(n-1){\chi'}^2_{k,\beta}\left(\frac{k}{n}\right)}{k\chi^2_{b,1-\gamma}}$$

(Modified Harmonic-Mean approximation),

where $a = \frac{k(b-2)}{n-k-2}$ and $b = \frac{k(n-k-1)(n-k-4)+4(n-2)}{n-2}$;

$$K_{\mathrm{V}} = \frac{(n-1){\chi'}^2_{k,\beta}\left(\frac{k}{n}\right)}{\chi^2_{n-k,1-\gamma}}$$

(Approximation based on $\Sigma_{11\cdot2}$);

$$K_{\mathrm{vhm}} = \frac{d(n-1){\chi'}^2_{k,\beta}\left(\frac{k}{n}\right)}{\chi^2_{e,1-\gamma}}$$

(Approximation based on Harmonic-Mean and $\Sigma_{11\cdot2}$),

where $d = \frac{e-2}{n-k-2}$ and $e = \frac{4k(n-k-1)(n-k)-12(k-1))n-k-2)}{3(n-2)+k(n-k-1)}$; and

$$K_{\mathrm{s}} = \frac{(n-1){\chi'}^2_{k,\beta}\left(\frac{k}{n}\right)}{h_\gamma} \quad \text{(Siotani's approximation)},$$

where h_γ denotes the $100(1-\gamma)$th percentile of $h(\boldsymbol{\ell})$ given by

$$h(\boldsymbol{\ell}) = \left(\prod_{i=1}^k \ell_i\right)^{1/k}\left\{\frac{\left(\prod_{i=1}^k \ell_i\right)^{1/k}}{\sum_{i=1}^k \ell_i/k}\right\}^2.$$

In Siotani's (1964) paper, the factor $n-1$ was inadvertently omitted. Calculation of h_γ presents substantial computational difficulties.

Based on an extensive numerical evaluation for different choices of n, k, β and γ, Krishnamoorthy and Mathew (1999) have observed that the approximation K_{vhm} for the tolerance factor performs quite satisfactorily except for the cases in which n is small and β is large ($\beta = 0.99$) in which case they recommend the use of the approximation K_{V}.

Fuchs and Kenett (1987) used multivariate tolerance regions in two practical examples: One deals with testing adulteration in citrus juice in which the data are six-dimensional, and the second one deals with the diagnosis of atopic diseases based on the levels of immunoglobulin in blood in which case the data are three-dimensional. In a subsequent paper, Fuchs and Kenett (1988) applied multivariate tolerance regions in a quality-control situation wherein a decision has to be made on whether ceramic substrate plates used in the microeletronics industry are conformal to a required standard.

10　TRUNCATED MULTIVARIATE NORMAL DISTRIBUTIONS

If the variables $X_1, \ldots, X_p$ ($p < k$) are truncated (in any way), but the remaining variables $X_{p+1}, \ldots, X_k$ are not, the conditional joint distribution of $X_{p+1}, \ldots, X_k$ given any set (or subset) $X_1, \ldots, X_p$ (or the specified subset) is as shown in Section 3. From this, it is possible to derive convenient formulas for the expected values, variances, and covariances of $X_{p+1}, \ldots, X_k$ in the truncated distribution.

Subject to the conditions that (i) regression of $X_1, \ldots, X_p$ on $X_{p+1}, \ldots, X_k$ is linear and (ii) the conditional distribution of $X_1, \ldots, X_p$ given $X_{p+1}, \ldots, X_k$ is of the same form (apart from a change in location), Aitken (1934) showed that for a *general* (not necessarily multivariate normal) distribution with expected value vector $\mathbf{0}$ and variance–covariance matrix

$$V = \begin{pmatrix} V_{11} & V_{12} \\ V_{21} & V_{22} \end{pmatrix}$$

(where V_{11} is the $p \times p$ variance–covariance matrix of $X_1, \ldots, X_p$), the expected value vector and variance–covariance matrix under a *general selection* on $X_1, X_2, \ldots, X_p$ can be expressed in the form

$$(\boldsymbol{\xi}_1^T, \boldsymbol{\xi}_1^T V_{11}^{-1} V_{12}) \tag{45.156}$$

and

$$\begin{pmatrix} U_{11} & U_{11}V_{11}^{-1}V_{12} \\ V_{21}V_{11}^{-1}U_{11} & V_{22} - V_{21}(V_{11}^{-1} - V_{11}^{-1}U_{11}V_{11}^{-1})V_{12} \end{pmatrix}, \quad (45.157)$$

where $\boldsymbol{\xi}_1^T, \boldsymbol{U}_{11}$ are the expected value vector and variance–covariance matrix of $X_1, \ldots, X_p$ after selection. These formulas had been obtained for special cases of selection from a multivariate normal distribution by Pearson (1903).

Lawley (1943) later pointed out that the *identity* of conditional distributions is not essential for formulas (45.156) and (45.157) to hold. All that is necessary is linearity of regression and homoscedasticity of the array variance–covariance matrices. Of course, it is necessary to evaluate $\boldsymbol{U}_{11}$ and $\boldsymbol{\xi}_1$ for the truncated variables.

For the special case when truncation is of type $X_j \geq h_j$ $(j = 1, \ldots, k)$— that is, values of X_j less than h_j are excluded—explicit, though complicated, formulas were obtained by Birnbaum, Paulson and Andrews (1950). Tallis (1961) gave an alternative derivation.

Note that truncation of functionally independent linear functions of the X's, such as

$$\sum_{j=1}^{k} a_j X_j \geq d,$$

can be reduced to cases of the form $X_j > h_j$ by appropriate transformation of variables.

An unusual type of truncation ("elliptical" truncation), which leads to remarkably simple formulas, has been described by Tallis (1963). Taking the standardized form of distribution, it is supposed that values of $\boldsymbol{X}$ are restricted by the condition

$$a \leq \boldsymbol{X}^T \boldsymbol{R}^{-1}\boldsymbol{X} \leq b \quad (0 \leq a < b). \quad (45.158)$$

Remembering that $\boldsymbol{X}^T \boldsymbol{R}^{-1}\boldsymbol{X}$ is distributed as χ^2 with k degrees of freedom, we obtain the following formula for the moment-generating function of $\boldsymbol{X}$:

$$\phi_{\boldsymbol{X}}(\boldsymbol{t})$$
$$= \{\Pr[a \leq \chi_k^2 \leq b]\}^{-1}(2\pi)^{-k/2}|\boldsymbol{R}|^{-1/2}\int \exp\left\{-\frac{1}{2}\boldsymbol{x}^T\boldsymbol{R}^{-1}\boldsymbol{x} + \boldsymbol{t}^T\boldsymbol{x}\right\}d\boldsymbol{x},$$

where the integral is over the region defined by (45.158). Making the transformation (see Section 2) $\boldsymbol{z}^T = \boldsymbol{x}^T\boldsymbol{H}^T$ with $\boldsymbol{H}^T\boldsymbol{H} = \boldsymbol{R}^{-1}$, we obtain

$$\phi_{\boldsymbol{X}}(\boldsymbol{t}) = \{\Pr[a \leq \chi_k^2 \leq b]\}^{-1}(2\pi)^{-k/2}$$

$$\times\ e^{(1/2)t^T Rt} \int \exp\left\{-\frac{1}{2}\ (z - Ht)^T(z - Ht)\right\} dz,$$

$$(45.159)$$

the integral now being over the region

$$a \leq z^T z \leq b$$

$\left(\text{note that } z^T z = \sum_{j=1}^{k} z_j^2\right).$

If the variables $Z^T = X^T H^T$ were to have joint density function

$$p_Z(z^T) = (2\pi)^{-k/2} \exp\left\{-\frac{1}{2}(z - Ht)^T(z - Ht)\right\},$$

then $Z^T Z$ would have a $\chi_k'^2(t^T Rt)$ distribution [see Chapter 29 of Johnson, Kotz, and Balakrishnan (1995)]. Hence, Eq. (45.159) can be expressed as

$$\begin{aligned}
\phi_X(t) &= \{\Pr[a < \chi_k^2 < b]\}^{-1} e^{(1/2)t^T Rt} \Pr[a \leq \chi_k'^2(t^T Rt) \leq b] \\
&= \frac{1}{\Pr[a \leq \chi_k^2 \leq b]} \sum_{j=0}^{\infty} \frac{\left(\frac{1}{2}t^T Rt\right)^j}{j!} \Pr[a \leq \chi_{k+2j}^2 \leq b];
\end{aligned}$$

$$(45.160)$$

see Chapter 29 of Johnson, Kotz, and Balakrishnan (1995).

From (45.160), it follows that

(i) the expected value vector, and indeed all moments and product-moments of odd order, of X are zero,

(ii) the variance–covariance vector of X is cR with

$$c = \{\Pr[a \leq \chi_{k+2}^2 \leq b]\}/\{\Pr[a \leq \chi_k^2 \leq b]\},$$

(iii) moments and product-moments of even order $(2m)$ are obtained from the corresponding values for the complete (untruncated) standardized multivariate normal distribution by multiplying by the factor

$$\{\Pr[a \leq \chi_{k+2m}^2 \leq b]\}/\{\Pr[a \leq \chi_k^2 \leq b]\}.$$

From (ii), we note that if a and b are chosen so that

$$\Pr[a \leq \chi_{k+2}^2 \leq b] = \Pr[a \leq \chi_k^2 \leq b],$$

then the truncated distribution has the *same* expected value vector and variance–covariance matrix as the untruncated distribution.

Tallis (1963) also discusses combination of elliptical truncation with "radial" truncation in which the angles made by the radius vector to the origin with coordinate axes are truncated. Tallis (1965) also considered truncation by sets of inequalities of form $\sum_{j=1}^{k} \alpha_{t_j} X_j > \alpha_t$; see also Yoneda (1961).

A special kind of truncation is considered by Beattie (1962). It is desired to truncate two variables (X_1, X_2) out of k from below in such a way that

(i) a specified proportion, P, of the original distribution is retained and

(ii) a certain linear function $\sum_{j=1}^{k} w_j E[X_j]$ is maximized.

Supposing that only values $X_1 \geq a_1$, $X_2 \geq a_2$ are retained, Beattie finds the following results. Put

$$A_1 = (1 - \rho_{12}^2)^{-1/2}(a_2 - \rho_{12}a_1) \text{ and } A_2 = (1 - \rho^2)^{-1/2}(a_1 - \rho_{12}a_2)$$

so that for the truncated distribution we obtain

$$E[X_1] = \{Z(a_1)[1 - \Phi(A_1)] + \rho_{12}Z(a_2)[1 - \Phi(A_2)]\}P^{-1}, \tag{45.161}$$

$$E[X_2] = \{\rho_{12}Z(a_1)[1 - \Phi(A_1)] + Z(a_2)[1 - \Phi(A_2)]\}P^{-1}, \tag{45.162}$$

$$E[X_j] = \rho_{j1\cdot 2}E[X_1] + \rho_{j2\cdot 1}E[X_2] \qquad (j = 3, \dots, k). \tag{45.163}$$

From the above equations, it is clear that

$$\sum_{j=1}^{k} w_j E[X_j] = \sum_{j=1}^{2} \alpha_j Z(a_j)[1 - \Phi(A_j)]$$

with

$$\alpha_1 = w_1 + \rho_{12}w_2 + (\rho_{j1\cdot 2} + \rho_{12}\rho_{j2\cdot 1}) \sum_{j=3}^{k} w_j,$$

$$\alpha_2 = \rho_{12}w_1 + w_2 + (\rho_{j2\cdot 1} + \rho_{12}\rho_{j1\cdot 2}) \sum_{j=3}^{k} w_j.$$

Marginal density function derived from a truncated multivariate normal distribution are not truncated normal in general. Cartinhour (1990a,b) has discussed the determination of one-dimensional marginal density functions. Let $\boldsymbol{X} = (X_1, \ldots, X_{k-1}, X_k)^T$ have a truncated multivariate normal distribution with density function

$$p(\boldsymbol{x}) = \frac{1}{K(2\pi)^{k/2}|\boldsymbol{V}|^{1/2}} \exp\left\{-\frac{1}{2}(\boldsymbol{x} - \boldsymbol{\xi})^T \boldsymbol{A}(\boldsymbol{x} - \boldsymbol{\xi})\right\}, \qquad \boldsymbol{x} \in R,$$

$$(45.164)$$

where R is a rectangle in k-dimensional space given by

$$R = \left\{(x_1, \ldots, x_k)^T : b_i \leq x_i \leq a_i, \ i = 1, \ldots, k\right\}$$

and K is the normalizing constant given by

$$K = \int \cdots \int_{\boldsymbol{x} \in R} \frac{1}{(2\pi)^{k/2}|\boldsymbol{V}|^{1/2}} \exp\left\{-\frac{1}{2}(\boldsymbol{x} - \boldsymbol{\xi})^T \boldsymbol{A}(\boldsymbol{x} - \boldsymbol{\xi})\right\} d\boldsymbol{x}.$$

In order to derive the marginal density function $p_k(x_k)$ of X_k, partition $\boldsymbol{y}^T \boldsymbol{A} \boldsymbol{y}$, where $\boldsymbol{y} = \boldsymbol{x} - \boldsymbol{\xi}$, in the form

$$\begin{aligned}
\boldsymbol{y}^T \boldsymbol{A} \boldsymbol{y} &= (\boldsymbol{y}_1^T \ y_k) \begin{pmatrix} \boldsymbol{A}_1 & \boldsymbol{a} \\ \boldsymbol{a}^T & a_{kk} \end{pmatrix} \begin{pmatrix} \boldsymbol{y}_1 \\ y_k \end{pmatrix} \\
&= \boldsymbol{y}_1^T \boldsymbol{A}_1 \boldsymbol{y}_1 + 2\boldsymbol{a}^T \boldsymbol{y}_1 y_k + a_{kk} y_k^2 \\
&= (\boldsymbol{y}_1 + y_k \boldsymbol{A}_1^{-1} \boldsymbol{a})^T \boldsymbol{A}_1 (\boldsymbol{y}_1 + y_k \boldsymbol{A}_1^{-1} \boldsymbol{a}) \\
&\quad - y_k^2 (\boldsymbol{a}^T \boldsymbol{A}_1^{-1} \boldsymbol{a} - a_{kk}).
\end{aligned}$$

Noting that

$$\begin{pmatrix} \boldsymbol{A}_1 & \boldsymbol{a} \\ \boldsymbol{a}^T & a_{kk} \end{pmatrix} \begin{pmatrix} \boldsymbol{V}_1 & \boldsymbol{v} \\ \boldsymbol{v}^T & v_{kk} \end{pmatrix} = \begin{pmatrix} \boldsymbol{I} & \boldsymbol{0} \\ \boldsymbol{0}^T & 1 \end{pmatrix},$$

we have

$$\boldsymbol{a} = -\frac{\boldsymbol{A}_1 \boldsymbol{v}}{v_{kk}} \qquad \text{and} \qquad a_{kk} - \boldsymbol{a}^T \boldsymbol{A}^{-1} \boldsymbol{a} = \frac{1}{v_{kk}} ;$$

consequently, we have

$$\begin{aligned}
&(\boldsymbol{x} - \boldsymbol{\xi})^T \boldsymbol{A}(\boldsymbol{x} - \boldsymbol{\xi}) \\
&= \left[\boldsymbol{x}_1 - \left\{\boldsymbol{\xi}_1 + \left(\frac{x_k - \xi_k}{v_{kk}}\right)\boldsymbol{v}\right\}\right]^T \boldsymbol{A}_1 \left[\boldsymbol{x}_1 - \left\{\boldsymbol{\xi}_1 + \left(\frac{x_k - \xi_k}{v_{kk}}\right)\boldsymbol{v}\right\}\right] \\
&\quad + \frac{(x_k - \xi_k)^2}{v_{kk}},
\end{aligned}$$

where $\boldsymbol{x}$ and $\boldsymbol{\xi}$ are partitioned as $(\boldsymbol{x}_1^T, x_k)^T$ and $(\boldsymbol{\xi}_1^T, \xi_k)^T$, respectively. Then, Cartinhour (1990a,b) has shown that the marginal density function of X_k can be written as

$$
\begin{aligned}
p_k(x_k) &= \frac{1}{K\{2\pi v_{kk}\}^{1/2}} \exp\left\{-\frac{1}{2v_{kk}}(x_k - \xi_k)^2\right\} \\
&\quad \times \int_{b_{k-1}}^{a_{k-1}} \cdots \int_{b_1}^{a_1} \frac{1}{(2\pi)^{(k-1)/2}|\boldsymbol{A}_1^{-1}|} \\
&\quad \times \exp\left\{-\frac{1}{2}(\boldsymbol{x}_1 - \boldsymbol{m}(x_k))^T \boldsymbol{A}_1 (\boldsymbol{x}_1 - \boldsymbol{m}(x_k))\right\} \, dx_1 \cdots dx_{k-1}
\end{aligned}
\tag{45.165}
$$

when $b_k \leq x_k \leq a_k$, where

$$
\boldsymbol{m}(x_k) = \boldsymbol{\xi}_1 + \left(\frac{x_k - \xi_k}{v_{kk}}\right) \boldsymbol{v}.
$$

The multivariate normal integral in (45.165) can be evaluated using MULNOR algorithm of Schervish (1984), for example, or by any of the methods described in Section 5. Thus, we can express the marginal density function of X_k from (45.165) as

$$
p_k(x_k) = \frac{S(x_k)}{\sqrt{2\pi\, v_{kk}}} \exp\left\{-\frac{1}{2v_{kk}}(x_k - \xi_k)^2\right\}, \quad b_k \leq x_k \leq a_k, \tag{45.166}
$$

where $S(x_k)$ can be viewed as a "skew function," as pointed out by Cartinhour (1990a,b).

Generalizing this expression, Sungur and Kovacevic (1990, 1991) have given the marginal joint cumulative distribution function of $(X_1, \ldots, X_\ell)^T$, when $\boldsymbol{X}$ is truncated to be in the rectangle R defined earlier, as

$$
\begin{aligned}
F_{X_1,\ldots,X_\ell}&(x_1, \ldots, x_\ell) \\
&= \frac{1}{L} \int_{a_1}^{x_1} \cdots \int_{a_\ell}^{x_\ell} p_{X_1,\ldots,X_\ell}(t_1, \ldots, t_\ell) K(t_1, \ldots, t_\ell; \boldsymbol{a}, \boldsymbol{b}) \\
&\qquad\qquad\qquad\qquad\qquad\qquad\qquad\qquad \cdot \, dt_\ell \cdots dt_\ell, \tag{45.167}
\end{aligned}
$$

where $L = \Pr(a_i \leq X_i^* \leq b_i \; \forall \; i = 1, \ldots, k)$ with $\boldsymbol{X}^* = (X_1^*, \ldots, X_k^*)$ being the original untruncated variable,

$$
\begin{aligned}
K(t_1, &\ldots, t_\ell; \boldsymbol{a}, \boldsymbol{b}) \\
&= \int_{a_{\ell+1}}^{b_{\ell+1}} \cdots \int_{a_k}^{b_k} p(t_{\ell+1}, \ldots, t_k | t_1, \ldots, t_\ell) dt_k \cdots dt_{\ell+1},
\end{aligned}
$$

$a = (a_{\ell+1}, \ldots, a_k)^T$, $b = (b_{\ell+1}, \ldots, b_k)^T$, and $p(t_{\ell+1}, \ldots, t_k | t_1, \ldots, t_\ell)$ is the conditional joint density function of $(X_{\ell+1}, \ldots, X_k)^T$, given $(X_1, \ldots, X_\ell)^T$. Note that the expression in (45.167) is valid for any truncated multivariate distribution.

Let the vector $(X, X_1^T, X_2^T)^T$ of dimension $(k_1 + k_2 + 1) \times 1$ possess a standard multivariate normal distribution with correlation matrix R. Let $Y = (X, X_1^T)^T$ and R be partitioned as

$$R = \begin{pmatrix} \Sigma_{YY} & \Sigma_{YX_2} \\ \Sigma_{X_2Y} & \Sigma_{X_2X_2} \end{pmatrix}.$$

The expectation $E[X | X_1 > x_1, \ X_2 = x_2]$, which can be interpreted as the mean of conditional truncated multivariate normal [observe that $E[X_i | X_j > x_j, \ j = 1, \ldots, k]$, where $X = (X_1, \ldots, X_k)^T \overset{d}{=} N_k(0, R)$, which was discussed by Tallis (1961), can be viewed as unconditional expectation of truncated multivariate normal variable], has been expressed by Waldman (1984) in the form

$$
\begin{aligned}
E[X | X_1 > x_1, \ X_2 &= x_2] \\
= \ b_1^T x_2 + d_1 E\Big[&(X - b_1^T x_2)d_1 \Big| D_{(1)}^{-1/2}(X_1 - B_{(1)}^T X_2) \\
&> D_{(1)}^{-1/2}(x_1 - B_{(1)}^T x_2), \ X_2 = x_2\Big],
\end{aligned}
$$

where $B = (b_1, B_{(1)}) = \Sigma_{X_2X_2}^{-1}\Sigma_{X_2Y}$, and

$$D = \begin{pmatrix} d_1^2 & 0^T \\ 0^T & D_{(1)} \end{pmatrix} = \text{diag}(d_1^2, \ldots, d_{k_1+1}^2)$$

with d_i being the diagonal elements of

$$\Sigma_{YY.X_2} = \Sigma_{YY} - \Sigma_{YX_2}\Sigma_{X_2X_2}^{-1}\Sigma_{X_2Y}.$$

Waldman (1984) also presented an alternate formula that unfortunately involves repeated evaluation of $(k_1 - 1)$-dimensional multivariate normal cumulative distribution functions and, hence, becomes computationally burdensome.

For the general truncated multivariate normal distribution, Gupta and Tracy (1976) have established some recurrence relations for the moments.

11 RELATED DISTRIBUTIONS

Sarabia (1995) studied the multivariate normal distribution with centered normal conditional that has a joint density function

$$p_X(x) = \beta_k(c) \cdot \frac{1}{(2\pi)^{k/2}} \sqrt{a_1 \cdots a_k}$$

$$\exp\left\{-\frac{1}{2}\left(\sum_{i=1}^{k} a_i x_i^2 + c \prod_{i=1}^{k} a_i x_i^2\right)\right\},$$

where $a_i \geq 0$ $(i = 1, \ldots, k)$, $c \geq 0$, and $\beta_k(c)$ is the normalizing constant. It is evident that when $c = 0$, this becomes the joint density function of k independent univariate normal variables (with mean 0 and variance $1/a_i$, $i = 1, 2, \ldots, k$). The marginal density function of $X_{(i)}$, where $X_{(i)}$ is the vector X with X_i been removed, is

$$p_{X_{(i)}}(x_{(i)}) = \frac{\beta_k(c)}{(2\pi)^{(k-1)/2}} \left\{\frac{\Pi_{(i)} a_j}{1 + c\Pi_{(i)} a_j x_j^2}\right\}^2$$

$$\times \exp\left\{-\frac{1}{2}\Sigma_{(i)} a_j x_j^2\right\};$$

here, $\Sigma_{(i)}$ and $\Pi_{(i)}$ denotes the sum and product over $j = 1, \ldots, k$ with $j = i$ excluded, respectively.

The distribution obtained by mixing multivariate normal distributions $N_k(\mathbf{0}, \Theta V)$ by ascribing the gamma distribution [see Chapter 17 of Johnson, Kotz and Balakrishnan (1998)]

$$p_\Theta(\theta) = \frac{\alpha^\alpha \theta^{\alpha-1}}{\Gamma(\alpha)} e^{-\alpha\theta}, \qquad \theta > 0, \ \alpha > 0$$

to Θ is called a *multivariate K-distribution*. The corresponding joint density function is

$$p_X(x) = \frac{1}{(2\pi)^{k/2}} \int_0^\infty \frac{1}{|\theta V|^{1/2}} \exp\left\{-\frac{1}{2} x^T (\theta V)^{-1} x\right\}$$

$$\times \frac{\alpha^\alpha \theta^{\alpha-1}}{\Gamma(\alpha)} e^{-\alpha\theta} d\theta$$

$$= \frac{1}{(2\pi)^{k/2} \Gamma(\alpha)|V|^{1/2}} \int_0^\infty \theta^{\alpha-1-k/2}$$

$$\times \exp\left\{-\frac{1}{2\theta} x^T V^{-1} x - \alpha\theta\right\} d\theta$$

$$= \frac{2^{\frac{k}{4}-\frac{\alpha}{2}+1}}{(2\pi)^{k/2}\Gamma(\alpha)|V|^{1/2}} (x^T V^{-1} x)^{\frac{\alpha}{2}-\frac{k}{4}}$$

$$\times K_{\frac{k}{2}-1}\left(\sqrt{2\alpha}(x^T V^{-1} x)^{1/2}\right),$$

where $K_v(\cdot)$ is a Bessel function of order v.

The univariate $(k = 1)$ K-distribution was introduced by Jakeman and Pusey (1976) to model "the non-Gaussian statistical properties of radiation scattered by objects as diverse as land and sea surfaces and extended and localized regions of turbulence"; see Jakeman and Tough (1987, 1988), Novak, Sechtin, and Cardullo (1989), and Yueh *et al.* (1991). The multivariate K-distribution was obtained by Barakat (1986), who termed it a "generalized" K-distribution.

The distribution obtained by mixing multivariate normal distributions $N_k(\boldsymbol{\xi} + \boldsymbol{W}\boldsymbol{\beta}\boldsymbol{V}, \boldsymbol{W}\boldsymbol{V})$ by ascribing the generalized inverse Gaussian distribution [see Chapter 15 of Johnson, Kotz, and Balakrishnan (1994)]

$$p_W(w) = \frac{(\eta/\delta)^\lambda}{2K_\lambda^*(\delta\eta)}\, w^{\lambda-1}\exp\left\{-\frac{1}{2}\left(\frac{\delta^2}{w}+\eta^2 w\right)\right\},\ w > 0$$

to W, where $\eta^2 = \alpha^2 - \boldsymbol{\beta}^T\boldsymbol{V}\boldsymbol{\beta}$ and $K_\lambda^*(\cdot)$ denotes the modified Bessel function of the third kind with index λ, is called a *generalized multivariate hyperbolic distribution* with index parameter λ; it is usually denoted by $H_k(\lambda, \alpha, \boldsymbol{\beta}, \delta, \boldsymbol{\xi}, \boldsymbol{V})$. The corresponding joint density function is

$$
\begin{aligned}
p_{\boldsymbol{X}}(\boldsymbol{x}) \ =\ & \frac{(\eta/\delta)^\lambda}{(2\pi)^{k/2}K_\lambda^*(\delta\eta)}\, \alpha^{(k/2)-\lambda}\{\delta^2+(\boldsymbol{x}-\boldsymbol{\xi})^T\boldsymbol{V}^{-1}(\boldsymbol{x}-\boldsymbol{\xi})\}^{-(k-2\lambda)/4} \\
& \times K_{\lambda-k/2}^*\left(\alpha\left\{\delta^2+(\boldsymbol{x}-\boldsymbol{\xi})^T\boldsymbol{V}^{-1}(\boldsymbol{x}-\boldsymbol{\xi})\right\}^{1/2}\right) \\
& \times \exp\{\boldsymbol{\beta}^T(\boldsymbol{x}-\boldsymbol{\xi})\},\quad \boldsymbol{x}\in\mathbb{R}^k;
\end{aligned}
$$

see, for example, Blaesild and Jensen (1980). Because of the simple formula

$$K_{1/2}^*(x) = \sqrt{\frac{\pi}{2x}}\, e^{-x},$$

the above density function reduces, in the special case of $\lambda = (k + 1)/2$, to

$$
\begin{aligned}
p_{\boldsymbol{X}}(\boldsymbol{x}) \ =\ & \frac{(\eta/\delta)^{(k+1)/2}}{(2\pi)^{(k-1)/2}2\alpha\, K_{(k+1)/2}^*(\delta\eta)} \\
& \times \exp\left[-\alpha\left\{\delta^2+(\boldsymbol{x}-\boldsymbol{\xi})^T\boldsymbol{V}^{-1}(\boldsymbol{x}-\boldsymbol{\xi})\right\}^{1/2}+\boldsymbol{\beta}^T(\boldsymbol{x}-\boldsymbol{\xi})\right]
\end{aligned}
$$

for $\boldsymbol{x}\in\mathbb{R}^k$, where $\eta^2 = \alpha^2 - \boldsymbol{\beta}^T\boldsymbol{V}\boldsymbol{\beta}$ as before. Due to the fact that the graph of the logarithm of this density function is a hyperboloid, this distribution is called a *multivariate hyperbolic distribution*. Also, due to this fact, it is evident that this distribution is log-concave and unimodal with the mode being $\boldsymbol{\xi} + \left(\frac{\delta}{\eta}\right)\boldsymbol{\beta}^T\boldsymbol{V}$. Furthermore, this family of distributions constitutes a regular exponential family of order $k+1$, for fixed values of δ, $\boldsymbol{\xi}$ and $\boldsymbol{V}$.

Muirhead (1982, Theorem 1.5.5) considered the following natural multivariate construction:

$$
\begin{aligned}
X_1 &= R\sin\theta_1\sin\theta_2\cdots\sin\theta_{k-3}\sin\theta_{k-2}\sin\theta_{k-1}, \\
X_2 &= R\sin\theta_1\sin\theta_2\cdots\sin\theta_{k-3}\sin\theta_{k-2}\cos\theta_{k-1}, \\
X_3 &= R\sin\theta_1\sin\theta_2\cdots\sin\theta_{k-3}\cos\theta_{k-2}, \\
&\cdots \qquad \cdots \\
&\cdots \qquad \cdots \\
X_{k-2} &= R\sin\theta_1\sin\theta_2\cos\theta_3, \\
X_{k-1} &= R\sin\theta_1\cos\theta_2, \\
X_k &= R\cos\theta_1,
\end{aligned}
$$

where $0 < \theta_i \le \pi$ $(i = 1, \ldots, k-2)$ and $0 < \theta_{k-1} < 2\pi$. The random vector $\boldsymbol{X} = (X_1, \ldots, X_k)^T$ is determined by $(k-1)$ random angles $\Theta_1, \ldots, \Theta_{k-1}$ and the random variable R. In two dimensions, for example, we have

$$X_1 = R\sin\theta \quad \text{and} \quad X_2 = R\cos\theta, \quad 0 < \theta < 2\pi,$$

and in three dimensions we have

$$X_1 = R\sin\theta_1\sin\theta_2, \qquad X_2 = R\sin\theta_1\cos\theta_2, \qquad X_3 = R\cos\theta_1,$$
$$0 < \theta_1 < \pi, \ 0 < \theta_2 < 2\pi.$$

In the case of multivariate normal distribution with independent standard normal components, $R, \Theta_1, \ldots, \Theta_{k-1}$ are independent and R is distributed as $\sqrt{\chi_k^2}$, and the densities of the angles are

$$f(\theta_i) \propto \sin^{k-1-i}(\theta_i), \quad 0 < \theta_i < \pi \ (i = 1, 2, \ldots, k-2)$$

and

$$f(\theta_{k-1}) = \frac{1}{2\pi}, \quad 0 < \theta_{k-1} < 2\pi \quad \text{(uniform)}.$$

In the bivariate case, let $f(\theta, r)$ be an arbitrary density function of (Θ, R) having the support $0 < \theta < 2\pi$ and $r > 0$. Then, with the transformation $X_1 = R\sin\theta$ and $X_2 = R\cos\theta$, we arrive at the joint density function of $(X_1, X_2)^T$ as

$$p_{X_1, X_2}(x_1, x_2) = \frac{1}{\sqrt{x_1^2 + x_2^2}} f\left(\tan^{-1}\left(\frac{x_2}{x_1}\right), \sqrt{x_1^2 + x_2^2}\right),$$
$$(x_1, x_2)^T \in \mathbb{R}^2.$$

Instead, if we assume Θ and R to be independent with R distributed as $\sqrt{\chi_2^2}$ and Θ having a density $g(\cdot)$, then we arrive at

$$p_{X_1,X_2}(x_1, x_2) = e^{-(x_1^2+x_2^2)/2} \, g\left(\tan^{-1}\left(\frac{x_2}{x_1}\right)\right).$$

The specific choice of $g(\theta) \propto \sin^{\alpha}(\theta/2)$, for example, yields

$$p_{X_1,X_2}(x_1, x_2) \propto e^{-(x_1^2+x_2^2)/2} \left(\frac{x_2}{\sqrt{x_1^2 + x_2^2}}\right)^{\alpha}, \qquad (x_1, x_2)^T \in \mathbb{R}^2.$$

Nachtsheim and Johnson (1988) have presented a variety of such densities derived in this manner.

Similarly, in the trivariate case, if we choose the joint density of $(\theta_1, \theta_2)^T$ as $\sin^{\alpha}(\theta_1)$ (i.e., using uniform distribution for θ_2), we obtain the trivariate density function

$$p_{X_1,X_2,X_3}(x_1, x_2, x_3) \propto e^{-(x_1^2+x_2^2+x_3^2)/2} \left(\frac{x_1^2 + x_2^2}{x_1^2 + x_2^2 + x_3^2}\right)^{\alpha-1},$$

which, when $\alpha = 1$, coincides with the trivariate distribution with independent standard normal components.

With $\boldsymbol{x} = (x_1, \ldots, x_k)^T$, let $Q(\boldsymbol{x})$ be a polynomial of degree ℓ in the k variables, namely, $Q(\boldsymbol{x}) = \sum_{q=0}^{\ell} Q^{(q)}(\boldsymbol{x})$, where $Q^{(q)}(\boldsymbol{x})$ is a homogeneous polynomial of degree q given by

$$Q^{(q)}(\boldsymbol{x}) = \sum c_{j_1 \cdots j_k}^{(q)} \prod_{i=1}^{k} x_i^{j_i}$$

with the summation taken over all nonnegative integer k-tuples $(j_1, \ldots, j_k)$ such that $j_1 + \cdots + j_k = q$. The polynomial $Q(\boldsymbol{x})$ is said to be admissible if $p(\boldsymbol{x}) = e^{-Q(\boldsymbol{x})}$ is integrable on $\mathbb{R}^k$. Then, as defined by Urzúa (1988), a random vector $\boldsymbol{X} = (X_1, \ldots, X_k)^T$ is said to have a *multivariate Q-exponential distribution* if the joint density function is of the form

$$p(\boldsymbol{x}) = \theta(\boldsymbol{c})e^{-Q(\boldsymbol{x})}, \qquad \boldsymbol{x} \in \mathbb{R}^k,$$

where $\theta(\boldsymbol{c})$ is the normalizing constant. Suppose now that the polynomial $Q(\boldsymbol{x})$ is of degree ℓ relative to each of its components x_i's. Then, the corresponding Q-exponential distribution is simply the multivariate normal distribution when $\ell = 2$. Urzúa (1988) has discussed many issues relating to the multivariate Q-exponential distribution including some characterization results and estimation methods.

Ernst (1997) studied the *multivariate generalized Laplace distribution* with joint density function

$$p_{\boldsymbol{X}}(\boldsymbol{x}) = \frac{\lambda \Gamma(k/2)}{2\pi^{k/2}\Gamma(k/\lambda)} |\boldsymbol{V}|^{-1/2}$$
$$\times \exp\left\{-[(\boldsymbol{x}-\boldsymbol{\xi})^T \boldsymbol{V}^{-1}(\boldsymbol{x}-\boldsymbol{\xi})]^{\lambda/2}\right\}.$$

It is denoted by $\text{MGL}_k(\boldsymbol{\xi}, \boldsymbol{V}, \lambda)$. Evidently, when $\lambda = 2$, this distribution reduces to the multivariate normal distribution $N_k(\boldsymbol{\xi}, \boldsymbol{V})$. The above multivariate generalized Laplace distribution has been used by Kuwana and Kariya (1991) while developing a test of multivariate normality ($\lambda = 2$ vs. $\lambda < 2$ or $\lambda > 2$) which allows for a variety of elliptically contoured alternatives. The marginal distribution of X_i ($1 \le i \le k$) is

$$p_{X_i}(x_i) = \frac{\lambda \Gamma(1/2)}{2\sqrt{\pi}\,\Gamma(1/\lambda)} |V_{ii}|^{-1/2} \exp\left\{-\left[\left(\frac{x_i-\mu_i}{V_{ii}}\right)^2\right]^{1/2}\right\}$$
$$= \frac{\lambda}{2V_{ii}\Gamma(1/\lambda)} \exp\left\{-\left|\frac{x_i-\xi_i}{V_{ii}}\right|^{\lambda}\right\},$$

which is the generalized Laplace density or the exponential power family or the error distribution [see Chapter 24 of Johnson, Kotz, and Balakrishnan (1995)]. Observe that, in general, when $\lambda = \infty$, the distribution corresponds to an uniform distribution on a k-dimensional ellipsoid.

A random variable Z is said to be skew-normal (with parameter λ) if its density function is

$$p(z; \lambda) = 2\varphi(z)\Phi(\lambda z), \qquad z \in \mathbb{R},$$

where $\varphi(z)$ and $\Phi(z)$ are standard normal density and cumulative distribution functions, respectively. This distribution is usually denoted by $\text{SN}(\lambda)$, and the parameter $\lambda \in \mathbb{R}$ regulates the skewness, with $\lambda = 0$ corresponding to the standard normal density; see, for example, Chapter 13 of Johnson, Kotz and Balakrishnan (1994). It has been shown by Azzalini (1986) and Henze (1986) that $Z \overset{d}{=} \delta|Y_0| + \sqrt{1-\delta^2}\,Y_1$, where $\delta = \lambda/\sqrt{1+\lambda^2}$ and Y_0 and Y_1 are independent standard normal variables. A multivariate extension of this distribution has been proposed by Azzalini and Dalla Valle (1996) as follows.

Consider a k-dimensional normal random variable $\boldsymbol{Y} = (Y_1, \ldots, Y_k)^T$ and an independent standard normal variable Y_0; that is,

$$\begin{pmatrix} Y_0 \\ \boldsymbol{Y} \end{pmatrix} \overset{d}{=} N_{k+1}\left(\boldsymbol{0}, \begin{pmatrix} 1 & \boldsymbol{0} \\ \boldsymbol{0}^T & \boldsymbol{\Psi} \end{pmatrix}\right),$$

where $\boldsymbol{\Psi}$ is a $k \times k$ correlation matrix. For $\delta_i \in (-1, 1)$, $i = 1, \ldots, k$, let us define $Z_j = \delta_j |Y_0| + \sqrt{1 - \delta_j^2}\, Y_j$ for $j = 1, 2, \ldots, k$, and $\lambda(\delta_j) = \frac{\delta_j}{\sqrt{1 - \delta_j^2}}$. Evidently, Z_j's are dependent and are marginally distributed as $\mathrm{SN}(\lambda(\delta_j))$. The joint density function of $\boldsymbol{Z} = (Z_1, \ldots, Z_k)^T$ is the *multivariate skew-normal density* given by

$$p_{\boldsymbol{Z}}(\boldsymbol{z}) = 2\varphi_k(\boldsymbol{z}, \boldsymbol{\Omega})\Phi(\boldsymbol{\alpha}^T \boldsymbol{z}), \qquad \boldsymbol{z} \in \mathbb{R}^k,$$

where

$$\boldsymbol{\alpha}^T = \frac{\boldsymbol{\lambda}^T \boldsymbol{\Psi}^{-1} \boldsymbol{\Delta}^{-1}}{\sqrt{1 + \boldsymbol{\lambda}^T \boldsymbol{\Psi}^{-1} \boldsymbol{\lambda}}}, \quad \boldsymbol{\Delta} = \mathrm{diag}\left(\sqrt{1 - \delta_1^2}, \ldots, \sqrt{1 - \delta_k^2}\right),$$

$$\boldsymbol{\lambda} = \left(\frac{\delta_1}{\sqrt{1 - \delta_1^2}}, \ldots, \frac{\delta_k}{\sqrt{1 - \delta_k^2}}\right)^T, \qquad \boldsymbol{\Omega} = \boldsymbol{\Delta}(\boldsymbol{\Psi} + \boldsymbol{\lambda}\boldsymbol{\lambda}^T)\boldsymbol{\Delta},$$

and $\varphi_k(\boldsymbol{z}, \boldsymbol{\Omega})$ denotes the density function of the k-dimensional multivariate normal distribution with standardized marginals and correlation matrix $\boldsymbol{\Omega}$. An alternate derivation of this distribution has been given by Arnold *et al.* (1993). These authors have noted that if $\boldsymbol{X} = (X_0, X_1, \ldots, X_k)^T \overset{d}{=} N_{k+1}(\boldsymbol{0}, \boldsymbol{\Omega}^*)$, where $\boldsymbol{\Omega}^*$ is a positive definite matrix given by

$$\boldsymbol{\Omega}^* = \begin{pmatrix} 1 & \delta_1 & \cdots & \delta_k \\ \delta_1 & & & \\ \vdots & & \boldsymbol{\Omega} & \\ \delta_k & & & \end{pmatrix},$$

then the joint density of $(X_1, \ldots, X_k)^T$, given $X_0 > 0$, is precisely the above-given multivariate skew-normal density. In fact, Azzalini and Dalla Valle (1996) have shown that these two definitions are equivalent.

The joint cumulative distribution function of $\boldsymbol{Z}$ can be expressed as

$$F_{\boldsymbol{Z}}(\boldsymbol{z}) = \Pr\left[\bigcap_{i=1}^{k}(Z_i \le z_i)\right]$$
$$= 2\Pr[Y_0^* \le 0, \; Y_1 \le z_1, \ldots, Y_k \le z_k],$$

where $\begin{pmatrix} Y_0 \\ \boldsymbol{Y} \end{pmatrix}$ is as described above and $Y_0^* = Y_0 - \boldsymbol{\alpha}^T \boldsymbol{Y}$. Similarly, the joint moment-generating function of $\boldsymbol{Z}$ is

$$M_{\boldsymbol{Z}}(\boldsymbol{t}) = E\left[e^{\boldsymbol{t}^T \boldsymbol{Z}}\right] = 2\int_{\mathbb{R}^k} \exp\{\boldsymbol{t}^T \boldsymbol{z}\}\varphi_k(\boldsymbol{z}; \boldsymbol{\Omega})\Phi(\boldsymbol{\alpha}^T \boldsymbol{z})\, d\boldsymbol{z}$$
$$= 2\exp\left(\frac{\boldsymbol{t}^T \boldsymbol{\Omega} \boldsymbol{t})}{2}\right)\Phi\left(\frac{\boldsymbol{\alpha}^T \boldsymbol{\Omega} \boldsymbol{t}}{\sqrt{1 + \boldsymbol{\alpha}^T \boldsymbol{\Omega} \boldsymbol{\alpha}}}\right).$$

From this, we obtain the elements of the correlation matrix of Z as

$$\rho_{ij} = \frac{\Omega_{ij} - \frac{2}{\pi}\delta_i\delta_j}{\sqrt{\left(1 - \frac{2}{\pi}\delta_i^2\right)\left(1 - \frac{2}{\pi}\delta_j^2\right)}} \ , \quad 1 \le i, j \le k.$$

Note that $\Omega_{ij} = \frac{2}{\pi}\delta_i\delta_j$ implies $\rho_{ij} = 0$. If $Z \stackrel{d}{=} \mathrm{SN}_k(\boldsymbol{\lambda}, \boldsymbol{\Psi})$ (the multivariate skew-normal distribution given above) and $\boldsymbol{D} = \mathrm{diag}(d_1, \ldots, d_k)$ where d_i's are either $+1$ or -1, then it can be shown that $\boldsymbol{DZ} \stackrel{d}{=} \mathrm{SN}_k(\boldsymbol{D\lambda}, \boldsymbol{D\Psi D})$. In particular $-\boldsymbol{Z} \stackrel{d}{=} \mathrm{SN}_k(-\boldsymbol{\lambda}, \boldsymbol{\Psi})$. Another interesting property of multivariate skew-normal distributions is that if $\boldsymbol{Z} \stackrel{d}{=} \mathrm{SN}_k(\boldsymbol{\lambda}, \boldsymbol{\Psi})$ and $\boldsymbol{\Omega} = \boldsymbol{\Delta}(\boldsymbol{\Psi} + \boldsymbol{\lambda}\boldsymbol{\lambda}^T)\boldsymbol{\Delta}$ (as given above), then $\boldsymbol{Z}^T\boldsymbol{\Omega}^{-1}\boldsymbol{Z} \stackrel{d}{=} \chi_k^2$, which may be compared with the corresponding property of a multivariate normal random variable. Azzalini and Capitanio (1999) showed that the parametrizations of multivariate skew-normal distribution via $(\boldsymbol{\Omega}, \boldsymbol{\alpha})$ and $(\boldsymbol{\Psi}, \boldsymbol{\lambda})$ are equivalent.

Filus and Filus (1994) derived *multivariate pseudonormal distributions* as generalizations of multivariate normal distributions in the following manner. Let $T_1, \ldots, T_k$ be k independent normal random variables. Then, consider the transformation

$$
\begin{aligned}
x_1 &= at_1, \\
x_2 &= \phi_1(t_1)t_2 + \theta_1(t_1) \\
x_3 &= \phi_2(t_1, t_2)t_3 + \theta_2(t_1, t_2), \\
&\cdots \quad \cdots \\
x_k &= \phi_{k-1}(t_1, \ldots, t_{k-1})t_k + \theta_{k-1}(t_1, \ldots, t_{k-1}),
\end{aligned}
$$

where a is a nonzero constant, and $\phi_i(\cdot)$ and $\theta_i(\cdot)$ $(i = 1, 2, \ldots, k - 1)$ are real continuous parameter functions that are assumed to be positive and nondecreasing with respect to each argument. The inverse of this transformation is

$$
\begin{aligned}
t_1 &= \frac{x_1}{a}, \\
t_2 &= \frac{x_2 - \theta_1\left(\frac{x_1}{a}\right)}{\phi_1\left(\frac{x_1}{a}\right)}, \\
t_3 &= \frac{x_3 - \theta_2\left(\frac{x_1}{a}, t_2(x_1, x_2)\right)}{\phi_2\left(\frac{x_1}{a}, t_2(x_1, x_2)\right)}, \\
&\cdots \quad \cdots
\end{aligned}
$$

$$t_k = \frac{x_k - \theta_{k-1}\left(\frac{x_1}{a}, t_2(x_1, x_2), \ldots, t_{k-1}(x_1, \ldots, x_{k-1})\right)}{\phi_{k-1}\left(\frac{x_1}{a}, t_2(x_1, x_2), \ldots, t_{k-1}(x_1, \ldots, x_{k-1})\right)}.$$

Then, from the above, we can express the joint density function of $(X_1, \ldots, X_k)^T$ as

$$
\begin{aligned}
& p_{\boldsymbol{X}}(\boldsymbol{x}) \\
&= p_1(x_1) \prod_{i=2}^{k} p_i(x_i | x_1, \ldots, x_{i-1}) \\
&= \frac{C}{(2\pi)^{k/2}\sigma_1 \cdots \sigma_k} \exp\left\{ -\frac{(x_1 - \xi_1)^2}{2\sigma_1^2 a^2} \right\} \\
&\quad \times \exp\left[-\sum_{i=2}^{k} \frac{\left\{ x_i - \xi_i \phi_{i-1}\left(\frac{x_1}{a}, t_2, \ldots, t_{i-1}\right) - \theta_{i-1}\left(\frac{x_1}{a}, t_2, \ldots, t_{i-1}\right) \right\}^2}{2\sigma_i^2 \phi_{i-1}^2\left(\frac{x_1}{a}, t_2, \ldots, t_{i-1}\right)} \right],
\end{aligned}
$$

where

$$\frac{1}{C} = \left| a\phi_1\left(\frac{x_1}{a}\right) \phi_2\left(\frac{x_1}{a}, t_2\right) \cdots \phi_{k-1}\left(\frac{x_1}{a}, t_2, \ldots, t_{k-1}\right) \right|.$$

The conditional density of X_i, given $X_1, \ldots, X_{i-1}$, is simply

$$
\begin{aligned}
& p_i(x_i | x_1, \ldots, x_{i-1}) \\
&= \frac{1}{\sqrt{2\pi}\, \sigma_i} \cdot \frac{1}{\phi_{i-1}\left(\frac{x_1}{a}, t_2, \ldots, t_{i-1}\right)} \\
&\quad \times \exp\left[-\frac{\left\{ x_i - \xi_i \phi_{i-1}\left(\frac{x_1}{a}, t_2, \ldots, t_{i-1}\right) - \theta_{i-1}\left(\frac{x_1}{a}, t_2, \ldots, t_{i-1}\right) \right\}^2}{2\sigma_i^2 \phi_{i-1}^2\left(\frac{x_1}{a}, t_2, \ldots, t_{i-1}\right)} \right];
\end{aligned}
$$

that is, it is univariate normal with mean

$$\xi_i \phi_{i-1}\left(\frac{x_1}{a}, t_2, \ldots, t_{i-1}\right) + \theta_{i-1}\left(\frac{x_1}{a}, t_2, \ldots, t_{i-1}\right)$$

and variance $\sigma_i^2 \phi_{i-1}^2\left(\frac{x_1}{a}, t_2, \ldots, t_{i-1}\right)$. Any multivariate normal density function is a special case of the above multivariate pseudonormal density when we set $\phi_1, \ldots, \phi_{k-1}$ to be constant and $\theta_1, \ldots, \theta_{k-1}$ to be linear. Filus and Filus (1994) have also given an alternate genesis for the distribution in terms of a reliability setting.

Let $(\boldsymbol{U}, \boldsymbol{V}) = (U_1, \ldots, U_k, V_1, \ldots, V_\ell) \stackrel{d}{=} N_{k+\ell}(\boldsymbol{\xi}, \boldsymbol{V})$. Let T be the function that associates one element of $\boldsymbol{V}$ with each element of $\boldsymbol{U}$, that is,

$$T : (1, \ldots, k) \rightarrow (1, \ldots, \ell),$$

and each V_i is associated with some U_j (hence, $\ell \leq k$). Let us consider the transformation from $(\boldsymbol{U}, \boldsymbol{V})$ to $(\boldsymbol{X}, \boldsymbol{Y})$ given by

$$X_1 = \frac{U_1}{V_{T(1)}}, \quad \ldots, X_k = \frac{U_k}{V_{T(k)}},$$
$$Y_1 = V_1, \quad \ldots, Y_\ell = V_\ell,$$

where the ordering of $V_{T(1)}, \ldots, V_{T(k)}$ is fixed. Yatchev (1986) has derived an explicit formula for the distribution of $\boldsymbol{X}$ by integrating the joint density of $(\boldsymbol{X}, \boldsymbol{Y})$:

$$f_{\boldsymbol{X}, \boldsymbol{Y}}(\boldsymbol{x}, \boldsymbol{y}) = \left| \prod_{i=1}^{k} y_{T(i)} \right| p(x_1 y_{T(1)}, \ldots, x_k y_{T(k)}, y_1, \ldots, y_\ell),$$

where p is the density of $N_{k+\ell}(\boldsymbol{\xi}, \boldsymbol{V})$. The expression derived by Yatchev is quite complicated. The main feature is that the density of $\boldsymbol{X}$ (consisting of ratios of normals) involves the evaluation of absolute moments (around some point $\boldsymbol{a}$) of a standardized normal vector with a specified correlation matrix $\boldsymbol{R}$. Special cases include Fieller's (1932) distribution (case $k = \ell = 1$) and the multivariate Cauchy distribution (case $k = k$, $\ell = 1$, $\boldsymbol{\xi} = \boldsymbol{0}$ and $\boldsymbol{V} = \boldsymbol{I}$) with joint density function

$$\Gamma\left(\frac{k+1}{2}\right) \Big/ \left\{ \pi \left(1 + \sum_{i=1}^{k} x_i^2\right) \right\}^{(k+1)/2}.$$

Final mention should be made here to the *multivariate lognormal distribution*; see also Chapter 44 for some details on this and other similar transformed normal distributions. Simply stated, if $\boldsymbol{X}$ has a multivariate normal distribution with mean vector $\boldsymbol{\xi}$ and variance–covariance matrix $\boldsymbol{V}$, then we say that $\boldsymbol{Y}$ has a multivariate lognormal distribution if $\log(\boldsymbol{Y}) \stackrel{d}{=} \boldsymbol{X}$. Realizing that this construction is the same as in the univariate case, we can proceed very much along the lines of Chapter 14 of Johnson, Kotz, and Balakrishnan (1994) to examine the properties of multivariate lognormal distributions. As an example, we have the joint moment of $\boldsymbol{Y}$ as

$$\mu_{\boldsymbol{r}}'(\boldsymbol{Y}) = E\left[Y_1^{r_1} \cdots Y_k^{r_k}\right] = E\left[e^{r_1 X_1} e^{r_2 X_2} \cdots e^{r_k X_k}\right]$$
$$= E\left[e^{\boldsymbol{r}^T \boldsymbol{X}}\right] = \exp\left\{\boldsymbol{r}^T \boldsymbol{\xi} + \frac{1}{2} \boldsymbol{r}^T \boldsymbol{V} \boldsymbol{r}\right\},$$

from which expressions for covariances and correlations can all be derived easily. Sampson and Siegel (1985), for example, derived a unique measure of "size" that is statistically independent of "shape" for random vectors of measurements following a multivariate lognormal distribution, and then they applied it to analyze data on length of antler and height of shoulder (taken as a measure of body size) in different species of cervine deer, studied earlier by Gould (1974, 1977).

12 MIXTURES OF MULTIVARIATE NORMAL DISTRIBUTIONS

If the joint distribution of $X_1, \ldots, X_k$ is a mixture of m multivariate normal distributions with weights $\{a_j\}$, then the joint distribution of any subset of the X's is a mixture of m multivariate normal distributions with the same weights; see Section 1 of Chapter 44.

For mixtures of univariate normal distributions, methods of estimation based on moments have been described in Chapter 13 of Johnson, Kotz and Balakrishnan (1994). Day (1969) has found that these methods are not greatly inferior to maximum likelihood, at least for the case of two components with equal variances. The situation is quite different for two-component mixtures of bivariate (and even more for multivariate) normal distributions with common variance–covariance matrix.

For the population density

$$
\begin{aligned}
p_{\boldsymbol{X}}(\boldsymbol{x}) = {} & (2\pi)^{-k/2}|\boldsymbol{V}|^{-1/2}\Big\{\omega \exp\Big[-\frac{1}{2}(\boldsymbol{x} - \boldsymbol{\xi}_1)^T\boldsymbol{V}^{-1}(\boldsymbol{x} - \boldsymbol{\xi}_1)\Big] \\
& + (1 - \omega)\exp\Big[-\frac{1}{2}(\boldsymbol{x} - \boldsymbol{\xi}_2)^T\boldsymbol{V}^{-1}(\boldsymbol{x} - \boldsymbol{\xi}_2)\Big]\Big\},
\end{aligned}
$$

the expected value vector is

$$
\bar{\boldsymbol{\xi}} = \omega\boldsymbol{\xi}_1 + (1 - \omega)\boldsymbol{\xi}_2 \tag{45.168}
$$

and the variance–covariance matrix is

$$
\begin{aligned}
\boldsymbol{V} + {} & \omega(1 - \omega)(\boldsymbol{\xi}_1 - \boldsymbol{\xi}_2)(\boldsymbol{\xi}_1 - \boldsymbol{\xi}_2)^T \\
& = \boldsymbol{V} + \omega(1 - \omega)^{-1}(\boldsymbol{\xi}_1 - \bar{\boldsymbol{\xi}})(\boldsymbol{\xi}_1 - \bar{\boldsymbol{\xi}})^T. \tag{45.169}
\end{aligned}
$$

The total number of parameters ω, $\boldsymbol{\xi}_1$, $\boldsymbol{\xi}_2$, and $\boldsymbol{V}$ is

$$
1 + 2k + \frac{1}{2}k(k + 1) = \frac{1}{2}(k^2 + 5k + 2).
$$

Equating sample first and second moments to corresponding population values gives only

$$k + \frac{1}{2}k(k+1) = \frac{1}{2}(k^2 + 3k)$$

equations, so that—due to the symmetry of V—further $(k+1)$ equations are needed. Equating third sample moments of each marginal distribution to the corresponding population values

$$E[(X_j - \bar{\xi}_j)^3] = \omega(1 - 2\omega)(1 - \omega)^{-2}(\xi_j - \bar{\xi}_j)^3, \quad j = 1, 2, \ldots, k,$$

$$(45.170)$$

provides k equations.

The choice of the final equation appears to be somewhat arbitrary. Day (1969) gives an equation based on a certain symmetrical function of third and fourth moments and product moments. We shall not present this here, as he found the moment estimators to be much inferior to maximum likelihood estimators. Day (1969) gave an ingenious method of organizing the iterative calculation of maximum likelihood estimators. The equations for these estimators can be arranged to give (i) the $\frac{1}{2}(k^2 + 3k)$ equations obtained by equating first and second moments with the corresponding population values, and (ii)

$$\begin{cases} \hat{a} = \dfrac{\hat{V}^{-1}(\hat{\xi}_1 - \hat{\xi}_2)}{1 - \hat{\omega}(1 - \hat{\omega})(\hat{\xi}_1 - \hat{\xi}_2)^T \hat{V}^{-1}(\hat{\xi}_1 - \hat{\xi}_2)}, \\[2mm] \hat{b} = -\frac{1}{2}\hat{a}^T(\hat{\xi}_1 + \hat{\xi}_2) + \log[(1 - \hat{\omega})/\hat{\omega}], \end{cases} \qquad (45.171)$$

where $a = V^{-1}(\xi_2 - \xi_1)$, $b = \frac{1}{2}(\xi_1 V^{-1}\xi_1^T - \xi_2 V^{-1}\xi_2^T) + \log[(1 - \omega)/\omega]$, and $\hat{V}$ is the matrix of sample variances and covariances.

A computer program for maximum likelihood fitting of multivariate normal mixtures with common variance–covariance matrix is described by Wolfe (1970). He has also given a program for the case when there is no common variance–covariance matrix; see, however, the remarks about mixtures with common mean vector in Chapter 44.

Note that if $(X, Y)^T$ be distributed as a mixture of standardized bivariate normals (with different ρ's), X and Y each have marginal unit normal distributions. If there are two components with $\rho_1 = -\rho_2$, X and Y are also uncorrelated, but they are not independent.

For a detailed account on mixtures of multivariate normal distributions, one may refer to Titterington, Smith, and Makov (1985) and McLachlan and Basford (1987).

13   COMPLEX MULTIVARIATE NORMAL DISTRIBUTIONS

It is possible to regard the joint distribution of $2k$ (real) random variables $\boldsymbol{X}^T = (X_1, \ldots, X_k)$, $\boldsymbol{Y}^T = (Y_1, \ldots, Y_k)$ as representing the joint distribution of k complex variables

$$Z_j = X_j + iY_j, \qquad (j = 1, 2, \ldots, k; \; i = \sqrt{-1}).$$

While such a concept may suggest new problems and enable some results to be expressed in a new form, it is clear that any distributional properties of the Z's are equivalent to properties of the real random variables $\boldsymbol{X}^T$ and $\boldsymbol{Y}^T$. There is nothing essentially new involved.

If $\boldsymbol{X}^T$ and $\boldsymbol{Y}^T$ have a joint multivariate normal distribution, then the Z's may be said to have a joint *complex multivariate normal distribution*. Wooding (1956) showed that in the special case when

$$\mathrm{var}(X_j) \;=\; \mathrm{var}(Y_j) = \sigma_j^2; \quad \mathrm{corr}(X_j, Y_j) = 0 \qquad (j = 1, \ldots, k),$$

$$\mathrm{corr}(X_j, X_\ell) \;=\; \mathrm{corr}(Y_j, Y_\ell) = \frac{1}{2}\alpha_{j\ell} \qquad (j \neq \ell),$$

$$\mathrm{corr}(X_\ell, Y_j) \;=\; -\mathrm{corr}(X_j, Y_\ell) = \frac{1}{2}\beta_{j\ell} \qquad (j \neq \ell), \tag{45.172}$$

the joint density of $\boldsymbol{X}^T$ and $\boldsymbol{Y}^T$ can be written as

$$\begin{aligned}
p_{\boldsymbol{X}^T, \boldsymbol{Y}^T}(\boldsymbol{x}^T, \boldsymbol{y}^T) \\
= \; \pi^{-k}|\boldsymbol{V}|^{-1} \exp[-(\boldsymbol{x} - i\boldsymbol{y} - \boldsymbol{\xi} + i\boldsymbol{\eta})^T \boldsymbol{V}^{-1}(\boldsymbol{x} + i\boldsymbol{y} - \boldsymbol{\xi} - i\boldsymbol{\eta})],
\end{aligned}$$
$$\tag{45.173}$$

where $\boldsymbol{\xi} = E[\boldsymbol{X}]$, $\boldsymbol{\eta} = E[\boldsymbol{Y}]$, and

$$\boldsymbol{V} = E[(\boldsymbol{X} + i\boldsymbol{Y})(\boldsymbol{X} - i\boldsymbol{Y})^T].$$

Writing $\boldsymbol{z} = \boldsymbol{x} + i\boldsymbol{y}$, the right-hand side of (45.173) can be expressed in the form

$$\pi^{-k}|\boldsymbol{V}|^{-1} \exp[-(\tilde{\boldsymbol{z}} - \tilde{\boldsymbol{\zeta}})^T \boldsymbol{V}^{-1}(\boldsymbol{z} - \boldsymbol{\zeta})], \tag{45.174}$$

where $\boldsymbol{\zeta} = \boldsymbol{\xi} + i\boldsymbol{\eta}$ and a tilde over a symbol means "conjugate complement of" (e.g., $\tilde{\boldsymbol{z}} = \boldsymbol{x} - i\boldsymbol{y}$). (The $\boldsymbol{\zeta}$ may be regarded as the "expected value" of $\boldsymbol{Z}$.) Since (45.174) depends on $\boldsymbol{x}$ and $\boldsymbol{y}$ only through $\boldsymbol{z}$ and $\tilde{\boldsymbol{z}}$, it may be regarded (in a formal sense) as the joint density function $p_{\boldsymbol{Z}}(\boldsymbol{z})$ of the complex variables $Z_1, Z_2, \ldots, Z_k$. It is not, in general, possible to obtain an expression for $p_{\boldsymbol{X},\boldsymbol{Y}}(\boldsymbol{x}, \boldsymbol{y})$ depending only on $\boldsymbol{z}$ and $\tilde{\boldsymbol{z}}$; this is a consequence of the special form assumed for the covariance matrix.

Goodman (1963) initiated further developments of the theory of this special kind of complex multivariate normal distribution; see also Khatri (1965a). Useful survey of the theory have been given by Miller (1964) and Young (1971).

To the best of our knowledge, no significant advance in the theory of complex multivariate normal distributions has been reported in the main-stream statistical literature during the last 25 years; however, numerous applications have been indicated in the engineering literature, particularly in communication theory.

BIBLIOGRAPHY

(Some bibliographical items not mentioned in the text are included here for completeness.)

Abbe, E. N. (1964). Experimental comparison of Monte Carlo sampling techniques to evaluate the multivariate normal integral, Technical Research Note No. 28, U. S. Army Behavioral Science Research Laboratory.

Abrahamson, I. G. (1964). Orthant probabilities for the quadrivariate normal distribution, *Annals of Mathematical Statistics*, **35**, 1685–1703.

Afifi, A. A., and Elashoff, R. M. (1966–1969). Missing observations in multivariate statistics, *Journal of the American Statistical Association*, **61**, 595–604; **62**, 10–29; **64**, 337–358; **64**, 359–365.

Ahmed, S. E., and Badahdah, S. O. (1992). On the estimation of the mean vector of a multivariate normal distribution under symmetry, *Communications in Statistics—Theory and Methods*, **21**, 1759–1778.

Ahsanullah, M. (1985). Some characerizations of the bivariate normal distribution, *Metrika*, **32**, 215–218.

Ahsanullah, M., and Sinha, B. K. (1986). On normality via conditional normality, *Calcutta Statistical Association Bulletin*, **35**, 193–202.

Ahsanullah, M., and Wesolowski, J. (1994). Multivariate normality via conditional normality, *Statistics & Probability Letters*, **20**, 235–238.

Aitchison, J., and Dunsmore, I. R. (1975). *Statistical Prediction Analysis*, Cambridge, England: Cambridge University Press.

Aitken, A. C. (1934). Note on selection from a multivariate normal population, *Proceedings of the Edinburgh Mathematical Society, Series 2*, **4**, 106–110.

Ali, A. M., and Saleh, A. K. Md. E. (1990). Estimation of the mean vector of a multivariate normal distribution under symmetry, *Journal of Statistical Computation and Simulation*, **35**, 209–226.

Anderson, D. E. (1968). The characterization of multivariate normal integrals and the distribution of linear combinations of order statistics from the multivariate normal distribution, Ph.D. thesis, Southern Methodist University, Dallas, Texas.

Anderson, D. E. (1970). A technique for reducing multivariate normal integrals, *Proceedings of the Symposium on Empirical Bayes Estimation, Lubbock, Texas*, pp. 212–217.

Anderson, M. R. (1971). A characterization of the multivariate normal distribution, *Annals of Mathematical Statistics*, **42**, 824–827.

Anderson, T. W. (1955). The integral of a symmetric unimodal function over a symmetric convex set and some probability inqualities, *Proceedings of the American Mathematical Society*, **6**, 170–176.

Anderson, T. W. (1957). Maximum likelihood estimates for a multivariate normal distribution when some observations are missing, *Journal of the American Statistical Association*, **52**, 200–203.

Anderson, T. W. (1984). *An Introduction to Multivariate Statistical Analysis*, second edition, New York: John Wiley & Sons.

Anderson, T. W., and Olkin, I. (1985). Maximum-likelihood estimation of the parameters of a multivariate normal distribution, *Linear Algebra and Its Applications*, **70**, 147–171.

Anis, A. A., and Lloyd, E. H. (1953). On the range of partial sums of a finite number of independent normal variates, *Biometrika*, **40**, 35–42.

Arnold, B. C. (1997). Characterizations involving conditional specification, *Journal of Statistical Planning and Inference*, **63**, 117–131.

Arnold, B. C., Beaver, R. J., Groeneveld, R. A., and Meeker, W. Q. (1993). The nontruncated marginal of a truncated bivariate normal distribution, *Psychometrika*, **58**, 471–478.

Arnold, B. C., Castillo, E., and Sarabia, J. M. (1992). *Conditionally Specified Distributions*, Lecture Notes in Statistics–**73**, New York: Springer-Verlag.

Arnold, B. C., Castillo, E., and Sarabia, J. M. (1994a). A conditional characterization of the multivariate normal distribution, *Statistics & Probability Letters*, **19**, 313–315.

Arnold, B. C., Castillo, E., and Sarabia, J. M. (1994b). Multivariate normality via conditional specification, *Statistics & Probability Letters*, **20**, 353–354.

Arnold, B. C., and Pourahmadi, M. (1988). Conditional characterizations of multivariate distributions, *Metrika*, **35**, 99–108.

Azzalini, A. (1986). Further results on a class of distributions which includes the normal ones, *Statistics*, **46**, 199–208.

Azzalini, A., and Capitanio, A. (1999). Some properties of the multivariate skew-normal distribution, *Journal of the Royal Statistical Society, Series B*, **61**, 579-602.

Azzalini, A., and Dalla Valle, A. (1996). The multivariate skew-normal distribution, *Biometrika*, **83**, 715–726.

Bacon, R. H. (1963). Approximations to multivariate normal orthant probabilities, *Annals of Mathematical Statistics*, **34**, 191–198.

Bairamov, I. G., and Gebizlioglu, O. L. (1997). On the ordering of random vectors in a norm sense, *Journal of Applied Statistical Science*, **6**, 77–86.

Balakrishnan, N. (1993). Multivariate normal distribution and multivariate order statistics induced by ordering linear combinations, *Statistics & Probability Letters*, **17**, 343–350.

Balakrishnan, N., and Cohen, A. C. (1991) *Order Statistics and Inference: Estimation Methods*, San Diego: Academic Press.

Barakat, R. (1986). Weak-scatterer generalization of the K-density function with application to laser scattering in atmospheric turbulence, *Journal of the Optical Society of America*, **73**, 269–276.

Baranchik, A. J. (1970). A family of minimax estimators of the mean of a multivariate normal distribution, *Annals of Mathematical Statistics*, **41**, 642–645.

Baranchik, A. J. (1973). Inadmissibility of maximum likelihood estimators in some multiple regression problems with three or more independent variables, *Annals of Statistics*, **1**, 312–321.

Basu, D. (1956). A note on the multivariate extension of some theorems related to the univariate normal distribution, *Sankhyā*, **17**, 221–224.

Beattie, A. W. (1962). Truncation in two variables to maximize a function of the means of a normal multivariate distribution, *Australian Journal of Statistics*, **4**, 1–3.

Becker, A., and Roux, J. J. J. (1995). Bayesian multivariate normal analysis with a Wishart prior, *Communications in Statistics—Theory and Methods*, **24**, 2485–2497.

Berger, J. (1976). Admissible minimax estimation of a multivariate normal mean with arbitrary quadratic loss, *Annals of Statistics*, **4**, 223–226.

Berger, J. (1979). Multivariate estimation with nonsymmetric loss functions, in *Optimizing Methods in Statistics* (J. S. Rustagi, ed.), New York: Academic Press.

Berger, J. (1980). A robust generalized Bayes estimator and confidence region for a multivariate normal mean, *Annals of Statistics*, **8**, 716–761.

Berger, J. (1982). Selecting a minimax estimator of a multivariate normal mean, *Annals of Statistics*, **10**, 81–92.

Berger, J. O. (1985). *Statistical Decision Theory and Bayesian Analysis*, New York: Springer-Verlag.

Bhargava, R. P. (1975). Some one-sample hypothesis testing problems when there is a monotone sample from a multivariate normal populaiton, *Annals of the Institute of Statistical Mathematics*, **27**, 327–339.

Bhattacharya, P. K. (1966). Estimating the mean of a multivariate normal population with general quadratic loss function, *Annals of Mathematical Statistics*, **37**, 1819–1824.

Bickel, P. J. (1967). Some contributions to the theory of order statistics, in *Proceedings of the Fifth Berkeley Symposium on Mathematical Statistics and Probability*, Vol. 1, pp. 575–591, Berkeley, California: University of California Press.

Bildikar, S., and Patil, G. P. (1968). Multivariate exponential-type distributions, *Annals of Mathematical Statistics*, **39**, 1316–1326.

Birnbaum, Z. W. (1950). Effect of linear truncation on a multivariate population, *Annals of Mathematical Statistics*, **21**, 272–279.

Birnbaum, Z. W., and Meyer, P. L. (1953). On the effect of truncation in some or all coordinates of a multinormal population, *Journal of the Indian Society for Agricultural Statistics*, **5**, 17–28.

Birnbaum, Z. W., Paulson, E., and Andrews, F. C. (1950). On the effect of selection performed on some coordinates of a multi-dimensional population, *Psychometrika*, **15**, 191–204.

Bischoff, W., and Fieger, W. (1991). Characterization of the multivariate normal distribution by conditional normal distributions, *Metrika*, **38**, 239–248.

Blaesild, P., and Jensen, J. L. (1980). Multivariate distributions of hyperbolic type, Research Report No. 67, Department of Theoretical Statistics, University of Aarhus, Aarhus, Denmark.

Bland, R. P., and Owen, D. B. (1966). A note on singular normal distributions, *Annals of the Institute of Statistical Mathematics*, **18**, 113–116.

Brewster, J. F., and Zidek, J. V. (1974). Improving on equivariant estimators, *Annals of Statistics*, **2**, 21–38.

Broffitt, J. D. (1986). Zero correlation, independence and normality, *The American Statistician*, **40**, 276–277.

Brown, J. L. (1968). On the expansion of the bivariate Gaussian probability density using results of nonlinear theorems, *IEEE Transactions on Information Theory*, **14**, 158–159.

Brown, L. D. (1968). Inadmissibility of the usual estimators of scale parameters in problems with unknown location and scale parameters, *Annals of Mathematical Statistics*, **39**, 29–48.

Butler, J., and Moffit, R. (1982). A computationally efficient quadrature for the one factor multinomial probit model, *Econometrica*, **50**, 761–764.

Campbell, L. L. (1970). Equivalence of Gauss's principle and minimum discrimination information estimation of probabilities, *Annals of Mathematical Statistics*, **41**, 1011–1015.

Carter, G. M., and Rolph, J. E. (1974). Empirical Bayes methods applied to estimating fire alarm probabilities, *Journal of the American Statistical Association*, **69**, 880–885.

Cartinhour, J. (1990a). One-dimensional marginal density functions of a truncated multivariate normal density function, *Communications in Statistics —Theory and Methods*, **19**, 197–203.

Cartinhour, J. (1990b). Lower dimensional marginal density functions of absolutely continuous truncated multivariate distributions, *Communications in Statistics—Theory and Methods*, **20**, 1569–1578.

Casella, G. (1977). Minimax ridge regression estimation, *Annals of Statistics*, **8**, 1036–1056.

Castillo, E., and Galambos, J. (1989). Conditional distributions and the bivariate normal distribution, *Metrika*, **36**, 209–214.

Chen, L., Eichenauer-Hermann, J., and Lehn, J. (1990). Gamma-minimax estimation of a multivariate normal mean, *Metrika*, **37**, 1–6.

Chen, S. Y. (1983). Restricted risk Bayes estimator, Ph.D. thesis, Department of Statistics, Purdue University, West Lafayette, Indiana.

Chen, S. Y. (1988). Restricted risk Bayes estimation for the mean of the multivariate normal distribution, *Journal of Multivariate Analysis*, **24**, 207–217.

Cheng, M. C. (1969). The orthant probabilities of four Gaussian variates, *Annals of Mathematical Statistics*, **40**, 152–161.

Chew, V. (1966). Confidence, prediction, and tolerance regions for the multivariate normal distribution, *Journal of the American Statistical Association*, **61**, 605–617.

Childs, D. R. (1967). Reduction of the multivariate normal integral to characteristic form, *Biometrika*, **54**, 293–299.

Chiou, W., and Cohen, A. (1985). On estimating a common multivariate normal mean vector, *Annals of the Institute of Statistical Mathematics*, **37**, 499–506.

Chow, Y. S., and Robbins, H. (1965). Asymptotic theory of fixed width confidence intervals for the mean, *Annals of Mathematical Statistics*, **36**, 457–462.

Cirel'son, B. S., Ibragimov, I. A., and Sudakov, V. N. (1976). Norms of Gaussian sample functions, in *Proceedings of the Third Japan–U.S.S.R. Symposium on Probability Theory*, Lecture Notes in Mathematics—**550**, pp. 20–41, New York: Springer-Verlag.

Cohen, A. C. (1957). Restriction and selection in multinormal distributions, *Annals of Mathematical Statistics*, **28**, 731–741.

Coxeter, H. S. M. (1935). The functions of Schläfli and Lobatschefsky, *Quarterly Journal of Mathematics*, **6**, 13–29.

Csörgö, M., and Seshadri, V. (1971). Characterizing the Gaussian and exponential laws by mappings onto the unit interval, *Zeitschrift Wahrscheinlichkeitstheorie und Verwandte Gebiete*, **18**, 333–339.

Curnow, R. N., and Dunnett, C. W. (1962). The numerical evaluation of certain multivariate normal integrals, *Annals of Mathematical Statistics*, **33**, 571–579.

Dahel, S. (1987). Minimaxity of maximum likelihood estimator, *Communications in Statistics—Theory and Methods*, **16**, 1289–1296.

Dahel, S., Giri, N., and Lepage, Y. (1985). Optimum properties of the m.l.e. of the mean of a multinormal population with additional information, Annual Meeting of the Institute of Mathematical Statistics, Paper No. 193-96.

Dahel, S., Giri, N., and Lepage, Y. (1994). Locally minimax tests for a multinormal data problem, *Metrika*, **4**, 363–374.

Das, S. C. (1956). The numerical evaluation of a class of integrals, II, *Proceedings of the Cambridge Philosophical Society*, **52**, 442–448.

DaSilva, A. G., and Han, C. P. (1984). Pooling means in a multivariate normal population, *Estadistica*, **36**, 63–75.

David, F. N. (1953). A note on the evaluation of the multivariate normal integral, *Biometrika*, **40**, 458–459.

David, F. N., and Mallows, C. L. (1961). The variance of Spearman's rho in normal samples, *Biometrika*, **48**, 19–28.

David, H. A. (1981). *Order Statistics*, second edition, New York: John Wiley & Sons.

David, H. A., and Six, F. B. (1971). Sign distribution of standard multinormal variables with equal positive correlation, *Review of the International Statistical Institute*, **39**, 1–3.

Day, N. E. (1969). Estimating the components of a mixture of normal distributions, *Biometrika*, **56**, 463–474.

Deàk, I. (1980a). Three digit accurate multiple normal probabilities, *Numerische Mathematik*, **35**, 369–380.

Deàk, I. (1980b). Computation of multiple normal probabilities, in *Recent Advances in Stochastic Programming* (P. Koll and A. Predorr, eds.), New York: Springer-Verlag.

DeGroot, M. H. (1970). *Optimal Statistical Decisions*, New York: McGraw-Hill.

Dey, D. K., and Gelfand, A. E. (1989). Improved estimation of a patterned covariance matrix, *Journal of Multivariate Analysis*, **31**, 107–116.

Dickey, J. M., Lindley, D. V., and Press, S. J. (1985). Bayesian estimation of the dispersion matrix of a multivariate normal distribution, *Communcations in Statistics—Theory and Methods*, **14**, 1019–1034.

DiDonato, A. R., and Hageman, R. K. (1982). A method for computing the integral of the bivariate normal distribution over an arbitrary polygon, *SIAM Journal on Scientific and Statistical Computing*, **3**, 434–446.

DiDonato, A. R., Jarnagin, M. P., and Hageman, R. K. (1980). Computation of the integral of the bivariate normal distribution over convex polygons, *SIAM Journal on Scientific and Statistical Computing*, **1**, 179–186.

Doktorov, B. Z. (1969). On some estimates of a multivariate normal distribution, *Teoriya Veroyatnostei i ee Primeneniya*, **14**, 552–554 (in Russian). English translation, pp. 526–528.

Donnelly, T. G. (1973). Algorithm 462: Bivariate normal distribution, *Communications of the Association of Computing Machinery*, **16**, 638.

Dowson, D. C., and Landau, B. V. (1982). The Fréchet distance between multivariate normal distributions, *Journal of Multivariate Analysis*, **12**, 450–455.

Drezner, Z. (1990). Approximations to the multivariate normal integral, *Communcations in Statistics—Simulation and Computation*, **19**, 527–534.

Dunn, O. J. (1958). Estimation of the means of dependent variables, *Annals of Mathematical Statistics*, **29**, 1095–1111.

Dunnett, C. W. (1955). Statistical need for new tables involving the multivariate normal distribution, *Meeting of the Institute of Mathematical Statistics*, New York.

Dunnett, C. W. (1989). Algorithm AS251: Multivariate normal probability integrals with product correlation structure, *Applied Statistics*, **38**, 564–579.

Dunnett, C. W., and Lamm, R. A. (1960). Some tables of the multivariate normal probability integral with correlation coefficients 1/3, *Mathematical Tables and Aids to Computation*, **14**, 290–291 (Abstract).

Dunnett, C. W., and Sobel, M. (1955). Approximations of the probability integral and certain percentage points of a multivariate analogue of Student's *t*-distribution, *Biometrika*, **42**, 258–260.

Eaton, M. L. (1966). Characterization of distributions by the identical distribution of linear forms, *Journal of Applied Probability*, **3**, 481–494.

Eaton, M. L., and Perlman, M. D. (1991). Concentration inequalities for multivariate distributions: I. Multivariate normal distributions, *Statistics & Probability Letters*, **12**, 487–504.

Efron, B., and Morris, C. (1971). Limiting the risk of Bayes and empirical Bayes estimators—Part 1: The Bayes case, *Journal of the American Statistical Associaiton*, **66**,

Efron, B., and Morris, C. (1973). Stein's estimation rule and its competitors. An empirical Bayes approach, *Journal of the American Statistical Association*, **68**, 117–130.

Ernst, M. D. (1997). A multivariate generalized Laplace distribution, Preprint, Department of Statistical Science, Southern Methodist University, Dallas, Texas.

Escoufier, Y. (1967). Calculs de probabilités par une methode de Monte Carlo pour une variable p-normale, *Revue de Statistique Appliquée*, **15**, No. 4, 5–15.

Evans, M., and Swartz, T. (1988). Monte Carlo computation of some multivariate normal probabilities, *Journal of Statistical Computation and Simulation*, **30**, 117–128.

Fairweather, W. R. (1973). A test for multivariate normality based on a characterization, Ph.D. thesis, University of Washington, Seattle, Washington.

Fang, K. T., Kotz, S., and Ng, K. W. (1989). *Multivariate Symmetric Distributions*, London: Chapman and Hall.

Fang, K. T., and Xu, J.-L. (1987). The Mills' ratio of multivariate normal distributions and spherical distributions, *Acta Mathematica Sinica*, **31**, 248–257. Reprinted in *Statistical Inference in Elliptically Contoured and Related Distributions* (K. T. Fang and T. W. Anderson, eds.), New York: Allerton Press.

Fieller, E. C. (1932). The distribution of the index in a normal bivariate population, *Biometrika*, **24**, 438–440.

Filus, J. K., and Filus, L. (1994). "Pseudonormal" probability densities in R^n as a generalization of the multivariate normals, Report, Department of Mathematics, Oakton College, Des Plaines, Illinois.

Findeisen, P. (1982). Die charakterisierung der Normal verteilung nach Gauss, *Metrika*, **29**, 55–64.

Finney, D. J. (1962). Cumulants of truncated multinormal distributions, *Journal of the Royal Statistical Society, Series B*, **24**, 535–536.

Fisk, P. R. (1970). A note on a characterization of the multivariate normal distribution, *Annals of Mathematical Statistics*, **41**, 486–494.

Fraser, D. A. S., and Guttman, I. (1956). Tolerance regions, *Annals of Mathematical Statistics*, **27**, 162–179.

Fréchet, M. (1951). Généralizations de la loi de probabilité de Laplace, *Annales de l'Institut Henri Poincaré*, **13**, 1–29.

Fréchet, M. (1957). Sur la distance de deux lois de probabilité, *Comptes Rendus, Academy of Sciences, Paris*, **244**, 689–692.

Fuchs, C., and Kenett, R. S. (1987). Multivariate tolerance regions and F-tests, *Journal of Quality Technology*, **19**, 122–131.

Fuchs, C., and Kenett, R. S. (1988). Appraisal of ceramic substrates of multivariate tolerance regions, *The Statistician*, **37**, 401–411.

Fujisawa, H. (1995). A note on the maximum likelihood estimators for multivariate normal distribution with monotone data, *Communications in Statistics—Theory and Methods*, **24**, 1377–1382.

Geary, R. C. (1933). A general expression for the moments of certain symmetrical functions of normal samples, *Biometrika*, **25**, 184–186.

Gehrlein, W. V. (1979). A representation for quadrivariate normal positive orthant probabilities, *Communications in Statistics—Simulation and Computation*, **8**, 349–358.

Geisser, S. (1964). Estimation in the uniform covariance case, *Journal of the Royal Statistical Society, Series B*, **26**, 477–483.

Genz, A. (1992). Numerical computation of multivariate normal probabilities, *Journal of Computational and Graphical Statistics*, **1**, 141–150.

Genz, A. (1993). Comparison of methods for the computation of multivariate normal probabilities, *Computing Science and Statistics*, **25**, 400–405.

Ghurye, S. G., and Olkin, I. (1962). A characterization of the multivariate normal distribution, *Annals of Mathematical Statistics*, **33**, 533–541.

Ghurye, S. G., and Olkin, I. (1969). Unbiased estimation of some multivariate probability densities and related functions, *Annals of Mathematical Statistics*, **40**, 1261–1271.

Gleser, L. J. (1987). Improved estimation of mean response in simulation when control variates are used, Technical Report No. 87-34, Department of Statistics, Purdue University, West Lafayette, Indiana.

Goodman, N. R. (1963). Statistical analysis based on a certain multivariate complex Gaussian distribution (an introduction), *Annals of Mathematical Statistics*, **34**, 152–177.

Gould, S. J. (1974). The evolutionary significance of 'bizarre' structures: Antler size and skull size in the 'Irish elk,' *Magaloceros giganteus, Evolution*, **28**, 191–200.

Gould, S. J. (1977). *Ontogeny and Phylogeny*, Cambridge, Massachusetts: Harvard University Press.

Gupta, A. K., and Tracy, D. S. (1976). Recurrence relations for the moments of truncated multinormal distribution, *Communications in Statistics— Theory and Methods*, **5**, 855–865.

Gupta, P. L., and Gupta, R. C. (1997). On the multivariate normal hazard, *Journal of Multivariate Analysis*, **62**, 64–73.

Gupta, P. L., and Gupta, R. C. (1998). Failure rate of the minimum and maximum of a multivariate normal distribution, *The IMS Bulletin*, **27**, 14–15 (Abstract).

Gupta, S. D. (1969). A note on some inequalities for multivariate normal distribution, *Bulletin of the Calcutta Statistical Association*, **18**, 179–180.

Gupta, S. S. (1963). Bibliography on the multivariate normal integrals and related topics, *Annals of Mathematical Statistics*, **34**, 829–838.

Guttman, I. (1970a). *Statistical Tolerance Regions, Classical and Bayesian*, London: Griffin.

Guttman, I. (1970b). Construction of β-content tolerance regions at confidence level γ for large samples from the k-variate normal distribution, *Annals of Mathematical Statistics*, **41**, 376–400.

Haff, L. R. (1977). Minimax estimators for a multivariate precision matrix, *Journal of Multivariate Analysis*, **7**, 374–385.

Haff, L. R. (1978). The multivariate normal mean with intraclass correlated components: Estimation of urban fire alarm probabilities, *Journal of the American Statistical Association*, **73**, 767–774.

Haff, L. R. (1979a). Estimation of the inverse covariance matrix: random mixtures of the inverse Wishart matrix and the identity, *Annals of Statistics*, **7**, 1264–1276.

Haff, L. R. (1979b). An identity for the Wishart distribution with application, *Journal of Multivariate Analysis*, **8**, 536–542.

Haff, L. R. (1980). Empirical Bayes estimation of the multivariate normal covariance matrix, *Annals of Statistics*, **8**, 586–597.

Haider, A. M. (1991). The estimation problem and incomplete data from multivariate populations, *ASA Proceedings of Business and Economic Statistics Section*, Annual Meeting at Atlanta, pp. 87–92.

Hall, I. J., and Sheldon, D. D. (1979). Improved bivariate normal tolerance regions with some applications, *Journal of Quality Technology*, **11**, 13–19.

Hamedani, G. G. (1984). Nonnormality of linear combinations of normal random variables, *The American Statistician*, **38**, 295–296.

Hamedani, G. G. (1992). Bivariate and multivariate normal characterizations: a brief survey, *Communications in Statistics—Theory and Methods*, **21**, 2665–2688.

Hannan, J. F., and Tate, R. F. (1965). Estimation of the parameters for a multivariate normal distribution when one variable is dichotomized, *Biometrika*, **52**, 664–668.

Harris, B., and Soms, A. P. (1980). The use of the tetrachoric series for evaluating multivariate normal probabilities, *Journal of Multivariate Analysis*, **10**, 252–267.

Henze, N. (1986). A probabilistic representation of the 'skew-normal' distribution, *Scandinavian Journal of Statistics*, **13**, 271–275.

Hickernell, F. J., and Hong, H. S. (1997). Computing multivariate normal probabilities using rank-1 lattice sequences, in *Proceedings of the Workshop on Scientific Computing*, pp. 209–215, New York: Springer-Verlag.

Hocking, R. R., and Smith, W. B. (1968). Estimation of parameters in the multivariate normal distribution with missing observations, *Journal of the American Statistical Association*, **63**, 159–173.

Holmquist, B. (1988). Moments and cumulants of the multivariate normal distribution, *Stochastic Analysis and Applications*, **6**, 273–278.

Hotelling, H. (1948). Fitting generalized truncated normal distributions, *Annals of Mathematical Statistics*, **19**, 597 (Abstract).

Houdré, C. (1995). Some applications of covariance identities and inequalities to functions of multivariate normal variables, *Journal of the American Statistical Association*, **90**, 965–968.

Hudson, H. M. (1974). Empirical Bayes estimation, Ph.D. thesis, Department of Statistics, Stanford University, Stanford, California.

Hyakutake, H., Takada, Y., and Aoshima, M. (1995). Fixed-size confidence regions for the multinormal mean in an intraclass correlation model, *American Journal of Mathematical and Management Sciences*, **15**, 291–308.

Ihm, P. (1959). Numerical evaluation of certain multivariate normal integrals, *Sankhyā*, **21**, 363–366.

Ihm, P. (1961). A further contribution to the numerical evaluation of certain multivariate integrals, *Sankhyā, Series A*, **23**, 205–206.

Ivshin, V. V., and Lumel'skii, Ya. P. (1995). *Statistical Problems of Estimation in the Model "Stress–Strength,"* Perm, Russia: Perm University Press (in Russian).

Jakeman, E., and Pusey, P. N. (1976). A model for non-Rayleigh sea echo, *IEEE Transactions on Antennas and Propagation*, **AP-24**, 806–814.

Jakeman, E., and Tough, R. J. A. (1987). Generalized K distribution: A statistical model for weak scattering, *Journal of the Optical Society of America*, **74**, 1764–1772.

Jakeman, E., and Tough, R. J. A. (1988). Non-Gaussian models for the statistics of scattering waves, *Advances in Physics*, **37**, 471–529.

James, A. T. (1954). Normal multivariate analysis and the orthogonal group, *Annals of Mathematical Statistics*, **25**, 40–75.

James, W., and Stein, C. (1955). Estimation with quadratic loss, *Proceedings of the Third Berkeley Symposium on Mathematical Statistics and Probability*, **1**, 361–379.

James, W., and Stein, C. (1961). Estimation with quadratic loss, in *Proceedings of the Fourth Berkeley Symposium on Mathematical Statistics and Probability* – **1**, pp. 361–380, Berkeley, California: University of California Press.

Jinadasa, K. G., and Tracy, D. S. (1992). Maximum likelihood estimation for multivariate normal distribution with monotone sample, *Communications in Statistics—Theory and Methods*, **21**, 41–50.

Joe, H. (1994). Multivariate extreme-value distributions with applications to environmental data, *Canadian Journal of Statistics*, **22**, 47–64.

Joe, H. (1995). Approximations to multivariate normal rectangle probabilities based on conditional expectations, *Journal of the American Statistical Association*, **90**, 957–964.

Jogdeo, K. (1970). A simple proof of an inequality for multivariate normal probabilities of rectangles, *Annals of Mathematical Statistics*, **41**, 1357–1359.

John, S. (1959). On the evaluation of the probability integral of a multivariate normal distribution, *Sankhyā*, **21**, 366–370.

John, S. (1963). A tolerance region for multivariate normal distributions, *Sankhyā, Series A*, **25**, 363–368.

John, S. (1966). On the evaluation of probabilities of convex polyhedra under multivariate normal and *t*-distributions, *Journal of the Royal Statistical Society, Series B*, **28**, 366-369.

John, S. (1968). A central tolerance region for the multivariate normal distribution, *Journal of the Royal Statistical Society, Series B*, **30**, 599–601.

Johnson, N. L., and Kotz, S. (1975). A vector multivariate hazard rate, *Journal of Multivariate Analysis*, **5**, 53–66.

Johnson, N. L., and Kotz, S. (1999). Square tray distributions, *Statistics & Probability Letters*, **42**, 157–165.

Johnson, N. L., Kotz, S., and Balakrishnan, N. (1994). *Continuous Univariate Distributions*, Vol. 1, second edition, New York: John Wiley & Sons.

Johnson, N. L., Kotz, S., and Balakrishnan, N. (1995). *Continuous Univariate Distributions*, Vol. 2, second edition, New York: John Wiley & Sons.

Johnson, N. L., Kotz, S., and Balakrishnan, N. (1997). *Discrete Multivariate Distributions*, New York: John Wiley & Sons.

Johnson, N. L., Kotz, S., and Kemp, A. W. (1992). *Univariate Discrete Distributions*, second edition, New York: John Wiley & Sons.

Judge, G. G., Hill, R. C., and Bock, M. E. (1990). An adaptive empirical Bayes estimator of the multivariate normal mean under quadratic loss, *Journal of Econometrics*, **44**, 189–213.

Kabe, D. G. (1968). Minimum variance unbiased estimate of a coverage probability, *Operations Research*, **16**, 1016–1020.

Kagan, A. (1998). Uncorrelatedness of linear forms implies their independence only for Gaussian random vectors, Preprint.

Kagan, A. M., Linnik, Yu. V., and Rao, C. R. (1965). On a characterization of the normal law based on a property of the sample average, *Sankhyā, Series A*, **27**, 405–406.

Karunamuni, R. J., and Schmuland, B. (1995). A robust generalized Bayes estimator of a multivariate normal mean, *Mathematical Methods in Statistics*, **4**, 472–482.

Kendall, M. G. (1941). Proof of relations connected with tetrachoric series and its generalization, *Biometrika*, **32**, 196–198.

Kendall, M. G. (1954). Two problems in sets of measurements, *Biometrika*, **41**, 560-564.

Kendall, M. G., and Stuart, A. (1963). *The Advanced Theory of Statistics*, Vol. 1, second edition, London: Griffin.

Kendall, M. G., and Stuart, A. (1966). *The Advanced Theory of Statistics*, Vol. 3, first edition, London: Griffin.

Kent, J., Briden, J., and Mardia, K. V. (1983). Linear and planar structure in ordered multivariate data as applied to progressive demagnetization of palaeomagnetic remanence, *Geophysical Journal of the Royal Astronomical Society*, **75**, 593–662.

Khan, R. A. (1968). Sequential estimation of the mean vector of multivariate normal distribution, *Sankhyā, Series B*, **30**, 331–334.

Khatri, C. G. (1965a). Classical statistical analysis based on a certain multivariate complex Gaussian distribution, *Annals of Mathematical Statistics*, **36**, 98–114.

Khatri, C. G. (1965b). Joint estimation of the parameters of multivariate normal distributions, *Journal of the Indian Statistical Association*, **1**, 125–133.

Khatri, C. G. (1967). On certain inequalities for normal distributions and their applications to simultaneous confidence bounds, *Annals of Mathematical Statistics*, **38**, 1853–1867.

Khatri, C. G. (1971). On characterization of gamma and multivariate normal distributions by solving some functional equations in vector variables, *Journal of Multivariate Analysis*, **1**, 70–89.

Kibble, W. F. (1945). An extension of a theorem of Mehler on Hermite polynomials, *Proceedings of the Cambridge Philosophical Society*, **41**, 12–15.

Kingman, A., and Graybill, F. A. (1970). A non-linear characterization of the normal distribution, *Annals of Mathematical Statistics*, **41**, 1889–1895.

Kowalski, C. J. (1973). Non-normal bivariate distributions with normal marginals, *The American Statistician*, **27**, 103–106.

Krishnamoorthy, K. (1991). Estimation of a common multivariate normal mean vector, *Annals of the Institute of Statistical Mathematics*, **43**, 761–771.

Krishnamoorthy, K., and Mathew, T. (1999). Comparison of approximation methods for computing tolerance factors for a multivariate normal population, *Technometrics*, **41**, 234–249.

Krishnamoorthy, K., and Pannala, M. K. (1999). Confidence estimation of a normal mean vector with incomplete data, *Canadian Journal of Statistics*, **27**, 395–407.

Krishnamoorthy, K., and Rohatgi, V. K. (1990). Unbiased estimation of the common mean of a multivariate normal distribution, *Communications in Statistics—Theory and Methods*, **19**, 1803–1810.

Kubokawa, T. (1989). Improved estimation of a covariance matrix under quadratic loss, *Statistics & Probability Letters*, **8**, 69–71.

Kudô, A. (1958). On the distribution of the maximum value of an equally correlated sample from a normal population, *Sankhyā*, **20**, 309–316.

Kunte, S., and Rattihalli, R. N. (1989). Bayes set estimators for the mean of a multivariate normal distribution and rate of convergence of their posterior risk, *Sankhyā, Series A*, **51**, 94–105.

Kusunori, K. (1967). On the estimates of the unknown parameters in the multivariate distribution with the intraclass correlation, Technical Report, Kansas University 1967, No. 9, 101–108.

Kuwana, Y., and Kariya, T. (1991). LBI tests for multivariate normality in exponential power distributions, *Journal of Multivariate Analysis*, **39**, 117–134.

Kwong, K. S. (1995). Evaluation of one-sided percentage points of the singular multivariate normal distribution, *Journal of Statistical Computation and Simulation*, **51**, 121–135.

Kwong, K. S., and Iglewicz, B. (1996). On singular multivariate normal distribution and its applications, *Computational Statistics & Data Analysis*, **22**, 271–285.

Laha, R. G. (1953). On an extension of Geary's theorem, *Biometrika*, **40**, 228–229.

Laha, R. G. (1955). On a characterization of the multivariate normal distribution, *Sankhyā*, **14**, 376–368.

Lancaster, H. O. (1959). Zero correlation and independence, *Australian Journal of Statistics*, **21**, 53–56.

Lawley, D. N. (1943). A note on Karl Pearson's selection formulae, *Proceedings of the Royal Society of Edinburgh*, **62**, 28–30.

Leonard, T. (1974). A Bayesian method for the simultaneous estimation of several parameters, Technical Report, University of Warwick, Warwick, U.K.

Leonard, T. (1976). Some alternative approaches to multiparameter estimation, *Biometrika*, **63**, 69–76.

Letac, G. (1981). Isotropy and sphericity: Some characterizations of the normal distribution, *Annals of Statistics*, **9**, 408–417.

Lewin, L. (1958). *Dilogarithms and Associated Functions*, London: Macdonald.

Li, H. C. (1978). A characterization of the multivariate normal distribution, *Journal of Multivariate Analysis*, **8**, 255–261.

Li, T. F., and Bhoj, D. S. (1986). A family of admissible minimax estimators of the mean of a multivariate normal distribution, *Canadian Journal of Statistics*, **14**, 245–250.

Li, T. F., and Bhoj, D. S. (1991). Bayes minimax estimators of a multivariate normal mean, *Statistics & Probability Letters*, **11**, 373–377.

Lin, P. E., and Tsai, H. L. (1973). Generalized Bayes minimax estimators of the multivariate normal mean with unknown covariance matrix, *Annals of Statistics*, **1**, 142–145.

Little, R. J. A., and Rubin, D. B. (1987). *Statistical Analysis with Missing Data*, New York: John Wiley & Sons.

Lohr, S. L. (1993). Algorithm AS285: Multivariate normal probabilities of star-shaped regions, *Applied Statistics*, **42**, 576–582.

Lord, F. M. (1955). Estimation of parameters from incomplete data, *Journal of the American Statistical Association*, **50**, 870–876.

Lukacs, E. (1942). A characterization of the normal distribution, *Annals of Mathematical Statistics*, **13**, 91–93.

Lukacs, E. (1956). Characterization of populations by properties of suitable statistics, *Proceedings of the Third Berkeley Symposium on Mathematical Statistics and Probability*, **2**, 195–214.

Lukacs, E., and Laha, R. G. (1964). *Applications of Characteristic Functions*, London: Griffin.

Lumel'skii, Ya. P. (1968). Unbiased sufficient estimations of probability for the multivariate normal distribution, *Vestnik Moskovskogo Universiteta*, No. 6, 14–17 (in Russian).

Lumel'skii, Ya. P., and Sapozhnikov, P. N. (1969). Unbiased estimates of density functions, *Theory of Probability and its Applications*, **14**, 357–364.

Ma, C. (1997). A note on the multivariate normal hazard, Preprint.

Marsaglia, G. (1963). Expressing the normal distribution with covariance matrix $A + B$ in terms of one with covariance matrix A, *Biometrika*, **50**, 535–538.

Martynov, G. V. (1981). Evaluation of the normal distribution function, *Journal of Soviet Mathematics*, **17**, 1857–1875.

Maruyama, Y. (1998). A unified and broadened class of admissible minimax estimators of a multivariate normal mean, *Journal of Multivariate Analysis*, **64**, 196–205.

Masuda, K. (1980). The unbiased estimation of the proportion on a given domain of the multivariate normal distribution based on the sufficient statistics, *TRU Mathematics*, **16**, 155–162.

McFadden, J. A. (1956). An approximation for the symmetric, quadrivariate normal integral, *Biometrika*, **43**, 206–207.

McFadden, J. A. (1960). Two expansions for the quadrivariate normal integral, *Biometrika*, **47**, 325–333.

McLachlan, G. J., and Basford, K. E. (1987). *Mixture Models: Inference and Applications to Clustering*, New York: Marcel Dekker.

Mehler, F. G. (1866). Über die Entwicklung einer Funktion von beliebig vielen Variablen nach Laplace'schen Funktionen höherer Ordnung, *Journal für die Reine und Angewandte Mathematik*, **66**, 161–176.

Melnick, E. L., and Tenenbein, A. (1982). Misspecification of the normal distribution, *The American Statistician*, **36**, 372–373.

Miller, K. S. (1964). *Multidimensional Gaussian Distributions*, New York: John Wiley & Sons.

Miller, K. S., and Sackrowitz, H. (1965). Distributions associated with the quadrivariate normal, *Journal of the Industrial Mathematics Society*, **15**, No. 2, 1–15.

Milton, R. C. (1972). Computer evaluation of the multivariate normal integral, *Technometrics*, **14**, 881–889.

Mises, R. von (1954). Numerische Berechnung mehrdimensionaler Integrale, *Zeitschrift für Angewandte Mathematik und Mechanik*, **34**, 201–210.

Moran, P. A. P. (1948). Rank correlation and product moment correlation, *Biometrika*, **35**, 203–206.

Moran, P. A. P. (1956). The numerical evaluation of a class of integrals, *Proceedings of the Cambridge Philosophical Society*, **52**, 230–233.

Moran, P. A. P. (1983). A new expansion for the multivariate normal distribution, *Australian Journal of Statistics*, **25**, 339–344.

Moran, P. A. P. (1984). The Monte Carlo evaluation of orthant probabilities for multivariate normal distribution, *Australian Journal of Statistics*, **26**, 39–44.

Moran, P. A. P. (1985). Calculations of multivariate normal probabilities— another special case, *Australian Journal of Statistics*, **27**, 60–67.

Muirhead, R. J. (1982). *Aspects of Multivariate Statistical Theory*, New York: John Wiley & Sons.

Mukhopadhyay, N., and Al-Mousawi, J. S. (1986). Fixed-size confidence regions for the mean vector of a multinormal distribution, *Sequential Analysis*, **5**, 139–168.

Murray, G. D. (1979). The estimation of multivariate normal density functions using incomplete data, *Biometrika*, **66**, 375–380.

Nachtsheim, C. J., and Johnson, M. E. (1988). A new family of multivariate distributions with applications to Monte Carlo studies, *Journal of the American Statistical Association*, **83**, 984–990.

Nagao, H. (1996). On fixed width confidence regions for multivariate normal mean when the covariance matrix has some structure, *Sequential Analysis*, **15**, 37–46.

Nelson, P. R. (1991). Numerical evaluation of multivariate normal integrals with correlations $\rho_{ij} = -\alpha_i\alpha_j$, in *The Frontiers of Statistical Scientific Theory and Industrial Applications, Proceedings of ICOSCO-I*, Vol. II, pp. 97–114, Columbus, Ohio: American Sciences Press.

Nguyen, T. T., and Sampson, A. R. (1991). A note on characterizations of multivariate stable distributions, *Annals of the Institute of Statistical Mathematics*, **43**, 793–801.

Novak, L. M., Sechtin, M. B., and Cardullo, M. J. (1989). Studies of target detection algorithms that use polarmetric radar data, *IEEE Transactions on Aerospace and Electronic Systems*, **AES–25**, 150–164.

Olkin, I., and Pratt, J. W. (1958). Unbiased estimation of certain correlation coefficients, *Annals of Mathematical Statistics*, **29**, 201–211.

Olkin, I., and Roy, S. N. (1954). On multivariate distribution theory, *Annals of Mathematical Statistics*, **25**, 329–339.

Olkin, I., and Viana, M. (1995). Correlation analysis of extreme observations from a multivariate normal distribution, *Journal of the American Statistical Association*, **90**, 1373–1379.

Pal, N. (1993). Estimating the normal dispersion matrix and the precision matrix from a decision-theoretic point of view: A review, *Statistische Hefte*, **34**, 1–26.

Pal, N., and Elfessi, A. (1995). Improved estimation of a multivariate normal mean vector and the dispersion matrix: How one affects the other, *Sankhyā, Series A*, **57**, 267–286.

Pal, N., and Ling, C. (1995). Improved minimax estimation of powers of the variance of a multivariate normal distribution under the entropy loss function, *Statistics & Probability Letters*, **24**, 205–211.

Pal, N., Sinha, B. K., Chaudhuri, G., and Chang, C.-H. (1995). Estimation of multivariate normal mean vector and local improvements, *Statistics*, **26**, 1–17.

Patil, G. P., and Boswell, M. T. (1970). A characteristic property of the multivariate normal distribution and some of its applications, *Annals of Mathematical Statistics*, **41**, 1970–1977.

Patil, G. P., and Seshadri, V. (1964). Characterization theorems for some univariate probability distributions, *Journal of the Royal Statistical Society, Series B*, **26**, 286–292.

Pearson, K. (1903). Mathematical contributions to the theory of evolution—XI. On the influence of natural selection on the variability and correlation of organs, *Philosophical Transactions of the Royal Society of London, Series A*, **200**, 1–66.

Peristiani, S. (1991). The $\mathcal{F}$-system distribution as an alternative to multivariate normality: An application in multivariate models with qualitative dependent variables, *Communications in Statistics—Theory and Methods*, **20**, 147–163.

Perron, F. (1990). Equivariant estimators of the covariance matrix, *Canadian Journal of Statistics*, **18**, 179–182.

Perron, F., and Giri, N. (1990). On the best equivariant estimator of mean of a multivariate normal population, *Journal of Multivariate Analysis*, **32**, 1–6.

Pierce, D. A., and Dykstra, R. L. (1969). Independence and the normal distribution, *The American Statistician*, **23**, 39.

Plackett, R. L. (1954). A reduction formula for normal multivariate integrals, *Biometrika*, **41**, 351–360.

Poznyakov, V. V. (1971). On one representation of the multidimensional normal distribution function, *Ukrainian Mathematical Journal*, **23**, 562–66.

Press, S. J. (1975). Simultaneous Bayesian estimation of multivariate normal parameters, Report No. P-5438, The Rand Corporation, Santa Monica, California.

Press, S. J., and Rolph, J. E. (1986). Empirical Bayes estimation of the mean of a multivariate normal distribution, *Communications in Statistics—Theory and Methods*, **15**, 2201–2228.

Provost, S. B. (1990). Estimators for the parameters of a multivariate normal random vector with incomplete data on two subvectors and test of independence, *Computational Statistics & Data Analysis*, **9**, 37–46.

Quenouille, M. H. (1950). The evaluation of probabilities in a normal multivariate distribution with special reference to correlation ratio, *Proceedings of the Edinburgh Mathematical Society*, **8**, 95–100.

Rao, C. R. (1965). *Linear Statistical Inference and Its Applications*, New York: John Wiley & Sons.

Rao, C. R. (1969a). On vector variables with a linear structure and a characterization of the multivariate normal distribution, *Bulletin of the International Statistical Institute*, **42**, 1207–1212.

Rao, C. R. (1969b). Some characterizations of the multivariate normal distribution, in *Multivariate Analysis-II* (P. R. Krishnaiah, eds.), pp. 321–328, New York: Academic Press.

Rattihalli, R. N. (1994). Bayes set estimators and rate of convergence of risk for the mean of multivariate normal distibution when precision matrix is unknown, *Sankhyā, Series A*, **56**, 516–523.

Rinott, Y., and Samuel-Cahn, E. (1994). Covariance between variables and their order statistics for multivariate normal variables, *Statistics & Probability Letters*, **21**, 153–155.

Ruben, H. (1961). On the numerical evaluation of a class of multivariate normal integrals, *Proceedings of the Royal Society of Edinburgh*, **65**, 272–281.

Ruben, H. (1962). An asymptotic expansion for a class of multivariate normal integrals, *Journal of the Australian Mathematical Society*, **2**, 253–264.

Ruben, H. (1964). An asymptotic expansion for the multivariate normal distribution and Mill's ratio, *Journal of Research, National Bureau of Standards*, **68B**, 3–11.

Rukhin, A. L. (1967). The complex normal law and the admissibility of the sample mean as an estimator of a location parameter, *Teoriya Veroyatnostei i ee Primeneniya*, **12**, 762–764 (in Russian). English translation, 695–697.

Rukhin, A. L., and Ananda, M. M. A. (1992). Risk behavior of variance estimators in multivariate normal distribution, *Statistics & Probability Letters*, **13**, 159–166.

Sampson, P. D., and Siegel, A. F. (1985). The measure of "size" independent of "shape" for multivariate lognormal populations, *Journal of the American Statistical Association*, **80**, 910–914.

Sarabia, J.-M. (1995). The centered normal conditionals distribution, *Communications in Statistics—Theory and Methods*, **24**, 2889–2900.

Savage, R. (1962). Mill's ratio for multivariate normal distributions, *Journal of Research, National Bureau of Standards*, **66B**, 93–96.

Schervish, M. J. (1984). Multivariate normal probabilities with error bound, *Applied Statistics*, **33**, 81–94. Correction, **34** (1985), 103–104.

Schläfli, L. (1858). On the multiple integral..., *Quarterly Journal of Pure and Applied Mathematics*, **2**, 269–301; **3**, 54–68; **3**, 97–108.

Scott, A. (1967). A note on conservative confidence regions for the means of a multivariate normal, *Annals of Mathematical Statistics*, **38**, 278–280.

Seneta, E. (1987). Multivariate probability in terms of marginal probability and correlation coefficient, *Biometrical Journal*, **29**, 375–380.

Seshadri, V. (1966). A characteristic property of the multivariate normal distribution, *Annals of Mathematical Statistics*, **37**, 1829–1831.

Shimizu, R. (1962). Characterization of the normal distribution, II, *Annals of the Institute of Statistical Mathematics*, **14**, 173–178.

Shinozaki, N. (1974). A note on estimating the mean vector of a multivariate normal distribution with general quadratic loss function, *Keio Engineering Reports*, **27**, 105–112.

Šidák, Z. (1967). Rectangular confidence regions for the means of multivariate normal distributions, *Journal of the American Statistical Association*, **62**, 626–633.

Šidák, Z. (1968). On multivariate normal probabilities of rectangles: Their dependence on correlations, *Annals of Mathematical Statistics*, **39**, 1425–1434.

Siegel, A. F. (1993). A surprising covariance involving the minimum of multivariate normal variables, *Journal of the American Statistical Association*, **88**, 77–80.

Singh, N. (1960). Estimation of parameters of a multivariate population from truncated and censored samples, *Journal of the Royal Statistical Society, Series B*, **22**, 307–311.

Sinha, B. K. (1987). Inadmissibility of the best equivariant estimators of the variance–covariance matrix, the precision matrix and the generalized variance: A survey, *Proceedings of the International Symposium on Advances in Multivariate Statistical Analysis*, Indian Statistical Institute, Calcutta.

Sinha, B. K., and Ghosh, M. (1987). Inadmissibility of the best equivariant estimators of the variance-covariance matrix, the precision matrix and the generalized variance under entropy loss, *Statistics and Decisions*, **5**, 201–227.

Siotani, M. (1964). Tolerance regions for a multivariate normal population, *Annals of the Institute of Statistical Mathematics*, **16**, 135–153.

Skitovič, V. P. (1954). Linear combinations of independent random variables and the normal distribution law, *Izvestiya Akademii Nauk SSSR, Series Mathematics*, **18**, 185–200. English translation in *Selected Translations in Mathematical Statistics and Probability*, **2**, 211–228 (1962).

Slepian, D. (1962). The one sided barrier problem for Gaussian noise, *Bell System Technical Journal*, **41**, 463–501.

Smith, W. B., and Hocking, R. R. (1968). A simple method for obtaining the information matrix for a multivariate normal distribution, *The American Statistician*, **22**, No. 1, 18–19.

Solow, A. R. (1990). A method for approximating multivariate normal orthant probabilities, *Journal of Statistical Computation and Simulation*, **37**, 225–229.

Sondhi, M. M. (1961). A note on the quadrivariate normal integral, *Biometrika*, **48**, 201–203.

Soong, W. C., and Hsu, J. C. (1997). Using complex integration to compute multivariate normal probabilities, *Journal of Computational and Graphical Statistics*, **6**, 397–415.

Srivastava, M. S. (1967). On fixed width confidence bounds for regression parameters and mean vector, *Journal of the Royal Statistical Society, Series B*, **29**, 132–140.

Srivastava, M. S., and Bhargava, R. P. (1979). On fixed-width confidence region for the mean, *Metron*, **27**, 163–174.

Srivastava, M. S., and Carter, E. M. (1983). *An Introduction to Applied Multivariate Statistics*, Amsterdam, The Netherlands: North-Holland.

Stadje, W. (1993). ML characterization of the multivariate normal distribution, *Journal of Multivariate Analysis*, **46**, 131–138.

Steck, G. P. (1958). A table for computing trivariate normal probabilities, *Annals of Mathematical Statistics*, **29**, 780–800.

Steck, G. P. (1979). Lower bounds for the multivariate normal Mills' ratio, *Annals of Probability*, **7**, 547–551.

Steck, G. P., and Owen, D. B. (1962). A note on the equicorrelated multivariate normal distribution, *Biometrika*, **49**, 269–271.

Stein, C. (1956). Inadmissibility of the usual estimator for the mean of a multivariate normal distribution, in *Proceedings of the Third Berkeley Symposium on Mathematical Statistics and Probability*—**1**, pp. 197–206, Berkeley, California: University of California Press.

Stein, C. (1964). Inadmissibility of the usual estimator for the variance of a normal distribution with unknown mean, *Annals of the Institute of Statistical Mathematics*, **16**, 155–160.

Stein, C. (1981). Estimation of the mean of a multivariate normal distribution, *Annals of Statistics*, **9**, 1135–1151.

Strawderman, W. E. (1971). Proper Bayes minimax estimators of the multivariate normal mean, *Annals of Mathematical Statistics*, **42**, 385–388.

Strawderman, W. E. (1974). Minimax estimation of powers of the variance of a normal population under squared error loss, *Annals of Statistics*, **2**, 190–198.

Stuart, A. (1958). Equally correlated variates and the multinormal integral, *Journal of the Royal Statistical Society, Series B*, **20**, 373–378.

Stuart, A., and Ord, J. K. (1994). *Kendall's Advanced Theory of Statistics*, Vol. 1, Sixth edition, London: Arnold.

Styan, G. P. H. (1968). Inference in multivariate normal populations with structure–Part I: Inference on variance when correlations are known, Technical Report No. 1, School of Aerospace Medicine, Brooks Air Force Base, Texas.

Sun, H.-J. (1988a). A Fortran subroutine for computing normal orthant probabilities of dimensions up to nine, *Communications in Statistics— Simulation and Computation*, **17**, 1097–1111.

Sun, H.-J. (1988b). A general reduction method for n-variate normal orthant probability, *Communications in Statistics—Theory and Methods*, **17**, 3913–3921.

Sungur, E. A., and Kovacevic, M. S. (1990). One dimensional marginal density functions of a truncated multivariate normal density function, *Communications in Statistics—Theory and Methods*, **19**, 197–203.

Sungur, E. A., and Kovacevic, M. S. (1991). Lower dimensional marginal density functions of absolutely continuous truncated multivariate distributions, *Communications in Statistics—Theory and Methods*, **20**, 1569–1578.

Tallis, G. M. (1961). The moment generating function of the truncated multinormal distribution, *Journal of the Royal Statistical Society, Series B*, **23**, 233–239.

Tallis, G. M. (1963). Elliptical and radial truncation in normal populations, *Annals of Mathematical Statistics*, **34**, 940–944.

Tallis, G. M. (1965). Plane truncation in normal populations, *Journal of the Royal Statistical Society, Series B*, **27**, 301–307.

Tallis, G. M. (1967). Approximate maximum likelihood estimates from grouped data, *Technometrics*, **9**, 599–606.

Teicher, H. (1961). Maximum likelihood characterization of normal distribution, *Annals of Mathematical Statistics*, **32**, 1214–1222.

Thomas, D. H. (1970). A Cauchy type functional equation and a characterization of the multivariate normal distribution, Report GMR-1003, General Motors Corporation.

Titterington, D. M., Smith, A. F. M., and Makov, U. E. (1985). *Statistical Analysis of Finite Mixture Distributions*, New York: John Wiley & Sons.

Todhunter, J. (1869). On the method of least squares, *Transactions of the Cambridge Philosophical Society*, **11**, 219–238.

Tong, Y. L. (1970). Some probability inequalities of multivariate normal and multivariate *t*, *Journal of the American Statistical Association*, **65**, 1243–1247.

Tong, Y. L. (1990). *The Multivariate Normal Distribution*, New York: Springer-Verlag.

Trawinski, I., and Bargmann, R. E. (1964). Maximum likelihood estimation with incomplete multivariate data, *Annals of Mathematical Statistics*, **35**, 647–658.

Urzúa, C. M. (1988). A class of maximum-entropy multivariate distributions, *Communications in Statistics—Theory and Methods*, **17**, 4039–4057.

van der Vaart, H. R. (1953). The content of certain spherical polyhedra for any number of dimensions, *Experientia*, **9**, 88–89.

van der Vaart, H. R. (1955). The content of some classes of non-Euclidean polyhedra for any number of dimensions, with several applications, I, II, *Proceedings of the Royal Academy of Sciences, Amsterdam, Series A*, **58**, 199–221.

Vijverberg, W. P. M. (1997). Monte-Carlo evaluation of multivariate normal probabilities, *Journal of Economics*, **76**, 281–307.

Vitale, R. A. (1996). Covariance identities for normal variables via convex polytopes, *Statistics & Probability Letters*, **30**, 363–368.

Wald, A. (1942). Setting of tolerance limits when the sample is large, *Annals of Mathematical Statistics*, **13**, 389–399.

Waldman, D. M. (1984). The mean of the conditional truncated multinormal distribution, *Communications in Statistics—Theory and Methods*, **13**, 2679–2689.

Wang, L. (1991). Admissible linear estimators of the multivariate normal mean without extra information, *Statistical Papers*, **32**, 155–165.

Wang, Y. J. (1987). The probability integrals for the bivariate normal distributions: A contingency-table approach, *Biometrika*, **74**, 185–190.

Wang, Y. J. (1997). Multivariate normal integrals and contingency tables with ordered categories, *Psychometrika*, **62**, 267–284.

Watterson, G. A. (1959). Linear estimation in censored samples from multivariate normal populations, *Annals of Mathematical Statistics*, **30**, 814–825.

Webster, J. T. (1970). On the application of the method of Das in normal integral, *Biometrika*, **57**, 657–660.

Weiss, M. C. (1966). Determination d'une variable de Gauss a plusieurs dimensions a l'aide de la function caractéristique, *Journal de la Société de Statistique de Paris*, **107**, 135–136.

Wolfe, J. H. (1970). Pattern clustering by multivariate mixture analysis, *Multivariate Behavioral Research*, **5**, 329–350.

Wooding, R. A. (1956). The multivariate distribution of complex normal variables, *Biometrika*, **43**, 212–215.

Woodroofe, M. (1982). *Nonlinear Renewal Theory in Sequential Analysis*, Philadelphia: Society for Industrial and Applied Mathematics.

Yatchev, A. (1986). Multivariate distributions involving ratios of normal variables, *Communications in Statistics—Theory and Methods*, **15**, 1905–1926.

Yoneda, K. (1961). Some estimations of the parameters of multinormal populations from linearly truncated samples, *Yokohama Mathematical Journal*, **9**, 149–161.

Young, J. C. (1971). Some inference problems associated with the complex multivariate normal distribution, Technical Report No. 102, Department of Statistics, Southern Methodist University, Dallas, Texas.

Yueh, S. H., Kong, J. A., Jao, J. K., Shin, R. T., Zebker, H. A., and Le Toan, T. (1991). K-distribution and multifrequency polarmetric terrain, *Journal of Electromagnetic Waves and Applications*, **5**, 1–15.

Zinger, A. A., and Linnik, Yu. V. (1964). A characteristic property of the normal distribution, *Theory of Probability and Its Applications*, **9**, 624–626.

CHAPTER 46

Bivariate and Trivariate Normal Distributions

1 DEFINITIONS AND APPLICATIONS

In this chapter we make frequent use of results presented in Chapter 45. Our attention is concentrated on the details of work that has been done on multivariate normal distributions with two or three variables, and not on general multivariate normal distributions.

When there are just two variables, X_1, and X_2, Eq. (45.1) becomes

$$
\begin{aligned}
p(x_1, x_2; \xi_1, \xi_2, \sigma_1, \sigma_2, \rho) \\
= \left[2\pi\sqrt{1-\rho^2}\right]^{-1} \exp\left[-\frac{1}{2(1-\rho^2)}\left\{\left(\frac{x_1-\xi_1}{\sigma_1}\right)^2\right.\right. \\
\left.\left. - 2\rho\left(\frac{x_1-\xi_1}{\sigma_1}\right)\left(\frac{x_2-\xi_2}{\sigma_2}\right) + \left(\frac{x_2-\xi_2}{\sigma_2}\right)^2\right\}\right],
\end{aligned} \tag{46.1}
$$

where $E[X_j] = \xi_j$, $\mathrm{var}(X_j) = \sigma_j^2$ $(j = 1, 2)$, and the correlation between X_1 and X_2 is ρ. This is the *bivariate normal* distribution. Other names are *Gaussian, Laplace-Gauss*, and *Bravais* (1846).

It is possible to regard the bivariate normal as a "univariate" complex normal distribution (Chapter 45), but this does not possess any advantages for our present interest.

For many purposes, it is sufficient to study the standardized distribution, obtained by putting $\xi_1 = \xi_2 = 0$ and $\sigma_1 = \sigma_2 = 1$ in (46.1):

$$
p(x_1, x_2; \rho) = \left[2\pi\sqrt{1-\rho^2}\right]^{-1} \exp\left\{-\frac{1}{2(1-\rho^2)}\left(x_1^2 - 2\rho x_1 x_2 + x_2^2\right)\right\}. \tag{46.2}
$$

If in (46.1) we have $\sigma_1 = \sigma_2$ and $\rho - 0$, it is called a *circular normal* density function. This should not be confused with the univariate circular normal distribution. If $\rho = 0$ but $\sigma_1 \neq \sigma_2$, the name *elliptical normal* is used sometimes.

The *standardized trivariate normal* probability density function of three random variables X_1, X_2, X_3 depends on the correlation coefficients $\rho_{23}, \rho_{13}, \rho_{12}$ and can be written as

$$p_{X_1,X_2,X_3}(x_1, x_2, x_3) = (2\pi)^{-3/2}\Delta^{-1/2}\exp\left\{-\frac{1}{2}\sum_{i=1}^{3}\sum_{j=1}^{3}A_{ij}x_ix_j\right\}, \quad (46.3)$$

where

$$\begin{aligned}
\Delta &= 1 - \rho_{23}^2 - \rho_{13}^2 - \rho_{12}^2 + 2\rho_{23}\rho_{13}\rho_{12}, \\
A_{11} &= (1 - \rho_{23}^2)/\Delta, \quad A_{22} = (1 - \rho_{13}^2)/\Delta, \quad A_{33} = (1 - \rho_{12}^2)/\Delta, \\
A_{12} &= A_{21} = (\rho_{13}\rho_{23} - \rho_{12})/\Delta, \quad A_{13} = A_{31} = (\rho_{12}\rho_{23} - \rho_{13})/\Delta, \\
A_{23} &= A_{32} = (\rho_{12}\rho_{13} - \rho_{23})/\Delta.
\end{aligned} \qquad (46.4)$$

If all ρ's are zero and all σ's are equal, the distribution is sometimes called *spherical normal*; if all ρ's are zero but σ's are not equal, the name *ellipsoidal normal* has been used.

Bivariate and trivariate normal distributions are used in a wide variety of applications. Among the oldest are applications in artillery fire control. To a first approximation, deviations from a target on a plane (for land artillery) are described by bivariate normal distributions. For aerial targets, trivariate normal distributions are appropriate.

Multivariate normal distributions are very commonly employed as approximations to joint distributions, even when the marginal distributions are not exactly normal, as already pointed out in Chapter 44. Although the theoretical framework thus constructed is useful as a basis for construction of tests and estimation procedures, it is only for the bivariate and (though to a lesser extent) trivariate normal distributions that it is easy to form a picture of the distribution. Study of these distributions is of special value in forming ideas of the effect of truncation, applied to one or two of a number of multivariate normal variables.

2   HISTORICAL REMARKS

Although the joint distribution of normal variables was considered occasionally as early as the beginning of the nineteenth century [Adrian (1808),

Bravais (1846), Plana (1813), and Helmert (1868)], it was not until the last quarter of that century that it became a subject of systematic study. The main impetus came from the work of Schols (1875) and especially of Galton (1877, 1888) at whose suggestion Dickson (1886) demonstrated a possible genesis for the bivariate normal distribution as the vector combination of independent normally distributed components on *oblique* axes. Subsequently, Pearson (1901, 1903) applied the bivariate normal distribution to biometric data. He also initiated work on tabulation of values of integral probabilities for bivariate normal distributions. Later work on tabulation of special values has been associated with development of techniques for selection among (univariate) normal populations with regard to their expected values; see, for example, Dunnett (1960) and Somerville (1954).

Accounts of the earlier history of the bivariate normal distribution have been given by Czuber (1891) and Pearson (1920). A briefer, but broader account, has been given by Anderson (1958).

3   PROPERTIES AND MOMENTS

For the standardized bivariate normal distribution in (46.2), the conditional distribution of X_2, given X_1, is normal with expected value ρX_1 and variance $(1 - \rho^2)$ [conversely, that of X_1, given X_2, is normal with expected value ρX_2 and variance $(1 - \rho^2)$]. This is reflected in the fact that $p(x_1, x_2)$ in (46.2) can be written as

$$Z(x_1)\, Z\left(\frac{x_2 - \rho x_1}{\sqrt{1 - \rho^2}}\right) \Big/ \sqrt{1 - \rho^2} \tag{46.5}$$

or

$$Z(x_2)\, Z\left(\frac{x_1 - \rho x_2}{\sqrt{1 - \rho^2}}\right) \Big/ \sqrt{1 - \rho^2}, \tag{46.6}$$

where $Z(x) = \frac{1}{\sqrt{2\pi}} \exp\left(-\frac{1}{2} x^2\right)$.

We may note here the characterization of the bivariate normal distributions, in terms of exponential-type distribution (see Chapter 44), obtained by Bildikar and Patil (1968). If X_1 and X_2 have a bivariate exponential-type distribution, that distribution is bivariate normal if and only if

(a) the regression of one variable or the other is linear, and

(b) the marginal distribution of one of the variables is normal.

Condition (b) may be replaced by the requirement that $(X_1 + X_2)$ have a normal distribution.

For the standardized trivariate normal distribution in (46.3), the regression of any variate on the other two is linear, with constant array variance. The distribution of X_3, given X_1 and X_2 for example, is normal with expected value $\rho_{13.2} X_1 + \rho_{23.1} X_2$ and variance $(1 - R_{3.12}^2)$, where $\rho_{ij.k}$ means partial correlation between X_i and X_j given X_k and $R_{3.12}^2$ is the multiple correlation of X_3 on X_1 and X_2. The joint distribution of X_1 and X_2, given X_3, is bivariate normal with marginal expected values $\rho_{13} X_3$ and $\rho_{23} X_3$, with variances $(1 - \rho_{13}^2)$ and $(1 - \rho_{23}^2)$, and with the correlation coefficient

$$\rho_{12.3} = \frac{\rho_{12} - \rho_{12}\rho_{23}}{\sqrt{(1 - \rho_{13}^2)(1 - \rho_{23}^2)}} . \tag{46.7}$$

The parameter $\rho_{12.3}$ is the *partial correlation* between X_1 and X_2, given X_3. The partial correlations $\rho_{23.1}$ and $\rho_{13.2}$ are defined similarly.

As special cases of the result stated earlier in Chapter 45, the statistics

$$\frac{1}{1 - \rho^2} (X_1^2 - 2\rho X_1 X_2 + X_2^2) \qquad \text{[for distribution (46.2)]} \tag{46.8}$$

and

$$\sum_{i=1}^{3} \sum_{j=1}^{3} A_{ij} X_i X_j \qquad \text{[for distribution (46.3)]} \tag{46.9}$$

have chi-square distributions with 2 and 3 degrees of freedom, respectively. This makes it easy to construct elliptical or ellipsoidal contours, within which specified proportions of the distributions lie. In the bivariate case, since $\Pr[\chi_2^2 < K] = 1 - e^{-K/2}$, the ellipse containing $100\alpha\%$ of the distribution has the simple equation

$$x_1^2 - 2\rho x_1 x_2 + x_2^2 = -2(1 - \rho^2) \log(1 - \alpha). \tag{46.10}$$

For the general bivariate distribution in (46.1), the corresponding ellipse has equation

$$\left(\frac{x_1 - \xi_1}{\sigma_1}\right)^2 - 2\rho \left(\frac{x_1 - \xi_1}{\sigma_1}\right)\left(\frac{x_2 - \xi_2}{\sigma_2}\right) + \left(\frac{x_2 - \xi_2}{\sigma_2}\right)^2$$
$$= -2(1 - \rho^2) \log(1 - \alpha). \tag{46.11}$$

Several such ellipses are shown in Figure 46.1. The major axis of the ellipse makes an angle $\theta = \frac{1}{2} \tan^{-1}[2\rho\sigma_1\sigma_2/(\sigma_1^2 - \sigma_2^2)]$ with the x_1 axis. Note that this angle is $45°$ if $\sigma_1 = \sigma_2$ and $\rho > 0$, whatever the numerical value of ρ. If $\rho = 0$ and $\sigma_1 = \sigma_2$ (i.e., the variables are independent and have equal variances), then (46.8) is the equation of a circle. The corresponding distribution is called *circular normal*. Some perspective drawings of the density function in (46.2) are shown in Figure 46.2. The values of $p(x_1, x_2)$ are measured in the vertical direction.

For bivariate normal distributions, zero correlation implies independence. This is, of course, not so in general. Examples of dependent normal variables with zero correlation are numerous; see, for example, Pitman (1939).

Mukherjea, Nakassis, and Miyashita (1986) wrote the joint cumulative distribution function corresponding to (46.2) in the form

$$F(x_1, x_2; \sigma_1, \sigma_2, \rho) = \frac{ab\sqrt{1-r^2}}{2\pi} \int_{-\infty}^{x_1} \int_{-\infty}^{x_2} e^{-\frac{1}{2}(a^2 u^2 - 2abruv + b^2 v^2)} \, du \, dv,$$

$$(46.12)$$

where $\sigma_1\sqrt{1-\rho^2} = 1/a$, $\sigma_2\sqrt{1-\rho^2} = 1/b$, and presented the following properties for the partial derivatives of $F(x_1, x_2)$ in (46.12):

$$\frac{\partial^2 F(x_1, x_2)}{\partial x_1 \partial x_2} = \frac{ab\sqrt{1-\rho^2}}{2\pi} e^{-\frac{1}{2}(a^2 x_1^2 - 2ab\rho x_1 x_2 + b^2 x_2^2)}, \qquad (46.13)$$

$$\frac{\partial F(x_1, x_2)}{\partial x_1} = \frac{a\sqrt{1-\rho^2}}{\sqrt{2\pi}} e^{-\frac{1}{2}a^2(1-\rho^2)x_1^2} \Phi(bx_2 - a\rho x_1)$$

$$= \frac{a\sqrt{1-\rho^2}}{\sqrt{2\pi}} e^{-\frac{1}{2}a^2(1-\rho^2)x_1^2} \{1 - \Phi(a\rho x_1 - bx_2)\},$$

$$(46.14)$$

and

$$\frac{\partial F(x_1, x_2)}{\partial x_1} = \frac{b\sqrt{1-\rho^2}}{\sqrt{2\pi}} e^{-\frac{1}{2}b^2(1-\rho^2)x_2^2} \Phi(ax_1 - b\rho x_2)$$

$$= \frac{b\sqrt{1-\rho^2}}{\sqrt{2\pi}} e^{-\frac{1}{2}b^2(1-\rho^2)x_2^2} \{1 - \Phi(b\rho x_2 - ax_1)\},$$

$$(46.15)$$

where $\Phi(\cdot)$ denotes the univariate standard normal cumulative distribution function.

Sungur (1990) has noted the property that

$$\frac{dF(x_1, x_2; \rho)}{d\rho} = p(x_1, x_2; \rho), \qquad (46.16)$$

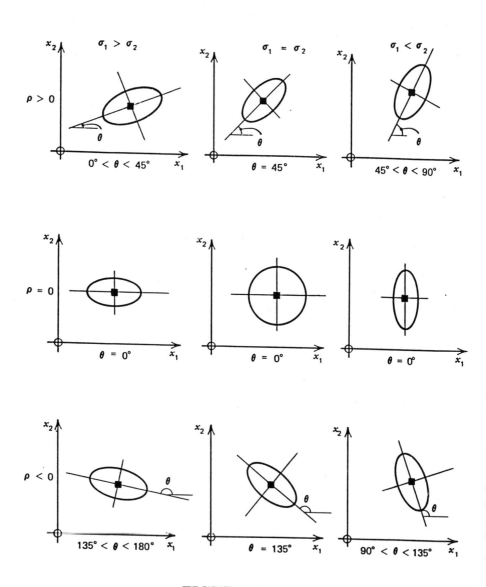

FIGURE 46.1

Contours of Equal Density of Bivariate Normal Distributions. (The Expected Value Point, or Centroid (ξ_1, ξ_2), is Denoted by ■.)

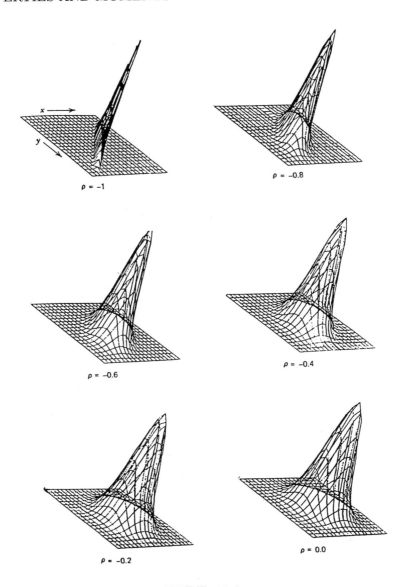

FIGURE 46.2
Plots of Standardized Bivariate Normal Density Function.

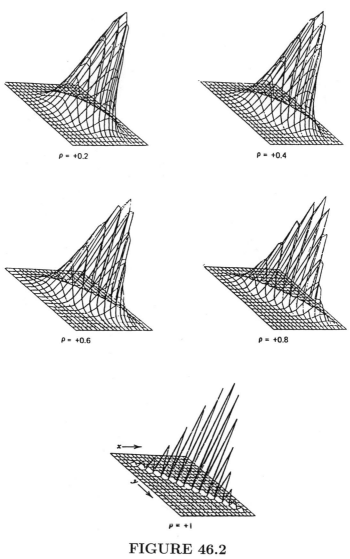

FIGURE 46.2
(*Continued*)

where $p(x_1, x_2; \rho)$ is the standard bivariate normal density function given (46.2) and $F(x_1, x_2; \rho)$ is the corresponding joint cumulative distribution

$$F(x_1, x_2; \rho) = \int_{-\infty}^{x_1} \int_{-\infty}^{x_2} p(t_1, t_2; \rho) \, dt_2 dt_1.$$

This property was pointed out earlier by Sibuya (1960). Sungur (1990) has also proposed a first-order approximation to the standard bivariate normal density function of the form

$$p(x_1, x_2; \rho) = \phi(x_1)\phi(x_2)(1 + \rho x_1 x_2) \qquad (46.17)$$

obtained by using the copula representation and the first-order Taylor expansion

$$C(u, v; \theta) = uv + (\theta - \theta_0)\phi(\Phi^{-1}(u))\phi(\Phi^{-1}(v)), \qquad (46.18)$$

where $\phi(\cdot)$ and $\Phi(\cdot)$ are the univariate standard normal density and cumulative distribution functions, respectively. Note that when we eliminate the influence of marginals on the bivariate normal distribution by using copulas, elliptical appearance disappears; hence, elliptical symmetry is not a property of the dependence structure of a bivariate normal distribution but that of its marginals. Sungur (1990) has then claimed that the best way of evaluating bivariate normality is to (i) verify that the univariate marginals are normal, (ii) standardize the observations, (iii) convert them to uniform variates by using the cumulative distribution function of normal, and (iv) plot the obtained variates against each other and compare them with the contours of the joint density function

$$\frac{1}{\sqrt{1-\rho^2}} \exp\left[-\frac{1}{2}\left\{\frac{(\rho\Phi^{-1}(u) - \Phi^{-1}(v))^2}{1-\rho^2} - (\Phi^{-1}(v))^2\right\}\right].$$

It is well-known that the standardized k-dimensional multivariate normal density function $\phi_k(\boldsymbol{x}; \boldsymbol{\Sigma})$, with correlation matrix $\boldsymbol{\Sigma}$, satisfies the identity

$$\frac{\partial}{\partial \rho_{ij}} \phi_k(\boldsymbol{x}; \boldsymbol{\Sigma}) = \frac{\partial^2}{\partial x_i \partial x_j} \phi_k(\boldsymbol{x}; \boldsymbol{\Sigma}); \qquad (46.19)$$

see Plackett (1954). Hence,

$$\frac{\partial}{\partial \rho_{ij}} \int_A \phi_k(\boldsymbol{x}; \boldsymbol{\Sigma}) \, d\boldsymbol{x} = \int_A \frac{\partial}{\partial \rho_{ij}} \phi_k(\boldsymbol{x}; \boldsymbol{\Sigma}) \, d\boldsymbol{x}$$

$$= \int_A \frac{\partial^2}{\partial x_i \partial x_j} \phi_k(\boldsymbol{x}; \boldsymbol{\Sigma}) \, d\boldsymbol{x} \qquad (46.20)$$

for many sets A of interest (such as orthants and rectangles). Let us now consider the bivariate case with density function $\phi(x_1, x_2; \rho)$. Moreover, let

$$
\begin{aligned}
F(x_1, x_2; \rho) &= \Pr[X_1 \le x_1, X_2 \le x_2] \text{ for } (x_1, x_2)^T \in \mathbb{R}^2, \\
G(x_1, x_2; \rho) &= \Pr[|X_1| \le x_1, |X_2| \le x_2] \text{ for } (x_1, x_2)^T \in \mathbb{R}_+^2,
\end{aligned}
$$

and

$$
h(x; \rho) = \rho(1 - \rho^2) + (\rho x_1 - x_2)(\rho x_2 - x_1).
$$

Applying Plackett's identity twice, we readily obtain

$$
\frac{\partial^2}{\partial \rho^2} F(x_1, x_2; \rho) = \frac{\phi(x_1, x_2; \rho) h(x; \rho)}{(1 - \rho^2)^2} \tag{46.21}
$$

and

$$
\begin{aligned}
&\frac{\partial^2}{\partial \rho^2} G(x_1, x_2; \rho) \\
&= \frac{2}{(1 - \rho^2)^2} \{\phi(x_1, x_2; \rho) h(x; \rho) + \phi(x_1, x_2; -\rho) h(x; -\rho)\}.
\end{aligned} \tag{46.22}
$$

Thus, $F(x_1, x_2; \rho)$ is convex in ρ for $h(x; \rho) \ge 0$, and $G(x_1, x_2; \rho)$ has the same property whenever

$$
\phi(x_1, x_2; \rho) h(x; \rho) + \phi(x_1, x_2; -\rho) h(x; -\rho) \ge 0.
$$

Iyengar and Tong (1989) have shown that: if $x_1 x_2 > 0$, then $F(x_1, x_2; \rho)$ is convex in $\rho \in [0, m]$; if $x_1 x_2 < 0$, then $F(x_1, x_2; \rho)$ is concave in $\rho \in [M, 0]$; and $G(x_1, x_2; \rho)$ is convex in $\rho \in [-m, m]$, where $m = \min\left(\frac{x_1}{x_2}, \frac{x_2}{x_1}\right)$ and $M = \max\left(\frac{x_1}{x_2}, \frac{x_2}{x_1}\right)$. In particular, when $x_1 = x_2 = c$ (i.e., on the diagonal), $F(x_1, x_2; \rho)$ is convex for all $\rho \ge 0$, and if $c \ge \sqrt{2} - 1$, $F(x_1, x_2; \rho)$ is convex in $\rho \in (-1, 1)$; also, $G(x_1, x_2; \rho)$ is convex for all $\rho \in (-1, 1)$ provided $c \ge 0$.

Azzalini and Dalla Valle (1996) have shown that if $(X_1, X_2)^T$ has a standard bivariate normal distribution with correlation coefficient ρ, then the conditional distribution of X_2 given $X_1 > 0$ is a skew-normal distribution with parameter $\lambda(\rho) = \frac{\rho}{\sqrt{1-\rho^2}}$; see Chapter 13 of Johnson, Kotz, and Balakrishnan (1994).

We have already noted in Section 2 of Chapter 45 that all cumulants, of order higher than 2, of *any* multivariate normal distribution are zero.

This is true, in particular, of bivariate normal and trivariate normal distributions, and it is therefore easy to evaluate the moments of these distributions. For the standardized distribution in (46.2), we have

$$
\begin{aligned}
\mu_{21} &= \mu_{12} = 0; \\
\mu_{31} &= \mu_{13} = 3\rho; \quad \mu_{22} = 1 + 2\rho^2; \\
\mu_{41} &= \mu_{14} = 0; \quad \mu_{32} = \mu_{23} = 0; \\
\mu_{51} &= \mu_{15} = 15\rho; \quad \mu_{42} = \mu_{24} = 3(1 + 4\rho^2); \quad \mu_{33} = 3\rho(3 + 2\rho^2).
\end{aligned}
\tag{46.23}
$$

In general,

$$
\mu_{2r,2s} = \frac{(2r)!(2s)!}{2^{r+s}} \sum_{j=0}^{t} \frac{(2\rho)^{2j}}{(r-j)!(s-j)!(2j)!}, \tag{46.24}
$$

$$
\mu_{2r+1,2s+1} = \frac{(2r+1)!(2s+1)!}{2^{r+s}}
$$
$$
\times \sum_{j=0}^{t} \frac{(2\rho)^{2j}}{(r-j)!(s-j)!(2j+1)!}, \tag{46.25}
$$

$$
\mu_{r,s} = 0 \quad \text{if } r+s \text{ is odd}, \tag{46.26}
$$

where $t = \min(r, s)$. The following recurrence relation exists [Kendall and Stuart (1963)]:

$$
\mu_{rs} = (r+s-1)\rho\mu_{r-1,s-1} + (r-1)(s-1)(1-\rho^2)\mu_{r-2,s-2}. \tag{46.27}
$$

Pearson and Young (1918) give tables of μ_{rs} to nine decimal places for $r, s \leq 10$ and $\rho = 0.00(0.05)1.00$.

The joint moment-generating function of X_1 and X_2 is

$$
E[e^{t_1 X_1 + t_2 X_2}] = \exp\left\{-\frac{1}{2}\left(t_1^2 + 2\rho t_1 t_2 + t_2^2\right)\right\}. \tag{46.28}
$$

Absolute moments are not so easily evaluated. It can be shown [Kamat (1953) and Nabeya (1951)] that

$$
\begin{aligned}
\nu_{rs} &= E[|X_1^r X_2^s|] \\
&= \pi^{-1} 2^{(r+s)/2} \Gamma\left(\frac{r+1}{2}\right) \Gamma\left(\frac{s+1}{2}\right) F\left(-\frac{1}{2}r, -\frac{1}{2}s; \frac{1}{2}; \rho^2\right),
\end{aligned}
\tag{46.29}
$$

where $F(\cdot)$ is the hypergeometric function

$$
F(\alpha, \beta; \gamma; z) = 1 + \frac{\alpha\beta}{1!\gamma} z + \frac{\alpha(\alpha+1)\beta(\beta+1)}{2!\gamma(\gamma+1)} z^2 + \cdots ;
$$

see Chapter 1. For small r and s, the hypergeometric and gamma functions can be evaluated in terms of more elementary functions, giving

$$
\left\{
\begin{aligned}
\nu_{11} &= \tfrac{2}{\pi}\left(\sqrt{1-\rho^2}+\rho\sin^{-1}\rho\right), \\[2mm]
\nu_{12} &= \nu_{21} = \sqrt{\tfrac{2}{\pi}}(1+\rho^2), \\[2mm]
\nu_{13} &= \nu_{31} = \tfrac{2}{\pi}\left\{\sqrt{1-\rho^2}(2+\rho^2)+3\rho\sin^{-1}\rho\right\}, \\[2mm]
\nu_{22} &= 1+2\rho^2, \\[2mm]
\nu_{14} &= \nu_{41} = \sqrt{\tfrac{2}{\pi}}(3+6\rho^2-\rho^4), \\[2mm]
\nu_{23} &= \nu_{32} = 2\sqrt{\tfrac{2}{\pi}}(1+3\rho^2), \\[2mm]
\nu_{15} &= \nu_{51} = \tfrac{2}{\pi}\left\{\sqrt{1-\rho^2}(8+9\rho^2-2\rho^4)+15\rho\sin^{-1}\rho\right\}, \\[2mm]
\nu_{24} &= \nu_{42} = 3(1+4\rho^2), \\[2mm]
\nu_{33} &= \tfrac{2}{\pi}\left\{\sqrt{1-\rho^2}(4+11\rho^2)+3\rho(3+2\rho^2)\sin^{-1}\rho\right\}.
\end{aligned}
\right.
\tag{46.30}
$$

Nabeya gives values of ν_{rs} for $r+s \leq 12$. The incomplete moments

$$
\left[[r,s] = \int_0^\infty \int_0^\infty x_1^r x_2^s p(x_1,x_2)dx_1\,dx_2\right]
$$

have been evaluated by Kamat (1958a). We have

$$
\begin{aligned}
[r,s] \;=\; & 2^{\frac{1}{2}(r+s)-2}\frac{1}{\pi}\Gamma\left(\frac{r+1}{2}\right)\Gamma\left(\frac{s+1}{2}\right)F\left(-\frac{1}{2}r,-\frac{1}{2}s;\frac{1}{2};\rho^2\right) \\[2mm]
& + 2\rho\Gamma\left(\frac{r}{2}+1\right)\Gamma\left(\frac{s}{2}+1\right)F\left(-\frac{1}{2}(r-1),-\frac{1}{2}(s-1);\frac{3}{2};\rho\right).
\end{aligned}
\tag{46.31}
$$

The value of $[0,0] = \Pr[(X_1 > 0) \cap (X_2 > 0)]$ was shown to be

$$
\frac{1}{4} + \frac{1}{2\pi}\sin^{-1}\rho
$$

by Sheppard (1899, 1900). Other special values are

$$
\left\{
\begin{aligned}
&[1,0] = \tfrac{1}{4}\sqrt{\tfrac{2}{\pi}}(1+\rho), \\[4pt]
&[2,0] = \tfrac{1}{4} + \tfrac{1}{2\pi}[\sin^{-1}\rho + \rho\sqrt{1-\rho^2}], \\[4pt]
&[1,1] = \tfrac{1}{2\pi}[\rho(\tfrac{1}{2}\pi + \sin^{-1}\rho) + \sqrt{1-\rho^2}], \\[4pt]
&[3,0] = \tfrac{1}{4}\sqrt{\tfrac{2}{\pi}}(1+\rho)^2(2-\rho), \\[4pt]
&[2,1] = \tfrac{1}{4}\sqrt{\tfrac{2}{\pi}}(1+\rho)^2, \\[4pt]
&[4,0] = \tfrac{1}{2\pi}[3(\tfrac{1}{2}\pi + \sin^{-1}\rho) + (5\rho - 2\rho^3)\sqrt{1-\rho^2}], \\[4pt]
&[3,1] = \tfrac{1}{2\pi}[3\rho(\tfrac{1}{2}\pi + \sin^{-1}\rho) + (2+\rho^2)\sqrt{1-\rho^2}], \\[4pt]
&[2,2] = \tfrac{1}{2\pi}[(1+2\rho^2)(\tfrac{1}{2}\pi + \sin^{-1}\rho) + 3\rho\sqrt{1-\rho^2}].
\end{aligned}
\right.
\tag{46.32}
$$

Kamat (1958a) gives tables of $[r, s]$ to six decimal places for $r+s \leq 4$ and $\rho = -0.9(0.1)1.0$.

For the standardized trivariate normal distributions, Nabeya (1952) gives the following values for absolute product moments, $\nu_{rst} = E[|X_1^r X_2^s X_3^t|]$:

$$
\nu_{111} = (2/\pi)^{3/2}(\Delta^{1/2} + \Sigma^*(\rho_{23} + \rho_{12}\rho_{13})\sin^{-1}\rho_{23.1})
\tag{46.33}
$$

(where Σ^* stands for a cyclic sum),

$$
\nu_{211} = \frac{2}{\pi}\left[(\rho_{23} + 2\rho_{12}\rho_{13})\sin^{-1}\rho_{23} + (1+\rho_{12}^2 + \rho_{13}^2)\sqrt{1-\rho_{23}^2}\right].
\tag{46.34}
$$

Kamat (1958a,b) gave the following values for trivariate incomplete moments:

$$
\begin{aligned}
[r,s,t] \;=\; &(2\pi)^{-3/2}\Delta^{-1/2}\int_0^\infty \int_0^\infty \int_0^\infty x_1^r x_2^s x_3^t \\
&\times \exp\left\{-\frac{1}{2}\sum_{i=1}^3 \sum_{j=1}^3 A_{ij}x_i x_j\right\} dx_1 dx_2 dx_3;
\end{aligned}
$$

$$
\left\{
\begin{aligned}
[1,0,0] &= (2\pi)^{-3/2}\{\tfrac{1}{2}\pi + \sin^{-1}\rho_{23.1} + \rho_{12}(\tfrac{1}{2}\pi + \sin^{-1}\rho_{13.2}) \\
&\quad + \rho_{13}(\tfrac{1}{2}\pi + \sin^{-1}\rho_{12.3})\}, \\[2mm]
[2,0,0] &= (4\pi)^{-1}\Big\{\tfrac{1}{2}\pi + \textstyle\sum_{i<j}^{3}\sin^{-1}\rho_{ij} + \Delta\rho_{23}\sqrt{1-\rho_{23}^2} + (2\rho_{12}\rho_{13} - \rho_{23}) \\
&\quad \times\sqrt{1-\rho_{23}^2} + \rho_{12}\sqrt{1-\rho_{12}^2} + \rho_{13}\sqrt{1-\rho_{13}^2}\Big\}, \\[2mm]
[1,1,0] &= (4\pi)^{-1}\Big\{\rho_{12}\big(\tfrac{1}{2}\pi + \textstyle\sum_{i<j}^{3}\sin^{-1}\rho_{ij}\big) \\
&\quad + \sqrt{1-\rho_{12}^2} + \rho_{13}\sqrt{1-\rho_{13}^2} + \rho_{23}\sqrt{1-\rho_{23}^2}\Big\}, \\[2mm]
[1,1,1] &= (2\pi)^{-3/2}[\Delta^{1/2} + \Sigma^{*}(\rho_{23} + \rho_{12}\rho_{13})(\tfrac{1}{2}\pi + \sin^{-1}\rho_{23.1})].
\end{aligned}
\right.
$$

$$(46.35)$$

Together with (46.32), this provides formulas for $[r, s, t]$ for all r, s, t with $r + s + t \le 3$. Further formulas will be found in Haldane (1942).

In an interesting article, Puente and Klebanoff (1994) constructed bivariate Gaussian distributions as transformations of diffuse probability distributions via space-filling fractal interpolating functions; see also Puente (1997).

4   BIVARIATE NORMAL INTEGRAL— TABLES AND APPROXIMATIONS

The joint cumulative distribution of random variables Y_1, Y_2 having joint standardized bivariate normal density in (46.2) is

$$
\begin{aligned}
\Phi(h, k; \rho) &= (2\pi\sqrt{1-\rho^2})^{-1} \\
&\quad \times \int_{-\infty}^{h}\int_{-\infty}^{k}\exp\left\{-\frac{1}{2(1-\rho^2)}(x_1^2 - 2\rho x_1 x_2 + x_2^2)\right\}dx_2\,dx_1.
\end{aligned}
$$

$$(46.36)$$

A more commonly tabulated quantity is

$$
\begin{aligned}
L(h, k; \rho) &= (2\pi\sqrt{1-\rho^2})^{-1} \\
&\quad \times \int_{h}^{\infty}\int_{k}^{\infty}\exp\left\{-\frac{1}{2(1-\rho^2)}(x_1^2 - 2\rho x_1 x_2 + x_2^2)\right\}dx_2\,dx_1.
\end{aligned}
$$

$$(46.37)$$

Note that $L(0,0;\rho) = [0,0]$ for standard bivariate normal distribution. The functions $\Phi(\cdot)$ and $L(\cdot)$ are related by the equation

$$\Phi(h,k;\rho) = 1 - L(h,-\infty;\rho) - L(-\infty,k;\rho) + L(h,k;\rho). \qquad (46.38)$$

Note also that

$$\Phi(h,\infty;\rho) = \Phi(h) \quad \text{and} \quad \Phi(\infty,k;\rho) = \Phi(k), \qquad (46.39)$$

where

$$\Phi(y) \equiv \frac{1}{\sqrt{2\pi}} \int_{-\infty}^{y} e^{-t^2/2}\, dt.$$

Further relations between the L and Φ functions are

$$L(h,k;\rho) = L(k,h;\rho), \qquad (46.40)$$

$$L(-h,k;\rho) + L(h,k;-\rho) = 1 - \Phi(k), \qquad (46.41)$$

$$L(-h,-k;\rho) - L(h,k;\rho) = 1 - \Phi(h) - \Phi(k), \qquad (46.42)$$

$$L(h,k;0) = \{1 - \Phi(h)\}\{1 - \Phi(k)\}, \qquad (46.43)$$

$$L(h,k;1) = \Phi(\max(h,k)), \qquad (46.44)$$

$$L(h,k;-1) = \begin{cases} 0 & (h+k \geq 0), \\ 1 - \Phi(h) - \Phi(k) & (h+k \leq 0). \end{cases} \qquad (46.45)$$

Using (46.5) and (46.6),

$$L(h,k;\rho) = \frac{1}{2\pi} \int_{h}^{\infty} Z(x_1) \int_{(k-\rho x_1)/\sqrt{(1-\rho^2)}}^{\infty} Z(x_2)\, dx_2 dx_1. \qquad (46.46)$$

For $h > 0$, $k > 0$, $\rho > 0$, this is the integral of the circular normal probability density $\frac{1}{2\pi}\exp[-\frac{1}{2}(x_1^2 + x_2^2)]$ over the shaded region shown in Figure 46.3. The slope of the line AB is $-\rho/\sqrt{1-\rho^2}$; the angle between AB and the x_2-axis is

$$\cot^{-1}\left(\frac{-\rho}{\sqrt{1-\rho^2}}\right) = \frac{\pi}{2} + \sin^{-1}\rho.$$

If $h = k = 0$, then A coincides with O and, from the symmetry of the circular normal distribution,

$$L(0,0;\rho) = \frac{1}{2\pi}\left(\frac{\pi}{2} + \sin^{-1}\rho\right) = \frac{1}{4} + \frac{1}{2\pi}\sin^{-1}\rho, \qquad (46.47)$$

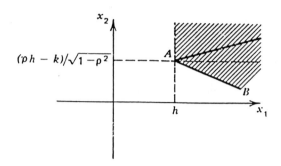

FIGURE 46.3
Shaded Region for the Integral of the Circular Normal Probability
Density Function for $L(h, k; \rho)$ when $h > 0$, $k > 0$, and $\rho > 0$.

a result obtained by Sheppard (1899). Sheppard (1900) gave values to
seven decimal places of $L(U_{0.75}, U_{0.75}; \rho)$ for $\pi^{-1} \cos^{-1} \rho = 0.00(0.01)0.80$.
($U_{0.75} = 0.67749$.)

An extensive set of tables of $L(h, k; \rho)$ were published by Karl Pearson
(1931). These collected together tables were published at various times
from 1910 onward by Everitt (1912), Elderton *et al.* (1930), and Lee
(1917, 1927). They give $L(h, k; \rho)$ for $h, k = 0.0(0.1)2.6$ to six decimal
places for $\rho = 0(0.05)1$ and to seven decimal places for $-\rho = 0(0.05)1$.
In 1959, these tables were extended by the National Bureau of Stan-
dards to give $L(h, k; \rho)$ for $h, k = 0(0.1)4.0$ to six decimal places for $\rho = 0(0.05)0.95(0.01)1.00$ and to seven places for $-\rho = 0(0.05)0.95(0.01)1.00$.
Tables for the special cases $\rho = 1/\sqrt{2}$ and $\rho = \frac{1}{3}$ have been given by
Dunnett (1958) and Dunnett and Lamm (1960), respectively.

Pólya (1949) has obtained the inequalities

$$1 - \Phi(h) - \frac{1-\rho^2}{\rho h - k} Z(k) \left\{ 1 - \Phi\left(\frac{h - \rho k}{\sqrt{1 - \rho^2}} \right) \right\} < L(h, k; \rho) < 1 - \Phi(h)$$

$$(46.48)$$

for $0 < \rho < 1$ and $\rho h - k > 0$. Since $\rho h - k > 0$, it follows that $h - \rho k > 0$
if $h > 0$ and so (46.48) is of the form $1 - \Phi(h) - \Delta < L(h, k; \rho) < 1 - \Phi(h)$
with $0 < \Delta < (2\sqrt{2\pi})^{-1}(1 - \rho^2)(\rho h - k)^{-1}$. Inequalities in (46.48) were
used for checking purposes on the calculations carried out by the National
Bureau of Standards.

Tables of $L(h, k; \rho)$ are of necessity rather bulky, since there are three arguments. Zelen and Severo (1960) pointed out that since

$$L(h, k; \rho) = L(h, 0; \rho(h, k)) + L(k, 0; \rho(k, h)) - \frac{1}{2}(1 - \delta_{hk}), \quad (46.49)$$

where

$$\rho(h, k) = \frac{(\rho h - k)f(h)}{\sqrt{h^2 - 2\rho h k + k^2}},$$

$$f(h) = \begin{cases} 1 & \text{if } h > 0, \\ -1 & \text{if } h < 0, \end{cases}$$

$$\delta_{hk} = \begin{cases} 0 & \text{if } \text{sgn}(h)\text{sgn}(k) = 1, \\ 1 & \text{otherwise}, \end{cases}$$

with

$$\begin{aligned} \text{sgn}(h) &= 1 \quad \text{if } h \geq 0, \\ \text{sgn}(h) &= -1 \quad \text{if } h < 0, \end{aligned}$$

it is possible to evaluate $L(h, k; \rho)$ from a table with $k = 0$, hence having only two arguments. Zelen and Severo (1960, 1964) presented charts from which values of $L(h, 0; \rho)$ can be read off, thus giving a rapid way of obtaining approximate values of $L(h, k; \rho)$.

We have already noted [see (46.46)] that $L(h, k; \rho)$ can be expressed as an integral of the standardized *circular normal* distribution over a certain region. Sheppard (1900) suggested that tabulation of the two-argument function

$$V(h, k) = \int_0^h Z(x_1) \int_0^{kx_1/h} Z(x_2)dx_2dx_1 \qquad (46.50)$$

would be useful, since

$$\begin{aligned} L(h, k; \rho) &= V\left(h, \frac{k - \rho h}{\sqrt{1 - \rho^2}}\right) + V\left(k, \frac{k - \rho k}{\sqrt{1 - \rho^2}}\right) \\ &\quad + 1 - \frac{1}{2}\{\Phi(h) + \Phi(k)\} - \frac{\cos^{-1}\rho}{2\pi}. \qquad (46.51) \end{aligned}$$

The quantity $V(h, k)$ is (for $h > 0$, $k > 0$) the integral of the standardized circular normal distribution over a triangle with vertices at the origin (O) and the points H, P with coordinates $(h, 0), (h, k)$, respectively (see Figure 46.4).

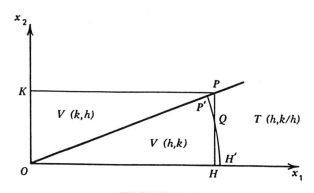

FIGURE 46.4

The Triangle for the Integral of the Standardized Circular Normal
Distribution of $V(h, k)$ when $h > 0$, and $k > 0$.

Similarly $V(k, h)$ is the integral over the triangle OKP. Since the integral over the rectangle $OHPK$ is simply $\Pr[(0 < X_1 < h) \cap (0 < X_2 < k)]$, we have

$$V(h, k) + V(k, h) = \left\{ \Phi(h) - \frac{1}{2} \right\} \left\{ \Phi(k) - \frac{1}{2} \right\}. \tag{46.52}$$

For negative values of h and/or k, $V(h, k)$ is defined by

$$V(h, -k) = -V(h, k) = V(-h, k); \tag{46.53}$$

hence $V(-h, -k) = V(h, k)$.

It was Nicholson (1943) who put Sheppard's suggestion into effect. He gave tables of $V(h, k)$ to six decimal places for $h, k = 0.0(0.1)3.1, \infty$. (Note that $V(h, \infty) = \Phi(h) - \frac{1}{2}$, $V(\infty, k) = 0$.) He used the formula

$$
\begin{aligned}
V(h, k) &= \frac{1}{2\pi} \left[\lambda(1 - e^{-m}) - \frac{1}{3}\lambda^3 (1 - e^{-m} - me^{-m}) \right. \\
&\quad \left. + \frac{1}{5}\lambda^5 \left(1 - e^{-m} - me^{-m} - \frac{m^2}{2!}e^{-m} \right) - \cdots \right]
\end{aligned}
\tag{46.54}
$$

with $\lambda = k/h \leq 1$ and $m = \frac{1}{2} h^2$, in his calculation. For $\lambda > 1$, (46.52) can be used.

Interpolation is facilitated if the variables are taken to be h and λ ($= k/h$) instead of h and k. The National Bureau of Standards (1959) tables published giving $V(h, \lambda h)$ to seven decimal places for $\lambda = 0.1(0.1)1.0$,

$h = 0.00(0.01)4.00(0.2)4.60(0.1)5.6, \infty$; and also $V(\lambda h, h)$ to seven decimal places for $\lambda = 0.1(0.1)1.0$, $h = 0.00(0.01)4.00(0.02)5.60, \infty$. Note
that $V(\infty, \lambda \cdot \infty) = \lim_{h \to \infty} V(h, \lambda h) = \frac{1}{2} \pi \tan^{-1} \lambda$. For $h \geq 5.6$ and
$0.1 \leq \lambda \leq 1$, $V(h, \lambda h)$ agrees with $V(\infty, \lambda \cdot \infty)$ to seven decimal places.
The same agreement is not found between $V(\lambda h, h)$ and $V(\lambda \cdot \infty, \infty) = \lim_{h \to \infty} V(\lambda h, h) = 2\pi \cot^{-1} \lambda$, but the approximation

$$V(\lambda h, h) \doteq \frac{1}{2\pi} \cot^{-1} \lambda - \frac{1}{2} \{1 - \Phi(\lambda h)\} \tag{46.55}$$

holds with an error less than $\frac{1}{2} \times 10^{-7}$ for $h \geq 5.6$. For small values of h
(up to about 0.8) and $\lambda \leq 1$, the simple approximation

$$V(h, \lambda h) \doteq \frac{\lambda h^2}{4\pi} \left\{ 1 - \frac{1}{4} h^2 \left(1 + \frac{1}{3} \lambda^2 \right) \right\} \tag{46.56}$$

gives useful results.

Cadwell (1951) has obtained another useful approximation to $V(h, k)$
for $k/h = \lambda$ small by replacing the boundary PH by the circular arc $P'H'$
(see Figure 46.4) with center at O and radius so chosen that $P'QP$ and
$H'QH$ have equal areas (Q is the point of intersection of $P'H'$ and PH).
The resulting approximation is

$$V(h, \lambda h) \doteq \frac{1}{2\pi} \tan^{-1} \lambda \left\{ 1 - \exp \left(-\frac{\frac{1}{2} h^2 \lambda}{\tan^{-1} \lambda} \right) \right\}. \tag{46.57}$$

This is always less than $V(h, \lambda h)$; the maximum error (for $\lambda < 1$) is 0.0015;
generally accuracy is much better. If the correction $0.04h^4(\lambda - \frac{3}{4})$ is added
to the exponent, the error is always less than 0.0005. For $\lambda > 1$, the
relation (46.52) can be used.

Owen (1956) has tabulated values of the integral of (46.2) over the remainder of the sector POH (i.e., the part to the right of the line PH in Figure 46.4). For $h > 0$, $k > 0$, we see that the integral over the whole sector is
(by the symmetry of the circular normal distribution) $(1/2\pi) \tan^{-1}(k/h)$,
hence the quantity tabulated by Owen is $(1/2\pi) \tan^{-1}(k/h) - V(h, k)$. He
regards this as a function $T(h, k/h)$ of h and k/h. The function thus is

$$T(h, \lambda) = \frac{1}{2\pi} \int_h^\infty Z(x_1) \int_{\lambda x_1}^\infty Z(x_2) \, dx_2 dx_1. \tag{46.58}$$

This can be expressed as the single integral

$$T(h, \lambda) = \frac{1}{2\pi} \int_0^\lambda (1 + x^2)^{-1} \exp \left\{ -\frac{1}{2} h^2 (1 + x^2) \right\} dx. \tag{46.59}$$

Owen showed that

$$T(h, \lambda) = \frac{1}{2\pi} \left\{ \tan^{-1} \lambda - \sum_{j=0}^{\infty} c_j \lambda^{2j+1} \right\} \tag{46.60}$$

with

$$c_j = \frac{(-1)^j}{2j+1} \left\{ 1 - e^{-(1/2)h^2} \sum_{i=0}^{j} \frac{(\frac{1}{2} h^2)^i}{i!} \right\}.$$

For small values of h and λ, convergence is rapid, and the formula is useful for computing $T(h, \lambda)$. Owen (1956) has given values of $T(h, \lambda)$ to six decimal places for $h = 0(0.01)2.00(0.02)3.00$, $\lambda = 0.25(0.25)1.00$,[1] and for $h = 0(0.05)4.75$, $\lambda = 0(0.25)1.00, \infty$ by Owen (1957); see also Owen and Wiesen (1959). Smirnov and Bol'shev (1962) have given tables of $T(h, \lambda)$ to seven decimal places for

(i) $h = 0(0.01)3.00$, $\lambda = 0(0.01)1.00$,
(ii) $h = 3.00(0.05)4.00$, $\lambda = 0.05(0.05)1.00$,
(iii) $h = 4.0(0.1)5.2$, $\lambda = 0.1(0.1)1.0$,

and they have also given tables of $T(h, 1)$ for $h = 0(0.001)3.000(0.005)4.000$ $(0.01)5.00(0.1)6.0$ and of $T(0, \lambda) = \frac{1}{2\pi} \tan^{-1} \lambda$ for $\lambda = 0.000(0.001)1.000$.

Amos (1969) and Sowden and Ashford (1969) have given instructive comparisons of time taken to compute the bivariate normal integral $L(h, k; \rho)$ on computers, using various formulas. Amos recommends formula (46.51) [with (46.59)] as being generally preferable. Sowden and Ashford agree with this conclusion for $0.2 \le \rho \le 0.95$, and also $-0.7 \le \rho \le -0.4$ if $|h - k| < 1$ or either of h, k exceeds 1. For other situations, they recommend direct quadrature using the formula

$$L(h, k; \rho) = \int_{-\infty}^{\infty} Z(t) \Phi\left(\frac{c_1 t - h}{\sqrt{1 - c_1^2}} \right) \Phi\left(\frac{c_2 t - k}{\sqrt{1 - c_2^2}} \right) dt, \tag{46.61}$$

which can be obtained by noting that X_1 and X_2 with distribution (46.2) can be represented as

$$X_i = c_i T + \sqrt{1 - c_i^2} \, U_i$$

with T, U_1 and U_2 being mutually independent unit normal variables, $0 \le c_i^2 \le 1$ $(i = 1, 2)$, $c_1 c_2 = \rho$. Subject to these limitations, c_1 and c_2 can

---

[1] Also for $h = 0(0.25)3.00$ and $\lambda = 0(0.01)1.00$ and for $h = 3.00(0.05)4.00(0.1)4.7$ and $\lambda = 0.1(0.1)1.0$.

be chosen arbitrarily. A good choice, suggested by Sowden and Ashford (1969), is to take $c_1 = c_2 = \sqrt{\rho}$ if $\rho > 0$ and $c_1 = -c_2 = \sqrt{-\rho}$ if $\rho < 0$. The first few terms of the series (46.190) below may be used as an approximation to $L(h_1, k_1; \rho)$, when ρ is small.

Expansions for $\Phi(h_1, h_2; \rho)$, of similar form to (46.190) have been used by Bofinger and Bofinger (1965) to obtain series expansions (in powers of ρ) for the correlation between $\max(X_{11}, \ldots, X_{1n})$ and $\max(X_{21}, \ldots, X_{2n})$, where (X_{1j}, X_{2j}) $(j = 1, \ldots, n)$ are independent vectors with a common bivariate normal distribution. They give a table of coefficients of $\rho, \rho^2, \rho^3, \rho^4$, and ρ^5 in this series, to five decimal places for $n = 2(1)50$.

The integral of (46.2) over any convex polygon can be expressed as the sum or difference of integrals over a number of triangles, each having one vertex at the origin. Any one of these can be expressed as the integral of the joint distribution of two *independent* unit normal variables over a triangular region of the same kind (see Figure 46.5). By suitable further transformation, the region of integration can be arranged to be as in Figure 46.6. Such integrals can be evaluated from tables of $V(h, k)$. In the case shown, the required integral is $V(h, k_2) - V(h, k_1)$. In general, some care is needed to keep the signs of the various terms correct.

Integrals over circles and ellipses are, in fact, probability integrals of positive definite quadratic forms in normal variables and, as such, will be discussed in a separate Chapter.

Tihansky (1970) has described construction of "equidistributional contours"–loci of points (x, y) such that $L(x, y; \rho) = \alpha$. Figure 46.7, taken from Tihansky (1970), shows such contours for $\alpha = 0.25$, $\rho = 0, \pm0.75$, ±0.99.

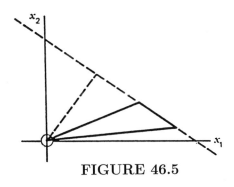

FIGURE 46.5

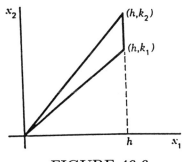

FIGURE 46.6

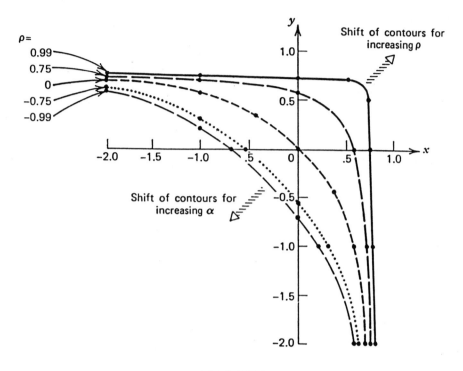

FIGURE 46.7
Equidistributional Contours $(L(x, y; \rho) = \alpha)$ for the Standard Bivariate
Normal Distribution with $\alpha = 0.25$. [From Tihansky (1970), with
permission.]

Burington and May (1970) have shown how two-way normal probability paper (with each scale as described earlier in Chapter 13) can be used as a basis for graphical quadrature of circular normal distributions over arbitrary regions.

Tables of random normal deviates (Chapter 13) can be used to construct tables of correlated random normal deviates. One way of doing this, described by Wold (1948), is to use *independent* unit normal variates U_1, U_2 and calculate $Y_1 = U_1, Y_2 = \rho U_1 + \sqrt{1 - \rho^2}\, U_2$. Then, Y_1, Y_2 are unit normal variates with correlation ρ. Fieller, Lewis, and Pearson (1956) and Iyer and Simha (1967) have published tables constructed by this method.

Following Tihansky (1970), Drezner and Wesolowsky (1989) discussed approximations to bivariate normal contours; however, unlike Tihansky (1970), they constructed contours for the probability of union of events rather than intersection. Mee and Owen (1983) have also presented a simple approximation for bivariate normal probabilities.

Let $(X_1, X_2)^T$ be a standard bivariate normal random variable with correlation coefficient ρ. From (46.37) we have

$$L(h_1, h_2; \rho) = \Pr[(X_1 \geq h_1) \cap (X_2 \geq h_2)]$$

and from (46.36) we obtain

$$1 - \Phi(h_1, h_2; \rho) = \Pr[(X_1 \geq h_1) \cup (X_2 \geq h_2)].$$

We define $\gamma(\alpha_1, \alpha_2; \rho) = 1 - \Phi(h_1, h_2; \rho)$, where $\alpha_1 = \Pr[X_1 \geq h_1]$ and $\alpha_2 = \Pr[X_2 \geq h_2]$. Note that $\gamma(\alpha_1, \alpha_2; \rho) = 1 - L(-h_1, -h_2; \rho)$. With the aim of approximating the contour $\gamma(\alpha_1, \alpha_2; \rho) = \alpha$, Drezner and Wesolowsky (1989) suggested the approximation

$$\gamma(\alpha_1, \alpha_2; \rho) \simeq \left(\alpha_1^{q(\alpha,\rho)} + \alpha_2^{q(\alpha,\rho)} \right)^{1/q(\alpha,\rho)}, \qquad (46.62)$$

where

$$q(\alpha, \rho) = \left[1 - \ln 2 \left\{ 1 + \left(\frac{2}{\alpha} - \frac{2\sqrt{1-\alpha}}{\alpha} \right)^{\frac{\sqrt{1-\rho}}{1+\rho} (1+\rho\psi)} \right\} \right]^{-1} ; \qquad (46.63)$$

and ψ is determined as a function of α and ρ through linear regression yielding

$$\begin{aligned}
\psi(\alpha, \rho) &= 0.173 + 0.968\alpha - 0.128\rho - 0.0756\alpha\sqrt{1-\alpha} - 0.417\rho\alpha^2 \\
&\quad + 0.215\alpha\rho^2 - 0.0657\alpha|\rho| - 0.789\alpha^2. \qquad (46.64)
\end{aligned}$$

Drezner and Wesolowsky (1989) calculated the approximate contours in (46.62) using $\alpha_2 = (\alpha - \alpha_1^q)^{1/q}$ with the values of q determined from (46.63) and (46.64), and observed that the approximations are better for small values of α. These authors have also discussed extension of this approximation to dimension more than 2.

Owen's (1956) method of computing the bivariate normal distribution function involves computing the integral

$$\begin{aligned}
T(x, a) &= T(-x, a) = -T(x, -a) \\
&= \frac{1}{2\pi} \int_0^a \frac{1}{1+u^2} e^{-x^2(1+u^2)/2} \, du \\
&= T\left(ax, -\frac{1}{a} \right) + \text{sgn}(a) \left[\frac{1}{4} - \{1 - \Phi(|x|)\} \{1 - \Phi(|ax|)\} \right]
\end{aligned}$$
$$(46.65)$$

in a power series expansion that is convergent for all x. But Daley (1974) has claimed that Simpson's rule for numerical integration is better, in that no problem of slow convergence (near $|a| = 1$) will be encountered. Daley (1974) adopted this numerical approach, along with the identity

$$\Pr[X_1 \leq x_1, X_2 \leq x_2]$$
$$= \frac{1}{2}\{\Phi(x_1) + \Phi(x_2) - \delta(x_1, x_2)\} - T(x_1, a_1) - T(x_2, a_2),$$

(46.66)

where
$$\delta(x, y) = 0 \quad \text{if } xy > 0 \text{ or } xy = 0 \text{ and } x + y \geq 0$$
$$= 1 \quad \text{otherwise,}$$

and $a_1 = \frac{(x_2/x_1) - \rho}{\sqrt{1 - \rho^2}}$ and $a_2 = \frac{(x_1/x_2) - \rho}{\sqrt{1 - \rho^2}}$. Daley (1974) has presented an adaptive quadrature technique that is an effective method of computing the T-function in (46.65) and has also presented a table giving the maximum error in calculating $T(x, a)$ using Simpson's rule with 10 sub-intervals.

Borth (1973) recommended the usage of Owen's series for computing $T(x, a)$ whenever $|x| \leq 1.6$ or $|a| \leq 3$ and otherwise approximating the integral in (46.65) by replacing the term $(1 + u^2)^{-1}$ in the integrand by an approximating polynomial. However, the constants arising in the approximation step need to be computed in order to make the error less than 10^{-7}. While Donnelly (1973) made a direct coding of Owen's formula, Cooper (1968a,b) used the function $T(x, a)$ in evaluating the noncentral t distribution; see Chapter 29 of Johnson, Kotz, and Balakrishnan (1995).

Young and Minder (1974) proposed the computation of $T(x, a)$ by means of 10-point Gaussian quadrature for integration over the whole range of a. Though they avoid using any algorithm for the univariate normal distribution function, it is not a simplification when the computation of the bivariate normal distribution function is done using $T(x, a)$. Daley (1974) pointed out that the number of points required to evaluate $T(x, a)$ via Gaussian quadrature with error at most 10^{-7} is less than 10 for small a and is about 6 or 7 for a close to 1; Young and Minder (1974) did agree with this observation. Daley (1974) also noted that an early approximation due to Cadwell (1951), given by

$$T(x, a) \simeq \frac{\theta}{2\pi} e^{-\frac{1}{2\theta}x^2 a}(1 + 0.00868\, x^4 a^4),$$

(46.67)

where $\theta = \tan^{-1} a$, has a maximum error of 5.1×10^{-5} (for $|a| \leq 1$ and all x).

Let $(X_1, X_2)^T$ be a bivariate normal random variable with mean $(\xi_1, \xi_2)^T$ and variance-covariance matrix $\begin{pmatrix} \sigma_1^2 & \rho\sigma_1\sigma_2 \\ \rho\sigma_1\sigma_2 & \sigma_2^2 \end{pmatrix}$. Let $R_1 \times R_2 = (h_1, \infty) \times (h_2, \infty)$ be the region of integration; let us denote $\Pr[X_2 \geq h_2]$ by $S(h_2)$ and $\frac{\phi(h_2)}{S(h_2)}$ by a_2. Then,

$$m_2 \equiv E[X_2 | X_2 \geq h_2] = \xi_2 + \sigma_2 a_2,$$
$$m_{1|2} \equiv E[X_1 | X_2 \geq h_2] = \xi_1 + (m_2 - \xi_2)\rho\sigma_1/\sigma_2$$

and

$$S_{1|2}^2 \equiv \mathrm{var}(X_1 | X_2 \geq h_2) = \sigma_1^2 \{1 - \rho^2 a_2 (a_2 - z_2)\},$$

where $z_2 = (h_2 - \xi_2)/\sigma_2$. Even though the conditional distribution of X_1 given $X_2 \geq h_2$ is not normal, Mendell and Elston (1974) assumed it to be normal so that $\Pr[X_1 \geq h_1, X_2 \geq h_2]$ can be approximated by $S(h_2)S(z_1)$, where $z_1 = (h_1 - m_{1|2})/S_{1|2}$; see also Rice, Reich, and Cloninger (1979). This approximation works especially well for small values of ρ.

Olson and Weissfeld (1991) suggested an approximation based on Taylor series expansion. Let $T[h(\boldsymbol{x})]$ denote the integral $\int_{\boldsymbol{R}} h(\boldsymbol{x})p(\boldsymbol{x})\, d\boldsymbol{x}$ where $\boldsymbol{R}$ is a rectangular region and $p(\boldsymbol{x})$ is the given bivariate density function. Let us take $h(\boldsymbol{x}) = 1$, $p(\boldsymbol{x})$ to be the bivariate normal density function in (46.1), and $\boldsymbol{R} = R_1 \times R_2$; furthermore, let E_2 denote the expected value of X_2 conditional on $X_2 \in R_2$, and let V_2 denote the second central moment (around E_2) of X_2 conditional on $X_2 \in R_2$. Note that E_2 and V_2 take the form of a ratio of univariate integrals; for example,

$$V_2 = \int_{R_2} (x_2 - E_2)^2 \phi(x_2)\, dx_2 \Big/ \int_{R_2} \phi(x_2)\, dx_2.$$

The Taylor series expansion of $T[h(\boldsymbol{x})]$ in this case (up to second order moment) is given by

$$\left\{1 - \frac{1}{2}\frac{\rho\sigma_1 V_2}{\sigma_2}\left(\frac{\rho}{\sigma_1\sigma_2(1-\rho^2)}\right)\right\} \int_{R_1} \phi(x_1 | E_2)\, dx_1$$
$$+ \frac{1}{2}V_2 \left(\frac{\rho}{\sigma_1\sigma_2(1-\rho^2)}\right)^2 \int_{R_1} \left\{x_1 - \xi_1 - \frac{\rho\sigma_1}{\sigma_2}(E_2 - \xi_2)\right\}^2 \phi(x_1 | E_2)\, dx_1.$$

$$(46.68)$$

A better approximation can be achieved by conditioning on the variable with the smaller marginal tail probability. In fact, if the marginal tail probability is very small, an excellent approximation can be achieved simply by using the term $\int_{R_1} \phi(x_1 | E_2)\, dx_1$. The comparative values reported in the following table, taken from Olson and Weissfeld (1991), are obtained by conditioning on X_2 as the formula in (46.68). Of course, the Taylor series approximation can be improved by including more terms involving higher-order moments.

TABLE 46.1

Approximation of Bivariate Normal Probabilities[a]

Lower Bounds		ρ			
X_1	X_2	−0.5	−0.1	0.1	0.5
−2.0	−2.0	0.954503	0.954780	0.955372	0.958553
		0.956324	0.954786	0.955367	0.959154
		0.955069	0.954785	0.955367	0.957860
−1.0	−1.0	0.686472	0.702300	0.714009	0.745203
		0.684176	0.702297	0.714007	0.743404
		0.686222	0.702299	0.714009	0.744651
0.0	0.0	0.166667	0.234058	0.265942	0.333333
		0.166266	0.234050	0.265950	0.333734
		0.165880	0.234050	0.265950	0.334120
1.0	1.0	0.003782	0.019610	0.031320	0.062514
		0.003857	0.019611	0.031320	0.062744
		0.003866	0.019610	0.031320	0.062719
2.0	2.0	0.000003	0.000280	0.000872	0.004053
		0.000003	0.000280	0.000872	0.004059
		0.000004	0.000280	0.000872	0.004057
−2.0	−1.0	0.818715	0.821028	0.823641	0.831861
		0.820255	0.821035	0.823635	0.831674
		0.819746	0.821035	0.823634	0.831073
−2.0	0.0	0.479276	0.486482	0.490769	0.497974
		0.479832	0.486485	0.490766	0.497877
		0.479798	0.486485	0.490766	0.497777
−2.0	1.0	0.145389	0.153609	0.156222	0.158508
		0.145433	0.153610	0.156221	0.158501
		0.145451	0.153610	0.156221	0.158496
−1.0	1.0	0.096141	0.127335	0.139045	0.154873
		0.095911	0.127335	0.139045	0.154798
		0.095936	0.127335	0.139045	0.154789
0.0	1.0	0.031257	0.069674	0.088981	0.127398
		0.031286	0.069674	0.088982	0.127369
		0.031241	0.069673	0.088982	0.127414
1.0	2.0	0.000147	0.002433	0.005046	0.013266
		0.000150	0.002433	0.005046	0.013279
		0.000150	0.002433	0.005046	0.013280

[a]First entry is table value, second entry is Taylor series approximation, and third entry is Mendell-Elston approximation. [From Olson and Weissfeld (1991), with permission.]

Albers and Kallenberg (1994) and Drezner and Wesolowsky (1990) have discussed simple approximations to the bivariate tail probability $L(h, k; \rho) = \Pr[X_1 \geq h, X_2 \geq k]$ for large values of the correlation coefficient ρ. For example, Drezner and Wesolowsky (1990) have given the following expression, which is particularly suitable for the calculation of the bivariate normal probability for large values of ρ:

$$L(h, k; \rho) = \frac{1}{2\pi} \int_0^{\sqrt{1-\rho^2}} \frac{1}{\sqrt{1-x^2}} e^{-(h^2 - 2\sqrt{1-x^2}\, hk + k^2)/2x^2} \, dx$$
$$+ \Phi(-\max\{h, k\}). \qquad (46.69)$$

As these authors have suggested, the integral in (46.69) can be evaluated using Gaussian quadrature formulas with only two or three points; and, in fact, use of five points (with some adjustment to the formula) gives the results to an accuracy of 2×10^{-7}.

Lin (1995) has proposed the following simple approximation for $L(h, 0; \rho)$:

$$L(h, 0; \rho) \simeq \frac{1}{\sqrt{8a}} e^{b^2/(4a)} \left\{ 1 - \Phi\left(\sqrt{2a}\left(h + \frac{b}{2a} \right) \right) \right\}, \qquad (46.70)$$

where $a = 0.5 + 0.416\, \rho^2/(1 - \rho^2)$ and $b = -0.717\, \rho/\sqrt{1 - \rho^2}$. By using another approximation for $S(\cdot)$ itself, Lin (1995) has suggested an even simpler approximation as

$$L(h, 0; \rho) \simeq \frac{1}{\sqrt{8a}} e^{b^2/(4a)}$$
$$\cdot \frac{1}{2} \frac{e^{-a^2(h + \frac{b}{2a})^2}}{1 + 0.91\{\sqrt{2a}(h + \frac{b}{2a})\}^{1.12}} . \qquad (46.71)$$

Lin has shown that the accuracy of these approximations are quite sufficient for many practical situations. With the use of (46.49), these approximations can be utilized to develop approximations for $L(h, k; \rho)$. Terza and Welland (1991) have provided a comparison of several bivariate normal algorithms.

5 CHARACTERIZATIONS

Brucker (1979) has presented sufficient conditions for the joint normality of the distribution of a bivariate random vector in terms of the conditional distributions of each component given the value of the other component. His specific conditions are

(a) $X_1 | (X_2 = x_2) \overset{d}{=} N(a + bx_2, g) \qquad \forall\, x_2 \in \mathbb{R}$

and

(b) $X_2|(X_1 = x_1) \stackrel{d}{=} N(c + dx_1, h)$ $\forall\, x_1 \in \mathbb{R},$

where $a, b, c, d, g(> 0)$, and $h(> 0)$ are all real numbers. Fraser and Streit (1980) have pointed out that Brucker's (1979) conditions could be relaxed by keeping (b) as above and replacing (a) by either

(a)' X_1 has a nonsingular marginal normal distribution

or

(a)'' The condition (a) above is satisfied for just one single value x_2^0 of X_2 (having nonsingular marginal density). If the existence of the joint density is not assumed, then the restriction $g < \frac{h}{d^2}$ needs to be imposed on $g = \mathrm{var}(X_1|X_2 = x_2^0)$ to ensure negative definiteness of the quadratic exponent that emerges in $f_{\mathbf{X}}(\mathbf{x})$.

In a survey article, Hamedani (1992) has presented eighteen different characterizations of the bivariate normal distribution, many of which do not possess straightforward generalizations to the multivariate case. Ahsanullah and Wesolowski (1992) have discussed a characterization of the bivariate normal distribution by normality of one conditional distribution and some properties of conditional moments of the other variable. Let $(X_1, X_2)^T$ be a bivariate random vector. If $E|X_1| < \infty$, $X_2|X_1 \stackrel{d}{=} N(\alpha X_1 + \beta, \sigma^2)$ (linear conditional mean and nonrandom conditional variance), and $E[X_1|X_2] = \gamma X_2 + \delta$ for some real numbers $\alpha, \beta, \gamma, \delta$, and σ with $\alpha \neq 0$, $\gamma \neq 0$, and $\sigma > 0$, then $(X_1, X_2)^T$ is distributed as bivariate normal.

A slight extension of this result is when $(X_1, X_2)^T$ is a bivariate random vector such that

$$X_2|X_1 \stackrel{d}{=} N(\alpha X_1 + \beta, \sigma^2(X_1)) \tag{46.72}$$

and

$$E[\sigma^2(X_1)|X_2 - \alpha X_1] = c \tag{46.73}$$

for real $\alpha \neq 0$, β and $c > 0$. If $E[X_1|X_2] = \gamma X_2 + \delta$, as above, then $(X_1, X_2)^T$ is distributed as bivariate normal.

In a somewhat obscure Swedish report, Wrigge (1971) has presented an extensive discussion on distributions that have normal marginal densities but are not jointly normally distributed. The classic example, due to Patil and Joshi (1970), is the bivariate density function

$$p(x_1, x_2) = \frac{1}{2\pi} e^{-(x_1^2 + x_2^2)/2} \left\{ 1 - \frac{x_1 x_2}{(1 + x_1^2)(1 + x_2^2)} \right\},$$
$$-\infty < x_1, x_2 < \infty. \tag{46.74}$$

Now, let us take X_1 and X_2 to be independent standard normal variables, and let

$$g_2(X_1, X_2) = \frac{X_1^2 - X_2^2}{\sqrt{X_1^2 + X_2^2}} \quad \text{and} \quad g_5(X_1, X_2) = \frac{4X_1 X_2(X_1^2 - X_2^2)}{(X_1^2 + X_2^2)^{3/2}}.$$

$$(46.75)$$

The characteristic function of the ratio $\frac{g_5(X_1, X_2)}{g_2(X_1, X_2)} = \frac{4X_1 X_2}{X_1^2 + X_2^2}$ is an ordinary Bessel function of order 0, namely, $J_0(t)$. Hence, the variables $g_2(X_1, X_2)$ and $g_5(X_1, X_2)$ are not jointly normal since the ratio would then be distributed as Cauchy whose characteristic function is not $J_0(t)$; see Chapter 16 of Johnson, Kotz, and Balakrishnan (1994). In general, if we define

$$g_{2n-2}(X_1, X_2) = \sum_{j=0}^{n/2} (-1)^j \binom{n}{2j} X_1^{n-2j} X_2^{2j} \Big/ (X_1^2 + X_2^2)^{(n-1)/2}$$

and

$$g_{2n-3}(X_1, X_2) = \sum_{j=0}^{(n-2)/2} (-1)^j \binom{n}{2j+1} X_1^{n-2j-1} X_2^{2j+1} \Big/ (X_1^2 + X_2^2)^{(n-1)/2},$$

then $g_{2n-3}(X_1, X_2)$ and $g_{2m-2}(X_1, X_2)$ $(m \neq n)$ are not jointly normal; also, $g_{2n-3}(X_1, X_2)$ and $g_{2m-3}(X_1, X_2)$ $(m \neq n)$ are not jointly normal.

It should be noted that, as indicated by Stoyanov (1987) and reproduced by Hamedani (1992), the bivariate distribution

$$p(x_1, x_2) = c \, e^{-(x_1^2 x_2^2 + x_1^2 + x_2^2)/2}$$

provides an example for the fact that the bivariate normality of $(X_1, X_2)^T$ is not determined by the properties that X_1 and X_2 are identically distributed and $X_1 | (X_2 = x_2)$ is distributed as normal for all $x_2 \in \mathbb{R}$ [Normal$(0, \frac{1}{1+x_2^2})$ in this case] and $X_2 | (X_1 = x_1)$ is also distributed as normal for all $x_1 \in \mathbb{R}$ [Normal$(0, \frac{1}{1+x_1^2})$ in this case]. However, as shown by Ahsanullah and Wesolowski (1992), conditions (46.72) and (46.73) and the fact that X and Y are identically distributed will ensure bivariate normality.

Ahsanullah et al. (1996) have presented a bivariate nonnormal random vector $(X_1, X_2)^T$ with normal marginal distributions, correlation coefficient ρ, and with corr$(X_1^2, X_2^2) = \rho^2$. Note that if $(X_1, X_2)^T$ is distributed as bivariate normal with correlation coefficient ρ, then X_1^2 and X_2^2 will have correlation ρ^2, but this fourth-moment relation is too weak to characterize bivariate normal distribution with zero mean vector. However, with additional conditions on X_1 and X_2 such as finiteness of the second and fourth moments, the conditional distribution of X_2 given $X_1 = x$ is normal with linear mean, and $E[X_2^2|X_1 = x] = b + cx^2$ for constants b and c, $\mathrm{corr}(X_1^2, X_2^2) = \rho^2$ is sufficient to characterize bivariate normality.

A recent result on the conditional specification of bivariate normal distribution is due to Kagan and Wesolowski (1996). Consider a bivariate random vector $(X_1, X_2)^T$ such that $X_1|X_2$ is distributed as $\mathrm{Normal}(\mu(X_2), \sigma^2(X_2))$ and that $X_2|X_1$ is distributed as $\mathrm{Normal}(\tilde{\mu}(X_1), \tilde{\sigma}^2(X_1))$, where $\mu, \tilde{\mu}, \sigma > 0$ and $\tilde{\sigma} > 0$ are real Borel functions. Then, the joint distribution of $(X_1, X_2)^T$ need not be bivariate normal [see Bhattacharyya (1943)] and it can even be bimodal, as pointed out by Gelman and Meng (1991). However, as mentioned earlier, if any of the conditional means is linear or any of the conditional variances is constant, the distribution of $(X_1, X_2)^T$ is bivariate normal. The linearity of $E[X_2|X_1]$ or the constancy of $\mathrm{var}(X_2|X_1)$ as a supplementary condition will ensure bivariate normality, as has been shown by Ahsanullah and Wesolowski (1992).

Now, let X_1 and X_2 be independent random variables and let $U = \alpha X_1 + \beta X_2$ and $V = \gamma X_1 + \delta X_2$, where α, β, γ, and δ are some real numbers such that $\alpha\delta - \beta\gamma \neq 0$. Then, Kagan and Wesolowski (1996) have shown that X_1 and X_2 are normal random variables if the conditional distribution of U given V is normal (with probability 1). In other words, these authors have shown that if U and V are linear functions of a pair of independent (but not necessarily identically distributed) random variables X_1 and X_2, then the conditional normality of U given V without any additional conditions on the structure of the parameters of this distribution implies the normality of both X_1 and X_2. Their proof of this result is based on solving a Cauchy-like functional equation.

Holland and Wang (1987) have shown that, for a bivariate function $p(x_1, x_2)$ defined on $\mathbb{R}^2$, if

(a) $\frac{\partial^2 \ln p(x_1, x_2)}{\partial x_1 \partial x_2} = \lambda$ (constant),

(b) $\int_{-\infty}^{\infty} p(x_1, x_2) \, dx_2 = \frac{1}{\sqrt{2\pi}} e^{-x_1^2/2}$,

and

(c) $\int_{-\infty}^{\infty} p(x_1, x_2) \, dx_1 = \frac{1}{\sqrt{2\pi}} e^{-x_2^2/2}$,

then $p(x_1, x_2)$ is the standard bivariate normal density function with correlation coefficient $\rho = \frac{\sqrt{1+4\lambda^2}-1}{2\lambda}$. In fact, this is a direct consequence of the following more general result of these authors. For any given integrable function $\gamma(x_1, x_2)$, defined over $S = \{(x_1, x_2) : a < x_1 < b, c < x_2 < d\}$, and any given continuous density functions $p_1(x_1)$ and $p_2(x_2)$ defined on (a, b) and (c, d), respectively, there exists an unique bivariate density function $p(x_1, x_2)$ defined on S such that

(a) $\frac{\partial^2 \ln p(x_1, x_2)}{\partial x_1 \partial x_2} = \gamma(x_1, x_2)$ $\forall (x_1, x_2) \in S$,

(b) $\int_c^d p(x_1, x_2)dx_2 = p_1(x_1)$ $\forall x_1 \in (a, b)$,

and

(c) $\int_a^b p(x_1, x_2)dx_1 = p_2(x_2)$ $\forall x_2 \in (c, d)$.

Extending Braverman's (1985) characterization of the univariate normal distribution [see Chapter 13 of Johnson, Kotz, and Balakrishnan (1994)], Szabłowski (1990) has shown that if X_1 and X_2 are independent and identically distributed random variables with zero means and if they satisfy the conditions

(i) $\exists\ \lambda > 0$ such that $E[e^{\lambda X_1^2}]$ and $E[e^{\lambda X_2^2}]$ are both finite

and

(ii) for all integers j, there exist $C_j > 0$ such that

$$E\left[(\alpha X_1 + \beta X_2)^{2j}\right] = C_j(\alpha^2 + \beta^2)^j$$

for all real-valued α and β,

then $(X_1, X_2)^T$ is distributed as bivariate normal. In fact, the independent and identically distributed assumption on X_1 and X_2 above can be relaxed by assuming symmetry for X_1 and X_2 (and retaining independence). Szabłowski (1992) has also established polynomial regression type characterizations of the bivariate normal distribution.

6 ORDER STATISTICS

Let $\boldsymbol{X} = (X_1, X_2)^T$ have a bivariate normal distribution with mean vector $\boldsymbol{\xi} = (\xi_1, \xi_2)^T$ and variance–covariance matrix $\Sigma = \begin{pmatrix} \sigma_1^2 & \rho\sigma_1\sigma_2 \\ \rho\sigma_1\sigma_2 & \sigma_2^2 \end{pmatrix}$, where $\rho^2 \neq 1$. Let $X_{(1)} = \min(X_1, X_2)$ and $X_{(2)} = \max(X_1, X_2)$. Then, as Cain (1994) has shown, the survival function of $X_{(1)}$ can be written as

$$
\begin{aligned}
1 - F_{X_{(1)}}(x) &= \Pr[X_1 > x, X_2 > x] \\
&= \int_{(x-\xi_1)/\sigma_1}^{\infty} \left\{ 1 - \Phi\left(\frac{x - \xi_2 - \rho\sigma_2 u}{\sigma_2\sqrt{1 - \rho^2}} \right) \right\} \phi(u) \, du,
\end{aligned}
$$

$$(46.76)$$

where $\phi(\cdot)$ and $\Phi(\cdot)$ are the probability density function and the cumulative distribution function of the standard normal distribution, respectively. From (46.76), the cumulative distribution function of $X_{(1)}$ readily becomes

$$
F_{X_{(1)}}(x) = \Phi\left(\frac{x - \xi_1}{\sigma_1} \right) + \int_{(x-\xi_1)/\sigma_1}^{\infty} \Phi\left(\frac{x - \xi_2 - \rho\sigma_2 u}{\sigma_2\sqrt{1 - \rho^2}} \right) \phi(u) \, du.
$$

$$(46.77)$$

From (46.77), the probability density function of $X_{(1)}$ can be expressed as

$$
f_{X_{(1)}}(x) = f_1(x) + f_2(x),
$$

$$(46.78)$$

where

$$
f_1(x) = \frac{1}{\sigma_1} \, \Phi\left\{ \frac{-\left(\frac{x - \xi_2}{\sigma_2} \right) + \rho\left(\frac{x - \xi_1}{\sigma_1} \right)}{\sqrt{1 - \rho^2}} \right\} \phi\left(\frac{x - \xi_1}{\sigma_1} \right)
$$

and

$$
f_2(x) = \frac{1}{\sigma_2} \, \Phi\left\{ \frac{-\left(\frac{x - \xi_1}{\sigma_1} \right) + \rho\left(\frac{x - \xi_2}{\sigma_2} \right)}{\sqrt{1 - \rho^2}} \right\} \phi\left(\frac{x - \xi_2}{\sigma_2} \right);
$$

note that $\int_{-\infty}^{\infty} f_2(x) \, dx = \Pr[X_1 > X_2]$.

From (46.78), the moment-generating function of $X_{(1)}$ is

$$
M_{X_{(1)}}(t) = M_1(t) + M_2(t),
$$

$$(46.79)$$

where

$$
M_1(t) = e^{t\xi_1 + \frac{1}{2}t^2\sigma_1^2} \, \Phi\left\{ \frac{\xi_2 - \xi_1 - t(\sigma_1^2 - \rho\sigma_1\sigma_2)}{\sqrt{\sigma_2^2 - 2\rho\sigma_1\sigma_2 + \sigma_1^2}} \right\}
$$

and

$$
M_2(t) = e^{t\xi_2 + \frac{1}{2}t^2\sigma_2^2} \, \Phi\left\{ \frac{\xi_1 - \xi_2 - t(\sigma_2^2 - \rho\sigma_1\sigma_2)}{\sqrt{\sigma_2^2 - 2\rho\sigma_1\sigma_2 + \sigma_1^2}} \right\}.
$$

Basu and Ghosh (1978) had earlier provided an expression for the moment-generating function of $X_{(1)}$, but their formula seems to be in error. From (46.79), we obtain the mean of $X_{(1)}$ to be

$$
\begin{aligned}
E[X_{(1)}] &= M_1'(0) + M_2'(0) \\
&= \xi_1 \Phi\left(\frac{\xi_2 - \xi_1}{\sqrt{\sigma_2^2 - 2\rho\sigma_1\sigma_2 + \sigma_1^2}}\right) + \xi_2 \Phi\left(\frac{\xi_1 - \xi_2}{\sqrt{\sigma_2^2 - 2\rho\sigma_1\sigma_2 + \sigma_1^2}}\right) \\
&\quad - \sqrt{\sigma_2^2 - 2\rho\sigma_1\sigma_2 + \sigma_1^2}\ \phi\left(\frac{\xi_2 - \xi_1}{\sqrt{\sigma_2^2 - 2\rho\sigma_1\sigma_2 + \sigma_1^2}}\right); \quad (46.80)
\end{aligned}
$$

similarly, we obtain the second raw moment of $X_{(1)}$ to be

$$
\begin{aligned}
E[X_{(1)}^2] &= (\xi_1^2 + \sigma_1^2)\Phi\left(\frac{\xi_2 - \xi_1}{\sqrt{\sigma_2^2 - 2\rho\sigma_1\sigma_2 + \sigma_1^2}}\right) \\
&\quad + (\xi_2^2 + \sigma_2^2)\Phi\left(\frac{\xi_1 - \xi_2}{\sqrt{\sigma_2^2 - 2\rho\sigma_1\sigma_2 + \sigma_1^2}}\right) \\
&\quad - (\xi_1 + \xi_2)\sqrt{\sigma_2^2 - 2\rho\sigma_1\sigma_2 + \sigma_1^2}\ \phi\left(\frac{\xi_2 - \xi_1}{\sqrt{\sigma_2^2 - 2\rho\sigma_1\sigma_2 + \sigma_1^2}}\right);
\end{aligned}
$$
$$(46.81)$$

see also David (1981, pp. 51–52).

Cain and Pan (1995) have extended Cain's (1994) results by establishing a recurrence relation for

$$
\mu_r' = E[X_{(1)}^r] = \int_{-\infty}^{\infty} x^r f_1(x)dx + \int_{-\infty}^{\infty} x^r f_2(x)\ dx. \quad (46.82)
$$

Specifically, they have shown that

$$
\begin{aligned}
&\int_{-\infty}^{\infty} x^r f_1(x)\ dx - \xi_1 \int_{-\infty}^{\infty} x^{r-1} f_1(x)\ dx - (r-1)\sigma_1^2 \int_{-\infty}^{\infty} x^{r-2} f_1(x)\ dx \\
&= -\frac{\sigma_1^2 - \rho\sigma_1\sigma_2}{\sigma_1\sigma_2\sqrt{1 - \rho^2}}\ \exp\left\{-\frac{(\xi_1 - \xi_2)^2}{2(\sigma_2^2 - 2\rho\sigma_1\sigma_2 + \sigma_1^2)}\right\} \cdot \int_{-\infty}^{\infty} \frac{x^{r-1}}{2\pi} \\
&\quad \times \exp\left[-\frac{\sigma_2^2 - 2\rho\sigma_1\sigma_2 + \sigma_1^2}{2\sigma_1^2\sigma_2^2(1 - \rho^2)}\left\{x - \frac{\xi_2\sigma_1^2 - \rho\sigma_1\sigma_2(\xi_1 + \xi_2) + \xi_1\sigma_2^2}{\sigma_2^2 - 2\rho\sigma_1\sigma_2 + \sigma_1^2}\right\}^2\right] dx \\
&= -\frac{\sigma_1^2 - \rho\sigma_1\sigma_2}{\sqrt{\sigma_2^2 - 2\rho\sigma_1\sigma_2 + \sigma_1^2}}\ \phi\left(\frac{\xi_1 - \xi_2}{\sqrt{\sigma_2^2 - 2\rho\sigma_1\sigma_2 + \sigma_1^2}}\right)
\end{aligned}
$$

$$
\times \int_{-\infty}^{\infty} \left\{ \frac{\sigma_1 \sigma_2 \sqrt{1-\rho^2}}{\sqrt{\sigma_2^2 - 2\rho\sigma_1\sigma_2 + \sigma_1^2}} \, u \right.
$$
$$
\left. + \frac{\xi_2 \sigma_1^2 - \rho\sigma_1\sigma_2(\xi_1 + \xi_2) + \xi_1 \sigma_2^2}{\sigma_2^2 - 2\rho\sigma_1\sigma_2 + \sigma_1^2} \right\}^{r-1} \phi(u) \, du.
$$

$$(46.83)$$

(When $r = 1$, the third term on the left-hand side will not appear.) Equation (46.83) provides a direct recursive relation for the required integral, involving two lags rather than one. The second integral in (46.82) can be obtained from the recurrence relation in (46.83) upon replacing $\xi_1(\xi_2)$ by $\xi_2(\xi_1)$ and $\sigma_1(\sigma_2)$ by $\sigma_2(\sigma_1)$.

Let us consider the case when $(X_1, X_2)^T$ has a standard bivariate normal distribution with correlation coefficient ρ. Let $U = a_1 X_{(1)} + a_2 X_{(2)}$, where a_1 and a_2 are constants. Let $b_i = 1/a_i$ for $i = 1, 2$. Then, using direct methods, Nagaraja (1982) has shown that the density function of U can be written as

$$
f_U(u) = \frac{|b_1 b_2|}{\pi \sqrt{1-\rho^2}} \exp\left\{ -\frac{(b_1 b_2 u)^2}{2\delta} \right\}
$$
$$
\times \int_{-\infty}^{b_2 u/(b_1+b_2)} \exp\left[-\frac{\delta}{2(1-\rho^2)} \left\{ v - b_2(b_2 + \rho b_1)\frac{u}{\delta} \right\}^2 \right] dv,
$$

$$(46.84)$$

where $\delta = b_1^2 + b_2^2 + 2\rho b_1 b_2$. The integral can be expressed as $\Phi(\eta u)$, where

$$
\eta = \frac{b_1 b_2 (b_1 - b_2)}{b_1 + b_2} \sqrt{\frac{1-\rho}{(1+\rho)\delta}}
$$

in the case when $b_1 + b_2 > 0$.

If a_1, a_2 are nonzero but $a_1 + a_2 = 0$, then $U = a_2|X_2 - X_1|$; since $X_2 - X_1 \overset{d}{=} N(0, 2(1-\rho))$, we simply have the density function of U to be

$$
f_U(u) = \begin{cases} f_2(u) & \text{if } a_2 > 0, \\ f_2(-u) & \text{if } a_2 < 0, \end{cases}
$$

where

$$
f_2(u) = \begin{cases} \frac{1}{\sqrt{2}|a_2|\sqrt{1-\rho}} \, \phi\left(\frac{u}{a_2\sqrt{2(1-\rho)}} \right) & \text{if } u \geq 0, \\ 0 & \text{if } u < 0. \end{cases}
$$

Finally, when a_1, a_2 are nonzero, the density function of U can be expressed as

$$
f_U(u) = \begin{cases} f_1(u) & \text{if } b_1 + b_2 > 0, \\ f_1(-u) & \text{if } b_1 + b_2 < 0, \end{cases}
$$

where

$$f_1(u) = \frac{2}{\sqrt{\zeta}} \, \phi\left(\frac{u}{\sqrt{\zeta}}\right) \Phi(\eta u)$$

with $\zeta = a_1^2 + 2\rho a_1 a_2 + a_2^2$. These results of Nagaraja (1982) correct the earlier erroneous derivation of Gupta and Pillai (1965).

Let $(X_{1i}, X_{2i})^T$, $i = 1, 2, \ldots, n$, be a random sample from a bivariate distribution with joint cumulative distribution function $F(x_1, x_2)$. If the sample is ordered by the X_1-values, then the X_2 value associated with the ith order statistic $X_{1(i)}$ is called the *concomitant of the ith order statistic* and is denoted by $X_{2[i]}$; see David (1973, 1981) and Bhattacharya (1974). Suppose that X_{1i} and X_{2i} $(i = 1, 2, \ldots, n)$ have means ξ_1 and ξ_2, variances σ_1^2 and σ_2^2, respectively, and are linked by the linear regression model

$$X_{2i} = \xi_2 + \rho \frac{\sigma_2}{\sigma_1}(X_{1i} - \xi_1) + \varepsilon_i, \tag{46.85}$$

where $|\rho| < 1$ and X_{1i} and ε_i are mutually independent. Then, from (46.85) it follows that $E[\varepsilon_i] = 0$, $\text{var}(\varepsilon_i) = \sigma_2^2(1-\rho^2)$ and $\rho = \text{corr}(X_1, X_2)$. In the special case when X_{1i} and ε_i are normal, $(X_{1i}, X_{2i})^T$ are distributed as bivariate normal. From (46.85), we have

$$X_{2[r]} = \xi_2 + \rho \frac{\sigma_2}{\sigma_1}(X_{1(r)} - \xi_1) + \varepsilon_{[r]}, \quad r = 1, \ldots, n, \tag{46.86}$$

where $\varepsilon_{[r]}$ denotes the specific ε_i associated with $X_{1(r)}$. Due to the independence of X_{1i} and ε_i, we have $X_{1(r)}$ to be independent of $\varepsilon_{[r]}$, with $\varepsilon_{[r]}$ being mutually independent having the same distribution as ε_i. Denoting now

$$\alpha_r = E\left[\frac{X_{1(r)} - \xi_1}{\sigma_1}\right] \quad \text{and} \quad \beta_{r,s} = \text{cov}\left(\frac{X_{1(r)} - \xi_1}{\sigma_1}, \frac{X_{1(s)} - \xi_1}{\sigma_1}\right),$$

we observe from (46.86) that

$$
\begin{aligned}
E[X_{2[r]}] &= \xi_2 + \rho\sigma_2\alpha_r, \\
\text{var}(X_{2[r]}) &= \sigma_2^2(\rho^2\beta_{rr} + 1 - \rho^2), \\
\text{cov}(X_{1(r)}, X_{2[s]}) &= \rho\sigma_1\sigma_2\beta_{rs}
\end{aligned}
$$

and

$$\text{cov}(X_{2[r]}, X_{2[s]}) = \rho^2\sigma_2^2\beta_{rs} \quad \text{for } r \neq s.$$

For the case of the bivariate normal distribution, these formulas were derived by Watterson (1959). As pointed out by Sondhauss (1994), these may be expressed alternatively as

$$E[X_{2[r]}] - \xi_2 = \rho(E[X_{2(r)}] - \xi_2),$$
$$\text{var}(X_{2[r]}) - \sigma_2^2 = \rho^2\{\text{var}(X_{2(r)}) - \sigma_2^2\}$$

and

$$\text{cov}(X_{2[r]}, X_{2[s]}) = \rho^2\text{cov}(X_{2(r)}, X_{2(s)}) \quad \text{for } r \neq s.$$

These formulas can be extended easily to the multivariate case. Asymptotic results on these concomitant order statistics have been established by David and Galambos (1974), David (1994), and Nagaraja and David (1994). An extensive review on this topic has been prepared recently by David and Nagaraja (1998).

Instead of just considering the concomitants of order statistics arising from ordering one of the components of $(X_{1i}, X_{2i})^T$, $i = 1, 2, \ldots, n$, one may consider ordering through a linear combination

$$S_i = aX_{1i} + bX_{2i} \quad \text{for } i = 1, 2, \ldots, n,$$

where a and b are two nonzero constants. Let $S_{(1)} \leq S_{(2)} \leq \cdots \leq S_{(n)}$ be the order statistics of $S_1, S_2, \ldots, S_n$, and let $X_{1[r]}$ and $X_{2[r]}$ be the concomitants associated with $S_{(r)}$. For the bivariate normal case, Balakrishnan (1993) has derived the following expressions:

$$E[X_{1[r]}] = \xi_1 + \left(\frac{a\sigma_1^2 + b\rho\sigma_1\sigma_2}{\sqrt{\Delta}}\right)\alpha_r$$

and

$$E[X_{2[r]}] = \xi_2 + \left(\frac{b\sigma_2^2 + a\rho\sigma_1\sigma_2}{\sqrt{\Delta}}\right)\alpha_r,$$

where $\Delta = a^2\sigma_1^2 + b^2\sigma_2^2 + 2ab\rho\sigma_1\sigma_2$;

$$\text{var}\left(X_{1[r]}\right) = \frac{\sigma_1^2}{\Delta}\left\{(a\sigma_1 + b\rho\sigma_2)^2\beta_{rr} + b^2(1 - \rho^2)\sigma_2^2\right\},$$

$$\text{var}\left(X_{2[r]}\right) = \frac{\sigma_2^2}{\Delta}\left\{(b\sigma_2 + a\rho\sigma_1)^2\beta_{rr} + b^2(1 - \rho^2)\sigma_1^2\right\},$$

$$\text{cov}\left(X_{1[r]}, X_{2[r]}\right) = \frac{\sigma_1\sigma_2}{\Delta}\left\{(a\sigma_1 + b\rho\sigma_2)(b\sigma_2 + a\rho\sigma_1)\beta_{rr} - ab(1 - \rho^2)\sigma_1\sigma_2\right\},$$

$$\text{cov}\left(X_{1[r]}, X_{1[s]}\right) = \frac{\sigma_1^2}{\Delta}(a\sigma_1 + b\rho\sigma_2)^2\beta_{rs}, \quad 1 \leq r < s \leq n,$$

$$\text{cov}\left(X_{2[r]}, X_{2[s]}\right) = \frac{\sigma_2^2}{\Delta}(b\sigma_2 + a\rho\sigma_1)^2\beta_{rs}, \quad 1 \leq r < s \leq n,$$

and

$$\text{cov}\left(X_{1[r]}, X_{2[s]}\right) = \frac{\sigma_1 \sigma_2}{\Delta}(a\sigma_1 + b\rho\sigma_2)(b\sigma_2 + a\rho\sigma_1)\beta_{rs}, \quad 1 \le r < s \le n.$$

These results are extensions of the corresponding ones for the independent normal variable case discussed earlier by Song, Buchberger and Deddens (1992). These authors have also given an interesting example in hydrology where this model is of great interest in extreme lake levels. The above-given formulas were also independently derived by Song and Deddens (1993) through conditioning arguments. The results have also been generalized to the multivariate normal case; see, for example, Balakrishnan (1993).

7   TRIVARIATE NORMAL INTEGRAL

As in the case of the bivariate normal integral, we confine ourselves to the consideration of standardized distributions, with expected value vector $(0,0,0)$ and variance–covariance matrix

$$\begin{pmatrix} 1 & \rho_{12} & \rho_{13} \\ \rho_{12} & 1 & \rho_{23} \\ \rho_{13} & \rho_{23} & 1 \end{pmatrix}.$$

Analogously to $\Phi(h, k; \rho)$ and $L(h, k; \rho)$, we define

$$\Phi(h_1, h_2, h_3; \rho_{23}, \rho_{13}, \rho_{12}) = \Pr[(X_1 < h_1) \cap (X_2 < h_2) \cap (X_3 < h_3)]$$
$$(46.87)$$

and

$$L(h_1, h_2, h_3; \rho_{23}, \rho_{13}, \rho_{12}) = \Pr[(X_1 > h_1) \cap (X_2 > h_2) \cap (X_3 > h_3)].$$
$$(46.88)$$

Equation (46.3) generalizes to

$$\Phi(0, 0, 0; \rho_{23}, \rho_{13}, \rho_{12}) = L(0, 0, 0; \rho_{23}, \rho_{13}, \rho_{12}).$$
$$(46.89)$$

Ruben (1954) has given a table of $\frac{1}{2} - \frac{3}{4}\pi \cos^{-1}\rho$ [equal to $\Phi(0, 0, 0; \rho, \rho, \rho)$] to eight decimal places for $\rho^{-1} = 2(1)11$ (his $\bar{V}_{n,n}(x) = \bar{u}_n(x)$ for $n = 3$). Tables of $\Phi(h, h, h; \frac{1}{2}, \frac{1}{2}, \frac{1}{2})$ have been published by (i) Teichroew (1955) to five decimal places for $h\sqrt{2} = 0(0.01)6.09$, and (ii) Somerville (1954) to four decimal places for $h = 0(0.1)2.0(0.5)3.0$. There are also unpublished tables by Owen [reported by Steck (1958)] of $\Phi(h, h, h; \rho, \rho, \rho)$ for (a) $\rho = (1+\sqrt{3})^{-1}, \frac{1}{4}, h = 0.(0.1)3.0(0.5)8.0$, and (b) $\rho = 0(0.1)0.9, h = 0(0.2)1.0$.

Steck (1958) has expressed the trivariate integral Φ in terms of a function (for $h, a, b > 0$)

$$
\begin{aligned}
S(h, a, b) = &\frac{1}{4\pi} \tan^{-1} \frac{b}{\sqrt{1+a^2+a^2b^2}} \\
&+ \Pr[(0 < U_1 < U_2 + bU_3) \cap (0 < U_2 < h) \cap (U_3 > aU_2)],
\end{aligned}
$$
(46.90)

where U_1, U_2, and U_3 are independent unit normal variables. For negative values of h, a, or b, $S(h, a, b)$ is defined through the formulas

$$
S(h, -a, b) = S(h, a, b),
$$
(46.91)

$$
S(h, a, -b) = -S(h, a, b),
$$
(46.92)

$$
S(-h, a, b) = \frac{1}{2\pi} \tan^{-1} \frac{b}{\sqrt{1+a^2+a^2b^2}} - S(h, a, b).
$$
(46.93)

Also,

$$
S(0, a, b) = \frac{1}{2} S(\infty, a, b) = \frac{1}{4\pi} \tan^{-1} \frac{b}{\sqrt{1+a^2+a^2b^2}},
$$
(46.94)

$$
S(h, 0, b) = \frac{1}{2\pi} \Phi(h) \tan^{-1} b,
$$
(46.95)

$$
S(h, a, 0) = 0 = S(h, \infty, b),
$$
(46.96)

$$
S(h, a, \infty) = \begin{cases} \frac{1}{2} \left[\frac{1}{2} \Phi(h) + T(h, |a|) \right] - \frac{1}{2\pi} \tan^{-1} |a| & (h \geq 0), \\[2mm] \frac{1}{2} \left[\frac{1}{2} \Phi(h) - T(h, |a|) \right] & (h < 0). \end{cases}
$$
(46.97)

The formulas for $\Phi(h_1, h_2, h_3; \rho_{23}, \rho_{13}, \rho_{12})$ are as follows:

1. For $h_1, h_2, h_3 \geq 0$ (or $h_1, h_2, h_3 \leq 0$),

$$
\begin{aligned}
&\Phi(h_1, h_2, h_3; \rho_{23}, \rho_{13}, \rho_{12}) \\
&= \sum_{j=1}^{3} \left(1 - \frac{1}{2} \delta_{a_j c_j} \right) \Phi(h_j) + \frac{1}{4} \sum_{i<j}^{3} \sum^{3} \delta_{h_i h_j}
\end{aligned}
$$

$$+ \frac{1}{2}\left[\sum_{i<j}^{3}\sum^{3} L(h_i, h_j; \rho_{ij}) - 3\right]$$

$$- \sum_{j=1}^{3}[S(h_j, a_j, b_j) + S(h_j, c_j, d_j)], \qquad (46.98)$$

where

$$a_1 = \frac{h_2 - h_1\rho_{12}}{h_1\sqrt{1-\rho_{12}^2}}, \quad a_2 = \frac{h_3 - h_2\rho_{23}}{h_2\sqrt{1-\rho_{23}^2}}, \quad a_3 = \frac{h_1 - h_3\rho_{13}}{h_3\sqrt{1-\rho_{13}^2}},$$

$$c_1 = \frac{h_3 - h_1\rho_{13}}{h_1\sqrt{1-\rho_{13}^2}}, \quad c_2 = \frac{h_1 - h_2\rho_{12}}{h_2\sqrt{1-\rho_{12}^2}}, \quad c_3 = \frac{h_2 - h_3\rho_{23}}{h_3\sqrt{1-\rho_{23}^2}},$$

$$b_1 = \sqrt{(1-\rho_{12}^2)(1-\rho_{13}^2)}(c_1 a_1^{-1} - \rho_{23.1})\Delta^{-1/2},$$

$$d_1 = \sqrt{(1-\rho_{12}^2)(1-\rho_{13}^2)}(a_1 c_1^{-1} - \rho_{23.1})\Delta^{-1/2},$$

and so on; and δ_{hk} is as defined after (46.49);

(ii) For $h_1, h_2 \geq 0$, $h_3 < 0$ (or $h_1, h_2 \leq 0$, $h_3 > 0$),

$$\Phi(h_1, h_2, h_3; \rho_{23}, \rho_{13}, \rho_{12}) = L(h_1, h_2; \rho_{12}) + \Phi(h_1) + \Phi(h_2) - 1$$
$$- \Phi(h_1, h_2, -h_3; -\rho_{23}, \rho_{13}, \rho_{12}). \qquad (46.99)$$

Two other similar formulas can be obtained by permuting the variates.

Steck (1958) gives tables of $S(h, a, b)$ to seven decimal places for $a = 0.0(0.1)2.0(0.2)5.0(0.5)8.0$, $b = 0.1(0.1)1.0$, and h increasing from zero by intervals of 0.1 to an upper limit such that beyond this limit

$$S(h, a, b) \doteq \frac{1}{2\pi}\,\Phi\left(h\sqrt{1 + a^2 + \frac{1}{4}\,a^2 b_2}\right)\tan^{-1}\left(\frac{b}{\sqrt{1 + a^2 + a^2 b^2}}\right) \qquad (46.100)$$

with an error less than 5×10^{-5}. For values of b greater than 1, the formulas

$$S(h, a, b) = \left[\Phi(h) - \frac{1}{2}\right]T(ah, b) - \left[\Phi(hab) - \frac{1}{2}\right]T(ah, a^{-1})$$
$$+ S(hab, b^{-1}, a^{-1}) \qquad \text{(for } a > 1\text{)} \qquad (46.101)$$

or

$$S(h, a, b) = \frac{1}{4}\,\Phi(h) + \left[\Phi(hab) - \frac{1}{2}\right]T(h, a) - S(hab, (ab)^{-1}, a)$$
$$- S(h, ab, b^{-1}) \qquad (\text{for } a < 1) \qquad\qquad (46.102)$$

may be used.

Mukerjea and Stephens (1990a) discussed various properties of trivariate normal distribution and also provided a solution to an identification problem. Starting with the trivariate normal density function written in the form

$$p(x_1, x_2, x_3) = (2\pi)^{-3/2}\sqrt{|M|} \cdot \exp\left\{-\frac{1}{2}(s_1^2 x_1^2 + s_2^2 x_2^2 + s_3^2 x_3^2\right.$$
$$\left. + 2s_1 s_2 r_{12} x_1 x_2 + 2s_1 s_3 r_{13} x_1 x_3 + 2s_2 s_3 r_{23} x_2 x_3)\right\},$$
$$(46.103)$$

where

$$|M| = s_1^2 s_2^2 s_3^2 \Delta^2 \quad \text{and} \quad \Delta^2 = 1 + 2r_{12} r_{13} r_{23} - r_{12}^2 - r_{13}^2 - r_{23}^2,$$

and denoting the variance of X_i by σ_i^2 ($i = 1, 2, 3$) and the correlation coefficient between X_i and X_j by ρ_{ij}, we have

$$\sigma_1^2 = \frac{1 - r_{23}^2}{s_1^2 \Delta^2}, \quad \sigma_2^2 = \frac{1 - r_{13}^2}{s_2^2 \Delta^2}, \quad \sigma_3^2 = \frac{1 - r_{12}^2}{s_3^2 \Delta^2},$$

$$\rho_{12} = \frac{r_{13} r_{23} - r_{12}}{\sqrt{(1 - r_{23}^2)(1 - r_{13}^2)}} \qquad \rho_{13} = \frac{r_{12} r_{23} - r_{13}}{\sqrt{(1 - r_{12}^2)(1 - r_{23}^2)}},$$

$$\text{and } \rho_{23} = \frac{r_{12} r_{13} - r_{23}}{\sqrt{(1 - r_{13}^2)(1 - r_{12}^2)}}.$$

The bivariate marginal density functions of (46.103) are as follows:

$$p_{12}(x_1, x_2) = \frac{s_1 s_2 \Delta}{2\pi}\exp\left\{-\frac{1}{2}[s_1^2(1 - r_{13}^2)x_1^2 + s_2^2(1 - r_{23}^2)x_2^2\right.$$
$$\left. + 2s_1 s_2(r_{12} - r_{13} r_{23})x_1 x_2]\right\},$$

$$p_{13}(x_1, x_3) = \frac{s_1 s_3 \Delta}{2\pi}\exp\left\{-\frac{1}{2}[s_1^2(1 - r_{12}^2)x_1^2 + s_3^2(1 - r_{23}^2)x_3^2\right.$$
$$\left. + 2s_1 s_3(r_{13} - r_{12} r_{23})x_1 x_3]\right\},$$

and

$$p_{23}(x_2, x_3) = \frac{s_2 s_3 \Delta}{2\pi} \exp\left\{-\frac{1}{2}[s_2^2(1 - r_{12}^2)x_2^2 + s_3^2(1 - r_{13}^2)x_3^2\right.$$
$$\left. +2s_2 s_3(r_{22} - r_{12}r_{13})x_2 x_3]\right\}.$$

If $F(x_1, x_2, x_3)$ denotes a nonsingular trivariate normal distribution function corresponding to (46.103), then the partial derivative F_{x_1} is

$$F_{x_1}(x_1, x_2, x_3) = \frac{\partial F(x_1, x_2, x_3)}{\partial x_1}$$

$$= \frac{s_1 \sqrt{1 + 2r_{12}r_{13}r_{23} - r_{12}^2 - r_{13}^2 - r_{23}^2}}{\sqrt{2\pi}\sqrt{1 - r_{23}^2}}$$

$$\times \exp\left\{-\frac{1}{2}\frac{s_1^2(1 + 2r_{12}r_{13}r_{23} - r_{12}^2 - r_{13}^2 - r_{23}^2)}{1 - r_{23}^2}x_1^2\right\}$$

$$\times N\left(\begin{pmatrix} m_1 \\ m_2 \end{pmatrix}, \begin{pmatrix} V_1^2 & V_{12} \\ V_{12} & V_2^2 \end{pmatrix}\right),$$

where

$$m_1 = x_2 + x_1\frac{s_1(r_{12} - r_{13}r_{23})}{s_2(1 - r_{23}^2)}, \quad m_2 = x_3 + x_1\frac{s_1(r_{13} - r_{12}r_{23})}{s_3(1 - r_{23}^2)},$$

$$V_1^2 = \frac{1}{s_2^2(1 - r_{23}^2)}, \quad V_2^2 = \frac{1}{s_3^2(1 - r_{23}^2)} \quad \text{and} \quad V_{12} = -\frac{r_{23}}{s_2 s_3(1 - r_{23}^2)}.$$

Similar expressions can be given for the partial derivatives F_{x_2} and F_{x_3}. The mixed partial derivatives of $F(x_1, x_2, x_3)$ are

$$F_{x_1,x_2}(x_1, x_2, x_3) = \frac{\partial^2 F(x_1, x_2, x_3)}{\partial x_1 \partial x_2}$$
$$= p_{12}(x_1, x_2)\Phi(s_1 r_{13}x_1 + s_2 r_{23}x_2 + s_3 x_3),$$

$$F_{x_1,x_3}(x_1, x_2, x_3) = \frac{\partial^2 F(x_1, x_2, x_3)}{\partial x_1 \partial x_3}$$
$$= p_{13}(x_1, x_3)\Phi(s_1 r_{12}x_1 + s_2 x_2 + s_3 r_{23}x_3),$$

and

$$F_{x_2,x_3}(x_1, x_2, x_3) = \frac{\partial^2 F(x_1, x_2, x_3)}{\partial x_2 \partial x_3}$$
$$= p_{23}(x_2, x_3)\Phi(s_1 x_1 + s_2 r_{12}x_2 + s_3 r_{13}x_3).$$

Mukherjea and Stephens (1990a) have also established the following factorization theorem for nonsingular trivariate normal distribution with zero means and nonzero correlations: Let $F_1, F_2, \ldots, F_m$ and $G_1, G_2, \ldots, G_n$ be nonsingular trivariate normal distribution functions with zero means and nonzero correlations. If the product of the F_i's and the product of the G_i's are identical, then $m = n$ and $\{F_1, \ldots, F_m\} = \{G_1, \ldots, G_m\}$.

They also noted that if this factorization holds for the product of two nonsingular 4-variate normal distributions with nonzero correlations, then the same is valid for k-variate ($k \geq 5$) such distributions. Suppose

$$F_1 F_2 = G_1 G_2 \Rightarrow \{F_1, F_2\} = \{G_1, G_2\}$$

whenever F_1, F_2, G_1, and G_2 are 4-variate nonsingular normal distribution functions with zero means and nonzero correlations. Then the same factorization holds for all such k-variate distributions with $k \geq 5$. The reason is the following. Suppose that $n = 5$. Suppose that $F_1 F_2 = G_1 G_2$ and F_1, F_2, G_1, and G_2 have covariance matrices M_1, M_2, N_1, and N_2, respectively. Let iM_j be the covariance matrix of the marginal of F_j as $x_i \to \infty$, and let iN_j be the same corresponding to G_j. Thus, with the assumption that factorization holds in the 4-variate case, we must have either $iM_1 = iN_1$ for at least three distinct i's or $iM_1 = iN_2$ for at least three distinct i's. It is then true that either $M_1 = N_1$ or $M_1 = N_2$. Consequently, the factorization holds for $n = 5$ and the case $n > 5$ can then be handled by induction.

The additional assumption that all the distributions have positive partial correlation allows to derive the above result in the general k-variate case; see Mukherjea and Stephens (1990b). The bivariate case for nonsingular normal distributions with positive correlation was proved originally by Anderson and Ghurye (1978), and the general bivariate case was solved by Mukherjea, Nakassis, and Miyashita (1986).

Arnold, Castillo, and Sarabia (1995) established the following conditional characterization of extended trivariate normal distributions. They examined all possible k-variate distributions for $\boldsymbol{X} = (X_1, \ldots, X_k)^T$ such that for any i and subvector $\tilde{\boldsymbol{X}}_{(i)}$ of $\boldsymbol{X}_{(i)}$, the conditional distribution of X_i given $\tilde{\boldsymbol{X}}_{(i)} = \tilde{\boldsymbol{x}}_{(i)}$ is normal; here, $\boldsymbol{X}_{(i)}$ denotes vector $\boldsymbol{X}$ with the i-th coordinate deleted. The classical k-variate normal distribution has this property, but the class includes other distributions as well. For example, in the case of $k = 3$, Arnold, Castillo, and Sarabia (1995) have pointed out that the joint density function

$$p(x_1, x_2, x_3) = \exp\left\{ -(a_{000} + a_{100}x_1 + a_{010}x_2 + a_{001}x_3 \right.$$

$$+a_{110}x_1x_2 + a_{101}x_1x_3 + a_{011}x_2x_3$$
$$+a_{111}x_1x_2x_3 + a_{200}x_1^2 + a_{020}x_2^2 + a_{002}x_3^2)\Big\}$$

$$(46.104)$$

has $X_1|(X_2, X_3)$, $X_2|(X_1, X_3)$, $X_3|(X_1, X_2)$, $X_1|X_2$, $X_2|X_1$, $X_1|X_3$, $X_3|X_1$, $X_2|X_3$, and $X_3|X_2$ all of the univariate normal form. The presence of a nonzero value for the parameter a_{111} in (46.104) identifies a distribution which is not the classical trivariate normal.

Considering the trivariate normal variable $\boldsymbol{X}$ with mean and variance–covariance matrix as

$$\boldsymbol{\xi} = \begin{pmatrix} 0 \\ 0 \\ 0 \end{pmatrix} \quad \text{and} \quad \boldsymbol{\Sigma} = \begin{pmatrix} 1 & \rho_{12} & \rho_{13} \\ \rho_{12} & 1 & \rho_{23} \\ \rho_{13} & \rho_{23} & 1 \end{pmatrix}$$

and denoting the conditional distribution and density function of $\binom{X_i}{X_j}|X_k$ by $\Phi(x_i, x_j|x_k)$ and $\phi(x_i, x_j|x_k)$, respectively, Sungur (1990) has established the following properties:

(i) $\frac{\partial \Phi(x_i,x_j|x_k)}{\partial \rho_{ij.k}} = \sqrt{1 - \rho_{ik}^2} \sqrt{1 - \rho_{jk}^2}\, \phi(x_i, x_j|x_k)$,
   where $\rho_{ij.k}$ is the partial correlation between X_i and X_j, given $X_k = x_k$,

(ii) $\frac{\partial \Phi(x_i,x_j|x_k)}{\partial \rho_{ij}} = \phi(x_i, x_j|x_k)$,

and so

(iii) $\frac{\partial \Phi(x_i,x_j|x_k)}{\partial \rho_{ij}}\, \phi(x_k) = \phi(x_i, x_j, x_k)$.

8 ESTIMATION

8.1 Bivariate Normal Distribution

The likelihood function of n independent pairs of random variables $(X_{11}, X_{21}), (X_{12}, X_{22}), \ldots, (X_{1n}, X_{2n})$, each having the same joint bivariate normal distribution with parameters $\xi_1, \xi_2, \sigma_1, \sigma_2$, and ρ, is

$$\left(\frac{1}{2\pi\sqrt{1 - \rho^2}}\right)^n \exp\left[-\frac{1}{2(1-\rho^2)}\left\{\frac{\sum_{j=1}^n (X_{1j} - \xi_1)^2}{\sigma_1^2}\right.\right.$$
$$\left.\left. -2\rho\frac{\sum_{j=1}^n (X_{1j} - \xi_1)(X_{2j} - \xi_2)}{\sigma_1\sigma_2} + \frac{\sum_{j=1}^n (X_{2j} - \xi_2)^2}{\sigma_2^2}\right\}\right]$$

$$
= \left(\frac{1}{2\pi\sqrt{1-\rho^2}}\right)^n \exp\left[-\frac{n}{2(1-\rho^2)}\right.
$$

$$
\times \left\{\frac{(\bar{X}_1 - \xi_1)^2}{\sigma_1^2} - 2\rho\,\frac{(\bar{X}_1 - \xi_1)(\bar{X}_2 - \xi_2)}{\sigma_1\sigma_2} + \frac{(\bar{X}_2 - \xi_2)^2}{\sigma_2^2}\right\}\right]
$$

$$
\times \exp\left[-\frac{n}{2(1-\rho^2)}\left\{\frac{S_1^2}{\sigma_1^2} - 2R\,\frac{S_1 S_2}{\sigma_1\sigma_2} + \frac{S_2^2}{\sigma_2^2}\right\}\right], \qquad (46.105)
$$

where

$$
\bar{X}_t = \frac{1}{n}\sum_{j=1}^{n} X_{tj}, \quad S_t^2 = \frac{1}{n}\sum_{j=1}^{n}(X_{tj} - \bar{X}_t)^2 \quad (t = 1, 2),
$$

$$
RS_1 S_2 = \frac{1}{n}\sum_{j=1}^{n}(X_{1j} - \bar{X}_1)(X_{2j} - \bar{X}_2).
$$

The maximum likelihood estimators of the parameters are

$$
\hat{\xi}_1 = \bar{X}_1, \quad \hat{\xi}_2 = \bar{X}_2, \quad \hat{\sigma}_1 = S_1, \quad \hat{\sigma}_2 = S_2, \quad \hat{\rho} = R. \qquad (46.106)
$$

The two sets of variables $(\hat{\xi}_1, \hat{\xi}_2)$ and $(\hat{\sigma}_1, \hat{\sigma}_2, \hat{\rho})$ are mutually independent. The set $\hat{\xi}_1, \hat{\xi}_2, \hat{\sigma}_1, \hat{\sigma}_2, \hat{\rho}$ is jointly sufficient for $\xi_1, \xi_2, \sigma_1, \sigma_2, \rho$. We have

$$
n\,\mathrm{var}(\hat{\xi}_t) = \sigma_t^2 \qquad (t = 1, 2), \qquad (46.107)
$$

and

$$
n\,\mathrm{var}(\hat{\sigma}_t^2) = 2\sigma_t^4\left(1 - \frac{1}{n}\right) \qquad (t = 1, 2). \qquad (46.108)
$$

For large n, we have

$$
n\,\mathrm{var}(\hat{\rho}) \doteq (1-\rho^2)^2, \qquad (46.109)
$$

$$
\mathrm{corr}(\hat{\sigma}_1^2, \hat{\sigma}_2^2) \doteq \rho^2, \qquad (46.110)
$$

$$
\mathrm{corr}(\hat{\sigma}_t^2, \hat{\rho}) \doteq \rho/\sqrt{2} \qquad (t = 1, 2). \qquad (46.111)
$$

Formulas (46.107) and (46.109) have already been encountered in Chapters 13 and 32 of Johnson, Kotz, and Balakrishnan (1994, 1995).

Note that the estimators of ξ_1, ξ_2, σ_1, and σ_2 are those that would be obtained by using separately the observed values of the appropriate variables. We have already described the distributions of $\hat{\xi}_t$ (in Chapter 13), $\hat{\sigma}_t$ (in Chapter 18), and $\hat{\rho}$ (in Chapter 32).

If the values of some of the parameters are known, different estimators of the remaining parameters are obtained. We now set out, briefly, some special cases.

(i) *One mean, say ξ_1, known.* The maximum likelihood estimators are now

$$
\begin{cases}
\hat{\xi}_2 = \bar{X}_2 - (\hat{\rho}\hat{\sigma}_2/\hat{\sigma}_1)(\bar{X}_1 - \xi_1), \\[2mm]
\hat{\sigma}_1 = \left[\frac{1}{n}\sum_{j=1}^n (X_{1j} - \xi_1)^2\right]^{1/2}, \quad \hat{\sigma}_2 = \left[\frac{1}{n}\sum_{j=1}^n (X_{2j} - \bar{X}_2)^2\right]^{1/2}, \\[2mm]
\hat{\rho} = \left[\frac{1}{n}\sum_{j=1}^n (X_{1j} - \xi_1)(X_{2j} - \bar{X}_2)\right]\Big/(\hat{\sigma}_1\hat{\sigma}_2).
\end{cases}
$$

$$(46.112)$$

The estimators $\hat{\sigma}_1, \hat{\sigma}_2, \hat{\rho}$ may be obtained from (46.106) by replacing $\bar{X}_1$ by ξ_1, but $\hat{\xi}_2$ cannot be obtained in this way.

(ii) *ξ_1 and ξ_2 known.* Maximum likelihood estimators of σ_1, σ_2, and ρ are obtained by replacing $\bar{X}_1$ by ξ_1 and $\bar{X}_2$ by ξ_2 in the corresponding formulas in (46.106). The three estimators so obtained are jointly sufficient for σ_1, σ_2, and ρ. For large n, the variances and correlations of $\hat{\sigma}_1^2, \hat{\sigma}_2^2$, and $\hat{\rho}$ are the same as in the case when no parameters are known; see Eqs. (46.107)–(46.111).

(iii) *ξ_1 and σ_1 known.* The maximum likelihood estimators are now

$$\hat{\xi}_2 = \bar{X}_2 - (RS_2/S_1)(\bar{X}_1 - \xi_1), \tag{46.113}$$

$$\hat{\sigma}_2^2 = S_2^2(1 - R^2 + R^2\sigma_1^2/S_1^2), \tag{46.114}$$

$$\hat{\rho} = (R\sigma_1/S_1)(1 - R^2 + R^2\sigma_1^2/S_1^2)^{-1/2}. \tag{46.115}$$

(iv) *ξ_1, σ_1, and ρ known.* The maximum likelihood estimator of σ_2 is (with $\rho > 0$)

$$\hat{\sigma}_2 = \frac{[(\rho S_{12}'/\sigma_1)^2 + 4(1 - \rho^2)S_{22}]^{1/2} - \rho S_{12}'/\sigma_1}{2(1 - \rho^2)}, \tag{46.116}$$

where

$$S_{12}' = \frac{1}{n}\sum_{j=1}^n (X_{1j} - \xi_1)(X_{2j} - \bar{X}_2),$$

$$S_{22} = \frac{1}{n}\sum_{j=1}^n (X_{2j} - \bar{X})^2.$$

The maximum likelihood estimator of ξ_2 is

$$\hat{\xi}_2 = \bar{X}_2 - (\rho\hat{\sigma}_2/\sigma_1)(\bar{X}_1 - \xi_1). \tag{46.117}$$

Note that if $\rho = 0$, the estimators $\hat{\xi}_2, \hat{\sigma}_2$ become $\hat{X}_2$, $\sqrt{S_{22}}$, respectively, based on observations of X_2 only.

(v) $\xi_1, \sigma_1, \xi_2, \sigma_2$ *known*. It is convenient to introduce the symbols

$$r = \frac{\frac{1}{n}\sum_{j=1}^{n}(X_{1j} - \xi_1)(X_{2j} - \xi_2)}{\sigma_1\sigma_2},$$

$$r_t = \frac{1}{n}\sum_{j=1}^{n}(X_{tj} - \xi_t)^2/\sigma_t^2 \quad (t = 1, 2).$$

The maximum likelihood estimator of ρ is a solution of the cubic equation

$$\sigma_1^2\sigma_2^2\hat{\rho}(1 - \hat{\rho}^2) + \sigma_1\sigma_2(1 + \hat{\rho}^2) - \hat{\rho}\left(\sigma_2^2 r_1 + \sigma_1^2 r_2\right) = 0. \tag{46.118}$$

Kendall and Stuart (1963) show that the probability that this equation has only one real root between -1 and $+1$ tends to 1 as n tends to infinity. Since the left-hand side of (46.118) is positive for $\hat{\rho} = -1$ and negative for $\hat{\rho} = 1$, there must always be a real root between -1 and $+1$.

In any particular case, however, there may be three roots between -1 and $+1$. That root should be chosen for which the likelihood function in (46.105) is greatest.

Madansky (1958) has shown that provided

$$r^2 < 3(r_1 + r_2 - 1)$$

(note that this will usually be true, since we expect r_1, r_2 to be about 1 and r^2 to be less than 1, on the average), we have

$$\hat{\rho} = \frac{2}{3}\{3(r_1 + r_2 - 1) - r^2\}^{1/2} \sinh\left(\frac{1}{3}\sinh^{-1}C\right) + \frac{1}{3}r, \tag{46.119}$$

where

$$C = \frac{36r + 2\hat{r}^3 - 9(r_1 + r_2)r}{2[3(r_1 + r_2 - 1) - r^2]^{1/2}}.$$

In the (unusual) case $r^2 > 3(r_1 + r_2 - 1)$, we have the following: If $|C| \geq 1$,

$$\hat{\rho} = \frac{2}{3}\{r^2 - 3(r_1 + r_2 - 1)\}^{1/2} \cosh\left(\frac{1}{3}\cosh^{-1}C\right) + \frac{1}{3}r; \tag{46.120}$$

If $|C| \leq 1$,

$$\hat{\rho} = \frac{2}{3}\{r^2 - 3(r_1 + r_2 - 1)\}^{1/2} \cos\left(\frac{4\pi}{3} + \frac{1}{3}\cos^{-1}C\right) + \frac{1}{3}r;$$

(46.121)

see Nadler (1967).

For many practical purposes, the following method of calculation can be used. Put $\hat{\rho}_1 = r$. Then

$$\hat{\rho}_1 = \rho_1 + \hat{\varepsilon},$$

where

$$\hat{\varepsilon} = \frac{\hat{\rho}_1}{1 + \hat{\rho}_1^2}\left[2 - \frac{\tilde{\sigma}_1^2}{\sigma_1^2} - \frac{\tilde{\sigma}_2^2}{\sigma_2^2}\right]$$

with

$$\tilde{\sigma}_t^2 = r_t \qquad (t = 1, 2).$$

(vi) $\sigma_1 = \sigma_2$ *(but common value unknown)*. DeLury (1938) has shown that the intraclass correlation coefficient $2RS_1S_2(S_1^2 + S_2^2)^{-1}$ is slightly more efficient than R as an estimator of ρ in this case. The common value of σ_1 and σ_2 is estimated by $[\frac{1}{2}(S_1^2 + S_2^2)]^{1/2}$; ξ_1, ξ_2 are estimated by $\bar{X}_1, \bar{X}_2$, respectively.

Ahsanullah (1970) has studied the properties of an estimation procedure in which it is first tested whether $\xi_1 = \xi_2$. The estimate of ξ_1 employed is $\bar{X}_1$ or that appropriate to item (vii) according to the result of the test.

(vii) $\xi_1 = \xi_2$, $\sigma_1 = \sigma_2$ *(common values unknown)*. The ρ can be estimated by

$$\left\{2\sum_{j=1}^{n}(X_{1j} - \hat{\xi})(X_{2j} - \hat{\xi})\right\}\left\{\sum_{j=1}^{n}(X_{1j} - \hat{\xi})^2 + \sum_{j=1}^{n}(X_{2j} - \hat{\xi})^2\right\}^{-1}.$$

(46.122)

The common value of ξ_1 and ξ_2 is estimated by $\frac{1}{2}(\bar{X}_1 + \bar{X}_2) = \hat{\xi}$; that of σ_1 and σ_2 by

$$\left[\frac{1}{2n}\left\{\sum_{t=1}^{2}\sum_{j=1}^{n}(X_{tj} - \hat{\xi})^2\right\}\right]^{1/2}.$$

(viii) $\sigma_1^2\sigma_2^2(1-\rho^2)^2 = \theta^2$ (*known*). Press (1965) has considered this case. He shows that the maximum likelihood estimator of ρ is

$$\hat{\rho} = -\frac{1}{2} \, \theta S_{12}^{-1} + \text{sgn}(S_{12})\sqrt{1+\frac{1}{4}(\theta S_{12}^{-1})^2} \,, \qquad (46.123)$$

where

$$S_{12} = \frac{1}{n}\sum_{j=1}^{n}(X_{1j}-\bar{X}_1)(X_{2j}-\bar{X}_2),$$

$$\text{sgn}(S_{12}) = \begin{cases} 1 & \text{if } S_{12} > 0, \\ 0 & \text{if } S_{12} = 0, \\ -1 & \text{if } S_{12} < 0. \end{cases}$$

(ix) $\xi_1 = \xi_2$ (*common value unknown*). Rastogi and Rohatgi (1972) show that the weighted mean of $\bar{X}_1$ and $\bar{X}_2$, given by

$$\alpha \bar{X}_1 + (1-\alpha)\bar{X}_2, \qquad (46.124)$$

with

$$\alpha = (S_2^2 - S_{12})/(S_1^2 + S_2^2 - 2S_{12}),$$

is an unbiased estimator of the common value of ξ_1 and ξ_2, with variance

$$\frac{n-1}{n(n-3)}\frac{\sigma_1^2\sigma_2^2(1-\rho^2)}{\sigma_1^2 + 2\rho\sigma_1\sigma_2 + \sigma_2^2}. \qquad (46.125)$$

Similarly, many other special cases can also be discussed.

(x) *Missing observations.* We have already noted in Chapter 45 that multivariate data may be lacking values of certain variates for some individuals in a random sample. Many different patterns of data are possible, and methods of estimation need to be related to the actual data pattern. General methods have been developed by Afifi and Elashoff (1966–1969) (who also give a useful list of references), Smith (1968), Hocking and Smith (1968), Smith and Hocking (1968), Trawinski and Bargmann (1964), and Kleinbaum (1969). Here we consider in detail only the bivariate case, where the variety of possible patterns is limited. We will also describe the general method proposed by Hocking and Smith (1968).

Suppose there are n_{12} observations on both X_1 and X_2, n_1 on X_1 alone, and n_2 on X_2 alone. A rational method of solution is to start by estimating ρ from the n_{12} observations on both X_1 and X_2, because these are the only data providing information on ρ. Then, $\xi_1, \sigma_1^2, \xi_2, \sigma_2^2$ are estimated by the sample means and variances of the whole available data $[(n_1 + n_{12})$ observations for $\xi_1, \sigma_1; (n_2 + n_{12})$ for $\xi_2, \sigma_2]$. Although unsophisticated, this procedure is good enough for many practical problems. However, it does suffer from the defect that one does not use the information available on ξ_1, σ_1, ξ_2, and σ_2 in the $(n_1 + n_2)$ observations where one variate is missing, in the estimation of ρ. If one attempts to do this using the estimator

$$\rho^* = \frac{\Sigma(X_{1j} - \tilde{\xi}_1)(X_{2j} - \tilde{\xi}_2)}{[\Sigma(X_{1j} - \tilde{\xi}_1)^2]^{1/2}[\Sigma(X_{2j} - \tilde{\xi}_2)^2]^{1/2}} \qquad (46.126)$$

(where summations are over the n_{12} observations on both X_1 and X_2 and $\tilde{\xi}_j$ is the arithmetic mean of all $(n_j + n_{12})$ observations on X_j), it is possible that a value of $|\rho^*|$ in excess of 1 may be obtained. The possible increase of accuracy in the estimator of ρ must be judged against the possibility of obtaining such a value of ρ^*. We are inclined not to use ρ^* unless n_1, n_2 are large compared with n_{12}. Then one can use the estimators of ξ_1, σ_1 and ξ_2, σ_2 from the sets of n_1, n_2 observations, respectively, as if they were the actual values, and one can estimate ρ from the n_{12} observations using the method of Madansky (1958); see item (v) above.

The maximum likelihood equations for estimators of $\xi_1, \sigma_1, \xi_2, \sigma_2$, and ρ are

$$\frac{n_j(\bar{X}_j - \hat{\xi}_j) + (1 + \hat{\rho}^2)^{-1} n_{12}(\bar{X}_j' - \hat{\xi}_j)}{\hat{\sigma}_j^2} = \frac{n_{12}\hat{\rho}}{1 - \hat{\rho}^2} \frac{\bar{X}_{j'}' - \hat{\xi}_{j'}}{\hat{\sigma}_1\hat{\sigma}_2}, \qquad (46.127)$$

$$n_j + n_{12} = \frac{\hat{S}_{jj} + (1 - \hat{\rho}^2)^{-1}\hat{S}_{jj}'}{\hat{\sigma}_j^2} - \frac{\hat{\rho}}{(1 - \hat{\rho}^2)} \frac{\hat{P}'}{\hat{\sigma}_1\hat{\sigma}_2}$$
$$(j = 1, 2; \; j' = 3 - j), \qquad (46.128)$$

$$n_{12}\hat{\rho} = \frac{\hat{\rho}}{1 - \hat{\rho}^2}\left(\frac{\hat{S}_{11}'}{\hat{\sigma}_1^2} + \frac{\hat{S}_{22}'}{\hat{\sigma}_2^2}\right) - \frac{1 + \hat{\rho}^2}{1 - \hat{\rho}^2}\frac{\hat{P}'}{\hat{\sigma}_1\hat{\sigma}_2}, \qquad (46.129)$$

where $\bar{X}_j, \hat{S}_{jj} = $ mean X_j, and sum of squares $(X_j - \hat{\xi}_j)^2$, for n_j observations on X_j alone; $\bar{X}_j', \hat{S}_{jj}' = $ mean X_j, and sum of squares $(X_j - \hat{\xi}_j)^2$, for n_{12} observations on both X_1 and X_2; and $\hat{P}' = $ sum of products $(X_1 - \hat{\xi}_1)(X_2 - \hat{\xi}_2)$ for n_{12} observations on both X_1 and X_2.

In deriving Eqs. (46.127)–(46.129), we have used the likelihood function

$$\left(2\pi\sigma_1\sigma_2\sqrt{1 - \rho^2}\right)^{-n_{12}} (\sqrt{2\pi}\sigma_1)^{-n_1}(\sqrt{2\pi}\sigma_2)^{-n_2}$$

$$\exp\left\{-\frac{1}{2(1-\rho^2)}\left(\frac{S'_{11}}{\sigma_1^2} - \frac{2\rho P'}{\sigma_1\sigma_2} + \frac{S'_{22}}{\sigma_2^2}\right) - \frac{S_{11}}{2\sigma_1^2} - \frac{S_{22}}{2\sigma_2^2}\right\}$$

(46.130)

in an obvious notation.

Solution of these five simultaneous equations is not simple. For the special case $n_2 = 0$, Anderson (1957) has obtained explicit formulas for the maximum likelihood estimators:

$$
\begin{aligned}
\hat{\xi}_1 &= (n_1\bar{X}_1 + n_{12}\bar{X}'_1)/(n_1 + n_{12}), \\
\hat{\sigma}_1^2 &= (S_{11} + S'_{11})/(n_1 + n_{12}), \\
\hat{\xi}_2 &= \bar{X}'_2 + (P'/S'_{11})(\bar{X}_1 - \bar{X}'_1), \\
\hat{\sigma}_2^2 &= n_{12}^{-1}S'_{22} + (P'/S'_{11})^2[\hat{\sigma}_1^2 - S'_{11}/n_{12}], \\
\hat{\rho} &= (P'/S'_{11})(\hat{\sigma}_1/\hat{\sigma}_2).
\end{aligned}
$$

(46.131)

These estimators are obtained expeditiously by writing the likelihood function as a product of the likelihood of the X_1's with the conditional likelihood function of the X_2's given the X_1's. A similar method can be used when the pattern of data allows, but this is not always the case.

If n_1, n_2, and n_{12} are large with $p_1 = n_1/n$, $p_2 = n_2/n$, $p_{12} = n_{12}/n$, where $n = n_1+n_2+n_{12}$, then $\sqrt{n}$ times the maximum likelihood estimators of $\sigma_1^2, \rho\sigma_1\sigma_2, \sigma_2^2$ should have limiting variance–covariance matrix

$$
cp_{12}\begin{pmatrix}
2\sigma_1^4 & 2\rho\sigma_1^3\sigma_2 & 2\rho^2\sigma_1^2\sigma_2^2 \\
2\rho\sigma_1^3\sigma_2 & (1+\rho^2)\sigma_1^2\sigma_2^2 & 2\rho\sigma_1\sigma_2^3 \\
2\rho^2\sigma_1^2\sigma_2^2 & 2\rho\sigma_1\sigma_2^3 & 2\sigma_2^4
\end{pmatrix} + c(1-\rho^2)
$$

$$
\times\begin{pmatrix}
2p_2(1+\rho^2)\sigma_1^4 & 2p_2\rho\sigma_1^3\sigma_2 & 0 \\
2p_2\rho\sigma_1^3\sigma_2 & [1 - p_{12} + p_1p_2(1-\rho^2)/p_{12}]\sigma_1^2\sigma_2^2 & 2p_1\sigma_1\sigma_2^3 \\
0 & 2p_1\sigma_1\sigma_2^3 & 2p_1(1+\rho^2)\sigma_2^4
\end{pmatrix},
$$

(46.132)

where $c = [p_{12} + p_1p_2(1 - \rho^4)]^{-1}$; see Nadler (1967) and Wilks (1932).

Hocking and Smith (1968) and Smith and Hocking (1968) have proposed a system of estimators that can be applied in many general situations to estimation of parameters of m-variate normal populations from data with missing observations. The steps in this method are as follows:

(i) Group the data according to the sets of variables for which values are available.

(ii) For each set of data, estimate all the parameters (means, variances, and covariances of the available variables). It is suggested that "best unbiased estimation" be used.

(iii) Starting with the group for which full data are available (if any; otherwise, the fullest data), modify the estimators by adding linear functions of difference between estimators of the same parameter from
this group and all other groups where such estimators are available.
(Means are treated separately from variances and covariances.)

TABLE 46.2
Smith's Suggestion for Estimators in the Bivariate Case

Set of Observations	ξ_1	σ_1^2	ξ_2	σ_2^2	$\rho\sigma_1\sigma_2$
			Estimators of		
n_{12} (on X_1 and X_2)	$\bar{X}_1'$	$\frac{S_{11}'}{n_{12}-1}$	$\bar{X}_2'$	$\frac{S_{22}'}{n_{12}-1}$	$\frac{S_{12}'}{n_{12}-1}$
n_1 (on X_1)	$\bar{X}_1$	$\frac{S_{11}}{n_1-1}$	—	—	—
n_2 (on X_2)	—	—	$\bar{X}_2$	$\frac{S_{22}}{n_2-1}$	—

Application of this method to the bivariate case has been described
by Smith (1968). The first estimators are shown in Table 46.2. The estimators in the first line are then "improved" by adding linear multipliers
of $(\bar{X}_1' - \bar{X}_1)$ (for ξ_1 and ξ_2) and of $((n_{12} - 1)^{-1}S_{11}' - (n_1 - 1)^{-1}S_{11})$ (for
$\sigma_1^2, \sigma_2^2,$ and $\rho\sigma_1\sigma_2$). These in turn are "improved" by introducing linear
terms involving $\bar{X}_2$ and $(n_2-1)^{-1}S_{22}$. The coefficients of the linear adjust-
ing functions should be chosen to minimize the variance of the resultant
estimators. They would be functions of the unknown parameters, and we
have to use estimators of these parameters to calculate the coefficients.
Srivastava and Zaatar (1972) [Report ARL 72-0032, Aerospace Research
Laboratories, U.S. Air Force] have found empirical evidence in favor of the
simple estimators $\tilde{\sigma}_j^2 = S_{jj}n_j^{-1}$, $\tilde{\rho} = P'(S_{11}S_{22}')^{-\frac{1}{2}}$.

It is also possible to estimate ρ from broadly grouped data according
to the scheme shown in Figure 46.8. n_1, n_2, n_3, n_4 denote the number of
observations falling into the cells where they are placed. Given these data
(only), the maximum likelihood estimator, $\hat{\rho},$ of ρ is the solution of the
equation

$$\frac{n_1 + n_3}{n_1 + n_2 + n_3 + n_4} = \frac{L(h, 0; \hat{\rho})}{1 - \Phi(h)}.$$

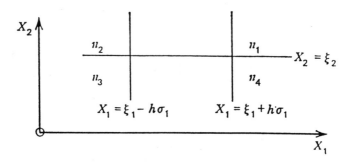

FIGURE 46.8
Scheme of Grouped Data.

For large n,

$$n \operatorname{var}(\hat{\rho}) \doteq \frac{L(h,0;\rho)[1-\Phi(h)-L(h,0;\rho)]}{2(1-\Phi(h))[\partial L(h,0;\rho)/\partial \rho]^2}$$

$(\partial L(h,0;\rho)/\partial \rho = [2\pi\sqrt{1-\rho^2}]^{-1}\exp[-\frac{1}{2}h^2(1-\rho^2)^{-1}])$; see Mosteller (1946). Table 46.3 gives values of h which minimize this function for a few values of ρ, along with corresponding approximate values of $n \operatorname{var}(\hat{\rho})$.

Assume that n observations are made on both X_1 and X_2, that n_1 observations are made only on X_1, and that n_2 observations are made only on X_2. Then the likelihood function is given by

$$\begin{aligned}
L = \ & C \cdot \sigma_1^{-(n+n_1)}\sigma_2^{-(n+n_2)}(1-\rho^2)^{-n/2}\exp\left[-\frac{1}{2(1-\rho^2)}\left\{\sum\left(\frac{X_{1i}-\xi_1}{\sigma_1}\right)^2\right.\right. \\
& \left.-2\rho\sum\left(\frac{X_{1i}-\xi_1}{\sigma_1}\right)\left(\frac{X_{2i}-\xi_2}{\sigma_2}\right)+\sum\left(\frac{X_{2i}-\xi_2}{\sigma_2}\right)^2\right\} \\
& \left.-\frac{1}{2}\sum_1\left(\frac{X_{1i}-\xi_1}{\sigma_1}\right)^2-\frac{1}{2}\sum_2\left(\frac{X_{2i}-\xi_2}{\sigma_2}\right)^2\right].
\end{aligned} \qquad (46.133)$$

Here, $\sum$ denotes summation over the common set of n observations, $\sum_1$ denotes summation over the n_1 observations on X_1 alone, and $\sum_2$ denotes

TABLE 46.3
Optimal Values of h

	Optimal Values		$n\text{var}(\hat{\rho})$
ρ	h	$\Phi(h)$	(approximate)
0.0	0.61	0.73	1.94
0.2	0.61	0.73	1.82
0.4	0.60	0.73	1.47
0.6	0.58	0.72	0.96
0.8	0.48	0.68	0.39

Note. The maximum value is not very critical.
For example, if we take $h = 0.60$ when $\rho = 0.8$,
then $n\,\text{var}(\hat{\rho}) = 0.40$.

summation over the n_2 observations on X_2 alone. It is not possible to solve the likelihood equations obtained from (46.133) to yield explicit expressions for each estimator solely in terms of the observations, but one can write each estimator as a function of the remaining estimators and the observations. The maximum likelihood equations for the five parameters in this case are given by Ratkowsky (1974) and are as follows:

$$\hat{\xi}_1 = \frac{n(n+n_2)\bar{X}_1}{\Delta} + \frac{n_1\{n+n_2(1-\rho^2)\}\bar{X}_1^*}{\Delta} - \frac{nn_2\rho\sigma_1(\bar{X}_2 - \bar{X}_2^*)}{\Delta\hat{\sigma}_2},$$

$$\hat{\xi}_2 = \frac{n(n+n_1)\bar{X}_2}{\Delta} + \frac{n_2\{n+n_1(1-\rho^2)\}\bar{X}_2^*}{\Delta} - \frac{nn_1\rho\sigma_2(\bar{X}_1 - \bar{X}_1^*)}{\Delta\hat{\sigma}_1},$$

$$\hat{\sigma}_1^2 = \frac{\sum(X_{1i} - \hat{\xi}_1)^2 + \sum_1(X_{1i} - \hat{\xi}_1)^2}{n + n_1 + \hat{\rho}^2\left\{n_2 - \sum_2\left(\frac{X_{2i}-\hat{\xi}_2}{\hat{\sigma}_2}\right)^2\right\}},$$

$$\hat{\sigma}_2^2 = \frac{\sum(X_{2i} - \hat{\xi}_2)^2 + \sum_2(X_{2i} - \hat{\xi}_2)^2}{n + n_2 + \hat{\rho}^2\left\{n_1 - \sum_1\left(\frac{X_{1i}-\hat{\xi}_1}{\hat{\sigma}_1}\right)^2\right\}},$$

and

$$\hat{\rho} = \frac{\sum\left(\frac{X_{1i}-\hat{\xi}_1}{\hat{\sigma}_1}\right)\left(\frac{X_{2i}-\hat{\xi}_2}{\hat{\sigma}_2}\right)}{n + \left\{n_1 - \sum_1\left(\frac{X_{1i}-\hat{\xi}_1}{\hat{\sigma}_1}\right)^2\right\} + \left\{n_2 - \sum_2\left(\frac{X_{2i}-\hat{\xi}_2}{\hat{\sigma}_2}\right)^2\right\}}, \qquad (46.134)$$

where

$$\bar{X}_1 = \frac{1}{n}\sum X_{1i}, \ \bar{X}_2 = \frac{1}{n}\sum X_{2i}, \ \bar{X}_1^* = \frac{1}{n_1}\sum{}_1 X_{1i}, \ \bar{X}_2^* = \frac{1}{n_2}\sum{}_2 X_{2i}$$

and $\Delta = n^2 + n(n_1 + n_2) + n_1 n_2(1 - \hat{\rho}^2)$.

When σ_1^2, σ_2^2 and ρ are known, then $\hat{\xi}_1$ and $\hat{\xi}_2$ are seen to be linear combinations of the observations and hence are normally distributed. By taking expectations, it can be easily shown that these estimators are unbiased.

The estimator $\hat{\rho}$ in (46.134) is very similar to the expression for the sample correlation coefficient r obtained from n complete observations, differing only in the two terms in braces in the denominator. In large samples, these terms will tend to approach 0 and, consequently, $\hat{\rho}$ will approach the sample correlation coefficient r. Since r is known to be a biased estimator of ρ, underestimating it in absolute value for all ρ except when $\rho = 0$ or $\rho = \pm 1$, one might expect $\hat{\rho}$ to exhibit a similar bias.

Based on an extensive simulation study wherein the likelihood equations for each data set were solved by using the Newton-Raphson iterative procedure, Ratkowsky (1974) concluded that the MLEs for "fragmentary" data are similar in their properties to the MLEs $\bar{x}_1, \bar{x}_2, s_1^2, s_2^2$ and r from complete data. Estimators $\hat{\sigma}_1^2$ and $\hat{\sigma}_2^2$ are similar to the sample variances s_1^2 and s_2^2 from complete data. There are, however, significant departures of the coefficient of kurtosis from their expected values. Asymptotic formulae for the MLEs of the parameters in the case of fragmented data seem to provide good approximations, even in small samples.

If instead we consider the simpler case when there are n observations made on both X_1 and X_2 and a further $N - n$ observations are made only on X_1, the likelihood function is given by

$$L = (2\pi\sigma^2)^{-(N+n)/2}(1 - \rho^2)^{-n/2} \exp\left\{-\frac{1}{2\sigma^2}\sum_{i=1}^{N}(X_{1i} - \xi_1)^2\right.$$

$$\left. - \frac{1}{2\sigma^2(1 - \rho^2)}\sum_{i=1}^{n}[X_{2i} - \xi_2 - \rho(X_{1i} - \xi_1)]^2\right\}. \quad (46.135)$$

Equation (46.135) yields the following four maximum likelihood estimators:

$$\hat{\xi}_1 = \bar{X}_1^*, \ \hat{\xi}_2 = \bar{X}_2 - \hat{\rho}(\bar{X}_1 - \bar{X}_1^*),$$

$$\hat{\rho} = \frac{s_{12}}{N\hat{\sigma}^2 - (s_1^{*2} - s_1^2)}$$

and

$$\hat{\sigma}^2 = \frac{1}{N+n}\left\{ s_1^{*2} - s_1^2 + \frac{s_1^2 + s_2^2 - 2\hat{\rho}s_{12}}{1 - \hat{\rho}^2} \right\}, \tag{46.136}$$

where

$$\bar{S}_1^* = \frac{1}{N}\sum_{i=1}^{N} X_{1i}, \ \bar{X}_1 = \frac{1}{n}\sum_{i=1}^{n} X_{1i}, \ \bar{X}_2 = \frac{1}{n}\sum_{i=1}^{n} X_{2i},$$

$$s_1^2 = \sum_{i=1}^{n}(X_{1i} - \bar{X}_1)^2, \ s_2^2 = \sum_{i=1}^{n}(X_{2i} - \bar{X}_2)^2,$$

$$s_1^{*2} = \sum_{i=1}^{N}(X_{1i} - \bar{X}_1^*)^2, \ \text{and} \ s_{12} = \sum_{i=1}^{n}(X_{1i} - \bar{X}_1)(X_{2i} - \bar{X}_2).$$

The equations for $\hat{\rho}$ and $\hat{\sigma}^2$ need to be solved numerically and they may have multiple roots. We may combine the equations and consider the equation

$$\begin{aligned} f(\hat{\rho}) &= n(s_1^{*2} - s_1^2)\hat{\rho}^3 - (N - n)s_{12}\hat{\rho}^2 \\ &\quad + \{N(s_1^2 + s_2^2) - n(s_1^{*2} - s_1^2)\}\hat{\rho} - (N+n)s_{12} = 0. \end{aligned} \tag{46.137}$$

Dahiya and Korwar (1980) have shown that (46.137) has exactly one real root in $[-1, 1]$ which has the same sign as s_{12}. This root is the unique MLE $\hat{\rho}$ of ρ. Dahiya and Korwar (1980) have suggested solving the cubic equation by the Newton–Raphson method with $\frac{2s_{12}}{s_1^2 + s_2^2}$ as the starting value, and they have observed that this numerical procedure converges after 3 or 4 iterations for most of the examples considered. These authors have also noted that $\hat{\xi}_1$ and $\hat{\xi}_2$ are asymptotically independent of $\hat{\sigma}^2$ and $\hat{\rho}$; see also Morrison (1971) for a study on the expectations and variances of the MLEs.

(xi) *Type II censored data.* Let $(X_{1i}, X_{2i})^T$, $i = 1, 2, \ldots, n$, be a random sample from the bivariate normal distribution in (46.1). Suppose $r_1 \geq 0$ smallest and $r_2 \geq 0$ largest X_2 observations are not available. Let $X_{2(r_1+1)}, \ldots, X_{2(n-r_2)}$ denote the available order statistics from X_2, and let the corresponding concomitant order statistics from X_1 be $X_{1[r_1+1]}, \ldots, X_{1[n-r_2]}$. Since the random variables

$$\frac{X_1 - \xi_1 - \rho\frac{\sigma_1}{\sigma_2}(X_2 - \xi_2)}{\sigma_1\sqrt{1 - \rho^2}} \quad \text{and} \quad \frac{X_2 - \xi_2}{\sigma_2}$$

are independent standard normal variables, the likelihood function based on the censored sample can be written as

$$L = L_1 L_2,$$

where

$$L_1 = \{2\pi\sigma_1^2(1-\rho^2)\}^{-(n-r_1-r_2)/2} \exp\left\{-\frac{1}{2\sigma_1^2(1-\rho^2)}\sum_{i=r_1+1}^{n-r_2}[X_{1[i]}-\xi_1\right.$$
$$\left.-\rho\frac{\sigma_1}{\sigma_2}(X_{2(i)}-\xi_2)]^2\right\}$$

and

$$L_2 = \frac{n!}{r_1!r_2!}(2\pi\sigma_2^2)^{-(n-r_1-r_2)/2}\exp\left\{-\frac{1}{2\sigma_2^2}\sum_{i=r_1+1}^{n-r_2}(X_{2(i)}-\xi_2)^2\right\}$$
$$\times\left\{\Phi\left(\frac{X_{2(r_1+1)}-\xi_2}{\sigma_2}\right)\right\}^{r_1}\left\{1-\Phi\left(\frac{X_{2(n-r_2)}-\xi_2}{\sigma_2}\right)\right\}^{r_2}.$$

Harrell and Sen (1979) have discussed the maximum likelihood estimation of all the five parameters. In addition to pointing out that the maximum likelihood equations have no explicit solutions, they have also mentioned that some elements of the expected information matrix are difficult to evaluate. Tiku and Gill (1989) have, therefore, derived modified maximum likelihood estimators by employing some suitable linear approximations in the likelihood equations; see Tiku, Tan and Balakrishnan (1986) for a detailed discussion of these estimators. Tiku and Gill (1989) have also shown that their estimators are efficient, that $\hat{\sigma}_1$ is always positive, and that $\hat{\rho}$ is between -1 and $+1$. These authors have also presented expressions for the asymptotic variances and covariances of their estimators.

(xii) *Estimation with preliminary testing.* Let $(X_{1i}, X_{2i})^T$, $i = 1, 2, \ldots,$ n, be a random sample from the bivariate normal distribution in (46.1). Suppose ξ_1 and ξ_2 are unknown, but σ_1^2, σ_2^2, and ρ are all known. Without loss of any generality, let us take $\sigma_1^2 = \sigma_2^2 = 1$. Then, if also ξ_1 is known, the regression estimator of ξ_2 is $\bar{y}+\rho(\xi_1-\bar{x})$; the variance of this estimator is $(1-\rho^2)/n$. If ξ_1 is unknown, the "preliminary test" estimator of ξ_2 (denoted by $\bar{x}_2^*$) depends on the outcome of the test for $H_0 : \xi_1 = 0$. (Note that it can be assumed without loss of generality that the hypothesized value of ξ_1 is 0.) Thus, we get

$$\bar{x}_2^* = \begin{cases} \bar{x}_2 - \rho\bar{x}_1 & \text{if } |\sqrt{n}\,\bar{x}_1| \le z_{\alpha/2}, \\ \bar{x}_2 & \text{if } |\sqrt{n}\,\bar{x}_1| > z_{\alpha/2}, \end{cases} \qquad (46.138)$$

where $z_{\alpha/2}$ is the upper $\alpha/2$ percentage point of the standard normal distribution. (If $H_0 : \xi_1 = 0$ is accepted, the regression estimator is used.) Since

$$E[\bar{X}_2^*] = E[\bar{X}_2] - \rho E\left[\frac{\bar{X}_1}{|\sqrt{n}\,\bar{X}_1|} \leq z_{\alpha/2}\right] \Pr\left[|\sqrt{n}\,\bar{X}_1| \leq z_{\alpha/2}\right],$$

$$(46.139)$$

the bias of the estimator $\bar{X}_2^*$ in (46.138) is obtained from (46.139) to be

$$
\begin{aligned}
B &= \int_{-z_{\alpha/2}/\sqrt{n}}^{z_{\alpha/2}/\sqrt{n}} \frac{-\sqrt{n}\,\bar{x}_1\rho}{\sqrt{2\pi}} e^{-n(\bar{x}_1-\xi_1)^2/2} \, d\bar{x}_1 \\
&= \frac{\sqrt{2}\,\rho}{\sqrt{\pi}\,n} e^{-(z_{\alpha/2}^2+a^2)/2} \sinh(az_{\alpha/2}) \\
&\quad -\frac{\rho a}{\sqrt{n}} \left\{\Phi\left(z_{\alpha/2}-a\right) - \Phi\left(-z_{\alpha/2}-a\right)\right\}, \quad (46.140)
\end{aligned}
$$

where $a = \sqrt{n}\,\xi_1$ and $\Phi(\cdot)$ is the standard normal distribution function. When $\alpha = 0$, $B = -\rho\xi_1$ (which is the bias of the regression estimator); when $\alpha = 1$, the estimator is $\bar{x}_2$ in which case $B = 0$. The bias changes sign when ρ or a change sign. The function $\sqrt{n}\,B$ is a function of a, ρ, and α. Han (1973) has presented a table of values of $-\sqrt{n}\,B$ for different choices of a, ρ, and α. For fixed n, α, and ρ, the magnitude of bias first increases and then decreases to 0 as ξ_1 increases. In general, the magnitude of bias is an increasing function of ρ and a decreasing function of α.

Han (1973) has derived the mean square error of $\bar{x}_2^*$ to be

$$\text{MSE}(\bar{x}_2^*) = \frac{1}{n}\{1 + f(a)\}, \quad (46.141)$$

where

$$
\begin{aligned}
f(a) &= \frac{\sqrt{2}\,\rho^2}{\sqrt{\pi}} e^{-(z_{\alpha/2}^2+a^2)}\{z_{\alpha/2}\cosh(az_{\alpha/2}) - a\sinh(az_{\alpha/2})\} \\
&\quad -\rho^2(1-a^2)\{\Phi(z_{\alpha/2}-a) - \Phi(-z_{\alpha/2}-a)\}.
\end{aligned}
$$

When $\alpha = 0$, $\text{MSE}(\bar{x}_2^*) = \frac{1}{n} - \rho^2(\frac{1}{n}-\xi_1^2)$, which is the MSE of the regression estimator; when $\alpha = 1$, $\text{MSE}(\bar{x}_2^*) = \frac{1}{n}$, which is the variance of $\bar{x}_2$. The relative efficiency of $\bar{x}_2^*$ to the usual estimator $\bar{x}_2$ is given by $e = \frac{1}{1+f(a)}$, which is a function of a and α. $f(a)$ is a symmetric function of ρ. Han (1973) noted that when ρ is small (say ≤ 0.2), the relative efficiency of $\bar{x}_2^*$ is close to 1. The estimator $\bar{x}_2^*$ does not change significantly for small

ρ and it is immaterial whether one chooses $\alpha = 0.05$ or $\alpha = 0.50$. The relative efficiency, however, fluctuates for large ρ.

When σ_1^2, σ_2^2 and ρ are unknown, the regression estimator is

$$\bar{x}_2' = \begin{cases} \bar{x}_2 - \frac{s_{12}}{s_1^2}\, \bar{x}_1 & \text{if } |t| \le t_{\alpha/2}, \\ \bar{x}_2 & \text{if } |t| > t_{\alpha/2}, \end{cases} \tag{46.142}$$

where

$$t = \frac{\bar{x}_1}{\sqrt{\frac{s_1^2}{n(n-1)}}}, \quad s_1^2 = \sum_{i=1}^{n}(x_{1i} - \bar{x}_1)^2, \quad s_2^2 = \sum_{i=1}^{n}(x_{2i} - \bar{x}_2)^2,$$

$$s_{12} = \sum_{i=1}^{n}(x_{1i} - \bar{x}_1)(x_{2i} - \bar{x}_2),$$

and $t_{\alpha/2}$ is the upper $\alpha/2$ percentage point of Student's t distribution with $n - 1$ degrees of freedom. The bias of this estimator is

$$\text{Bias} = \sigma_2 \sum_{i=0}^{((n-1)/2)-1} H_i\, \rho\, \mu_{2i+1}', \tag{46.143}$$

where μ_r' is the rth raw moment of $N\left(\frac{\xi_1}{c\sigma_1}, \frac{n}{c}\right)$ and

$$H_i = \frac{1}{i!} \left\{ \frac{n(n-1)}{2t_{\alpha/2}^2} \right\}^i \frac{1}{c} \exp\left\{ -\frac{n(n-1)}{2(t_{\alpha/2}^2 + n - 1)} \left(\frac{\xi_1}{\sigma_1} \right)^2 \right\}$$

with $c = 1 + \frac{n-1}{t_{\alpha/2}^2}$. (Here, it is assumed that n is odd and at least 3 so that $\frac{n-1}{2}$ is an integer.) The magnitude of the bias first increases and then decreases to 0 as $\frac{\xi_1}{\sigma_1}$ increases; it is an increasing function of ρ and a decreasing function of α, and hence the behavior of the bias is similar to the case when the covariance matrix is known.

Han (1973) has provided quite a complicated expression for $\text{MSE}(\bar{x}_2')$. For $\alpha = 0$, when the regression estimator is used, we have

$$\frac{\text{MSE}(\bar{x}_2')}{\sigma_2^2} = \frac{1 - \rho^2}{n} + \rho^2 \left(\frac{\xi_1}{\sigma_1} \right)^2 + \frac{1 - \rho^2}{n - 3} \left\{ \frac{1}{n} + \left(\frac{\xi_1}{\sigma_1} \right)^2 \right\}$$

which may be compared with $\text{MSE}(\bar{x}_2) = \sigma_2^2/n$. Han (1973) also investigated the relative efficiency of $\bar{x}_2'$ to the usual estimator $\bar{x}_2$. This relative efficiency has a minimum larger than 1 at $\xi_1 = 0$, provided that $\rho < 1/\sqrt{n-2}$. Han (1973) has made the recommendation that if $\rho \le 0.3$

and $n \leq 10$, then one should use $\bar{x}_2$; and for $\rho \geq 0.5$, the preliminary test estimate should be used because it results in higher precision.

Proceeding in a similar way, Lakshminarayan and Han (1997) have discussed the problem of estimating the mean ξ_1 of one of the components with equal variances or unequal variances. When the mean of the other component ξ_2 is equal to ξ_1, it is better to pool the two sample means as an estimator of ξ_1. When one is uncertain whether $\xi_1 = \xi_2$, they suggest a preliminary test of significance be used at level α to test $\xi_1 = \xi_2$. Lakshminarayan and Han (1997) have then examined the performance of preliminary test estimator, adaptive preliminary test estimator, and weighting function estimator (which is a linear combination of the two sample means with the weight obtained by minimizing the mean square error).

(xiii) *Estimation based on ranks.* Let $(\zeta, \eta)^T$ be a bivariate absolutely continuous random variable with joint density function $p(x, y)$ and marginal density functions $p_1(x)$ and $p_2(y)$. Let us assume that $(\zeta, \eta)^T$ has finite mean-square contingency, namely;

$$\int_{\mathbb{R}^2} \frac{p^2(x, y)}{p_1(x) p_2(y)} \, dx dy - 1 < \infty. \tag{46.144}$$

The joint density function of the random variables

$$U = F_1(\zeta) \quad \text{and} \quad V = F_2(\eta), \tag{46.145}$$

where $F_1(\cdot)$ and $F_2(\cdot)$ are the marginal cumulative distribution functions of ζ and η, respectively, has the form

$$h(u, v) = p(F_1^{-1}(\zeta), F_2^{-1}(\eta)) \frac{dF_1^{-1}(u)}{du} \frac{dF_2^{-1}(v)}{dv}, \quad 0 \leq u, v \leq 1; \tag{46.146}$$

Hoeffding (1940/41) has termed this the *normed bivariate density function*.

Let $\{P_i(u), \ i = 0, 1, 2, \ldots\}$ be a system of normed Legendre polynomials on $[0,1]$ and the Fourier coefficients ρ_{ij} be defined as

$$\rho_{ij} = \int_{[0,1] \times [0,1]} P_i(u) P_j(v) h(u, v) \, du dv = E[P_i(U) P_j(V)]. \tag{46.147}$$

The joint density $h(u, v)$ is then uniquely representable as

$$h(u, v) = \sum_{i=0}^{\infty} \sum_{j=0}^{\infty} \rho_{ij} P_i(u) P_j(v) = 1 + \sum_{i=1}^{\infty} \sum_{j=1}^{\infty} \rho_{ij} P_i(u) P_j(v) \tag{46.148}$$

since $\rho_{00} = 1$ and $\rho_{i0} = \rho_{0i} = 0$ for $i \geq 1$.

The usual estimator of $h(u, v)$ [based on (46.148)] suggested by Mirzahmedov and Hasimov (1972) is

$$\tilde{h}_n(u, v) = 1 + \sum_{i=1}^{m} \sum_{j=1}^{m} \tilde{\rho}_{ij} P_i(u) P_j(v), \qquad (46.149)$$

where

$$\tilde{\rho}_{ij} = \frac{1}{n} \sum_{\ell=1}^{n} P_i(u_\ell) P_j(v_\ell) \qquad \text{for } i, j = 1, 2, \ldots, m,$$

and $(u_\ell, v_\ell)^T$ $(\ell = 1, 2, \ldots, n)$ denotes the sample.

Rödel (1987) proposed an estimator based on ranks as

$$\hat{h}_n(u, v) = 1 + \sum_{i=1}^{m} \sum_{j=1}^{m} \hat{\rho}_{ij} P_i(u) P_j(v), \qquad (46.150)$$

where

$$\hat{\rho}_{ij} = \frac{1}{n} \sum_{\ell=1}^{n} P_i\left(\frac{R_\ell}{n+1}\right) P_j\left(\frac{S_\ell}{n+1}\right) \qquad \text{for } i, j = 1, 2, \ldots, m$$

with $(R_\ell, S_\ell)^T$ denoting the bivariate ranks. Note that this estimator, unlike the usual estimator in (46.149), does not require the calculation of the sample $(u_i, v_i)^T = (F_1(\zeta_i), F_2(\eta_i))^T$ for $i = 1, 2, \ldots, n$. Recall that the marginal distribution functions are both unknown. Rödel (1987) has discussed the order of the error involved in this approximation, paying particular attention to the case of positive dependence; see, for example, Lehmann (1966) and Chapter 33 of Johnson, Kotz, and Balakrishnan (1995).

8.2   Trivariate Normal Distributions

We shall not discuss a wide range of special situations, as we did in the case of bivariate normal distributions. There are, in general, nine parameters to be estimated, three means, three standard deviations, and three correlation coefficients. When no parameters are known, estimators are the same as in case (i) of Section 46.8.1. The maximum likelihood estimator of the multiple correlation of X_1 with X_2 and X_3,

$$R^2 = (\rho_{12}^2 + \rho_{13}^2 - 2\rho_{12}\rho_{13}\rho_{23})/(1 - \rho_{23}^2)$$

is, of course,

$$\hat{R}^2 = (\hat{\rho}_{12}^2 + \hat{\rho}_{13}^2 - 2\hat{\rho}_{12}\hat{\rho}_{13}\hat{\rho}_{23})/(1 - \hat{\rho}_{23}^2). \qquad (46.151)$$

Olkin and Pratt (1958) suggested the adjusted estimator

$$\hat{R}^2 - 2(n-1)^{-1}(1-\hat{R}^2)^{-2}, \tag{46.152}$$

which has bias of order n^{-2} (while $\hat{R}^2$ has bias of order n^{-1}).

In cases when values of some of the expected values and variances are known, formulas for maximum likelihood estimators of correlation coefficients are the same as for the bivariate distribution of the two variates concerned.

If $E[X_1] = \xi_1$ and $E[X_2] = \xi_2$ (but no other parameters are known), then the maximum likelihood estimator of ξ_3 is

$$\hat{\xi}_3 = \bar{X}_3 - (\hat{\rho}_{13.2}\hat{\sigma}_3/\hat{\sigma}_1')(\bar{X}_1 - \hat{\xi}_1) - (\hat{\rho}_{23.1}\hat{\sigma}_3/\hat{\sigma}_2')(\bar{X}_2 - \hat{\xi}_2), \quad (46.153)$$

where $\hat{\rho}_{13.2} = (\hat{\rho}_{13}' - \hat{\rho}_{12}'\hat{\rho}_{23}')(1 - \hat{\rho}_{12}'^2)^{-1/2}(1 - \hat{\rho}_{23}'^2)^{-1/2}$, with a similar formula for $\rho_{23.1}$, and primes denote that X_1, X_2 are replaced by ξ_1, ξ_2, respectively, in $\hat{\sigma}_1, \hat{\sigma}_2, \hat{\rho}_{12}, \hat{\rho}_{13}, \hat{\rho}_{23}$. Cases when additional parameters are known give modifications of (46.153) analogous to those for the bivariate case.

9   TRUNCATED BIVARIATE AND TRIVARIATE NORMAL DISTRIBUTIONS

In the following discussion, we will use the standardized distributions in (46.2) or (46.3). Extension to the general case (by using new variables $\xi_t + \sigma_t X_t$) is straightforward.

The most common form of truncation of a bivariate normal distribution is single truncation (from above or below) with respect to one of the variables. If we select so that only values of X_1 that exceed h are used, the resulting joint distribution has density function

$$p_{X_1,X_2}(x_1, x_2)$$
$$= \frac{1}{2\pi\sqrt{1-\rho^2}\{1-\Phi(h)\}} \exp\left\{-\frac{1}{2(1-\rho^2)}(x_1^2 - 2\rho x_1 x_2 + x_2^2)\right\},$$
$$x_1 > h. \tag{46.154}$$

Using the fact that the conditional distribution of X_2, given X_1, is normal with expected value ρX_1 and standard deviation $\sqrt{1-\rho^2}$, we have

$$E[X_2] = E[E[X_2 \mid X_1]] = \rho E[X_1], \tag{46.155}$$

$$E[X_1 X_2] = E[X_1 E[X_2 \mid X_1]] = \rho E[X_1^2], \qquad (46.156)$$

$$E[X_2^2] = E[E[X_2^2 \mid X_1]] = E[\rho^2 X_1^2 + 1 - \rho^2]$$
$$= \rho^2 E[X_1^2] + 1 - \rho^2. \qquad (46.157)$$

Hence,

$$\mathrm{var}(X_2) = \rho^2 \mathrm{var}(X_1) + 1 - \rho^2, \qquad (46.158)$$

$$\mathrm{cov}(X_1, X_2) = \rho \mathrm{var}(X_1), \qquad (46.159)$$

and

$$\mathrm{corr}(X_1, X_2) = \rho \sqrt{\mathrm{var}(X_1)/\mathrm{var}(X_2)} = \rho \left[\rho^2 + \frac{1 - \rho^2}{\mathrm{var}(X_1)} \right]^{-1/2}. \qquad (46.160)$$

Note that (46.160) applies to *any* form of truncation of X_1, provided X_2 is not truncated. From Chapter 13, we have

$$\mathrm{var}(X_1) = 1 + \frac{hZ(h)}{1 - \Phi(h)} - \left\{ \frac{Z(h)}{1 - \Phi(h)} \right\}^2. \qquad (46.161)$$

Some values of $\mathrm{var}(X_1)$ are presented in Chapter 13. Since $\mathrm{var}(X_1) \leq 1$, it follows that

$$|\mathrm{corr}(X_1, X_2)| \leq |\rho|.$$

Thus, we would expect the correlation in the truncated population to be numerically less than that in the original population. Table 46.4 gives a few numerical values; see also Aitkin (1964).

Thus, if individuals are chosen on the basis of their X_1 values, with a view to controlling their X_2 values, the observed results tend to be "disappointing" in the sense that the observed correlation between X_1 and X_2 is less than that in the original population. This does not mean (though it is sometimes taken to do so) that the accuracy of prediction of X_2, given X_1, is any less; it is, in fact, the same. It is also worth noting that, while the regression of X_2 on X_1 is linear, that of X_1 on X_2 is, in the truncated population,

$$E[X_1 \mid X_2] = \rho X_2 + \frac{\phi[(h - \rho X_2)/\sqrt{1 - \rho^2}]}{1 - \Phi[(h - \rho X_2)/\sqrt{1 - \rho^2}]} \sqrt{1 - \rho^2}. \qquad (46.162)$$

TABLE 46.4

Correlation in Truncated Bivariate Normal Population

Degree of Truncation $\Phi(h)$	Original Correlation ρ			
	0.25	0.5	0.75	0.9
0.1	0.213	0.438	0.691	0.867
0.2	0.193	0.403	0.655	0.845
0.3	0.178	0.376	0.623	0.823
0.4	0.165	0.351	0.593	0.802
0.5	0.154	0.329	0.564	0.780
0.6	0.143	0.307	0.535	0.755
0.7	0.132	0.285	0.504	0.728
0.8	0.120	0.261	0.468	0.695
0.9	0.106	0.231	0.423	0.647

If both X_1 and X_2 are truncated from below ($X_1 > h, X_2 > k$), then the moments (μ'_{rs}) are given by

$$L(h, k; \rho)\mu'_{10} = \phi(h)\{1 - \Phi(A)\} + \rho\phi(k)\{1 - \Phi(B)\}, \quad (46.163)$$

$$L(h, k; \rho)\mu'_{20} = h\phi(h)\{1 - \Phi(A)\} + \rho^2 k\phi(k)\{1 - \Phi(B)\} + \rho(1 - \rho^2)\frac{1}{2} \phi(h, k; \rho) + L(h, k; \rho), \quad (46.164)$$

$$L(h, k; \rho)\mu'_{11} = \rho[h\phi(h)\{1 - \Phi(A)\} + k\phi(k)\{1 - \Phi(B)\} + L(h, k; \rho)] + (1 - \rho^2)\phi(h, k; \rho), \quad (46.165)$$

where

$$A = \frac{k - \rho h^2}{\sqrt{1 - \rho^2}}, \qquad B = \frac{h - \rho k^2}{\sqrt{1 - \rho^2}},$$

$$\phi(h) = \frac{1}{2\sqrt{\pi}} \exp\{-Z^2/2\},$$

$$\phi(h, k; \rho) = \frac{1}{2\pi\sqrt{1 - \rho^2}} \exp\left\{-\frac{1}{2(1 - \rho^2)}(h^2 - 2\rho hk + k^2)\right\}.$$

Formulas for μ'_{0r} are obtained from those for μ'_{r0} by interchanging h and k and interchanging A and B.

From Eqs. (46.163)–(46.165),

$$(h+k)\rho^2 - \{(h+k)\mu'_{11} - hk(\mu'_{10} + \mu'_{01})\}\rho$$
$$- (h+k) - hk(\mu'_{10} + \mu'_{01}) + k\mu'_{20} + h\mu'_{02} = 0; \quad (46.166)$$

see Rosenbaum (1961).

Cases where both variables are truncated (either singly or doubly) have been considered by Des Raj (1953), Shah and Parikh (1964), and Regier and Hamdan (1971). Supposing the retained ranges of values are

$$h_1 < X_1 < k_1 \qquad \text{and} \qquad h_2 < X_2 < k_2,$$

Shah and Parikh (1964) obtained several recurrence relations among the product moments $\mu'_{r,s} = E[X_1^r X_2^s]$. Among these, we quote

$$\mu'_{r,s} - (r-1)(1-\rho^2)\mu'_{r-2,s} - \rho\mu'_{r-1,s+1}$$
$$= P^{-1}(1-\rho^2)[h_1^{r-1}\phi(h_1)G_s(h_2, k_2, h_1\rho; \sqrt{1-\rho^2})$$
$$- k_1^{r-1}Z(k_1)G_s(h_2, k_2, k_1\rho; \sqrt{1-\rho^2})], \qquad r \geq 2, s \geq 2,$$
$$(46.167)$$

where

$$P = \int_{h_1}^{h_2} \int_{k_1}^{k_2} Z(0,0;1,1;\rho) \, dx_1 dx_2$$

and

$$G_s(a_1, a_2, b; c) = \frac{1}{c} \int_{a_1}^{a_2} x^s \phi\left(\frac{x-b}{c}\right) dx.$$

Also,

$$\mu'_{r-1,s+1} - s(1-\rho^2)\mu'_{r-1,s-1} - \rho\mu'_{r,s}$$
$$= P^{-1}(1-\rho^2)[h_2^s\phi(h_2)G_{r-1}(h_1, k_1, h_2\rho; \sqrt{1-\rho^2})$$
$$- k_2^s\phi(k_2)G_{r-1}(h_1, k_1, k_2\rho; \sqrt{1-\rho^2})]. \qquad (46.168)$$

For the case when both variates are singly truncated from below at the *same* (standardized) point (so that the retained values are $X_1 > a$, $X_2 > a$), Regier and Hamdan (1971) give values of the correlation coefficient (ρ') as a function of a and the pretruncation correlation coefficient (ρ) to three decimal places for $\pm a = 0.0(0.1)1.1(0.2)1.5(0.5)2.5$, $\rho = 0.05(0.05)0.95$.

Use of tables of $L(h, k; \rho)$ to evaluate $\Pr[X_1 - X_2 > 0]$ when X_1, X_2 are independent and one is truncated has been described by Lipow and Eidemiller (1964) for single truncation and (in a very similar manner) by Parikh and Sheth (1966) for double truncation.

Let us assume that $(X_1, X_2)^T$ has a standard bivariate normal distribution with correlation coefficient ρ and density function $\phi(x_1, x_2; \rho)$. If X_1 is truncated below c, then the joint density function of such a singly truncated standard bivariate normal distribution is given by

$$p(x_1, x_2; \rho) = \frac{\phi(x_1, x_2; \rho)}{\Phi(-c)} \qquad \text{for } c \le x_1 < \infty,\ -\infty < x_2 < \infty,$$

(46.169)

where $\Phi(\cdot)$ denotes the univariate standard normal cumulative distribution function. The joint moment-generating function of this distribution is

$$M(t_1, t_2) = e^{\frac{1}{2}(t_1^2 + 2\rho t_1 t_2 + t_2^2)} \frac{\Phi(t_1 + \rho t_2 - c)}{\Phi(-c)}, \qquad (46.170)$$

and hence the joint cumulant-generating function is

$$
\begin{aligned}
K(t_1, t_2) &= \ln M(t_1, t_2) \\
&= -\ln \Phi(-c) + \ln \Phi(t_1 + \rho t_2 - c) + \frac{1}{2} t_1^2 + t_1 t_2 + \frac{1}{2} t_2^2.
\end{aligned}
$$

(46.171)

From (46.171), we obtain

$$\kappa_{10} = \frac{\phi(c)}{\Phi(-c)}, \qquad \kappa_{01} = \rho \kappa_{10},$$

$$\kappa_{20} = \frac{c\phi(c)}{\Phi(-c)} - \left\{ \frac{\phi(c)}{\Phi(-c)} \right\}^2 + 1 = c\kappa_{10} - \kappa_{10}^2 + 1,$$

$$\kappa_{02} = \rho^2(\kappa_{20} - 1) + 1, \qquad \kappa_{11} = \rho \kappa_{20},$$

and

$$\kappa_{ij} = \rho^j \left. \frac{\partial^{i+j-1}}{\partial x^{i+j-1}} \left\{ \frac{\phi(x)}{\Phi(x)} \right\} \right|_{x=-c} \qquad \text{for } i + j \ge 3,$$

where $\phi(\cdot)$ denotes the univariate standard normal density function. Note that κ_{i0} is free of ρ. Chou and Owen (1984) derived these explicit formulas and also presented a table of κ_{i0} $(i = 1, \ldots, 8)$ for $c = -3.0(0.2)3.0$. These values are useful for obtaining [via the Cornish–Fisher expansion;

see Chapter 12 of Johnson, Kotz, and Balakrishnan (1994)] approximations to percentage points of a variable of the bivariate normal distribution when the other variable is truncated below.

Arnold et al. (1993) have considered the following truncated bivariate normal model. Let X_1 and X_2 jointly have a truncated bivariate normal distribution wherein both lower and upper truncation are permitted on X_2. It is assumed that X_1 values are available only for the nontruncated X_2 values, while the values of X_2 (truncated or not) are not available. Estimation of the population parameters for the marginal distribution of X_1 has been considered by Arnold et al. (1993) in a specific truncated bivariate normal case, in which X_1 values are available only for those X_2 values exceeding the expectation of X_2. This has been shown to be equivalent to estimation of the parameters in the skew-normal distribution of Azzalini (1985); see also Chapter 13 of Johnson, Kotz, and Balakrishnan (1994). Specifically, let us consider the joint density function of $(X_1, X_2)^T$ to be

$$
p_{X_1,X_2}(x_1, x_2) = \begin{cases} \dfrac{p(x_1,x_2)}{\Phi\left(\frac{b-\xi_2}{\sigma_2}\right)-\Phi\left(\frac{a-\xi_2}{\sigma_2}\right)}, & -\infty < x_1 < \infty, \ a < x_2 < b \\ \qquad 0, & \text{otherwise}, \end{cases}
$$

$$(46.172)$$

where $p(x_1, x_2)$ is the bivariate normal density function in (46.1), $\Phi(\cdot)$ denotes the cumulative distribution function of the univariate standard normal distribution, and a and b are real constants that are the lower and upper truncation points for X_2, respectively. Direct integration then yields

$$
p_{X_1}(x_1) = \frac{\frac{1}{\sigma_1}\phi\left(\frac{x_1-\xi_1}{\sigma_1}\right)\left\{\Phi\left(\frac{\frac{b-\xi_2}{\sigma_2}-\frac{\rho(x_1-\xi_1)}{\sigma_1}}{\sqrt{1-\rho^2}}\right) - \Phi\left(\frac{\frac{a-\xi_2}{\sigma_2}-\frac{\rho(x_1-\xi_1)}{\sigma_1}}{\sqrt{1-\rho^2}}\right)\right\}}{\Phi\left(\frac{b-\xi_2}{\sigma_2}\right) - \Phi\left(\frac{a-\xi_2}{\sigma_2}\right)},
$$
$$-\infty < x_1 < \infty, \qquad (46.173)$$

where $\phi(\cdot)$ is the univariate standard normal density function. Thus, denoting $\beta = \frac{b-\xi_2}{\sigma_2}$ and $\alpha = \frac{a-\xi_2}{\sigma_2}$, we can rewrite (46.173) as

$$
p_{X_1}(x_1) = \frac{1}{\sigma_1}\, g\left(\frac{x_1 - \xi_1}{\sigma_1}\right), \qquad -\infty < x_1 < \infty, \qquad (46.174)
$$

where

$$
g(y) = \frac{\phi(y)\left\{\Phi\left(\frac{\beta-\rho y}{\sqrt{1-\rho^2}}\right) - \Phi\left(\frac{\alpha-\rho y}{\sqrt{1-\rho^2}}\right)\right\}}{\Phi(\beta) - \Phi(\alpha)}. \qquad (46.175)
$$

This expression coincides with the one given by Chou and Owen (1984) for the case when $\beta = \infty$. Note that $Y = (X_1 - \xi_1)/\sigma_1$ has the density function $g(y)$ in (46.175). For the case when $\alpha = 0$ and $\beta = \infty$, the density in (46.174) becomes

$$p_{X_1}(x_1) = \frac{1}{\sigma_1} \, g_1 \left(\frac{x_1 - \xi_1}{\sigma_1} \right), \tag{46.176}$$

where

$$g_1(y) = 2\phi(y)\Phi\left(\frac{\rho y}{\sqrt{1-\rho^2}}\right) = 2\phi(y)\Phi(\lambda y), \tag{46.177}$$

which is Azzalini's (1985) *skew-normal* distribution.

Arnold *et al.* (1993) have also discussed the maximum likelihood estimation of ξ_1, σ_1^2 and ρ, and as in Azzalini (1985) they use profile maximum likelihood to estimate λ. These authors have also dealt with the case when $\alpha \neq 0$ and $\beta = \infty$, in which case the likelihood function for X_1 becomes

$$
\begin{aligned}
&L(x_1, \xi_1, \sigma_1, \lambda, \alpha) \\
&= \prod_{i=1}^{n} \left[\frac{\Phi\left(\frac{x_{1i}-\xi_1}{\sigma_1}\right) \Phi\left\{\lambda\left(\frac{x_{1i}-\xi_1}{\sigma_1}\right) - \alpha\sqrt{1+\lambda^2}\right\}}{\sigma_1\{1 - \Phi(\alpha)\}} \right].
\end{aligned}
\tag{46.178}
$$

Arnold *et al.* (1993) have observed, in this specific case, that the log-likelihood does not change very much for all values of α considered. Profiles with respect to α, therefore, would be very flat. In short, if α is known, it should be used; if, however, α is unknown, this fourth parameter will cause severe identifiability problems in determining the complete model of X_1.

If truncation of a single variable (selection of $X_1 \geq h$) is applied to a trivariate normal population, then since the conditional joint distribution of X_2 and X_3, given X_1, is bivariate normal with parameters $\rho_{12}X_1, \rho_{13}X_1, \sqrt{1-\rho_{12}^2}, \sqrt{1-\rho_{13}^2}, \rho_{23.1}$, we have the following results. As in the bivariate case, for $t = 2, 3$,

$$
\begin{aligned}
E[X_t] &= \rho_{1t}E[X_1], \qquad E[X_t^2] = \rho_{1t}^2 E[X_1^2] + 1 - \rho_{1t}^2, \\
\mathrm{var}(X_t) &= \rho_{1t}^2 \mathrm{var}(X_1) + 1 - \rho_{1t}^2,
\end{aligned}
$$

and

$$\mathrm{corr}(X_1, X_t) = \rho_{1t} \left[\rho_{1t}^2 + \frac{1 - \rho_{1t}^2}{\mathrm{var}(X_1)} \right]^{-1/2}.$$

Also,

$$
\begin{aligned}
E[X_2 X_3] &= E[E[X_2 X_3 \mid X_1]] \\
&= E[\rho_{23.1}\sqrt{(1 - \rho_{12}^2)(1 - \rho_{13}^2)} + \rho_{12}\rho_{13}X_1^2] \\
&= \rho_{12}\rho_{13}E[X_1^2] + \rho_{23} - \rho_{12}\rho_{13}.
\end{aligned}
\tag{46.179}
$$

Hence,

$$
\text{cov}(X_2, X_3) = \rho_{12}\rho_{13}\text{var}(X_1) + \rho_{23} - \rho_{12}\rho_{13}
\tag{46.180}
$$

and

$$
\text{corr}(X_2, X_3) = \frac{\rho_{23} - \rho_{12}\rho_{13}\left(\frac{1}{\text{var}(X_1)} - 1\right)}{\sqrt{\left(\rho_{12}^2 + \frac{1 - \rho_{12}^2}{\text{var}(X_1)}\right)\left(\rho_{13}^2 + \frac{1 - \rho_{13}^2}{\text{var}(X_1)}\right)}}.
\tag{46.181}
$$

If truncation is by selection only of values X_1, X_2 such that $\alpha_1 X_1 + \alpha_2 X_2 > h$, the problem reduces to that of bivariate normal with single truncation of one variable, discussed at the beginning of this section. This is because $(\alpha_1 X_1 + \alpha_2 X_2)$ and X_3 have (before truncation) a bivariate normal distribution. If there is a further truncation, $\alpha_1' X_1 + \alpha_2' X_2 > h'$ (say), then by transformation to new variables we obtain truncations of the form $Z > h$, $Z' > h'$; see also Birnbaum (1950a,b) and Young and Weiler (1960).

We shall now no longer suppose that we are discussing standardized distributions and shall discuss, briefly, the estimation of parameters for truncated bivariate and trivariate normal distributions. For the case when each of two variates X_1, X_2 is truncated from below ($X_1 \geq h_1, X_2 \geq h_2$), Rosenbaum (1961) has given a method of estimating $h_1, h_2, \xi_1, \xi_2, \sigma_1, \sigma_2$, and ρ by using equations (46.163)–(46.166), which relate to the standardized variables $(X_1 - \xi_1)/\sigma_1$, $(X_2 - \xi_2)/\sigma_2$ with $h = (h_1 - \xi_1)/\sigma_1$, $k = (h_2 - \xi_2)/\sigma_2$. Approximate values of $\hat{h}$, $\hat{k}$, and $\hat{\rho}$ are used to evaluate $\hat{\mu}_{rs}'$. These, in turn, are used to calculate

$$
\hat{\sigma}_1 = [S_{11}(\hat{\mu}_{20}' - \hat{\mu}_{10}'^2)^{-1}]^{1/2},
\tag{46.182}
$$

$$
\hat{\sigma}_2 = [S_{22}(\hat{\mu}_{02}' - \hat{\mu}_{01}'^2)^{-1}]^{1/2},
\tag{46.183}
$$

$$
\hat{\xi}_1 = \bar{X}_1 - \hat{\mu}_{10}'\hat{\sigma}_1,
\tag{46.184}
$$

$$
\hat{\xi}_2 = \bar{X}_2 - \hat{\mu}_{01}'\hat{\sigma}_2.
\tag{46.185}
$$

From these values, new values of $\hat{h}$ and $\hat{k}$ can be obtained. Finally, a new value for $\hat{\rho}$ is obtained by solving (46.166) (with all quantities replaced by estimates) and a new cycle of calculation started.

Maximum likelihood equations for the cases when both variables are singly or doubly truncated (and for linear truncation) are given by Nath (1971), who also gives formulas from which the asymptotic variances and covariances of the maximum likelihood estimators can be calculated.

Jaiswal and Khatri (1967) have given moment estimators applicable when only one of the two variables is truncated. As high moments are needed, the estimators are likely to be variable.

Votaw, Rafferty, and Deemer (1950) have described a method of calculating maximum likelihood estimators for some parameters of a trivariate normal distribution with one variable truncated from below $(X_1 \leq h_1)$ when the parameters $\xi_1, \xi_2, \sigma_1, \sigma_2, \sigma_{12}$, and h_1 are known. The parameters to be estimated are $\xi_3, \sigma_3, \rho_{13}$, and ρ_{23}—that is, those relating to the third variable, X_3.

Elliptical truncation [Tallis (1963)] has been discussed in Chapter 45. For the case of standardized bivariate normal distribution, Tallis (1963) included a table that enabled one to choose a region

$$a_1 < \frac{1}{1 - \rho^2} (X_1^2 - 2\rho X_1 X_2 + X_2^2) < a_2$$

such that the variance–covariance matrix of the truncated distribution equals that of the original distribution. This table, reproduced in Table 46.5, gives appropriate pairs of values a_1 and a_2 for each of a number of different degrees of truncation (q). Values of a_1 and a_2 are obtained as solutions of the equations

$$\Pr[a_1 < \chi_4^2 < a_2] = \Pr[a_1 < \chi_2^2 < a_2] = 1 - q, \qquad (46.186)$$

that is,

$$a_1 e^{-(1/2)a_1} = a_2 e^{-(1/2)a_2} \qquad \text{and} \qquad e^{-(1/2)a_1} - e^{-(1/2)a_2} = 1 - q.$$

TABLE 46.5
Constants for Elliptical Truncation of Bivariate
Normal

q	a_1	a_2	q	a_1	a_2
0.1	0.171	8.632	0.6	1.068	3.361
0.2	0.335	6.161	0.7	1.277	2.956
0.3	0.506	5.144	0.8	1.500	2.601
0.4	0.684	4.411	0.9	1.740	2.285
0.5	0.871	3.836			

10   DICHOTOMIZED VARIABLES

When data are grouped, calculations can be made as if all observations in a group were at one specific point in the group. Subsequently, corrections may be applied. In many practical cases, these corrections are of relatively small importance.

When the grouping is very coarse, however, the situation is different. In this section, we consider the coarsest possible grouping in which one, or both, of the variables is dichotomized—that is, divided into two groups $X \leq x_0$, $X > x_0$.

10.1   Tetrachoric Correlation

When both X_1 and X_2 are dichotomized, the available data can be represented in the form of a 2×2 table; see Figure 46.9. The symbols $a, b, c,$ and d stand for the frequencies of observations, in a sample of size n, of the events

$$
\begin{aligned}
a &: \quad (X_1 \leq x_{10}) \cap (X_2 \leq x_{20}), \\
b &: \quad (X_1 > x_{10}) \cap (X_2 \leq x_{20}), \\
c &: \quad (X_1 \leq x_{10}) \cap (X_2 > x_{20}), \\
d &: \quad (X_1 > x_{10}) \cap (X_2 > x_{20}).
\end{aligned}
$$

Evidently, $a + b + c + d = n$. Although there are only three distinct observations and at least five unknown parameters (seven if, as is commonly the case, x_{10} and x_{20} are unknown), it is possible to construct useful estimators of ρ. The method now to be described was originally constructed by Pearson (1901).

	$X_1 \leq x_{10}$	$X_2 > x_{10}$
$X_2 > x_{20}$	c	d
$X_2 \leq x_{20}$	a	b

FIGURE 46.9

The observed proportion of X_1's less than x_{10} is $(a + c)/n$, and we estimate $h_1 = (x_{10} - \xi_1)/\sigma_1$ by $\tilde{h}_1$, which satisfies the equation

$$\Phi(\tilde{h}_1) = (a + c)/n. \tag{46.187}$$

Similarly, $h_2 = (x_{20} - \xi_2)/\sigma_2$ is estimated by $\tilde{h}_2$, where

$$\Phi(\tilde{h}_2) = (a + b)/n. \tag{46.188}$$

Then ρ is estimated by $\tilde{\rho}$, where

$$L(\tilde{h}_1, \tilde{h}_2; \tilde{\rho}) = d/n. \tag{46.189}$$

The resulting estimator is called the *tetrachoric correlation*, because it is based on the *tetrachoric* (four-entry) table in Figure 46.9. Hamdan (1970) has shown that this is the maximum likelihood estimator of ρ, for the given data. Equation (46.189) may be solved by an iterative process, using the tables of $L(h_1, h_2; \rho)$ described earlier in Section 46.4.

Before these tables were available, approximate analytic methods of solution were devised. Pearson (1901) obtained the expansion

$$
\begin{aligned}
L(h_1, h_2; \rho) \\
&= \Phi(h_1)\Phi(h_2) + \phi(h_1)\phi(h_2)\left\{\rho + \frac{\rho^2}{2!} h_1 h_2 + \frac{\rho^3}{3!} (h_1^2 - 1)(h_2^2 - 1) + \cdots\right\} \\
&= \sum_{j=0}^{\infty} \tau_j(h_1)\tau_j(h_2)\rho^j,
\end{aligned}
\tag{46.190}
$$

where

$$\tau_j(h) = \frac{(-1)^{j-1}}{\sqrt{j!}} \frac{d^{j-1}\phi(h)}{dh^{j-1}}, \qquad j \geq 1,$$

and
$$\tau_0(h) = \Phi(h).$$

The $\tau_j(h)$ is called the jth *tetrachoric function*. It is a multiple of the $(j-1)$th Hermite polynomial (for $j \geq 1$). Lee (1917) presented tables of $\tau_j(h)$ to seven decimal places for $j = 0(1)19$, $h = 0.0(0.1)4.0$. These tables are included in the tables of Pearson (1931). Note that (46.190) can be obtained by integration of Mehler's series expansion (see Chapter 45):

$$\phi(x_1, x_2; \rho) = \phi(x_1)\phi(x_2)\left\{1 + \rho H_1(x_1)H_1(x_2) + \frac{\rho^2}{2!}H_2(x_1)H_2(x_2) + \cdots\right\}.$$

The fact that $|\tau_j(h)| < 1$ makes possible a rough assessment of the convergence of the series in (46.190). For values of $|h_1|$ and $|h_2|$ less than 1, another series, also obtained by Pearson, gives rather better convergence. This series is an expansion in powers of $\theta = \sin^{-1}\rho$ and starts

$$
\begin{aligned}
L(h_1, h_2; \rho) &= \Phi(h_1)\Phi(h_2) \\
&+ \phi(h_1)\phi(h_2)\Big[\theta + \frac{\theta^2}{2!}h_1 h_2 + \frac{\theta^3}{3!}(h_1^2 + h_2^2 - h_1^2 h_2^2) \\
&+ \frac{\theta^4}{4!}h_1 h_2\{5 - 3(h_1^2 + h_2^2) + h_1^2 h_2^2\} + \cdots\Big].
\end{aligned}
\tag{46.191}
$$

It will be appreciated that solution of either (46.190) or (46.191) for ρ is usually troublesome. A number of approximations, which are simple to compute, have been suggested.

Using the values of $\Phi(\tilde{h}_1)$ and $\Phi(\tilde{h}_2)$ from (46.187) and (46.188) in (46.190), we find

$$R_1 = \frac{ad - bc}{n^2\phi(\tilde{h}_1)\phi(\tilde{h}_2)} = \tilde{\rho} + \frac{\tilde{\rho}^2}{2!}\tilde{h}_1\tilde{h}_2 + \frac{\tilde{\rho}^3}{3!}(\tilde{h}_1^2 - 1)(\tilde{h}_2^2 - 1) + \cdots.$$

$$\tag{46.192}$$

The jth term in the series is $(\tilde{\rho}^j/j!)H_{j-1}(\tilde{h}_1)H_{j-1}(\tilde{h}_2)$. This suggests that if $\tilde{\rho}$ is small, then $\tilde{\rho} \doteq R_1$; this approximate formula was given by Pearson (1901). Yule (1897) had previously suggested the estimator

$$R_2 = (ad - bc)/(ad + bc).\tag{46.193}$$

Further estimators suggested by Pearson (1903) include

$$R_3 = \cos[\pi(1 + ad/bc)^{-1}],\tag{46.194}$$

$$R_4 = \sin\left[\frac{\pi}{2}\left\{1 + \frac{2bc}{ad - bc}\frac{n}{b + c}\right\}^{-1}\right] \quad (ad \geq bc),$$

$$\tag{46.195}$$

and

$$R_5 = \sin\left[\frac{\pi}{2}\left\{1 + \frac{4abcdn^2}{(ad - bc)^2(a + d)(b + c)}\right\}^{-1/2}\right].$$ (46.196)

Pearson made a number of numerical comparisons among these formulas. His work was extended by Castellan (1966), who found that an estimator proposed by Camp (1931) gave results generally considerably closer to $\hat{\rho}$. This estimator is constructed as follows. First arrange (by changing signs of variables, if necessary) that $a+c \geq b+d$. Then calculate $\tilde{h}_2^-$, $\tilde{h}_2^+$ from

$$\Phi(\tilde{h}_2^-) = a/(a + c), \qquad \Phi(\tilde{h}_2^+) = b/(b + d)$$ (46.197)

and

$$M = \frac{(a + c)(b + d)}{n^2}\frac{\tilde{h}_2^- + \tilde{h}_2^+}{\phi(\tilde{h}_1)}.$$ (46.198)

The estimator is then

$$R_6 = \frac{M}{1 + \theta^2 M},$$ (46.199)

where θ can be found by interpolation in the following table:

$$(a + c)/n = 0.50\ 0.55\ 0.60\ 0.65\ 0.70\ 0.75\ 0.80\ 0.85\ 0.90$$
$$\theta = 0.64\ 0.63\ 0.63\ 0.63\ 0.62\ 0.61\ 0.60\ 0.58\ 0.56.$$

For many purposes, it suffices to take $\theta = \frac{5}{8}$. Note that a different value of R_6 is obtained, in general, if X_1 and X_2 are interchanged.

The approximate standard deviation of $\hat{\rho}$ (and of each of the R's insofar as they approximate to $\hat{\rho}$) for large n is

$$\frac{1}{\sqrt{n}}\frac{\sqrt{\Phi(h_1)[1 - \Phi(h_1)]\Phi(h_2)[1 - \Phi(h_2)]}}{\phi(h_1)\phi(h_2)};$$ (46.200)

see Pearson (1901). This is often estimated by

$$\frac{1}{n^{5/2}}\frac{\sqrt{(a + b)(c + d)(a + c)(b + d)}}{\phi(\hat{h}_1)\phi(\hat{h}_2)}.$$ (46.201)

10.2   Biserial Correlation

If only one of the two variables, say X_2, is dichotomized, we may regard the available data as represented by n independent pairs of random variables (X_{1j}, Y_j), where

$$Y_j = \begin{cases} 1 & \text{if } X_{2j} > x_{20}, \\ 0 & \text{if } X_{2j} \leq x_{20}. \end{cases}$$

The correlation between X_{1j} and Y_j, say ρ', is related to ρ by the formula

$$\rho' = \rho \phi(h_2)[\Phi(h_2)\{1 - \Phi(h_2)\}]^{-1/2}, \tag{46.202}$$

where $h_2 = (x_{20} - \xi_2)/\sigma_2$.

It is natural to take, as an estimator of ρ, the statistic

$$\rho^* = (\text{sample correlation between } X_1 \text{ and } Y)\frac{\sqrt{\Phi(\hat{h}_2)[1 - \Phi(\hat{h}_2)]}}{\phi(\hat{h}_2)},$$

where $\Phi(\hat{h}_2) = \bar{Y} = \frac{1}{n}\sum_{j=1}^{n} Y_j$ = proportion of X_2's greater than x_{20}. Since

$$\sum_{j=1}^{n}(Y_j - \bar{Y})^2 = n\Phi(\hat{h}_2)[1 - \Phi(\hat{h}_2)],$$

it follows that

$$\rho^* = \frac{\frac{1}{n}\sum_{j=1}^{n}(X_{1j} - \bar{X}_1)(Y_j - \bar{Y})}{\phi(\hat{h}_2)\left[\frac{1}{n}\sum_{j=1}^{n}(X_{1j} - \bar{X}_1)^2\right]^{1/2}}. \tag{46.203}$$

This formula was obtained by Pearson (1903) and termed by him the *biserial correlation coefficient*. As $n \to \infty$, $E[\rho^*] \to \rho$. Soper (1915) showed that for n large

$$n\text{var}(\rho^*) \doteq \rho^4 + (h_2^2\rho^2 - 1)\frac{\Phi(h_2)[1 - \Phi(h_2)]}{[\phi(h_2)]^2}$$
$$+ \rho^2\left\{\frac{(2\Phi(h_2) - 1)h_2}{\phi(h_2)} - \frac{5}{2}\right\}. \tag{46.204}$$

Tate (1955) gives values of the square root of the right-hand side of (46.204) to three decimal places for $\rho = 0.0(0.1)1.0$, $\Phi(h_2) = 0.05(0.05)0.50$. Note that the value is unchanged if the sign of ρ or h_2 is reversed.

Maximum likelihood estimation of the parameters ρ, h_2, ξ_1, and σ_1 has been studied by Tate (1955). The ξ_1 and σ_1^2 are estimated by $\bar{X}_1$

and $\frac{1}{n}\sum_{j=1}^{n}(X_{1j} - \bar{X}_1)^2 = S^2$, respectively. Then $\hat{\rho}$ and $\hat{h}_2$, the maximum likelihood estimators of ρ and h_2, respectively, have to satisfy the equations

$$\Sigma^+(X'_{1j} - \hat{\rho}\hat{h}_2)\left[\Re\left(\frac{\hat{h}_2 - \hat{\rho}X'_{1j}}{\sqrt{1-\hat{\rho}^2}}\right)\right]^{-1}$$
$$= \Sigma^-(X'_{1j} - \hat{\rho}\hat{h}_2)\left[\Re\left(-\frac{\hat{h}_2 - \hat{\rho}X'_{1j}}{\sqrt{1-\hat{\rho}^2}}\right)\right]^{-1}, \qquad (46.205)$$

$$\Sigma^+\left[\Re\left(\frac{\hat{h}_2 - \hat{\rho}X'_{1j}}{\sqrt{1-\hat{\rho}^2}}\right)\right]^{-1} = \Sigma^-\left[\Re\left(-\frac{\hat{h}_2 - \hat{\rho}X'_{1j}}{\sqrt{1-\hat{\rho}^2}}\right)\right]^{-1}, \qquad (46.206)$$

where Σ^+ denotes summation over j for $Y_j = 1$ and Σ^- denotes summation over j for $Y_j = 0$, $X'_{1j} = \frac{X_{1j} - \bar{X}_1}{S}$, and $\Re(u) = \frac{1-\Phi(u)}{Z(u)}$ is Mills' ratio.

Tate (1955) has constructed an iterative method of solving (46.205) and (46.206) for $\hat{\rho}$ and $\hat{h}_2$. He suggests taking the biserial correlation ρ^* and h^* as initial values. For large n,

$$n\mathrm{var}(\hat{h}_2) \doteq \frac{(1-\rho^2)(\psi_2 - 2\rho h_2\psi_1 + \rho^2 h_2\psi_0)}{\psi_0\psi_2 - \psi_1^2} + \rho^2(\rho^2 h_2^2 + 2),$$
$$\qquad (46.207)$$

$$n\mathrm{var}(\hat{\rho}) \doteq \frac{(1-\rho^2)^3\psi_0}{\psi_0\psi_2 - \psi_1^2} + \rho^2(1-\rho^2)^2, \qquad (46.208)$$

where

$$\psi_r = \int_{-\infty}^{\infty} t^r\phi(t)\left[\Re\left(\frac{h_2 - \rho t}{\sqrt{1-\rho^2}}\right)\Re\left(-\frac{h_2 - \rho t}{\sqrt{1-\rho^2}}\right)\right]^{-1} dt. \qquad (46.209)$$

Prince and Tate (1966) give values of the right-hand sides of (46.207) and (46.208) and of ψ_0, ψ_1, and ψ_2 to five decimal places for $h_2 = 0(0.1)0.8(0.05)$ 1.60(0.025)1.65 and various values of ρ. Note that (46.207) and (46.208) are unchanged by reversal of sign of ρ or h_2; ψ_r is unchanged by reversal of sign of ρ, but is multiplied by $(-1)^r$ if the sign of h_2 is changed.

Birnbaum (1950a,b) has discussed the situation arising when X_1 is truncated. Many recent works on point biserial correlation coefficient may be found in the Psychology literature (for example, in *Psychometrika*).

11   RELATED DISTRIBUTIONS

Distributions obtained by simple transformations of multivariate normal variables have been discussed earlier in Chapter 44. For example, Greenland (1996) has studied the exponentiation of bivariate normal random variables (essentially, pairs of lognormal random variables) and their correlation properties.

11.1   Mixtures of Bivariate Normal Distributions

Mixtures of bivariate normal distributions were described by Akesson (1916) and by Charlier and Wicksell (1924), but little further work has been published using such distributions. Hyrenius (1952) used a mixture of bivariate normal distributions as a specimen nonnormal distribution in assessing the effects of nonnormality on distributions of sample arithmetic means, variances, and covariances. A special case was described by Charnley (1941).

The relative accuracy of moment estimators for mixtures of univariate distributions is exploited in the following practical method of fitting a two-component mixture of bivariate normal distributions with common variance-covariance matrix [Day (1969)]. Denoting the variates by X and Y, each marginal distribution is fitted separately by a two-component mixture of normal distributions, each with common variance, using the method of moments. Call the fitted values of ω (the proportion of first component) for the two cases $\hat{\omega}_x$, $\hat{\omega}_y$. These should be estimators of the same value ω. In fact, substantial difference between $\hat{\omega}_x$ and $\hat{\omega}_y$ can be taken as indication that the mixture of two bivariate normal distributions is inappropriate.

Take $\tilde{\omega} = \frac{1}{2}(\hat{\omega}_x + \hat{\omega}_y)$ and then fit each marginal distribution to make first three sample and population moments agree. Denote the fitted parameters for $X, \tilde{\xi}_1, \tilde{\xi}_2, \tilde{\sigma}_x$; and for $Y, \tilde{\eta}_1, \tilde{\eta}_2, \tilde{\sigma}_y$. Then, ρ is estimated by equating the sample covariance between X and Y to

$$\tilde{\rho}\tilde{\sigma}_x\tilde{\sigma}_y + \tilde{\omega}(1 - \tilde{\omega})(\tilde{\xi}_1 - \tilde{\xi}_2)(\tilde{\eta}_1 - \tilde{\eta}_2).$$

11.2   Bivariate Half-Normal Distribution

Some, or all, of the variates in a multivariate normal distribution may be replaced by their absolute values. This produces a joint distribution with some, or all, of the marginal distributions folded normal (see Chapter 13). If the expected values of the original variables are zero, the marginal distributions are half-normal. In particular, we have the *bivariate half-normal* distribution

$$
\begin{aligned}
p_{X_1,X_2}(x_1,x_2) \;=\;& 2[\pi\sigma_1\sigma_2\sqrt{1-\rho^2}]^{-1} \\
&\times \left[\exp\left\{-\frac{(x_1/\sigma_1)^2+(x_2/\sigma_2)^2}{2(1-\rho^2)}\right\}\right]\cosh\left[\frac{\rho x_1 x_2}{(1-\rho^2)\sigma_1\sigma_2}\right],
\end{aligned}
$$

$$
0 < x_1,\ 0 < x_2. \tag{46.210}
$$

The X_1 and X_2 each have half normal distributions. The conditional distribution of X_2, given X_1, is folded normal, being the distribution of the absolute value of a normal variable with expected value $|\rho|X_1$ and variance $(1-\rho^2)$. The regression function of X_2 on X_1 is

$$
E[X_2 \mid X_1] = |\rho|X_1 + 2Z\left(\frac{\rho X_1}{\sqrt{1-\rho^2}}\right). \tag{46.211}
$$

11.3 Distribution of Ratios

The distribution of ratios of variables having a joint bivariate normal distribution has attracted some attention. If X_1, X_2 have joint density in (46.1), then

$$
\begin{aligned}
\Pr[X_1/X_2 \le g] \;=\;& \Pr[(U_1\sigma_1+\xi_1)/(U_2\sigma_2+\xi_2)\le g] \\
=\;& \Pr[(U_1+\xi_1\sigma_1^{-1})/(U_2+\xi_2\sigma_2^{-1})\le g\sigma_2\sigma_1^{-1}],
\end{aligned}
$$

$$
\tag{46.212}
$$

where U_1, U_2 have a joint standardized bivariate normal distribution. We, therefore, need to consider only the distribution of $(U_1+\delta_1)/(U_2+\delta_2)$, and it can always be arranged to have δ_1 and δ_2 nonnegative. For a contrary opinion, see Hinkley (1969). Nicholson (1943) has provided convenient formulas for computations.

Fieller (1932) obtained an expression for the cumulative distribution of the ratio, effectively in the form

$$
\Pr[(U_1+\delta_1)/(U_2+\delta_2)\le g] = L(\varepsilon,-\delta_2;\rho') + L(-\varepsilon,\delta_2;\rho') \tag{46.213}
$$

with

$$
\begin{aligned}
\varepsilon &= (\delta_1-\delta_2 g)(g^2-2\rho g+1)^{-1/2}, \\
\rho' &= (g-\rho)(g^2-2\rho g+1)^{-1/2}.
\end{aligned}
$$

An equivalent expression is

$$1 - \frac{\cos^{-1}\rho'}{\pi} - 2\left\{ V\left(\varepsilon, \frac{\delta_2 + \rho'\varepsilon}{\sqrt{1-\rho'^2}}\right) + V\left(\delta_2, \frac{\varepsilon + \rho'\delta_2}{\sqrt{1-\rho'^2}}\right)\right\}.$$

(46.214)

The $L(\cdot), V(\cdot)$ are as defined earlier [see (46.37) and (46.50)]. If δ_2 is large, so that $\Pr[U_2 + \delta_2 < 0]$ is negligible, an approximate formula is easily obtained. We have, from Geary (1930),

$$\begin{aligned} \Pr[(U_1 + \delta_1)/(U_2 + \delta_2) \le g] &= \Pr[U_1 + \delta_1 \le g(U_2 + \delta_2)] \\ &= \Pr[U_1 - gU_2 \le g\delta_2 - \delta_1] \\ &= \Phi((g\delta_2 - \delta_1)(g^2 - 2\rho g + 1)^{-1/2}). \end{aligned}$$

(46.215)

Hinkley (1969) has investigated the accuracy of this approximation. He suggests adding the correction $\pm\Phi(-\delta_2)$; the sign being the same as that of $(\delta_1 - \rho\delta_2)$.

A paper by Marsaglia (1965) includes a number of graphs of the density function, which can be bimodal. Further details on this distribution have been provided by Shanmugalingam (1982). In the special case $\delta_1 = \delta_2 = 0$, we have a Cauchy distribution; see Chapter 16 of Johnson, Kotz, and Balakrishnan (1994).

11.4   "Bivariate Normal" Distribution with Centered Normal Conditionals

Sarabia (1995) has studied the "bivariate normal" distribution with joint density function

$$p_{X_1,X_2}(x_1, x_2) = K(c)\frac{\sqrt{ab}}{2\pi} e^{-(ax_1^2 + bx_2^2 + abcx_1^2 x_2^2)/2},$$

$$x_1, x_2 \in \mathbb{R}, \qquad (46.216)$$

where $a, b > 0$, $c \ge 0$, and $K(c)$ is a normalizing constant. Equation (46.216) corresponds to the most general density function that has normal conditional distributions with zero means and is the so-called *bivariate normal distribution with centered normal conditionals*.

The marginal density functions are given by

$$p_{X_1}(x_1) = K(c)\sqrt{\frac{a}{2\pi}} \frac{1}{\sqrt{1 + acx_1^2}} e^{-ax_1^2/2}, \qquad x_1 \in \mathbb{R} \qquad (46.217)$$

and

$$p_{X_2}(x_2) = K(c)\sqrt{\frac{b}{2\pi}}\,\frac{1}{\sqrt{1 + bcx_2^2}}\,e^{-bx_2^2/2}, \qquad x_2 \in \mathbb{R}. \qquad (46.218)$$

Except when $c = 0$, (46.217) and (46.218) are not normal density functions. The normalizing constant $K(c)$ is

$$K(c) = \frac{\sqrt{2c}}{U(\frac{1}{2}, 1, \frac{1}{2})}, \qquad (46.219)$$

where

$$U(\alpha, \beta, z) = \frac{1}{\Gamma(\alpha)} \int_0^\infty e^{-tz} t^{\alpha-1} (1 + t)^{\beta-\alpha-1}\,dt \qquad (\alpha > 0,\ z > 0) \qquad (46.220)$$

is the confluent hypergeometric function. $K(c)$ is a monotonic increasing function of c, though it increases very slowly (for example, $K(1000) = 9.42$).

From the joint density function of X_1 and X_2 in (46.216), it can be shown easily that

$$Z_1 = \sqrt{a}\,X_1\sqrt{1 + bcX_2^2} \quad \text{and} \quad Z_2 = \sqrt{b}\,X_1\sqrt{1 + acX_1^2} \qquad (46.221)$$

are univariate standard normal variables, with Z_1 being independent of X_2. The joint-moment generating function of (X_1^2, X_2^2) is given by

$$
\begin{aligned}
M_{X_1^2, X_2^2}(t_1, t_2) &= E\left[e^{t_1 X_1^2 + t_2 X_2^2}\right] \\
&= \frac{\left(1 - \frac{2t_1}{a}\right)^{-1/2}\left(1 - \frac{2t_2}{b}\right)^{-1/2} K(c)}{K\left(c\left(1 - \frac{2t_1}{a}\right)^{-1}\left(1 - \frac{2t_2}{b}\right)^{-1}\right)} \qquad (46.222)
\end{aligned}
$$

for $t_1 < \frac{a}{2}$ and $t_2 < \frac{b}{2}$, where $K(\cdot)$ is as given in (46.219). From (46.222), we obtain

$$\text{corr}(X_1^2, X_2^2) = \frac{1 - 2\ell(c) - 4c\ell(c) + 4c^2\ell^2(c)}{-1 - 2\ell(c) + 4c^2\ell^2(c)}, \qquad (46.223)$$

where

$$\ell(c) = \frac{d\log K(c)}{dc} = \frac{K'(c)}{K(c)}. \qquad (46.224)$$

The variables X_1 and X_2 are symmetric unimodal random variables whose tails are lighter than those of the standard normal distribution and have their coefficient of kurtosis to be less than 3 corresponding to the standard normal distribution; see Chapter 13 of Johnson, Kotz, and Balakrishnan (1994). The even moments of X_1, for example, are

$$
\begin{aligned}
E[X^m] \\
= K(c) \frac{1}{\sqrt{2\pi}} a^{-m/2} c^{-(m+1)/2} \Gamma\left(\frac{m+1}{2}\right) U\left(\frac{m+1}{2}, \frac{m}{2}+1, \frac{1}{2c}\right),
\end{aligned}
$$

$$(46.225)$$

where $U(\alpha, \beta, z)$ is the confluent hypergeometric function defined earlier in (46.220).

A natural generalization of the marginal density function of X_1 in (46.217) is a symmetric distribution with density function

$$
\frac{x^{2d}(1+\beta\gamma x^2)^{b-d-3/2} e^{-\beta x^2/2}}{\Gamma\left(d+\frac{1}{2}\right)(\beta\gamma)^{-d-1/2} U\left(d+\frac{1}{2}, b, \frac{1}{2\gamma}\right)}
$$

$$(46.226)$$

which reduces to the density in (46.217) when $d = 0$ and $b = 1$. This density function is bimodal for $d > 0$ with modes at

$$
\pm \left\{\frac{-(1+3\gamma-2b\gamma)+[(1+3\gamma-2b\gamma)^2+8\gamma\delta]^{1/2}}{2b\gamma}\right\}^{1/2}.
$$

Based on a random sample of size n from the bivariate distribution in (46.216), the maximum likelihood estimators of a, b, and c are obtained by solving the equations

$$
\left.
\begin{aligned}
\frac{1}{a} &= \overline{x_1^2} + bc, \overline{x_1^2 x_2^2} \\
\frac{1}{b} &= \overline{x_2^2} + ac, \overline{x_1^2 x_2^2} \\
\ell(c) &= ab/2\overline{x_1^2 x_2^2},
\end{aligned}
\right\}
$$

$$(46.227)$$

where $\ell(c)$ is as defined in (46.224), and

$$
\overline{x_1^2} = \frac{1}{n}\sum_{i=1}^{n} x_{1i}^2, \quad \overline{x_2^2} = \frac{1}{n}\sum_{i=1}^{n} x_{2i}^2 \text{ and } \overline{x_1^2 x_2^2} = \frac{1}{n}\sum_{i=1}^{n} x_{1i}^2 x_{2i}^2.
$$

Noting that the last equation of (46.227) can be rewritten as

$$
\frac{1-2c\ell(c)}{\sqrt{\ell(c)}} = \sqrt{\frac{2\overline{x_1^2}\ \overline{x_2^2}}{\overline{x_1^2 x_2^2}}},
$$

Sarabia (1995) has presented a table of values of the function $\frac{\{1-2c\ell(c)\}}{\sqrt{\ell(c)}}$ using which the maximum likelihood estimate of c can be easily determined.
The 'strongly consistent' estimators in this case are

$$\left. \begin{array}{l} \hat{c} = \frac{1}{4}\left(1 + rt - 2t - \frac{1-t^2}{1+rt}\right), \\[2ex] \hat{\ell} = \frac{2(1+rt)}{(1-r)^2 t^2}, \\[2ex] \hat{a} = \frac{1-2\hat{c}\hat{\ell}}{x_1^2}, \\[2ex] \hat{b} = \frac{1-2\hat{c}\hat{\ell}}{x_2^2}, \end{array} \right\} \qquad (46.228)$$

where (in an obvious notation)

$$r = \frac{S_{x_1^2, x_2^2}}{S_{x_1^2} S_{x_2^2}} \quad \text{and} \quad t = \frac{S_{x_1^2} S_{x_2^2}}{x_1^2 \, x_2^2}.$$

The uncentered form of the bivariate distribution in (46.216) corresponds to the joint density function

$$p(x_1, x_2) = \exp\left\{\sum_{i=0}^{2} \sum_{j=0}^{2} a_{ij} x_1^i x_2^j\right\}. \qquad (46.229)$$

which belongs to the exponential family of distributions. The choice of
the parameters $a_{12} = a_{21} = a_{22} = 0$ yields the classic bivariate normal
density function provided that $a_{02} < 0$, $a_{20} < 0$ and $a_{11}^2 < 4a_{02}a_{20}$. The
joint density in (46.216) corresponds to the case $a_{22} \neq 0$. Though a simple
expression of moments is not available in this case, the regression function
can be shown to be

$$E[X_1|X_2 = x_2] = -\frac{a_{12}x_2^2 + a_{11}x_2 + a_{10}}{2(a_{22}x_2^2 + a_{21}x_2 + a_{20})},$$

which is clearly nonlinear but bounded.

11.5   Bivariate Skew-Normal Distribution

The density function of the bivariate skew-normal distribution, as first
given by Azzalini and Dalla Valle (1996), is

$$p_{X_1, X_2}(x_1, x_2) = 2\phi(x_1, x_2; \omega)\Phi(\lambda_1 x_1 + \lambda_2 x_2), \qquad (46.230)$$

where $\phi(x_1, x_2; \omega)$ denotes the bivariate standard normal density function with correlation coefficient ω, $\Phi(\cdot)$ denotes the univariate standard normal cumulative distribution function, and

$$\lambda_1 = \frac{\delta_1 - \delta_2\omega}{\sqrt{(1-\omega^2)(1-\omega^2-\delta_1^2-\delta_2^2+2\delta_1\delta_2\omega)}} \quad \text{and} \quad \lambda_2 = \frac{\delta_2 - \delta_1\omega}{\sqrt{(1-\omega^2)(1-\omega^2-\delta_1^2-\delta_2^2+2\delta_1\delta_2\omega)}} .$$

$$(46.231)$$

The moment-generating function of $X = (X_1, X_2)^T$ is

$$M_X(t) = 2e^{(t_1^2+2\omega t_1 t_2+t_2^2)/2}\Phi(\delta_1 t_1 + \delta_2 t_2) \qquad (46.232)$$

and the marginal distributions of X are

$$X_i \stackrel{d}{=} \delta_i|Z_0| + \sqrt{1-\delta_i^2}\, Z_i, \qquad i = 1, 2, \qquad (46.233)$$

where Z_0, Z_1, and Z_2 are independent standard normal variables and ω is restricted by the condition that

$$\delta_1\delta_2 - \sqrt{(1-\delta_1^2)(1-\delta_2^2)} < \omega < \delta_1\delta_2 + \sqrt{(1-\delta_1^2)(1-\delta_2^2)} . \quad (46.234)$$

The regression and the conditional variances are

$$E[X_2|X_1 = x_1] = \omega x_1 + \left(\frac{\delta_2 - \omega\delta_1}{\sqrt{1-\delta_1^2}}\right) H(-\tau_1 x_1) \qquad (46.235)$$

and

$$\begin{aligned}
&\text{var}(X_2|X_1 = x_1) \\
&= 1 - \omega^2 - \left(\frac{\delta_2 - \omega\delta_1}{\sqrt{1-\delta_1^2}}\right)^2 H(-\tau_1 x_1)\{\tau_1 x_1 + H(-\tau_1 x_1)\},
\end{aligned}$$

$$(46.236)$$

where $H(x)$ denotes the hazard function of the standard normal distribution which is $H(x) = \phi(x)/\{1 - \Phi(x)\}$.

The correlation curve [a local analog of the classical correlation; see, for example, Bjerve and Doksum (1993)] is

$$\rho(x_1) = \frac{\sigma_1\beta(x_1)}{[\{\sigma_1\beta(x_1)\}^2 + \sigma^2(x_1)]^{1/2}} , \qquad (46.237)$$

where

$$\begin{aligned}
\beta(x_1) &= \frac{d}{dx_1} E[X_2|X_1 = x_1], \\
\sigma^2(x_1) &= \text{var}(X_2|X_1 = x_1)
\end{aligned}$$

and
$$\sigma_1^2 = \text{var}(X_1).$$

For the bivariate skew-normal distribution, we have $\sigma_1^2 = \text{var}(X_1) = 1 - \frac{2}{\pi}\delta_1^2$, $\sigma^2(x_1) = \text{var}(X_2|X_1 = x_1)$ to be as given above in (46.236), and

$$\beta(x_1) = \omega - \left(\frac{\delta_2 - \omega\delta_1}{\sqrt{1-\delta_1^2}}\right)\{H(-\tau_1 x_1)\tau_1^2 x_1 + H^2(-\tau_1 x_1)\tau_1\}.$$

The statistical dependence on the value of x_1 is noteworthy.

BIBLIOGRAPHY

(Some bibliographical items not mentioned in the text are included here for completeness.)

Adrian, R. (1808). Research concerning the probabilities of errors which happen in making observations, etc., *The Analyst; or Mathematical Museum*, **1**(4), 93–109.

Afifi, A. A., and Elashoff, R. M. (1966–1969). Missing observations in multivariate analysis, *Journal of the American Statistical Association*, **61**, 595–604; **62**, 10–29; **62**, 337–358; **64**, 359–365.

Ahsanullah, M. (1970). On the estimation of means in a bivariate normal distribution with equal marginal variance, *Annals of Mathematical Statistics*, **41**, 1155–1156 (Abstract).

Ahsanullah, M., Bansal, N., Hamedani, G. G., and Zhang, H. (1996). A note on bivariate normal distribution, Report, Rider University, Lawrenceville, New Jersey.

Ahsanullah, M., and Wesolowski, J. (1992). Bivariate normality via Gaussian conditional structure, Report, Rider College, Lawrenceville, New Jersey.

Ahsanullah, M., and Wesolowski, J. (1994). Multivariate normality via conditional normality, *Statistics & Probability Letters*, **20**, 235–238.

Aitkin, M. A. (1964). Correlation in a singly truncated bivariate normal distribution, *Psychometrika*, **29**, 263–270.

Aitkin, M. A., and Hume, M. W. (1966). Correlation in a singly truncated bivariate normal distribution, III. Correlation between ranks and variate-values, *Biometrika*, **53**, 278–281.

Akesson, O. A. (1916). On the dissection of correlation surfaces, *Arkiv für Matematik, Astronomi och Fysik*, **11**, No. 16, 1–18.

Albers, W., and Kallenberg, W. C. M. (1994). A simple approximation to the bivariate normal distribution with large correlation coefficient, *Journal of Multivariate Analysis*, **49**, 87–96.

Amos, D. E. (1969). On computation of the bivariate normal distribution, *Mathematics of Computation*, **23**, 655–659.

Anderson, T. W. (1957). Maximum likelihood estimates for a multivariate normal distribution when some observations are missing, *Journal of the American Statistical Association*, **52**, 200–203.

Anderson, T. W. (1958). *An Introduction to Multivariate Statistical Analysis*, New York: John Wiley & Sons.

Anderson, T. W., and Ghurye, S. G. (1978). Unique factorization of products of bivariate normal cumulative distribution functions, *Annals of the Institute of Statistical Mathematics*, **30**, 63–69.

Arnold, B. C., Beaver, R. J., Groeneveld, R. A., and Meeker, W. Q. (1993). The nontruncated marginal of a truncated bivariate normal distribution, *Psychometrika*, **58**, 471–488.

Arnold, B. C., Castillo, E., and Sarabia, J. M. (1995). General conditional specification models, *Communications in Statistics—Theory and Methods*, **24**, 1–11.

Azzalini, A. (1985). A class of distributions which includes the normal ones, *Scandinavian Journal of Statistics*, **12**, 171–178.

Azzalini, A., and Dalla Valle, A. (1996). The multivariate skew-normal distribution, *Biometrika*, **83**, 715–726.

Balakrishnan, N. (1993). Multivariate normal distribution and multivariate order statistics induced by ordering linear combinations, *Statistics & Probability Letters*, **17**, 343–350.

Basu, A. P., and Ghosh, J. K. (1978). Identifiability of the multinormal and other distributions under competing risks model, *Journal of Multivariate Analysis*, **8**, 413–429.

Bhattacharya, P. K. (1974). Convergence of sample paths of normalized sums of induced order statistics, *Annals of Statistics*, **2**, 1034–1039.

Bhattacharyya, A. (1943). On some sets of sufficient conditions leading to the normal bivariate distributions, *Sankhyā*, **6**, 399–406.

Bildikar, S., and Patil, G. P. (1968). Multivariate exponential-type distributions, *Annals of Mathematical Statistics*, , 1316–1326.

Birnbaum, Z. W. (1950a). On the effect of cutting score when selection is performed against a dichotomized criterion, *Psychometrika*, **15**, 385–389.

Birnbaum, Z. W. (1950b). Effect of linear truncation on a multinormal population, *Annals of Mathematical Statistics*, **21**, 272–279.

Bjerve, S., and Doksum, K. (1993). Correlation curves: measures of association as functions of covariate values, *Annals of Statistics*, **21**, 890–902.

Bofinger, E., and Bofinger, V. J. (1965). The correlation of maxima in samples drawn from a bivariate normal distribution, *Australian Journal of Statistics*, **7**, 57–61.

Borth, D. M. (1973). A modification of Owen's method for computing the bivariate normal integral, *Applied Statistics*, **22**, 82–85.

Bravais, A. (1846). Analyse mathématique sur la probabilité des erreurs de situation d'un point, *Mémoires presentés à l'Académie Royale des Sciences, Paris*, **9**, 255–332 (English translation, 1958; White Sands Proving Ground, New Mexico).

Braverman, M. S. (1985). Characteristic properties of normal and stable distributions, *Theory of Probability and its Applications*, **30**, 465–474.

Brucker, J. (1979). A note on the bivariate normal distribution, *Communications in Statistics—Theory and Methods*, **8**, 175–177.

Burington, R. S., and May, D. C. (1970). *Handbook of Probability and Statistics with Tables*, second edition, pp. 144–147, New York: McGraw-Hill.

Cadwell, J. H. (1951). The bivariate normal integral, *Biometrika*, **38**, 475–479.

Cain, M. (1994). The moment-generating function of the minimum of bivariate normal random variables, *The American Statistician*, **48**, 124–125.

Cain, M., and Pan, E. (1995). Moments of the minimum of bivariate normal random variables, *The Mathematical Scientist*, **20**, 119–122.

Camp, B. H. (1931). *The Mathematical Parts of Elementary Statistics*, New York: D. C. Heath.

Castellan, N. J. (1966). On the estimation of the tetrachoric correlation coefficient, *Psychometrika*, **31**, 67–73.

Charlier, C. V. L., and Wicksell, S. D. (1924). On the dissection of frequency functions, *Arkiv für Matematik, Astronomi och Fysik*, **18**, No. 6, 1–64.

Charnley, F. (1941). Some properties of a composite, bivariate distribution in which the means of the component normal distributions are linearly related, *Canadian Journal of Research*, **19**, 139–151.

Chew, V. (1964). Confidence, prediction and tolerance regions, *Technical Memo No. 64-9*, RCA System Analysis, U.S. Air Force, Patrick Air Force Base, Florida.

Chou, Y.-M., and Owen, D. B. (1984). An approximation to percentiles of a variable of the bivariate normal distribution when the other variable is truncated, with applications, *Communications in Statistics—Theory and Methods*, **13**, 2535–2547.

Cohen, A. C. (1955). Restriction and selection in samples from bivariate normal distributions, *Journal of the American Statistical Association*, **50**, 884–893.

Cohen, A. C. (1959). Simplified estimators for the normal distribution when samples are singly censored or truncated, *Technometrics*, **1**, 217–237.

Cooper, B. E. (1968a). Algorithm AS4. An auxiliary function for distribution integrals, *Applied Statistics*, **17**, 190–192.

Cooper, B. E. (1968b). Algorithm AS5. The integral of the non-central t-distribution, *Applied Statistics*, **17**, 193–194.

Czuber, E. (1891). *Theorie der Beobachtungsfehler*, Leipzig: Teubner.

Dahiya, R. C., and Korwar, R. M. (1980). Maximum likelihood estimates for a bivariate normal distribution with missing data, *Annals of Statistics*, **8**, 687–692.

Daley, D. J. (1974). Computation of bi- and tri-variate normal integrals, *Applied Statistics*, **23**, 435–438.

David, H. A. (1973). Concomitants of order statistics, *Bulletin of the International Statistical Institute*, **45**, 295–300.

David, H. A. (1981). *Order Statistics*, Second edition, New York: John Wiley & Sons.

David, H. A. (1994). Concomitants of extreme order statistics, in *Extreme Value Theory and Applications* (J. Galambos *et al.*, eds.), pp. 211–224, Dordrecht, The Netherlands: Kluwer Academic Publishers.

David, H. A., and Galambos, J. (1974). The asymptotic theory of concomitants of order statistics, *Journal of Applied Probability*, **11**, 762–770.

David, H. A., and Nagaraja, H. N. (1998). Concomitants of order statistics, in *Handbook of Statistics-16: Order Statistics: Theory and Methods* (N. Balakrishnan and C. R. Rao, eds.), pp. 487–513, Amsterdam, The Netherlands: North-Holland.

Day, N. E. (1969). Estimating the components of a mixture of normal distributions, *Biometrika*, **56**, 463–474.

DeLury, D. B. (1938). Note on correlations, *Annals of Mathematical Statistics*, **9**, 149–151.

Des Raj, S. (1953). On estimating the parameters of bivariate normal populations from doubly and singly linearly truncated samples, *Sankhyā*, **12**, 277–290.

Dickson, I. D. H. (1886). Appendix to "Family likeness in stature," by F. Galton, *Proceedings of the Royal Society of London*, **40**, 63–73.

Donnelly, T. G. (1973). Algorithm 462. Bivariate normal distribution, *Communications of the Association of Computing Machinery*, **16**, 638.

Drezner, Z., and Wesolowsky, G. O. (1989). An approximation method for bivariate and multivariate normal equiprobability contours, *Communications in Statistics—Theory and Methods*, **18**, 2331–2344.

Drezner, Z., and Wesolowsky, G. O. (1990). On the computation of the bivariate normal integral, *Journal of Statistical Computation and Simulation*, **35**, 101–107.

Dunnett, C. W. (1958). *Tables of the Bivariate Normal Distribution with Correlation* $1/\sqrt{2}$, Deposited in UMT file [abstract in *Mathematics of Computation*, **14**, (1960), 79].

Dunnett, C. W. (1960). On selecting the largest of k normal population means, *Journal of the Royal Statistical Society, Series B*, **22**, 1–30.

Dunnett, C. W. and Lamm, R. A. (1960). *Some tables of the multivariate normal probability integral with correlation coefficient 1/3*, deposited in UMF file [abstract in *Mathematics of Computation*, **14**, (1960), 290–291].

Elderton, E. M., Moul, M., Fieller, E. C., Pretorius, S. J., and Church, A. E. R. (1930). On the remaining tables for determining the volumes of a bivariate normal surface, *Biometrika*, **22**, 13–35; Introduction by K. Pearson, 1–12.

Everitt, P. F. (1912). Supplementary tables for finding the correlation coefficient from tetrachoric groupings, *Biometrika*, **8**, 385–395.

Fieller, E. C. (1932). The distribution of the index in a normal bivariate population, *Biometrika*, **24**, 428–440.

Fieller, E. C., Lewis, T., and Pearson, E. S. (1956). *Correlated Random Normal Deviates*, Tracts for Computers, **26**, Cambridge University Press: London.

Fisher, R. A. (1929). Moments and product moments of sampling distributions, *Proceedings of the London Mathematical Society, Series 2*, **30**, 199–238.

Fraser, D. A. S., and Streit, F. (1980). A further note on the bivariate normal distribution, *Communications in Statistics—Theory and Methods*, **9**, 1097–1099.

Galton, F. (1877). *Typical Laws of Heredity in Man*, Address to Royal Institution of Great Britain.

Galton, F. (1888). Co-relations and their measurement chiefly from anthropometric data, *Proceedings of the Royal Society of London*, **45**, 134–145.

Gauss, C. F. (1823). *Theoria Combinationis Observationum Erroribus Minimis Obnoxiae*, Göttingen.

Geary, R. C. (1930). The frequency distribution of the quotient of two normal variates, *Journal of the Royal Statistical Society, Series A*, **93**, 442–446.

Gelman, A., and Meng, X.-L. (1991). A note on bivariate distributions that are conditionally normal, *The American Statistician*, **45**, 125–126.

Greenland, S. (1996). A lower bound for the correlation of exponentiated bivariate normal pairs, *The American Statistician*, **50**, 163–164.

Grundy, P. M., Healy, M. J. R., and Rees, D. H. (1956). Economic choice of the amount of experimentation, *Journal of the Royal Statistical Society, Series B*, **18**, 32–49.

Gupta, R. C., and Subramanian, S. (1998). Estimation of reliability in bivariate normal with equal coefficient of variation, *Communications in Statistics—Simulation and Computation*, **27**, 675–697.

Gupta, S. S. (1963). Bibliography on the multivariate normal integrals and related topics, *Annals of Mathematical Statistics*, **34**, 829–838.

Gupta, S. S., and Pillai, K. C. S. (1965). On linear functions of ordered correlated normal random variables, *Biometrika*, **52**, 367–379.

Haldane, J. B. S. (1942). Moments of the distributions of powers and products of normal variates, *Biometrika*, **32**, 226–242.

Hamdan, M. A. (1970). The equivalence of tetrachoric and maximum likelihood estimates of ρ in 2×2 tables, *Biometrika*, **57**, 212–215.

Hamedani, G. G. (1992). Bivariate and multivariate normal characterizations: A brief survey, *Communications in Statistics: Theory and Methods*, **21**, 2665–2688.

Han, C.-P. (1973). Regression estimation for bivariate normal distributions, *Annals of the Institute of Statistical Mathematics*, **23**, 335–343.

Harrell, F. E., and Sen, P. K. (1979). Statistical inference for censored bivariate normal distributions based on induced order statistics, *Biometrika*, **66**, 293–298.

Helmert, F. R. (1868). Studien über rationelle Vermessungen, im Gebiete der höheren Geodäsie, *Zeitschrift für Mathematik und Physik*, **13**, 73–129.

Hinkley, D. V. (1969). On the ratio of two correlated normal random variables, *Biometrika*, **56**, 635–639.

Hocking, R. R., and Smith, W. B. (1968). Estimation of parameters in the multivariate normal distribution with missing observations, *Journal of the American Statistical Association*, **63**, 159–173.

Hoeffding, W. (1940/41). Maßstabsinvariante Korrelationstheorie. Schriften des Mathem, *Seminars und des Instituts für Angewandte Mathematik der Universität Berlin*, **5**, 179–233.

Holland, P. W., and Wang, Y. J. (1987). Dependence function for continuous bivariate densities, *Communications in Statistics—Theory and Methods*, **16**, 863–876.

Hughes, H. M. (1949). Estimation of the variance of the bivariate normal distribution, *University of California at Berkeley, Publications in Statistics*, **1**, 37–52.

Hyrenius, H. (1952). Sampling from bivariate non-normal universes by means of compound normal distributions, *Biometrika*, **39**, 238–246.

Iyengar, S., and Tong, Y. L. (1989). Convexity properties of elliptically contoured distributions with applications, *Sankhyā, Series A*, **51**, 13–29.

Iyer, P. V. K., and Simha, P. S. (1967). *Tables of Bivariate Random Normal Deviates*, Defense Science Laboratory, Delhi, India.

Jaiswal, M. C., and Khatri, C. G. (1967). Estimation of parameters for the selected samples from bivariate normal populations, *Metron*, **26**, Nos. 3–4, 1–8.

Johnson, N. L., Kotz, S., and Balakrishnan, N. (1994). *Continuous Univariate Distributions*, Vol. 1, second edition, New York: John Wiley & Sons.

Johnson, N. L., Kotz, S., and Balakrishnan, N. (1995). *Continuous Univariate Distributions*, Vol. 2, second edition, New York: John Wiley & Sons.

Kagan, A., and Wesolowski, J. (1996). Normality via conditional normality of linear forms, *Statistics & Probability Letters*, **29**, 229–232.

Kamat, A. R. (1953). Incomplete and absolute moments of the multivariate normal distribution with some applications, *Biometrika*, **40**, 20–34.

Kamat, A. R. (1958a). Hypergeometric expansions for incomplete moments of the bivariate normal distribution, *Sankhyā*, **20**, 317–320.

Kamat, A. R. (1958b). Incomplete moments of the trivariate normal distribution, *Sankhyā*, **20**, 321–322.

Kendall, M. G. (1941). Proof of relations connected with the tetrachoric series and its generalization, *Biometrika*, **32**, 196–198.

Kendall, M. G., and Stuart, A. (1963). *The Advanced Theory of Statistics*, **1**, London: Griffin.

Khatri, C. G., and Jaiswal, M. C. (1963). Estimation of parameters of a truncated bivariate normal distribution, *Journal of the American Statistical Association*, **58**, 519–526.

Kleinbaum, D. (1969). Estimation and hypothesis testing for generalized multivariate linear models, Ph.D. thesis, University of North Carolina, Chapel Hill.

Lakshminarayan, C. K., and Han, C.-P. (1997). On estimating the mean in a bivariate normal distribution with equal or unequal variances, *Journal of Statistical Computation and Simulation*, **58**, 155–170.

Lancaster, H. O. (1959). Zero correlation and independence, *Australian Journal of Statistics*, **1**, 53–56.

Laplace, P. S. (1810). Mémoire sur les intégrales définies, et leur application aux probabilités, *Mémoires de l'Institut Impérial de France*, 279–347.

Lee, A. (1917). Further supplementary tables for determining high (tetrachoric) correlations from tetrachoric groupings, *Biometrika*, **11**, 287–291.

Lee, A. (1927). Supplementary tables for determining correlation from tetrachoric groupings (tetrachoric correlations), *Biometrika*, **19**, 354–404.

Lehmann, E. L. (1966). Some concepts of dependence, *Annals of Mathematical Statistics*, **37**, 1137–1153.

Lin, J.-T. (1995). A simple approximation for the bivariate normal integral, *Probability in the Engineering and Informational Sciences*, **9**, 317–321.

Lipow, M., and Eidemiller, R. L. (1964). Application of the bivariate normal distribution to a stress vs strength problem in reliability analysis, *Technometrics*, **6**, 325–328.

Lukatzkaya, M. L. (1965). Some properties of random variables with a generalized normal distribution, *Nauchnye Trudy Novosibirsk. U-ta*, **5**, 57–71 (in Russian).

Madansky, A. (1958). On the maximum likelihood estimate of the correlation coefficient, Report No. P-1355, RAND Corporation, Santa Monica, California.

Mallows, C. L. (1958). An approximate formula for bivariate normal probabilities, Technical Report No. 30, Statistical Techniques Research Group, Princeton University.

Maritz, J. S. (1953). Estimation of the correlation coefficient in the case of a bivariate normal population when one of the variables is dichotomized, *Psychometrika*, **18**, 97–110.

Marsaglia, G. (1965). Ratios of normal variables and ratios of sums of uniform variables, *Journal of the American Statistical Association*, **60**, 193–204.

Mee, R. W., and Owen, D. B. (1983). A simple approximation for bivariate normal probabilities, *Journal of Quality Technology*, **15**, 72–75.

Melnick, E. L., and Tenenbein, A. (1982). Misspecification of the normal distribution, *The American Statistician*, **36**, 372–373.

Mendell, N. R., and Elston, R. C. (1974). Multifactorial qualitative traits: Genetic analysis and prediction of recurrence risks, *Biometrics*, **30**, 41–47.

Milton, R. C. (1972). Computer evaluation of the multivariate normal integral, *Technometrics*, **14**, 881–889.

Morrison, D. (1971). Expectations and variances of maximum likelihood estimates of the multivariate normal distribution parameters with missing data, *Journal of the American Statistical Association*, **66**, 602–604.

Mosteller, F. (1946). On some useful "inefficient" statistics, *Annals of Mathematical Statistics*, **17**, 377–408.

Mukherjea, A., Nakassis, A., and Miyashita, J. (1986). Identification of parameters by the distribution of the maximum random variable: The Anderson-Ghurye theorems, *Journal of Multivariate Analysis*, **18**, 178–186.

Mukherjea, A., and Stephens, R. (1990a). The problem of identification of parameters by the distribution of the maximum random variable: Solution for the trivariate normal case, *Journal of Multivariate Analysis*, **34**, 95–115.

Mukherjea, A., and Stephens, R. (1990b). Identification of parameters by the distribution of the maximum random variable: The general multivariate normal case, *Probability Theory & Related Fields*, **84**, 289–296.

Nabeya, S. (1951). Absolute moments in 2-dimensional normal distribution, *Annals of the Institute of Statistical Mathematics*, **3**, 2–6.

Nabeya, S. (1952). Absolute moments in 3-dimensional normal distribution, *Annals of the Institute of Statistical Mathematics*, **4**, 15–20.

Nadler, J. (1967). Bivariate samples with missing values, *Technometrics*, **9**, 679–682.

Nagaraja, H. N. (1982). A note on linear functions of ordered correlated normal random variables, *Biometrika*, **69**, 284–285.

Nagaraja, H. N., and David, H. A. (1994). Distribution of the maximum of concomitants of selected order statistics, *Annals of Statistics*, **22**, 478–494.

Nath, G. B. (1971). Estimation in truncated bivariate normal distributions, *Applied Statistics*, **20**, 313–319.

National Bureau of Standards (1959). *Tables of the Bivariate Normal Distribution Function and Related Functions*, Applied Mathematics Series, **50**.

Nicholson, C. (1943). The probability integral for two variables, *Biometrika*, **33**, 59–72.

Olkin, I., and Pratt, J. W. (1958). Unbiased estimation of certain correlation coefficients, *Annals of Mathematical Statistics*, **29**, 201–211.

Olson, J. M., and Weissfeld, L. A. (1991). Approximation of certain multivariate integrals, *Statistics & Probability Letters*, **11**, 309–317.

Owen, D. B. (1956). Tables for computing bivariate normal probabilities, *Annals of Mathematical Statistics*, **27**, 1075–1090.

Owen, D. B. (1957). The bivariate normal probability distribution, Research Report SC 3831-TR, Sandia Corporation.

Owen, D. B., and Wiesen, J. M. (1959). A method of computing bivariate normal probabilities with an application to handling errors in testing and measuring, *Bell System Technical Journal*, **38**, 553–572.

Parikh, N. T., and Sheth, R. J. (1966). Applications of bivariate normal distribution; to a stress vs. strength problem in reliability analysis, *Journal of the Indian Statistical Association*, **4**, 105–107.

Patil, G. P., and Joshi, S. N. (1970). Further results on minimum variance unbiased estimation and additive number theory, *Annals of Mathematical Statistics*, **41**, 561–575.

Paulson, E. (1942). A note on the estimation of some mean values for a bivariate distribution, *Annals of Mathematical Statistics*, **13**, 440–445.

Pearson, K. (1901). Mathematical contributions to the theory of evolution–VII. On the correlation of characters not quantitatively measurable, *Philosophical Transactions of the Royal Society of London, Series A*, **195**, 1–47.

Pearson, K. (1903). Mathematical contributions to the theory of evolution–XI. On the influence of natural selection on the variability and correlation of organs, *Philosophical Transactions of the Royal Society of London, Series A*, **200**, 1–66.

Pearson, K. (1920). Notes on the history of correlation, *Biometrika*, **13**, 25–45.

Pearson, K. (1931). *Tables for Statisticians and Biometricians*, Vol. 2, London: Cambridge University Press.

Pearson, K., and Young, A. W. (1918). On the product-moments of various orders of the normal correlation surface of two variates, *Biometrika*, **12**, 86–92.

Pitman, E. J. G. (1939). A note on normal correlation, *Biometrika*, **31**, 9–12.

Plackett, R. L. (1954). A reduction formula for normal multivariate integrals, *Biometrika*, **41**, 351–360.

Plana, G. A. A. (1813). Mémoire sur divers problèmes de probabilité, *Mémoires de l'Académie Impériale de Turin*, **20**, 355–408.

Pólya, G. (1949). Remarks on computing the probability integral in one and two dimensions, *Proceedings of the 1st Berkeley Symposium in Mathematical Statistics and Probability*, pp. 63–78.

Press, S. J. (1965). Correlation in a bivariate normal distribution when the conditional variances are known, RAND Corporation Report.

Prince, J., and Tate, R. F. (1966). Accuracy of maximum likelihood estimates of correlation for a biserial model, *Psychometrika*, **31**, 85–92.

Puente, C. E. (1997). The remarkable kaleidoscopic decompositions of the bivariate Gaussian distribution, *Fractals*, **5**, 47–61.

Puente, C. E., and Klebanoff, A. D. (1994). Gaussians everywhere, *Fractals*, **2**, 65–79.

Rastogi, S. C., and Rohatgi, V. K. (1972). On unbiased estimation of the common mean of a bivariate normal distribution, *Biometrische Zeitung*, **16**, 155–166.

Ratkowsky, D. A. (1974). Maximum likelihood estimation in small incomplete samples from the bivariate normal distribution, *Applied Statistics*, **23**, 180–189.

Regier, M. H., and Hamdan, M. A. (1971). Correlation in a bivariate normal distribution with truncation in both variables, *Australian Journal of Statistics*, **13**, 77–82.

Rice, J., Reich, T., and Cloninger, C. R. (1979). An approximation to the multivariate normal integral: Its application to multifactorial qualitative traits, *Biometrics*, **35**, 451–459.

Rödel, E. (1987). R-estimation of normed bivariate density functions, *Statistics*, **18**, 573–585.

Rosenbaum, S. (1961). Moments of a truncated bivariate normal distribution, *Journal of the Royal Statistical Society, Series B*, **23**, 405–408.

Ruben, H. (1954). On the moments of order statistics in samples from normal populations, *Biometrika*, **41**, 200–227; Correction: **41**, ix.

Ruben, H. (1961). Probability content of regions under spherical normal distributions. III: The bivariate normal integral, *Annals of Mathematical Statistics*, **32**, 171–186.

Ruben, H. (1960). On the numerical evaluation of a class of multivariate normal integrals, *Proceedings of the Royal Society of Edinburgh*, **65A**, 272–281.

Sapra, S. K. (1995). Comment on "The moment-generating function of the minimum of bivariate normal random variables" by M. Cain, *The American Statistician*, **49**, 240.

Sarabia, J.-M. (1995). The centered normal conditionals distribution, *Communications in Statistics—Theory and Methods*, **24**, 2889–2900.

Schols, C. M. (1875). Over de theorie der fouten in de ruimte en in het platte vlak, *Verhandelingen van de Koninklijke Akademie van Wetenschappen*, **15** (English translation, 1958; White Sands Proving Ground, New Mexico).

Shah, S. M., and Parikh, N. T. (1964). Moments of singly and doubly truncated standard bivariate normal distribution, *Vidya (Gujarat University)*, **7**, 82–91.

Shanmugalingam, S. (1982). On the analysis of the ratio of two correlated normal variables, *The Statistician*, **31**, 251–258.

Sheppard, W. F. (1898). On the geometric treatment of the 'normal curve' of statistics with special reference to correlation and to the theory of errors, *Proceedings of the Royal Society of London*, **62**, 170–173.

Sheppard, W. F. (1899). On the application of the theory of error to cases of normal distribution and normal correlation, *Philosophical Transactions of the Royal Society of London, Series A*, **192**, 101–167.

Sheppard, W. F. (1900). On the calculation of the double integral expressing normal correlation, *Transactions of the Cambridge Philosophical Society*, **19**, 23–66.

Sibuya, M. (1960). Bivariate extreme statistics, I, *Annals of the Institute of Statistical Mathematics*, **11**, 195–210.

Singh, N. (1960). Estimation of parameters of a multivariate normal population from truncated and censored samples, *Journal of the Royal Statistical Society, Series B*, **22**, 307–311.

Smirnov, N. V., and Bol'shev, L. N. (1962). *Tables for Evaluating a Function of a Bivariate Normal Distribution*, Izdatel'stvo Akademii Nauk SSSR, Moscow.

Smith, W. B. (1968). Bivariate samples with missing values, *Technometrics*, **10**, 867–868.

Smith, W. B., and Hocking, R. R. (1968). A simple method for obtaining the information matrix for a multivariate normal distribution, *American Statistician*, **22**, 18–20.

Somerville, P. N. (1954). Some problems of optimum sampling, *Biometrika*, **41**, 420–429.

Sondhauss, U. (1994). Asymptotische Eigenschaften intermediärer Ordnungs-statistiken und ihrer Konkomitanten, *Diplomarbeit*, Department of Statistics, Dortmund University, Germany.

Song, R., Buchberger, S. G., and Deddens, J. A. (1992). Moments of variables summing to normal order statistics, *Statistics & Probability Letters*, **15**, 203–208.

Song, R., and Deddens, J. A. (1993). A note on moments of variables summing to normal order statistics, *Statistics & Probability Letters*, **17**, 337–341.

Soper, H. E. (1915). On the probable error for the bi-serial expression for the correlation coefficient, *Biometrika*, **10**, 384–390.

Sowden, R. R., and Ashford, J. R. (1969). Computation of the bivariate normal integral, *Applied Statistics*, **18**, 169–180.

Steck, G. P. (1958). A table for computing trivariate normal probabilities, *Annals of Mathematical Statistics*, **29**, 780–800.

Stoyanov, J. (1987). *Counter Examples in Probability*, New York: John Wiley & Sons.

Sungur, E. A. (1990). Dependence information in parametrized copulas, *Communications in Statistics—Simulation and Computation*, **19**, 1339–1360.

Szabłowski, P. J. (1990). Moment rotation invariant multivariate distributions, Preprint No. 13, Politechnika Warszawska, Instytut Matematyki, Pl. Jedności Robotniczej 1, Warsaw, Poland.

Szabłowski, P. J. (1992). Binormal law characterized by polynomial conditional moments, *Demonstratio Mathematica*, **25**, 385–402.

Tallis, G. M. (1963). Elliptical and radial truncation in normal populations, *Annals of Mathematical Statistics*, **34**, 940–944.

Tate, R. F. (1955). The theory of correlation between two continuous variables when one variable is dichotomized, *Biometrika*, **42**, 205–216.

Teichroew, D. (1955). Probabilities associated with order statistics in samples from two normal populations with equal variances, *Chemical Corps Engineering Agency*, Army Chemical Center, Maryland.

Terza, J. V., and Welland, U. (1991). A comparison of bivariate normal algorithms, *Journal of Statistical Computation and Simulation*, **39**, 115–127.

Tihansky, D. P. (1970). Properties of the bivariate normal cumulative distribution, RAND Corporation Report P4400, Santa Monica, California.

Tiku, M. L., and Gill, P. S. (1989). Modified maximum likelihood estimators for the bivariate normal based on Type II censored samples, *Communications in Statistics—Theory and Methods*, **18**, 3505–3518.

Tiku, M. L., Tan, W. Y., and Balakrishnan, N. (1986). *Robust Inference*, New York: Marcel Dekker.

Trawinski, I., and Bargmann, R. E. (1964). Maximum likelihood estimation with incomplete multivariate data, *Annals of Mathematical Statistics*, **35**, 647–657.

Votaw, D.F., Rafferty, J. A., and Deemer, W. L. (1950). Estimation of parameters in a truncated trivariate normal distribution, *Psychometrika*, **15**, 339–347.

Watterson, G. A. (1959). Linear estimation in censored samples from multivariate normal populations, *Annals of Mathematical Statistics*, **30**, 814–824.

Weiler, H. (1959). Mean and standard deviations of a truncated normal bivariate distribution, *Australian Journal of Statistics*, **1**, 73–81.

Welsh, G. S. (1955). A tabular method of obtaining tetrachoric r with median-cut variables, *Psychometrika*, **20**, 83–85.

Wilks, S. S. (1932). Moments and distributions of estimates of population parameters from fragmentary samples, *Annals of Mathematical Statistics*, **3**, 163–195.

Williams, J. M., and Weiler, H. (1964). Further charts for the means of truncated normal bivariate distributions, *Australian Journal of Statistics*, **6**, 117–129.

Wold, H. (1948). *Random Normal Deviates*, Tracts for Computers, **35**, Cambridge University Press: London.

Wrigge, S. (1971). On distributions with normal marginal densities which are not jointly normal, FOAP Report C 8298-MI, Stockholm, Sweden.

Young, J. C., and Minder, C. E. (1974). Algorithm AS76. An integral useful in calculating non-central t and bivariate normal probabilities, *Applied Statistics*, **23**, 455–457.

Young, S. S. Y., and Weiler, H. (1960). Selection for two correlated traits by independent culling levels, *Journal of Genetics*, **57**, 329–338.

Yule, G. U. (1897). On the theory of correlation, *Journal of the Royal Statistical Society, Series A*, **60**, 812–854.

Zelen, M., and Severo, N. C. (1960). Graphs for bivariate normal probabilities, *Annals of Mathematical Statistics*, **31**, 619–624.

Zelen, M., and Severo, N. C. (1964). Probability function, Applied Mathematics Series **55**, National Bureau of Standards, Washington, D.C. [Also *Handbook of Mathematical Functions with Formulas, Graphs and Mathematical Tables* (M. Abramowitz and I. A. Stegun, eds.) (1965), pp. 936–940, New York: Dover Publications.]

CHAPTER 47

Multivariate Exponential Distributions

1   INTRODUCTION

It is clearly evident from Chapters 45 and 46 that considerable amount of work carried out in multivariate distribution theory has been based on multivariate (or bivariate) normality. Yet, as in the case of univariate distribution theory [see, for example, Johnson, Kotz, and Balakrishnan (1994, 1995a)], bivariate and multivariate exponential distributions have served as friendly "alternative arena" for those involved in theoretical and/or applied aspects of multivariate distributions. The volume on the exponential distribution prepared by Balakrishnan and Basu (1995) provides ample testimony to this fact.

In this chapter, which is a substantial extension of Sections 3 and 4 of Chapter 41 in the first edition of this volume, we first present a detailed discussion on bivariate exponential distributions, describing many different forms that have been proposed in the literature, their properties and applications, and inferential issues. Next, we summarize various developments on multivariate exponential distributions. It should be mentioned that although this chapter includes numerous results from the voluminous literature on this topic, it can by no means be regarded as an exhaustive coverage of this active area of research. We hope that this discussion will nonetheless provide a reader with a clear picture of the present status of this topic.

2 BIVARIATE EXPONENTIAL DISTRIBUTIONS

2.1 Introduction

The term *bivariate exponential* usually refers to bivariate distributions with both marginal distributions being exponential (BEDs). It is mostly the case that these are standard exponential distributions, but location and scale parameters can be easily introduced, if needed, through appropriate linear transformations.

First, we mention a simple special case of the bivariate gamma distributions discussed in Chapter 48. The special case of $\alpha = 2$, with a reparameterization, yields the joint density function

$$p_{X_1,X_2}(x_1, x_2) \;=\; \frac{\theta_1 \theta_2}{1 - \rho} \, I_0 \left(\frac{2\sqrt{\rho \theta_1 \theta_2 x_1 x_2}}{1 - \rho} \right) \exp \left(- \frac{\theta_1 x_1 + \theta_2 x_2}{1 - \rho} \right),$$
$$x_1, x_2 > 0, \qquad (47.1)$$

where $I_0(z) = \sum_{j=0}^{\infty} \left(\frac{z}{2\,j!} \right)^{2j}$ is the well-known modified Bessel function of the first kind of order zero; see Chapter 1 of Johnson, Kotz and Kemp (1992). Note that X_1 and X_2 are mutually independent if and only if $\rho = 0$. This is the so-called *Moran–Downton bivariate exponential distribution* (discussed in Section 2.7). Nagao and Kadoya (1971) have studied this distribution in detail in a rather obscure publication.

There are now a number of different kinds of bivariate exponential distributions. However, it was only in 1960 that a pioneering paper, specifically devoted to bivariate exponential distributions, was published in a journal with a wide circulation. In that paper, Gumbel (1960) [almost immediately followed by Freund (1961)] introduced a number of bivariate exponential distributions—mainly as a warning against undue reliance on multivariate normal techniques. He stressed the fact that many properties of bivariate exponential distributions differ markedly from those of bivariate normal distributions.

We now list some systems of bivariate exponential distributions, starting with three systems adumbrated by Gumbel (1960).

2.2 Gumbel's Bivariate Exponentials

Model I: The joint cumulative distribution function is

$$F_{X_1,X_2}(x_1, x_2) \;=\; 1 - e^{-x_1} - e^{-x_2} + e^{-(x_1 + x_2 + \theta x_1 x_2)},$$
$$x_1, x_2 > 0, \ 0 \le \theta \le 1. \quad (47.2)$$

The joint survival function is

$$\bar{F}_{X_1,X_2}(x_1, x_2) = e^{-(x_1+x_2+\theta x_1 x_2)}, \qquad x_1, x_2 > 0, \qquad (47.3)$$

and the joint density function is

$$p_{X_1,X_2}(x_1, x_2) = e^{-(x_1+x_2+\theta x_1 x_2)}\{(1+\theta x_1)(1+\theta x_2) - \theta\},$$
$$x_1, x_2 > 0. \quad (47.4)$$

The marginal distributions of each of X_1 and X_2 are standard exponential. The conditional density of X_2, given $X_1 = x_1$, is

$$p_{X_2|X_1}(x_2|x_1) = e^{-(1+\theta x_1)x_2}\{(1+\theta x_1)(1+\theta x_2) - \theta\}, \qquad x_2 > 0. \quad (47.5)$$

If $\theta = 0$, X_1 and X_2 are mutually independent. The conditional sth moment of X_2 about zero is

$$E[X_2^s|X_1 = x_1] = \frac{s!(1 + s\theta + x_1\theta)}{(1 + x_1\theta)^{s+1}}. \qquad (47.6)$$

In particular, the conditional mean and variance of X_2 are

$$E[X_2|X_1 = x_1] = \frac{1 + \theta + x_1\theta}{(1 + x_1\theta)^2} \qquad (47.7)$$

and

$$\text{var}(X_2|X_1 = x_1) = \frac{(1 + \theta + x_1\theta)^2 - 2\theta^2}{(1 + x_1\theta)^4}. \qquad (47.8)$$

Also,

$$E[X_1 X_2] = \frac{1}{\theta} e^{1/\theta} Ei\left(\frac{1}{\theta}\right), \qquad (47.9)$$

where

$$Ei(z) = \int_1^\infty e^{-tz} \frac{1}{t}\, dt.$$

Hence, the correlation coefficient between X_1 and X_2 is

$$\text{corr}(X_1, X_2) = 1 - \frac{1}{\theta} e^{1/\theta} Ei\left(\frac{1}{\theta}\right). \qquad (47.10)$$

[Remember that $E[X_i] = \text{var}(X_i) = 1$ for $i = 1, 2$.] The correlation is, of course, zero for $\theta = 0$, and it decreases to -0.40365 as θ increases to 1.

Barnett (1985) has considered maximum likelihood and moment estimation of θ. Based on a random sample $(X_{1i}, X_{2i})^T$, $i = 1, 2, \ldots, n$, he has shown that the maximum likelihood estimator (MLE) of θ is a solution of the equation

$$\sum_{i=1}^{n} \frac{X_{1i} + X_{2i} - 1 + 2\theta X_{1i} X_{2i}}{1 + (X_{1i} + X_{2i} - 1)\theta + X_{1i} X_{2i}\theta^2} = \sum_{i=1}^{n} X_{1i} X_{2i}. \qquad (47.11)$$

A moment estimator of θ can be obtained as the solution of the equation

$$\frac{1}{\theta} e^{1/\theta} Ei\left(\frac{1}{\theta}\right) = 1 - \text{ (sample correlation coefficient)}.$$

Barnett (1985) has presented formulas for asymptotic variances of these two estimators.

This distribution is characterized by the properties

$$E[X_i - x_i | (X_i > x_i) \cap (X_{3-i} > x_{3-i})] = E[X_i - x_i | X_{3-i} > x_{3-i}],$$
$$i = 1, 2, \qquad (47.12)$$

which can be viewed as a form of lack of memory property, but is more commonly referred to as *bivariate mean residual* (or *remaining*) *life constancy*; see Nair and Nair (1988) and Ebrahimi and Zahedi (1989).

Castillo, Sarabia, and Hadi (1997) have discussed an estimation method for the BED with joint survival function

$$\bar{F}(x_1, x_2) = \exp\left\{-\frac{x_1}{\theta_1} - \frac{x_2}{\theta_2} - \frac{\alpha x_1 x_2}{\theta_1 \theta_2}\right\}, \quad x_1, x_2 \geq 0, \ \theta_1, \theta_2 > 0, \ 0 < \alpha < 1$$

and marginal distributions

$$\bar{F}_{X_1}(x_1) = e^{-x_1/\theta_1} \ (x_1 > 0) \text{ and } \bar{F}_{X_2}(x_2) = e^{-x_2/\theta_2} \ (x_2 > 0).$$

Writing $\bar{F}_{X_1}(x_1) = \exp(-x_1/\theta_1) = q^{x_1}$ (proportion of points in the sample where $X_1 \leq x_1$) and

$$\bar{F}(x_1, x_2) = \exp\left(-\frac{x_1}{\theta_1} - \frac{x_2}{\theta_2} - \frac{\alpha x_1 x_2}{\theta_1 \theta_2}\right) = q^{x_1 x_2}$$

(proportion of points in the sample where $X_1 \leq x_1$ and $X_2 \leq x_2$), they used the estimators

$$\hat{x}_1 = -\theta_1 \log q^{x_1}, \quad \hat{x}_2 = -\theta_2 \left(\frac{\log q^{x_1 x_2} - \log q^{x_1}}{1 - \alpha q^{x_1}}\right)$$

and their symmetric counterparts

$$\hat{x}_1 = -\theta_1 \left(\frac{\log q^{x_1 x_2} - \log q^{x_2}}{1 - \alpha q^{x_2}} \right) \quad \text{and} \quad \hat{x}_2 = -\theta_2 \log q^{x_2}.$$

Replacing $\hat{x}_1$ and $\hat{x}_2$ by $\hat{x}_{1,i}$ and $\hat{x}_{2,i}$ $(i = 1, \ldots, n)$ and taking the averages of the above two sets of estimates, we obtain

$$\hat{x}_{1,i} = \theta_1 r_i \quad \text{and} \quad \hat{x}_{2,i} = \theta_2 s_i,$$

where

$$r_i = \frac{1}{2} \left(-\log q_i^{x_1} - \frac{\log q_i^{x_1 x_2} - \log q_i^{x_2}}{1 - \alpha q_i^{x_2}} \right)$$

and

$$s_i = \frac{1}{2} \left(-\log q_i^{x_2} - \frac{\log q_i^{x_1 x_2} - \log q_i^{x_1}}{1 - \alpha q_i^{x_1}} \right).$$

Minimizing the sum of squares

$$\sum_{i=1}^{n} \left\{ (x_{1,i} - \theta_1 r_i)^2 + (x_{2,i} - \theta_2 s_i)^2 \right\}$$

with respect to θ_1, θ_2 and α, we can obtain estimates of θ_1, θ_2 and α. Castillo, Sarabia, and Hadi (1997) have claimed that this method gives quite reliable estimates. These authors have also shown that, if U and V are independent Uniform(0,1) random variables, then $(X_1, X_2)^T$ having this BED can be obtained by the transformation

$$X_1 = -\theta_1 \log U \quad \text{and} \quad \exp\left\{ -\frac{X_2}{\theta_2} - \frac{\alpha X_1 X_2}{\theta_1 \theta_2} \right\} \left(1 + \frac{\alpha X_2}{\theta_2} \right) = V.$$

Model II: This is just a special form of Farlie–Gumbel–Morgenstern's bivariate distributions (see Chapter 44) with marginal distributions which are both standard exponential. The joint cumulative distribution function is

$$\begin{aligned} F_{X_1,X_2}(x_1, x_2) &= (1 - e^{-x_1})(1 - e^{-x_2})(1 + \alpha e^{-x_1 - x_2}), \\ &\quad x_1, x_2 > 0, \ |\alpha| < 1, \end{aligned} \tag{47.13}$$

and the joint density function is

$$p_{X_1,X_2}(x_1, x_2) = e^{-x_1 - x_2} \{ 1 + \alpha(2e^{-x_1} - 1)(2e^{-x_2} - 1) \}. \tag{47.14}$$

The conditional density of X_2, given $X_1 = x_1$, is

$$p_{X_2|X_1}(x_2|x_1) = e^{-x_2}\{1 + \alpha(2e^{-x_1} - 1)(2e^{-x_2} - 1)\} \qquad (47.15)$$

and the conditional cumulative distribution function of X_2, given $X_1 = x_1$, is

$$
\begin{aligned}
F_{X_2|X_1}(x_2|x_1) &= \{1 - \alpha(2e^{-x_1} - 1)\}(1 - e^{-x_2}) \\
&\quad + \alpha(2e^{-x_1} - 1)(1 - e^{-2x_2}).
\end{aligned} \qquad (47.16)
$$

Provided that

$$0 < 1 - \alpha(2e^{-x_1} - 1) < 1, \qquad \text{i.e., } \alpha(\log 2 - x_1) > 0,$$

this is a mixture of two exponential distributions with means 1 and $\frac{1}{2}$, respectively. The conditional sth moment of X_2 about zero is

$$E[X_2^s|X_1 = x_1] = s!\{1 - \alpha(1 - 2^{-s})(2e^{-x_1} - 1)\}. \qquad (47.17)$$

In particular, the conditional mean and variance of X_2 are

$$E[X_2|X_1 = x_1] = 1 + \frac{1}{2}\alpha - \alpha e^{-x_1} \qquad (47.18)$$

and

$$\text{var}(X_2|X_1 = x_1) = 1 + \frac{1}{2}\alpha - \frac{1}{4}\alpha^2 - \alpha(1 - \alpha)e^{-x_1} - \alpha^2 e^{-2x_1}. \quad (47.19)$$

Also,

$$E[X_1 X_2] = 1 + \frac{1}{4}\alpha, \qquad (47.20)$$

whence

$$\text{corr}(X_1, X_2) = \frac{1}{4}\alpha. \qquad (47.21)$$

Note that since $|\alpha| \le 1$, the correlation cannot exceed $\frac{1}{4}$ or be less than $-\frac{1}{4}$.

Bilodeau and Kariya (1993) observed that the densities of both Models I and II are of the form

$$p_{X_1,X_2}(x_1, x_2) = \lambda_1 \lambda_2 \, g(\lambda_1 x_1, \lambda_2 x_2; \theta)e^{-\lambda_1 x_1 - \lambda_2 x_2}. \qquad (47.22)$$

Model III: The joint cumulative distribution function is

$$F_{X_1,X_2}(x_1, x_2) = 1 - e^{-x_1} - e^{-x_2} + \exp\left\{-(x_1^m + x_2^m)^{1/m}\right\},$$
$$x_1, x_2 > 0, \ m \geq 1, \qquad (47.23)$$

and the joint density function is

$$p_{X_1,X_2}(x_1, x_2) = (x_1^m + x_2^m)^{-2+(1/m)} x_1^{m-1} x_2^{m-1} \left\{(x_1^m + x_2^m)^{1/m} + m - 1\right\}$$
$$\times \exp\left\{-(x_1^m + x_2^m)^{1/m}\right\},$$
$$x_1, x_2 > 0, \ m \geq 1. \qquad (47.24)$$

If $m = 1$, X_1 and X_2 are mutually independent. Gumbel (1960) considered this model only briefly, but Lu and Bhattacharyya (1991a,b) have discussed it in detail.

2.3 Freund's Bivariate Exponential (Bivariate Exponential Mixture Distributions)

Freund (1961) constructed a model representing the following situation. An instrument has two components C_1 and C_2, with lifetimes having independent density functions (when both are in operation)

$$p_{X_i}(x_i) = \alpha_i \, e^{-\alpha_i x_i}, \qquad x_i > 0, \ \alpha_i > 0 \ (i = 1, 2)$$

—symbolically, $X_i \overset{d}{=} \exp(\alpha_i)$—but X_1 and X_2 are dependent because a failure of either component changes the parameter of the life distribution of the other component. When C_i (with lifetime X_i) fails $(i = 1, 2)$, the parameter for X_{3-i} changes from α_{3-i} to α'_{3-i}. There is no other dependence.

The time to first failure is, of course, distributed as $\exp(\alpha_1 + \alpha_2)$. The probability that component C_i is the first to fail is $\alpha_i/(\alpha_1 + \alpha_2)$, $i = 1, 2$, whenever the first failure occurs. The distribution of the time from first failure to failure of the other component is thus a mixture of $\exp(\alpha'_1)$ and $\exp(\alpha'_2)$ in proportions $\frac{\alpha_2}{\alpha_1+\alpha_2}$ and $\frac{\alpha_1}{\alpha_1+\alpha_2}$, respectively. The joint density function of X_1 and X_2 is

$$p_{X_1,X_2}(x_1, x_2) = \begin{cases} \alpha_1 \alpha'_2 \, e^{-\alpha'_2 x_2 - \gamma_2 x_1} & \text{for } 0 \leq x_1 < x_2 \\ \alpha'_1 \alpha_2 \, e^{-\alpha'_1 x_1 - \gamma_1 x_2} & \text{for } 0 \leq x_2 < x_1, \end{cases} \qquad (47.25)$$

where $\gamma_i = \alpha_1 + \alpha_2 - \alpha'_i$ $(i = 1, 2)$. The joint survival function is

$$\bar{F}_{X_1,X_2}(x_1, x_2) = \Pr[X_1 > x_1, X_2 > x_2]$$

$$= \begin{cases} \frac{1}{\gamma_2} \left\{ \alpha_1 \, e^{-\gamma_2 x_1 - \alpha_2' x_2} + (\alpha_2 - \alpha_2') e^{-(\alpha_1 + \alpha_2) x_2} \right\} \\ \qquad\qquad\qquad\qquad \text{for } 0 \le x_1 < x_2, \\ \frac{1}{\gamma_1} \left\{ \alpha_2 \, e^{-\gamma_1 x_2 - \alpha_1' x_1} + (\alpha_1 - \alpha_1') e^{-(\alpha_1 + \alpha_2) x_1} \right\} \\ \qquad\qquad\qquad\qquad \text{for } 0 \le x_2 < x_1. \end{cases} \tag{47.26}$$

Provided $\alpha_1 + \alpha_2 \ne \alpha_i'$ $(i = 1, 2)$, the marginal density function of X_i $(i = 1, 2)$ is

$$p_{X_i}(x_i) = \frac{1}{\alpha_1 + \alpha_2 - \alpha_i'} \left\{ (\alpha_i - \alpha_i')(\alpha_1 + \alpha_2) e^{-(\alpha_1 + \alpha_2) x_i} + \alpha_i' \alpha_{3-i} e^{-\alpha_i' x_i} \right\},$$

$$x_i \ge 0. \tag{47.27}$$

These are not exponential but rather mixtures of exponential distributions if $\alpha_i > \alpha_i'$; otherwise, they are weighted averages. For this reason (as indicated in the title of this section), this system of distributions should be termed *bivariate mixture exponential distributions* rather than simply BEDs. The term *bivariate exponential extension distributions* (BEEs) is often used.

It was noted by Leurgans, Tsai, and Crowley (1982) that these distributions form an exponential family with natural parameters $\alpha_1 + \alpha_2$, α_1', α_2' and $\log \left(\frac{\alpha_1 + \alpha_2 - \alpha_1'}{\alpha_1 + \alpha_2 - \alpha_2'} \right)$.

The random variables X_1 and X_2 are independent if and only if $\alpha_i = \alpha_i'$ $(i = 1, 2)$. If $\alpha_2 > \alpha_2'$ (or $\alpha_1 > \alpha_1'$), the expected life of component C_2 (C_1) improves when component C_1 (C_2) fails. The joint-moment generating function of X_1 and X_2 is

$$E \left[e^{t_1 X_1 + t_2 X_2} \right]$$

$$= \frac{1}{\alpha_1 + \alpha_2 - t_1 - t_2} \left\{ \alpha_2 \left(1 - \frac{t_1}{\alpha_1'} \right)^{-1} + \alpha_1 \left(1 - \frac{t_2}{\alpha_2'} \right)^{-1} \right\},$$

$$\tag{47.28}$$

whence

$$E[X_i] = \frac{\alpha_i' + \alpha_{3-i}}{\alpha_i'(\alpha_1 + \alpha_2)}, \qquad i = 1, 2, \tag{47.29}$$

$$\mathrm{var}(X_i) = \frac{\alpha_i'^2 + 2\alpha_1 \alpha_2 + \alpha_{3-i}^2}{\{\alpha_i'(\alpha_1 + \alpha_2)\}^2}, \qquad i = 1, 2, \tag{47.30}$$

and

$$\text{corr}(X_1, X_2) = \frac{\alpha_1' \alpha_2' - \alpha_1 \alpha_2}{\sqrt{(\alpha_1'^2 + 2\alpha_1 \alpha_2 + \alpha_2^2)(\alpha_2'^2 + 2\alpha_1 \alpha_2 + \alpha_1^2)}} . \qquad (47.31)$$

Note that $-\frac{1}{3} < \text{corr}(X_1, X_2) < 1$. In many applications, $\alpha_i' > \alpha_i$ ($i = 1, 2$); in such cases, the correlation is positive. (The future lifetime tends to be shorter when the other component is out of action, in these cases.) The conditional density of X_2, given X_1, is

$$p_{X_2|X_1}(x_2|x_1) = \begin{cases} \alpha_1 \alpha_2'(\alpha_1 + \alpha_2 - \alpha_1')\frac{1}{h(x_1)} \, e^{-(\alpha_1+\alpha_2-\alpha_2')x_1 - \alpha_2' x_2} \\ \qquad\qquad \text{for } 0 \leq x_1 < x_2, \\ \alpha_1' \alpha_2(\alpha_1 + \alpha_2 - \alpha_1')\frac{1}{h(x_1)} \, e^{-\alpha_1' x_1 - (\alpha_1+\alpha_2-\alpha_1')x_2} \\ \qquad\qquad \text{for } 0 \leq x_2 < x_1, \end{cases}$$
$$(47.32)$$

where $h(x_1) = (\alpha_1 - \alpha_1')(\alpha_1 + \alpha_2)e^{-(\alpha_1+\alpha_2)x_1} + \alpha_1' \alpha_2 \, e^{-\alpha_1' x_1}$. The regression of X_2 on X_1 is

$$E[X_2|X_1 = x_1] = \frac{1}{(\alpha_1 + \alpha_2 - \alpha_1')h(x_1)} \left[\frac{\alpha_1}{\alpha_2'} (\alpha_1 + \alpha_2 - \alpha_1')(1 + \alpha_2' x_1) \right.$$
$$- \frac{\alpha_1' \alpha_2}{\alpha_1 + \alpha_2 - \alpha_1'} \{1 + (\alpha_1 + \alpha_2 - \alpha_1')x_1\} e^{-(\alpha_1+\alpha_2)x_1}$$
$$\left. + \frac{\alpha_1' \alpha_2}{\alpha_1 + \alpha_2 - \alpha_1'} \, e^{-\alpha_1' x_1} \right]. \qquad (47.33)$$

In the special case when $\alpha_1 + \alpha_2 = \alpha_1' = \alpha_2'$, the joint density is

$$p_{X_1,X_2}(x_1, x_2) = \begin{cases} \alpha_1(\alpha_1 + \alpha_2)e^{-(\alpha_1+\alpha_2)x_2} & \text{for } 0 \leq x_1 < x_2, \\ \alpha_2(\alpha_1 + \alpha_2)e^{-(\alpha_1+\alpha_2)x_1} & \text{for } 0 \leq x_2 < x_1. \end{cases} \qquad (47.34)$$

The marginal density function of X_i ($i = 1, 2$) is

$$p_{X_i}(x_i) = \{\alpha_i + (\alpha_1 + \alpha_2)\alpha_{3-i}x_i\} e^{-(\alpha_1+\alpha_2)x_i}, \qquad x_i \geq 0. \qquad (47.35)$$

This is a mixture of $\exp(\alpha_1 + \alpha_2)$ and $\star^2\text{Exp}(\alpha_1 + \alpha_2)$ in proportions $\frac{\alpha_i}{\alpha_1+\alpha_2}$ and $\frac{\alpha_{3-i}}{\alpha_1+\alpha_2}$, respectively. This fact can also be noted directly as follows. As mentioned earlier, the first failure time is distributed as $\exp(\alpha_1 + \alpha_2)$ and the probability that it is from C_i is $\frac{\alpha_i}{\alpha_1+\alpha_2}$ ($i = 1, 2$). When the component C_i is first to fail, X_i is distributed as $\exp(\alpha_1 + \alpha_2)$, but when C_i is second to fail (with probability $\frac{\alpha_{3-i}}{\alpha_1+\alpha_2}$), X_i is distributed as $\star^2\exp(\alpha_1 + \alpha_2)$ (since

$\alpha_i' = \alpha_1 + \alpha_2$). The conditional densities are

$$p_{X_{3-i}|X_i}(x_{3-i}|x_i)$$
$$= \begin{cases} \dfrac{\alpha_{3-i}(\alpha_1 + \alpha_2)}{\alpha_i + (\alpha_1 + \alpha_2)\alpha_{3-i}x_i} & \text{for } 0 \le x_{3-i} \le x_i, \\[3mm] \dfrac{\alpha_i(\alpha_1 + \alpha_2)}{\alpha_i + (\alpha_1 + \alpha_2)\alpha_{3-i}x_i} \, e^{-(\alpha_1+\alpha_2)(x_{3-i}-x_i)} & \text{for } 0 \le x_i \le x_{3-i}. \end{cases}$$

$$(47.36)$$

Note the remarkable result that the conditional distribution of X_{3-i}, given $X_i = x_i$, is uniform over the interval $[0, x_i]$ and then exponential over (x_i, ∞). Also, since $M = \max(X_1, X_2)$ is distributed as the sum of two mutually independent random variables—$\min(X_1, X_2)$ and $|X_1 - X_2|$— each distributed as $\exp(\alpha_1 + \alpha_2)$, we have

$$E[M] = E[\max(X_1, X_2)] = \frac{2}{\alpha_1 + \alpha_2}. \qquad (47.37)$$

With $X_{(1)}$ denoting $\min(X_1, X_2)$ and $X_{(2)}$ denoting $\max(X_{(1)}, X_{(2)})$, Nagaraja and Baggs (1996) have shown that their joint density function is

$$p_{X_{(1)},X_{(2)}}(x_1, x_2) = \alpha_1\alpha_2' e^{-\alpha_2'x_2 - \gamma_2 x_1} + \alpha_1'\alpha_2 e^{-\alpha_1'x_2 - \gamma_1 x_1}$$
$$\text{for } 0 < x_1 < x_2,$$

and have given an expression for the joint survival function of $(X_{(1)}, X_{(2)})^T$; see also Baggs (1994). While the marginal distribution of $X_{(1)}$ is Exponential($\alpha_1 + \alpha_2$), the distribution of $X_{(2)}$ takes on four forms as given in Nagaraja and Baggs (1996) which agree with those provided by Klein and Moeschberger (1986).

Papadoulos (1981) carried out a Bayesian analysis in this case, ascribing mutually independent uniform prior distributions to α_1 and α_2 over intervals $[a_1, b_1]$ and $[a_2, b_2]$, respectively. The Bayesian estimator of α_i is then

$$\tilde{\alpha}_i = \frac{\Gamma(R_i + 2, b_iZ_i) - \Gamma(R_i + 2, a_iZ_i)}{Z_i\{\Gamma(R_i + 1, b_iZ_i) - \Gamma(R_i + 1, a_iZ_i)\}} \qquad (i = 1, 2), \qquad (47.38)$$

where $Z_i = \sum_{j=1}^{n} X_{ij} + \Sigma^* X_{3-i,j}$, R_i is the number of items for which C_i failed first ($R_1 + R_2 = n$), Σ^* denotes summation over those R_i items, and $\Gamma(u, v) = \int_0^v e^{-t}t^{u-1}\, du$ is the incomplete gamma function.

Another special case of interest is when $\alpha_1'(\alpha_2') \to \infty$ so that if the component $C_1(C_2)$ has not failed before $C_2(C_1)$, then it must fail simultaneously with $C_2(C_1)$, but $C_2(C_1)$ need not fail at the same time as $C_1(C_2)$ if $C_1(C_2)$ fails first [since $\alpha_2'(\alpha_1') < \infty$].

Returning now to the general case, we turn to the estimation of the parameters $\alpha_1, \alpha_2, \alpha_1'$, and α_2', using lifetimes X_{ij} $(j = 1, \ldots, n;\ i = 1, 2)$ for component C_i of the jth item in a random sample of size n. The likelihood function in this case is

$$(\alpha_1 \alpha_2')^{R_1} (\alpha_1' \alpha_2)^{R_2} \exp\left\{ -\alpha_1' \sum_{j=1}^{n} X_{1j} - \alpha_2' \sum_{j=1}^{n} X_{2j} \right.$$
$$\left. -(\alpha_1 + \alpha_2 - \alpha_1')\Sigma^* X_{1j} - (\alpha_1 + \alpha_2 - \alpha_2')\Sigma^* X_{2j} \right\}, \quad (47.39)$$

where R_1, R_2, and Σ^* are as defined in (47.38). The maximum likelihood estimators are

$$\hat{\alpha}_i = \frac{R_i}{\Sigma^* X_{1j} + \Sigma^* X_{2j}} \quad \text{and} \quad \hat{\alpha}_i' = R_{3-i} \left(\sum_{j=1}^{n} X_{ij} - \Sigma^* X_{3-i,j} \right), \quad i = 1, 2.$$
$$(47.40)$$

Note that $\Sigma^* X_{1j} + \Sigma^* X_{2j}$ is the sum of all times to first failure. Furthermore,

$$E[\hat{\alpha}_i] = \frac{n}{n-1} \alpha_i, \ \operatorname{var}(\hat{\alpha}_i) = \frac{n}{(n-1)^2(n-2)} \{n\alpha_i + (n-1)\alpha_{3-i}\},$$

and

$$E\left[\frac{1}{\hat{\alpha}_i'}\right] = \frac{1}{\alpha_i'}.$$

Hanagal and Kale (1991a,b,c) showed that the asymptotic (as $n \to \infty$) joint distribution of $\sqrt{n}(\hat{\alpha}_1 - \alpha_1, \hat{\alpha}_2 - \alpha_2, \hat{\alpha}_1' - \alpha_1', \hat{\alpha}_2' - \alpha_2')^T$ is multivariate normal with mean vector $\mathbf{0}$ and diagonal variance–covariance matrix with diagonal elements

$$\alpha_1(\alpha_1 + \alpha_2), \quad \alpha_2(\alpha_1 + \alpha_2), \quad \alpha_1'^2 \frac{(\alpha_1 + \alpha_2)}{\alpha_2}, \quad \alpha_2'^2 \frac{(\alpha_1 + \alpha_2)}{\alpha_1},$$

respectively. In the symmetric case when $\alpha_1 = \alpha_2 = \alpha$ and $\alpha_1' = \alpha_2' = \alpha'$, the MLEs of α and α' are

$$\hat{\alpha} = \frac{n}{2 \sum_{j=1}^{n} \min(X_{1j}, X_{2j})} \quad \text{and} \quad \hat{\alpha}' = \frac{n}{\sum_{j=1}^{n} |X_{1j} - X_{2j}|}.$$

The asymptotic (as $n \to \infty$) joint distribution of $(\hat{\alpha}, \hat{\alpha}')^T$ is bivariate normal with mean vector $(0, 0)^T$ and variance–covariance matrix $\text{Diag}(\alpha^2/n, \alpha'^2/n)$; also see SenGupta (1991). The statistics $\sum_{j=1}^{n} \min(X_{1j}, X_{2j})$ and $\sum_{j=1}^{n} |X_{1j} - X_{2j}|$ jointly form a complete sufficient statistic in this case. Weier (1981) studied this case using the formulation $\alpha' = \theta\alpha$ and estimating α. The MLE of θ is, of course, $\hat{\theta} = \hat{\alpha}'/\hat{\alpha}$.

Estimation based on right censored data has been discussed by Leurgans, Tsai, and Crowley (1982). If S_i is the censoring time for X_i $(i = 1, 2)$, then the observed data are

$$T_{ij} = \min(X_{ij}, S_i), \ \delta_{ij} = I(X_{ij} \leq S_i), \ i = 1, 2; \ j = 1, \dots, n.$$

If $S_1 = S_2$ ("univariate" censoring), the maximum likelihood estimators of $\alpha_1, \alpha_2, \alpha_1'$ and α_2' are

$$
\hat{\alpha}_i = \frac{1}{\sum_{j=1}^{n} \min(T_{1j}, T_{2j})} \sum_{j=1}^{n} [\delta_{ij} \{1 - \delta_{3-i,j} I(T_{ij} > T_{3-i,j})\}],
$$

$$
\hat{\alpha}_i' = \frac{1}{\sum_{j=1}^{n} (T_{ij} - T_{3-i,j}) I(T_{ij} > T_{3-i,j})} \sum_{j=1}^{n} \delta_{1j} \delta_{2j} I(T_{ij} > T_{3-i,j}),
$$

$$i = 1, 2. \qquad (47.41)$$

If $S_1 \neq S_2$ ("bivariate" censoring), the maximum likelihood estimators of $\alpha_1, \alpha_2, \alpha_1'$ and α_2' have to be determined by numerical methods. Details are presented by Leurgans, Tsai, and Crowley (1982).

O'Neill (1985) constructed a likelihood ratio test of symmetry (H_0 : $\alpha_i = \alpha_i'$, $i = 1, 2$) based on complete data from a random sample of size n. This uses the test statistic

$$
L = \frac{n^{2n} U_i^R U_{3-i}^{n-R}}{2^n R^{2R} (n - R)^{2(n-R)} (U_1 + U_2)^n},
$$

where $U_i = \sum_{j=1}^{n} \{X_{ij} - \min(X_{1j}, X_{2j})\}$ $(i = 1, 2)$, and R_i is the number of items for which $X_{ij} > X_{3-i,j}$. If n is large and H_0 is valid, L has approximately a χ_2^2 distribution. O'Neill has found the small-sample distribution of L and has provided exact critical values for $n = 2(1)20(5)40$.

Hanagal and Kale (1991b) constructed a test based on Wald's statistic using the criterion

$$
W = \frac{n(\hat{\alpha}_1 - \hat{\alpha}_2)}{(\hat{\alpha}_1 + \hat{\alpha}_2)^2} + \frac{n(\hat{\alpha}_1' - \hat{\alpha}_2')^2}{(\hat{\alpha}_1 + \hat{\alpha}_2) \left(\frac{\hat{\alpha}_1'^2}{\hat{\alpha}_2} + \frac{\hat{\alpha}_2'^2}{\hat{\alpha}_1} \right)},
$$

where $\hat{\alpha}_i$ and $\hat{\alpha}'_i$ are as in (47.40). If n is large and H_0 is valid, W also has approximately a χ_2^2 distribution.

There has been a considerable number of other modifications and/or specializations of the Freund (1961) model. We now describe, fairly briefly, a few of them. We do not include here variations which are simply reparameterizations of the original model. Interested readers may consult Weier (1981) and Hashino (1985) for additional details on this subject.

Tosch and Holmes (1980) proposed a bivariate failure model, as a generalization of Freund's BED, in which the mean residual lifetime of one component is dependent on the working status of the other. Suppose the lifetimes of the components of a system are denoted by X_1 and X_2. Let W_1, W_2, U_1, and U_2 be nonnegative mutually independent random variables with W_1 and W_2 being absolutely continuous. Furthermore, let

$$X_1 = \min(W_1, W_2) + U_1 I_{[W_1 > W_2]} \quad \text{and} \quad X_2 = \min(W_1, W_2) + U_2 I_{[W_1 \le W_2]}.$$

Then, the joint survival function of X_1 and X_2 is

$$
\bar{F}_{X_1, X_2}(x_1, x_2)
$$
$$
= \begin{cases}
\bar{F}_{W_1}(x_2)\bar{F}_{W_2}(x_2) + \int_{x_1}^{x_2} \bar{F}_{U_2}(x_2 - t)\bar{F}_{W_2}(t)\, dF_{W_1}(t) & \text{if } x_1 < x_2 \\
\bar{F}_{W_1}(x_1)\bar{F}_{W_2}(x_1) & \text{if } x_1 = x_2 \\
\bar{F}_{W_1}(x_1)\bar{F}_{W_2}(x_1) + \int_{x_2}^{x_1} \bar{F}_{U_1}(x_1 - t)\bar{F}_{W_1}(t)\, dF_{W_2}(t) & \text{if } x_1 > x_2.
\end{cases}
$$

In the special case when W_1 is $\text{Exp}(\alpha)$, W_2 is $\text{Exp}(\beta)$, $P(U_1 > t) = qe^{-\alpha' t}$, $P(U_2 > t) = qe^{-\beta' t}$, where $\alpha, \beta, \alpha', \beta' > 0$ and $0 \le q \le 1$, $t > 0$, the above distribution yields a generalization of Freund's BED which corresponds to the case $q = 1$. Note that $P(U_1 = 0)$ and $P(U_2 = 0)$ are in general positive. Tosch and Holmes (1980) have discussed estimation of parameters in the exponential case of this model.

Heinrich and Jensen (1995) developed a more general approach to the problem of constructing bivariate models of this kind, without immediate specialization to marginal exponential distributions. Let Z_1 and Z_2 be mutually independent positive random variables with cumulative distribution functions $F_{Z_1}(z_1)$ and $F_{Z_2}(z_2)$ and density functions $p_{Z_1}(z_1)$ and $p_{Z_2}(z_2)$, respectively. Define

$$X_i = \min(Z_1, Z_2) + Y_i\, I(Z_{3-i} < Z_i), \qquad i = 1, 2, \qquad (47.42)$$

where Y_1 and Y_2 are independent of Z_1 and Z_2 and of each other. The joint survival function of X_1 and X_2 is

$$\bar{F}_{X_1,X_2}(x_1, x_2)$$
$$= \begin{cases} \bar{F}_{Z_1}(x_2)\bar{F}_{Z_2}(x_2) + \int_{x_1}^{x_2} \Pr[Y_2 > x_2 - u]\bar{F}_{Z_2}(u)p_{Z_1}(u)\, du \\ \hspace{6cm} \text{for } x_1 < x_2, \\ \bar{F}_{Z_1}(x)\bar{F}_{Z_2}(x) \hspace{3.5cm} \text{for } x_1 = x_2 = x, \\ \\ \bar{F}_{Z_1}(x_1)\bar{F}_{Z_2}(x_1) + \int_{x_2}^{x_1} \Pr[Y_1 > x_1 - u]\bar{F}_{Z_1}(u)p_{Z_2}(u)\, du \\ \hspace{6cm} \text{for } x_1 > x_2. \end{cases}$$

$$(47.43)$$

If $\bar{F}_{Z_1}$ and $\bar{F}_{Z_2}$ are chosen to be Weibull distributions, (47.43) becomes a bivariate Weibull distribution which is different from Lu's (1989) bivariate Weibull family; for more details on bivariate Weibull distributions, see Section 4.

Block and Basu (1974) constructed a system of BEDs by modifying Marshall and Olkin's BEDs (see Section 2.4); but Block and Basu's system is, in fact, just a reparameterization of a special case of Freund's BEDs, with

$$\alpha_i = \frac{\lambda_i \lambda}{\lambda_1 + \lambda_2} \quad \text{and} \quad \alpha_i' = \lambda_i + \lambda_{12} \quad (i = 1, 2),$$

or conversely

$$\lambda_i = \alpha_1 + \alpha_2 - \alpha_{3-i}' \quad (i = 1, 2) \quad \text{and} \quad \lambda_{12} = \alpha_1' + \alpha_2' - \alpha_1 - \alpha_2,$$

where $\lambda = \lambda_1 + \lambda_2 + \lambda_{12}$ [see Eq. (47.45) below]. Block and Basu (1974) termed these ACBVE distributions to emphasize that they are absolutely continuous. Since $\alpha_1' + \alpha_2' - \alpha_1 - \alpha_2 > 0$, $\alpha_1' + \alpha_2'$ must be greater than $\alpha_1 + \alpha_2$.

Ebrahimi (1987) and Achcar (1995) have discussed accelerated life tests based on bivariate exponential distributions. For Block and Basu's ACBVE, Achcar and Santander (1993) and Achcar and Leandro (1998) have discussed Bayesian inferential methods, with the latter proposing Metropolis algorithms and Gibbs sampling.

2.4 Marshall and Olkin's Bivariate Exponential

A system of BEDs developed by Marshall and Olkin (1967a,b), denoted by MOBEDs, has become a widely used BED system. Numerous papers have been written on its properties, extensions, and applications during the subsequent thirty years. We first describe its background and definition.

The univariate exponential distribution has derived considerable importance from its role as the distribution of the waiting time in a Poisson process; see, for example, Balakrishnan and Basu (1995) and Chapter 19 of Johnson, Kotz, and Balakrishnan (1994). It is, therefore, natural to inquire whether a similar relationship exists between some BEDs and the waiting times in a suitably defined two-dimensional Poisson process.

We first assume a two-component system, subjected to "shocks" that are always "fatal." These shocks are assumed to be governed by independent Poisson processes with parameters λ_1, λ_2, and λ_{12}, according as the shock applies to component 1 only, component 2 only, or both components, respectively. Then, the joint survival function of the lifetimes X_1 and X_2 of the two components is

$$
\begin{aligned}
\bar{F}_{X_1,X_2}(x_1, x_2) &= \Pr[X_1 > x_1, X_2 > x_2] \\
&= e^{-\lambda_1 x_1 - \lambda_2 x_2 - \lambda_{12} \max(x_1, x_2)}, \quad x_1, x_2 > 0 \quad (47.44)
\end{aligned}
$$

—that is,

$$
\bar{F}_{X_1,X_2}(x_1, x_2) = \begin{cases} \exp\{-\lambda_1 x_1 - (\lambda_2 + \lambda_{12})x_2\}, & 0 \leq x_1 \leq x_2, \\ \exp\{-(\lambda_1 + \lambda_{12})x_1 - \lambda_2 x_2\}, & 0 \leq x_2 \leq x_1. \end{cases}
$$

This means that times between shocks are independently exponentially distributed with means $\frac{1}{\lambda_1}, \frac{1}{\lambda_2}$ and $\frac{1}{\lambda_{12}}$, respectively. (A similar distribution is obtained when the shocks are "fatal" not always, but with fixed probabilities, though the values of λ's are changed.) A formal construction for the system is through $X_i = \min(Z_i, Z_{12})$, $i = 1, 2$, where Z_1, Z_2 and Z_{12} are mutually independent exponential random variables with parameters λ_1, λ_2 and λ_{12}, respectively. The marginal distribution of X_i is $\exp(\lambda_i + \lambda_{12})$, $i = 1, 2$. The joint moment-generating function of $(X_1, X_2)^T$ is

$$
E\left[e^{t_1 X_1 + t_2 X_2}\right] = \frac{(\lambda + t_1 + t_2)(\lambda_1 + \lambda_{12})(\lambda_2 + \lambda_{12}) + \lambda_{12} t_1 t_2}{(\lambda_1 + \lambda_{12} - t_1)(\lambda_2 + \lambda_{12} - t_2)}, \quad (47.45)
$$

where $\lambda = \lambda_1 + \lambda_2 + \lambda_{12}$. The correlation coefficient between X_1 and X_2 is

$$
\text{corr}(X_1, X_2) = \frac{\lambda_{12}}{\lambda}.
$$

The non-negativity of $\text{corr}(X_1, X_2)$ also follows from the fact that the variables X_1 and X_2 are "associated" in the sense of Esary, Proschan and Walkup (1972). The distribution is singular on the line $x_1 = x_2$ since

$$
\begin{aligned}
\Pr[X_1 = X_2] &= \Pr[\text{first "fatal shock" affects both components}] \\
&= \frac{\lambda_{12}}{\lambda} = \frac{\lambda_{12}}{\lambda_1 + \lambda_2 + \lambda_{12}} > 0. \quad (47.46)
\end{aligned}
$$

Note that $\Pr[X_1 = X_2] = \text{corr}(X_1, X_2) > 0$. The joint survival function can be written as a mixture of an absolutely continuous survival function

$$\bar{F}_a(x_1, x_2) = \frac{\lambda}{\lambda_1 + \lambda_2} e^{-\lambda_1 x_1 - \lambda_2 x_2 - \lambda_{12} \max(x_1, x_2)} - \frac{\lambda_{12}}{\lambda_1 + \lambda_2} e^{-\lambda \max(x_1, x_2)},$$

$$(47.47)$$

and a singular survival function

$$\bar{F}_s(x_1, x_2) = e^{-\lambda \max(x_1, x_2)} \qquad (47.48)$$

in the form

$$\bar{F}_{X_1, X_2}(x_1, x_2) = \frac{1}{\lambda} \left\{ (\lambda_1 + \lambda_2) \bar{F}_a(x_1, x_2) + \lambda_{12} \bar{F}_s(x_1, x_2) \right\}. \qquad (47.49)$$

Note that $\min(X_1, X_2)$ has an $\exp(\lambda)$ distribution.

The joint density function of $(X_1, X_2)^T$ is

$$p_{X_1, X_2}(x_1, x_2) = \begin{cases} \lambda_2(\lambda_1 + \lambda_{12}) e^{-\lambda_1 x_1 - \lambda_2 x_2 - \lambda_{12} x_1} & \text{for } 0 < x_2 < x_1, \\ \lambda_1(\lambda_2 + \lambda_{12}) e^{-\lambda_1 x_1 - \lambda_2 x_2 - \lambda_{12} x_2} & \text{for } 0 < x_1 < x_2 \end{cases}$$

$$(47.50)$$

or, equivalently,

$$p_{X_1, X_2}(x_1, x_2) = \begin{cases} \lambda_2(\lambda_1 + \lambda_{12}) \bar{F}_{X_1, X_2}(x_1, x_2) & \text{for } 0 < x_2 < x_1, \\ \lambda_1(\lambda_2 + \lambda_{12}) \bar{F}_{X_1, X_2}(x_1, x_2) & \text{for } 0 < x_1 < x_2, \\ \lambda_{12} \bar{F}_{X_1, X_2}(x, x) & \text{for } x_1 = x_2 = x > 0. \end{cases}$$

$$(47.51)$$

Although the joint distribution is singular, the marginal distributions are continuous, as noted above.

The conditional density function of X_2, given X_1, is

$$p_{X_2 | X_1}(x_2 | x_1) = \begin{cases} \frac{\lambda_1(\lambda_2 + \lambda_{12})}{\lambda_1 + \lambda_{12}} e^{-\lambda_2 x_2 - \lambda_{12}(x_2 - x_1)} & \text{for } x_2 > x_1, \\ \lambda_2 e^{-\lambda_2 x_2} & \text{for } x_2 < x_1. \end{cases} \qquad (47.52)$$

The distribution is not infinitely divisible except in the degenerate case when $\lambda_1 = 0$ (or $\lambda_2 = 0$) or when $\lambda_{12} = 0$ (in the latter case, the distribution has independent exponential marginals); see Block, Paulson and Kohberger (1975).

For the MOBED, by denoting $\theta_i = 1/\lambda_i$ ($i = 1, 2$), Boland (1998) has shown that $c_1 X_1 + c_2 X_2$ is *stochastically arrangement increasing* in $c = (c_1, c_2)^T$ and $\theta = (\theta_1, \theta_2)^T$. In this context, a real-valued function g of two vector arguments $\boldsymbol{u}$ and $\boldsymbol{v}$ is said to be "arrangement increasing" if it increases in value as the components of $\boldsymbol{u}$ and $\boldsymbol{v}$ become more similarly arranged.

Marshall and Olkin's BEDs possess *bivariate lack of memory* (BLOM) property

$$\bar{F}_{X_1,X_2}(x_1 - t, x_2 - t) = \bar{F}_{X_1,X_2}(x_1, x_2)\bar{F}_{X_1,X_2}(t, t),$$
$$\min(x_1, x_2) > t > 0. \qquad (47.53)$$

As a matter of fact, these distributions are the only ones with exponential marginal distributions that satisfy the functional equation (47.53). However, Downton (1970) pointed out that this preservation of the LOM property in two (or more) dimensions may sometimes limit the applicability of the system. For example, there may be correlation arising if one component possesses, in some sense, a memory of the time to failure of the other. Finally, note that $\min(X_1, X_2)$ is distributed as exponential, a result similar to the case when X_1 and X_2 have independent exponential distributions. Furthermore, Nagaraja and Baggs (1996) have shown that the survival function of $X_{(2)} = \max(X_1, X_2)$ is

$$\bar{F}_{X_{(2)}}(x) = e^{-(\lambda_1+\lambda_{12})x} + e^{-(\lambda_2+\lambda_{12})x} - e^{-(\lambda_1+\lambda_2+\lambda_{12})x}, \quad x > 0,$$

which was given earlier by Downton (1970).

Given values of n independent pairs $(X_{1j}, X_{2j})^T$, $j = 1, \ldots, n$, of random variables, each having the distribution (47.44), the following consistent estimators of λ_1, λ_2, and λ_{12} were proposed by Arnold (1968):

$$\lambda_i^* = \frac{N_i}{n(n-1)T} \quad (i = 1, 2) \quad \text{and} \quad \lambda_{12}^* = \frac{N_{12}}{n(n-1)T}, \qquad (47.54)$$

where N_i $(i = 1, 2)$ is the number of pairs for which $X_{ij} < X_{3-i,j}$, N_{12} is the number of pairs for which $X_{1j} = X_{2j}$ and $T = \sum_{j=1}^n \min(X_{1j}, X_{2j})$. Defining, in addition,

$$T' = \sum_{j=1}^n \max(X_{1j}, X_{2j}) \quad \text{and} \quad S_i = \sum_{j=1}^n X_{ij} \quad (i = 1, 2), \qquad (47.55)$$

we note that (N_1, N_2, N_{12}) have a Multinomial$\left(n; \frac{\lambda_1}{\lambda}, \frac{\lambda_2}{\lambda}, \frac{\lambda_{12}}{\lambda}\right)$ distribution [see Chapter 35 of Johnson, Kotz, and Balakrishnan (1997)] and S_1, S_2 and T' have gamma distributions with scale parameters $\lambda_1 + \lambda_{12}$, $\lambda_2 + \lambda_{12}$ and λ, respectively, and common shape parameter n [see Chapter 17 of Johnson, Kotz, and Balakrishnan (1994)]. The log-likelihood function is

$$\log L = -\lambda_1 S_1 - \lambda_2 S_2 - \lambda_{12} T' + N_1 \log\{\lambda_1(\lambda_2 + \lambda_{12})\}$$
$$+ N_2 \log\{\lambda_2(\lambda_1 + \lambda_{12})\} + N_{12} \log \lambda_{12}; \qquad (47.56)$$

see, for example, Bemis, Bain, and Higgins (1972), Proschan and Sullo (1974, 1976), and Bhattacharyya and Johnson (1971, 1973). The MLEs satisfy the equations

$$\frac{N_i}{\hat{\lambda}_i} + \frac{N_{3-i}}{\hat{\lambda}_i + \hat{\lambda}_{12}} = S_i \; (i = 1, 2) \quad \text{and} \quad \frac{N_1}{\hat{\lambda}_2 + \hat{\lambda}_{12}} + \frac{N_2}{\hat{\lambda}_1 + \hat{\lambda}_{12}} + \frac{N_{12}}{\hat{\lambda}_{12}} = T'.$$

$$(47.57)$$

If all N's are positive, the MLEs are the unique solutions of (47.57). If $N_{12} = 0$, then $(\hat{\lambda}_1, \hat{\lambda}_2, \hat{\lambda}_{12}) = \left(\frac{n}{S_1}, \frac{n}{S_2}, 0\right)$ provided that $N_1 N_2 > 0$, but if $N_{12} = 0$ and either $N_1 = 0$ or $N_2 = 0$, the MLE exists but there are multiple solutions of (47.57). If $N_{12} > 0$ and either $N_1 = 0$ or $N_2 = 0$, the MLE cannot be obtained as the solution of (47.57).

Proschan and Sullo (1976) proposed estimators based on the first iteration to maximization of $\log L$ in (47.56). These are

$$\hat{\lambda}'_i = \frac{nN_i}{(N_1 + N_{12})S_i} \quad (i = 1, 2) \quad \text{and}$$

$$(47.58)$$

$$\hat{\lambda}'_{12} = \frac{n}{T'}\left\{1 + \frac{N_1}{N_2 + N_{12}} + \frac{N_2}{N_1 + N_{12}}\right\}.$$

Bemis, Bain, and Higgins (1972) constructed estimators based on method of moments, namely,

$$\tilde{\lambda}_i = \frac{\frac{n}{S_i} - \frac{N_{12}}{S_{3-i}}}{1 + \frac{N_{12}}{n}} \; (i = 1, 2) \quad \text{and} \quad \tilde{\lambda}_{12} = \frac{N_{12}\left(\frac{1}{S_1} + \frac{1}{S_2}\right)}{1 + \frac{N_{12}}{n}}. \quad (47.59)$$

The asymptotic (as $n \to \infty$) joint distributions of each of these sets of estimators are trivariate normal. Proschan and Sullo (1976) provided formulas for the asymptotic relative efficiency of $(\hat{\lambda}'_1, \hat{\lambda}'_2, \hat{\lambda}'_{12})$.

Hanagal and Kale (1991a) constructed consistent moment-type estimators of the form

$$\tilde{\lambda}^*_i = \frac{n}{T} + \frac{N_1}{T - S_{3-i}} \; (i = 1, 2) \quad \text{and} \quad \tilde{\lambda}^*_{12} = \frac{N_1}{T - S_2} + \frac{N_2}{T - S_1} - \frac{n}{T'}.$$

$$(47.60)$$

Awad, Azzam, and Hamdan (1981) recommended "partial maximum likelihood estimators" of the form

$$\hat{\hat{\lambda}}_i = \frac{N_i\left(\bar{X}_1^{-1} - \bar{X}_2^{-1}\right)}{N_1 - N_2} \; (i = 1, 2) \quad \text{and} \quad \hat{\hat{\lambda}}_{12} = \frac{N_{12}\left(\bar{X}_1^{-1} - \bar{X}_2^{-1}\right)}{N_1 - N_2}.$$

$$(47.61)$$

While these estimators may be useful in some circumstances, it should be borne in mind that all three estimators will be negative if $\left(\bar{X}_1^{-1} - \bar{X}_2^{-1}\right)$ and $(N_1 - N_2)$ are of opposite sign.

Note that

$$\Pr[X_{1j} < X_{2j}] = \frac{\lambda_1}{\lambda}, \quad \Pr[X_{1j} > X_{2j}] = \frac{\lambda_2}{\lambda}, \quad \Pr[X_{1j} = X_{2j}] = \frac{\lambda_{12}}{\lambda}.$$
$$(47.62)$$

A special case of the MOBEDs is the symmetric MOBEDs, with $\lambda_1 = \lambda_2 = \theta$ (say). Bhattacharyya and Johnson (1971, 1973) showed that, in this case, the MLEs of θ and λ_{12} are uniquely determined if $N_{12} < n$ and are given by

$$\hat{\theta} = \begin{cases} \dfrac{2n}{S_1 + S_2} & \text{if } N_{12} = 0, \\[2ex] \dfrac{(n - N_{12})\hat{\lambda}_{12}}{N_{12} + \hat{\lambda}_{12}S_1} & \text{if } N_{12} > 0 \end{cases}$$

and

$$\hat{\lambda}_{12} = \begin{cases} 0 & \text{if } N_{12} = 0, \\[2ex] \dfrac{\{n^2(S_1 - S_2)^2 + 4N_{12}S_1S_2\}^{1/2} + n(S_1 - S_2)}{2S_1S_2} & \text{if } N_{12} > 0. \end{cases}$$

The MLEs of the parameters if only $\min(X_{1j}, X_{2j})$ $(j = 1, \ldots, n)$, N_1 and N_2 (and thus N_{12}) are recorded can be similarly discussed.

For the MOBED, Chen et al. (1998) have investigated the asymptotic properties of the MLEs of the parameters based on a mixed censored data. Specifically, they have assumed the available data to be $(X_{1(1)}, X_{2[1]}^*)^T, (X_{1(2)}, X_{2[2]}^*)^T, \ldots, (X_{1(r)}, X_{2[r]}^*)^T, (X_{1(r+1)}^*, X_{2[r+1]}^*)^T, \ldots,$ $(X_{1(n)}^*, X_{2[n]}^*)^T$, where $X_{1(1)} \leq X_{1(2)} \leq \cdots \leq X_{1(r)}$ are ordered lifetimes from component C_1, $X_{2[i]}$ is the concomitant order statistic corresponding to $X_{1(i)}$ from component C_2, and $X_{1(r+1)}^*, \ldots, X_{1(n)}^*$ and $X_{2[i]}^*$ $(i = 1, 2, \ldots, n)$ are all censored at time $X_{1(r)} = x_{1(r)}$.

Among many applications of MOBEDs, we note especially the fields of nuclear reactor safety, competing risks, and reliability. These fields have in common the need to study the operation of multiple causes of failure. For references on competing risks and on lengths of life, one may refer to Gail (1975), Prentice et al. (1978), Tolley, Manton, and Poss (1978), and Langberg, Proschan, and Quinzi (1981); for references on nuclear risks, one may refer to Vesely (1977) and Hagen (1980); for references in the context of reliability, one may refer to Apostolakis (1976) and Sarkar (1971).

Beg and Balasubramanian (1996) have studied the concomitants of order statistics arising from the MOBED.

Saw (1969) generalized MOBEDs by replacing $\lambda_{12} \max(x_1, x_2)$ in (47.44) by

$$\lambda_{12} \int_0^{\max(x_1, x_2)} \frac{t}{\gamma + t} \, dt$$
$$= \lambda_{12} \left[\max(x_1, x_2) - \gamma \log\{\gamma + \max(x_1, x_2)\} \right], \quad \gamma > 0. \quad (47.63)$$

The corresponding joint survival function is

$$\bar{F}_{X_1, X_2}(x_1, x_2) = \{\gamma + \max(x_1, x_2)\}^{\lambda_{12}\gamma} e^{-\lambda_1 x_1 - \lambda_2 x_2 - \lambda_{12} \max(x_1, x_2)},$$
$$x_1, x_2 > 0. \quad (47.64)$$

Hyakutake (1990) suggested incorporating location parameters ξ_1 and ξ_2 in the MOBED system. The joint survival function and the joint density function are

$$\bar{F}_{X_1, X_2}(x_1, x_2) = e^{-\lambda_1(x_1 - \xi_1) - \lambda_2(x_2 - \xi_2) - \lambda_{12} \max(x_1 - \xi_1, x_2 - \xi_2)},$$
$$x_1 > \xi_1, \ x_2 > \xi_2 \quad (47.65)$$

and

$$p_{X_1, X_2}(x_1, x_2)$$
$$= \begin{cases} \lambda_1(\lambda_2 + \lambda_{12})\bar{F}_{X_1, X_2}(x_1, x_2) & \text{for } 0 < x_1 - \xi_1 < x_2 - \xi_2, \\ \lambda_2(\lambda_1 + \lambda_{12})\bar{F}_{X_1, X_2}(x_1, x_2) & \text{for } 0 < x_2 - \xi_2 < x_1 - \xi_1, \\ \lambda_{12}\bar{F}_{X_1, X_2}(x, x) & \text{for } x_1 - \xi_1 = x_2 - \xi_2 = x. \end{cases}$$
$$(47.66)$$

Sarkar (1987) modified the MOBED model to ensure absolute continuity of the joint survival function and proposed

$$\bar{F}_{X_1, X_2}(x_1, x_2) = \left[1 - \left\{ 1 - e^{-\lambda_i x_{3-i}} \right\}^{-\lambda_{12}/(\lambda_1 + \lambda_2)} \left\{ 1 - e^{-\lambda_i x_i} \right\}^{-\lambda/(\lambda_1 + \lambda_2)} \right]$$
$$\times e^{-(\lambda_{3-i} + \lambda_{12})x_{3-i}}, \quad 0 < x_i \leq x_{3-i}. \quad (47.67)$$

It is sometimes denoted as ACBVE$_2$. X_1 and X_2 are mutually independent if $\lambda_{12} = 0$. The marginal distributions are $\exp(\lambda_1 + \lambda_{12})$ and $\exp(\lambda_2 + \lambda_{12})$, respectively, and $T = \min(X_1, X_2)$ has an $\exp(\lambda)$ distribution. The hazard function for component C_i in the presence of component C_{3-i} is

$$h_i(t) = \frac{\lambda \lambda_i}{\lambda_1 + \lambda_2}, \quad i = 1, 2$$

and the joint survival function of T and the failing component I is

$$\Pr[(T > t) \cap (I = i)] = \frac{\lambda_i}{\lambda_1 + \lambda_2} e^{-\lambda t}, \qquad i = 1, 2.$$

Knowing T and I, the parameters $\frac{\lambda_i}{\lambda_1 + \lambda_2}$ and λ are identifiable; however, the parameters λ_1, λ_2, and λ_{12} can not be identified. The formula for the density is rather cumbersome; see, for example, Hutchinson and Lai (1990). Unlike the MOBED model, this distribution does not possess the bivariate lack of memory property. Wada, Sen, and Shimakura (1996) used Sarkar's BED in a competing risk model with two causes of failure. Due to the nonidentifiability of the parameters of this distribution under competing risk, these authors suggested a reparameterization and the covariates are related to the reparameterized parameters through log-linear and logistic models.

Brockett (1984) has discussed bivariate Makeham distribution with joint survival function of the form

$$\bar{F}_M(x_1, x_2) = \bar{F}_G(x_1, x_2) \exp\{-d_1 x_1 - d_2 x_2 - d_3 \max(x_1, x_2)\},$$

where $\bar{F}_G(x_1, x_2)$ is the bivariate Gompertz survival function corresponding to the bivariate hazard function

$$a_1 \exp\{c_1 x_1 + c_2 x_2 + c_3 x_1 x_2\},$$

and the second component is the independent MOBED. Brockett (1984) has provided a rationale for this model in terms of Poisson processes.

2.5   Friday and Patil's Bivariate Exponential

These distributions are defined as mixtures of Freund's BED in (47.26) and a singular part with survival function

$$_s\bar{F}_{X_1,X_2}(x_1, x_2) = e^{-(\alpha_1 + \alpha_2) \max(x_1, x_2)}. \qquad (47.68)$$

The joint survival function of $(X_1, X_2)^T$ is thus

$$\bar{F}_{X_1,X_2}(x_1, x_2) = p\,{}_a\bar{F}_{X_1,X_2}(x_1, x_2) + (1 - p)\,{}_s\bar{F}_{X_1,X_2}(x_1, x_2),$$
$$0 < p < 1, \quad (47.69)$$

where $_a\bar{F}_{X_1,X_2}(x_1, x_2)$ is defined in (47.26); see Friday and Patil (1977). Freund's BEDs in (47.26) are obtained simply by putting $p = 1$. MOBEDs are obtained by replacing α_i and α_i' $(i = 1, 2)$ in ${}_a\bar{F}_{X_1,X_2}(x_1, x_2)$ by $\lambda_i\lambda(\lambda_1 + \lambda_2)$ and $\lambda_i + \lambda_{12}$, respectively, and setting $p = \frac{\lambda_1+\lambda_2}{\lambda}$.

Another generalization of both Freund's BEDs and MOBEDs was proposed by Proschan and Sullo (1974). This system was originally introduced "as an example for some sampling and inference results." The joint survival function is

$$\bar{F}_{X_1,X_2}(x_1, x_2) = \frac{1}{\phi_1 + \phi_2 - \phi_{3-i}'}\left[\phi_i\, e^{-(\phi_1+\phi_2-\phi_{3-i}')x_i-(\phi_0+\phi_{3-i}')x_{3-i}}\right.$$
$$\left. +(\phi_{3-i} - \phi_{3-i}')e^{-\phi x_{3-i}}\right], \quad 0 < x_i \le x_{3-i}\ (i = 1, 2),$$
$$(47.70)$$

where the parameters $\phi_0, \phi_1, \phi_2, \phi_1'$, and ϕ_2' are all positive, and $\phi = \phi_0+\phi_1+\phi_2$. Friday and Patil (1977) denoted this system by PSE (Proschan and Sullo's Extension). This distribution possesses the bivariate lack of memory property. For Proschan and Sullo's (1974) BED, Hanagal (1997) has discussed inference procedures based on hybrid randomly censored samples. Specifically, the bivariate lifetimes $(X_1, X_2)^T$ are recorded until the rth $\min(X_1, X_2) = Y$ (say) is observed—that is, $Y_{(r)}$. Then, the sampling scheme considered by Hanagal terminates the experiment at random time $T^* = \min(Y_{(r)}, T_i)$, where the censoring times T_i's are exponentially distributed with parameter γ and are independent of $(X_{1i}, X_{2i})^T$. Hanagal (1997) obtained the MLEs of the parameters and developed large-sample tests. He has also considered some other BEDs such as MOBED, Freund's BED, and Block and Basu's model.

2.6 Arnold and Strauss' Bivariate Exponential

Arnold and Strauss (1988) introduced a model based on the requirement that the conditional distributions of X_i, given X_{3-i} $(i = 1, 2)$, be exponential. The joint density is given by

$$p_{X_1,X_2}(x_1, x_2) = C(\beta_3)\beta_1\beta_2 e^{-\beta_1 x_1-\beta_2 x_2-\beta_1\beta_2\beta_3 x_1 x_2},$$
$$x_i > 0,\ \beta_i > 0\ (i = 1, 2),\ \beta_3 \ge 0, \quad (47.71)$$

where

$$C(\beta_3) = \int_0^\infty \frac{e^{-u}}{1 + \beta_3 u}\, du.$$

The statistics $\sum_{j=1}^n X_{ij}$ $(i = 1, 2)$ and $\sum_{j=1}^n X_{1j}X_{2j}$ form jointly a sufficient statistic for $(\beta_1, \beta_2, \beta_3)^T$. In this form, β_1 and β_2 are the scale (intensity) parameters while β_3 is the dependence parameter. Arnold and Strauss have given an equivalent parametrization as

$$p_{X_1, X_2}(x_1, x_2) = \exp\{m x_1 x_2 - a x_1 - b x_2 + c\},$$

where, for convergence, we must have $a, b > 0$ and $m \leq 0$, and c is a normalizing constant.

Arnold and Strauss (1988) motivated the use of this model by observing that it often happens that a researcher has better insight into the forms of conditional distributions rather than the joint distribution.

The joint survival function is

$$\bar{F}_{X_1, X_2}(x_1, x_2) = \frac{C(\beta_3) e^{-\beta_1 x_1 - \beta_2 x_2 - \beta_1 \beta_2 \beta_3 x_1 x_2}}{(1 + \beta_1 \beta_3 x_1)(1 + \beta_2 \beta_3 x_2) C\left(\frac{\beta_3}{(1 + \beta_1 \beta_3 x_1)(1 + \beta_2 \beta_3 x_2)}\right)}.$$

$$(47.72)$$

For small values of β_3 we have

$$C(\beta_3) = 1 + \beta_3 - \beta_3^2 + 3\beta_3^3 + o(\beta_3^3).$$

The moment estimators of β_1, β_2, and β_3 satisfy

$$\tilde{\beta}_i = \frac{C(\tilde{\beta}_3) - 1}{\tilde{\beta}_3 \bar{X}_i} \ (i = 1, 2) \quad \text{and} \quad \tilde{\beta}_1 \tilde{\beta}_2 = \frac{1 + \tilde{\beta}_3 - C(\tilde{\beta}_3)}{\tilde{\beta}_3 \bar{X}_1 \bar{X}_2}, \qquad (47.73)$$

where $\bar{X}_i = \frac{1}{n} \sum_{j=1}^{n} X_{ij}$ $(i = 1, 2)$, which coincide with the maximum likelihood equations. Since these equations cannot be solved in closed form, Arnold and Strauss (1988) have recommended an alternate approach by regarding the distribution as having four distinct parameters, ignoring the fact that $C(\cdot)$ is a function of β_3. This approach enables the derivation of consistent estimators presented by Arnold and Strauss (1988).

2.7 Moran and Downton's Bivariate Exponential

This system was mentioned earlier, in Section 1. It was introduced by Moran (1967a), popularized by Downton (1970), and studied extensively by Nagao and Kadoya (1971). The joint density function of $(X_1, X_2)^T$ is

$$p_{X_1, X_2}(x_1, x_2) = \frac{\theta_1 \theta_2}{1 - \rho} I_0 \left(\frac{2\sqrt{\rho \theta_1 \theta_2 x_1 x_2}}{1 - \rho}\right) \exp\left(-\frac{\theta_1 x_1 + \theta_2 x_2}{1 - \rho}\right),$$

$$x_1, x_2 > 0, \ \theta_1, \theta_2 > 0, \ 0 \leq \rho \leq 1,$$

as given in (47.1). For this density, even the first order derivatives do not exist. Independence corresponds to $\rho = 0$, and the correlation between X_1 and X_2 is ρ. The regression function of X_2 on X_1 is

$$E[X_2|X_1 = x_1] = \frac{1 - \rho(1 - \theta_1 x_1)}{\theta_2}.$$ (47.74)

These distributions can arise from "shocks" causing various types of failure to components which have geometric distributions for lifetimes. An early genesis was to set

$$X_1 = \frac{1}{2\theta_1}(U_1^2 + U_2^2) \quad \text{and} \quad X_2 = \frac{1}{2\theta_2}(U_3^2 + U_4^2),$$

where each U_i $(i = 1, 2, 3, 4)$ is a standard normal variable, and (U_1, U_3) and (U_2, U_4) each have a bivariate normal distribution with correlation coefficient ρ. Note that the joint characteristic function of X_1 and X_2 is $\{(1 - it_1)(1 - it_2) + \theta t_1 t_2\}^{-1}$.

Downton (1970) presented an alternate construction. He assumed that the two components C_1 and C_2 receive shocks occurring in independent Poisson streams at rates λ_1 and λ_2, respectively, and that the numbers N_1 and N_2 of shocks needed to cause failure of C_1 and C_2, respectively, have a joint distribution with a joint probability-generating function $\pi(t_1, t_2)$. The times to failure $(X_1, X_2)^T$ of the two components have a joint distribution with Laplace transform

$$\phi(t_1, t_2) = E\left[e^{-t_1 X_1 - t_2 X_2}\right] = \pi\left(\frac{\lambda_1}{\lambda_1 + t_1}, \frac{\lambda_2}{\lambda_2 + t_2}\right).$$ (47.75)

Downton assigned a bivariate geometric distribution to $(N_1, N_2)^T$ with joint probability-generating function [see Johnson, Kotz, and Balakrishnan (1997)]

$$\pi(t_1, t_2) = t_1 t_2 \{1 + \beta_1(1 - t_1) + \beta_2(1 - t_2) + \beta_3(1 - t_1 t_2)\}$$ (47.76)

leading to

$$\phi(t_1, t_2) = \frac{\theta_1 \theta_2}{(\theta_1 + t_1)(\theta_2 + t_2) - \rho t_1 t_2},$$ (47.77)

where

$$\theta_i = \frac{\lambda_i}{1 + \beta_i + \beta_3} \quad (i = 1, 2) \quad \text{and} \quad \rho = \frac{\beta_1 \beta_2 + \beta_1 \beta_3 + \beta_2 \beta_3 + \beta_2 \beta_3^2}{(1 + \beta_1 + \beta_3)(1 + \beta_2 + \beta_3)}.$$

Although five parameters $(\beta_1, \beta_2, \beta_3, \lambda_1, \lambda_2)$ have been used in this construction, these distributions depend only on the parameters $(\theta_1, \theta_2, \rho)$.

Nagao and Kadoya (1971) suggested

$$\tilde{\rho} = \frac{\sum_{i=1}^{n}(X_{1i} - \bar{X}_1)(X_{2i} - \bar{X}_2)}{n\bar{X}_1\bar{X}_2}, \tag{47.78}$$

where $\bar{X}_i = \frac{1}{n}\sum_{j=1}^{n} X_{ij}$ $(i = 1, 2)$, as a moment-based estimator of ρ. Al-Saadi and Young (1980) modified this estimator to

$$\tilde{\rho}_1 = \begin{cases} 0 & \text{if } \tilde{\rho} \leq 0 \\ \tilde{\rho} & \text{if } 0 < \tilde{\rho} < 1 \\ 1 & \text{if } \tilde{\rho} \geq 1 \end{cases} \tag{47.79}$$

and used $\tilde{\rho}_1$ as a test statistic for the hypothesis $\rho = 0$ against the alternative $\rho \neq 0$; see also Al-Saadi, Scrimshaw and Young (1979), and Al-Saaadi and Young (1982). For large n,

$$E[\tilde{\rho}] \doteq \rho\left\{1 - \frac{1}{n}(3 - \rho)\right\}. \tag{47.80}$$

An approximate bias reduction may, therefore, be effected by using

$$\tilde{\rho}_2 = \tilde{\rho}\left(1 + \frac{3}{n}\right) - \frac{1}{n}\tilde{\rho}^2, \tag{47.81}$$

provided it takes values in the interval $(0, 1)$. The reduction in bias is quite marked for $\rho \geq 0.5$.

Al-Saadi and Young (1980) also proposed an estimator based on the sample correlation coefficient r as

$$\tilde{\rho}_3 = \begin{cases} 0 & \text{if } -1 \leq r \leq 0 \\ r & \text{if } 0 < r \leq 1. \end{cases}$$

Using the approximation

$$E[r] \simeq \rho\left\{1 - \frac{1}{n}(2 - \rho - \rho^2)\right\}, \tag{47.82}$$

Al-Saadi and Young (1980) suggested an alternate estimator

$$\tilde{\rho}_4 = \begin{cases} 0 & \text{if } r\left(1 + \frac{2}{n}\right) - \frac{1}{n}(r^2 + r^3) < 0 \\ r\left(1 + \frac{2}{n}\right) - \frac{1}{n}(r^2 + r^3) & \text{if } 0 \leq r\left(1 + \frac{2}{n}\right) - \frac{1}{n}(r^2 + r^3) \leq 1 \\ 1 & \text{if } r\left(1 + \frac{2}{n}\right) - \frac{1}{n}(r^2 + r^3) > 1. \end{cases}$$

They compared the bias and mean squared error of the estimators $\tilde{\rho}_1$, $\tilde{\rho}_2$, $\tilde{\rho}_3$, and $\tilde{\rho}_4$ through a Monte Carlo simulation study using the choices $\rho = 0.1(0.1)0.9$ and $n = 10, 20$. They observed that in small samples, $\tilde{\rho}_2$ and $\tilde{\rho}_4$ have a much smaller bias than $\tilde{\rho}_1$ and $\tilde{\rho}_3$, respectively, except for very small values of ρ.

From (47.80), Balakrishnan and Ng (2000) proposed an estimator

$$\tilde{\rho}_5 = \begin{cases} 0 & \text{if } \frac{-(n-3)+\sqrt{(n-3)^2+4n\tilde{\rho}}}{2} < 0 \\ \frac{-(n-3)+\sqrt{(n-3)^2+4n\tilde{\rho}}}{2} & \text{if } 0 \leq \frac{-(n-3)+\sqrt{(n-3)^2+4n\tilde{\rho}}}{2} \leq 1 \\ 1 & \text{if } \frac{-(n-3)+\sqrt{(n-3)^2+4n\tilde{\rho}}}{2} > 1. \end{cases}$$

From (47.82), these authors proposed another estimator

$$\tilde{\rho}_6 = \begin{cases} 0 & \text{if } \tilde{\rho}^* < 0 \\ \tilde{\rho}^* & \text{if } 0 \leq \tilde{\rho}^* \leq 1 \\ 1 & \text{if } \tilde{\rho}^* > 1, \end{cases}$$

where $\tilde{\rho}^* = C + D - \frac{1}{3}$,

$$C = \left\{ -\frac{B}{2} + \sqrt{\frac{A^3}{27} + \frac{B^2}{4}} \right\}^{1/3}, \quad D = \left\{ -\frac{B}{2} - \sqrt{\frac{A^3}{27} + \frac{B^2}{4}} \right\}^{1/3},$$

$$A = n - \frac{7}{3} \quad \text{and} \quad B = \frac{20}{27} - n\left(r - \frac{1}{3}\right).$$

In addition, Balakrishnan and Ng (2000) suggested the jackknifed version of the estimators $\tilde{\rho}_5$ and $\tilde{\rho}_6$ given by

$$\tilde{\rho}_{5,J} = \begin{cases} 0 & \text{if } n\tilde{\rho}_5 - (n-1)\tilde{\rho}_{5(\cdot)} < 0 \\ n\tilde{\rho}_5 - (n-1)\tilde{\rho}_{5(\cdot)} & \text{if } 0 \leq n\tilde{\rho}_5 - (n-1)\tilde{\rho}_{5(\cdot)} \leq 1 \\ 1 & \text{if } n\tilde{\rho}_5 - (n-1)\tilde{\rho}_{5(\cdot)} > 1 \end{cases}$$

and

$$\tilde{\rho}_{6,J} = \begin{cases} 0 & \text{if } n\tilde{\rho}_6 - (n-1)\tilde{\rho}_{6(\cdot)} < 0 \\ n\tilde{\rho}_6 - (n-1)\tilde{\rho}_{6(\cdot)} & \text{if } 0 \leq n\tilde{\rho}_6 - (n-1)\tilde{\rho}_{6(\cdot)} \leq 1 \\ 1 & \text{if } n\tilde{\rho}_6 - (n-1)\tilde{\rho}_{6(\cdot)} > 1, \end{cases}$$

where

$$\tilde{\rho}_{5(\cdot)} = \frac{1}{n}\sum_{i=1}^{n} \tilde{\rho}_{5(i)} \quad \text{and} \quad \tilde{\rho}_{6(\cdot)} = \frac{1}{n}\sum_{i=1}^{n} \tilde{\rho}_{6(i)},$$

with $\tilde{\rho}_{5(i)}$ and $\tilde{\rho}_{6(i)}$ being the estimators $\tilde{\rho}_5$ and $\tilde{\rho}_6$ determined by leaving out the i-th observation $(x_{1i}, x_{2i})^T$.

Balakrishnan and Ng (2000) carried out an extensive Monte Carlo simulation study and examined the bias and mean squared error of all these estimators of ρ for $n = 10, 20, 50, 100, 200$, and $\rho = 0.1(0.1)0.9$. For example, Tables 47.1 and 47.2, taken from Balakrishnan and Ng (2000), present the bias and mean squared error values of all the estimators for sample sizes 20 and 50. From these simulational results, it is observed that the jackknife estimators $\tilde{\rho}_{5,J}$ and $\tilde{\rho}_{6,J}$ both reduce bias substantially. Though $\tilde{\rho}_{6,J}$ seems to be the best estimator in terms of bias, it has a large mean squared error. Overall, $\tilde{\rho}_6$ seems to be the best estimator as it possesses a small bias as well as a much smaller mean squared error than that of $\tilde{\rho}_{6,J}$.

Hawkes (1972) replaced the bivariate geometric distribution in (47.76) by a "natural" generalization. Let E_1 and E_2 be two events with

$$\Pr[E_1 \cap E_2] = p_{11}, \ \Pr[E_1 \cap \bar{E}_2] = p_{10}, \ \Pr[\bar{E}_1 \cap E_2] = p_{01}, \ \Pr[\bar{E}_1 \cap \bar{E}_2] = p_{00},$$

and let N_1 and N_2 be the number of observations up to the first occurrence of E_1 and E_2, respectively. The joint probability generating function of N_1 and N_2 is then

$$\pi(t_1, t_2)$$
$$= \frac{t_1 t_2 \{p_{11} - (p_{11}Q_1 - p_{01}P_1)t_1 - (p_{11}Q_2 - p_{10}P_2)t_2 - (p_{00}P_2Q_1 + p_{01}P_1Q_2 - p_{11}Q_1Q_2)\}}{(1 - Q_1 t_1)(1 - Q_2 t_2)(1 - p_{00}t_1 t_2)},$$

where $P_1 = p_{11} + p_{10} = 1 - Q_1$ and $P_2 = p_{11} + p_{01} = 1 - Q_2$. This corresponds to a BED with five parameters; this model was also derived by Paulson and Kohberger (1974) [see also Block, Paulson, and Kohberger (1975)]. The regression of X_2 on X_1 is

$$E[X_2 | X_1 = x_1] = \frac{1}{P_2}\left[1 + \frac{p_{00}p_{11} - p_{01}p_{10}}{p_{01}}\left\{1 - \frac{(1 - p_{00})}{P_1} e^{-P_1 p_{01} x_1}\right\}\right].$$
$$(47.83)$$

Compare this with the regression for the MOBED model which is also of an exponential form.

Nagao and Kadoya (1971) presented detailed tables of the conditional cumulative distribution function

$$F(\eta | \xi) = \int_0^\eta f(\eta | \xi) \, d\eta = \frac{e^{-\rho \xi/(1-\rho)}}{1 - \rho} \int_0^\eta e^{-\eta/(1-\rho)} I_0\left(\frac{2\sqrt{\rho \xi \eta}}{1 - \rho}\right) d\eta$$

by providing the values of η for which $F(\eta | \xi) = 0.001(0.001)0.01(0.01)0.20$ $(0.05)0.80(0.01)0.99(0.001)0.999$, $\xi = 0(0.25)3.0(0.5)5(1)10(2)18$, and $\rho = 0.1(0.1)0.9$.

TABLE 47.1

Simulated Bias of the Estimators $\tilde{\rho}_1, \tilde{\rho}_2, \tilde{\rho}_3, \tilde{\rho}_4, \tilde{\rho}_5, \tilde{\rho}_6, \tilde{\rho}_{5,J}$, and $\tilde{\rho}_{6,J}$

$$n = 20$$

ρ	$\tilde{\rho}_1$	$\tilde{\rho}_2$	$\tilde{\rho}_3$	$\tilde{\rho}_4$	$\tilde{\rho}_5$	$\tilde{\rho}_6$	$\tilde{\rho}_{5,J}$	$\tilde{\rho}_{6,J}$
0.1	0.03636	0.05380	0.05142	0.06250	0.05604	0.06302	0.01300	0.00865
0.2	0.00380	0.02881	0.02228	0.03752	0.03186	0.03809	0.00298	−0.00559
0.3	−0.02483	0.00792	0.00081	0.02014	0.01174	0.02066	−0.00221	−0.00891
0.4	−0.04740	−0.00755	−0.01377	0.00901	−0.00313	0.00937	0.00143	−0.00697
0.5	−0.06993	−0.02359	−0.02385	0.00150	−0.01870	0.00157	−0.00151	−0.00498
0.6	−0.08628	−0.03505	−0.02378	0.00266	−0.02995	0.00228	0.00150	0.00213
0.7	−0.10583	−0.05150	−0.02253	0.00283	−0.04640	0.00197	−0.00678	0.00321
0.8	−0.13113	−0.07516	−0.01960	0.00181	−0.07030	0.00055	−0.02594	0.00135
0.9	−0.15871	−0.10464	−0.01062	0.00276	−0.10029	0.00155	−0.05227	0.00156

$$n = 50$$

ρ	$\tilde{\rho}_1$	$\tilde{\rho}_2$	$\tilde{\rho}_3$	$\tilde{\rho}_4$	$\tilde{\rho}_5$	$\tilde{\rho}_6$	$\tilde{\rho}_{5,J}$	$\tilde{\rho}_{6,J}$
0.1	0.01867	0.02515	0.02259	0.02668	0.02549	0.02678	0.00157	0.00044
0.2	−0.00184	0.00869	0.00378	0.01017	0.00921	0.01031	−0.00363	−0.00474
0.3	−0.01361	0.00115	−0.00608	0.00250	0.00184	0.00265	−0.00135	−0.00390
0.4	−0.01753	0.00136	−0.00767	0.00269	0.00218	0.00280	0.00687	0.00138
0.5	−0.02566	−0.00341	−0.00991	0.00153	−0.00252	0.00157	0.00706	0.00100
0.6	−0.03422	−0.00941	−0.01107	0.00058	−0.00848	0.00051	0.00811	0.00028
0.7	−0.04632	−0.01992	−0.01100	−0.00012	−0.01902	−0.00029	0.00682	−0.00039
0.8	−0.06025	−0.03379	−0.00793	0.00089	−0.03296	0.00065	−0.00025	0.00004
0.9	0.08574	−0.06144	−0.00489	0.00043	−0.06074	0.00021	−0.02031	−0.00017

TABLE 47.2

Simulated Mean Squared Error of the Estimators $\tilde{\rho}_1, \tilde{\rho}_2, \tilde{\rho}_3, \tilde{\rho}_4,$ $\tilde{\rho}_5, \tilde{\rho}_6, \tilde{\rho}_{5,J}$, and $\tilde{\rho}_{6,J}$

$$n = 20$$

ρ	$\tilde{\rho}_1$	$\tilde{\rho}_2$	$\tilde{\rho}_3$	$\tilde{\rho}_4$	$\tilde{\rho}_5$	$\tilde{\rho}_6$	$\tilde{\rho}_{5,J}$	$\tilde{\rho}_{6,J}$
0.1	0.03404	0.04317	0.03566	0.04105	0.04424	0.04117	0.07374	0.06743
0.2	0.04765	0.05828	0.04428	0.04994	0.05942	0.04994	0.10816	0.08599
0.3	0.05736	0.06734	0.04951	0.05448	0.06829	0.05434	0.12899	0.09259
0.4	0.06653	0.07394	0.05125	0.05470	0.07450	0.05441	0.14641	0.09033
0.5	0.07092	0.07415	0.04735	0.04879	0.07421	0.04842	0.14777	0.07648
0.6	0.07331	0.07121	0.03904	0.03874	0.07072	0.03837	0.15165	0.06078
0.7	0.07222	0.06390	0.02725	0.02562	0.06289	0.02536	0.14088	0.03996
0.8	0.07099	0.05675	0.01548	0.01366	0.05532	0.01357	0.12522	0.02197
0.9	0.07083	0.05104	0.00478	0.00387	0.04931	0.00389	0.10740	0.00661

$$n = 50$$

ρ	$\tilde{\rho}_1$	$\tilde{\rho}_2$	$\tilde{\rho}_3$	$\tilde{\rho}_4$	$\tilde{\rho}_5$	$\tilde{\rho}_6$	$\tilde{\rho}_{5,J}$	$\tilde{\rho}_{6,J}$
0.1	0.01778	0.01990	0.01676	0.01794	0.02000	0.01797	0.02825	0.02737
0.2	0.02643	0.02906	0.02256	0.02383	0.02917	0.02384	0.04019	0.03461
0.3	0.03264	0.03530	0.02485	0.02587	0.03539	0.02587	0.04703	0.03414
0.4	0.03795	0.04041	0.02432	0.02498	0.04048	0.02495	0.05715	0.03125
0.5	0.03955	0.04100	0.02069	0.02088	0.04101	0.02085	0.06214	0.02528
0.6	0.04027	0.04037	0.01648	0.01628	0.04033	0.01633	0.06829	0.02033
0.7	0.03879	0.03712	0.01086	0.01047	0.03703	0.01080	0.07168	0.01340
0.8	0.03537	0.03169	0.00562	0.00528	0.03155	0.00640	0.06457	0.00683
0.9	0.03340	0.02756	0.00168	0.00152	0.02737	0.00524	0.05611	0.00205

By taking $(X_1, X_2)^T$ and $(Y_1, Y_2)^T$ to be independently and identically distributed with bivariate density function

$$p(x_1, x_2) = \frac{e^{-\frac{x_1 + x_2}{1 - \rho}}}{1 - \rho} I_0\left(\frac{2\sqrt{\rho x_1 x_2}}{1 - \rho}\right), \qquad x_1, x_2 \geq 0,$$

Ulrich and Chen (1987) have discussed a *bivariate double exponential distribution* as the joint distribution of $Z_1 = X_1 - Y_1$ and $Z_2 = X_2 - Y_2$. For example, the joint moment generating function of $(Z_1, Z_2)^T$ can be shown to be

$$M_{Z_1, Z_2}(t_1, t_2) = E\left[e^{t_1 Z_1 + t_2 Z_2}\right]$$

$$= \frac{1}{\{t_1 t_2 (1-\rho) - t_1 - t_2 + 1\}\{t_1 t_2 (1-\rho) + t_1 + t_2 + 1\}}.$$

2.8 Singpurwalla and Youngren's Bivariate Exponential

Singpurwalla and Youngren (1993) introduced a bivariate exponential distribution with joint survival function

$$\bar{F}_{X_1, X_2}(x_1, x_2)$$

$$= \sqrt{\frac{1 - m\min(x_1, x_2) + m\max(x_1, x_2)}{1 + m(x_1 + x_2)}} \, \exp\{-m\max(x_1, x_2)\},$$

$$x_1, x_2 \geq 0.$$

This distribution has exponential marginals and is indexed by a single parameter m. It arises naturally in a *shot-noise process environment*.

The joint density of X_1 and X_2 is

$$m^2 e^{-mx_1} \frac{(1+mx_1)\{(1+mx_1)^2 - m^2 x_2^2\} + \{1 + m(x_1 - x_2)\}^2 - mx_2(1+mx_1)}{\{1 + m(x_1 - x_2)\}^{3/2}\{1 + m(x_1 + x_2)\}^{5/2}}$$

on the set of points $x_1 > x_2$; on the set of points $x_2 > x_1$, x_1 is replaced by x_2 and vice versa in the above expression. The joint density is undefined on the set of points $x_1 = x_2$ which is similar to the behavior of the MOBED model. However, this model cannot be decomposed into an absolutely continuous and a singular part as in the case of MOBED. This distribution has been further discussed by Kotz and Singpurwalla (1999).

2.9 Raftery's Bivariate Exponential

Raftery (1984) defined a bivariate exponential distribution with joint survival function

$$\bar{F}(x_1, x_2)$$

$$= \begin{cases} e^{-x_1} + \frac{1-\delta}{1+\delta} e^{-x_1/(1-\delta)} \left\{ e^{x_2 \delta/(1-\delta)} - e^{-x_2/(1-\delta)} \right\} & \text{for } x_1 \geq x_2 \\ e^{-x_2} - \frac{1-\delta}{1+\delta} e^{-x_2/(1-\delta)} \left\{ e^{x_1 \delta/(1-\delta)} - e^{-x_1/(1-\delta)} \right\} & \text{for } x_1 \leq x_2. \end{cases}$$

Similar to many of the bivariate exponential models discussed earlier, this distribution also arises from a shock model. This corresponds to a system that has two components, C_1 and C_2, each of which can be functioning normally, unsatisfactorily, or may have failed. The system is subject to three kinds of shock, governed by independent Poisson processes. These kinds of shock cause normal components to become unsatisfactory, an unsatisfactory C_1 to fail, and an unsatisfactory C_2 to fail, respectively. Then, the model underlying this distribution is based on the stochastic representation

$$X_1 = (1 - \delta)Y_1 + I Y_{12} \quad \text{and} \quad X_2 = (1 - \delta)Y_2 + I Y_{12},$$

where I, Y_1, Y_2, and Y_{12} are independent random variables with Y's being standard exponential and I being a Bernoulli(δ) random variable with probability mass function

$$P(I = 1) = 1 - P(I = 0) = \delta.$$

Here, X_1 and X_2 are positively correlated with an exchangeable joint distribution.

A slightly more general model is as follows. Let Y_1, Y_2, and Y_{12} be independent Exponential(λ) random variables, I_1 and I_2 be binary random variables with joint distribution

$$p_{ij} = P(I_1 = i, I_2 = j) \qquad \text{for } i, j = 0, 1,$$

and

$$\delta_i = P(I_i = 1) = 1 - P(I_i = 0) \qquad \text{for } i = 1, 2.$$

Then, the model for $(X_1, X_2)^T$ is

$$X_i = (1 - \delta_i)Y_i + I_i Y_{12} \qquad \text{for } i = 1, 2.$$

A converse model is readily obtained by interchanging the roles of Y_i and Y_{12}.

Alternatively, we may construct a bivariate exponential distribution using

$$X_1 = (1 - \delta_1)Y_1 + I_1 Y_{12} \quad \text{and} \quad X_2 = (1 - \delta_2)Y_2 + I_2 Y'_{12},$$

where Y'_{12} has maximum negative correlation with Y_{12} given by $e^{-\lambda Y_{12}} + e^{-\lambda Y'_{12}} = 1$. Here, dependence between X_1 and X_2 is specified by three parameters δ_1, δ_2, and p_{11}, subject to $p_{11} \geq 0$, $\delta_1 \geq p_{11}$, $\delta_2 \geq p_{11}$ and $1 - \delta_1 - \delta_2 + p_{11} \geq 0$. The most asymmetric case is when $\delta_1 = 1$ in which case the data points are concentrated in a triangle. The marginals are Exponential(λ) and the correlations are

$$\rho = \text{corr}(X_1, X_2) = \begin{cases} 2p_{11} - \delta_1\delta_2 \\ (1-c)p_{11} - \delta_1\delta_2 \end{cases}$$

for the first and the second general models, respectively, where $c = -\text{corr}(Y_{12}, Y'_{12}) = \frac{\pi^2}{6} - 1 = 0.6449$. The Fréchet bound is attained when $\delta_1 = \delta_2 = p_{11} = 1$.

Bhattacharyya (1997) adopted Raftery's bivariate exponential construction to propose an absolutely continuous bivariate model for modeling survival data with random censoring and when the censoring pattern and the failure pattern are dependent and follow exponential distributions with different means. Bhattacharyya (1997) has proved the identifiability of this model and the asymptotic normality of the MLEs of the parameters of this model.

2.10   Hayakawa's Bivariate Exponential

Using a finite population of exchangeable two-component systems based on indifference principle, Hayakawa (1994) derived a class of bivariate exponential distributions which includes the Freund, Marshall and Olkin, and Block and Basu models as special cases.

Let $\{C_j, \, j \geq 1\}$ and $\{(X_{1i}, X_{2i})^T, \, i \geq 1\}$ be two infinite sequences of non-negative random variables. Let

$$\begin{cases} C_{2i-1} &= \kappa_1 X_{1i} - \kappa_{12}\min(X_{1i}, X_{2i}) + \Delta \\ C_{2i} &= \kappa_2 X_{2i} - \kappa_{21}\min(X_{1i}, X_{2i}) \end{cases} \quad \text{when } X_{1i} \neq X_{2i}$$

and

$$\begin{cases} C_{2i-1} &< \Delta \\ C_{2i} &= \kappa X_{1i} = \kappa X_{2i} \end{cases} \quad \text{when } X_{1i} = X_{2i},$$

where $\kappa_1, \kappa_2, \kappa_{12}, \kappa_{21} > 0$, $\kappa_1 > \kappa_{12}$, $\kappa_2 > \kappa_{21}$, $\kappa = (\kappa_1 - \kappa_{12}) + (\kappa_2 - \kappa_{21})$, and $\Delta \geq 0$.

Now, let $\{C_j, j \geq 1\}$ be exchangeable and a sequence of generalized exponentials in the sense that for all $n \geq 1$

$$g_n(c_1, \ldots, c_n) = \int_0^\infty \theta^{-n} \exp\left\{-\sum_{i=1}^n c_i/\theta\right\} \, dH(\theta),$$

where g_n is the n-dimensional marginal density of $\{C_j, \, j \geq 1\}$ and $H(\theta)$ is the distribution function of the parameter θ—that is, any finite sequence of $\{C_j, \ j \geq 1\}$ can be represented as a mixture of i.i.d. exponentials. Then, the joint survival function of the pair $\{(X_{1i}, X_{2i})^T, \ i = 1, 2, \ldots\}$ can be represented as a mixture of Friday and Patil distributions given by

$$\bar{F}(x_{1i}, x_{2i}) = \int \bar{F}(x_{1i}, x_{2i} \mid \phi) \ dH(\phi),$$

where the conditional survival function $\bar{F}(x_{1i}, x_{2i} \mid \phi)$ can be decomposed into an absolutely continuous part $\bar{F}_a$ and a singular part $\bar{F}_s$ as

$$\bar{F}(x_{1i}, x_{2i} \mid \phi) = e^{-\Delta/\phi}\bar{F}_a(x_{1i}, x_{2i} \mid \phi) + (1 - e^{-\Delta/\phi})\bar{F}_s(x_{1i}, x_{2i} \mid \phi)$$

with

$$\bar{F}_s(x_{1i}, x_{2i} \mid \phi) = \exp\left\{-\frac{\kappa}{\phi}\max(x_{1i}, x_{2i})\right\}$$

and $\bar{F}_a(x_{1i}, x_{2i} \mid \phi)$ has a density function

$$p(x_{1i}, x_{2i} \mid \phi)$$
$$= \begin{cases} \{\kappa_2(\kappa_1 - \kappa_{12})/\phi^2\} \exp\{-[(\kappa - \kappa_2)x_{1i} + \kappa_2 x_{2i}]/\phi\} & \text{if } x_{1i} < x_{2i} \\ \{\kappa_1(\kappa_2 - \kappa_{21})/\phi^2\} \exp\{-[\kappa_1 x_{1i} + (\kappa - \kappa_1)x_{2i}]/\phi\} & \text{if } x_{2i} > x_{1i}. \end{cases}$$

This class of distributions includes mixtures of Freund's, Marshall and Olkin's, and Block and Basu's distributions as special cases.

2.11   Lindley and Singpurwalla's Bivariate Exponential Mixture

Consider a two-component system in which, for a given environment η, the component lifetimes X_1 and X_2 are independently exponentially distributed with failure rates $\eta\lambda_1$ and $\eta\lambda_2$, respectively. λ_1 and λ_2 are the failures under the test (*laboratory*) environment. The unconditional joint density of X_1 and X_2 is then given by

$$p(x_1, x_2) = \int \eta\lambda_1 e^{-\eta\lambda_1 x_1}\eta\lambda_2 e^{-\eta\lambda_2 x_2} \ dG(\eta),$$

where $G(\eta)$ is the distribution function of η. By assuming $G(\cdot)$ to be a gamma distribution with density function

$$\frac{dG(\eta)}{d\eta} = b^{a+1}\eta^a e^{-\eta b}/a! \qquad \text{for } \eta > 0,$$

Lindley and Singpurwalla (1986) derived a bivariate exponential mixture distribution with joint density

$$p(x_1, x_2 \mid \lambda_1, \lambda_2, a, b) = \frac{\lambda_1 \lambda_2 (a+1)(a+2) b^{a+1}}{(\lambda_1 x_1 + \lambda_2 x_2 + b)^{a+3}}$$

and with the bivariate failure rate

$$r(x_1, x_2 \mid \lambda_1, \lambda_2, a, b) = \frac{(a+1)(a+2)\lambda_1 \lambda_2}{(\lambda_1 x_1 + \lambda_2 x_2 + b)^2}$$

which is a decreasing function of the argument.

Note that for $a = 0$ and $b = 1$ (that is, a standard exponential density for the environment η), we have $P(\eta > 1) = 0.3679$ implying that the conditions under exponential environment is likely to be more "gentle" than the test environment. However, for $a = 1$ and $b = 1$, we have $P(\eta > 1) = 0.7358$ in which case the situation is reversed. Currit and Singpurwalla (1988) have made a detailed study of the reliability features of Lindley and Singpurwalla's bivariate model.

Lefevre and Malice (1987) considered mixing with exponential distributions for the components and analyzed them using partial ordering. Bhattacharya and Kumar (1986) considered a compound exponential with inverse Gaussian distributed mean parameter. Whitmore and Lee (1991) also considered the case of mixing with inverse Gaussian distribution. Dey and Liu (1990) studied the case when the life distribution for the components is a generalized life model and the environment is modeled by an inverse gamma distribution.

In the construction of the Lindley and Singpurwalla distribution, Al-Mutairi (1997) used a MOBED with parameters $\eta\lambda_1, \eta\lambda_2$, and $\eta\lambda_{12}$ (instead of independent exponentials) and an inverse Gaussian distribution for η with density function

$$p(\eta) = \frac{1}{\sqrt{2\pi\nu\eta^3}} e^{-(\delta\eta-1)^2/(2\nu\eta)}, \qquad \delta, \nu > 0.$$

The resulting joint density of $(X_1, X_2)^T$ turns out to be

$$p(x_1, x_2)$$

$$= \begin{cases} \lambda_i \gamma_j \exp\left\{\frac{1}{\nu}\left(\delta - \sqrt{\delta^2 + 2\nu(\lambda_i x_i + \gamma_j x_j)}\right)\right\} \frac{\sqrt{\delta^2 + 2\nu(\lambda_i x_i + \gamma_j x_j)} + \nu}{\{\delta^2 + 2\nu(\lambda_i x_i + \gamma_j x_j)\}^{3/2}} \\ \qquad\qquad \text{if } 0 < x_i < x_j, \ i \neq j = 1, 2 \\[2ex] \dfrac{\lambda_{12} \exp\left\{\frac{1}{\nu}(\delta - \sqrt{\delta^2 + 2\nu\lambda x})\right\}}{\sqrt{\delta^2 + 2\nu\lambda x}} \qquad \text{if } x_1 = x_2 = x, \end{cases}$$

where $\gamma_i = \lambda_i + \lambda_{12}$ $(i = 1, 2)$ and $\lambda = \lambda_1 + \lambda_2 + \lambda_{12}$. The marginal survival distributions are

$$\bar{F}_{X_1}(x_1) = \exp\left\{\frac{1}{\nu}\left(\delta - \sqrt{\delta^2 + 2\nu\gamma_1 x_1}\right)\right\} \qquad \text{for } x_1 > 0$$

and

$$\bar{F}_{X_2}(x_2) = \exp\left\{\frac{1}{\nu}\left(\delta - \sqrt{\delta^2 + 2\nu\gamma_2 x_2}\right)\right\} \qquad \text{for } x_2 > 0.$$

In face, the random variables $\delta^2 + 2\nu\gamma_1 X_1$ and $\delta^2 + 2\nu\gamma_2 X_2$ are truncated Weibull over (δ^2, ∞) with parameters $\frac{1}{2}$ and $\frac{1}{\nu}$.

Nayak (1987) extended Lindley and Singpurwalla's bivariate distribution to the multivariate case by assuming that the k component lifetimes $X_1, \ldots, X_k$ of the system to be independently exponentially distributed with failure rates $\eta\lambda_1, \ldots, \eta\lambda_k$, respectively. He further assumed that the environment η is distributed as gamma with density function

$$b^a \eta^{a-1} e^{-\eta b}/\Gamma(a), \qquad \eta > 0, \ a, b > 0.$$

(Note the slight change in the parameterization.) This yields the unconditional joint density function as

$$p(x_1, \ldots, x_k) = a(a+1)\cdots(a+k-1)\left(\prod_{i=1}^{k} \theta_i\right)\bigg/\left(1 + \sum_{i=1}^{k} \theta_i x_i\right)^{a+k},$$

where $\theta_i = \frac{\lambda_i}{b}$ for $i = 1, \ldots, k$, and the joint survival function as

$$\bar{F}(x_1, \ldots, x_k) = \left(1 + \sum_{i=1}^{k} \theta_i x_i\right)^{-a}, \qquad x_1, \ldots, x_k > 0.$$

Nayak (1987) referred to this distribution as *multivariate Lomax* (Pareto Type 2) distribution in order to distinguish it from Mardia's (1962) multivariate Pareto Type 2 distribution; see Chapter 52 for details. If we make the transformation $V_i = (1 + \theta_i X_i)^{-a}$ $(i = 1, \ldots, k)$, we obtain the joint survival function of $(V_1, \ldots, V_k)^T$ as

$$\bar{F}(v_1, \ldots, v_k) = \left(\sum_{i=1}^{k} v_i^{-1/a} - k + 1\right)^{-a}, \qquad 0 < v_i \leq 1,$$

which is Cook and Johnson's (1981) multivariate non-elliptically symmetric distribution. Instead, if we make the transformation $W_i = (\theta_i X_i/d_i)^{1/c_i}$ $(i = 1, \ldots, k)$ with $c_i, d_i > 0$, we obtain the multivariate Burr distribution.

2.12 Ghurye's Extended Bivariate Lack of Memory Distributions

Ghurye (1987) studied bivariate and multivariate distributions possessing an extended version of lack of memory (LOM) property which could provide "more realistic" models than bivariate exponentials. In the bivariate case, the joint survival function is

$$\bar{F}_{X_1,X_2}(x_1, x_2) = \bar{A}(\min(x_1, x_2))\bar{K}(x_1 - x_2), \qquad x_1, x_2 > 0,$$

where

$$\bar{K}(\omega) = \begin{cases} \bar{G}(\omega) & \text{for } \omega > 0 \\ \bar{H}(|\omega|) & \text{for } \omega < 0, \end{cases}$$

and $\bar{A}$, $\bar{G}$, and $\bar{H}$ are survival functions on $[0, \infty)$ (being equal to 1 at 0) of $\min(X_1, X_2)$, X_1, and X_2, respectively. Ghurye (1987) provided rather involved conditions on $\bar{A}$, $\bar{G}$, and $\bar{H}$ which will assure that $\bar{F}(x_1, x_2)$ is a valid bivariate survival function on $\mathbb{R}_+^2$. Note that the bivariate survival function with LOM property can be represented as

$$\bar{F}(x_1, x_2) = \begin{cases} e^{-\theta x_2}\bar{G}(x_1 - x_2) & \text{for } x_1 \geq x_2 \geq 0 \\ e^{-\theta x_1}\bar{H}(x_2 - x_1) & \text{for } x_2 \geq x_1 \geq 0, \end{cases}$$

where $\bar{G}$ and $\bar{H}$ are univariate survival functions and have unique right derivatives

$$g(\omega) = \lim_{\delta \downarrow 0} \frac{\bar{G}(\omega) - \bar{G}(\omega + \delta)}{\delta} \quad \text{and} \quad h(\omega) = \lim_{\delta \downarrow 0} \frac{\bar{H}(\omega) - \bar{H}(\omega + \delta)}{\delta}$$

which are right-continuous, of bounded variation and have at most a countable number of discontinuities; further, $e^{\theta\omega}g(\omega)$ and $e^{\theta\omega}h(\omega)$ are nondecreasing in ω, and

$$\bar{G}(\omega) + \bar{H}(\omega) \leq 1 + e^{-\theta\omega} \qquad \text{for all } \omega \geq 0.$$

This $\bar{F}(x_1, x_2)$ corresponds to the distribution of the first failure of a two-component system and the marginal distributions correspond to the lifetimes of the individual components.

Note that the generic definition of the LOM class is given by the joint survival function satisfying

$$\bar{F}(x_1 + t, x_2 + t) = \bar{F}(x_1, x_2)\bar{F}(t, t)$$

for all $x_1, x_2, t > 0$; see, for example, Ghurye and Marshall (1984). This is closely associated with many forms of BEDs. Ghurye (1987) provided a different extension of the LOM property by generalizing the above equation by introducing an ageing factor as

$$\bar{F}(x_1 + t, x_2 + t) = \bar{F}(x_1, x_2)\bar{F}(t, t)\bar{B}(t; x_1, x_2),$$

where $\bar{B}(0; x_1, x_2) = 1 = \bar{B}(t; 0, 0)$ and $\bar{B}$ is decreasing in t and $(x_1, x_2)^T$. Taking $\bar{B}(t; x_1, x_2) = \exp\{-2t(\alpha x_1 + \beta x_2)\}$, it can be shown that $\bar{F}(x_1, x_2)$ satisfying the above equation can be represented as

$$\bar{F}(x_1, x_2) = \bar{F}_0(x_1, x_2)e^{-(\alpha x_1^2 + \beta x_2^2)},$$

where $\bar{F}_0(x_1, x_2)$ is a survival function belonging to the extended LOM class—that is,

$$\bar{F}(x_1, x_2) = \exp\{-\theta_0 \min(x_1, x_2)\}\bar{K}_0(x_1 - x_2)e^{-(\alpha x_1^2 + \beta x_2^2)}.$$

Moreover, the marginal survival function of X_1 satisfies

$$\bar{G}(\omega) = \bar{G}_0(\omega)e^{-\alpha \omega^2},$$

the first failure has the survival function

$$\bar{F}(t, t) = e^{-\theta_0 t - (\alpha + \beta)t^2},$$

and the conditional survival function of X_1, subject to both components surviving beyond time t, is given by

$$\frac{\bar{F}(t + \omega, t)}{\bar{F}(t, t)} = \bar{G}(\omega)e^{-2\alpha t \omega} = \bar{G}_0(\omega)e^{-2\alpha t \omega - \alpha \omega^2}.$$

Another extension of the LOM property is due to Raja Rao, Damaraju, and Alhumound (1993) and is as follows. A class of bivariate life distributions $\{\bar{F}(x_1, x_2, \omega); x_1 \geq 0, x_2 \geq 0, \omega \in \Omega\}$ is said to have the *setting the clock back to zero property* if, for each $\omega \in \Omega$ and $x_0 > 0$, the following two equations are satisfied:

$$\frac{\bar{F}(x_1 - x_0, x_0; \omega)}{\bar{F}(x_0, x_0; \omega)} = \bar{F}(x_1, x_0; \omega^*)$$

and

$$\frac{\bar{F}(x_0, x_2 + x_0; \omega)}{\bar{F}(x_0, x_0; \omega)} = \bar{F}(x_0, x_2; \omega^{**}),$$

where $\omega^* = \omega^*(x_0) \in \Omega \cup \Omega_0$ and $\omega^{**} = \omega^{**}(x_0) \in \Omega \cup \Omega_0$, with Ω_0 being the boundary of Ω.

The conditional distribution of the additional survival time of an individual due to any one risk R_1 (assuming that the risk R_2 has not killed the individual first) given that the individual has survived both the risks for time x_0 remains in the family.

Model I of Gumbel as well as the MOBED model possesses this "setting the clock back to zero property."

2.13 Cowan's Bivariate Exponential

Cowan (1987) derived a bivariate exponential distribution, using the theory of Poisson line processes, with joint cumulative distribution function

$$
\begin{aligned}
F(x_1, x_2) &= 1 - e^{-\lambda x_1} - e^{-\lambda x_2} \\
&+ \exp\left\{ -\frac{1}{2}\left(x_1 + x_2 + \sqrt{x_1^2 + x_2^2 - 2x_1 x_2 \cos a} \right) \right\}, \\
&\qquad x_1, x_2 \geq 0, \ 0 \leq a \leq \pi.
\end{aligned}
$$

This joint distribution function is absolutely continuous with respect to (x_1, x_2) and, therefore, has a density which is given by

$$
\begin{aligned}
p(x_1, x_2) &= \frac{\lambda(1 - \eta)}{2s^3} \left[4\eta x_1 x_2 + s\{ s(x_1 + x_2) + x_1^2 + x_2^2 + 2x_1 x_2 \eta \} \right] \\
&\quad \times \exp\left\{ -\frac{1}{2}\lambda(x_1 + x_2 + s) \right\}, \quad x_1, x_2 \geq 0,
\end{aligned}
$$

where $s^2 = x_1^2 + x_2^2 - 2x_1 x_2 \cos a = (x_1 + x_2)^2 - 4x_1 x_2 \eta$. The correlation between X_1 and X_2 is

$$
\begin{aligned}
&\text{corr}(X_1, X_2) \\
&= \begin{cases}
1 & \text{if } a = 0 \\
-1 + \frac{4}{1+\cos a}\left\{ 1 - \frac{1-\cos a}{1+\cos a} \log\left(\frac{2}{1-\cos a} \right) \right\} & \text{if } 0 < a < \pi \\
0 & \text{if } a = \pi.
\end{cases}
\end{aligned}
$$

In the above case, X_1 and X_2 have identical marginal exponential distributions with mean $1/\lambda$; however, the variables can be scaled to have different means.

2.14 Infinite Divisibility

Rvaceva (1962) has shown that a bivariate distribution is *infinitely divisible* if its joint characteristic function $\phi(t_1, t_2)$ can be represented uniquely as

$$
\begin{aligned}
\log \phi(t_1, t_2) &= i(\gamma_1 t_1 + \gamma_2 t_2) + \iint_{\mathbb{R}^2} \left\{ e^{i(t_1 x_1 + t_2 x_2)} - 1 - \frac{i(t_1 x_1 + t_2 x_2)}{1 + x_1^2 + x_2^2} \right\} \\
&\quad \times \frac{1 + x_1^2 + x_2^2}{x_1^2 + x_2^2} \, dG(x_1, x_2),
\end{aligned}
$$

where $G(x_1, x_2)$ is a finite non-negative measure on $\mathbb{R}^2$ and γ_1 and γ_2 are constants. An alternative form for infinite divisibility, applicable to many bivariate exponential distributions, is

$$\log \phi(t_1, t_2) = \iint_{\mathbb{R}_+^2} \left\{ e^{i(t_1 x_1 + t_2 x_2)} - 1 \right\} \frac{1}{(x_1^2 + x_2^2)} \, dK(x_1, x_2),$$

where $K(x_1, x_2)$ is a finite non-negative measure on $\mathbb{R}_+^2$.

Clearly, bivariate exponential distribution with independent marginals for which

$$\phi(t_1, t_2) = \{(1 - i\theta_1 t_1)(1 - i\theta_2 t_2)\}^{-1} \qquad (\theta_i > 0, \; i = 1, 2)$$

admits the above representation with

$$K_1(x_1, x_2) = \int_0^{x_1} e^{-y_1/\theta_1} y_1 \, dy_1 \quad \text{and} \quad K_2(x_1, x_2) = \int_0^{x_2} e^{-y_2/\theta_2} y_2 \, dy_2$$

for $x_1, x_2 \geq 0$ and $K(x_1, x_2) = K_1(x_1, x_2) + K_2(x_1, x_2)$ concentrated on the positive axes $x_i = 0$ $(i = 1, 2)$. Also, singular bivariate characteristic function

$$\phi(t_1, t_2) = \{1 - i(\theta_1 t_1 + \theta_2 t_2)\}^{-1} \qquad (\theta_i > 0, \; i = 1, 2)$$

admits the above representation with

$$K(x_1, x_2) = (\theta_1^2 + \theta_2^2) \int_0^{\min(x_1/\theta_1, x_2/\theta_2)} e^{-y} y \, dy$$

concentrated on the line $x_1/\theta_1 = x_2/\theta_2$ in the first quadrant; see Block (1975b).

2.15   Characterizations

Characterization of the MOBED by the LOM property was mentioned earlier. Recall that the MOBED contains a singular component. Then, Block and Basu (1974) established that if $(X_1, X_2)^T$ is a nonnegative bivariate random vector which is absolutely continuous and possesses the LOM property, then the exponentiality of the marginal distributions implies that X_1 and X_2 are independent random variables.

Block (1977) proved that $(X_1, X_2)^T$ has a MOBED iff one of the following three equivalent conditions hold:
  (a) $(X_1, X_2)^T$ has exponential marginals,
  (b) $\min(X_1, X_2)$ is exponential,
  (c) $\min(X_1, X_2)$ is independent of $X_1 - X_2$.

A similar result due to Basu involves the concept of bivariate failure rate

$$h(x_1, x_2) = \frac{p(x_1, x_2)}{\overline{F}(x_1, x_2)} .$$

For bivariate distributions with a finite Laplace transform, if the function $h(x_1, x_2)$ is constant in x_1 and x_2 and possesses exponential marginals, then this bivariate distribution is the product of independent exponential marginals. Thus, there is no absolutely continuous bivariate exponential distribution with a constant failure rate other than the one with independent exponential marginals.

An alternate vector-valued definition of failure rate leads to a different conclusion. Seshadri and Patil (1964) characterized Gumbel's Model I distribution by stipulating the marginal and conditional distributions of the same variable.

Asha and Nair (1999) have augmented characterizations of MOBEDs used in reliability modeling as discussed in Barlow and Proschan (1981) and Galambos and Kotz (1978).

Wu (1997) has characterized the MOBED by using bivariate stopping $\mathbf{Y} = (Y_1, Y_2)^T$. Specifically, let $(Y_1, Y_2)^T$ have a general bivariate geometric distribution with joint probability mass function

$$P(Y_1 = m, Y_2 = n)$$
$$= \begin{cases} p_{11}^{n-1}(p_{10} + p_{11})^{m-n-1} p_{10}(p_{01} + p_{00}) & \text{if } m > n \\ p_{11}^{m-1} p_{00} & \text{if } m = n \\ p_{11}^{m-1}(p_{01} + p_{11})^{n-m-1} p_{01}(p_{10} + p_{00}) & \text{if } m < n. \end{cases}$$

Let $\{X_{1i}\}$ and $\{X_{2i}\}$ be sequences of random variables such that $E[X_{1i}] = \frac{1}{\lambda_1 + \lambda_{12}}$ and $E[X_{2i}] = \frac{1}{\lambda_2 + \lambda_{12}}$. Let $(Y_1, Y_2)^T$ have a general bivariate geometric distribution with $p_{01} = \lambda_1 \theta$, $p_{10} = \lambda_2 \theta$, $p_{00} = \lambda_{12} \theta$ $(\theta > 0)$ and $p_{00} + p_{01} + p_{10} + p_{11} = 1$, $p_{10} + p_{11} < 1$, $p_{01} + p_{11} < 1$. Then, the joint distribution of

$$\left((p_{00} + p_{01}) \sum_{i=1}^{Y_1} X_{1i}, (p_{00} + p_{10}) \sum_{i=1}^{Y_2} X_{2i} \right)^T$$

converges weakly, as $\theta \to 0$, to a MOBED.

3   MULTIVARIATE EXPONENTIAL DISTRIBUTIONS

In this section, we will present various forms of multivariate exponential distributions and their generalizations. These are natural extensions of the corresponding bivariate forms discussed in Section 2.

First of all, by considering n independent and identically distributed k-variate exponential random vectors with independent $\text{Exp}(\mu, \theta_i)$ ($i = 1, \ldots, k$) components, Bordes, Nikulin, and Voinov (1997) have derived an UMVUE of the joint density function from the UMVUE of the joint distribution function. They have also illustrated the usefulness of the UMVUE of the joint density function in developing a chi-square goodness of fit for this model.

3.1   Freund's Multivariate Exponential

Weinman (1966) extended Freund's BED (presented earlier in Section 2.3) to the multivariate setting in the following way. Suppose a system has k identical components with times to failure $X_1, \ldots, X_k$. They all have the exponential density function

$$p_X(x) = \frac{1}{\theta_0} \, e^{-x/\theta_0}, \qquad x \ge 0, \ \theta_0 > 0.$$

It is further supposed that if ℓ components have failed (and not been replaced), the conditional joint distribution of the lifetimes of the remaining $k - \ell$ components is that of independent random variables, each having the density function

$$p_X(x) = \frac{1}{\theta_\ell} \, e^{-x/\theta_\ell}, \qquad x \ge 0, \ \theta_\ell > 0.$$

In this case, clearly, $0 \le X_1 \le X_2 \le \cdots \le X_k$. Then, Weinman has shown that the joint density of $X_1, \ldots, X_k$ is then

$$p_{X_1, \ldots, X_k}(x_1, \ldots, x_k) = \prod_{i=0}^{k} \frac{1}{\theta_i} \, e^{-(k-i)(x_{i+1}-x_i)/\theta_i},$$
$$x_0 = 0, \ 0 \le x_1 \le x_2 \le \cdots \le x_k. \quad (47.84)$$

[It is of interest ot mention that the joint density function of progressively Type II right censored order statistics from an exponential distribution is a member of the family in (47.84); see, for example, Balakrishnan and Aggarwala (2000) and Kamps (1995).] The joint moment-generating function is

$$E\left[e^{t_1 X_1 + \cdots + t_k X_k}\right] = \frac{1}{k!} \sum_{P}^{*} \prod_{i=0}^{k-1} \left\{ 1 - \frac{\theta_i}{k-i} \sum_{j=i+1}^{k} t_{P(j)} \right\}^{-1}, \quad (47.85)$$

where $\{t_{P(1)}, \ldots, t_{P(k)}\}$ is one of the $k!$ possible permutations of $t_1, \ldots, t_k$, and $\sum_P^*$ denotes the summation over all such permutations. The distribution is symmetrical in $X_1, \ldots, X_k$. For each i $(= 1, 2, \ldots, k)$, we have

$$E[X_i] = \frac{1}{k} \sum_{i=0}^{k-1} \theta_i,$$

$$\text{var}(X_i) = \frac{1}{k^2} \left[\sum_{i=0}^{k-1} \frac{k+1}{k-i} \theta_i^2 + 2 \sum_{i<j} \sum i(k-i)\theta_i \theta_j \right],$$

and

$$\text{cov}(X_i, X_j) = \frac{1}{k^2(k-1)} \left[\sum_{i=0}^{k-1} \left(k - \frac{k+i}{k-i} \right) \theta_i^2 - 2 \sum_{i<j} \sum i(k-i)\theta_i \theta_j \right].$$

$$(47.86)$$

The joint moment generating function of the *ordered* variables $X_{(1)} \leq X_{(2)} \leq \cdots \leq X_{(k)}$ has the relatively simple form

$$\prod_{i=0}^{k-1} \left\{ 1 - \frac{\theta_i}{k-i} \sum_{j=i+1}^{k} t_j \right\}^{-1}.$$

[The joint density of $X_{(1)}, X_{(2)}, \ldots, X_{(k)}$ is, of course, $k!$ times the density in (47.84).] From the above expression, we have

$$E[X_{(i)}] = \sum_{j=0}^{i-1} \frac{\theta_j}{k-j}, \quad \text{var}(X_{(i)}) = \sum_{j=0}^{i-1} \frac{\theta_j^2}{(k-j)^2}, \quad \text{and}$$

$$(47.87)$$

$$\text{cov}(X_{(i)}, X_{(j)}) = \text{var}(X_{(i)}) \qquad \text{for } 1 \leq i < j \leq k.$$

Cramer and Kamps (1997) derived an UMVUE of $\Pr(X_1 < X_2)$ based on Type II censored samples from Weinman's multivariate exponential distribution. The UMVUE has been shown to have a Gauss's hypergeometric distribution. Explicit expressions for the variances of the estimators have been derived which have been used to calculate the relative efficiency.

Block (1977) extended the Freund–Weinman MED to the case of non-identical marginals as follows. Let $X_1, \ldots, X_k$ be the times to failure of k components. If at time $x > 0$ none of the components have failed, we assume that the components act independently with densities $p_i^{(0)}(x)$ for $i = 1, \ldots, k$. If there have been j failures up to time x and $1 \leq j \leq k-1$, and the failures have been to components $i_1, \ldots, i_j$ at times $0 \leq x_{i_1} < \cdots < x_{i_j}$, respectively, we assume that the remaining $k - j$ components act independently with densities $p_{i|i_1,\ldots,i_j}^{(j)}(x)$ (for $x \geq x_{i_j}$) and these densities do not depend on the order of $i_1, \ldots, i_j$. The joint density function of $(X_1, \ldots, X_k)^T$ is then

$$
p(x_1, \ldots, x_k) = p_{i_1}^{(0)}(x_{i_1}) \prod_{j=2}^{k} \int_{x_{i_1}}^{\infty} p_{i_j}^{(0)}(x_{i_j}) dx_{i_j}
$$

$$
\times \prod_{\ell=2}^{k} \left\{ p_{i_\ell | i_1, \ldots, i_{\ell-1}}^{(\ell-1)}(x_{i_\ell}) \right.
$$

$$
\left. \times \prod_{j=\ell+1}^{k} \int_{x_{i_\ell}}^{\infty} p_{i_j | i_1, \ldots, i_{\ell-1}}^{(\ell-1)}(x_{i_j}) dx_{i_j} \right\},
$$

$$
0 = x_{i_0} < x_{i_1} < \cdots < x_{i_k}, \qquad (47.88)
$$

where density $p_{i|i_1,\ldots,i_{\ell-1}}^{(\ell-1)}(x) = 0$ for $x \leq x_{i_{\ell-1}}$. If all the densities on the RHS of (47.88) are exponential and the times of failures are $0 < x_{i_1} < x_{i_2} < \cdots < x_{i_k}$ and $x_{i_0} = 0$, then

$$
p_{\ell|i_1,\ldots,i_j}^{(j)}(x) = \alpha_{\ell|i_1,\ldots,i_j}^{(j)} \exp\left\{ -\alpha_{\ell|i_1,\ldots,i_j}^{(j)} (x - x_{i_j}) \right\}, \quad x \geq x_{i_j}
$$

or (suppressing the indices $i_1, \ldots, i_j$)

$$
p_{\ell}^{(j)}(x) = \alpha_{\ell}^{(j)} \exp\left\{ -\alpha_{\ell}^{(j)} (x - x_{i_j}) \right\}, \qquad x \geq x_{i_j}.
$$

Thus, (47.88) becomes

$$
p(x_1, \ldots, x_k) = \prod_{\ell=1}^{k} \left[\alpha_{i_\ell}^{(\ell-1)} \prod_{j=\ell}^{k} \exp\left\{ -\alpha_{i_j}^{(\ell-1)} (x_{i_\ell} - x_{i_{\ell-1}}) \right\} \right]
$$

$$
\text{for } 0 = x_{i_0} < x_{i_1} < \cdots < x_{i_k}, \qquad (47.89)
$$

or, equivalently,

$$
p(x_1, \ldots, x_k) = \left\{ \prod_{\ell=1}^{k} \alpha_{i_\ell}^{(\ell-1)} \right\} \exp\left\{ -\sum_{\ell=1}^{k} \left(\sum_{j=\ell}^{k} \alpha_{i_j}^{(\ell-1)} \right) (x_{i_\ell} - x_{i_{\ell-1}}) \right\}
$$

$$
\text{for } 0 = x_{i_0} < x_{i_1} < \cdots < x_{i_k}, \qquad (47.90)
$$

where $\alpha_{i_j | i_1, \ldots, i_{\ell-1}}^{(\ell-1)} > 0$. This distribution has the multivariate lack of memory property, namely,

$$
p(x_1 + t, \ldots, x_k + t) = \Pr[X_1 > t, \ldots, X_k > t] \, p(x_1, \ldots, x_k).
$$

The joint moment generating function for the above generalized Freund–Weinman–Block MED is given by

$$E\left[e^{t_1 X_1 + \cdots + t_k X_k}\right]$$

$$= \int \cdots \int e^{t_1 x_1 + \cdots + t_k x_k} p(x_1, \ldots, x_k) dx_1 \cdots dx_k$$

$$= \sum_P^* \prod_{\ell=1}^k \alpha_{i_\ell}^{(\ell-1)} \int \cdots \int \exp\left\{-\sum_{\ell=1}^k \left(\sum_{j=\ell}^k \alpha_{ij}^{(\ell-1)}\right)(x_{i_\ell} - x_{i_{\ell-1}})\right.$$

$$\left. -t_{i_\ell} x_{i_\ell}\right\} dx_{i_k} \cdots d_{x_{i_1}}$$

$$= \sum_P^* \prod_{\ell=1}^k \left\{\alpha_{i_\ell}^{(\ell-1)} \Big/ \sum_{j=\ell}^k \left(\alpha_{ij}^{(\ell-1)} - t_{ij}\right)\right\}, \qquad (47.91)$$

where, as before, $\sum_P^*$ denotes summation over all permutations $(i_1, \ldots, i_k)$ of $(1, \ldots, k)$.

Basu and Sun (1997) pointed out that the distribution in (47.90) is a complete generalization of Freund's BED, which can be derived from a fatal shock model. Consider a k-component system with independent non-homogeneous Poisson processes governing the occurrence of fatal shocks. The Freund–Weinman–Block distribution is derived by assuming that there are $(n - j)\binom{n}{j}$ classes of the processes $\{Z_{\ell|i_1,\ldots,i_j}^{(j)}(t) : \ell, i_1, \ldots, i_j$ are $j + 1$ distinct elements of $1, \ldots, k\}$ for $j = 0, \ldots, k - 1$. However, it is possible in some cases that the processes are independent of not only the order of $i_1, \ldots, i_j$ but also the elements of $i_1, \ldots, i_j$; that is, there are just k classes of these processes for $j = 0, 1, \ldots, k - 1$. Then this distribution has k^2 parameters and has a somewhat simpler density function

$$p(x_1, \ldots, x_k) = \left\{\prod_{\ell=1}^k \alpha_{i_\ell}^{(\ell-1)}\right\} \exp\left\{-\sum_{\ell=1}^k \left(\sum_{j=\ell}^k \alpha_{ij}^{(\ell-1)}\right)(x_{i_\ell} - x_{i_{\ell-1}})\right\}$$

$$\text{for } 0 = x_{i_0} < x_{i_1} < \cdots < x_{i_k}. \qquad (47.92)$$

3.2 Marshall and Olkin's Multivariate Exponential

Marshall and Olkin (1967a) have generalized their MOBED described earlier in Section 2.4, denoted by MOMED, in the following manner. In a system of k components, the distribution of times between "fatal shocks" to the combination $\{a_1, \ldots, a_\ell\}$ of components is supposed to have an exponential distribution with mean $1/\lambda_{a_1,\ldots,a_\ell}$. The $2^{k-1} - 1$ different distributions of this kind are supposed to be a mutually independent set.

The resulting joint distribution of lifetimes $X_1, \ldots, X_k$ of the components is

$$
\begin{aligned}
\bar{F}_{X_1,\ldots,X_k}(x_1,\ldots,x_k) = \ &\exp\Big\{-\sum_{i=1}^{k}\lambda_i x_i - \sum\sum_{i_1<i_2}\lambda_{i_1,i_2}\max(x_{i_1},x_{i_2}) \\
&-\sum\sum\sum_{i_1<i_2<i_3}\lambda_{i_1,i_2,i_3}\max(x_{i_1},x_{i_2},x_{i_3}) \\
&-\cdots-\lambda_{1\,2\cdots k}\max(x_1,\ldots,x_k)\Big\}. \quad (47.93)
\end{aligned}
$$

This is also a mixed distribution, as in the bivariate case.

Arnold (1968) pointed out that estimation of the parameters λ's by standard maximum likelihood or moment methods is not simple. He suggested the following method of estimation which exploits the singular nature of the distribution. Let

$$
Z_{a_1,\ldots,a_\ell} = \begin{cases} 1 & \text{if } X_{a_1} = \cdots = X_{a_\ell} < X_i \text{ for all } i \neq a_1,\ldots,a_\ell \\ 0 & \text{otherwise.} \end{cases}
$$

Given n independent observations $\boldsymbol{X}_j = (X_{1j},\ldots,X_{kj})^T$ $(j = 1,\ldots,n)$, each having the joint MOMED in (47.93), the estimator of $\lambda_{a_1,\ldots,a_\ell}$ is, in an obvious notation,

$$
\frac{\frac{1}{n}\sum_{j=1}^{n} Z_{a_1,\ldots,a_\ell}(j)}{\frac{1}{n-1}\sum_{j=1}^{n}\min(X_{1j},\ldots,X_{kj})}. \quad (47.94)
$$

The numerator and the denominator of (47.94) are mutually independent. The estimator is unbiased and has variance

$$
\frac{1}{n(n-1)}\lambda_{a_1,\ldots,a_\ell}\left\{(n-1)\lambda + \lambda_{a_1,\ldots,a_\ell}\right\},
$$

where λ is the sum of $\lambda_{a_1,\ldots,a_\ell}$'s over all possible sets $\{a_1,\ldots,a_\ell\}$. However, if the sample size n is not large, many of the estimators in (47.94) will be 0. In fact, for each $\boldsymbol{X}_j$, only one Z (at most) will not be 0, so there must be at least $(2^k - 1 - n)$ estimators with 0 values.

The $(k-1)$-dimensional marginal distributions of (47.93) have the same structure, and the two-dimensional marginal distributions are MOBEDs of the form (47.44). Moreover, the functional equation

$$
\bar{F}(x_1 + t,\ldots,x_k + t) = \bar{F}(x_1,\ldots,x_k)\bar{F}(t,\ldots,t) \quad (47.95)
$$

is satisfied, and the only distributions with exponential marginal distributions that satisfy (47.95) are the MOMEDs in (47.93). For more details, see Section 3.9 on characterizations.

A simplified version of MOMED is given by the survival function

$$\bar{F}_{X_1,\ldots,X_k}(x_1,\ldots,x_k) = \exp\left\{-\sum_{i=1}^{k}\lambda_i x_i - \lambda_{k+1}\max(x_1,\ldots,x_k)\right\},$$

$$x_i, \lambda_i > 0, \quad \lambda_{k+1} \geq 0, \quad \sum_{i=1}^{k+1}\lambda_i = \lambda. \qquad (47.96)$$

Symmetry corresponds to $\lambda_1 = \cdots = \lambda_k$—that is, $\gamma_i = \lambda_i - \lambda_k = 0$ ($i = 1,\ldots,k-1$)—while mutual independence corresponds to $\lambda_{k+1} = 0$. Also, $\Pr[X_1 = \cdots = X_k] = \frac{\lambda_{k+1}}{\lambda}$. Proschan and Sullo (1976) considered the distribution in (47.96) and denoted it by $\boldsymbol{X} \overset{d}{=} \mathrm{MVE}(k+1,\boldsymbol{\lambda})$ in order to distinguish it from the MOMED in (47.93) which they denoted by $\boldsymbol{X} \overset{d}{=} \mathrm{MVE}(2^k - 1)$.

Proschan and Sullo (1976) showed that $\boldsymbol{X} \overset{d}{=} \mathrm{MVE}(k+1,\boldsymbol{\lambda})$ if and only if there exist $k+1$ mutually independent exponential random variables $Y_0, Y_1, \ldots, Y_k$ with corresponding failure rates λ_i such that $X_i = \min(Y_0, Y_i)$ for $i = 1,\ldots,k$ (see also Section 3.9 on characterizations). Assuming that $\boldsymbol{X}_j$ ($j = 1,\ldots,n$) is a random sample from $\mathrm{MVE}(k+1,\boldsymbol{\lambda})$, they used the notation

$$\begin{aligned} X_{(1)} &= \min(X_1,\ldots,X_k), \\ X_{(k)} &= \max(X_1,\ldots,X_k), \\ Z_i(\boldsymbol{X}) &= \begin{cases} 1 & \text{if } X_i < X_{(k)}, \\ 0 & \text{otherwise,} \end{cases} \end{aligned} \qquad (47.97)$$

and

$$W(\boldsymbol{X}) = \begin{cases} 1 & \text{if } X_i = X_j = X_{(k)} \text{ for any } i \neq j, \\ 0 & \text{otherwise.} \end{cases}$$

The arguments in the functions defined in (47.97) may be suppressed when no confusion arises. Let $Z_{ij} \equiv Z_i(\boldsymbol{X}_j)$ and $W_j \equiv W(\boldsymbol{X}_j)$ and let

$$\begin{aligned} n_0 &= \sum_{j=1}^{n} W_j, \quad n_i = \sum_{j=1}^{n} W_{ij}, \\ n_i^{(c)} &= \sum_{j=1}^{n}(1 - Z_{ij})(1 - W_j), \end{aligned} \qquad (47.98)$$

and

$$n_0(i) = \sum_{j=1}^{n}(1 - Z_{ij})W_j = n - n_i - n_i^{(c)}.$$

Then if $n_i > 0$ (for $i = 0, 1, \ldots, k$), the MLE of $\boldsymbol{\lambda}$ exists and is given by the unique solution in $\boldsymbol{\Lambda}^+ = \{\boldsymbol{\lambda} : 0 < \lambda_i < \infty \text{ for } i = 0, 1, \ldots, k\}$ of the system

$$\frac{n_i}{\lambda_i} + \frac{n_i^{(c)}}{\gamma_i} = \sum_{j=1}^{n} X_{ij} \qquad \text{and} \qquad \frac{n_0}{\lambda_0} + \sum_{i=1}^{k} \frac{n_i^{(c)}}{\gamma_i} = \sum_{j=1}^{n} X_{(k)j},$$

$$i = 1, \ldots, k, \quad (47.99)$$

where $\gamma_i = \lambda_0 + \lambda_i$ $(i = 1, 2, \ldots, k)$.

If $n_i = 0$ for some $i = 0, 1, \ldots, k$, then the MLE of $\boldsymbol{\lambda}$ is given explicitly by

$$\hat{\lambda}_i^{(L)} = \begin{cases} \dfrac{\left(\dfrac{n n_i}{n - n_i^{(c)}}\right)}{\sum_{j=1}^{n} X_{ij}} & \text{if } n_i^{(c)} < n \\[4mm] \dfrac{n}{\sum_{j=1}^{n} X_{ij}} & \text{if } n_i^{(c)} = n \end{cases} \qquad (i = 1, \ldots, k), \qquad (47.100)$$

$$\hat{\lambda}_0^{(L)} = \begin{cases} \dfrac{n - \sum_{i=1}^{k} \dfrac{n_i n_i^{(c)}}{n - n_i^{(c)}}}{\sum_{j=1}^{n} X_{(k)j}} & \text{if } n_i^{(c)} < n \text{ for all } i = 1, \ldots, k, \\[4mm] 0 & \text{if } n_i^{(c)} = n \text{ for some } i = 1, \ldots, k. \end{cases} \qquad (47.101)$$

In particular, if $\lambda_i = \lambda_1$ for all $i \neq 0$, there is a unique MLE, given as:

(i) For $0 < n_0 < n$,

$$\hat{\lambda}_0^{(L)} = \frac{1}{2a} \left\{ \sqrt{b^2 + 4ac} - b \right\} \qquad \text{and}$$

$$\hat{\lambda}_1^{(L)} = \frac{\hat{\lambda}_0^{(L)} \sum_{i=1}^{k} n_i}{\left\{ \sum_{i=1}^{k} \sum_{j=1}^{n} X_{ij} - \sum_{j=1}^{n} X_{(k)j} \right\} \hat{\lambda}_0^{(L)} + n_0}, \qquad (47.102)$$

where

$$a = \left(\sum_{i=1}^{k} \sum_{j=1}^{n} X_{ij} - \sum_{j=1}^{n} X_{(k)j} \right) \sum_{j=1}^{n} X_{(k)j},$$

$$b = \sum_{i=0}^{k} n_i \sum_{j=1}^{n} X_{(k)j} - n \left(\sum_{i=1}^{k} \sum_{j=1}^{n} X_{ij} - \sum_{j=1}^{n} X_{(k)j} \right),$$

and

$$c = n_0 \left(n + \sum_{i=1}^{k} n_i \right);$$

(ii) For $n_0 = 0$,

$$\hat{\lambda}_0^{(L)} = 0 \qquad \text{and} \qquad \hat{\lambda}_1^{(L)} = \frac{nk}{\sum_{i=1}^{k} \sum_{j=1}^{n} X_{ij}} ; \qquad (47.103)$$

(iii) For $n_0 = n$,

$$\hat{\lambda}_0^{(L)} = \frac{n}{\sum_{j=1}^{n} X_{(k)j}} \qquad \text{and} \qquad \hat{\lambda}_1^{(L)} = 0. \qquad (47.104)$$

The MLEs are strongly consistent, asymptotically efficient, and asymptotically distributed as $(k+1)$-variate normal.

Proschan and Sullo (1976) described an iterative procedure to solve the likelihood equations in (47.99). They recommended initial (intuitive) estimators

$$\hat{\lambda}_i^{(T)} = \frac{n_i + \frac{n_i n_i^{(c)}}{n_i + n_0(i)}}{\sum_{j=1}^{n} X_{ij}} = \frac{\left(\frac{nn_i}{n - n_i^{(c)}}\right)}{\sum_{j=1}^{n} X_{ij}}, \qquad i = 1, \ldots, k,$$

$$(47.105)$$

$$\hat{\lambda}_0^{(T)} = \frac{n_0 + \sum_{i=1}^{k} \frac{n_0(i) n_i^{(c)}}{n_i + n_0(i)}}{\sum_{j=1}^{n} X_{(k)j}} = \frac{n - \sum_{i=1}^{k} \frac{n_i n_i^{(c)}}{n - n_i^{(c)}}}{\sum_{j=1}^{n} X_{(k)j}}.$$

For the case when $n_i^{(c)} = n$ for some i, $\hat{\lambda}_i^{(T)}$ is taken as $\hat{\lambda}_i^{(L)}$ in (47.100). In fact, this initial estimator coincides with the MLE for the special case when $n_i = 0$ for some i. The bivariate version of the initial estimator for $n_1, n_2 \neq n$ is

$$\hat{\lambda}_i^{(T)} = \frac{\left(\frac{nn_i}{n_i + n_0}\right)}{\sum_{j=1}^{n} X_{ij}} \quad (i = 1, 2) \qquad \text{and} \qquad \hat{\lambda}_0^{(T)} = \frac{n_0\left(1 + \frac{n_2}{n_1 + n_0} + \frac{n_1}{n_2 + n_0}\right)}{\sum_{j=1}^{n} X_{(2)j}}.$$

$$(47.106)$$

Note here that $n_1^{(c)} = n_2$, $n_2^{(c)} = n_1$, and $n_0(1) = n_0(2) = n_0$. These may be compared with Arnold's (1968) estimators, mentioned earlier, given by

$$\hat{\lambda}_i^{(A)} = \frac{n_i\left(1 - \frac{1}{n}\right)}{\sum_{j=1}^{n} X_{(1)j}}, \qquad i = 0, 1, 2. \qquad (47.107)$$

Returning now to the MVE$(2^k - 1)$ distribution in (47.93), $\mathbf{X} \overset{d}{=}$ MVE$(2^k - 1)$ if and only if there exist $2^k - 1$ mutually independent exponential random variables $\{Y_s : s \in S_k\}$ with corresponding failure rates λ_s

such that $X_i = \min\{Y_s : s_i = 1\}$. See Section 3.9 on characterizations for more details.

Proschan and Sullo (1976) also provided maximum likelihood estimators as well as intuitive estimators of parameters λ_s for this multivariate exponential distribution. The expressions are rather complicated, and not all cases are covered. The estimators are strongly consistent, asymptotically efficient, and asymptotically distributed as $(2^k - 1)$-variate normal. They also presented an initial estimator in this case.

There have been a number of ingenious attempts to modify and extend the MOMEDs. Among these, the distributions proposed by Arnold (1975a,b), Langberg, Proschan, and Quinzi (1978), Esary and Marshall (1974), Marshall (1975), and Proschan and Sullo (1974) deserve to be mentioned.

Pickands (1977) defined a vector $\boldsymbol{X} = (X_1, \ldots, X_k)^T$ to be distributed as "exponential" if its joint survival function $\bar{F}(x_1, \ldots, x_k)$ satisfies

$$-t \log \bar{F}\left(\frac{x_1}{t}, \ldots, \frac{x_k}{t}\right) = -\log \bar{F}(x_1, \ldots, x_k) \qquad (47.108)$$

for any $\boldsymbol{x} = (x_1, \ldots, x_k)^T$ and $t > 0$. The MOMED is exponential in the Pickands' sense; see, for example, Galambos and Kotz (1978) for details.

3.3   Block and Basu's Multivariate Exponential

This model is an extension of the ACBED of Block and Basu (1974), described earlier in Section 2, to the multivariate case and constitutes the absolutely continuous part of the MOMED discussed in the preceding section. If $\boldsymbol{X} = (X_1, \ldots, X_k)^T$ represents the joint lifetime of k components, the corresponding $(k+1)$-parameter density function is

$$p_{\boldsymbol{X}}(\boldsymbol{x}) = \frac{\lambda_{i_1} + \lambda_{k+1}}{\alpha} \prod_{r=2}^{k} \lambda_{i_r} \bar{F}_M(\boldsymbol{x}), \quad x_{i_1} > \cdots > x_{i_k},$$
$$i_1 \neq i_2 \neq \cdots \neq i_k = 1, 2, \ldots, k, \qquad (47.109)$$

where

$$\bar{F}_M(\boldsymbol{x}) = \exp\left\{-\sum_{r=1}^{k} \lambda_{i_r} x_{i_r} - \lambda_{k+1} x_{\langle k \rangle}\right\},$$

$$\alpha = \sum_{i_1 \neq \cdots \neq i_k = 1}^{k} \cdots \sum \frac{\prod_{r=2}^{k} \lambda_{i_r}}{\prod_{r=2}^{k} \left(\sum_{j=1}^{r} \lambda_{i_j} + \lambda_{k+1}\right)},$$

and $x_{\langle k \rangle}$ is $\max(x_1, \ldots, x_k)$.

The failure times $X_1, \ldots, X_k$ are independent iff $\lambda_{k+1} = 0$. The condition $\lambda_1 = \cdots = \lambda_k$ implies symmetry and it is equivalent to identical marginals of all the k components. The model in (47.109) satisfies the lack of memory property, but the marginals are weighted combinations of exponentials. The marginals are exponential only in the independent case; see Block (1975a).

Let $\boldsymbol{X}_1, \ldots, \boldsymbol{X}_n$ be a random sample from (47.109). Let n_{i_1} denote the number of observations with $X_{i_1} > \max(X_{i_2}, \ldots, X_{i_k})$. The expected value of n_{i_1} is

$$E[n_{i_1}] = \frac{n}{\alpha} \sum_{i_2 \neq \cdots \neq i_k = 1}^{k} \cdots \sum^{k} \prod_{r=2}^{k} \frac{\lambda_{i_r}}{\sum_{j=1}^{r} \lambda_{i_j} + \lambda_{k+1}} . \qquad (47.110)$$

The likelihood equations are

$$\frac{\partial \log L}{\partial \lambda_{i_1}} = -n\,\alpha_{i_1} + \frac{n_{i_1}}{\lambda_{i_1} + \lambda_{k+1}} + \frac{n - n_{i_1}}{\lambda_{i_1}} - \sum_{j=1}^{n} X_{i_1 j} = 0,$$

$$i_1 = 1, 2, \ldots, k, \qquad (47.111)$$

$$\frac{\partial \log L}{\partial \lambda_{k+1}} = -n\,\alpha_{k+1} + \sum_{i_1=1}^{k} \frac{n_{i_1}}{\lambda_{i_1} + \lambda_{k+1}} - \sum_{j=1}^{n} X_{\langle k \rangle j} = 0,$$

where $\alpha_{i_1} = \frac{\partial \log \alpha}{\partial \lambda_{i_1}}$, $i_1 = 1, \ldots, k+1$.

Each pair $(X_{i_1}, X_{i_2})^T$ (for $i_1 \neq i_2 = 1, \ldots, k$) follows ACBED of Block and Basu (1974), in which case Hanagal and Kale (1991a) obtained consistent estimators $\tilde{\lambda}_{i_1}$, $\tilde{\lambda}_{i_2}$ and $\tilde{\lambda}_3$. Hanagal (1993) suggested to use $k-1$ different consistent estimators of λ_i ($i = 1, \ldots, k$) and $\binom{k}{2}$ different consistent estimators of λ_{k+1} by considering all $\binom{k}{2}$ different pairs of components. The average of these consistent estimators is also consistent for the corresponding parameters, and these averages $(\bar{\lambda}_1, \bar{\lambda}_2, \ldots, \bar{\lambda}_{k+1})^T$ are used as a trial solution for obtaining the MLE $\hat{\boldsymbol{\lambda}} = (\hat{\lambda}_1, \ldots, \hat{\lambda}_{k+1})$ by the Newton–Raphson method or Fisher's method of scoring. The Fisher information matrix

$$n\boldsymbol{I}(\boldsymbol{\lambda}) = ((n\,I_{ij})) = \left(\left(E\left[-\frac{\partial^2 \log L}{\partial \lambda_i \partial \lambda_j}\right]\right)\right), \quad i, j = 1, \ldots, k+1$$

is positive definite in this case, and $\sqrt{n}(\hat{\boldsymbol{\lambda}} - \boldsymbol{\lambda})$ has asymptotic multivariate normal distribution with mean vector $\boldsymbol{0}$ and variance–covariance matrix $\boldsymbol{I}^{-1}(\boldsymbol{\lambda})$.

Weier and Basu (1980) represented the density alternatively as

$$p(x_1, \ldots, x_k) = \left(\frac{1}{k!} \sum_{r=1}^{k} C_r \right) \exp \left\{ -\sum_{r=1}^{k} (C_r - C_{r+1}) x_{\langle r \rangle} \right\}, \quad (47.112)$$

where C_r's $(r = 1, \ldots, k)$ are parameters, $C_{k+1} \equiv 0$, and $x_{\langle r \rangle}$ is the r-th smallest component of the k-variate vector $(x_1, \ldots, x_k)^T$. This form is more appropriate for conducting tests of independence. Suppose $\theta_i = C_i/(k - i + 1)$ and

$$U_i = (k - i + 1) \sum_{j=1}^{n} \left(X_{\langle i \rangle j} - X_{\langle i-1 \rangle j} \right), \qquad X_{\langle 0 \rangle j} \equiv 0,$$

U_i's being independent Gamma(θ_i), with densities $p(u_i, \theta_i)$ for $i = 1, \ldots, k$. Then, the hypothesis regarding independence becomes $H_0 : \theta_1 = \cdots = \theta_k$ vs. $H_1 : \theta_1 < \cdots < \theta_k$, and the likelihood function in terms of U_i's becomes $\prod_{i=1}^{k} p(u_i, \theta_i)$. The testing problem thus reduces to that of homogeneity of ordered gamma distributions.

A trivariate Block–Basu model has been proposed by Weier and Basu (1980), viewing it as a special case of trivariate Marshall and Olkin's distribution, with joint density function

$$
\begin{aligned}
&p(x_1, x_2, x_3) \\
&= \frac{(3\lambda_0 + \lambda_4)(2\lambda_0 + \lambda_4)(\lambda_0 + \lambda_4)}{6} \cdot \exp\{-\lambda_0(x_1 + x_2 + x_3) \\
&\quad - \lambda_4 \max(x_1, x_2, x_3)\}, \quad x_1, x_2, x_3 > 0, \ \lambda_0 > 0, \ \lambda_4 \geq 0.
\end{aligned}
$$

All the marginal distributions in this case are $Exp(\lambda_0)$ and independence corresponds to the case $\lambda_4 = 0$.

3.4   Olkin and Tong's Multivariate Exponential

Olkin and Tong (1994) studied an important subclass of MOMEDs. Let $U_1, \ldots, U_k, V_1, \ldots, V_k$ and W be independent exponential random variables with $E[U_i] = 1/\lambda_1$, $E[V_i] = 1/\lambda_2$ $(i = 1, \ldots, k)$, and $E[W] = 1/\lambda_0$. Let $\boldsymbol{K} = (K_1, \ldots, K_k)^T$ be a vector of non-negative integers with

$$\sum_{s=1}^{k} K_s = k, \quad K_1 \geq \cdots \geq K_r \geq 1, \ K_{r+1} = \cdots = K_k = 0 \quad (47.113)$$

for some $r \leq k$. For a given $\boldsymbol{K}$, let $\boldsymbol{X}(\boldsymbol{K}) = (X_1, \ldots, X_k)^T$ be a k-dimensional multivariate exponential random variable defined by

$$X_j = \begin{cases} \min(U_j, V_1, W), & j = 1, \ldots, K_1 \\ \min(U_j, V_2, W), & j = K_1 + 1, \ldots, K_1 + K_2 \\ \quad \vdots \\ \min(U_j, V_r, W), & j = K_1 + \cdots + K_{r-1} + 1, \ldots, k. \end{cases} \qquad (47.114)$$

Note that the distribution of $(X_1, \ldots, X_k)^T$ belongs to a subclass of the MOMED family. The latter, as mentioned earlier in Section 3.2, requires $2^k - 1$ independent variables to generate a k-variate exponential distribution. The univariate marginal distributions of X_j's are exponential with mean $1/(\lambda_0 + \lambda_1 + \lambda_2)$.

The joint distribution of the X_i's is exchangeable when $\boldsymbol{K} = (k, 0, \ldots, 0)^T$ and also when $\boldsymbol{K} = (1, \ldots, 1)^T$. The components $X_j = \min(U_j, V_1, W)$, $j = 1, \ldots, n$, of $\boldsymbol{X}(k, 0, \ldots, 0)$ are more positively dependent than $X_j = \min(U_j, V_j, W)$, $j = 1, \ldots, n$, of $\boldsymbol{X}(1, \ldots, 1)$. (Note that the former depends on the same variable V_1, while the latter allows for different V_j's.)

For a fixed but arbitrary k, $\boldsymbol{\lambda} = (\lambda_0, \lambda_1, \lambda_2)^T$ and t, let $\boldsymbol{K}$ and $\boldsymbol{K}'$ be two vectors satisfying (47.113). If $\boldsymbol{K} > \boldsymbol{K}'$ where $>$ denotes majorization order, then Olkin and Tong (1994) have established that

$$\bar{F}_{\boldsymbol{K}, \boldsymbol{\lambda}}(t, \ldots, t) \geq \bar{F}_{\boldsymbol{K}', \boldsymbol{\lambda}}(t, \ldots, t).$$

Also, if $\boldsymbol{\lambda} \overset{t}{<} \boldsymbol{\lambda}^*$, namely, $\lambda_1 \leq \lambda_1^*$, $\lambda_1 + \lambda_2 \leq \lambda_1^* + \lambda_2^*$ and $\lambda_1 + \lambda_2 + \lambda_0 = \lambda_1^* + \lambda_2^* + \lambda_0^*$ (note that the ordering $\overset{t}{>}$, unlike majorization, does not require an ordering of elements), then for fixed $k, \boldsymbol{K}$ and t

$$\bar{F}_{\boldsymbol{K}, \boldsymbol{\lambda}}(x_1, \ldots, x_k) > \bar{F}_{\boldsymbol{K}, \boldsymbol{\lambda}^*}(x_1, \ldots, x_k)$$

for all $\boldsymbol{x} \in \mathbb{R}_+^k$, provided that $\boldsymbol{\lambda} \neq \boldsymbol{\lambda}^*$.

3.5 Marshall and Olkin's Multivariate Exponential with Limited Memory

Marshall and Olkin (1991, 1995) studied multivariate exponential distributions with limited memory—that is, having a property of the form

$$\bar{F}(\boldsymbol{x} + \boldsymbol{y}) = \bar{F}(\boldsymbol{x})\bar{F}(\boldsymbol{y}) \qquad \text{for all } (\boldsymbol{x}, \boldsymbol{y})^T \in \mathcal{S}, \qquad (47.115)$$

where $\mathcal{S}$ is a proper subset of $\mathbb{R}_+^{2k}$. They considered the following classes:

$$\mathcal{C}(A) = \{\text{Multivariate exponential with independent marginals}\},$$
$$\mathcal{C}(B) = \{\text{MOMEDs}\},$$
$$\mathcal{C}(C) = \{\text{Multivariate exponential with exponential scaled minima}\},$$
$$\mathcal{C}(D) = \{\text{Multivariate exponential with exponential minima}\},$$

and

$$\mathcal{C}(E) = \{\text{Distributions with exponential marginals}\}.$$

(In these, "multivariate exponential" is a distribution with exponential univariate marginals.) These classes are all characterized as a family of solutions of (47.115) for an appropriate choice of $\mathcal{S}$. For $\mathcal{C}(A)$, $\mathcal{S} \equiv \mathbb{R}_+^{2k}$; for $\mathcal{C}(B)$, $\mathcal{S} = \{(\boldsymbol{x}, \boldsymbol{y}) : \boldsymbol{x} \in \mathbb{R}_+^k$ and $\boldsymbol{x}$ and $\boldsymbol{y}$ are similarly ordered$\}$; for $\mathcal{C}(C)$, $\mathcal{S} = \{(\boldsymbol{x}, \boldsymbol{y}) : \boldsymbol{x} \in \mathbb{R}_+^k$ and $\boldsymbol{y} = a\boldsymbol{x}$ for some $a > 0\}$; for $\mathcal{C}(D)$, $\mathcal{S} = \{(\boldsymbol{x}, \boldsymbol{y}) : \boldsymbol{x} \in \mathbb{R}_+^k$ and all nonzero components of $\boldsymbol{x}$ are equal and $\boldsymbol{y} = a\boldsymbol{x}\}$; finally, for $\mathcal{C}(E)$, $\mathcal{S} = \{(\boldsymbol{x}, \boldsymbol{y}) : \boldsymbol{x} \in \mathbb{R}_+^k$ has only one non-zero component and $\boldsymbol{y} = a\boldsymbol{x}$ for $a > 0$, or $\boldsymbol{x} = \boldsymbol{0}$, or $\boldsymbol{y} = \boldsymbol{0}\}$. Evidently, $\mathcal{C}(A) \subset \mathcal{C}(B) \subset \mathcal{C}(C) \subset \mathcal{C}(D) \subset \mathcal{C}(E)$, and these classes are distinct.

3.6   Moran and Downton's Multivariate Exponential

Al-Saadi and Young (1982) generalized Moran and Downton's bivariate exponential distribution, discussed earlier in Section 2.7, to the equicorrelated multivariate case as follows. Let $X_i = \sum_{j=1}^M Y_{ij}$, where Y_{ij}'s are independent and identically distributed random variables with density function

$$p_{Y_i}(y) = \frac{\theta_i}{1 - \rho} \, e^{-\theta_i y/(1-\rho)}, \qquad y > 0, \; i = 1, \ldots, k; \qquad (47.116)$$

let M have a geometric distribution with probability mass function

$$\Pr[M = m] = (1 - \rho)\rho^{m-1}, \qquad 0 \le \rho < 1, \; m = 1, 2, \ldots . \qquad (47.117)$$

Then, conditional on $M = m$, the distribution of X_i is gamma with probability density function

$$f_i(x) = \left(\frac{\theta_i}{1 - \rho}\right)^m \frac{x^{m-1}}{(m-1)!} \, e^{-\theta_i x/(1-\rho)}, \qquad x > 0,$$

and the joint unconditional density function of $\boldsymbol{X} = (X_1, \ldots, X_k)^T$ is

$$p_{\boldsymbol{X}}(\boldsymbol{x}) = \sum_{m=1}^{\infty} \Pr[M = m] \prod_{i=1}^k f_i(x_i)$$

$$= \frac{\theta_1 \cdots \theta_k}{(1-\rho)^{k-1}} \exp\left\{-\frac{1}{1-\rho} \sum_{i=1}^{k} \theta_i x_i\right\} S_k\left(\frac{\rho\theta_1 x_1 \theta_2 x_2 \cdots \theta_k x_k}{(1-\rho)^k}\right),$$

$$x_i > 0, \quad i = 1, \ldots, k, \quad (47.118)$$

where $S_k(z) = \sum_{i=0}^{\infty} z^i/(i!)^k$.

The marginal distribution of X_i is exponential with parameter θ_i ($i = 1, \ldots, k$). Noting that $I_0(z)$—the modified Bessel function of the first kind of order zero—is $I_0(z) = S_2(z^2/4)$ [cf. (47.1)], we observe that (47.118) reduces readily to the bivariate Moran and Downton's density presented in Section 2.7. The mixed moment of order $(r_1, \ldots, r_k)$ is

$$E\left[X_1^{r_1} \cdots X_k^{r_k}\right] = \sum_{j_1=0}^{r_1} \cdots \sum_{j_k=0}^{r_k} \frac{(r-j)!\rho^{r-j}(1-\rho)^j}{\theta_1^{r_1} \theta_2^{r_2} \cdots \theta_k^{r_k}}$$

$$\times \prod_{i=1}^{k} \left[\binom{r_i}{j_i} \frac{\left\{r_i + \sum_{\ell=1}^{i-1}(r_\ell - j_\ell)\right\}!}{\left\{\sum_{\ell=1}^{i}(r_\ell - j_\ell)\right\}!}\right],$$

$$(47.119)$$

where $r = \sum_{\ell=1}^{k} r_\ell$, $j = \sum_{\ell=1}^{k} j_\ell$, and $r_0 = j_0 = 0$. In particular, setting $r_s = r_t = 1$ and $r_i = 0$, for $i \neq s, t$, we obtain

$$E[X_s X_t] = \frac{1+\rho}{\theta_s \theta_t}, \quad s = 1, \ldots, k-1; \ t = s+1, \ldots, k,$$

which shows that each pair of random variables has correlation coefficient equal to ρ.

3.7 Raftery's Multivariate Exponential

Raftery (1984) and O'Cinneide and Raftery (1989) studied a multivariate exponential distribution which is defined as follows. Suppose that $Y_1, \ldots, Y_k$ and $Z_1, \ldots, Z_\ell$ are independent exponential(λ) random variables and that $(J_1, \ldots, J_k)$ is a random vector taking on values in $\{0, 1, \ldots, \ell\}^k$ with marginals

$$\Pr[J_i = 0] = 1 - \pi_i \quad \text{and} \quad \Pr[J_i = j] = \pi_{ij},$$

$$i = 1, \ldots, k, \ j = 1, \ldots, \ell, \quad (47.120)$$

where $\pi_i = \sum_{j=1}^{\ell} \pi_{ij}$. Let $Z_0 \equiv 0$. Then, the model for $X_1, \ldots, X_k$ is

$$X_i = (1 - \pi_i)Y_i + Z_{J_i}, \quad i = 1, \ldots, k. \quad (47.121)$$

The main properties of this model are similar to those of the multivariate normal distribution in the sense that univariate marginals are exponential while bivariate marginals belong to Raftery's bivariate exponential distribution, given by

$$X_i = (1 - \pi_i)Y_i + I_i Z, \qquad i = 1, 2, \tag{47.122}$$

a linear combination of the underlying independent random variables. Here, $Y_1, Y_2,$ and Z are independent exponential(λ) random variables, and I_i's ($i = 1, 2$) are binary 0-1 random variables with

$$\Pr[I_i = 1] = \pi_i \quad i = 1, 2 \quad \text{and} \quad \Pr[I_1 = i, I_2 = j] = p_{ij} \quad i, j = 0, 1.$$

When $\ell = 1$, $p_{11} = \Pr[J_i = J_j = 1]$ and moreover

$$\rho_{ij} = \mathrm{corr}(X_i, X_j) = \alpha_{ij} + \beta_{ij} + \pi_i + \pi_j - \pi_i \pi_j - 1$$

with $\alpha_{ij} = \Pr[J_i = J_j = 0]$ and $\beta_{ij} = \Pr[J_i = J_j \neq 0]$ so that the correlation structure is independent of the marginal distributions. Unfortunately, the dependence structure involves $(\ell + 1)^k - 1$ parameters. Raftery (1984) has therefore recommended to constrain the bivariate marginal distributions to be exchangeable. For example, in the three-dimensional case, assume X_i ($i = 1, 2, 3$) are such that $0 \leq \rho_{12} \leq \rho_{23} \leq \rho_{31} \leq 1$. Taking $\ell = 1$ and $\pi_i = \pi$, we have from (47.121)

$$X_i = (1 - \pi)Y + I_i Z,$$

where $(I_1, I_2, I_3)^T$ is a vector of binary 0-1 random variables. Then, $p_{abc} = \Pr[I_1 = a, I_2 = b, I_3 = c]$ are expressed as follows (since $\Pr[I_i = 1] = \pi$ and $2\Pr[I_i = 1, I_j = 1] - \pi^2 = \rho_{ij}$, $i, j = 1, 2, 3$, provided that we search for a solution with the largest p_{111}):

$$p_{000} = 1 - 3\pi + \pi^2 + \tfrac{1}{2}(\rho_{23} + \rho_{31}), \qquad p_{100} = \pi - \tfrac{1}{2}\pi^2 - \tfrac{1}{2}\rho_{31},$$

$$p_{001} = \pi - \tfrac{1}{2}\pi^2 - \tfrac{1}{2}(\rho_{31} + \rho_{23} - \rho_{12}), \qquad p_{101} = \tfrac{1}{2}(\rho_{31} - \rho_{12}),$$

$$p_{010} = \pi - \tfrac{1}{2}\pi^2 - \tfrac{1}{2}\rho_{23}, \qquad p_{110} = 0,$$

$$p_{011} = \tfrac{1}{2}(\rho_{23} - \rho_{12}), \qquad p_{111} = \tfrac{1}{2}(\rho_{12} + \pi^2).$$

We thus have here only four parameters (one more than the trivariate normal distribution; see Chapter 46). The model covers full range of correlations and seems to be useful in asymmetric situations.

In the bivariate case, Nagaraja and Baggs (1996) have discussed the joint and marginal distributions of order statistics $X_{(1)} = \min(X_1, X_2)$ and $X_{(2)} = \max(X_1, X_2)$ as well as some reliability properties of these order statistics.

O'Cinneide and Raftery (1989) have shown that the multivariate exponential distribution discussion here is a *multivariate phase type (MPH) distribution* (a joint distribution of two or more finite hitting times in a regular finite-state continuous-time time-homogeneous Markov chain) introduced by Assaf *et al.* (1984).

3.8 Krishnamoorthy and Parthasarathy's Multivariate Exponential

A further example of a multivariate exponential distribution can be obtained by taking $\nu = 2$ in the multivariate gamma distribution of Krishnamoorthy and Parthasarathy (1951); see Chapter 48 for more details. The joint characteristic function is

$$E\left[e^{i(t_1 X_1 + \cdots + t_k X_k)}\right] = |I_k - 2i RD_t|^{-1}, \tag{47.123}$$

where R is a correlation matrix, I_k is an identity matrix of order k, and $D_t = \text{Diag}(t_1, \ldots, t_k)$. Since $|I_k - 2i RD_t|$ is a polynomial in $(1 - 2it_1), \ldots, (1 - 2it_k)$, the joint distribution of $(X_1, \ldots, X_k)^T$ can be expressed formally as a mixture of a finite number of χ^2-distributions.

By considering two independent copies of Krishnamoorthy and Parthasarathy's multivariate gamma variables of index $\frac{1}{2}$, and adding them, one could obtain a multivariate exponential distribution. Kent (1983) has shown the equivalence of the distribution so obtained and the distribution derived from considering the *sojourn time vector* of a birth-death process up to a first passage time. Recall that in the univariate case [see Chapter 18 of Johnson, Kotz, and Balakrishnan (1994)], we have two derivations of exponential distributions—one based on the lack of memory property which is equivalent to the waiting time spent in a given state of continuous-time Markov process before jumping into a new state, and the other, based on the normal distribution, as the distribution of $X_1^2 + X_2^2$ when X_1 and X_2 are independent normal random variables with zero mean and same variance.

3.9 Characterizations

Some bivariate exponential distributions were characterized earlier, in Section 2. Here, we will present several basic characterizations of the models discussed in this section.

1. A random vector $\boldsymbol{X}$ has MOMED with joint survival function in (47.93) if and only if there exists a collection H_J, $J \in \mathcal{J}$ (where $\mathcal{J}$ is the class of nonempty subsets of $\{1, 2, \ldots, n\}$) of independent exponential random variables such that $X_i = \min(H_J, J \in \mathcal{J}, i \in J)$, $i = 1, \ldots, k$. More explicitly, denote by V the set of vectors $\boldsymbol{v} = (v_1, \ldots, v_k)$ where each v_i is either 0 or 1, but $(v_1, \ldots, v_k) \neq (0, \ldots, 0)$. Eq. (47.93) can then be rewritten as

$$\bar{F}(x_1, \ldots, x_k) = \exp\left\{-\sum_V \lambda_{v_1, \ldots, v_k} \max(x_1 v_1, \ldots, x_k v_k)\right\},$$
$$x_i \geq 0 \ (i = 1, \ldots, k).$$

The characterization of (47.93) in terms of minima asserts the existence of $2^k - 1$ independent exponential random variables $Z_{\boldsymbol{v}}$, $\boldsymbol{v} \in V$, such that $X_i = \min_V \{Z_{\boldsymbol{v}} | v_i = 1\}$.

2. The basic characterization of the MOMED as a unique k-dimensional distribution satisfying the lack of memory property for $n = k$ with all $(k - 1)$-dimensional marginals being MOMED, can be rephrased by saying that the lack of memory property holds for any n-dimensional marginal for $n = 1, 2, \ldots, k$. (The lack of memory property for $n = 1$ yields the exponentiality of the univariate marginals.)

3. A generalization of the proof of Result **2** leads to the following result. The absolute continuity of the joint distribution coupled with the lack of memory property for $n = 2, \ldots, k$ and the exponentiality of the univariate marginals results in a MED with independent exponential components. This result served as a stimulus for deriving the ACMED by Block and Basu (1974).

4. A random vector $(X_1, \ldots, X_k)^T$ has the Freund–Weinman–Block MED in (47.90) if and only if [see Basu and Sun (1997)]

   (a) it has constant $r(t)$, and constant

   $$\Pr[\min(X_i : i \neq j) > X_j | \min(X_1, \ldots, X_n) = t],$$

   where the summation of these over $j = 1, \ldots, k$ is 1;

   (b) given $\min(X_i : i \neq j) > X_j = x_j$, the conditional distribution of $(X_i : i \neq j)$ is the $(k - 1)$-dimensional Freund–Weinman–Block MED for all j.

Here, $r(t) = -d \log \Pr[X_1 > t, \ldots, X_k > t]/dt$ is called the one-stage constant failure rate (in Basu-Sun sense). (Note that in the bivariate case absolute continuity and constant $r(t)$ and a condition on $\min(X_1, X_2)$ results in Freund's bivariate exponential distribution.)

5. Basu and Sun (1997) proposed a new concept of *total failure rate* (not to be confused with scalar failure rate and the vector-valued failure rate). If the joint survival function $\bar{F}(x_1, \ldots, x_k)$ is absolutely continuous on $x_i \neq x_j$ ($i \neq j$), the vector

$$(r_{R_L, D_\ell}(t|x_{D_\ell}) \text{ for } x_{R_\ell} > x_{D_\ell}, \ R_L = \{i_\ell, i_{\ell+1}, \ldots, i_k\}, \ \ell = 1, \ldots, k)$$

is called the *total failure rate* of $(X_1, \ldots, X_k)^T$, where

$$r_{R_1, D_1}(t|x_{D_1}) = -d \log \Pr[X_1 > t, \ldots, X_k > t]/dt$$

(one-stage total failure rate) and for $\ell = 2, \ldots, k$

$$r_{R_L, D_\ell}(t|x_{D_\ell}) = -d \log \Pr[X_{R_L} > t | X_{D_\ell} = x_{D_\ell}]/dt \text{ for } D_\ell \neq \emptyset.$$

This concept has been used by Basu and Sun (1997) to characterize the MOMEDs.

The vector $(X_1, \ldots, X_k)^T$ is distributed as MOMED if and only if $(X_1, \ldots, X_k)^T$ and all its n-dimensional marginals have constant total failure rates, $n = 1, \ldots, k-1$. Alternatively, the vector $(X_1, \ldots, X_k)^T$ is distributed as MOMED if and only if $(X_1, \ldots, X_k)^T$ has a constant total failure rate and all its $(k-1)$-dimensional marginals are MOMEDs.

6. As mentioned above, there are various definitions of failure (hazard) rate functions. The scalar quantity $r(\boldsymbol{x}) = p(\boldsymbol{x})/\bar{F}(\boldsymbol{x})$, due to Basu, is useful for characterizing bivariate exponential distributions. For the multivariate case, the concept of vector-valued multivariate hazard rate is useful. It is defined as

$$\begin{aligned}
h_{\boldsymbol{X}}(\boldsymbol{x}) &= \nabla H_{\boldsymbol{X}}(\boldsymbol{x}) \\
&= \left(\frac{\partial}{\partial x_1}, \ldots, \frac{\partial}{\partial x_k}\right)^T \{-\log \bar{F}_{\boldsymbol{X}}(\boldsymbol{x})\} \\
&= \left(-\frac{\partial}{\partial x_1} \log \bar{F}_{\boldsymbol{X}}(\boldsymbol{x}), \ldots, -\frac{\partial}{\partial x_k} \log \bar{F}_{\boldsymbol{X}}(\boldsymbol{x})\right)^T \\
&= \left(h_{\boldsymbol{X}}(\boldsymbol{x})_1, \ldots, h_{\boldsymbol{X}}(\boldsymbol{x})_k\right)^T ;
\end{aligned}$$

see Johnson and Kotz (1975). Evidently, the only multivariate distribution for which the multivariate hazard gradient is strictly constant (i.e., $h_{\boldsymbol{X}}(\boldsymbol{x}) = \boldsymbol{c}$, where $\boldsymbol{c} = (c_1, \ldots, c_k)^T$ is absolutely constant with respect to all variables) is the MED with independent exponential marginals.

7. The vector-valued multivariate hazard rate $h_{\boldsymbol{X}}(\boldsymbol{x})$ is continuous and locally constant (i.e., the ith component does not depend on x_i, $i = 1, \ldots, k$) if and only if the joint distribution of $\boldsymbol{X}$ is Gumbel's Type 1 multivariate distribution with survival function

$$\bar{F}_{\boldsymbol{X}}(\boldsymbol{x}) = \exp\left\{ -\sum_{i=1}^{k} \theta_i x_i - \sum_{i<j}\sum \theta_{ij} x_i x_j - \cdots - \theta_{1\ldots k} x_1 \cdots x_k \right\}$$

with θ's ≥ 0; see Johnson and Kotz (1975).

8. If the hazard components $h_{\boldsymbol{X}}(\boldsymbol{x})_i$, $i = 1, \ldots, k$, are stationary in $x_1, \ldots, x_k$ and $\bar{F}_{\boldsymbol{X}}(\boldsymbol{x})$ is absolutely continuous, then $\bar{F}_{\boldsymbol{X}}(\boldsymbol{x})$ possesses the lack of memory property and conversely.

9. Obretenov (1985) modified the characterization based on the lack of memory property and also characterized the MOMED by the integrated lack of memory property. Noting that the lack of memory property can be written as

$$\bar{F}(x_1 + t, \ldots, x_k + t) = \bar{F}(x_1, \ldots, x_k)\bar{F}(t, \ldots, t),$$

—that is, $\bar{F}(\boldsymbol{x} + t\boldsymbol{1}) = \bar{F}(\boldsymbol{x})\bar{F}(t\boldsymbol{1})$, where $\boldsymbol{x} = (x_1, \ldots, x_k)^T$ and $\boldsymbol{1} = (1, \ldots, 1)^T$, Obretenov defined a weak lack of memory property by

$$\bar{F}(t\boldsymbol{1} + \boldsymbol{x} \circ \boldsymbol{a}) = \bar{F}(t\boldsymbol{1})\bar{F}(\boldsymbol{x} \circ \boldsymbol{a}),$$

where $\boldsymbol{x} \circ \boldsymbol{a} = (x_1 a_1, \ldots, x_k a_k)^T$ and

$$\boldsymbol{a} \in \boldsymbol{E} = \{\boldsymbol{a} : \text{ only one of } a_i\text{'s is 0 and the others are 1}\}.$$

Now, if the joint survival function $\bar{F}$ has weak lack of memory property and $\bar{F}(t\boldsymbol{1})$ is an exponential function of t and all marginals of $\bar{F}$ are MOMEDs, then $\bar{F}$ is a MOMED. Moreover, if $\bar{F}$ has all its marginals of Marshall and Olkin's type and also satisfies the equation

$$\int_0^\infty G(x_1 + t, \ldots, x_k + t)\mu(dt) = G(x_1, \ldots, x_k)$$

for some Borel measure $\mu(t)$ on $\mathbb{R}^+$, then $\bar{F}$ is a k-dimensional MOMED. This result is based on the well-known Lau-Rao (1982) result on the integrated Cauchy functional equation; see Rao and Shanbhag (1994).

4 MULTIVARIATE WEIBULL DISTRIBUTIONS

Since the Weibull distribution can be obtained from an exponential distribution by power transformation [see Chapter 21 of Johnson, Kotz, and Balakrishnan (1994)], a multivariate Weibull distribution can in general be obtained from a multivariate exponential distribution by power transformations; see (e) below.

Marshall and Olkin (1967a) and Lee and Thompson (1974) discussed multivariate Weibull distributions of the form

$$
\begin{aligned}
\bar{F}_{X_1,\ldots,X_k}(x_1,\ldots,x_k) &= \Pr[X_1 > x_1, \ldots, X_k > x_k] \\
&= \exp\left\{-\sum_J \lambda_J \max(x_i^\alpha)\right\}, \\
&\quad x_i > 0 \ (i = 1, \ldots, k), \ \alpha > 0,
\end{aligned}
$$
(47.124)

$\lambda_J > 0$ for $J \in \mathcal{J}$, where the sets J are elements of the class $\mathcal{J}$ of nonempty subsets of $\{1, 2, \ldots, k\}$ having the property that for each i, $i \in J$ for some $J \in \mathcal{J}$. This is a generalization of the MOMED, MVE$(k+1, \boldsymbol{\lambda})$, which is the case when $\alpha = 1$ in (47.124).

Lee (1979) considered several classes of multivariate Weibull distributions as presented below:

(a) $X_1, \ldots, X_k$ are independent and X_i has a Weibull distribution of the form $\bar{F}_{X_i}(x_i) = e^{-\lambda_i x_i^\alpha}$, $x_i \geq 0$ $(i = 1, \ldots, k)$;

(b) $X_1, \ldots, X_k$ have a multivariate distribution generated from independent Weibull variables by setting

$$
X_i = \min(Z_J : i \in J), \qquad i = 1, \ldots, k,
$$

where the sets J are elements of class $\mathcal{J}$ of nonempty subsets of $\{1, \ldots, k\}$ having the property that for each i, $i \in J$ for some $J \in \mathcal{J}$, and the random variables Z_J, $J \in \mathcal{J}$, are independent having Weibull distributions of the form $\bar{F}_J(x) = \exp(-\lambda_J x^\alpha)$. This is equivalent to Marshall and Olkin's (1967) multivariate Weibull distribution in (47.124).

(c) $X_1, \ldots, X_k$ have a joint distribution that satisfies

$$\Pr\left[\min_i(a_i x_i) > x\right] = \exp\left\{-K(\boldsymbol{a})x^\alpha\right\}, \qquad x \geq 0 \quad (47.125)$$

for some $\alpha > 0$, where $a_i > 0$ $(i = 1, \ldots, k)$ are arbitrary.

(d) $X_1, \ldots, X_k$ have a joint distribution satisfying

$$\Pr\left[\min_{i \in S} X_i > x\right] = \exp\left(-\lambda_S \, x^\alpha\right)$$

for some $\lambda_S > 0$ and all nonempty subsets S of $\{1, \ldots, k\}$;

(e) Each X_i $(i = 1, \ldots, k)$ has a Weibull distribution of the form $\bar{F}_{X_i}(x_i) = \exp(-\lambda_i x_i^{\alpha_i})$, $x_i \geq 0$, $\alpha_i > 0$ $(i = 1, \ldots, k)$. In other words, by specifying $Y_1, \ldots, Y_k$ to have a multivariate distribution with exponential marginals, $X_i = Y_i^{1/\alpha_i}$ $(i = 1, \ldots, k)$ produce a multivariate distribution having Weibull marginals.

The class **(c)** contains class **(a)** and class **(b)**, but there are also other multivariate Weibull distributions belonging to class **(c)**.

Basically, two types of bivariate Weibull distributions emerge: one of the forms

A1 : $\quad \bar{F}(x_1, x_2) = \exp\left\{-\left(\lambda_1 c_1^\alpha x_1^\alpha + \lambda_2 c_2^\alpha x_2^\alpha + \lambda_{12}\max(c_1^\alpha x_1^\alpha, c_2^\alpha x_2^\alpha)\right)\right\}$

or

A2 : $\quad \bar{F}(x_1, x_2) = \exp\left\{-\left(\lambda_1 x_1^{\alpha_1} + \lambda_2 x_2^{\alpha_2} + \lambda_{12}\max(x_1^{\alpha_1}, x_2^{\alpha_2})\right)\right\}$

on one hand, and that of absolutely continuous form

B : $\quad \bar{F}(x_1, x_2) = \exp\left\{-(\lambda_1 x_1^\beta + \lambda_2 x_2^\beta)^\gamma\right\}$

and their mixtures. In the form **A2**, the cases $\alpha_1 \neq \alpha_2$ and $\alpha_1 = \alpha_2$ yield quite different distributions. In particular, the situation $\alpha_1 = \alpha_2$ arises from independent Weibull distributions $\Pr[Z_1 > x] = \exp(-\lambda_1 x^\alpha)$, $\Pr[Z_2 > x] = \exp(-\lambda_2 x^\alpha)$ and $\Pr[Z_{12} > x] = \exp(-\lambda_{12} x^\alpha)$ by the representation $X_1 = \min(Z_1, Z_{12})$ and $X_2 = \min(Z_2, Z_{12})$ as specified in **(b)**.

In the form **A1**, the distribution of $(X_1, X_2)^T$ has a singular component on $c_1 x_1 = c_2 x_2$ and thus differs (for $c_1 \neq c_2$) from the form **B**. The bivariate

Weibull distribution in form **B** has been used prominently in dependent failure-times analysis [see, for example, Hougaard (1986)] in the form

$$\Pr[T_1 > t_1, T_2 > t_2] = \exp\left\{-\frac{\delta}{\alpha}\,(t_1^\gamma + t_2^\gamma)^\alpha\right\}. \tag{47.126}$$

In this case, the marginal distributions are Weibull with shape parameter $\alpha\gamma$, and the distribution of the minimum is also Weibull with the same shape parameter. Thus, both individuals have equal probability $\left(\frac{1}{2}\right)$ of dying first independently of the time of the first death. Hence, if only the marginal distributions or the minimum lifetime is observed, it is impossible to identify the parameters.

Returning to the general form **B** now, the variables X_1 and X_2 can be represented in terms of independent random variables. Let $Z_i = \lambda_i X_i^\beta$ $(i = 1, 2)$. Consider now the transformation

$$U = \frac{Z_1}{Z_1 + Z_2} \qquad \text{and} \qquad S = (Z_1 + Z_2)^\gamma.$$

It is known that U and S are independent random variables with U uniformly distributed over $(0,1)$ and S has the density

$$f(s) = (1 - \gamma + \gamma s)e^{-s}, \qquad s > 0.$$

Thus, $Z_1 = US^{1/\gamma}$ and $Z_2 = (1 - U)S^{1/\gamma}$ thus provide representations of X_1 and X_2 in terms of independent random variables U and S.

Hougaard (1986, 1989) presented a multivariate Weibull distribution with joint survival function

$$\bar{F}(x_1, \ldots, x_k) = \exp\left\{-\left(\sum_{i=1}^k \theta_i x_i^p\right)^\ell\right\}, \quad p \geq 0,\ \ell > 0,\ x_i \geq 0. \tag{47.127}$$

This distribution has been used *inter alia* to test the hypothesis of independence of litter mates in the proportional hazards model. In the bivariate case, with a different parameterization, the joint density is given by

$$\begin{aligned}
p(x_1, x_2) &= \kappa^2 \varepsilon_1^\phi \varepsilon_2^\phi x_1^{\kappa\phi-1} x_2^{\kappa\phi-1} \\
&\quad \times \left\{ \left(\varepsilon_1^\phi x_1^{\kappa\phi} + \varepsilon_2^\phi x_2^{\kappa\phi}\right)^{2\alpha-2} + (\phi - 1)\left(\varepsilon_1^\phi x_1^{\kappa\phi} + \varepsilon_2^\phi x_2^{\kappa\phi}\right)^{\alpha-2}\right\} \\
&\quad \times \exp\left\{-\left(\varepsilon_1^\phi x_1^{\kappa\phi} + \varepsilon_2^\phi x_2^{\kappa\phi}\right)^\alpha\right\},
\end{aligned}$$

where $\phi \geq 1$ is the measure of dependence, $\alpha = 1/\phi$, and $\frac{1}{\varepsilon}$ and κ are the scale and shape parameters of the marginal Weibull distributions. This density can be derived as an accelerated life-test model as well.

For the multivariate Weibull distribution in (47.127), $\min\{\frac{\theta_1}{a_1} X_1, \ldots, \frac{\theta_k}{a_k} X_k\}$ is distributed as Weibull with shape parameter ℓp when a_i's ≥ 0 ($i = 1, \ldots, k$) are such that $\|a\| = (\sum_{i=1}^{k} a_i^p)^{1/p} = 1$.

Crowder (1989) extended Hougaard's distributions and proposed "multivariate distributions with Weibull connections" with

$$\bar{F}(x_1, \ldots, x_k) = \exp\left\{\nu^\ell - \left(\nu + \sum_{i=1}^{k} \theta_i x_i^{p_i}\right)^\ell\right\},$$

where $\ell > 0$, $\nu \geq 0$ and $p_i > 0$. In the special case when $p_1 = \cdots = p_k = p$, the marginals are all Weibull with the same parameter. In this case, $\min\left\{\frac{\theta_1}{a_1} X_1, \ldots, \frac{\theta_k}{a_k} X_k\right\}$ is a random variable with survival function

$$\bar{F}_0(x) = \exp\left\{\nu^\ell - (\nu + x^p)^\ell\right\}, \quad x \geq 0,$$

for all a_i's ≥ 0 such that $\|a\| = (\sum_{i=1}^{k} a_i^p)^{1/p} = 1$.

Suppose that $W_1, \ldots, W_k$ are independent and identically distributed as Weibull with shape parameter p and with density

$$f(w) = pw^{p-1} e^{-w^p}, \quad w > 0.$$

Let $\boldsymbol{W} = (W_1, \ldots, W_k)^T$ and $\boldsymbol{U} = (U_1, \ldots, U_k)^T = \frac{\boldsymbol{W}}{\|\boldsymbol{W}\|}$, where $\|x\| = (\sum_{i=1}^{k} x_i^p)^{1/p}$. Also, let

$$A_k(a) = \left\{\boldsymbol{x} \in \mathbb{R}_+^k : \left(\sum_{i=1}^{k} x_i^p\right)^{1/p} < a\right\} \quad \text{for } a > 0 \text{ and } A_k = A_k(1).$$

Yue and Ma (1995) have then shown that the joint density of $\boldsymbol{U}$ is

$$\Gamma(k)p^{k-1} \prod_{i=1}^{k-1} u_i^{k-1} I_{A_{k-1}}(u_1, \ldots, u_{k-1}), \tag{47.128}$$

where $I_{A(\cdot)}$ is the indicator function of set A. The corresponding joint survival function is

$$\Pr[U_1 > u_1, \ldots, U_k > u_k] = (1 - \|\boldsymbol{u}\|^p)^{k-1} I_{A_k}(\boldsymbol{u}).$$

It follows then that the joint density of $(U_1, \ldots, U_m)^T$ is (for $1 \leq m \leq k-1$)

$$\frac{\Gamma(k)}{\Gamma(k-m)} p^m \prod_{i=1}^{m} u_i^{p-1} \left(1 - \sum_{i=1}^{m} u_i^p\right)^{k-m-1} I_{A_m}(u_1, \ldots, u_m).$$

Writing the density $p(x_1, x_2)$ of $(X_1, X_2)^T$ as

$$p(x_1, x_2) = f(x_2|x_1)g(x_1) = f(x_1|x_2)h(x_2),$$

where $f(x_1|x_2)$, $f(x_2|x_1)$, $g(x_1)$, and $h(x_2)$ are conditional and marginal densities, respectively, and assuming that $f(x_1|x_2)$ and $f(x_2|x_1)$ are each Weibull, we obtain

$$p(x_1, x_2) = \begin{cases} m(x_1)e^{-a(x_1)(x_2-K)^{c(x_1)}}a(x_1)(x_2 - K)^{c(x_1)-1}, & x_2 > K, \\ m^*(x_2)e^{-d(x_2)(x_1-L)^{f(x_2)}}d(x_2)(x_1 - L)^{f(x_2)-1}, & x_1 > L, \end{cases}$$
$$(47.129)$$

where K and L are location parameters, and $a(x_1)$, $c(x_1)$, $d(x_2)$, and $f(x_2)$ are the scale and shape parameters of the Weibull distributions. Note also that

$$m(x_1) = a(x_1)c(x_1)g(x_1) > 0 \quad \text{and} \quad m^*(x_2) = d(x_2)f(x_2)h(x_2) > 0.$$
$$(47.130)$$

Thus, for $x_1 > L$ and $x_2 > K$, the functions $a(x_1)$, $c(x_1)$, $m(x_1)$, $d(x_2)$, $f(x_2)$ and $m^*(x_2)$ are all positive. Castillo and Galambos (1990) have provided a particular solution of (47.129) under the condition (47.130), which provides a "conditionally specified bivariate Weibull distribution."

Walker and Stephens (1998) have generalized the mixture approach of Hougaard and Crowder, wherein a single mixing variable is used for all k variables thus resulting in a single association parameter, by using k mixing variables thus resulting in one association parameter of each pair of the k-variables. These authors have also shown that the resulting multivariate family is *dimensionally coherent*, a notion introduced by Haro-López and Smith (1997), which basically says that the marginal distributions of the k-variate family are members of the r-variate family ($r < k$).

Patra and Dey (1999) have constructed a class of multivariate distributions in which each component has a mixture of Weibull distributions. Specifically, by taking

$$Y_i \overset{d}{=} \sum_{j=1}^{\ell} a_{ij}Y_{ij} \ (i = 1, \ldots, k) \quad \text{and} \quad Z \overset{d}{=} \text{Exp}(\theta_0),$$

where Y_{ij} is distributed as two-parameter Weibull with density

$$\alpha_{ij}\theta_{ij}y^{\alpha_{ij}-1}e^{-\theta_{ij}y^{\alpha_{ij}}}, \qquad y > 0, \ \theta_{ij}, \alpha_{ij} > 0$$

and Z is independent of Y_{ij}'s, and defining $X_i = \min(Y_i, Z)$, they considered the joint distribution of $(X_1, \ldots, X_k)^T$.

For Gumbel's form of bivariate Weibull distribution, Begum and Khan (1997) have discussed the marginal and joint distributions of concomitants of order statistics and their single moments.

5 BIVARIATE DISTRIBUTIONS INDUCED BY FRAILTIES

Let T be a positive survival time and let there be a positive random variable W such that

$$\Pr[T > t | W = w] = \{B(t)\}^w, \qquad (47.131)$$

where $B(t)$ is a continuous baseline survival function. The variable W is referred to as a *frailty* and (47.131) is called a *frailty model*. The terms were originally introduced by Vaupel, Manton, and Stallari (1979) and were later utilized by Hougaard (1984). The model (47.131) is equivalent to the classical proportional hazards model of Cox (1972).

From (47.131), the unconditional survival function of T is

$$\bar{F}(t) = \Pr[T > t] = p\left(-\log B(t)\right), \qquad (47.132)$$

where $p(u) = E[e^{-uW}]$ is the Laplace transform of W. The function $B(t)$ and $p(u)$ are unidentifiable from data only on T.

Oakes (1989) extended (47.132) to frailty models for bivariate distributions with joint survival function $\bar{F}(t_1, t_2) = \Pr[T_1 > t_1, T_2 > t_2]$. Here, T_1 and T_2 are conditionally independent given W, each satisfying (47.131); and furthermore,

$$\bar{F}(t_1, t_2) = \int \{B_1(t_1) B_2(t_2)\}^w \, dG(w) \qquad (47.133)$$

for some baseline survival functions B_1 and B_2, and some $G(\cdot)$, which is the frailty distribution of W. We have

$$
\begin{aligned}
\bar{F}(t_1, t_2) &= \int \{B_1(t_1) B_2(t_2)\}^w \, dG(w) \\
&= \int \exp\left[-\{-\log B_1(t_1) - \log B_2(t_2)\} w\right] dG(w) \\
&= p\left(-\log B_1(t_1) - \log B_2(t_2)\right).
\end{aligned}
$$

The above bivariate frailty distributions are a subclass of Archimedean copulas of Genest and MacKay (1986a,b) defined by (see Chapter 44)

$$\bar{F}(t_1, t_2) = p\left(q\{\bar{F}_1(t_1)\} + q\{\bar{F}_2(t_2)\}\right), \qquad (47.134)$$

where $p(u)$ is now any nonnegative decreasing function with $p(0) = 1$ and $p''(u) \geq 0$, and $q(v)$ is its inverse function. If $p(u)$ is a Laplace transform, (47.134) is equivalent to (47.133) provided $B_j(t_j) = \exp[-q\{\bar{F}_j(t_j)\}]$, $j = 1, 2$. Any bivariate frailty model leads to an Archimedean survival function, but not conversely (since $p(u)$ may not be a Laplace transform). The cross-ratio

$$\theta^*(\boldsymbol{t}) = \theta^*(t_1, t_2) = \frac{\bar{F}(\boldsymbol{t})D_1 D_2 \bar{F}(\boldsymbol{t})}{\{D_1 \bar{F}(\boldsymbol{t})\}\{D_2 \bar{F}(\boldsymbol{t})\}},$$

which was originally introduced by Clayton (1978), where D_j denotes the operator $-\frac{\partial}{\partial t_j}$, form the basis for construction of frailty distributions. Oakes (1989) proposed an empirical estimate of the cross-ratio $\theta^*(\boldsymbol{t})$ which can be calculated by counting the concordant and discordant pairs; he also proposed a diagnostic plot to assess the goodness of fit. Oakes (1989) has further shown that if the joint survival function is as in (47.134), then $\theta^*(\boldsymbol{t})$ depends on $\boldsymbol{t}$ only through some function $\theta(v)$ of $v = \bar{F}(t_1, t_2)$, namely,

$$\theta^*(t_1, t_2) = \theta\left(\bar{F}(t_1, t_2)\right)$$

and explicitly

$$\theta(v) = -\frac{vq''(v)}{q'(v)},$$

while the inverse function of $p(u)$ is determined in terms of $\theta(v)$ (up to a constant multiple specified by $k > 0$) as

$$q_k(v) = \int_{z=v}^1 \exp\left(\int_{y=z}^{1-k} \frac{\theta(y)}{y} \, dy\right) dz.$$

Some examples of frailty models include the following:

1. The case when

$$q(v) = \begin{cases} (1/v)^{c-1} - 1, & c > 1, \\ -\log v, & c = 1, \\ 1 - v^{1-c}, & 0 < c < 1, \end{cases}$$

that is,

$$p(u) = \begin{cases} \left(\frac{1}{1+u}\right)^{1/(c-1)}, & c > 1, \\ e^{-u}, & c = 1, \\ (1-u)^{1/(1-c)}, & 0 < c < 1 \end{cases}$$

for $0 < u < 1$. This corresponds to $\theta(v) = c$ — the original example of Clayton (1978).

Here, for $c > 1$, then $p(u)$ is the Laplace transform of a gamma distribution with index $1/(c-1)$. The joint survival function is

$$\bar{F}(t_1, t_2) = \left[\left\{ \frac{1}{\bar{F}_1(t_1)} \right\}^{c-1} + \left\{ \frac{1}{\bar{F}_2(t_2)} \right\}^{c-1} - 1 \right]^{-1/(c-1)},$$

which, as $c \to \infty$, becomes

$$\bar{F}(t_1, t_2) = \min \left\{ \bar{F}_1(t_1), \bar{F}_2(t_2) \right\},$$

the Fréchet upper bound (see Chapter 44).

If $c = 1$, then $\bar{F}(t_1, t_2) = \bar{F}_1(t_1)\bar{F}_2(t_2)$ (the independence case), corresponding to a frailty distribution degenerate at unity.

If $c < 1$, then

$$\bar{F}(t_1, t_2) = \max \left[\left\{ \bar{F}_1(t_1) \right\}^{1-c} + \left\{ \bar{F}_2(t_2) \right\}^{1-c} - 1, 0 \right].$$

Here, the support depends on c. As $c \to 0$, the distribution approaches the Fréchet lower bound, given by $\max\{\bar{F}_1(t_1) + \bar{F}_2(t_2) - 1, 0\}$ (see Chapter 44).

2. The case $p(u) = e^{-u^\alpha}$, $0 < \alpha \le 1$, corresponding to positive stable distributions with parameter α, has been popularized by Hougaard (1986) as frailty distributions. Taking the Oakes model in (47.133) with the corresponding $\theta(v) = 1 + \frac{1-\alpha}{(-\alpha \log v)}$, we have

$$\bar{F}(t_1, t_2) = \exp\left[-\left\{ \left\{ -\log \bar{F}_1(t_1) \right\}^{1/\alpha} + \left\{ -\log \bar{F}_2(t_2) \right\}^{1/\alpha} \right\}^\alpha \right].$$

$$(47.135)$$

Note that $\theta(v)$ decreases from ∞ to 1 as v decreases from 1 to 0. When $\alpha = 1$, the survival times are independent; when $\alpha \to 0$, we obtain maximal positive dependence. The simplest case in (47.135) is the bivariate model with a common parameter:

$$\Pr[T_1 > t_1, T_2 > t_2] = \exp\left\{ -\frac{\delta}{\alpha} \, (t_1^\gamma + t_2^\gamma)^\alpha \right\}$$

which is a bivariate Weibull distribution mentioned earlier in Section 4.

Bjarnason and Hougaard (1999) have derived the Fisher information for two gamma frailty bivariate Weibull models, one in which the survival distribution is of Weibull form conditional on the frailty and the other in which the marginal distribution is of Weibull form.

The Weibull model has also been discussed by Lee (1979), Lu (1989), and Lu and Bhattacharyya (1990, 1991a,b).

Let $(T_1, T_2)^T$ have a bivariate frailty distribution. Consider the conditional survival function of $(T_1, T_2)^T$, given $(T_1 > a_1, T_2 > a_2)$. From (47.134), using Bayes' theorem, we have

$$\Pr[T_1 > t_1, T_2 > t_2 \mid T_1 > a_1, T_2 > a_2] = \frac{p(s+u)}{p(u)} \, ,$$

where

$$u = -\log\{B_1(a_1) B_2(a_2)\} = q\{\bar{F}(a_1, a_2)\}$$

and

$$s = -\log\left[\left\{ \frac{B_1(t_1)}{B_1(a_1)} \right\} \left\{ \frac{B_2(t_2)}{B_2(a_2)} \right\} \right].$$

The terms in the braces are the conditional baseline survival functions of T_1 and T_2, given $T_1 > a_1$ and $T_2 > a_2$. Therefore, the conditional distribution of $(T_1, T_2)^T$ is also a bivariate frailty distribution with a new Laplace transform $\tilde{p}(s) = \frac{p(s+u)}{p(u)}$. Note that provided $t_1 > a_1$ and $t_2 > a_2$, the conditional survival function $\bar{F}(t_1, t_2)$ depends on the truncation point (a_1, a_2) only through $v = \bar{F}(a_1, a_2)$.

For estimating the dependence coefficient in (47.135), Manatunga and Oakes (1996) utilized the concordance coefficient (Kendall's τ). They noted that for bivariate frailty models, the variance of U can also be written in terms of the Laplace transform of the frailty distribution. For any bivariate frailty model,

$$\tau = E[U] = 4 \int_0^\infty s p(s) \, p''(s) \, ds - 1.$$

Thus, for the model in (47.135), we have $E[U] = 1 - \alpha$, giving the simple estimate $\hat{\alpha} = 1 - U$. The asymptotic variance of $\hat{\alpha}$, denoted by $\sigma_\alpha^2 = \lim_{n \to \infty} n \text{var}(\hat{\alpha})$, is

$$\sigma_\alpha^2 = 4 \left\{ -\frac{5}{9} (3 + 10\alpha) + 32 G_3 + 8 G_4 - (1 - \alpha)^2 \right\},$$

where

$$G_3(\alpha) = \left(\frac{1}{\alpha} - 1 \right) \int_0^1 \frac{v^{\frac{1}{\alpha} - 1}}{(v + 2)^2} \, dv \qquad \text{and}$$

$$G_4(\alpha) = \int_0^1 \left\{ \frac{1 - \alpha}{1 + v^\alpha + (1 - v)^\alpha} + \frac{\alpha}{\{1 + v^\alpha + (1 - v)^\alpha\}^2} \right\} \, dv.$$

The estimator $\hat{\alpha}$ is asymptotically normal, in view of the general result of Hoeffding (1948). When $\alpha = 1$, which corresponds to independence between T_1 and T_2, we obtain $\sigma_\alpha^2 = \frac{4}{9}$, which can also be verified directly. As $\alpha \to 0$, we have $\sigma_\alpha^2 \to 0$ (corresponding to the maximal dependence).

A final mention should be made to the work of Hougaard (1995) which provides a lucid survey of multivariate frailty models.

BIBLIOGRAPHY

(Some bibliographical items not mentioned in the text are included here for completeness.)

Achcar, J. A. (1995). Inferences for accelerated life tests considering a bivariate exponential distribution, *Statistics*, **26**, 269–283.

Achcar, J. A., and Leandro, R. A. (1998). Use of Markov chain Monte Carlo methods in a Bayesian analysis of the Block and Basu bivariate exponential distribution, *Annals of the Institute of Statistical Mathematics*, **50**, 403–416.

Achcar, J. A., and Santander, L. A. M. (1993). Use of approximate Bayesian methods for the Block and Basu bivariate exponential distribution, *Journal of the Italian Statistical Society*, **3**, 233–250.

Al-Mutairi, D. K. (1997). Properties of an inverse Gaussian mixture of bivariate exponential distribution and its genrealization, *Statistics & Probability Letters*, **33**, 359–365.

Al-Saadi, S. D., Scrimshaw, D. F., and Young, D. H. (1979). Tests for independence of exponential variables, *Journal of Statistical Computation and Simulation*, **9**, 217–233.

Al-Saadi, S. D., and Young, D. H. (1980). Estimators for the correlation coefficient in a bivariate exponential distribution, *Journal of Statistical Computation and Simulation*, **11**, 13–20.

Al-Saadi, S. D., and Young, D. H. (1982). A test for independence in a multivariate exponential distribution with equal correlation coefficient, *Journal of Statistical Computation and Simulation*, **14**, 219–227.

Apostolakis, G. E. (1976). The effest of a certain class of potential common model faliures on the reliability of redundant systems, *Nuclear Engineering Design*, **36**, 123–133.

Arnold, B. C. (1968). Parameter estimation for a multivariate exponential distribution, *Journal of the American Statistical Association*, **63**, 848–852.

Arnold, B. C. (1975a). Multivariate exponential distributions based on hierarchical successive damage, *Journal of Applied Probability*, **12**, 142–147.

Arnold, B. C. (1975b). A characterization of the exponential distribution by multivariate geometric compounding, *Sankhyā, Series A*, **37**, 164–173.

Arnold, B. C., and Strauss, D. (1988). Bivariate distributions with exponential conditionals, *Journal of the American Statistical Association*, **83**, 522–527.

Asha, G., and Nair, N. U. (1999). Characterizations of the Marshall–Olkin bivariate exponential distributions, Technical Report, Department of Statistics, Cochin University of Science, Cochin, India.

Assaf, D., Langberg, N. A., Savits, T. H., and Shaked, M. (1984). Multivariate phase-type distributions, *Operations Research*, **32**, 688–701.

Awad, A. M., Azzam, M. M., and Hamdan, M. A. (1981). Some inference results on $\Pr(X < Y)$ in the bivariate exponential model, *Communications in Statistics—Theory and Methods*, **10**, 2515–2525.

Baggs, G. E. (1994). Properties of order statistics from bivariate exponential distributions, Ph.D. dissertation, Department of Statistics, The Ohio State University, Columbus, Ohio.

Balakrishnan, N., and Aggarwala, R. (2000). *Progressive Censoring: Theory, Methods and Applications*, Boston, Massachusetts: Birkhäuser.

Balakrishnan, N., and Basu, A. P. (eds.) (1995). *The Exponential Distribution: Theory, Methods and Applications*, Amsterdam, The Netherlands: Gordon and Breach Science Publishers.

Balakrishnan, N., and Ng, H. K. T. (2000). Improved estimation of the correlation coefficient in a bivariate exponential distribution, submitted for publication, *Journal of Statistical Computation and Simulation* (to appear).

Barlow, R. E., and Proschan, F. (1981). *Statistical Theory of Reliability and Life Testing: Probability Models*, Silver Spring, Maryland: To Begin With.

Barnett, V. (1985). The bivariate exponential distribution: A review and some new results, *Statistica Neerlandica*, **39**, 343–356.

Basu, A. P. (1988). Multivariate exponential distribuitons and their applications in reliability, in *Handbook of Statistics—7* (P. R. Krishnaiah and P. K. Sen, eds.), pp. 467–477, New York: North-Holland.

Basu, A. P. (1995). Bivariate exponential distributions, in *The Exponential Distribution: Theory, Methods and Applications* (N. Balakrishnan and A. P. Basu, eds.), pp. 327–332, Amsterdam, The Netherlands: Gordon and Breach Science Publishers.

Basu, A. P., and Sun, K. (1997). Multivariate exponential distributions with constant failure rates, *Journal of Multivariate Analysis*, **61**, 159–169.

Beg, M. I., and Balasubramanian, K. (1996). Concomitants of order statistics in the bivariate exponential distributions of Marshall and Olkin, *Calcutta Statistical Association Bulletin*, **46**, 109–115.

Begum, A. A., and Khan, A. H. (1997). Concomitants of order statistics from Gumbel's bivariate Weibull distribution, *Calcutta Statistical Association Bulletin*, **47**, 132–138.

Bemis, B. M., Bain, L. J., and Higgins, J. J. (1972). Estimation and hypothesis testing for the parameters of a bivariate exponential distribution, *Journal of the American Statistical Association*, **67**, 927–929.

Bhattacharya, S. K., and Kumar, S. (1986). E-IG model in life testing, *Calcutta Statistical Association Bulletin*, **35**, 85–90.

Bhattacharyya, A. (1997). Modelling exponential survival data with dependent censoring, *Sankhyā, Series A*, **59**, 242–267.

Bhattacharyya, G. K., and Johnson, R. A. (1971). Maximum likelihood estimation and hypothesis testing in the bivariate exponential model of Marshall and Olkin, Technical Report No. 276, Department of Statistics, University of Wisconsin, Madison, Wisconsin.

Bhattacharyya, G. K., and Johnson, R. A. (1973). On a test of independence in a bivariate exponential distribution, *Journal of the American Statistical Association*, **68**, 704–706.

Bilodeau, M., and Kariya, T. (1993). LBI tests of independence in bivariate exponential distributions, *Seventh International Conference on Multivariate Analysis*, New Delhi, India.

Bjarnason, H., and Hougaard, P. (1999). Fisher information for two gamma frailty bivariate Weibull models, *Lifetime Data Analysis* (to appear).

Block, H. W. (1975a). Continuous multivariate exponential extensions, in *Reliability and Fault Tree Analysis* (R. E. Barlow, J. B. Fussell, and N. D. Singpurwalla, eds.), Philadephia, Pennsylvania: Society for Industrial and Applied Mathematics.

Block, H. W. (1975b). Infinite divisibility of a bivariate exponential extension and mixtures of bivariate exponential distributions, Research Report 75-21, Department of Mathematics, University of Pittsburgh, Pittsburgh, PA.

Block, H. W. (1977). A characterization of a bivariate exponential distribution, *Annals of Statistics*, **5**, 808–812.

Block, H. W., and Basu, A. P. (1974). A continuous bivariate exponential extension, *Journal of the American Statistical Association*, **69**, 1031–1037.

Block, H. W., Paulson, A. S., and Kohberger, R. C. (1975). A class of bivariate distributions, preprint.

Boland, P. J. (1998). An arrangement increasing property of the Marshall–Olkin bivariate exponential, *Statistics & Pobability Letters*, **37**, 167–170.

Bordes, L., Nikulin, M., and Voinov, V. (1997). Unbiased estimation for a multivariate exponential whose components have a common shift, *Journal of Multivariate Analysis*, **63**, 199–221.

Brockett, P. L. (1984). General bivariate Makeham laws, *Scandinavian Actuarial Journal*, 150–156.

Castillo, E., and Galambos, J. (1990). Bivariate distributions with Weibull conditionals, *Analysis Mathematica*, **16**, 3–9.

Castillo, E., Sarabia, J. M., and Hadi, A. S. (1997). Fitting continous bivariate distributions to data, *The Statistician*, **46**, 355–369.

Chen, D., Lu, J.-C., Hughes-Oliver, J. M., and Li, C.-S. (1998). Asymptotic properties of maximum likelihood estimates for a bivariate exponential distribution and mixed censored data, *Metrika*, **48**, 109–125.

Clayton, D. G. (1978). A model for association in bivariate life tables and its application in epidemiological studies of familial tendency in chronic disease incidence, *Biometrika*, **65**, 141–151.

Cook, R. D., and Johnson, M. E. (1981). A family of distributions for modelling non-elliptically symmetric multivariate data, *Journal of the Royal Statistical Society, Series B*, **43**, 210–218.

Cowan, R. (1987). A bivariate exponential distribution arising in random geometry, *Annals of the Institute of Statistical Mathematics*, **39**, 103–111.

Cox, D. R. (1972). Regression models and life tables (with discussion), *Journal of the Royal Statistical Society, Series B*, **34**, 187–220.

Cramer, E., and Kamps, U. (1997). The UMVUE of $P(X < Y)$ based on type-II censored samples from Weinman multivariate exponential distributions, *Metrika*, **46**, 93–121.

Crowder, M. (1989). A multivariate distribution with Weibull connections, *Journal of the Royal Statistical Society, Series B*, **51**, 93–107.

Currit, A., and Singpurwalla, N. D. (1988). On the reliability function of a system of components sharing a common environment, *Journal of Applied Probability*, **26**, 763–771.

Dey, D. K., and Liu, P.-S. L. (1990). Estimation of parameters and reliability from generalized life models, *Communications in Statistics—Theory and Methods*, **19**, 1073–1099.

Downton, F. (1970). Bivariate exponential distributions in reliability theory, *Journal of the Royal Statistical Society, Series B*, **32**, 408–417.

Ebrahimi, N. (1987). Analysis of bivariate accelerated life test data from bivariate exponential of Marshall and Olkin, *American Journal of Mathematical and Management Sciences*, **6**, 175–190.

Ebrahimi, N., and Zahedi, H. (1989). Testing for bivariate Gumbel against bivariate new better than used in expectation, *Communications in Statistics—Theory and Methods*, **18**, 1357–1371.

Esary, J. D., and Marshall, A. W. (1974). Multivariate distributions with exponential minimums, *Annals of Statistics*, **2**, 84–93.

Freund, J. (1961). A bivariate extension of the exponential distribution, *Journal of the American Statistical Association*, **56**, 971–977.

Friday, D. S. (1976). A new multivariate life distribution, Ph.D. dissertation, Department of Statistics, Pennsylvania State University, State College, Pennsylvania.

Friday, D. S., and Patil, G. P. (1977). A bivariate exponential model with applications to reliability and computer generation of random variables, in *Theory and Applications of Reliability*, Vol. 1 (C. P. Tsokos and I. N. Shimi, eds.), pp. 527–549, New York: Academic Press.

Galambos, J., and Kotz, S. (1978). *Characterizations of Distributions*, Lecture Notes in Mathematics—**675**, New York: Springer-Verlag.

Gail, M. (1975). A review and critique of some models used in competing risk analysis, *Biometrics*, **31**, 209–222.

Genest, C., and MacKay, J. (1986a). Copules Archimediennes et familles de lois bidimensionelles dont les marges sont données, *Canadian Journal of Statistics*, **14**, 145–159.

Genest, C., and MacKay, J. (1986b). The joy of copulas: Bivariate distributions with given marginals, *The American Statistician*, **40**, 280–283.

Ghurye, S. G. (1987). Some multivariate lifetime distributions, *Advances in Applied Probability*, **19**, 138–155.

Ghurye, S. G., and Marshall, A. W. (1989). Shock processes with after-effects and multivariate lack of memory, *Journal of Applied Probability*, **21**, 786–801.

Gumbel, E. J. (1960). Bivariate exponential distributions, *Journal of the American Statistical Association*, **55**, 698–707.

Gumbel, E. J. (1961). Multivariate exponential distributions, *Bulletin of the International Statistical Institute*, **39**, 469–475.

Gupta, P. L., and Gupta, R. C. (1990). A bivariate random environmental stress model, *Advances in Applied Probability*, **22**, 501–503.

Hagen, E. W. (1980). Common-mode/common-cause failure: A review, *Annals of Nuclear Energy*, **7**, 509–519.

Hanagal, D. D. (1992). Some inference results in bivariate exponential distributions based on censored samples, *Communications in Statistics— Theory and Methods*, **21**, 1273–1295.

Hanagal, D. D. (1993). Some inference results in an absolutely continuous multivariate exponential model of Block, *Statistics & Probability Letters*, **16**, 177–180.

Hanagal, D. D. (1997). Inference procedures in some bivariate exponential models under hybrid random censoring, *Statistical Papers*, **38**, 167–189.

Hanagal, D. D., and Kale, B. K. (1991a). Large sample tests of independence for an absolutely continuous bivariate exponential model, *Communications in Statistics—Theory and Methods*, **20**, 1301–1313.

Hanagal, D. D., and Kale, B. K. (1991b). Large sample tests of λ_3 in the bivariate exponential distribution, *Statistics & Probability Letters*, **12**, 311–313.

Hanagal, D. D., and Kale, B. K. (1991c). Large sample tests for testing symmetry and independence in bivariate exponential models, in *Proceedings of the Symposium in Honour of V. P. Godambe*, University of Waterloo, Waterloo, Ontario, Canada.

Haro-López, R. A., and Smith, A. F. M. (1997). Scale mixtures of exponential power distributions, preprint.

Hashino, M. (1985). Formulation of the joint return period of two hydrologic variables associated with a Poisson process, *Journal of Hydroscience and Hydraulic Engineering*, **3**, 73–84.

Hawkes, A. G. (1972). A bivariate exponential distribution with applications to reliability, *Journal of the Royal Statistical Society, Series B*, **34**, 129–131.

Hayakawa, Y. (1994). The construction of new bivariate exponential distributions from a Bayesian perspective, *Journal of the American Statistical Association*, **89**, 1044–1049.

Heinrich, G., and Jensen, U. (1995). Parameter estimation for a bivariate life-distribution in reliability with multivariate extensions, *Metrika*, **42**, 49–65.

Hoeffding, W. (1948). A class of statistics with asymptotically normal distributions, *Annals of Mathematical Statistics*, **19**, 293–325.

Hougaard, P. (1984). Life table methods for heterogeneous populations: Distributions describing the heterogeneity, *Biometrika*, **71**, 75–83.

Hougaard, P. (1986). A class of multivariate failure time distributions, *Biometrika*, **73**, 671–678.

Hougaard, P. (1989). Fitting a multivariate failure time distribution, *IEEE Transactions on Reliability*, **R-38**, 444–448.

Hougaard, P. (1995). Frailty models for survival data, *Lifetime Data Analysis*, **1**, 255–273.

Hutchinson, T. P., and Lai, C. D. (1990). *Continuous Bivariate Distributions, Emphasizing Applications*, Adelaide, Australia: Rumsby Scientific Publishers.

Hyakutake, H. (1990). Statistical inferrences on location parameters of bivariate exponential distributions, *Hiroshima Mathematical Journal*, **20**, 525–547.

Jaisingh, L. R., Dey, D. K., and Griffith, W. S. (1993). Properties of a multivariate survival distribution generated by a Weibull and inverse Gaussian mixture, *IEEE Transactions on Reliability*, **R-42**, 618–621.

Johnson, N. L., and Kotz, S. (1975). A vector multivariate hazard rate, *Journal of Multivariate Analysis*, **5**, 53–66.

Johnson, N. L., Kotz, S., and Balakrishnan, N. (1994). *Continuous Univariate Distributions*, Vol. 1, second edition, New York: John Wiley & Sons.

Johnson, N. L., Kotz, S., and Balakrishnan, N. (1995a). *Continuous Univariate Distributions*, Vol. 2, second edition, New York: John Wiley & Sons.

Johnson, N. L., Kotz, S., and Balakrishnan, N. (1995b). Related distributions and some generalizations, in *The Exponential Distribution: Theory, Methods and Applications* (N. Balakrishnan and A. P. Basu, eds.), pp. 297–306, Amsterdam, The Netherlands: Gordon and Breach Science Publishers.

Johnson, N. L., Kotz, S., and Balakrishnan, N. (1997). *Discrete Multivariate Distributions*, New York: John Wiley & Sons.

Johnson, N. L., Kotz, S., and Kemp, A. W. (1992). *Univariate Discrete Distributions*, second edition, New York: John Wiley & Sons.

Kamps, U. (1995). *A Concept of Generalized Order Statistics*, Stuttgart, Germany: Teubner.

Kendall, M. G. (1938). A new measure of rank correlation, *Biometrika*, **30**, 81–93.

Kent, J. T. (1983). The appearance of a multivariate exponential distribution in sojourn times for birth-death and diffusion processes, in *Probability, Statistics and Analysis* (J. F. C. Kingman and G. E. H. Reuter, eds.), pp. 161–179, Cambridge, England: Cambridge University Press.

Kibble, W. F. (1941). A two-variate gamma type distribution, *Sankhyā,* **5,** 137–150.

Klein, J. P. (1995). Inference for multivariate exponential distributions, in *The Exponential Distribution: Theory, Methods and Applications* (N. Balakrishnan and A. P. Basu, eds.), pp. 333–350, Amsterdam, The Netherlands: Gordon and Breach Science Publishers.

Klein, J. P., and Basu, A. P. (1985). Estimating reliability for bivariate exponential distributions, *Sankhyā, Series B,* **47,** 346–353.

Klein, J. P., and Moeschberger, M. L. (1986). The independence assumption for a series or parallel system when component lifetimes are exponential, *IEEE Transactions on Reliability,* **R-35,** 330–335.

Kotz, S., and Singpurwalla, N. D. (1999). On a bivariate distribution with exponential marginals, *Scandinavian Journal of Statistics,* **26,** 451–464.

Krishnamoorthy, A. S., and Parthasarathy, M. (1951). A multivariate gamma-type distribution, *Annals of Mathematical Statistics,* **22,** 549–557; Correction, **31,** 229.

Langberg, N., Proschan, F., and Quinzi, A. J. (1978). Converting dependent models into independent ones, with applications in reliability, *Annals of Probability,* **6,** 174–181.

Langberg, N., Proschan, F., and Quinzi, A. J. (1981). Estimating dependent life-lengths, with application to the theory of competing risks, *Annals of Statistics,* **9,** 157–167.

Lau, K.-S., and Rao, C. R. (1982). Integrated Cauchy functional equation and characterizations of the exponential law, *Sankhyā, Series A,* **44,** 72–90. Correction, **44,** 452.

Lee, L., and Thompson, W. A., Jr. (1974). Results on failure time and pattern for the series system, in *Reliability and Biometriy: Statistical Analysis of Lifelength* (F. Proschan and R. J. Serfling, eds.), pp. 291–302, Philadelphia, Pennsylvania: Society for Industrial and Applied Mathematics.

Lee, L. (1979). Multivariate distributions having Weibull properties, *Journal of Multivariate Analysis,* **9,** 267–277.

Lefevre, C., and Malice, M. P. (1987). On a system of components with joint lifetimes distributed as a mixture of independent exponential laws, Technical Report No. 263, Department of Statistics, University of Kentucky, Lexington, Kentucky.

Leurgans, S., Tsai, T. W.-Y., and Crowley, J. (1982). Freund's bivariate exponential distribution and censoring, in *Survival Analysis* (R. A. Johnson, eds.), IMS Lecture Notes, Hayward, California: Institute of Mathematical Statistics.

Lindley, D. V., and Singpurwalla, N. D. (1986). Multivariate distributions for the life lengths of components of a system sharing a common environment, *Journal of Applied Probability*, **23**, 418–431.

Lu, J. C. (1989). Weibull extensions of the Freund and Marshall-Olkin distributions, *IEEE Transactions on Reliability*, **R-38**, 615–619.

Lu, J., and Bhattacharyya, G. K. (1990). Some new constructions of bivariate Weibull models, *Annals of the Institute of Statistical Mathematics*, **42**, 543–559.

Lu, J., and Bhattacharyya, G. K. (1991a). Inference procedures for bivariate exponential distributions, *Statistics & Probability Letters*, **12**, 37–50.

Lu, J., and Bhattacharyya, G. K. (1991b). Inference procedures for a bivariate exponential model of Gumbel based on life test of component and system, *Journal of Statistical Planning and Inference*, **27**, 283–296.

Manatunga, A., and Oakes, D. (1996). A measure of association for bivariate frailty distributions, *Journal of Multivariate Analysis*, **56**, 60–74.

Mardia, K. V. (1962). Multivariate Pareto distributions, *Annals of Mathematical Statistics*, **33**, 1008–1015.

Marshall, A. W. (1975a). Some comments on the hazard gradient, *Stochastic Processes*, **3**, 293–300.

Marshall, A. W. (1975b). Multivariate distributions with monotone hazard rate, in *Reliability and Fault Tree Analysis*, pp. 259–284, Philadelphia, Pennsylvania: Society for Industrial and Applied Mathematics.

Marshall, A. W., and Olkin, I. (1967a). A multivariate exponential distribution, *Journal of the American Statistical Association*, **62**, 30–44.

Marshall, A. W., and Olkin, I. (1967b). A generalized bivariate exponential distribution, *Journal of Applied Probability*, **4**, 291–302.

Marshall, A. W., and Olkin, I. (1991). Functional equations for multivariate exponential distributions, *Journal of Multivariate Analysis*, **39**, 209–215.

Marshall, A. W., and Olkin, I. (1995). Multivariate exponential and geometric distributions with limited memory, *Journal of Multivariate Analysis*, **53**, 110–125.

McLachlan, G. J. (1995). Mixture models and applications, in *The Exponential Distribution: Theory, Methods and Applications* (N. Balakrishnan and A. P. Basu, eds.), pp. 333–350, Amsterdam, The Netherlands: Gordon and Breach Science Publishers.

Mehrotra, K. G., and Michalek, J. E. (1976). Estimation of parameters and tests of independence in a continuous bivariate exponential distribution, Technical Report, Syracuse University, Syracuse, New York.

Moran, P. A. P. (1967a). Testing for correlation between non-negative variates, *Biometrika*, **54**, 385–394.

Moran, P. A. P. (1967b). A non-Markovian quasi-Poisson process, *Studia Scientiarum Mathematicae Hungaricae*, **2**, 425–429.

Moran, P. A. P. (1969). Statistical inference with bivariate gamma distribution, *Biometrika*, **56**, 627–634.

Nagao, M., and Kadoya, M. (1971). Two-variate exponential distribution and its numerical table for engineering application, *Bulletin of the Disaster Prevention Research Institute*, Kyoto University, **20**, 183–215.

Nagaraja, H. N., and Baggs, G. E. (1996). Order statistics of bivariate exponential random variables, in *Statistical Theory and Applications: Papers in Honor of Herbert A. David* (H. N. Nagaraja, P. K. Sen, and D. F. Morrison, eds.), pp. 129–141, New York: Springer-Verlag.

Nair, K. R. M., and Nair, N. U. (1988). On characterizing the bivariate exponential and geometric distributions, *Annals of the Institute of Statistical Mathematics*, **40**, 267–271.

Nayak, T. K. (1987). Multivariate Lomax distribution: Properties and usefulness in reliability theory, *Journal of Applied Probability*, **24**, 170–171.

Oakes, D. (1989). Bivariate survival models induced by frailties, *Journal of the American Statistical Association*, **84**, 487–493.

O'Cinneide, C. A., and Raftery, A. E. (1989). A continuous multivariate exponential distribution that is multivariate phase type, *Statistics & Probability Letters*, **7**, 323–325.

Olkin, I., and Tong, Y. L. (1994). Positive dependence of a class of multivariate exponential distributions, *SIAM Journal on Control and Optimization*, **32**, 965–974.

O'Neill, T. J. (1985). Testing for symmetry and independence in a bivariate exponential distribution, *Statistics & Probability Letters*, **3**, 269–274.

Papadoulos, A. S. (1981). A bivariate exponential failure model for life testing, *Statistica*, **41**, 39–58.

Patra, K., and Dey, D. K. (1999). A multivariate mixture of Weibull distributions in reliability modeling, *Statistics & Probability Letters*, **45**, 225–235.

Paulson, A. S. (1969). Statistical inference with bivariate gamma distributions, *Biometrika*, **56**, 627–634.

Paulson, A. S. (1974). A characterization of the exponential distribution and a bivariate exponential distribution, *Sankhyā, Series A*, **35**, 69–78.

Paulson, A. S., and Kohberger, R. C. (1974). Unpublished Report, Research Report No. 37-74PJ, Rensselaer Polytechnic Institute, Troy, New York.

Pickands, J. (1980). Multivariate negative exponential and extreme value distributions, Research Report, University of Pennsylvania, Philadelphia, Pennsylvania.

Prentice, R. L., Kalbfleisch, J. D., Peterson, A. V., Flournoy, N., Farewell, V. T., and Breslow, N. E. (1978). The analysis of failure times in the presence of competing risks, *Biometrics*, **34**, 541–554.

Proschan, F., and Sullo, P. (1974). Estimating the parameters of a bivariate exponential distribution in several sampling situations, in *Reliability and Biometry: Statistical Analysis of Lifelengths* (F. Proschan and R. Serfling, eds.), pp. 423–440, Philadelphia, PA: Society for Industrial and Applied Mathematics.

Proschan, F., and Sullo, P. (1976). Estimating the parameters of a multivariate exponential distribution, *Journal of the American Statistical Association*, **71**, 465–472.

Raftery, A. E. (1984). A continuous multivariate exponential distribution, *Communications in Statistics—Theory and Methods*, **13**, 947–965.

Raftery, A. E. (1985). Some properties of a new continuous bivariate exponential distribution, *Statistics and Decisions*, Supplement **2**, 53–58.

Raja Rao, B., Damaraju, C. V., and Alhumound, J. M. (1993). Setting the clock back to zero property of a class of bivariate life distributions, *Communications in Statistics—Theory and Methods*, **22**, 2067–2080.

Rao, C. R., and Shanbhag, D. N. (1994). *Choquet-Deny Type Functional Equations with Applications to Stochastic Models*, Chichester, England: John Wiley & Sons.

Rvacheva, E. L. (1962). On domains of attraction of multi-dimensional distributions, in *Selected Translations in Mathematical Statistics and Probability*, **2**, 183–205, Providence, Rhode Island: American Mathematical Society.

Samanta, M. (1986). On asymptotic optimality of some tests for exponential distribution, *Australian Journal of Statistics*, **28**, 164–172.

Sarkar, S. K. (1987). A continuous bivariate exponential distribution, *Journal of the American Statistical Association*, **82**, 667–675.

Sarkar, T. K. (1971). An exact lower confidence bound for the reliability of a series system where each component has exponential time to failure distribution, *Technometrics*, **13**, 535–546.

Saw, J. G. (1969). A bivariate exponential density and a test that two identical components in parallel behave independently, Technical Report No. 22, Department of Industrial and Systems Engineering, University of Florida, Gainesville, Florida.

Scheaffer, R. L. (1975). Optimum age replacement in the bivariate exponential case, *IEEE Transactions on Reliability*, **R-24**, 214–215.

SenGupta, A. (1991). A review of optimality of multivariate tests, *Statistics & Probability Letters*, **12**, 527–535.

SenGupta, A. (1995). Optimal tests in multivariate exponential distributions, in *The Exponential Distribution: Theory, Methods and Applications* (N. Balakrishnan and A. P. Basu, eds.), pp. 351–376, Amsterdam, The Netherlands: Gordon and Breach Science Publishers.

Seshadri, V., and Patil, G. P. (1964). A characterization of a bivariate distribution by the marginal and the conditional distributions of the same component, *Annals of the Institute of Statistical Mathematics*, **15**, 215–221.

Shanubhogue, A., and Jani, P. N. (1993). Consequences of departure from independence in two component systems when component life times depend through FGM model, *Communications in Statistics—Theory and Methods*, **22**, 2969–2981.

Singpurwalla, N. D., and Youngren, M. A. (1993). Multivariate distributions induced by dynamic environments, *Scandinavian Journal of Statistics*, **20**, 251–261.

Tolley, H. D., Manton, K. G., and Poss, S. S. (1978). A linear models application of competing risks to multiple causes of death, *Biometrics*, **34**, 581–591.

Tosch, T. J., and Holmes, P. T. (1980). A bivariate failure model, *Jounal of the American Statistical Association*, **75**, 415–417.

Ulrich, G., and Chen, C.-C. (1987). A bivariate double exponential distribution and its generalizations, *ASA Proceedings of the Statistical Computing Section*, 127–129.

Vaupel, J. W., Manton, K. G., and Stallari, E. (1979). The impact of heterogeneity in individual frailty on the dynamics of mortality, *Demography*, **16**, 439–454.

Vesely, W. E. (1977). In *Nuclear Systems Reliability Engineering and Risk Assessment* (J. B. Fussell and G. R. Burdick, eds.), Philadelphia: Society for Industrial and Applied Mathematics.

Wada, C. Y., Sen, P. K., and Shimakura, S. E. (1996). A bivariate exponential model with covariates in competing risk data, *Calcutta Statistical Association Bulletin*, **46**, 197–210.

Walker, S. G., and Stephens, D. A. (1998). A multivariate family of distributions on $(0, \infty)^p$, preprint.

Weier, D. R. (1981). Bayes estimation for bivariate survival models, based on the exponential distribution, *Communications in Statistics—Theory and Methods*, **10**, 1415–1427.

Weier, D. R., and Basu, A. P. (1980). Testing for independence in multivariate exponential distributions, *Australian Journal of Statistics*, **22**, 276–288.

Weier, D. R., and Basu, A. P. (1981). On tests of independence under bivariate exponential models, in *Statistical Distributions in Scientific Work*, Vol. 5 (C. P. Taillie, G. P. Patil, and B. Baldessari, eds.), pp. 169–181, Dordrecht, The Netherlands: D. Reidel.

Weinman, D. G. (1966). A multivariate extension of the exponential distribution, Ph.D. dissertation, Arizona State Univesity, Tempe, Arizona.

Whitmore, G. A., and Lee, M.-L. T. (1991). A multivariate survival distribution generated by an exponential-inverse Gaussian mixture, *Technometrics*, **33**, 39–50.

Wu, C. (1997). New characterization of Marshall–Olkin-type distributions via bivariate random summation scheme, *Statistics & Probability Letters*, **34**, 171–178.

Yang, G. L. (1989). Indentifiability of some bivariate exponential and Weibull distributions in the competing risk study, Technical Report, Department of Mathematics, University of Maryland, College Park, MD.

Yue, X., and Ma, C. (1995). Multivariate ℓ_p-norm symmetric distributions, *Statistics & Probability Letters*, **24**, 281–288.

CHAPTER 48

Multivariate Gamma Distributions

1   INTRODUCTION

The distributions discussed in this chapter can be regarded as generalization of the univariate gamma distributions studied in Chapter 17 of Johnson, Kotz, and Balakrishnan (1994). As is often the case, generalizations can take on a number of different forms and we describe here their motivation, method of construction, and some properties.

We first present various forms of bivariate gamma distributions. Amongst the bivariate gamma distributions that have been studied in the Russian literature is a system introduced by Sarmanov (1970a) [see Lee (1996)], which has some points of similarity with (but differs from) a system described by Eagleson (1964) and the system studied by D'jachenko (1961, 1962a,b).

Occasionally, in the statistical distribution literature, Wishart distributions have been referred to as "multivariate gamma distributions"; see, for example, Tan (1968). We, however, restrict this term to those distributions for which the marginal distribution are of gamma form.

This chapter represents a substantial revision and extension of the first five sections of Chapter 40 of the first edition of this volume by Johnson and Kotz (1972).

2   BIVARIATE GAMMA DISTRIBUTIONS

In this section, we survey various forms of bivariate gamma distributions. Most of them exploit various properties of the univariate gamma distribution to construct bivariate families.

2.1   McKay's Bivariate Gamma

One of the earliest forms of the bivariate gamma distribution is due to McKay (1934), defined by the density function

$$p_{X_1,X_2}(x_1, x_2) \; = \; \frac{c^{a+b}}{\Gamma(a)\Gamma(b)} \, x_1^{a-1}(x_2 - x_1)^{a-1} e^{-cx_2},$$

$$x_2 > x_1 > 0, \; a, b, c > 0.$$

Plots of this density function for the three cases ($a = b = 0.5$, $c = 2.0$), ($a = b = c = 0.5$), and ($a = 0.2$, $b = 0.8$, $c = 1.0$) have been given by Kellogg and Barnes (1987).

    The marginal distributions are gamma with shape parameters a and $a + b$, respectively. The correlation coefficient is corr$(X_1, X_2) = \sqrt{\frac{a}{a+b}}$. It is easy to verify that the conditional density of $X_1 | (X_2 = x_2)$ is beta with parameters a and b. The original deviation of the distribution was based on the joint distribution of the sample variances $S_N^2 \, S_n^2$, where $X_1, \dots, X_N$ is a random sample from a normal population and S_n^2 is the variance of a subsample of size n. The distribution is the bivariate Pearson Type IVa distribution; see Table 44.1 in Chapter 44.

    A more general form of the distribution is given by the density function

$$p_{X_1,X_2}(x_1, x_2) \; = \; K \cdot (d_1 x_1 + d_2)^a (d_3 x_1 + d_4 x_2 + d_5)^b e^{cx_2},$$

$$x_1 > 0, \; x_2 > 0.$$

An interesting application of this distribution in hydrology as the joint distribution of annual streamflow and areal precipitation has been given by Clarke (1980).

2.2   Cheriyan and Ramabhadran's Bivariate Gamma

A bivariate generalization of the gamma distribution may be constructed by a method similar to the one due to Holgate used in constructing bivariate Poisson and Neyman Type A distributions; see, for example, Chapter

37 of Johnson, Kotz, and Balakrishnan (1997). Specifically, let Y_0, Y_1, and Y_2 be independent gamma random variables with probability density functions

$$p_{Y_i}(y_i) = \frac{1}{\Gamma(\theta_i)} \, e^{-y_i} y_i^{\theta_i - 1}, \qquad y_i > 0, \; \theta_i > 0 \; (i = 0, 1, 2). \qquad (48.1)$$

Let $X_i = Y_0 + Y_i$ for $i = 1, 2$. Then, from the joint density function of $(Y_0, Y_1, Y_2)^T$ given by

$$p_{Y_0, Y_1, Y_2}(y_0, y_1, y_2)$$
$$= \frac{1}{\Gamma(\theta_0) \Gamma(\theta_1) \Gamma(\theta_2)} \, e^{-(y_0 + y_1 + y_2)} y_0^{\theta_0 - 1} y_1^{\theta_1 - 1} y_2^{\theta_2 - 1},$$
$$y_i > 0, \; \theta_i > 0 \; (i = 0, 1, 2), \qquad (48.2)$$

we obtain the joint density function of $(Y_0, X_1, X_2)^T$ as

$$p_{Y_0, X_1, X_2}(y_0, x_1, x_2)$$
$$= \frac{1}{\Gamma(\theta_0) \Gamma(\theta_1) \Gamma(\theta_2)} \, y_0^{\theta_0 - 1} (x_1 - y_0)^{\theta_1 - 1} (x_2 - y_0)^{\theta_2 - 1} \, e^{y_0 - x_1 - x_2},$$
$$x_i \geq y_0 \geq 0 \; (i = 1, 2). \qquad (48.3)$$

In order to integrate out the variable Y_0, it is necessary to evaluate the integral

$$\int_0^{\tilde{x}} y_0^{\theta_0 - 1} (x_1 - y_0)^{\theta_1 - 1} (x_2 - y_0)^{\theta_2 - 1} \, e^{y_0} \, dy_0, \qquad (48.4)$$

where $\tilde{x} = \min(x_1, x_2)$. From (48.3), we then obtain the bivariate gamma density function as

$$p_{X_1, X_2}(x_1, x_2)$$
$$= \frac{e^{-(x_1 + x_2)}}{\Gamma(\theta_0) \Gamma(\theta_1) \Gamma(\theta_2)} \int_0^{\min(x_1, x_2)} y_0^{\theta_0 - 1} (x_1 - y_0)^{\theta_1 - 1} (x_2 - y_0)^{\theta_2 - 1} \, e^{y_0} \, dy_0; \qquad (48.5)$$

see Cheriyan (1941) (who considered the case $\theta_1 = \theta_2$) and Ramabhadran (1951). Independently, David and Fix (1961) obtained this distribution and derived several properties. They also presented explicit expressions for (48.5) for various combinations of integer values of θ_0, θ_1, and θ_2. In particular, if $\theta_1 = \theta_2 = 1$ and θ_0 is an integer, we obtain from (48.5)

$$p_{X_1,X_2}(x_1, x_2) = \frac{e^{-(x_1+x_2)}}{\Gamma(\theta_0)} \int_0^{\min(x_1,x_2)} y_0^{\theta_0-1} e^{y_0} dy_0$$

$$= \frac{e^{-(x_1+x_2)}}{\Gamma(\theta_0)} [e^{\tilde{x}} \{ \tilde{x}^{\theta_0-1} - (\theta_0 - 1)\tilde{x}^{\theta_0-2}$$
$$+ (\theta_0 - 1)(\theta_0 - 2)\tilde{x}^{\theta_0-3} + \cdots$$
$$+ (-1)^{\theta_0-1}(\theta_0 - 1)! \} + (-1)^{\theta_0}(\theta_0 - 1)!]$$

$$= e^{-(x_1+x_2)}(-1)^{\theta_0} \left[1 - e^{\tilde{x}} \left\{ 1 - \frac{\tilde{x}}{1!} + \frac{\tilde{x}}{2!} + \cdots \right. \right.$$
$$\left. \left. + (-1)^{\theta_0-1}\frac{\tilde{x}^{\theta_0-1}}{(\theta_0 - 1)!} \right\} \right], \qquad (48.6)$$

where $\tilde{x} = \min(x_1, x_2)$, as before. Moran (1967) observed that the density in (48.5) has different expressions for $x_1 < x_2$ and $x_1 > x_2$. Szántai (1986) rederived this distribution and provided an expression for the density funciton for arbitrary positive shape parameters in terms of Laguerre polynomials.

From (48.3), we note that the marginal distribution of X_i is a standard gamma distribution with parameter $\theta_0 + \theta_i$, and consequently, $\text{var}(X_i) = \theta_0 + \theta_i$ $(i = 1, 2)$. Also,

$$\text{cov}(X_1, X_2) = \text{cov}(Y_0 + Y_1, Y_0 + Y_2) = \text{var}(Y_0) = \theta_0.$$

Hence,

$$\text{corr}(X_1, X_2) = \frac{\theta_0}{\sqrt{(\theta_0 + \theta_1)(\theta_0 + \theta_2)}} \qquad (48.7)$$

which is nonnegative. Furthermore, the conditional distribution of Y_0, given $X_i = Y_0 + Y_i = x_i$, can be derived from the fact that the random variables X_i and Y_0/X_i are independent; see Chapter 17 of Johnson, Kotz and Balakrishnan (1994). It follows, therefore, that the conditional distribution of Y_0, given $X_i = x_i$, is that of $x_i \times$ (beta random variable with parameters θ_0 and θ_i). Specifically, we have

$$E[Y_0 \mid X_i = x_i] = x_i \frac{\theta_0}{\theta_0 + \theta_i}$$

and

$$\text{var}(Y_0 \mid X_i = x_i) = x_i^2 \frac{\theta_0\theta_i}{(\theta_0 + \theta_i)^2(\theta_0 + \theta_i + 1)}.$$

Hence,

$$E[X_2 \mid X_1 = x_1] = E[Y_0 \mid X_1 = x_1] + E[Y_2] = x_1 \, \frac{\theta_0}{\theta_0 + \theta_1} + \theta_2 \quad (48.8)$$

since Y_0, Y_1 and Y_2 are independent, and

$$
\begin{aligned}
\mathrm{var}(X_2 \mid X_1 = x_1) &= \mathrm{var}(Y_0 \mid X_1 = x_1) + \mathrm{var}(Y_2) \\
&= x_1^2 \, \frac{\theta_0 \theta_1}{(\theta_0 + \theta_1)^2 (\theta_0 + \theta_1 + 1)} + \theta_2. \quad (48.9)
\end{aligned}
$$

Thus, the regression of X_2 on X_1 is linear, but the variation about the regression is not homoscedastic. We further note that the conditional distribution of X_2, given $X_1 = x_1$, is that of the sum of two independent random variables, one distributed as $x_1 \times$ (beta random variable with parameters θ_0 and θ_1) and the other as a standard gamma random variable with parameter θ_2.

The joint moment generating function of X_1 and X_2 is

$$
\begin{aligned}
E\left[e^{t_1 X_1 + t_2 X_2}\right] &= E\left[e^{t_1(Y_0 + Y_1) + t_2(Y_0 + Y_2)}\right] \\
&= E\left[e^{(t_1 + t_2)Y_0}\right] E\left[e^{t_1 Y_1}\right] E\left[e^{t_2 Y_2}\right] \\
&= \frac{1}{(1 - t_1 - t_2)^{\theta_0} (1 - t_1)^{\theta_1} (1 - t_2)^{\theta_2}} \, ; \quad (48.10)
\end{aligned}
$$

see Cheriyan (1941) and Ramabhadran (1951).

More parameters can be introduced by considering the joint distribution of

$$X_i = \lambda_i (Y_0 + Y_i), \qquad i = 1, 2.$$

Ghirtis (1967) has referred to this distribution as the *double-gamma distribution*. He also studied some properties of estimators of parameters of this distribution.

Jensen (1969a) has shown that if X_1 and X_2 have a bivariate gamma distribution in (48.5), then we obtain

$$\Pr[c_1 \le X_1 \le c_2, c_1 \le X_2 \le c_2] \ge \Pr[c_1 \le X_1 \le c_2] \Pr[c_1 \le X_2 \le c_2]$$
$$(48.11)$$

for any $0 \le c_1 < c_2$. Another way of expressing (48.11) is

$$\Pr[c_1 \le X_2 \le c_2 \mid c_1 \le X_1 \le c_2] \ge \Pr[c_1 \le X_2 \le c_2]$$

which means that if it is known that X_1 is between c_1 and c_2, then it increases the probability that X_2 is between c_1 and c_2.

2.3   Kibble and Moran's Bivariate Gamma

Kibble (1941) and Moran (1967) discussed a symmetrical bivariate gamma distribution with joint characteristic function

$$\frac{1}{\{(1-it_1)(1-it_2)+\omega^2 t_1 t_2\}^\alpha}, \qquad \alpha > 0. \qquad (48.12)$$

The corresponding moment-generating function is due to Wicksell (1933). Vere-Jones (1967) has shown that this distribution is infinitely divisible. The marginal distributions of X_1 and X_2 are both standard gamma distributions with shape parameter α. The correlation coefficient, $\rho = \text{corr}(X_1, X_2)$, is equal to ω^2. An explicit expression for the joint density function of $(X_1, X_2)^T$ is

$$
\begin{aligned}
p_{X_1,X_2}(x_1, x_2) &= \sum_{j=0}^{\infty} \frac{\omega^{2j} \alpha^{[j]}}{j!} \left[\prod_{k=1}^{2} \left\{ \sum_{\ell=0}^{j} (-1)^j \binom{j}{\ell} \frac{1}{\ell!} x_k^\ell e^{-x_k} \right\} \right] \\
&= \left\{ \prod_{k=1}^{2} \frac{1}{\Gamma(\alpha)} x_k^{\alpha-1} e^{-x_k} \right\} \left[1 + \sum_{j=1}^{\infty} \rho^j L_j^{(\alpha-1)}(x_1) L_j^{(\alpha-1)}(x_2) \right],
\end{aligned}
$$
$$\alpha > 0, \ 0 \le \rho < 1, \ x_1 > 0, \ x_2 > 0, \qquad (48.13)$$

where $L_j^{(\alpha-1)}(x)$ is the Laguerre polynomial given by

$$L_j^{(\alpha-1)}(x) = \left\{ \frac{\Gamma(\alpha)\Gamma(\alpha+j)}{j!} \right\}^{1/2} \sum_{k=0}^{j} (-1)^k \binom{j}{k} \frac{1}{\Gamma(\alpha+k)} x^k; \quad (48.14)$$

also see Sarmanov (1968) and D'jachenko (1962a,b). Contours of the density function of Kibble distribution have been presented by Izawa (1965) for the cases ($\alpha = 1$, $\rho = 0.5$) and ($\alpha = 2$, $\rho = 0.5$). He also applied this distribution to model rainfall at two nearby rain gauges.

Moran (1969, 1970) has also discussed the use of a bivariate gamma distribution such that normalizing transformations on each variate produce a joint bivariate normal distribution (see Chapter 46) in the analysis of data obtained from rainmaking experiments.

A minor generalization of the bivariate gamma density function in (48.13) can be obtained by replacing $\{\rho^j\}$ by a sequence $\{c_j^2\}$ such that $\sum_{j=1}^{\infty} c_j^2 < \infty$.

The characteristic function of the standard gamma distribution with shape parameter α [see Chapter 17 of Johnson, Kotz, and Balakrishnan (1994)] is $(1-it)^{-\alpha}$. It follows that the compound gamma distribution

$$\text{Gamma}(\alpha + \theta) \bigwedge_{\theta} \text{Negative binomial}(\alpha, \beta)$$

has characteristic function

$$(1 - it)^{-\alpha}\{\beta + 1 - \beta(1 - it)^{-1}\}^{-\alpha} = \{(\beta + 1)(1 - it) - \beta\}^{-\alpha} \quad (48.15)$$

and therefore has a gamma distribution with shape parameter α, scale parameter $\beta + 1$, and location parameter 0.

This fact has been exploited by Gaver (1970) to construct the Kibble distribution by means of compounding. Let X_1 and X_2 be independent random variables (for given θ), each having the standard gamma distribution with parameter $\alpha + \theta$. Assuming that the parameter θ has a negative binomial distribution with probability generating function

$$\left(\frac{\beta}{1 + \beta - z}\right)^{\alpha} = \sum b_n(\alpha, \beta)z^n,$$

where $\alpha, \beta > 0$ are the parameters of the negative binomial distribution, we readily obtain the joint moment generating function of X_1 and X_2 as

$$M_{X_1,X_2}(t_1, t_2) = \left\{1 - \frac{\beta + 1}{\beta} t_1 - \frac{\beta + 1}{\beta} t_2 - \frac{\beta + 1}{\beta} t_1 t_2\right\}^{-\alpha}, \qquad \alpha > 0,$$

which is the moment generating function of the Kibble–Moran distribution with correlation coefficient $\omega^2 = 1/(1 + \beta)$.

Hamdan and Martinson (1971) have obtained, for Kibble's bivariate gamma distribution, a formula analogous to that of the bivariate normal distribution, with the left-hand side multiplied by $\alpha(\tilde{h}_1\tilde{h}_2)^{-1}$ and Laguerre polynomials replacing Hermite polynomials on the right-hand side.

2.4   Sarmanov's Bivariate Gamma

Sarmanov (1970a,b) introduced asymmetrical bivariate gamma distributions, which are extensions of Kibble–Moran's bivariate gamma in (48.13), with joint density function

$$p_{X_1,X_2}(x_1, x_2)$$

$$= \left\{\prod_{k=1}^{2} \frac{1}{\Gamma(\alpha_k)} x_k^{\alpha_k - 1} e^{-x_k}\right\} \left[1 + \sum_{j=1}^{\infty} a_j L_j^{(\alpha_1 - 1)}(x_1) L_j^{(\alpha_2 - 1)}(x_2)\right],$$

$$x_1, x_2 > 0, \qquad (48.16)$$

where $\alpha_1 > \alpha_2$ (and for some $0 < \lambda < 1$),

$$a_j = \lambda^j \left\{\frac{\Gamma(\alpha_1)\Gamma(\alpha_2 + j)}{\Gamma(\alpha_1 + j)\Gamma(\alpha_2)}\right\}^{1/2} = \lambda^j \left\{\frac{\alpha_2^{[j]}}{\alpha_1^{[j]}}\right\}^{1/2}, \qquad j = 1, 2, \ldots.$$

The correlation coefficient between X_1 and X_2 is

$$\text{corr}(X_1, X_2) = a_1 = \lambda\sqrt{\alpha_2/\alpha_1} \ .$$

Note that in the case when $\alpha_1 = \alpha_2$, the density in (48.16) reduces immediately to (48.13). The distribution in (48.16) has been analyzed and extended by Lee (1996).

2.5   Jensen's Bivariate Gamma

Another generalization of the Kibble–Moran distribution is due to Jensen (1970a). Consider $X_i = \sum_{k=1}^{v} Z_{ki}^2$ ($i = 1, 2$), where $\boldsymbol{Z}_k = (Z_{k1}, Z_{k2})^T$ are mutually independent (for $k = 1, 2, \ldots, v$) and $\boldsymbol{Z}_k$ has a standardized bivariate normal distribution with correlation coefficient ρ_k (for $k = 1, 2, \ldots, v$). Then, the joint characteristic function of X_1 and X_2 is [Jensen (1970a)]

$$\{(1 - 2it_1)(1 - 2it_2)\}^{-v/2} \sum_{j=0}^{\infty} C_j(\rho_1, \ldots, \rho_v) \left\{ -\frac{4t_1t_2}{(1 - 2it_1)(1 - 2it_2)} \right\}^j ,$$

where

$$C_j(\rho_1, \ldots, \rho_v) = \sum \cdots \sum_{j_1 + \cdots + j_v = j} a_{j_1}(\rho_1) \cdots a_{j_v}(\rho_v)$$

and

$$a_j(\rho_\ell) = \frac{\rho_\ell^{2j}\, \Gamma(j + \frac{1}{2})}{\sqrt{\pi}\, \Gamma(j + 1)} \ .$$

The joint density function of X_1 and X_2 in this case is

$$
\begin{aligned}
p_{X_1, X_2}&(x_1, x_2) \\
&= \left\{ \frac{e^{-x_1/2} x_1^{(v/2)-1}}{2^{v/2}\, \Gamma(v/2)} \right\} \left\{ \frac{e^{-x_2/2} x_2^{(v/2)-1}}{2^{v/2}\, \Gamma(v/2)} \right\} \\
&\quad \times \sum_{j=0}^{\infty} \left\{ \frac{j!\, \Gamma(v/2)}{\Gamma(j + v/2)} \right\}^2 C_j(\rho_1, \ldots, \rho_v) L_j^{(v/2)}\left(\frac{x_1}{2} \right) L_j^{(v/2)}\left(\frac{x_2}{2} \right),
\end{aligned}
$$

$$(48.17)$$

where $L_j^{(\alpha-1)}(x)$ is the Laguerre polynomial as defined in (48.14). Though the above form is a bivariate chi-square distribution, it becomes a form of bivariate gamma if we consider the joint density function of $\left(\frac{X_1}{2}, \frac{X_2}{2} \right)^T$.

The marginal distributions are again gamma but with different shape parameters. Note the similarity with Kibble–Moran's bivariate gamma density function in (48.13).

The correlation between X_1 and X_2 in (48.17) is

$$(2v)^{-1}\mathrm{cov}(X_1, X_2) = (2v)^{-1}\sum_{\ell=1}^{v}\mathrm{cov}(Z_{\ell 1}^2, Z_{\ell 2}^2)$$

$$= v^{-1}\sum_{\ell=1}^{v}\rho_\ell^2.$$

The density (48.17) was studied extensively by D'jachenko (1961, 1962a,b), who has also given tables and "prisomograms" for $\rho = 0.9$ and $\alpha = 2(1)5$.

Gunst and Webster (1973) considered Jensen's distribution in the equicorrelated case, when all $\rho's$ are either zero or a constant. Gunst (1973) provided tables of upper 5% critical points using the procedure presented by Gunst and Webster (1973).

If $\boldsymbol{X}^T = (\boldsymbol{X}_{(1)}^T, \boldsymbol{X}_{(2)}^T)$ has a multivariate normal distribution with zero expected value vector and variance–covariance matrix

$$\boldsymbol{V} = \begin{pmatrix} \boldsymbol{V}_{11} & \boldsymbol{V}_{12} \\ \boldsymbol{V}_{21} & \boldsymbol{V}_{22} \end{pmatrix},$$

where the $v_j \times v_j$ matrix $\boldsymbol{V}_{jj}$ is the variance–covariance matrix of $\boldsymbol{X}_{(j)}^T$ $(j = 1, 2)$, then the quadratic forms

$$\boldsymbol{Y}_j = \boldsymbol{X}_{(j)}^T \boldsymbol{V}_{jj}^{-1} \boldsymbol{X}_{(j)}$$

have χ^2 distributions with v_j degrees of freedom $(j = 1, 2)$. The joint distribution of $\boldsymbol{Y}_1$ and $\boldsymbol{Y}_2$ is thus a form of bivariate χ^2 distribution. It has been studied by Jensen (1970a), who has pointed out that the distribution is the same as that of

$$Y_1 = \sum_{j=1}^{v_1} Z_{1j}^2; \qquad Y_2 = \sum_{j=1}^{v_2} Z_{2j}^2 \qquad (v_1 \leq v_2),$$

where $(Z_{1j}, Z_{2j})^T$ $(j = 1, \ldots, v_1)$ are independent and have standardized bivariate normal distributions with correlations $\rho_1, \rho_2, \ldots, \rho_{v_1}$ (the canonical correlations between $\boldsymbol{X}_{(1)}$ and $\boldsymbol{X}_{(2)}$) and $Z_{2,v_1+1} \cdots Z_{2,v_2}$ are independent standard normal variables that are also independent of all other Z's.

Jensen (1970b) has also considered the joint distribution of

$$Y_j = \sum_{\ell=1}^{v} Z_{\ell j}^2 \qquad (j = 1, 2, \ldots, k),$$

where $\boldsymbol{Z}_\ell = (Z_{\ell 1}, \ldots, Z_{\ell k})$ are mutually independent ($\ell = 1, 2, \ldots, v$) and $\boldsymbol{Z}_\ell$ has a standardized multivariate normal distribution with variance–covariance matrix $\boldsymbol{V}_\ell$ (not necessarily the same for all ℓ).

He has also obtained the general joint distribution of Y_1, Y_2, and Y_3, and of $Y_1, Y_2, \ldots, Y_k$ for general k, when the variance–covariance matrices $\boldsymbol{V}_\ell$ are each of Jacobi form—that is, with all elements zero except those on the principal and its two adjacent diagonals.

2.6   Royen's Bivariate Gamma

Let $\boldsymbol{R} = \begin{pmatrix} 1 & \rho \\ \rho & 1 \end{pmatrix}$ be a nonsingular correlation matrix, let $\boldsymbol{Y}_1, \ldots, \boldsymbol{Y}_d$ be independent standard bivariate normal random variables with correlation matrix $\boldsymbol{R}$, and let $\boldsymbol{Y}$ be the $(2 \times d)$ matrix with columns $\boldsymbol{Y}_j$ ($j = 1, 2, \ldots, d$). Then, according to Royen (1991a), the joint distribution function of the squared Euclidean norms of the *row* vectors of $\boldsymbol{Y}$ is the distribution function of a bivariate gamma distribution of order $\alpha = d/2$ and accompanying correlation matrix $\boldsymbol{R}$. Let $G_{\alpha+n}(x)$ denote the cumulative distribution function of a standard gamma distribution with shape parameter $\alpha + n$ [see Chapter 17 of Johnson, Kotz, and Balakrishnan (1994)] and $G_{\alpha+n}^{(n)}(x)$ its n-th derivative with respect to x. Denoting the density function by $g_{\alpha+n}(x) = e^{-x} x^{\alpha+n-1} / \Gamma(\alpha + n)$, we have

$$g_{\alpha+n}^{(n)}(x) = \frac{g_\alpha(x) L_{n-1}^{(\alpha-1)}(x)}{\binom{\alpha-1+n}{n}} \quad \text{and} \quad G_{\alpha+n}^{(n)}(x) = \frac{g_{\alpha+1}(x) L_{n-1}^{(\alpha)}(x)}{\binom{\alpha-1+n}{n}},$$

where $L_n^{(\lambda)}(x) = (-1)^n x^{-\lambda} e^x \frac{d^n}{dx^n}\left(e^{-x} x^{\lambda+n}\right)$ is the nth generalized Laguerre polynomial. Then, the joint cumulative distribution function of this bivariate gamma distribution is

$$
\begin{aligned}
&\Pr[X_1 \leq x_1, X_2 \leq x_2] \\
&= \frac{1}{\Gamma(\alpha)} \sum_{n=0}^{\infty} \frac{\Gamma(\alpha+n)\rho^{2n}}{n!} G_{\alpha+n}^{(n)}\left(\frac{x_1}{2}\right) G_{\alpha+n}^{(n)}\left(\frac{x_2}{2}\right) \\
&= \frac{(1-\rho^2)^\alpha}{\Gamma(\alpha)} \sum_{n=0}^{\infty} \frac{\Gamma(\alpha+n)\rho^{2n}}{n!} G_{\alpha+n}\left(\frac{x_1}{2(1-\rho^2)}\right) G_{\alpha+n}\left(\frac{x_2}{2(1-\rho^2)}\right).
\end{aligned}
$$

Royen (1991a) has also discussed a trivariate gamma distribution.

2.7 Farlie–Gumbel–Morgenstern-Type Bivariate Gamma

The bivariate gamma distribution of the Farlie–Gumbel–Morgenstern type was studied by D'Este (1981) and Gupta and Wong (1989). Recall (from Chapter 44) that the probability density function of the Farlie-Gumbel-Morgenstern distribution in the general case is

$$
\begin{aligned}
p(x_1, x_2) &= g(x_1)h(x_2)[1 + \lambda\{2G(x_1) - 1\}\{2H(x_2) - 1\}] \\
&= g(x_1)h(x_2) + \lambda[g(x_1)\{2G(x_1) - 1\}][h(x_2)\{2H(x_2) - 1\}], \\
&\qquad\qquad\qquad\qquad |\lambda| \le 1, \qquad\qquad (48.18)
\end{aligned}
$$

where $G(x_1)$ and $H(x_2)$ are the marginal cumulative distribution functions, and $g(x_1)$ and $h(x_2)$ are the corresponding probability density functions. Representation (48.18) shows that the integrals on the joint space decompose into a product of integrals of a single variable. In the case when $g(x_1)$ and $h(x_2)$ are standard gamma density functions given by

$$
g(x_1) = \frac{1}{\Gamma(\alpha)}\, e^{-x_1} x_1^{\alpha-1}, \qquad x_1 \ge 0,\ \alpha > 0
$$

and

$$
h(x_2) = \frac{1}{\Gamma(\beta)}\, e^{-x_2} x_2^{\beta-1}, \qquad x_2 \ge 0,\ \beta > 0,
$$

we have

$$
\begin{aligned}
E[X_1^n X_2^m] &= E[X_1^n]E[X_2^m] \\
&\quad + \lambda\, E[X_1^n]E[X_2^m]\left\{\frac{2I(\alpha, n)}{B(\alpha, \alpha+n)} - 1\right\}\left\{\frac{2I(\beta, m)}{B(\beta, \beta+m)} - 1\right\},
\end{aligned}
$$

where

$$
I(a, k) = \int_0^1 \frac{z^{a-1}}{(1+z)^{2a+k}}\, dz, \quad I(a, k; x) = \int_0^x \frac{z^{a-1}}{(z+1)^{2a+k}}\, dz
$$

and

$$
B(a, b) = \frac{\Gamma(a)\Gamma(b)}{\Gamma(a+b)}\,.
$$

When k is a positive integer, as in this case, the integral $I(a, k)$ can be evaluated using the recurrence relation

$$
(2a + k)I(a, k + 1) = 2^{-(2a+k)} + (a + k)I(a, k)
$$

and the standard integral $I(a, 0) = \frac{1}{2} B(a, a)$.

Note that the above expression for $E[X_1^n X_2^m]$ can be readily rewritten in terms of the incomplete beta function ratio $I-p(a, b) = \int_0^p \frac{1}{B(a,b)} t^{a-1}(1- t)^{b-1} dt$ since we have $\frac{I(\alpha,n)}{B(\alpha,\alpha+n)} = I_{1/2}(\alpha, \alpha + n)$.

The joint moment generating function was derived by Gupta and Wong (1989) as

$$M_{X_1,X_2}(t_1, t_2) = \prod_{j=1}^{2} (1 - t_j)^{-alpha_j} \left[1 + \lambda \prod_{j=1}^{2} \left\{ \frac{2I\left(\alpha_j, 0; \frac{1}{1-t_j}\right)}{I(\alpha_j, 0; 1)} \right\} \right],$$
$$|t_j| < 1. \qquad (48.19)$$

X_1 and X_2 are independent iff $\lambda = 0$.

Since the probability structure has the property that the variables can be separated, conditional moments can be determined easily, Indeed,

$$E[X_1|X_2 = x_2] = \alpha + \frac{\lambda\alpha\Gamma(\alpha + 1/2)}{(\alpha + 1)\sqrt{\pi}} \{2G(x_2) - 1\}$$

for which an asymptotic expansion has been provided by Gupta and Wong (1989).

The correlation coefficient between X_1 and X_2 is

$$\text{corr}(X_1, X_2) = \rho = \lambda K(\alpha) K(\beta), \qquad (48.20)$$

where

$$K(\alpha) = 1/\left\{ 2^{2\alpha-1} B(\alpha, \alpha)\sqrt{\alpha} \right\}.$$

Since $|\lambda| \leq 1$, it is necessary that $K(\alpha) \geq 1$ in order that the bivariate gamma distribution of Farlie–Gumbel–Morgenstern type be applicable to all sets of marginals and correlations. However, $\lim_{\alpha\to\infty} K(\alpha) = \frac{1}{\sqrt{\pi}}$ and $K(\alpha)$ is monotonically increasing, and so the maximal admissible correlation coefficient between X_1 and X_2 is $\rho = \frac{1}{\pi} = 0.3183$.

2.8 Prékopa and Szántai's Bivariate Gamma

Prékopa and Szántai (1978) introduced a multivariate gamma distribution as the distribution of the random vector $\boldsymbol{X} = \boldsymbol{AY}$, where $\boldsymbol{Y}$ has independent standard gamma components and the matrix $\boldsymbol{A}$ consists of all possible nonzero column vectors having components 0 and 1.

Szántai (1986) investigated further the bivariate case of this multivariate gamma family. The bivariate structure is

$$X_1 = Y_1 + Y_2 \quad \text{and} \quad X_2 = Y_1 + Y_3,$$

where Y_1, Y_2 and Y_3 are independent standard gamma random variables with shape parameters α_1, α_2 and α_3, respectively. Note the similarity of this structure to Holgate's bivariate Poisson as described in Chapter 37 of Johnson, Kotz, and Balakrishnan (1997).

Direct calculations readily yield

$$F_{X_1,X_2}(x_1, x_2) = \int_0^{\min(x_1,x_2)} F_{\alpha_2}(x_1 - y) F_{\alpha_3}(x_2 - y) p_{\alpha_1}(y) \, dy. \quad (48.21)$$

The joint density function of X_1 and X_2 can be written explicitly as

$$
\begin{aligned}
p_{X_1,X_2}&(x_1, x_2)\\
&= p_{\alpha_1+\alpha_2}(x_1) p_{\alpha_1+\alpha_3}(x_2)\\
&\quad \times \left\{ 1 + \sum_{r=1}^{\infty} r! \, \frac{\Gamma(\alpha_1 + r)}{\Gamma(\alpha_1)} \frac{\Gamma(\alpha_1 + \alpha_2)}{\Gamma(\alpha_1 + \alpha_2 + r)} \frac{\Gamma(\alpha_1 + \alpha_3)}{\Gamma(\alpha_1 + \alpha_3 + r)} \right.\\
&\qquad \left. \cdot L_r^{(\alpha_1+\alpha_2-1)}(x_1) L_r^{(\alpha_1+\alpha_3-1)}(x_2) \right\}\\
&= p_{\alpha_1+\alpha_2}(x_1) p_{\alpha_1+\alpha_3}(x_2)\\
&\quad \times \left\{ 1 + \sum_{r=1}^{\infty} \frac{r! \alpha_1^{[r]}}{(\alpha_1 + \alpha_2)^{[r]} (\alpha_1 + \alpha_3)^{[r]}} L_r^{(\alpha_1+\alpha_2-1)}(x_1) L_r^{(\alpha_1+\alpha_3-1)}(x_2) \right\},
\end{aligned}
$$
$$(48.22)$$

where

$$p_\alpha(x) = \frac{1}{\Gamma(\alpha)} \, x^{\alpha-1} e^{-x}, \qquad x \geq 0, \; \alpha > 0,$$

and $L_r^{(\alpha)}(x)$ are the Laguerre polynomials as defined earlier in (48.14). Recall that $L_0^{(\alpha)}(x) \equiv 1$, $L_1^{(\alpha)}(x) = \alpha - x + 1$, and

$$(r+1)L_{r+1}^{(\alpha)}(x) - (2r+\alpha+1-x)L_r^{(\alpha)}(x) + (r+\alpha)L_{r-1}^{(\alpha)}(x) = 0 \text{ for } r = 1, 2, \ldots .$$

2.9 Smith, Adelfang, and Tubbs's Bivariate Gamma

Smith, Adelfang, and Tubbs (1982) discussed the bivariate gamma distribution with joint density function

$$
\begin{aligned}
p(x_1, &x_2; \gamma_1, \gamma_2, \eta)\\
&= \frac{x_1^{\gamma_1-1} x_2^{\gamma_2-1} e^{-(x_1+x_2)/(1-\eta)}}{(1-\eta)^{\gamma_1} \Gamma(\gamma_1) \Gamma(\gamma_2 - \gamma_1)} \sum_{k=0}^{\infty} a_k I_{\gamma_2+k-1}\left(\frac{2\sqrt{\eta x_1 x_2}}{1 - \eta} \right), \quad (48.23)
\end{aligned}
$$

where

$$a_k = \frac{(\eta x_2)^{k/2} \Gamma(\gamma_2 - \gamma_1 + k)}{x_1^{k/2} \, k!}, \qquad k = 0, 1, 2, \ldots,$$

$\gamma_2 > \gamma_1 > 1$ are shape parameters, $0 < \eta < 1$ is a dependency parameter satisfying $\eta = \rho\sqrt{\gamma_2/\gamma_1}$ where ρ is the correlation coefficient between X_1 and X_2, and $I_v(\cdot)$ is the modified Bessel function with index v [see Chapter 1 of Johnson, Kotz, and Kemp (1992)]. For the case of equal shape parameters (viz., $\gamma_1 = \gamma_2 = \gamma$), the joint density function in (48.23) reduces to

$$
\begin{aligned}
&p(x_1, x_2; \gamma, \eta) \\
&= \frac{(x_1 x_2)^{(\gamma-1)/2} \; e^{-(x_1+x_2)/(1-\eta)}}{\eta^{(\gamma-1)/2}(1-\eta)\Gamma(\gamma)} \; I_{\gamma-1}\left(\frac{2\sqrt{\eta x_1 x_2}}{1-\eta}\right)
\end{aligned}
\tag{48.24}
$$

which is Kibble's bivariate density function with $\rho^2 = \eta$ (see Section 2.2). Smith and Adelfang (1981) utilized this family for modeling wind gust data.

Brewer, Tubbs, and Smith (1987) examined the location of the mode in Kibble's and in generalized Kibble's bivariate gamma densities. They noted that the Kibble's density (48.24) attains its maximum in the region $\mathbb{R}_+^2 = \{(x_1, x_2) : x_1 \geq 0, \; x_2 \geq 0\}$ on the line $x_1 = x_2$. For a fixed $\gamma > 1$, denoting by $\tau(\eta)$ or by $\tau(\eta, \gamma)$ (the so-called modal location function) the value at which $p(\tau(\eta), \tau(\eta); \gamma, \eta)$ is maximum, we find τ to be continuously differentiable for $0 < \eta < 1$. Moreover, for $\gamma = 3/2$ we have

$$
\tau(\eta) = \frac{1-\eta}{4\sqrt{\eta}} \; \ln\left(\frac{1+\sqrt{\eta}}{1-\sqrt{\eta}}\right),
$$

and the function $\tau(\eta, \gamma)$ is a decreasing function of γ for fixed $\eta \in (0, 1)$ and $\gamma > 1$; furthermore,

$$
\lim_{\eta \to 1} \tau(\eta, \gamma) = \max\left(\gamma - \frac{3}{2}, 0\right) \quad \text{for} \quad \gamma > 1.
$$

The modal location function $\tau(\eta, \gamma)$ satisfies a nonlinear differential equation in ρ.

For unequal shape parameters with $\gamma_1 = 1$ and $\gamma_2 \geq 2$, the density function in (48.23) becomes

$$
p(x_1, x_2; 1, \gamma, \eta) = \frac{x_2^{\gamma-1} \; e^{-s_2}}{(1-\eta)\Gamma(\gamma-1)} \sum_{j=0}^{\infty} e^{-s_1} c_j \, s_1^j / j!,
\tag{48.25}
$$

where $s_1 = x_1/(1-\eta)$, $s_2 = x_2/(1-\eta)$, and

$$
c_j = \sum_{k=0}^{\infty} (\eta s_2)^{j+k} \frac{\Gamma(\gamma+k-1)}{k!\Gamma(\gamma+j+k)}, \quad j = 0, 1, 2, \dots .
$$

Let $\mu(\eta)$ (or $\mu(\eta, \gamma)$) denote the value for which $p(0, \mu(\eta, \gamma); 1, \gamma, \eta)$ is a maximum of $p(x_1, x_2; 1, \gamma, \eta)$ in (48.25). If $\gamma = 2$, then

$$\mu(\eta) = \frac{1 - \eta}{\eta} \ln \left(\frac{1}{1 - \eta} \right) \qquad \text{for } 0 < \eta < 1;$$

moreover, $\mu(\eta, \gamma)$ is a decreasing function of η for fixed $\gamma > 2$ and

$$\lim_{\eta \to 1} \mu(\eta, \gamma) = \gamma - 2 \qquad \text{for } \gamma \geq 2.$$

In the case of the general bivariate gamma distribution in (48.23) with unequal shape parameters, Brewer, Tubbs, and Smith (1987) have provided the following empirical approximations for the modal values:

$$x_1 = \frac{\gamma_1 - 1}{\gamma_2 - 1} \tau(\eta, \gamma_2) \quad \text{and} \quad x_2 = \mu(\eta, \gamma_2) + \frac{\gamma_1 - 1}{\gamma_2 - 1} \{\tau(\eta, \gamma_2) - \mu(\eta, \gamma_2)\},$$

$$(48.26)$$

where τ and μ are as defined above.

Brewer, Tubbs and Smith (1987) have presented graphs illustrating the behaviour of the modal location function.

2.10 Dussauchoy and Berland's Bivariate Gamma

Dussauchoy and Berland (1975) pointed out that the joint characteristic function

$$\phi(u, v) = (1 - iu)^{-\ell_1}(1 - iv)^{-\ell_2} \left(1 + \frac{zuv}{(1 - iu)(1 - iv)} \right)^{-\ell_3},$$

where $\ell_1, \ell_2, \ell_3 > 0$ and $0 \leq z \leq 1$, corresponds to bivariate distributions with gamma marginals, a result due to Griffiths (1969) based on series expansion of bivariate frequency functions.

On the other hand, using the approach of David and Fix (1961), linear combinations of two independent gamma random variables can be used to define two nonindependent gamma random variables. The joint characteristic function in this case is of the form

$$\phi(u, v) = (1 - iu)^{-\ell_1}(1 - iv)^{-\ell_2}\{1 - i(u + v)\}^{-\ell_3}.$$

Dussauchoy and Berland (1972) provided a joint distribution of two dependent gamma random variables X_1 and X_2 with the property that

$X_2 - \beta X_1$ and X_1 are statistically independent. The joint characteristic function of this distribution is

$$\phi(u,v) = \left(1 - \frac{iv}{a_2}\right)^{-\ell_2} \left(1 - \frac{i\beta v}{a_1}\right)^{\ell_1} \left\{1 - \frac{i(u+\beta v)}{a_1}\right\}^{-\ell_1}.$$

The corresponding density (with the support $x_2 > \beta x_1 > 0$) is

$$\frac{\beta a_2^{\ell_2}}{\Gamma(\ell_1)\Gamma(\ell_2 - \ell_1)} (\beta x_1)^{\ell_1 - 1} e^{-a_2 x_1} (x_2 - \beta x_1)^{\ell_2 - \ell_1 - 1} e^{-a_2(x_2 - \beta x_1)/\beta}$$

$$\times\, _1F_1\left[\ell_1, \ell_2 - \ell_1; \left(\frac{a_1}{\beta} - a_2\right)(x_2 - \beta x_1)\right],$$

$$\beta \geq 0,\ 0 < a_2 \leq \frac{a_1}{\beta},\ 0 < \ell_1 < \ell_2,$$

where $_1F_1$ is the confluent hypergeometric function. The correlation coefficient is

$$\mathrm{corr}(X_1, X_2) = \frac{\beta a_2}{a_1}\sqrt{\frac{\ell_1}{\ell_2}}.$$

These results have been extended by Dussauchoy and Berland (1975) to form a similar multivariate gamma-type distribution that is defined by means of a characteristic function (see Section 3.8).

2.11 Becker and Roux's and Steel and le Roux's Bivariate Gamma

Becker and Roux (1981) and Steel and le Roux (1987, 1989) studied bivariate gamma extensions of the gamma distribution which are based on plausible physical models; they also include Freund's (1961) bivariate exponential distribution as a special case.

The original model suggested by Becker and Roux (1981) was slightly reparametrized by Steel and le Roux (1987) to a form that seems to be more suitable for practical situations.

Consider two components, C_1 and C_2, operating in a system. These components are subject to shocks S_1 and S_2, respectively. Assume that C_1 fails after receiving h shocks and C_2 fails after receiving ℓ shocks. While both components are functioning, the occurrence of S_i is governed by a Poisson process with parameter $1/\alpha_i$, $i = 1, 2$. If C_1 fails first, a parameter shift occurs and the subsequent occurrence of S_2 is governed by a Poisson process with parameter λ_2/α_2. Similarly, if C_2 fails first, a parameter shift occurs and the subsequent occurrence of S_1 is governed by a Poisson process with parameter λ_1/α_1. The four possible Poisson processes governing the occurrence of shocks are assumed to be independent.

In this model, simple restrictions on the values of λ_1 and λ_2 are sufficient to identify models for various plausible physical situations. If $0 < \lambda_1, \lambda_2 < 1$, a "competition model" arises in which the components (or individuals) compete for the same limited resources, and failure (or death) of one leads to a decrease in the rate at which shocks subsequently strike the other component (or individual). If $\lambda_1, \lambda_2 > 1$, a "shared load model" arises in which the two components share a certain load, and failure of one leads to an increase in the rate at which shocks subsequently strike the other component. If $\lambda_1 > 1$ and $0 < \lambda_2 < 1$, or vice versa, an "asymmetrical model" arises in which C_1 can be regarded as a parasite and C_2 as a host. Finally, if $\lambda_1 = \lambda_2 = 1$, no parameter shift occurs and, in this case, the components act independently.

Let X_1 and X_2 denote the lifetimes of the components C_1 and C_2, respectively. Then, the bivariate density function of $(X_1, X_2)^T$ has been derived by Steel and le Roux (1987) as

$$p_{X_1,X_2}(x_1, x_2) = \begin{cases} \frac{\lambda_2 x_1^{h-1}}{\Gamma(h)\Gamma(\ell)\alpha_1^h \alpha_2^\ell} \{\lambda_2(x_2 - x_1) + x_1\}^{\ell-1} \\ \quad \times \exp\left\{-\left(\frac{1}{\alpha_1} + \frac{1}{\alpha_2} - \frac{\lambda_2}{\alpha_2}\right) x_1 - \frac{\lambda_2}{\alpha_2} x_2\right\} \\ \quad \text{if } 0 < x_1 < x_2 < \infty, \\[2mm] \frac{\lambda_1 x_2^{\ell-1}}{\Gamma(h)\Gamma(\ell)\alpha_1^h \alpha_2^\ell} \{\lambda_1(x_1 - x_2) + x_2\}^{h-1} \\ \quad \times \exp\left\{-\left(\frac{1}{\alpha_1} + \frac{1}{\alpha_2} - \frac{\lambda_1}{\alpha_1}\right) x_2 - \frac{\lambda_1}{\alpha_1} x_1\right\} \\ \quad \text{if } 0 < x_2 < x_1 < \infty. \end{cases}$$

(48.27)

In the special case when $h = \ell = 1$, the density in (48.27) reduces to

$$p_{X_1,X_2}(x_1, x_2) = \begin{cases} \frac{\lambda_2}{\alpha_1 \alpha_2} \exp\left\{-\left(\frac{1}{\alpha_1} + \frac{1}{\alpha_2} - \frac{\lambda_2}{\alpha_2}\right) x_1 - \frac{\lambda_2}{\alpha_2} x_2\right\} \\ \quad \text{if } 0 < x_1 < x_2 < \infty, \\[2mm] \frac{\lambda_1}{\alpha_1 \alpha_2} \exp\left\{-\left(\frac{1}{\alpha_1} + \frac{1}{\alpha_2} - \frac{\lambda_1}{\alpha_1}\right) x_2 - \frac{\lambda_1}{\alpha_1} x_1\right\} \\ \quad \text{if } 0 < x_2 < x_1 < \infty. \end{cases}$$

(48.28)

This is a reparametrized version of Freund's (1961) bivariate exponential distribution (see Chapter 47).

For the bivariate density function in (48.27), we have

$$E[X_1^r X_2^s]$$

$$
\begin{aligned}
= {} & \frac{1}{\Gamma(h)\Gamma(\ell)\alpha_1^h \alpha_2^\ell} \Bigg\{ \lambda_2^{-s}\alpha_2^{s+\ell} \left(\frac{1}{\alpha_1} + \frac{1}{\alpha_2} \right)^{-h-r} \sum_{i=0}^{s} \sum_{j=0}^{\ell-1} \binom{s}{i}\binom{\ell-1}{j} \\
& \cdot \left(\frac{\lambda_2}{\alpha_2} \right)^i \alpha_2^{-j} \left(\frac{1}{\alpha_1} + \frac{1}{\alpha_2} \right)^{-i-j} \Gamma(h+r+i+j)\Gamma(s+\ell-i-j) \\
& + \lambda_1^{-r}\alpha_1^{r+h} \left(\frac{1}{\alpha_1} + \frac{1}{\alpha_2} \right)^{-\ell-s} \sum_{i=0}^{r} \sum_{j=0}^{h-1} \binom{r}{i}\binom{h-1}{j} \\
& \cdot \left(\frac{\lambda_1}{\alpha_1} \right)^i \alpha_1^{-j} \left(\frac{1}{\alpha_1} + \frac{1}{\alpha_2} \right)^{-i-j} \Gamma(\ell+s+i+j)\Gamma(r+h-i-j) \Bigg\}
\end{aligned}
$$

(48.29)

for $r, s = 0, 1, 2, \ldots$.

Steel and le Roux (1989) studied compound distributions of the bivariate gamma distribution in (48.27). First, by considering the variables $Y_1 = \min(X_1, X_2)$ and $Y_2 = \max(X_1, X_2) - \min(X_1, X_2) = |X_1 - X_2|$, we obtain the joint density function of Y_1 and Y_2 from (48.27) as

$$
\begin{aligned}
p_{Y_1,Y_2}(y_1, y_2) = {} & \frac{1}{\Gamma(h)\Gamma(\ell)\alpha_1^h \alpha_2^\ell} \exp\left\{ -\left(\frac{1}{\alpha_1} + \frac{1}{\alpha_2} \right) y_1 \right\} \\
& \times \Big\{ \lambda_1 y_1^{\ell-1}(y_1 + \lambda_1 y_2)^{h-1} e^{-\lambda_1 y_2/\alpha_1} \\
& + \lambda_2 y_1^{h-1}(y_1 + \lambda_2 y_2)^{\ell-1} e^{-\lambda_2 y_2/\alpha_2} \Big\}, \quad y_1, y_2 > 0.
\end{aligned}
$$

(48.30)

Let us now view λ_1 and λ_2 as values of random variables Λ_1 and Λ_2 with density functions $p_{\Lambda_1}(\lambda_1)$ and $p_{\Lambda_2}(\lambda_2)$, respectively. Let us further assume that Λ_1 and Λ_2 are statistically independent with supports (possibly infinite) A_1 and A_2 that are subintervals of $\mathbb{R}^+$. Then, the compounded joint density function of Y_1 and Y_2 is

$$
\begin{aligned}
& p_{Y_1,Y_2}(y_1, y_2) \\
& = \frac{1}{\Gamma(h)\Gamma(\ell)\alpha_1^h \alpha_2^\ell} \exp\left\{ -\left(\frac{1}{\alpha_1} + \frac{1}{\alpha_2} \right) y_1 \right\} y_1^{h+\ell-2} \\
& \times \Bigg\{ \sum_{k=0}^{h-1} \binom{h-1}{k}(y_2/y_1)^k \int_{A_1} \lambda_1^{k+1} e^{-y_2\lambda_1/\alpha_1} p_{\Lambda_1}(\lambda_1)\, d\lambda_1 \\
& + \sum_{k=0}^{\ell-1} \binom{\ell-1}{k}(y_2/y_1)^k \int_{A_2} \lambda_2^{k+1} e^{-y_2\lambda_2/\alpha_2} p_{\Lambda_2}(\lambda_2)\, d\lambda_2 \Bigg\},
\end{aligned}
$$

$$y_1, y_2 > 0; \qquad (48.31)$$

the corresponding compounded joint survival function of Y_1 and Y_2 is

$$
\begin{aligned}
\bar{F}_{Y_1,Y_2}(y_1, y_2) &= \Pr[Y_1 > y_1, Y_2 > y_2] \\
&= \frac{\alpha_1^\ell \alpha_2^h}{\Gamma(h)\Gamma(\ell)} \left\{ \sum_{k=0}^{h-1} \binom{h-1}{k} \frac{\Gamma\left(h + \ell - k - 1; \left(\frac{1}{\alpha_1} + \frac{1}{\alpha_2}\right) y_1\right)}{\alpha_2^{k+1}(\alpha_1 + \alpha_2)^{h+\ell-k-1}} \right. \\
&\qquad \times \int_{A_1} p_{\Lambda_1}(\lambda_1) \Gamma\left(k + 1; \frac{\lambda_1 y_2}{\alpha_1}\right) d\lambda_1 \\
&\qquad + \sum_{k=0}^{\ell-1} \binom{\ell-1}{k} \frac{\Gamma\left(h + \ell - k - 1; \left(\frac{1}{\alpha_1} + \frac{1}{\alpha_2}\right) y_1\right)}{\alpha_1^{k+1}(\alpha_1 + \alpha_2)^{h+\ell-k-1}} \\
&\qquad \left. \times \int_{A_2} p_{\Lambda_2}(\lambda_2) \Gamma\left(k + 1; \frac{\lambda_2 y_2}{\alpha_2}\right) d\lambda_2 \right\}, \qquad y_1, y_2 > 0,
\end{aligned}
$$

$$(48.32)$$

where

$$\Gamma(k; z) = \int_z^\infty e^{-u} u^{k-1} \, du, \qquad z > 0.$$

In the special case when $h = \ell = 1$, the joint survival function in (48.32) reduces to

$$
\begin{aligned}
\bar{F}_{Y_1,Y_2}(y_1, y_2) \\
= \frac{1}{\alpha_1 + \alpha_2} \exp\left\{-\left(\frac{1}{\alpha_1} + \frac{1}{\alpha_2}\right) y_1\right\} \\
\times \left\{\alpha_1 \int_{A_1} p_{\Lambda_1}(\lambda_1) e^{-\lambda_1 y_2/\alpha_1} d\lambda_1 + \alpha_2 \int_{A_2} p_{\Lambda_2}(\lambda_2) e^{-\lambda_2 y_2/\alpha_2} d\lambda_2\right\},
\end{aligned}
$$

$$y_1, y_2 > 0. \qquad (48.33)$$

In this special case, it is evident that Y_1 and Y_2 are independently distributed with Y_1 distributed as $\exp\left(\frac{1}{\alpha_1} + \frac{1}{\alpha_2}\right)$ and Y_2 distributed as a (finite) mixture of two distributions with mixing coefficients $\alpha_1/(\alpha_1 + \alpha_2)$ and $\alpha_2/(\alpha_1 + \alpha_2)$. These two distributions are obtained by compounding $\exp(\lambda_i/\alpha_i)$ with prior distributions of Λ_i for $i = 1, 2$.

From the joint density function in (48.31), we have

$$E[Y_1^r Y_2^s]$$

$$= \frac{\alpha_1^{\ell+r} \alpha_2^{h+r}}{\Gamma(h)\Gamma(\ell)(\alpha_1 + \alpha_2)^{h+\ell+r}}$$

$$\times \left\{ \alpha_1^s \sum_{k=0}^{h-1} \binom{h-1}{k} \Gamma(h+\ell+r-k-1)\Gamma(k+s+1) \left(\frac{1}{\alpha_1} + \frac{1}{\alpha_2}\right)^{k+1} \right.$$

$$\cdot \int_{A_1} \lambda_1^{-s} p_{\Lambda_1}(\lambda_1) \, d\lambda_1$$

$$+ \alpha_2^s \sum_{k=0}^{\ell-1} \binom{\ell-1}{k} \Gamma(h+\ell+r-k-1)\Gamma(k+s+1) \left(\frac{1}{\alpha_1} + \frac{1}{\alpha_2}\right)^{k+1}$$

$$\left. \cdot \int_{A_2} \lambda_2^{-s} p_{\Lambda_2}(\lambda_2) d\lambda_2 \right\} \tag{48.34}$$

for $r, s = 0, 1, 2, \ldots$.

Steel and le Roux (1989) investigated two cases:

(i) the "competition model" with Λ_1 and Λ_2 both distributed as $Beta(a, b)$ with density $\frac{1}{B(a,b)} \lambda_j^{a-1}(1-\lambda_j)^{b-1}$, $0 < \lambda_j < 1$,

and

(ii) the "shared load model" with Λ_1 and Λ_2 both distributed as *exponential* with density $\theta \, e^{-\theta(\lambda_j-1)}$, $1 < \lambda_j < \infty$,

leading to compound bivariate gamma-beta and compound bivariate gamma-exponential distributions, respectively.

2.12   Schmeiser and Lal's Bivariate Gamma

Schmeiser and Lal (1982) developed an algorithm that enables, for any two given gamma marginal distributions with parameters (β_i, α_i), $i = 1, 2$, any associated correlation ρ, and any choice of linear or nonlinear regression curves, the construction of a bivariate gamma distribution with parameters $(\beta_1, \beta_2, \alpha_1, \alpha_2, \rho)$. Specifically, let Z, W_1, and W_2 be independent gamma random variables with unit scale parameter and shape parameters γ, δ_1, and δ_2, respectively, and let U be an independent Uniform$(0, 1)$ random variable. Let

$$X_1 = \frac{G_{\lambda_1}^{-1}(U) + Z + W_1}{\beta_1} \quad \text{and} \quad X_2 = \frac{G_{\lambda_2}^{-1}(V) + Z + W_2}{\beta_2},$$

where $G_\lambda(\cdot)$ is the distribution function of a gamma random variable with unit scale parameter and shape parameter λ, $G_\lambda^{-1}(\cdot)$ is the inverse function of $G_\lambda(\cdot)$, and either $V = U$ or $V = 1-U$. The variables X_i are distributed as gamma with scale parameters β_i and shape parameters $\alpha_i = \gamma + \lambda_i + \delta_i$, $i = 1, 2$, respectively, and have a correlation coefficient

$$\rho = \frac{E\left\{G_{\lambda_1}^{-1}(U)G_{\lambda_2}^{-1}(V) - \lambda_1\lambda_2 + \gamma\right\}}{\sqrt{\alpha_1\alpha_2}}.$$

The procedure of selecting the parameters such that

$$\begin{cases} \gamma + \lambda_i + \delta_i = \alpha_i, & i = 1, 2, \\ E\left\{G_{\lambda_1}^{-1}(U)G_{\lambda_2}^{-1}(V) - \lambda_1\lambda_2 + \gamma\right\} = \rho\sqrt{\alpha_1\alpha_2}, \\ \gamma \geq 0, \ \lambda_i \geq 0, \ \delta_i \geq 0 \end{cases}$$

(with β_i's being set directly) is equivalent to finding a *feasible solution* to a nonlinear programming problem. Compare this approach with that of Cheriyan and Ramabhadran presented in Section 2.2.

2.13 Bivariate Chi-Square Distributions

A distribution with marginal χ^2 distributions arises naturally in the following way. Consider a random sample of size n represented by n independent vectors $(X_{i1}, \ldots, X_{ik})$ $(i = 1, 2, \ldots, n)$, each vector having the same multivariate normal distribution with variance–covariance matrix $\boldsymbol{V}$ and having each diagonal element equal to 1 ($\boldsymbol{V}$ is, of course, also the correlation matrix). Then the statistics $S_j = \sum_{i=1}^n (X_{ij} - \bar{X}_{\cdot j})^2$ $(j = 1, 2, \ldots, k)$ each have a χ_{n-1}^2 distribution. Their joint distribution may, following Krishnaiah, Hagis, and Steinberg (1963), be called a *multivariate chi-square distribution*. It is also called *generalized Rayleigh distribution*; see, for example, Miller (1964).

The conditional distribution of $(X_{11}, \ldots, X_{n1})$, given $(X_{12}, \ldots, X_{n2})$, is that of n independent normal variables with expected values $(\rho X_{12}, \ldots, \rho X_{n2})$ and common variance $(1 - \rho^2)$, where $\rho = \text{corr}(X_{i1}, X_{i2})$. It follows that X_{i1} can be represented as $\rho X_{i2} + \sqrt{1 - \rho^2}\, U_i$, where $U_1, U_2, \ldots, U_n$ are independent standard normal variables. Hence, given $(X_{12}, \ldots, X_{n2})$, S_1 is distributed as

$$\sum_{i=1}^n \left\{\rho(X_{i2} - \bar{X}_2) + \sqrt{1 - \rho^2}\,(U_i - \bar{U})\right\}^2, \qquad (48.35)$$

that is, as $(1 - \rho^2) \times$ (noncentral χ^2 with $(n - 1)$ degrees of freedom and noncentrality parameter $\rho^2 S_2 (1 - \rho^2)^{-1}$). Since this depends on $\{X_{i2}\}$ only in S_2, this is also the conditional distribution of S_1 given S_2, so that

$$
\Pr[S_1 \leq s | S_2 = s_2] = \exp\left[-\frac{1}{2}\frac{\rho^2 s_2}{1 - \rho^2}\right] \sum_{j=0}^{\infty} \frac{1}{j!} \left[\frac{1}{2}\frac{\rho^2 s_2}{1 - \rho^2}\right]^j
$$
$$
\times \Pr\left[\chi^2_{n-1+2j} \leq (1 - \rho^2)^{-1} s\right].
$$

Noting that S_2 has a χ^2_{n-1} distribution, we calculate

$$
\Pr[(S_1 < s_1) \cap (S_2 < s_2)]
$$
$$
= \int_0^{s_2} \Pr[S_1 \leq s_1 | S_2 = s_2] p_{S_2}(s_2) \, ds_2
$$
$$
= \sum_{j=0}^{\infty} c_j \Pr[\chi^2_{n-1+2j} \leq (1 - \rho^2)^{-1} s_1]
$$
$$
\times \Pr[\chi^2_{n-1+2j} \leq (1 - \rho^2)^{-1} s_2], \tag{48.36}
$$

where

$$
c_j = \frac{\Gamma\left(\frac{1}{2}(n - 1) + j\right)(1 - \rho^2)^{\frac{1}{2}(n-1)}\rho^{2j}}{\Gamma\left(\frac{1}{2}(n - 1)\right)j!} ;
$$

see Bose (1935), Finney (1938), Johnson (1962), Vere-Jones (1967), and Moran and Vere-Jones (1969).

     This is a *standard* bivariate chi-square distribution. Additional parameters can be introduced by considering the variables $S_j' = S_j \sigma_j^2$ giving a *general* bivariate chi-square distribution. Note that $c_0, c_1, c_2, \ldots$ are terms in the expression of the negative binomial

$$
\left(\frac{1}{1 - \rho^2} - \frac{\rho^2}{1 - \rho^2}\right)^{-\frac{1}{2}(n-1)},
$$

so that $\sum_{j=0}^{\infty} c_j = 1$. The joint distribution of S_1 and S_2 can thus be regarded as a mixture of joint distributions, with weights c_j, in which S_1 and S_2 each have independent χ^2_{n-1+2j} distributions.

     It follows that S_1/S_2 is distributed as a mixture, in the same proportions as c_j, of $F_{n-1+2j, n-1+2j}$ distributions.

     The density function of $G = S_1/S_2$ can also be written as

$$
p_G(g) = \frac{(1 - \rho^2)^{\frac{1}{2}v}}{B\left(\frac{1}{2}, \frac{1}{2}v\right)} \frac{g^{(1/2)v-1}}{(1 + g)^v} \left(1 - \frac{4\rho^2 g}{(1 + g)^2}\right)^{-\frac{1}{2}(v+1)} \quad (g > 0); \tag{48.37}
$$

see Bose (1935) and Finney (1938). The sum $S_1 + S_2$ is distributed as a mixture of $\chi^2_{2(n-1)+4j}$ distributions with weights c_j.

Furthermore, the expected value of $S_1^{\alpha_1} S_2^{\alpha_2}$ is

$$
\begin{aligned}
\mu'_{\alpha_1,\alpha_2} &= \sum_{j=0}^{\infty} c_j \mu'_{\alpha_1}(\chi^2_{n-1+2j}) \mu'_{\alpha_2}(\chi^2_{n-1+2j}) \\
&= \sum_{j=0}^{\infty} c_j \frac{2^{\alpha_1+\alpha_2}\Gamma[\frac{1}{2}(n-1)+j+\alpha_1]\Gamma[\frac{1}{2}(n-1)+j+\alpha_2]}{\{\Gamma[\frac{1}{2}(n-1)+j]\}^2}.
\end{aligned}
$$
(48.38)

This formula applies for any values of α_1 and α_2, provided only that $\min(\alpha_1, \alpha_2) > -\frac{1}{2}(n-1)$.

From (48.35), it is clear that

$$
\begin{aligned}
E[S_1|S_2] &= (1-\rho^2)E\left[\chi'^2_{n-1}\left(\frac{\rho^2 S_2}{1-\rho^2}\right)\right] \\
&= (n-1)(1-\rho^2) + \rho^2 S_2.
\end{aligned}
$$
(48.39)

In addition,

$$
\begin{aligned}
\mathrm{var}(S_1|S_2) &= (1-\rho^2)^2[2(n-1) + 4\rho^2 S_2/(1-\rho^2)] \\
&= 2(n-1)(1-\rho^2)^2 + 4\rho^2(1-\rho^2)S_2.
\end{aligned}
$$
(48.40)

The regression of S_1 on S_2 is linear, but the array distributions are not homoscedastic.

The joint distribution of $\sqrt{S_1}$ and $\sqrt{S_2}$ (the *bivariate chi distribution*) has been studied by Krishnaiah, Hagis, and Steinberg (1963).

Probabilities associated with the distribution of S_1/S_2 can be evaluated from tables of the incomplete beta function ratio [see Eq. (1.91) in Chapter 1 of Johnson, Kotz, and Kemp (1992)] using the relation [Finney (1938)]

$$
\Pr\left[\max\left(\frac{S_1}{S_2}, \frac{S_2}{S_1}\right) > y^2\right] = I_\eta\left(\frac{1}{2}(n-1), \frac{1}{2}(n-1)\right),
$$
(48.41)

where $\eta = \frac{1}{2}[1 - (y - y^{-1})\{(y+y^{-1})^2 - 4\rho^2\}^{-1/2}]$, with $y > 1$. Johnson (1962) showed that a useful approximation to (48.39) is

$$
2\Pr[F_{v',v'} > y^2]
$$

with $v' = (v - 2\rho^2)/(1 - \rho^2)$.

Note that we can define a *general standard bivariate gamma distribution* by replacing $(n-1)$ in (48.36) by v, which should be positive, but need not be an integer. This distribution depends on the two parameters v, ρ. All the properties of (48.36) are also valid for the general case. Thus, for example, S_1/S_2 is distributed as a mixture of $F_{v+2j,v+2j}$ distributions with weights that are terms in the expansion of the negative binomial

$$
\left(\frac{1}{1-\rho^2} - \frac{\rho^2}{1-\rho^2} \right)^{-v/2}.
$$

Some additional models of bivariate gamma distributions have been discussed by Hutchinson and Lai (1990). We take this opportunity to recommend their book as an excellent factual source for continuous bivariate distributions.

3 MULTIVARIATE GAMMA DISTRIBUTIONS

It is a rather daunting task to attempt to present a coherent, properly classified and an organized description of multivariate gamma distributions, due to an abundance of isolated and disconnected results and substantial time stretches during which little research was carried out in this area followed by booming research activity. The pioneering paper by Krishnamoorthy and Parthasarathy (1951) served for many years as a guiding light along this uneven path.

In this section, we describe various forms of multivariate gamma distributions and their properties. This discussion, though not exhaustive, will hopefully provide an adequate coverage that is useful for applications and also for further theoretical investigations.

3.1 Cheriyan and Ramabhadran's Multivariate Gamma

Let $Y_0, Y_1, \ldots, Y_k$ be independent gamma random variables with probability density functions

$$
p_{Y_i}(y_i) = \frac{1}{\Gamma(\theta_i)}\, e^{-y_i}\, y_i^{\theta_i - 1}, \quad y_i > 0,\ \theta_i > 0\ (i = 0, 1, \ldots, k).
$$

Let $X_i = Y_0 + Y_i$ for $i = 1, 2, \ldots, k$. Then, from the joint density function of $(Y_0, Y_1, \ldots, Y_k)^T$ given by

$$
p_{Y_0, Y_1, \ldots, Y_k}(y_0, y_1, \ldots, y_k)
$$

$$= \frac{1}{\prod_{i=0}^{k} \Gamma(\theta_i)} \, e^{-\sum_{i=0}^{k} y_i} \prod_{i=0}^{k} y_i^{\theta_i-1}, \quad y_i > 0, \; \theta_i > 0 \; (i = 0, 1, \ldots, k),$$

we obtain the joint density function of $(Y_0, X_1, \ldots, X_k)^T$ as

$$p_{Y_0, X_1, \ldots, X_k}(y_0, x_1, \ldots, x_k)$$
$$= \frac{1}{\prod_{i=0}^{k} \Gamma(\theta_i)} \, y_0^{\theta_0-1} \left\{ \prod_{i=1}^{k} (x_i - y_0)^{\theta_i-1} \right\} \exp \left\{ (k-1)y_0 - \sum_{i=1}^{k} x_i \right\},$$
$$x_i \geq y_0 \geq 0 \; (i = 1, 2, \ldots, k). \qquad (48.42)$$

In order to integrate out the variable Y_0, it is necessary to evaluate the integral

$$\int_0^{\tilde{x}} y_0^{\theta_0-1} \left\{ \prod_{i=1}^{k} (x_i - y_0)^{\theta_i-1} \right\} e^{(k-1)y_0} \, dy_0, \qquad (48.43)$$

where $\tilde{x} = \min(x_1, \ldots, x_k)$. In the general case, (48.43) leads to very complicated expressions. Some special cases are, however, quite simple. For example, if $\theta_1 = \cdots = \theta_k = 1$ (that is, $Y_1, \ldots, Y_k$ each have an exponential distribution), then

$$p_{X_1, \ldots, X_k}(x_1, \ldots, x_k) = \frac{1}{\Gamma(\theta_0)} \, e^{-\sum_{i=1}^{k} x_i} g(\tilde{x}; \theta_0), \quad x_i > 0 \; (i = 1, \ldots, k),$$
$$(48.44)$$

where

$$g(\tilde{x}; \theta_0) = \int_0^{\tilde{x}} y_0^{\theta_0-1} e^{(k-1)y_0} \, dy_0.$$

Evidently, $\tilde{x} = \min(x_1, \ldots, x_k)$ is a sufficient statistic for θ_0. The maximum likelihood estimator, $\hat{\theta}_0$, of θ_0 satisfies the equation

$$\frac{\partial}{\partial \hat{\theta}_0} \log g(\tilde{x}; \hat{\theta}_0) = \psi(\hat{\theta}_0),$$

where $\psi(\cdot)$ is the digamma function; see Chapter 1 [Eq. (1.37)] of Johnson, Kotz, and Kemp (1992).

The marginal distribution of X_i is a standard gamma distribution with parameter $\theta_0 + \theta_i$ and, hence, $\text{var}(X_i) = \theta_0 + \theta_i$ $(i = 1, \ldots, k)$. Furthermore,

$$\text{cov}(X_i, X_j) = \text{var}(Y_0) = \theta_0$$

and

$$\text{corr}(X_i, X_j) = \frac{\theta_0}{\sqrt{(\theta_0 + \theta_i)(\theta_0 + \theta_j)}}$$

which is nonnegative. Next, proceeding as in Section 2.2, we can show that

$$E[X_i \mid X_j = x_j] = x_j \frac{\theta_0}{\theta_0 + \theta_j} + \theta_i$$

and

$$\mathrm{var}(X_i \mid X_j = x_j) = x_j^2 \frac{\theta_0 \theta_j}{(\theta_0 + \theta_j)^2 (\theta_0 + \theta_j + 1)} + \theta_i$$

which reveal that the regression of X_i on X_j is linear, but the variation about the regression is not homoscedastic.

The joint moment generating function of $\boldsymbol{X} = (X_1, \ldots, X_k)^T$ is

$$
\begin{aligned}
E[e^{\boldsymbol{t}^T \boldsymbol{X}}] &= E\left[e^{Y_0 \sum_{i=1}^k t_i} \prod_{i=1}^k e^{t_i Y_i} \right] \\
&= E\left[e^{Y_0 \sum_{i=1}^k t_i} \right] \prod_{i=1}^k E\left[e^{t_i Y_i} \right] \\
&= \left(1 - \sum_{i=1}^k t_i \right)^{-\theta_0} \prod_{i=1}^k (1 - t_i)^{-\theta_i};
\end{aligned}
\qquad (48.45)
$$

see Cheriyan (1941) and Ramabhadran (1951).

3.2   Gaver's Multivariate Gamma

An extension of the argument at the end of Section 2.3 due to Gaver (1970) leads to a general multivariate gamma distribution with joint characteristic function

$$\left\{ (\beta + 1) \prod_{j=1}^k (1 - it_j) - \beta \right\}^{-\alpha}, \qquad \alpha, \beta > 0. \qquad (48.46)$$

He has thus considered a mixture of gamma variable with negative binomial weights. This distribution is symmetrical in all k variates. The covariance between X_i and X_j is

$$E_\theta[(\alpha + \theta)^2] - [\alpha(\beta + 1)]^2 = \alpha\beta(\beta + 1)$$

for any $i \neq j$, and so the correlation coefficient between X_i and X_j is

$$\mathrm{corr}(X_i, X_j) = \frac{\beta}{\beta + 1} .$$

3.3   Krishnamoorthy and Parthasarathy's Multivariate Gamma and Its Extension

The most general class (without location parameters) consists of all continuous distributions on $\mathbb{R}_+^k$ with univariate gamma marginal distribution functions with scale parameters β_j and shape parameters α_j. Even with identical standard gamma (that is, with unit scale parameter) marginal densities $g_\alpha(x_j)$, this is a very broad class since it contains, at least for $2\alpha > k - 2$, all mixtures of k-variate standard gamma distributions in the sense of Krishnamoorthy and Parthasarathy (1951) belonging to a random nonsingular correlation matrix $\mathbf{R}$ with any distribution. Simple representations for such general gamma distributions seem to exist mainly for the bivariate case, using orthogonal expansions with Laguerre polynomials and canonical correlations, as presented earlier in Section 2; see also Griffiths (1969) and Sarmanov (1970a,b).

The joint distribution of the diagonal elements Y_{jj} in a $W_k(v, \Sigma)$ Wishart matrix $\mathbf{Y}$ [that is, the matrix $\mathbf{Y}$ with elements

$$Y_{j\ell} = \sum_{i=1}^{k}(X_{ij} - \bar{X}_{\cdot j})(X_{i\ell} - \bar{X}_{\cdot \ell}),$$

where $\mathbf{X} = (X_{i1}, \ldots, X_{ik})^T$, is a k-variate normal random variable with mean $\boldsymbol{\xi}$ and variance-covariance matrix Σ] is a k-variate chi-square distribution with v degrees of freedom, belonging to the covariance matrix Σ and scaled by $\text{Diag}(\Sigma)$, which is also called a k-variate gamma distribution in the sense of Krishnamoorthy and Parthasarathy (1951) with shape parameter $\alpha = v/2$ and scaled by $2\text{Diag}(\Sigma)$. For $\Sigma > 0$ (i.e., positive definite), an extension of this distribution to noninteger values $2\alpha > k - 1$ is always possible due to the existence of the corresponding Wishart density.

The k-variate standard gamma distribution of Krishnamoorthy and Parthasarathy (1951) is defined by its characteristic function

$$\phi_{\mathbf{X}}(t) = E\left[e^{it^T \mathbf{X}}\right] = |\mathbf{I} - i\mathbf{R}T|^{-\alpha}, \qquad (48.47)$$

where $\mathbf{I}$ is $k \times k$ identity matrix, $\mathbf{R} = ((r_{ij}))$ is any $k \times k$ correlation matrix, $\mathbf{T} = \text{Diag}(t_1, \ldots, t_k)$, and positive integer values 2α or real $2\alpha > k - 2 \geq 0$. For $k \geq 3$, the admissible noninteger values $0 < 2\alpha < k - 2$ depend on $\mathbf{R}$. In particular, every $\alpha > 0$ is admissible iff $|\mathbf{I} - i\mathbf{R}T|^{-1}$ is infinitely divisible, which holds iff the cofactors R_{ij} of $\mathbf{R}$ satisfy the conditions

$$(-1)^\ell R_{i_1 i_2} R_{i_2 i_3} \cdots R_{i_\ell i_1} \geq 0 \qquad (48.48)$$

for every subset $\{i_1, \ldots, i_\ell\} \subseteq \{1, 2, \ldots, k\}$ with $\ell \geq 3$.

There are three types of absolutely convergent series for the k-variate gamma distribution derived from (48.45) with nonsingular $\mathbf{R}$.

Royen (1991a, 1992) has provided a detailed and ingenious derivation of these expansions. The first type generalizes the orthogonal expansion with generalized Laguerre polynomials as given by Krishnamoorthy and Parthasarathy (1951). The second type involves univariate gamma distributions that converge to a multivariate distsribution. The third type, the so-called *tetrachoric expansion*, which is a modification of the second type, allows the calculation of probabilities over unbounded rectangular regions. Direct application of the Fourier inversion formula to (48.45) leads to a 'numerically unsuitable' integral. Royen (1992), after noting taht the classical orthogonal series expansion of Krishnamoorthy and Parthasarathy (1951) has rather poor conversion properties, has improved on their sufficient conditions for convergence [Royen (1991a)]. Simpler series and integrals are obtained for matrix $\mathbf{R}$ of the special form $\mathbf{R} = \mathbf{D} + \mathbf{a}\mathbf{a}^T$, where $\mathbf{D} > 0$ is a diagonal matrix and $a_i^2 < 1$ $(i = 1, 2, \ldots, k)$; see Section 3.6.

3.4   Prékopa and Szántai's Multivariate Gamma

Extending Ramabhadran's (1951) construction, Prékopa and Szántai (1978) defined the following multivariate gamma distribution. Let $X_1, \ldots, X_k$ be mutually independent gamma random variables with scale parameters α_i and shape parameters θ_i, $i = 1, \ldots, k$, respectively. Then, $Y_i = X_i/\alpha_i$ $(i = 1, \ldots, k)$ are standard gamma random variables with shape parameters θ_i, $i = 1, \ldots, k$. They suggested approximating the joint distribution of $X_1, \ldots, X_k$ by the joint distribution of the random vector $\mathbf{Z}$ of the form $\mathbf{Z} = \mathbf{A}\mathbf{W}$, where W_i are independent standard gamma random variables and $\mathbf{A}$ is a matrix with 0,1 entries (there are 2^k distinct column vectors with 0 or 1 as components; in fact, there are $2^k - 1$ if we disregard the column vector consisting of solely 0 as components). Since the covariance of partial sums of independent random variables is the sum of variances of the common terms, the covariances of two random vectors having components $Z_1, \ldots, Z_k$ coincide iff

$$\mathbf{A}\boldsymbol{\beta} = \boldsymbol{\theta}, \quad \tilde{\mathbf{A}}\boldsymbol{\beta} = \mathbf{c}, \quad \text{and} \quad \boldsymbol{\beta} \geq \mathbf{0}, \qquad (48.49)$$

where $\boldsymbol{\theta} = (\theta_1, \ldots, \theta_k)^T$, $\boldsymbol{\beta} = (\eta_1, \ldots, \eta_r)^T$ (where $r = 2^k 1-$) is an unknown vector of parameters of random variables $W_1, \ldots, W_r$, $\mathbf{c}$ is the vector containing all the covariances of W_i $(i = 1, \ldots, r)$ in an appropriate ordering, and $\tilde{\mathbf{A}}$ is a matrix of order $\frac{k(k+1)}{2} \times r$ constructed of the componentwise product of the rows of $\boldsymbol{A}$ which follow in the same order as the components of $\boldsymbol{c}$. [Actually, the first condition in (48.47) is superfluous.]

As an example, let us consider the case when $k = 4$. Here, $\boldsymbol{A}$ is the matrix

$$\boldsymbol{A} = \begin{bmatrix} 1 & 0 & 0 & 0 & 1 & 1 & 1 & 0 & 0 & 0 & 1 & 1 & 1 & 0 & 1 \\ 0 & 1 & 0 & 0 & 1 & 0 & 0 & 1 & 1 & 0 & 1 & 1 & 0 & 1 & 1 \\ 0 & 0 & 1 & 0 & 0 & 1 & 0 & 1 & 0 & 1 & 1 & 0 & 1 & 1 & 1 \\ 0 & 0 & 0 & 1 & 0 & 0 & 1 & 0 & 1 & 1 & 0 & 1 & 1 & 1 & 1 \end{bmatrix}$$

and

$$\begin{aligned}
Z_1 &= W_1 + W_5 + W_6 + W_7 + W_{11} + W_{12} + W_{13} + W_{15}, \\
Z_2 &= W_2 + W_5 + W_8 + W_9 + W_{11} + W_{12} + W_{14} + W_{15}, \\
Z_3 &= W_3 + W_6 + W_8 + W_{10} + W_{11} + W_{13} + W_{14} + W_{15}, \\
Z_4 &= W_4 + W_7 + W_9 + W_{10} + W_{12} + W_{13} + W_{14} + W_{15},
\end{aligned}$$

while

$$\begin{aligned}
\eta_1 + \eta_5 + \eta_6 + \eta_7 + \eta_{11} + \eta_{12} + \eta_{13} + \eta_{15} &= c_{11}, \\
\eta_2 + \eta_5 + \eta_8 + \eta_9 + \eta_{11} + \eta_{12} + \eta_{14} + \eta_{15} &= c_{22}, \\
\cdots\cdots\cdots\cdots\cdots\cdots\cdots\cdots\cdots &\quad \cdots \\
\eta_{10} + \eta_{13} + \eta_{14} + \eta_{15} &= c_{34}.
\end{aligned}$$

In this case,

$$\boldsymbol{c} = \left(c_{11}, c_{22}, c_{33}, c_{44}, c_{12}, c_{13}, c_{14}, c_{23}, c_{24}, c_{34} \right)^T.$$

Evidently, this construction is not restricted to the underlying gamma distribution.

Prékopa and Szántai (1978) have studied the conditional distributions of $\boldsymbol{Z}$ and noted that the components Z_i's are independent iff they are uncorrelated. The special feature of this construction in the gamma case is that the vector of components $\left(\dfrac{W_1}{\sum_{i=1}^{k} W_i}, \dfrac{W_2}{\sum_{i=1}^{k} W_i}, \cdots, \dfrac{W_k}{\sum_{i=1}^{k} W_i} \right)$ are independent of the variables $\sum_{i=1}^{k} W_i$; see Chapter 17 of Johnson, Kotz, and Balakrishnan (1994).

3.5   Kowalczyk and Tyrcha's Multivariate Gamma

The following family of multivariate gamma distributions was proposed by Kowalczyk and Tyrcha (1989). Firstly, let $X \stackrel{d}{=} G(\alpha, \mu, \sigma)$ with probability density function [see Chapter 17 of Johnson, Kotz, and Balakrishnan

(1994)]

$$p_X(x) = \frac{1}{\Gamma(\alpha)\sigma^\alpha} (x - \mu)^{\alpha-1} e^{-(x-\mu)/\sigma}, \qquad x \geq \mu, \ \sigma > 0, \ \alpha > 0.$$

Given $\boldsymbol{\alpha} = (\alpha_1, \ldots, \alpha_k)^T \in \mathrm{IR}_+^k \backslash \mathbf{0}$, $0 \leq \theta_0 \leq \min(\alpha_1, \ldots, \alpha_k)$, $\boldsymbol{\mu} = (\mu_1, \ldots, \mu_k)^T \in \mathrm{IR}^k$, and $\boldsymbol{\sigma} = (\sigma_1, \ldots, \sigma_k)^T \in \mathrm{IR}_+^k \backslash \mathbf{0}$, let $V_0, V_1, \ldots, V_k$ be a sequence of mutually independent gamma random variables such that

$$V_0 \overset{d}{=} G(\theta_0, 0, 1) \quad \text{and} \quad V_i \overset{d}{=} G(\alpha_i - \theta_0, 0, 1), \ i = 1, 2, \ldots, k.$$

Let $X_i = \mu_i + \sigma_i(V_0 + V_i - \alpha_i)/\sqrt{\alpha_i}$, $i = 1, 2, \ldots, k$. Then, the joint distribution of $\boldsymbol{X} = (X_1, \ldots, X_k)^T$ is said to be a k-dimensional multivariate gamma distribution with parameters θ_0, $\boldsymbol{\alpha}$, $\boldsymbol{\mu}$, and $\boldsymbol{\sigma}$ and is denoted by $\boldsymbol{X} \overset{d}{=} G_k(\theta_0, \boldsymbol{\alpha}, \boldsymbol{\mu}, \boldsymbol{\sigma})$. Evidently,

$$\mathrm{corr}(X_i, X_j) = \frac{\theta_0}{\sqrt{\alpha_i \alpha_j}} \qquad \text{for } i \neq j.$$

This family has its marginals (of any dimension) as gamma and is also closed with respect to linear transformation of components. The family is, in addition, closed relative to convolutions in the following sense. Let

$$\boldsymbol{X} \overset{d}{=} G_k(\theta_0, \boldsymbol{\alpha}, \boldsymbol{\mu}, \boldsymbol{\sigma}) \quad \text{and} \quad \boldsymbol{X}' \overset{d}{=} G_k(\theta_0', \boldsymbol{\alpha}', \boldsymbol{\mu}', \boldsymbol{\sigma}')$$

be two independent random vectors. Let $\sigma_i/\sqrt{\alpha_i} = \sigma_i'/\sqrt{\alpha_i'}$ for $i = 1, 2, \ldots, k$. Then, $\boldsymbol{X} + \boldsymbol{X}' \overset{d}{=} G_k(\theta_0'', \boldsymbol{\alpha}'', \boldsymbol{\mu}'', \boldsymbol{\sigma}'')$, where $\theta_0'' = \theta_0 + \theta_0'$, $\boldsymbol{\alpha}'' = \boldsymbol{\alpha} + \boldsymbol{\alpha}'$, $\boldsymbol{\mu}'' = \boldsymbol{\mu} + \boldsymbol{\mu}'$, and $\sigma_i'' = \sqrt{\sigma_i^2 + \sigma_i'^2}$ for $i = 1, 2, \ldots, k$.

The variables $X_1, \ldots, X_k$ are positively quadrant-dependent; that is,

$$\Pr[X_1 < x_1, \ldots, X_k < x_k] \geq \prod_{i=1}^{k} \Pr[X_i < x_i],$$

$$\Pr[X_1 \geq x_1, \ldots, X_k \geq x_k] \geq \prod_{i=1}^{k} \Pr[X_i \geq x_i].$$

This distribution appears in Karlin and Rinott (1980), where it is shown that if $\alpha_i - \theta_0 \geq 1$ for $i = 1, 2, \ldots, k$, then the variables $X_1, \ldots, X_k$ are totally positive of order two; that is,

$$p_X(\boldsymbol{x}) p_X(\boldsymbol{x}') \leq p_X(\boldsymbol{x} \vee \boldsymbol{x}') p_X(\boldsymbol{x} \wedge \boldsymbol{x}'),$$

where $p_X(\boldsymbol{x})$ is the probability density function of $\boldsymbol{X} = (X_1, \ldots, X_k)^T$,

$$\boldsymbol{x} \vee \boldsymbol{x}' = (\max(x_1, x_1'), \ldots, \max(x_k, x_k'))^T$$

and

$$\boldsymbol{x} \wedge \boldsymbol{x}' = (\min(x_1, x_1'), \ldots, \min(x_k, x_k'))^T.$$

Let $\theta_0^{(n)}$ and $\boldsymbol{\alpha}^{(n)} = (\alpha_1^{(n)}, \ldots, \alpha_k^{(n)})$ (for $n = 1, 2, \ldots$) be such that $\alpha_i^{(n)} \to \infty$ as $n \to \infty$ for $i = 1, 2, \ldots, k$ and $A_i = \lim_{n \to \infty} \theta_0^{(n)}/\alpha_i^{(n)}$ exists. Kowalczyk and Tyrcha (1989) have shown that, for any $\boldsymbol{\mu} \in \mathbb{R}^k$ and $\boldsymbol{\sigma} \in \mathbb{R}_+^k \setminus \boldsymbol{0}$, the sequence $G_k^{(n)} = G_k(\theta_0^{(n)}, \boldsymbol{\alpha}^{(n)}, \boldsymbol{\mu}, \boldsymbol{\sigma})$ ($n = 1, 2, \ldots$) converges weakly to a k-dimensional normal distribution with mean vector $\boldsymbol{\mu}$ and variance–covariance matrix Σ (see Chapter 45), where

$$\begin{aligned} \Sigma_{ij} &= \sigma_i \sigma_j \sqrt{A_i A_j} \quad \text{for } i \neq j, \\ &= \sigma_i^2 \qquad\qquad \text{for } i = j, \end{aligned} \qquad i, j = 1, 2, \ldots, k.$$

Kowalczyk and Tyrcha (1989) have also discussed the estimation of the shape parameter $\boldsymbol{\alpha}$. If all components of $\boldsymbol{\alpha}$ are assumed to be different, each component is estimated separately from the respective marginal data. If any ℓ of them are assumed to be equal, then these authors recommend averaging the separate estimates.

3.6   Royen's Multivariate Gammas

Royen (1991b, 1994) studied two forms of multivariate gamma distributions, one based on one-factorial correlation matrices and the other motivated by multivariate Rayleigh distribution.

A $k \times k$ correlation matrix $\mathbf{R} = ((r_{ij}))$ is said to be *one-factorial* if there are any numbers $a_1, \ldots, a_k$ with

$$r_{ij} = a_i a_j \ (i \neq j) \quad \text{and} \quad a_1, \ldots, a_k \in (-1, 1) \qquad (48.50)$$

or

$$r_{ij} = -a_i a_j \ (i \neq j) \quad \text{and} \quad \mathbf{R} \text{ is positive semidefinite.} \qquad (48.51)$$

Royen (1991b) has shown that if $\mathbf{R}$ is a one-factorial $k \times k$ correlation matrix, then for any positive integer 2α the multivariate gamma distribution with joint characteristic function

$$\phi_{\boldsymbol{X}}(\boldsymbol{t}) = E\left[e^{i t^T \boldsymbol{X}}\right] = |\boldsymbol{I} - 2i \text{Diag}(t_1, \ldots, t_k) \mathbf{R}|^{-\alpha}$$

is given by

$$\frac{1}{\Gamma(\alpha)} \int_0^\infty e^{-y} \, y^{\alpha-1} \prod_{j=1}^k \left\{ \exp\left(\frac{\mp a_j^2 y}{1 \mp a_j^2}\right) \sum_{n=0}^\infty G_{\alpha+n}\left(\frac{x_j/2}{1 \mp a_j^2}\right) \right.$$
$$\left. \left(\frac{\pm a_j^2 y}{1 \mp a^2 j}\right)^n \Big/ n! \right\} dy, \qquad (48.52)$$

where the upper signs hold for (48.48) and the lower ones hold for (48.49). Under condition (48.48), also *any* positive value α is admissible. In (48.50), $G_{\alpha+n}(\cdot)$ denotes the cumulative distribution of the standard gamma distribution with shape parameter $\alpha + n$. In the case of (48.48), the expression (48.50) is a mixture of products of noncentral gamma distribution functions, providing therefore a distribution function for every positive α.

Royen (1994) proposed the following multivariate gamma distribution motivated by multivariate Rayleigh density with a tridiagonal correlation matrix $\mathbf{R}$ with $\mathbf{R}^{-1} = ((r^{ij}))$ given by Blumenson and Miller (1963). As before, let $g_\alpha(x)$ denote the density function of a standard gamma distribution with shape parameter α. Let us define the modified Bessel function as

$$I_{\alpha-1}(x) = \left(\frac{x}{2}\right)^{\alpha-1} {}_0F_1\left(\alpha; \frac{x^2}{4}\right) \Big/ \Gamma(\alpha),$$

where

$$_0F_1(\alpha; x) = \Gamma(\alpha) \sum_{n=0}^{\infty} \frac{x^n}{\Gamma(\alpha+n)n!}$$

is the confluent hypergeometric function. Royen's k-variate gamma density function, which is a modification of Blumenson and Miller's (1963) Rayleigh density function, is given by

$$\begin{aligned}
p(x_1, &\ldots, x_k; \alpha, \mathbf{R}) \\
&= \left(|\mathbf{R}| \prod_{i=1}^{k} r^{ii}\right)^{-\alpha} \prod_{i=1}^{k} r^{ii} g_\alpha(r^{ii} x_i) \\
&\qquad \times \prod_{i<j} {}_0F_1\left(\alpha; (r^{ij})^2 x_i x_j\right),
\end{aligned} \tag{48.53}$$

where 2α is not restricted to integer values.

Royen (1994) has pointed out that no elementary formula is known for the corresponding density with a tridiagonal $\mathbf{R}$. Jensen (1970b) has given some series expansions for generalized multivariate Rayleigh distributions with several different tridiagonal correlation matrices (see Section 4), however, they are based on a formula for determinants of tridiagonal $k \times k$ matrices which does not hold for $k > 3$ and may lead to incorrect formulas for densities even in the case of identical correlation matrices.

Royen (1994) has discussed multivariate gamma distributions when the matrix $\mathbf{R}$ or $\mathbf{R}^{-1}$ is of a "tree type." Any covariance matrix $C_{k \times k} = ((c_{ij}))$ is said to be of a tree type if the graph $\mathcal{G}(C)$ with the vertices $1, 2, \ldots, k$ is a spanning tree containing the edge $[i, j]$ iff $c_{ij} \neq 0$. By definition, a spanning tree is connected and has no cycles. Thus, it contains exactly $k - 1$ edges and it holds for all "cyclic products" of C that

$$c_{i_1 i_2} c_{i_2 i_3} \cdots c_{i_\ell i_1} = 0 \quad (\{i_1, \ldots, i_\ell\} \subseteq \{1, 2, \ldots, k\}, \; 3 \le \ell \le k).$$

In particular, $\mathbf{R}$ belongs to this class if $\mathbf{R}^{-1}$ or $\mathbf{R}$ is tridiagonal.

Let us denote

$$
\begin{aligned}
G_\alpha(x, y) &= e^{-y} \int_0^x {}_0F_1(\alpha; \xi y) g_\alpha(\xi) d\xi = \sum G_{\alpha+n}^{(n)}(x)(-y)^n/n! \\
&= e^{-y} \sum_{n=0}^\infty {}_0F_1(\alpha + 1 + n; xy) g_{\alpha+1+n}(x) \\
&= \begin{cases} e^{-x-y} \sum_{n=0}^\infty (\sqrt{x/y})^{\alpha+n} I_{\alpha+n}(2\sqrt{xy}), & y > 0 \\ e^{-x-y} \sum_{n=0}^\infty (\sqrt{-x/y})^{\alpha+n} I_{\alpha+n}(2\sqrt{-xy}), & y < 0, \end{cases}
\end{aligned}
$$

where

$$G_{\alpha+n}^{(n)}(x) = \frac{d^n}{dx^n} G_{\alpha+n}(x) = \binom{\alpha+n-1}{n-1}^{-1} L_{n-1}^{(\alpha)}(x) g_{\alpha+1}(x),$$

and $L_{n-1}^{(\alpha)}(x)$ denotes the generalized Laguerre polynomial. Let $\mathbf{R}_{k \times k} = ((r_{ij}))$ or its standardized inverse $\mathbf{Q} = ((q_{ij}))$ be a correlation matrix $C = ((c_{ij}))$ of a tree type. In any spanning tree $\mathcal{G}(\mathbf{C})$, the degree d_i of i is the number of edges $[i, j]$ of $\mathcal{G}$. Let us define

$$L = \{\ell \mid d_\ell = 1\}, \quad I = \{i \mid d_i > 1\} = \{1, 2, \ldots, k\} \backslash L$$

$$(48.54)$$

and for any $i \in I$ the (possibly empty) set

$$L_i = \{\ell \in L \mid c_{i\ell} \ne 0\}. \tag{48.55}$$

Furthermore, let

$$
\begin{aligned}
I_1 &= \{i \in I \mid L_i \ne \emptyset\}, \quad I_2 = I \backslash I_1, \\
\mathcal{I} &= \{(i, j) \mid i, j \in I, \; i < j, \; c_{ij} \ne 0\}.
\end{aligned}
$$

Let $\sum_{(n)}$ denote the summation over all partitions $n = \sum_{(i,j) \in \mathcal{I}} n_{ij}$ or $n = \sum_{1 \le i < j \le k, \; c_{ij} \ne 0} n_{ij}$ with nonnegative integers n_{ij}, and let $N_i = \sum_{j=1, \; c_{ij} \ne 0}^k n_{ij}$, $n_i = \sum_{j \in I, \; c_{ij} \ne 0} n_{ij}$ (with $n_{ji} = n_{ij}$ and $n_{ii} = 0$). Let I be of size m. Then,

Royen (1994) has proved that if $\mathbf{R}$ is of a tree type and 2α is a positive integer or $\alpha > (k-1)/2$, then

$$
F(x_1, \ldots, x_k; \alpha, \mathbf{R})
$$
$$
= \frac{1}{\Gamma(\alpha)} \sum_{n=0}^{\infty} \sum_{(n)} \prod_{\substack{i<j \\ r_{ij} \neq 0}} \frac{r_{ij}^{2n_{ij}}}{\Gamma(\alpha + n_{ij})n_{ij}!} \prod_{i=1}^{k} \Gamma(\alpha + N_i) G_{\alpha+N_i}^{(N_i)}(x_i).
$$

$$(48.56)$$

The simplest situation in the integral representation arises if all $r_{ij} = 0$ for $i, j < k$ $(i \neq j)$ [see Royen (1994) for details]. The correlation matrix $\mathbf{R}$ belongs to this class iff the elements q_{ij} of the standardized inverse satisfy the relations $q_{ij} = q_{ik}q_{jk}$ $(i, j < k, \ i \neq j)$; that is, $\mathbf{Q}$ is the limit case of "one-factorial" correlation matrices $((q_{ij}))$ with $q_{ij} = a_i a_j$ $(i \neq j)$, $a_i^2 < 1$ $(i = 1, \ldots, k)$, and $a_k^2 \to 1$, discussed above.

Royen (1994) has also provided similar but somewhat more complicated expressions for $F(x_1, \ldots, x_k; \alpha, \mathbf{R})$ when the inverse $\mathbf{Q}$ is of a tree type without any restriction on the parameter $\alpha > 0$.

The Laplace transform of the k-variate gamma density function in (48.51) is

$$
|\mathbf{I} + \mathbf{RT}|^{-\alpha} =
\begin{cases}
\left(\prod_{i=1}^{k} z_i^{\alpha} \right) |\mathbf{I} + \dot{\mathbf{R}}\mathbf{U}|^{-\alpha} \\
\quad \left(\begin{array}{c} z_i = (1+t_i)^{-1}, \ u_i = 1 - z_i = t_i z_i, \\ \mathbf{U} = \mathrm{Diag}(u_1, \ldots, u_k) \end{array} \right), \\[2em]
\left(|\mathbf{Q}|^{\alpha} \prod_{i=1}^{k} z_i^{\alpha} \right) |\mathbf{I} + \dot{\mathbf{Q}}\mathbf{Z}|^{-\alpha} \\
\quad \left(\begin{array}{c} z_i = (1 + t_i/r^{ii})^{-1}, \ u_i = 1 - z_i, \\ \mathbf{Z} = \mathrm{Diag}(z_1, \ldots, z_k) \end{array} \right),
\end{cases}
$$

$$(48.57)$$

where $\dot{\mathbf{R}} = \mathbf{R} - \mathrm{Diag}(r_{11}, \ldots, r_{kk})$, $\mathbf{Q}$ is its standardized inverse with elements $q_{ij} = r^{ij}/\sqrt{r^{ii}r^{jj}}$, and $\mathbf{T} = \mathrm{Diag}(t_1, \ldots, t_k)$. Griffiths (1984) has shown that the Laplace transform $|\mathbf{I} + \mathbf{RT}|^{-1}$ $(k \geq 3)$ is infinitely divisible if the elements r^{ij} of $\mathbf{R}^{-1}$ satisfy the condition

$$
(-1)^{\ell} r^{i_1 i_2} r^{i_2 i_3} \cdots r^{i_{\ell} i_1} \geq 0
$$

$$(48.58)$$

for all subsets $\{i_1, \ldots, i_{\ell}\} \subseteq \{1, \ldots, k\}$ $(3 \leq \ell \leq k)$. Thus, $|\mathbf{I} + \mathbf{RT}|^{-1}$ is infinitely divisible for any nonsingular correlation matrix $\mathbf{R}_{k \times k}$ $(k \geq 3)$ if $\mathbf{R}^{-1}$ is of tree type due to the relation

$$
c_{i_1 i_2} c_{i_2 i_3} \cdots c_{i_{\ell} i_1} = 0 \ (\{i_1, \ldots, i_{\ell}\} \subseteq \{1, \ldots, k\}, \ 3 \leq \ell \leq k).
$$

However, $|I + RT|^{-1}$ is not infinitely divisible if R itself has a tree type. More details on infinite divisibility are presented in Section 5. Royen's expansions, and especially integral representations, are numerically efficient at least for small α and dimension k. For larger α, application of central limit theorem and multivariate Edgeworth expansion is recommended; see Khatri, Krishnaiah, and Sen (1977).

3.7 Mathai and Moschopoulos's Multivariate Gamma

Mathai and Moschopoulos (1991) introduced a new form of multivariate gamma distributions using three-parameter univariate gamma distributions as a building block. This family is especially useful for models in reliability theory and renewal processes.

Let us consider the three-parameter $G(\alpha, \gamma, \beta)$ random variable with density function

$$\frac{1}{\Gamma(\alpha)\beta^\alpha} (x - \gamma)^{\alpha-1} e^{-(x-\gamma)/\beta}, \qquad x > \gamma, \ \alpha > 0, \ \beta > 0. \qquad (48.59)$$

Now let $V_i \overset{d}{=} G(\alpha_i, \gamma_i, \beta_i)$, $i = 0, 1, \ldots, k$, be mutually independent random variables, and let

$$X_i = \frac{\beta_i}{\beta_0} V_0 + V_i, \qquad i = 1, 2, \ldots, k. \qquad (48.60)$$

Mathai and Moschopoulos (1991) have proposed the distribution of $X = (X_1, \ldots, X_k)^T$ as a multivariate gamma distribution. The moment-generating function of X is

$$M_X(t) = E\left[e^{t^T X}\right] = \frac{\exp\left\{\left(\gamma + \frac{\gamma_0}{\beta_0} \beta\right)^T t\right\}}{(1 - \beta^T t)^{\alpha_0} \prod_{i=1}^{k}(1 - \beta_i t_i)^{\alpha_i}}, \qquad (48.61)$$

where $\beta = (\beta_1, \ldots, \beta_k)^T$, $\gamma = (\gamma_1, \ldots, \gamma_k)^T$, $t = (t_1, \ldots, t_k)^T$, $|\beta_i t_i| < 1$ for $i = 1, \ldots, k$, and $|\beta^T t| = \left|\sum_{i=1}^{k} \beta_i t_i\right| < 1$.

This model was motivated by Mathai and Moschopoulos (1991) as follows: Consider $Y_1, \ldots, Y_k$ to be independent gamma random variables representing runoffs to a dam from k different streams. These variables are disturbed to form the new variables $X_i = Y_i + \delta_i Z$ $(i = 1, \ldots, k)$, where Z is another gamma variable (independent of Y_i's) and δ_i's are constants; for example, Z could be the contribution from a new rainfall in the region and δ_i's could be the coefficients representing the catchment areas of the different streams. In this case, $\boldsymbol{X} = (X_1, \ldots, X_k)^T$ possesses the above described multivariate gamma distribution of Mathai and Moschopoulos (1991). A stochastic routing problem also leads to this distribution.

Evidently, $X_i \overset{d}{=} G(\alpha_0 + \alpha_i, \frac{\gamma_0}{\beta_0} \beta_i + \gamma_i, \beta_i)$, $i = 1, \ldots, k$, and

$$
\begin{aligned}
E[X_i] &= (\alpha_0 + \alpha_i)\beta_i + \frac{\gamma_0}{\beta_0} \beta_i + \gamma_i, \\
\text{var}(X_i) &= (\alpha_0 + \alpha_i)\beta_i^2
\end{aligned}
$$

and

$$
\text{cov}(X_i, X_j) = \alpha_0 \beta_i \beta_j > 0 \qquad \text{for } i \neq j. \tag{48.62}
$$

This class of multivariate gamma distributions is closed under the shift transformation $\boldsymbol{W} = \boldsymbol{X} + \boldsymbol{d}$, where $\boldsymbol{d} = (d_1, \ldots, d_k)^T$, and also under the convolution of two independent $\boldsymbol{X}_1$ and $\boldsymbol{X}_2$. Clearly,

$$
M_{\boldsymbol{X}_1 + \boldsymbol{X}_2}(\boldsymbol{t}) = M_{\boldsymbol{X}_1}(\boldsymbol{t}) M_{\boldsymbol{X}_2}(\boldsymbol{t}),
$$

which is also of the form (48.59) provided that the scale parameters β_i are the same for both $\boldsymbol{X}_1$ and $\boldsymbol{X}_2$.

Denote

$$
E[V_i^m] = \sum_{\ell=0}^{m} \binom{m}{\ell} (\alpha_i)_\ell \beta_i^\ell \gamma_i^{m-\ell}
$$

by $M_i^{(m)}$, where $(\alpha)_\ell = \alpha(\alpha+1)\cdots(\alpha+\ell-1)$ and $(\alpha)_0 = 1$, and V_i's are the gamma variables defined after (48.57). Then,

$$
\begin{aligned}
&E[X_i^m X_j^n] \\
&= \sum_{r=0}^{m} \sum_{s=0}^{n} \binom{m}{r}\binom{n}{s} \left(\frac{\beta_i}{\beta_0}\right)^r \left(\frac{\beta_j}{\beta_0}\right)^s M_0^{(r+s)} M_i^{(m-r)} M_j^{(n-s)} \tag{48.63}
\end{aligned}
$$

and the cumulants are given by

$$
\kappa_{mn} = \frac{\partial^{m+n}}{\partial t_j^n \partial t_i^m} \left(\log M_{\boldsymbol{X}}(\boldsymbol{t})\right) |_{\boldsymbol{t}=0} = \alpha_0 (m+n-1)! \beta_i^m \beta_j^n, \tag{48.64}
$$

yielding

$$
\kappa_{20} = (\alpha_0 + \alpha_i)\beta_i^2 = \text{var}(X_i)
$$

and

$$
\kappa_{11} = \alpha_0 \beta_i \beta_j = \text{cov}(X_i, X_j) \qquad (i \neq j)
$$

[see (48.60)]. The covariance matrix of $\boldsymbol{X} = (X_1, \ldots, X_k)^T$ is

$$\boldsymbol{\Sigma} = ((\sigma_{ij})) = \begin{pmatrix} \sigma_{11} & \boldsymbol{\Sigma}_{12} \\ \boldsymbol{\Sigma}_{21} & \boldsymbol{\Sigma}_{22} \end{pmatrix},$$

where $\sigma_{ii} = (\alpha_0 + \alpha_i)\beta_i^2$ and $\sigma_{ij} = \alpha_0 \beta_i \beta_j$ $(i \neq j)$. The conditional expectation

$$E[X_1 | X_2 = x_2] = B_0 + B_1(x_2 - E[X_2])$$

is a linear function in x_2, where

$$B_0 = E[X_1] = (\alpha_0 + \alpha_1)\beta_1 + \gamma_1 + \frac{\gamma_0}{\beta_0}\beta_1$$

and

$$B_1 = \frac{\text{cov}(X_1, X_2)}{\text{var}(X_2)} = \frac{\alpha_0 \beta_1}{\beta_2(\alpha_0 + \alpha_2)}.$$

In general,

$$\begin{aligned} &E[X_1^r | X_2 = x_2] \\ &= \frac{\beta_1^r}{\Gamma(\alpha_1)} \sum_{i=0}^{r} \binom{r}{i} \frac{\Gamma(r - i + \alpha_1)}{\beta_0^i} \sum_{j=0}^{i} \binom{i}{j} \frac{(\alpha_0)^{(j)}}{(\alpha_0 + \alpha_2)^{(j)}} \delta^{i-j} \omega^j, \end{aligned}$$

where $\delta = \gamma_0 + \frac{\beta_0}{\beta_1}\gamma_1$, $\omega = \frac{\beta_0}{\beta_2}(x_2 - \gamma_2) - \gamma_0$, and $(\alpha_0)^{(j)} = \alpha_0(\alpha_0 - 1)\cdots(\alpha_0 - j + 1)$. From this expression, the conditional variance can be derived. Expressions for $E[X_1 X_2 | X_3 = x_3]$ and $E[X_1^r X_2^s | X_3 = x_3]$ can be derived by using direct but rather tedious calculations.

The joint density function of $X_1, \ldots, X_k$ and $X_{k+1} \equiv V_0$ is

$$\begin{aligned} &p_{k+1}(x_1, \ldots, x_k, x_{k+1}) \\ &= \left\{ \beta_0^{\alpha_0}\Gamma(\alpha_0) \prod_{i=1}^{k} \beta_i^{\alpha_i}\Gamma(\alpha_i) \right\}^{-1} (x_{k+1} - \gamma_0)^{\alpha_0 - 1} e^{-(x_{k+1} - \gamma_0)/\beta_0} \\ &\quad \times \prod_{i=1}^{k} \left(x_i - \frac{\beta_i}{\beta_0} x_{k+1} - \gamma_i \right)^{\alpha_i - 1} \exp\left\{ -\left(x_i - \frac{\beta_i}{\beta_0} x_{k+1} - \gamma_i \right) \middle/ \beta_i \right\}. \end{aligned}$$

By integrating out x_{k+1} in the above expression, Mathai and Moschopoulos (1991) have derived the joint density function of $\boldsymbol{U} = (U_1, \ldots, U_k)^T$ as

$$\begin{aligned} &p_{\boldsymbol{U}}(\boldsymbol{u}) \\ &= \frac{\prod_{i=1}^{k}(\beta_i/\beta_0)^{\alpha_i - 1}}{\prod_{i=0}^{k}(\beta_i^{\alpha_i}\Gamma(\alpha_i))} \left(\prod_{i=1}^{k} u_i^{\alpha_i - 1} \right) u_1^{\alpha_0} \end{aligned}$$

$$\times \sum_{r_0=0}^{\infty} \cdots \sum_{r_k=0}^{\infty} \frac{\left(-\frac{u_1}{\beta_0}\right)^{r_0}}{r_0!} \cdots \frac{\left(-\frac{u_k}{\beta_0}\right)^{r_k}}{r_k!} \frac{\Gamma(\alpha_0+r_0)\Gamma(\alpha_1+r_1)}{\Gamma(\alpha_0+r_0+\alpha_1+r_1)}$$

$$\times \int_0^1 y^{\alpha_0+r_0-1}(1-y)^{\alpha_1+r_1-1}\left(1-\frac{u_1}{u_2}y\right)^{\alpha_2+r_2-1}$$

$$\cdots \left(1-\frac{u_1}{u_k}y\right)^{\alpha_k+r_k-1} dy,$$

$$(48.65)$$

where $U_i = \frac{\beta_0}{\beta_i}(X_i - \gamma_i) - \gamma_0$ for $i = 1, 2, \ldots, k$. The last integral can be expressed in terms of the Lauricella function defined, for example, in Mathai and Saxena (1978).

The vector Z of standardized X_i's, namely, $Z_i = (X_i - E[X_i])/\sqrt{\text{var}(X_i)}$, is asymptotically standard normal. The estimation of parameters of this multivariate gamma distribution can be developed easily based on the method of moments. Denoting the sample cumulants of z_i $(i = 1, \ldots, k)$ by $m_1^{(i)}, m_2^{(i)}, \ldots$, we have, for example, $\hat{\beta}_i = \frac{m_3^{(i)}}{2m_2^{(i)}}$ or a different estimate $\hat{\beta}_i = \frac{m_4^{(i)}}{3m_3^{(i)}}$.

Mathai and Moschopoulos (1992) presented a simplified version of their earlier model. Let $V_i \overset{d}{=} G(\alpha_i, \gamma_i, \beta)$, $i = 1, 2, \ldots, k$, be independent three-parameter gamma variables with a *common* scale parameter β. Let

$$X_1 = V_1, \ X_2 = V_1 + V_2, \ldots, V_k = V_1 + \cdots + V_k.$$

Then, according to Mathai and Moschopoulos (1992), the joint distribution of $X = (X_1, \ldots, X_k)^T$ is a multivariate gamma with density function

$$p_X(x) = \frac{(x_1-\gamma_1)^{\alpha_1-1}}{\beta^{\alpha_k^*}\prod_{i=1}^k \Gamma(\alpha_i)}(x_2-x_1-\gamma_2)^{\alpha_2-1}\cdots(x_k-x_{k-1}-\gamma_k)^{\alpha_k-1}$$
$$\times e^{-\{x_k-(\gamma_1+\cdots\gamma_k)\}/\beta} \qquad (48.66)$$

for $\alpha_i > 0$, $\beta > 0$, γ_i real, $z_{i-1} < z_i - \gamma_i$ $(i = 2, \ldots, k)$, $z_k < \infty$, $\gamma_1 < z_1$, and $\alpha_k^* = \alpha_1 + \cdots + \alpha_k$.

A model motivating this distribution in reliability applications is as follows. An item is installed at time $X - 0 = 0$ and when it fails, it is replaced by an identical (or different item). Then, when the new item fails, it is replaced again by another item and the process continues. Here $X_i = X_{i-1} + V_i$, where V_i is the time of operation of the ith item, and X_i is time at which the ith replacement is needed and X_k denotes the time interval in which a total of k items need replacement.

The joint moment-generating function of $\boldsymbol{X}$ is

$$
\begin{aligned}
&M_{\boldsymbol{X}}(\boldsymbol{t}) \\
&= \frac{e^{\gamma_1(t_1+\cdots+t_k)}}{\{1-\beta(t_1+\cdots+t_k)\}^{\alpha_1}} \; \frac{e^{\gamma_2(t_2+\cdots+t_k)}}{\{1-\beta(t_2+\cdots+t_k)\}^{\alpha_2}} \; \cdots \; \frac{e^{\gamma_k t_k}}{(1-\beta t_k)^{\alpha_k}}
\end{aligned}
\tag{48.67}
$$

which exists if $|t_i+\cdots+t_k| < 1/\beta$ for $i = 1,2,\ldots,k$. The marginal distribution of X_i is $G(\alpha_i^*,\gamma_i^*,\beta)$ for $i = 1,2,\ldots,k$, where $\alpha_i^* = \sum_{j=1}^i \alpha_j$ and $\gamma_i^* = \sum_{j=1}^i \gamma_j$; also,

$$
\begin{aligned}
E[X_i] &= \beta\alpha_i^* + \gamma_i^*, \quad \text{var}(X_i) = \beta^2\alpha_i^*, \\
\text{cov}(X_i, X_j) &= \text{var}(X_i) = \beta^2\alpha_i^* \quad (\text{for } i < j)
\end{aligned}
$$

and

$$
\rho = \text{corr}(X_i, X_j) = \sqrt{\alpha_i^*/\alpha_j^*}.
$$

Compare these with (48.60). The correlation is, therefore, always positive and the variance–covariance matrix is of the interesting form

$$
\boldsymbol{\Sigma} =
\begin{pmatrix}
\sigma_1^2 & \sigma_1^2 & \cdots & \sigma_1^2 \\
 & \sigma_1^2 + \sigma_2^2 & \cdots & \sigma_1^2 + \sigma_2^2 \\
 & & \ddots & \\
 & & & \sum_{i=1}^k \sigma_1^2
\end{pmatrix}
\quad \text{with } \sigma_1^2 = \alpha_i^*\beta^2.
$$

Evidently, $|\boldsymbol{\Sigma}| = \prod_{i=1}^k \sigma_i^2$ and $\max_j \sum_{i=1}^k |\sigma_{ij}| = \text{trace}(\boldsymbol{\Sigma})$, where σ_{ij} is the (i,j)th element of $\boldsymbol{\Sigma}$. Also, the multiple correlation coefficient of X_1 on $X_2,\ldots,X_k$ is of the form

$$
R^2_{1(2\cdots k)} = \frac{\sigma_1^2}{\sigma_1^2 + \sigma_2^2},
$$

free of $\sigma_3,\ldots,\sigma_k$.

Furthermore, if $\boldsymbol{X}_1$ is distributed as in (48.66) with parameters $(\alpha_i, \gamma_i, \beta)$, $i = 1,\ldots,k$, and $\boldsymbol{X}_2$ is also distributed as in (48.66) with $(\alpha_i', \gamma_i', \beta)$, independently of $\boldsymbol{X}_1$, then $\boldsymbol{X}_1 + \boldsymbol{X}_2$ is distributed once again as in (48.66) with parameters $(\alpha_i + \alpha_i', \gamma_i + \gamma_i', \beta)$, $i = 1,\ldots,k$.

From (48.67), we obtain the joint cumulant-generating function $\boldsymbol{X}$ as

$$
\begin{aligned}
K_{\boldsymbol{X}}(\boldsymbol{t}) = \; & \gamma_1 \sum_{i=1}^k t_i + \gamma_2 \sum_{i=2}^k t_i + \cdots + \gamma_k t_k \\
& - \alpha_1 \ln\left(1 - \beta \sum_{i=1}^k t_i\right) - \alpha_2 \ln\left(1 - \beta \sum_{i=2}^k t_i\right) \\
& - \cdots - \alpha_k \ln(1 - \beta t_k)
\end{aligned}
$$

from which we get the mth cumulant of X_i as

$$\kappa_m(X_i) = \begin{cases} \gamma_i^* + \beta\alpha_i^*, & \text{for } m = 1 \\ (m-1)!\beta^m\alpha_i^*, & \text{for } m \geq 2 \end{cases}$$

and get the joint cumulant as

$$\kappa_{m_1,m_2} = (m_1 + m_2 - 1)!\beta^{m_1+m_2}\alpha^r, \qquad \text{where } r = \min(m_1, m_2).$$

The densities of $(X_1, \ldots, X_{k-1})^T$ and that of $(X_1, \ldots, X_{i-1}, X_{i+1}, \ldots, X_k)^T$, as well as the joint densities of all subsets of $(X_1, \ldots, X_k)^T$ are also of the same form as in (48.66). Also, the variables

$$Y_1 = \frac{X_1 - \gamma_1}{X_k - \gamma_k^*}, Y_2 = \frac{X_2 - X_1 - \gamma_2}{X_k - \gamma_k^*}, \ \cdots \ , Y_{k-1} = \frac{X_k - X_{k-1} - \gamma_k}{X_k - \gamma_k^*}$$

jointly have the Dirichlet density with parameters $\alpha_1, \ldots, \alpha_k$ (see Chapter 49) given by

$$\frac{\Gamma(\alpha_k^*)}{\prod_{\ell=1}^{k}\Gamma(\alpha_\ell)}\left(\prod_{\ell=1}^{k-1}y_\ell^{\alpha_\ell-1}\right)\left(1 - \sum_{\ell=1}^{k-1}y_\ell\right)^{\alpha_k-1}, \qquad 0 \leq y_\ell \leq 1, \ \sum_{\ell=1}^{k-1}y_\ell \leq 1.$$

Clearly, each Y_i is then a beta variable of Type 1; see Chapter 25 of Johnson, Kotz, and Balakrishnan (1995).

3.8   Dussauchoy and Berland's Multivariate Gamma

A multivariate extension of the characteristic function presented in Section 2.10 can be written as

$$\phi(u_1, \ldots, u_k) = \prod_{j=1}^{k}\left\{\frac{\phi_j\left(u_j + \sum_{b=j+1}^{k}\beta_{jb}u_b\right)}{\phi_j\left(\sum_{b=j+1}^{k}\beta_{jb}u_b\right)}\right\},$$

where

$$\phi_j(u_j) = (1 - i\,u_j/a - J)^{\ell_j} \qquad (j = 1, \ldots, k),$$
$$\beta_{jb} \geq 0, \ a_j \geq b_{jb}a_b > 0, \ j < b = 1, \ldots, k,$$

and

$$0 < \ell_1 \leq \ell_2 \leq \cdots \leq \ell_k.$$

An explicit form of the density function is not available except in the bivariate case (see Section 2.10)

4 MULTIVARIATE (JENSEN-TYPE) CHI-SQUARE DISTRIBUTIONS

Returning to Section 2.12, we note that the derivation of the joint distribution of $S_1, S_2, \ldots, S_k$ is more difficult. Using the methods employed in Section 2.12 (for $k = 2$), one can show, for example, that the conditional distribution of S_1 given

$$(X_{12}, \ldots, X_{n2})(X_{13}, \ldots, X_{n3}) \cdots (X_{1k}, \ldots, X_{nk})$$

is that of $(1 - R_{1 \cdot 2 3 \ldots k}^2) \times$ noncentral χ^2 with $(n - 1)$ degrees of freedom and noncentrality parameter

$$(1 - R_{1 \cdot 2 3 \ldots k}^2)^{-1} \left\{ \sum_{j=2}^{k} a_j^2 S_j + \sum_{\ell=2}^{k} \sum_{j \leq \ell} a_j a_\ell P_{j\ell} \right\},$$

where

$$a_j = \rho_{1j \cdot 2 \ldots (j-1),(j+1) \ldots k} \quad \text{(partial correlation coefficients)}$$

and

$$P_{j\ell} = \sum_{i=1}^{n} (X_{ij} - \bar{X}_j)(X_{i\ell} - \bar{X}_\ell),$$

and $R_{1 \cdot 2 3 \ldots k}^2$ is the multiple correlation of X_1 on $X_2, \ldots, X_k$. The joint distribution of $S_2, \ldots, S_n, P_{23}, \ldots, P_{k-1,k}$ is a Wishart distribution $W_{k-1}(n - 1; \boldsymbol{V}_{11})$, where $\boldsymbol{V}_{11}$ is the cofactor of the first diagonal element of $\boldsymbol{V}$ as defined in Section 2.12. It is thus straightforward to obtain the joint distribution of $S_1, S_2, \ldots, S_k, P_{23}, \ldots, P_{k-1,k}$, but elimination of the P's poses difficulties.

For the special case when

$$\boldsymbol{V} = \begin{pmatrix} 1 & \rho & \cdots & \rho \\ \rho & 1 & \cdots & \rho \\ \vdots & \vdots & & \vdots \\ \rho & \rho & \cdots & 1 \end{pmatrix},$$

Johnson (1962) has suggested the approximate formula

$$\Pr \left[\bigcap_{j=1}^{k} (S_j \leq s_j) \right] \doteq \sum_{j=0}^{\infty} c_j \prod_{\ell=1}^{k} \Pr[\chi_{n-1+2j}^2 < s_\ell] \qquad (48.68)$$

with c_j's as given in Section 2.12. This leads to the correct values for $\Pr[S_j \leq s_j]$ and $\Pr[(S_j \leq s_j) \cap (S_{j'} \leq s_{j'})]$ for any j, j' (i.e., all marginal univariate and bivariate distributions are correct). It would seem likely that (48.66) should give usefully accurate values for $k = 3$ or 4, but that the accuracy would decrease with increasing k.

Krishnamoorthy and Parthasarathy (1951) and Lukacs and Laha (1964) have shown that the joint characteristic function of $S_1, \ldots, S_k$ is

$$E\left[\exp\left(i\sum_{j=1}^{k} t_j S_j\right)\right] = |\boldsymbol{I} - 2i\boldsymbol{V}\boldsymbol{D}_t|^{-v/2}$$

where $\boldsymbol{D}_t = \text{diag}(t_1, \ldots, t_k)$.

This joint distribution could be used to construct simultaneous confidence intervals for the variances $\sigma_1^2, \ldots, \sigma_k^2$ (diagonal elements of $\boldsymbol{V}$). This would require a knowledge of the correlation coefficients (elements of $\boldsymbol{R}$). Jensen and Jones (1969) have shown, however, that very good approximations can be obtained without the need to use this distribution. *Bonferroni intervals* [Bens and Jensen (1967)] are formed by using ordinary univariate intervals for each σ_j^2 with confidence coefficients $1 - \gamma_j = 1 - \alpha/k$ ($1 - \alpha$ being the required joint confidence coefficient). These give satisfactory results with $k = 2$ for $\alpha = 0.01, 0.10$, over a wide range of values of v and ρ (the correlation coefficient).

Moran and Vere-Jones (1969) have shown that if $\rho_{ij} = \rho$ for all i, j, the joint distribution of $S_1, \ldots, S_k$ is infinitely divisible [i.e., for any α (> 0), $|\boldsymbol{I} - 2i\boldsymbol{V}\boldsymbol{D}_t|^{-\alpha}$ is a characteristic function]. They have also shown that this is true for $k = 3$ with

$$\boldsymbol{V} = \begin{pmatrix} 1 & \rho & \rho^2 \\ \rho & 1 & \rho \\ \rho^2 & \rho & 1 \end{pmatrix}.$$

The distribution of the maximum of $S_1, \ldots, S_k$ has been considered by Fomin (1970). He has also given an approximate formula for the cumulative distribution.

Krishnamoorthy and Parthasarathy (1951), Miller, Bernstein and Blumenson (1958), and Zaharov, Sarmanov, and Sevastjanov (1969) have extended distributions of type (48.75) to the multivariate case ($k > 2$). The last authors have considered the joint distribution of series of $\frac{1}{2}\chi_v^2$ statistics obtained sequentially by increasing the sample size, taking groups of observations at a time. They have shown that the limiting joint distribution of the first k variables $X_1, \ldots, X_k$, as $n_1, \ldots, n_k$ (the corresponding numbers of observations) tend to infinity in fixed ratios $(n_i/n_j = \rho_{ij})$, is

$$
p_{\boldsymbol{X}}(\boldsymbol{x}) = \frac{(x_1 x_k / \lambda_1 \lambda_k)^{(1/2)v-1} \exp\left[-\sum_{j=1}^{k} x_j / \lambda_j\right]}{K^{(1/2)v} \left(\prod_{i=1}^{k-1} b_i\right)^{(1/2)v} \Gamma\left(\frac{1}{2}v\right) \prod_{i=1}^{k} \lambda_i}
$$
$$
\times \prod_{i=1}^{k-1} I_{(1/2)v-1}\left(2\sqrt{\frac{b_i x_i x_{i+1}}{\lambda_i \lambda_{i+1}}}\right), \qquad (48.69)
$$

where

$$
K = \prod_{i=1}^{k-1} \left[\frac{1 - \rho_j^2 \rho_{j-1}^2}{1 - \rho_j^2}\right] \quad (\rho_j = \rho_{j,j+1};\ \rho_0 = 0),
$$
$$
b_j = \frac{\rho_j^2 (1 - \rho_{j-1}^2)(1 - \rho_{j+1}^2)}{(1 - \rho_{j-1}^2 \rho_j^2)(1 - \rho_j^2 \rho_{j+1}^2)},
$$
$$
\lambda_j = \frac{(1 - \rho_{j-1}^2)(1 - \rho_j^2)}{1 - \rho_{j-1}^2 \rho_j^2},
$$

and $I(\cdot)$ is a modified Bessel function of the first kind; see Chapter 1 of Johnson, Kotz, and Kemp (1992).

This distribution has been generalized by Jensen (1970c) in the following manner. If $\boldsymbol{X}^T = (X_1, \ldots, X_p)$ has a multivariate normal distribution function with zero expected value vector and variance–covariance matrix $\boldsymbol{V}$, then each of the disjoint subsets $\boldsymbol{X}_{(1)}^T = (X_1, \ldots, X_{p_1})$, $\boldsymbol{X}_{(2)}^T = (X_{p_1+1}, \ldots, X_{p_1+p_2}), \ldots, \boldsymbol{X}_{(k)}^T = (X_{p-p_k+1}, \ldots, X_p)$ has a multivariate normal distribution with zero expected value vector and variance-covariance matrices $\boldsymbol{V}_{11}, \boldsymbol{V}_{22}, \ldots, \boldsymbol{V}_{kk}$, respectively, with

$$
\boldsymbol{V} = \begin{pmatrix}
\boldsymbol{V}_{11} & \boldsymbol{V}_{12} & \cdots & \boldsymbol{V}_{1k} \\
\boldsymbol{V}_{21} & \boldsymbol{V}_{22} & \cdots & \boldsymbol{V}_{2k} \\
\vdots & \vdots & & \vdots \\
\boldsymbol{V}_{k1} & \boldsymbol{V}_{k2} & \cdots & \boldsymbol{V}_{kk}
\end{pmatrix},
$$

being partitioned into sets of $p_1, p_2, \ldots, p_k$ rows and columns. Given v independent sets of $\boldsymbol{X}$'s, $\boldsymbol{X}_j^T = (X_{1j}, \ldots, X_{pj})$ $(j = 1, \ldots, k)$, the Wishart matrix

$$
\boldsymbol{S} = \sum_{j=1}^{v} \boldsymbol{X}_j \boldsymbol{X}_j^T
$$

can be partitioned similarly, with elements $(\boldsymbol{S}_{\ell\ell'})$. From the theory of quadratic forms in normal variables, the variables $Y_j = \operatorname{tr} \boldsymbol{S}_{jj} \boldsymbol{V}_{jj}^{-1}$ $(j =$

$1, 2, \ldots, k$) each have a χ^2 distribution—and the number of degrees of freedom for Y_j is vp_j. The joint distribution of $Y_1, \ldots, Y_k$ can be regarded as a multivariate chi-squared or, more generally (allowing v and the p_j's to take fractional values), a multivariate gamma distribution. As might be expected [from (48.76)], the mathematical expression of this distribution is rather complicated. Jensen (1970c), however, has shown that the *structure* of the density function is easily comprehended. The joint density function is

$$
p_{\boldsymbol{Y}}(\boldsymbol{y}) = 2^k \sum_{h=0}^{\infty} \frac{\Gamma\left(\frac{1}{2}v + h\right)}{h! \Gamma\left(\frac{1}{2}v\right)}
$$
$$
\times \sum_{j_1=0}^{h} \sum_{j_2=0}^{h} \cdots \sum_{j_k=0}^{h} A_{\boldsymbol{j}} f_{\boldsymbol{j}}\left(\frac{1}{2}\boldsymbol{y}; \frac{1}{2}v\boldsymbol{p}\right), \qquad (48.70)
$$

where

(i) $\boldsymbol{j}^T = (j_1, \ldots, j_k)$, $\boldsymbol{p}^T = (p_1, \ldots, p_k)$,

(ii) $f_{\boldsymbol{j}}(\boldsymbol{s}; \boldsymbol{\theta}) = \prod_{h=1}^{k} \{\Gamma(\theta_h + j_h)\}^{-1} \left(-\frac{d}{ds_h}\right)^{j_h} [s_h^{\theta_h + j_h - 1} e^{-s_h}]$,

(iii) $A_{\boldsymbol{j}}$ is defined by the identities

$$
[B(z)]^h \equiv \sum_{j_1=0}^{h} \sum_{j_2=0}^{h} \cdots \sum_{j_k=0}^{h} \left\{ A_{\boldsymbol{j}} \prod_{h=1}^{k} z_j^{j_h} \right\},
$$

with

$$
B(z) = 1 - \begin{vmatrix} \boldsymbol{I}_{p_1} & -z_1\boldsymbol{R}_{12} & -z_1\boldsymbol{R}_{13} & \cdots & -z_1\boldsymbol{R}_{1k} \\ -z_2\boldsymbol{R}_{21} & \boldsymbol{I}_{p_2} & -z_2\boldsymbol{R}_{23} & \cdots & -z_2\boldsymbol{R}_{2k} \\ -z_3\boldsymbol{R}_{31} & -z_3\boldsymbol{R}_{32} & \boldsymbol{I}_{p_3} & \cdots & -z_3\boldsymbol{R}_{3k} \\ \vdots & \vdots & \vdots & & \vdots \\ -z_k\boldsymbol{R}_{k1} & -z_k\boldsymbol{R}_{k2} & -z_k\boldsymbol{R}_{k3} & \cdots & \boldsymbol{I}_{p_k} \end{vmatrix}
$$

and $\boldsymbol{R}_{gh} = \boldsymbol{V}_{gg}^{-1/2} \boldsymbol{V}_{gh} \boldsymbol{V}_{hh}^{-1/2}$ (symmetric positive definite square roots). The characteristic function of distribution in (48.68) is

$$
E[\exp(it^T \boldsymbol{Y})] = |\boldsymbol{I}_p - 2i\boldsymbol{D}(t)\boldsymbol{V}|^{-(1/2)v}, \qquad (48.71)
$$

where

$$
\boldsymbol{D}(t) = \operatorname{diag}(t_1 \boldsymbol{V}_{11}^{-1}, \ldots, t_k \boldsymbol{V}_{kk}^{-1})
$$

is a "block-diagonal" matrix.

An alternative form to (48.69) is

$$|I_p - 2iD(t)\mathbf{R}|^{-(1/2)v},\tag{48.72}$$

where

$$\mathbf{R} = \begin{pmatrix} I_{p_1} & \mathbf{R}_{12} & \cdots & \mathbf{R}_{1k} \\ \mathbf{R}_{21} & I_{p_2} & \cdots & \mathbf{R}_{2k} \\ \vdots & & & \\ \mathbf{R}_{k1} & \mathbf{R}_{k2} & \cdots & I_{p_k} \end{pmatrix}.$$

The cumulant generating function is

$$\frac{1}{2}v \sum_{j=0}^{\infty} j^{-1}(2i)^j \mathrm{tr}[D(t)\mathbf{R}]^j.$$

From this we can find, for example,

$$\mathrm{cov}(Y_j, Y_h) = 2v\ \mathrm{tr}(\mathbf{R}_{jh}\mathbf{R}_{hj}).\tag{48.73}$$

Jensen (1969a,b) has also shown that

$$\mathrm{Pr}\left[\bigcap_{j=1}^{k}(Y_j \le c_j)\right] \le \prod_{j=1}^{k} \mathrm{Pr}[Y_j \le c_j].\tag{48.74}$$

Compare with the corresponding inequalities in the bivariate case.

5   NONCENTRAL MULTIVARIATE CHI-SQUARE (GAMMA) DISTRIBUTIONS

Each of the types of multivariate gamma distributions discussed in Sections 2 and 3 can be extended to noncentral cases in a natural fashion. Distributions constructed by compounding can be generalized by supposing each Y_ℓ ($\ell = 0, 1, \ldots, k$) to have a noncentral gamma distribution, as defined in Chapter 17 of Johnson, Kotz, and Balakrishnan (1994). Since this means that the distribution of Y_ℓ is a mixture of (central) gamma distributions with Poisson weights, the joint distribution of $X_j = Y_0 + Y_j$ ($j = 1, \ldots, k$) will be a mixture of joint distributions of the kind described in Section 2, with weights that are products of Poisson weights.

For distributions based on χ^2 marginals involving multivariate normal distribution, we may suppose that the expected value vector of $\mathbf{X}_i^T =$

$(X_{i1}, \ldots, X_{ik})$ depends on i, though the other conditions (multivariate normality, homoscedasticity, independence of $\boldsymbol{X}_i^T$ and $\boldsymbol{X}_j^T$) remain satisfied. The resulting distribution is sometimes called a *biased generalized Rayleigh distribution* [see Section 2.12 of Blumenson and Miller (1963)]. The marginal distributions of $S_1, \ldots, S_k$ are then noncentral χ^2 distributions each with $(n-1)$ degrees of freedom and noncentrality parameters

$$\sum_{i=1}^n (\xi_{i1} - \bar{\xi}_1)^2, \ldots, \sum_{i=1}^n (\xi_{ik} - \bar{\xi}_k)^2,$$

respectively, where

$$E[X_{ij}] = \xi_{ij} \quad \text{and} \quad \bar{\xi}_j = n^{-1} \sum_{i=1}^n \xi_{ij}.$$

Derivation of explicit expressions for the joint distribution is difficult, even for $k = 2$. A particular case for general k with all elements v^{ij} of $\boldsymbol{V}^{-1}$ zero except for $|i - j| \leq 1$, has been worked out by Blumenson and Miller (1963). Miller and Sackrowitz (1967) have obtained a fairly simple form for the *ratio* of a noncentral distribution with k dimensions to the central distribution with $(k+1)$ dimensions. The conditional distribution of X_{i1}, given $(X_{12}, \ldots, X_{n2})$ is normal with expected value $\rho X_{i2} + \xi_{i1} - \rho \xi_{i2}$ and variance $(1 - \rho^2)$, where $\rho = \text{corr}(X_{i1}, X_{i2})$. Hence, the conditional distribution of S_1, given $(X_{12}, \ldots, X_{n2})$, is that of

$$\sum_{i=1}^n \left[\rho(X_{i2} - \bar{X}_{.2}) + (\xi_{i1} - \bar{\xi}_1) - \rho(\xi_{i2} - \bar{\xi}_2) + (U_i - \bar{U})\sqrt{1 - \rho^2} \right]^2,$$

where $U_1, \ldots, U_n$ are independent standard normal variables. This distribution is that of

$$(1 - \rho^2) \times \left(\text{noncentral } \chi^2 \text{ with } (n-1) \text{ degrees} \right.$$

$$\text{of freedom and noncentrality parameter}$$

$$\left. (1 - \rho^2)^{-1} \sum_{i=1}^n \left[\rho(X_{i2} - \bar{X}_{.2}) + (\xi_{i1} - \bar{\xi}_1) - \rho(\xi_{i2} - \bar{\xi}_2) \right]^2 \right).$$

$$(48.75)$$

Unfortunately, the noncentrality is now not a function of S_2 only, as it was in the central case.

Jensen (1969b) has shown that the limiting joint distribution, as the noncentrality parameters tend to infinity, is multivariate normal. Zaharov, Sarmanov, and Sevastjanov (1969) have obtained a noncentral form of (48.69), corresponding to a departure from specified values of multinomial cell probabilities $p_1, \ldots, p_{v+1}$.

6 INFINITE DIVISIBILITY OF MULTIVARIATE GAMMA

Vere-Jones' results on infinite divisibility have already been mentioned. Griffiths (1984) and Bapat (1989) studied characterization of matrices V for which the Laplace transform $\psi(t) = |I + VT|^{-1/2}$, where T is a diagonal matrix with diagonal elements $t_1, \ldots, t_k$, is infinitely divisible. Note that if $Y = (Y_1, \ldots, Y_k)^T \overset{d}{=} N_k(0, V)$, then $X = (X_1, \ldots, X_k)^T$, where $X_i = Y_i^2/2$ $(i = 1, \ldots, k)$, has the above form as its Laplace transform. Griffiths's (1984) necessary and sufficient condition for infinite divisibility involves the concept of "cycle product of a matrix." Actually, he has considered the infinite divisibility of $|I + VT|^{-1}$ rather than with power $-1/2$, but the two problems are clearly equivalent.

Bapat (1989) has used the concept of "M-matrices" in his result. A $k \times k$ matrix A is said to be an M-matrix if $a_{ij} \le 0$ for all $i \ne j$ and if any one of the following equivalent conditions is satisfied:

(a) A is nonsingular and $A^{-1} \ge 0$.

(b) $A = \lambda I - B$, where $B \ge 0$ and λ is greater than the absolute value of any eigenvalue of B.

(c) All principal minors of A are positive.

The results of Griffiths (1984) and Bapat (1989) can be summarized as follows. Let $X = (X_1, \ldots, X_k)^T$ have the Laplace transform $\psi(t) = |I + VT|^{-1/2}$, where V is a $k \times k$ positive definite matrix, $T = \mathrm{Diag}(t_1, \ldots, t_k)$, and let $W = V^{-1}$. Then, the following conditions are equivalent:

(i) $\psi(t)$ is infinitely divisible.

(ii) for any $\{i_1, \ldots, i_\ell\} \subset \{1, \ldots, k\}$, $\ell \ge 3$,

$$(-1)^\ell w_{i_1 i_2} w_{i_2 i_3} \cdots w_{i_\ell i_1} \ge 0.$$

(iii) There exists a signature matrix D such that DWD is an M-matrix.

(A matrix is a *signature matrix* if all its diagonal entries are either 1 or -1.) The proof that (i) $\Rightarrow$ (ii) is due to Griffiths (1984). The result that (ii) $\Rightarrow$ (iii) and (iii) $\Rightarrow$ (i) is due to Bapat (1989), wherein the latter part relies heavily on Griffiths's arguments. Bapat's (1989) proof led to the following necessary condition for infinite divisibility of $|I + VT|^{-1/2}$.

Let $\boldsymbol{X} = (X_1, \ldots, X_k)^T$ have the Laplace transform $\psi(\boldsymbol{t}) = |\boldsymbol{I} + \boldsymbol{VT}|^{-1/2}$, where $\boldsymbol{V}$ is a positive definite matrix, and suppose that $\psi(\boldsymbol{t})$ is infinitely divisible. Then there exists a signature matrix $\boldsymbol{D}$ such that $\boldsymbol{DVD} \geq 0$.

Paranjape (1978) has shown that a sufficient condition for $\psi(\boldsymbol{t})$ to be infinitely divisible is the existence of a set of positive constants $c_1, \ldots, c_k$ so that the principal minors of $\boldsymbol{V}^{-1}$ - $\mathrm{Diag}(c_1, \ldots, c_k)$ are nonpositive.

BIBLIOGRAPHY

(Some bibliographical items not mentioned in the text are included here for completeness.)

Bapat, R. B. (1989). Infinite divisibility of multivariate gamma distributions and M-matrices, *Sankhyā, Series A*, **51**, 73–78.

Becker, P. J., and Roux, J. J. J. (1981). A bivariate extension of the gamma distribution, *South African Statistical Journal*, **15**, 1–12.

Bens, G. B., and Jensen, D. R. (1967). Percentage points of the Bonferroni approximations chi-square statistics, Technical Report No. 3, Department of Statistics, Virginia Polytechnic Institute, Blacksburg, Virginia.

Blumenson, L. E., and Miller, K. S. (1963). Properties of generalized Rayleigh distributions, *Annals of Mathematical Statistics*, **34**, 903–910.

Bose, S. S. (1935). On the distribution of the ratio of variances of two samples drawn from a given normal bivariate correlated population, *Sankhyā*, **2**, 65–72.

Brewer, D. W., Tubbs, J. D., and Smith, O. E. (1987). A differential equations approach to the modal location for a family of bivariate gamma distributions, *Journal of Multivariate Analysis*, **21**, 53–66.

Cheriyan, K. C. (1941). A bivariate correlated gamma-type distribution function, *Journal of the Indian Mathematical Society*, **5**, 133–144.

Clarke, R. T. (1980). Bivariate gamma distributions for extending annual stream flow records from precipitation, *Water Resources Research*, **16**, 863–870.

David, F. N., and Fix, E. (1961). In *Proceedings of the Fourth Berkeley Symposium*, Vol. 1, pp. 177–197, Berkeley, California: University of California Press.

D'Este, G. M. (1981). A Morgenstern-type bivariate gamma distribution, *Biometrika*, **68**, 339–340.

D'jachenko, Z. N. (1961). On moments of bivariate γ-distribution, *Izvestiya Vysschych Uchebnych Zavedeniĭ Matematika*, **1**, 55–65 (in Russian).

D'jachenko, Z. N. (1962a). On a form of bivariate γ-distribution, *Nauchnye Trudy Leningradskoi Lesotekhnicheskoi Akademii*, **94**, 5–17 (in Russian).

D'jachenko, Z. N. (1962b). Distribution surfaces of γ type, *Trudy VI Vsesoyuznogo Soveshchaniya po Teorii Verojatnostei, Matematicheskoi Statistike, Vilnjus*, 389–395 (in Russian).

Dussauchoy, A., and Berland, R. (1972). Lois gamma à deux dimensions, *Comptes Rendus, de l'Academie des Sciences, Paris, Série A*, **274**, 1946–1949.

Dussauchoy, A., and Berland, R. (1975). A multivariate gamma distribution whose marginal laws are gamma, in *Statistical Distributions in Scientific Work*, Vol. 1 (G. P. Patil, S. Kotz, and J. K. Ord, eds.), pp. 319–328, Dordrecht, The Netherlands: D. Reidel.

Eagleson, G. K. (1964). Polynomial expansions of bivariate distributions, *Annals of Mathematical Statistics*, **35**, 1208–1215.

Finney, D. J. (1938). The distribution of the ratio of estimates of the two variances in a sample from a normal bivariate population, *Biometrika*, **30**, 190–192.

Fomin, Ye. A. (1970). Maximum value distribution for realization of discrete random time processes, *Problemy Peredachi Informatsii*, **6**, 99-103 (in Russian).

Freund, R. J. (1961). A bivariate extension of the exponential distribution, *Journal of the American Statistical Association*, **56**, 971–977.

Gaver, D. P. (1970). Multivariate gamma distributions generated by mixture, *Sankhyā, Series A*, **32**, 123–126.

Ghirtis, G. C. (1967). Some problems of statistical inference relating to double-gamma distribution, *Trabajos de Estadistica*, **18**, 67–87.

Griffiths, R. C. (1969). The canonical coefficients of bivariate gamma distributions, *Annals of Mathematical Statistics*, **40**, 1401–1408.

Griffiths, R. C. (1984). Characterization of infinitely divisible multivariate gamma distributions, *Journal of Multivariate Analysis*, **15**, 13–20.

Gunst, R. F. (1973). On computing critical points for a bivariate chi-square random variable, *Communications in Statistics*, **2**, 221–229.

Gunst, R. F., and Webster, J. T. (1973). Density functions of the bivariate shi-square distribution, *Journal of Statistical Computation and Simulation*, **2**, 275–288.

Gupta, A. K., and Wong, C. F. (1989). On a Morgenstern-type bivariate gamma distribution, *Metrika*, **31**, 327–332.

Hutchinson, T. P., and Lai, C. D. (1990). *Continuous Bivariate Distributions, Emphasising Applications*, Adelaide, Australia: Rumsby Scientific Publishers.

Jensen, D. R. (1969a). An inequality for a class of bivariate chi-square distributions, *Journal of the American Statistical Association*, **64**, 333–336.

Jensen, D. R. (1969b). Limit properties of noncentral multivariate Rayleigh and chi square distributions, *SIAM Journal on Applied Mathematics*, **17**, 802–814.

Jensen, D. R. (1970a). The joint distribution of quadratic forms and related distributions, *Australian Journal of Statistics*, **12**, 13–22.

Jensen, D. R. (1970b). A generalization of the multivariate Rayleigh distribution, *Sankhyā, Series A*, **32**, 193–206.

Jensen, D. R. (1970c). The joint distribution of traces of Wishart matrices and some applications, *Annals of Mathematical Statistics*, **41**, 133–145.

Jensen, D. R. (1976). Gaussian approximation to bivariate Rayleigh distributions, *Journal of Statistical Computation and Simulation*, **4**, 259–267.

Jensen, D. R., and Jones, M. Q. (1969). Simultaneous confidence intervals for variances, *Journal of the American Statistical Association*, **64**, 324–332.

Johnson, N. L. (1962). Some notes on the investigation of heterogeneity in interactions, *Trabajos de Estadistica*, **13**, 183–199.

Johnson, N. L., and Kotz, S. (1972). *Continuous Multivaruate Distributions*, first edition, New York: John Wiley & Sons.

Johnson, N. L., Kotz, S., and Balakrishnan, N. (1994). *Continuous Univariate Distributions*, Vol. 1, Second edition, New York: John Wiley & Sons.

Johnson, N. L., Kotz, S., and Balakrishnan, N. (1997). *Discrete Multivariate Distributions*, New York: John Wiley & Sons.

Johnson, N. L., Kotz, S., and Kemp, A. W. (1992). *Univariate Discrete Distributions*, second edition, New York: John Wiley & Sons.

Karlin, S., and Rinott, Y. (1980). Classes of orderings of measures and related correlation inequalities. I. Multivariate totally positive distributions, *Journal of Multivariate Analysis*, **10**, 467–498.

Kellogg, S. D., and Barnes, J. W. (1987). The distribution of producrs, quotients and powers of two dependent *H*-function variates, *Mathematics and Computers in Simulation*, **31**, 91–111.

Khatri, C. G., Krishnaiah, P. R., and Sen, P. K. (1977). A note on the joint distribution of correlated quadratic forms, *Journal of Statistical Planning and Inference*, **1**, 299–307.

Kibble, W. F. (1941). A two-variate gamma type distribution, *Sankhyā*, **5**, 137–150.

Kowalczyk, T., and Tyrcha, J. (1989). Multivariate gamma distributions, properties and shape estimation, *Statistics*, **20**, 465–474.

Krishnaiah, P. R., Hagis, P., and Steinberg, L. (1963). A note on the bivariate chi distribution, *SIAM Review*, **5**, 140–144.

Krishnaiah, P. R., and Rao, M. M. (1961). Remarks on a multivariate gamma distribution, *American Mathematical Monthly*, **68**, 342–346.

Krishnamoorthy, A. S., and Parthasarathy, M. (1951). A multivariate gamma-type distribution, *Annals of Mathematical Statistics*, **22**, 549–557. Correction (1960), **31**, 229.

Lee, M.-L. T. (1996). Properties and applications of Sarmanov family of bivariate distributions, *Communications in Statistics—Theory and Methods*, **25**, 1207–1222.

Lukacs, E., and Laha, R. G. (1964). *Applications of Characteristic Functions*, London: Griffin.

Mathai, A. M., and Moschopoulos, P. G. (1991). On a multivariate gamma, *Journal of Multivariate Analysis*, **39**, 135–153.

Mathai, A. M., and Moschopoulos, P. G. (1992). A form of multivariate gamma distribution, *Annals of the Institute of Statistical Mathematics*, **44**, 97–106.

Mathai, A. M., and Saxena, R. K. (1978). *The H-Function with Applications in Statistics and Other Disciplines*, New York: John Wiley & Sons.

McKay, A. T. (1934). Sampling from batches, *Journal of the Royal Statistical Society, Supplement,* **1**, 207–216.

Miller, K. S. (1964). *Multidimensional Gaussian Distributions,* New York: John Wiley & Sons.

Miller, K. S., Bernstein, R. I., and Blumenson, L. E. (1958). Generalized Rayleigh processes, *Quarterly of Applied Mathematics,* **16**, 137–145. Correction, **20**, 395.

Miller, K. S., and Sackrowitz, H. (1967). Relationships between biased and unbiased Rayleigh distributions, *SIAM Journal of Applied Mathematics,* **15**, 1490–1495.

Moran, P. A. P. (1967). Testing for correlation between non-negative variates, *Biometrika,* **54**, 385–394.

Moran, P. A. P. (1969). Statistical inference with bivariate gamma distributions, *Biometrika,* **56**, 627–634.

Moran, P. A. P. (1970). The methodology of rain making experiments, *Review of the International Statistical Institute,* **38**, 105–115; Discussion, 115–119.

Moran, P. A. P., and Vere-Jones, D. (1969). The infinite divisibility of multivariate gamma distributions, *Sankhyā, Series A,* **31**, 191–194.

Paranjape, S. R. (1978). Simpler proofs for the infinite divisibility of multivariate gamma distributions, *Sankhyā, Series A,* **40**, 393–398.

Prékopa, A., and Szántai, T. (1978). A new multivariate gamma distribution and its fitting to empirical stream flow data, *Water Resources Research,* **14**, 19–29.

Ramabhadran, V. R. (1951). A multivariate gamma-type distribution, *Sankhyā,* **11**, 45–46.

Royen, T. (1991a). Expansions for the multivariate chi-square distribution, *Journal of Multivariate Analysis,* **38**, 213–232.

Royen, T. (1991b). Multivariate gamma distributions with one-factorial accompanying correlation matrices and applications to the distribution of the multivariate range, *Metrika,* **38**, 299–315.

Royen, T. (1992). On representation and computation of multivariate gamma distributions, in *Data Analysis and Statistical Inference–Festschrift in Honour of F. Eicker* (S. Schach and G. Trenkler, eds.), pp. 201–216, Bergisch Gladbach: Eul-Verlag.

Royen, T. (1994). On some multivariate gamma-distributions connected with spanning trees, *Annals of the Institute of Statistical Mathematics*, **46**, 361–371.

Sarmanov, I. O. (1968). A generalized symmetric gamma correlation, *Doklady Akademii Nauk SSSR*, **179**, 1279–1285; *Soviet Mathematics Doklady*, **9**, 547–550.

Sarmanov, I. O. (1970a). Gamma correlation process and its properties, *Doklady Akademii Nauk, SSSR*, **191**, 30–32 (in Russian).

Sarmanov, I. O. (1970b). An approximate calculation of correlation coefficients between functions of dependent random variables, *Matematicheskie Zametki*, **7**, 617–625 (in Russian). English translation in *Mathematical Notes, Academy of Sciences of the USSR*, **7**, 373–377.

Schmeiser, G. B. W., and Lal, R. (1982). Bivariate gamma random vectors, *Operations Research*, **30**, 355–374.

Smith, O. E., and Adelfang, S. I. (1981). Gust model based on the bivariate gamma distribution, *Journal of Spacecraft*, **18**, 545–549.

Smith, O. E., Adelfang, S. I., and Tubbs, J. D. (1982). A bivariate gamma probability distribution with application to gust modelling, *NASA Technical Report TM-82483*.

Steel, S. J., and le Roux, N. J. (1987). A reparametrisation of a bivariate gamma extension, *Communications in Statistics—Theory and Methods*, **16**, 293–305.

Steel, S. J., and le Roux, N. J. (1989). A class of compound distributions of the reparametrised bivariate gamma extension, *South African Statistical Journal*, **23**, 131–141.

Szántai, T. (1986). Evaluation of special multivariate gamma distribution function, *Mathematical Programming Study*, **27**, 1–16.

Tan, W. Y. (1968). Some distribution theory associated with complex Gaussian distribution, *Tamkang Journal of Mathematics*, **7**, 263–302.

Vere-Jones, D. (1967). The infinite divisibility of a bivariate gamma distribution, *Sankhyā, Series A*, **27**, 421–422.

Wicksell, S. D. (1933). On correlation functions of type III, *Biometrika*, **25**, 121–133.

Zaharov, V. K., Sarmanov, O. V. and Sevastjanov, B. A. (1969). Sequential χ^2 tests, *Matematicheskii Sbornik*, **79**, 444–460 (in Russian).

CHAPTER 49

Dirichlet and Inverted Dirichlet Distributions

Before reading this chapter, the reader is strongly advised to review Chapter 35 of the book *Discrete Multivariate Distributions* by Johnson, Kotz, and Balakrishnan (1997) [especially Sections 4 and 5] dealing with Multinomial Distributions. In the first edition of this book *Multivariate Continuous Distributions*, only two sections of Chapter 40 were devoted to Dirichlet and inverted Dirichlet distributions. The expanded nature of our discussion is a good indication of the amount of work that has been carried out on these distributions during the past 25 years or so.

1 DIRICHLET DISTRIBUTION

Suppose that $X_0, X_1, \ldots, X_m$ are independent random variables, with X_j distributed as χ^2 with v_j degrees of freedom, for $j = 0, 1, \ldots, m$ (v_j need not be an integer, through it must be greater than zero.) We seek the joint distribution of $Y_1, Y_2, \ldots, Y_m$ where

$$Y_j = \frac{X_j}{\sum_{i=0}^m X_i} \qquad (j = 1, 2, \ldots, m).$$

The joint probability density function of $X_0, X_1, \ldots, X_m$ is

$$
\begin{aligned}
p_{X_0,\ldots,X_m}(x_0, \ldots, x_m) &= \left[\prod_{j=0}^m \Gamma\left(\frac{1}{2} v_j \right) \right]^{-1} 2^{-(1/2)\sum_{j=0}^m v_j} \left[\prod_{j=0}^m x_j^{(v_j/2)-1} \right] \\
&\quad \times \exp\left[-\frac{1}{2} \sum_{j=0}^m x_j \right] \qquad (0 \leq x_j; \ j = 1, \ldots, m).
\end{aligned}
$$

(49.1)

485

Making the transformation to new variables $Y_0 = \sum_{i=1}^m X_i, Y_1, Y_2, \ldots, Y_m$, we find

$$p_{Y_0,\ldots,Y_m}(y_0, \ldots, y_m)$$

$$= \left[\prod_{j=0}^m \Gamma\left(\frac{1}{2} v_j\right)\right]^{-1} 2^{-(1/2)\sum_{j=0}^m v_j}$$

$$\times \left[\left\{y_0\left(1 - \sum_{j=1}^m y_j\right)\right\}^{(v_0/2)-1} \prod_{j=1}^m (y_0 y_j)^{(v_j/2)-1} \exp\left[-\frac{1}{2}y_0\right] J\right.$$

$$\left(0 \le y_j;\ j = 0, 1, \ldots, m;\ \sum_{j=1}^m y_j \le 1\right), \qquad (49.2)$$

where J is the Jacobian

$$J = \frac{\partial(x_0, x_1, \ldots, x_m)}{\partial(y_0, y_1, \ldots, y_m)}$$

$$= \begin{vmatrix} 1 - \sum_{j=1}^m y_j & -y_0 & -y_0 & \cdots & -y_0 \\ y_1 & y_0 & 0 & \cdots & 0 \\ y_2 & 0 & y_0 & \cdots & 0 \\ \cdots & \cdot & \cdot & \cdots & \cdot \\ \cdots & \cdot & \cdot & \cdots & \cdot \\ \cdots & \cdot & \cdot & \cdots & \cdot \\ y_m & 0 & 0 & \cdots & y_0 \end{vmatrix} = y_0^{m-1}.$$

Formula (49.2) can be rearranged in the form

$$p_{Y_0,Y_1,\ldots,Y_m}(y_0, y_1, \ldots, y_m)$$

$$= \left[2^{(1/2)\sum_{j=0}^m v_j} \prod_{j=0}^m \Gamma\left(\frac{1}{2} v_j\right)\right]^{-1} \left[\prod_{j=1}^m y_j^{(v_j/2)-1}\right]$$

$$\times y_0^{(1/2)\sum_{j=0}^m v_j - 1}\left(1 - \sum_{j=1}^m y_j\right)^{(v_0/2)-1} e^{-(y_0/2)} \qquad (49.3)$$

defined over $w(\boldsymbol{y}) = \left\{(y_1, \ldots, y_m) \mid y_j \ge 0;\ j = 0, 1, \ldots, m;\ \sum_{j=1}^m y_j \le 1\right\}$.
Integrating out the variable y_0, we obtain the joint density of $Y_1, Y_2, \ldots, Y_m$ as

$$p_{Y_1,\ldots,Y_m}(y_1, \ldots, y_m)$$

$$= \frac{\Gamma\left(\frac{1}{2}\sum_{j=0}^m v_j\right)}{\prod_{j=0}^m \Gamma\left(\frac{1}{2} v_j\right)}\left[\prod_{j=1}^m y_j^{(v_j/2)-1}\right]\left(1 - \sum_{j=1}^m y_j\right)^{(v_0/2)-1} \qquad (49.4)$$

defined over $w(\boldsymbol{y})$.

Since (49.4) is a density function, we obtain

$$
\int\int\cdots\int\left[\prod_{j=1}^{m} y_j^{(v_j/2)-1}\right]\left(1-\sum_{j=1}^{m} y_j\right)^{(v_0/2)-1} dy_1\cdots dy_m
$$
$$
= \frac{\prod_{j=0}^{m}\Gamma\left(\frac{1}{2}\,v_j\right)}{\Gamma\left(\frac{1}{2}\sum_{j=0}^{m} v_j\right)} \tag{49.5}
$$

[integration being over the region $w(\boldsymbol{y})$ defined after (49.4)]. Formula (49.5) is a particular case of a multiple integral evaluated by Dirichlet (1839). An integral similar to the one in (49.5) is mentioned to be *Dirichlet integral* in Cramér (1946). The name *Dirichlet distribution* has been given to the class of distributions (49.4). It is usual to replace $\frac{1}{2}v_j$ by θ_j $(j=0,1,\ldots,m)$; the *standard Dirichlet* distribution with parameters $\theta_1,\ldots,\theta_m,\theta_0$ has density function

$$
p_{Y_1,\ldots,Y_m}(y_1,\ldots,y_m)
$$
$$
= \frac{\Gamma\left(\sum_{j=0}^{m}\theta_j\right)}{\prod_{j=0}^{m}\Gamma(\theta_j)}\left(1-\sum_{j=1}^{m} y_j\right)^{\theta_0-1}\prod_{j=1}^{m} y_j^{\theta_j-1}
$$
$$
\left(0\le y_j;\ j=1,\ldots,m;\ \sum_{j=1}^{m} y_j\le 1\right). \tag{49.6}
$$

Tiao and Afonja (1969) obtained approximations to the probability integral

$$
\Pr\left[\bigcap_{j=1}^{m}(Y_j\le a_j)\right] = \int_0^{a_m}\cdots\int_0^{a_1} p_{Y_1,\ldots,Y_m}(y_1,\ldots,y_m)dy_1\cdots dy_m
$$

with $p_{Y_1,\ldots,Y_m}(y_1,\ldots,y_m)$ given by (49.4). It is clear, from our derivation of the Dirichlet distribution, that Y_j has a standard beta distribution with parameters θ_j, $\sum_{i=0}^{m}\theta_i-\theta_j$; see Chapter 25 of Johnson, Kotz and Balakrishnan (1995). It is thus reasonable to regard the distribution as a *multivariate generalization* of the beta distribution. The mixed moments can be easily evaluated using (49.5). We have

$$
\begin{aligned}
\mu'_{r_1,\ldots,r_m} &= E\left[\prod_{j=1}^{m} Y_j^{r_j}\right] \\
&= \frac{\Gamma\left(\sum_{j=0}^{m}\theta_j\right)}{\prod_{j=0}^{m}\Gamma(\theta_j)} \int\int\cdots\int_{\omega(\boldsymbol{y})} \left(1 - \sum_{j=1}^{m} y_j\right)^{\theta_0-1} \prod_{j=1}^{m} y_j^{\theta_j+r_j-1} d\boldsymbol{y} \\
&= \frac{\Gamma\left(\sum_{j=0}^{m}\theta_j\right)}{\prod_{j=0}^{m}\Gamma(\theta_j)} \frac{\prod_{j=0}^{m}\Gamma(\theta_j+r_j)}{\Gamma\left(\sum_{j=0}^{m}(\theta_j+r_j)\right)} \\
&= \frac{\prod_{j=0}^{m}\theta_j^{[r_j]}}{\left\{\sum_{j=0}^{m}\theta_j\right\}^{[\sum_{j=0}^{m} r_j]}} \cdot
\end{aligned}
\tag{49.7}
$$

In particular, the covariance between X_i and X_j is

$$
\frac{\theta_i\theta_j}{\Theta(\Theta+1)} - \frac{\theta_i\theta_j}{\Theta^2} = -\frac{\theta_i\theta_j}{\Theta^2(\Theta+1)}
\tag{49.8}
$$

where $\Theta = \sum_{j=0}^{m}\theta_j$. Since

$$
E[Y_i] = \theta_i/\Theta \quad \text{and} \quad \mathrm{var}(Y_i) = \frac{\theta_i(\Theta-\theta_i)}{\Theta^2(\Theta+1)},
\tag{49.9}
$$

then

$$
\mathrm{corr}(Y_i, Y_j) = -\sqrt{\frac{\theta_i\theta_j}{(\Theta-\theta_i)(\Theta-\theta_j)}}.
\tag{49.10}
$$

Note that all pairwise correlations are negative. [Compare this with the corresponding formula for the multinomial distribution presented in Eq. (35.9) of Chapter 35 of Johnson, Kotz, and Balakrishnan (1997).] It can be seen, either by integrating out $y_{s+1},\ldots,y_m$ from (49.6) or by considering the derivation of the distribution given at the beginning of this section, that the variables $Y_1, Y_2,\ldots, Y_s$ ($s < m$) have a standard joint Dirichlet distribution with parameters $\theta_1,\theta_2,\ldots,\theta_s; \Theta - \sum_{j=1}^{s}\theta_j$.

It immediately follows that the conditional joint distribution of

$$
Y'_j = \frac{Y_j}{1 - \sum_{i=1}^{s} Y_i} \qquad (j = s+1,\ldots,m),
$$

given $Y_1,\ldots,Y_s$, is a standard Dirichlet distribution with parameters $\theta_{s+1},$ $\ldots,\theta_m,\theta_0$. (In particular, the distribution of Y'_j, given $Y_1,\ldots,Y_s$, is standard beta with parameters $\theta_j, \sum_{i=s+1}^{m}\theta_i + \theta_0 - \theta_j$.)

This is in fact a *characterization* of the Dirichlet distribution (provided that the Y's are positive random variables with continuous density functions, and $\sum_{j=1}^{m} Y_j \leq 1$), given by Darroch and Ratcliff (1971). See Section 3 for more details.

An orthonormal expansion of a two-dimensional Dirichlet distribution, with Jacobi polynomials as the orthonormal functions, has been described by Lee (1971). Specifically, the two-variable $(X_1, X_2)^T$ Dirichlet distribution [denoted $D(v_1, v_2; v_3)$] with density

$$p(x_1, x_2) = \frac{\Gamma(v_1 + v_2 + v_3)}{\Gamma(v_1)\Gamma(v_2)\Gamma(v_3)} \, x_1^{v_1-1} x_2^{v_2-1} (1 - x_1 - x_2)^{v_3-1},$$

$$(49.11)$$

where $v_i > 0$, $i = 1, 2, 3$, $x_j \geq 0$, $j = 1, 2$ and $x_1 + x_2 \leq 1$, admits the following diagonal expansion in terms of orthonormal polynomials.

Since the marginals are

$$
\begin{aligned}
p_1(x_1) &= Beta(v_1, v_2 + v_3), \\
p_2(x_2) &= Beta(v_2, v_1 + v_3),
\end{aligned}
$$

$E[X_1^n | X_2 = x_2] = (1 - x_2)^n$, $E[X_2^n | X_1 = x_1] = (1 - x_1)^n$, and the marginal beta density is the weight function for the shifted Jacobi polynomials $R_n^{(\alpha,\beta)}(x) = \frac{(1+\alpha)_n}{n!} \, {}_2F_1[-n, n + \alpha + \beta + 1; \alpha + 1; x]$ where $Re(\alpha) > -1$, $Re(\beta) > -1$ [see Eq. (1.175) in Chapter 1 of Johnson, Kotz, and Kemp (1992)], the expansion is

$$
\begin{aligned}
p(x_1, x_2) &= p_1(x_1) p_2(x_2) \sum_{n=0}^{\infty} (-1)^n n! \, \frac{v_1 + v_2 + v_3 + 2n - 1}{v_1 + v_2 + v_3 + n - 1} \\
&\quad \times \frac{(v_1 + v_2 + v_3)_n}{(v_1 + v_3)_n (v_2 + v_3)_n} \, R_n^{(v_1-1, v_2+v_3-1)}(x_1) R_n^{(v_2-1, v_1+v_3-1)}(x_2), \\
&\quad v_i > 0, \ i = 1, 2, 3; \ x_j \geq 0, \ j = 1, 2; \ x_1 + x_2 \leq 1.
\end{aligned}
$$

The use of a Dirichlet distribution as an approximation to a *multinomial* distribution [Johnson [1960]] has been discussed in Chapter 35 of Johnson, Kotz, and Balakrishnan (1997).

From the structure of the Dirichlet distribution, as described in this section, it follows [see Chapter 25 of Johnson, Kotz, and Balakrishnan (1995)] that the random variables

$$Z_j = \frac{Y_j}{\sum_{i=j}^{m} Y_i} \qquad (j = 1, 2, \ldots, m)$$

are mutually independent standard beta variables, with parameters θ_j, $\sum_{i=j+1}^{m} \theta_i$, respectively. Provost and Cheong (2000) have discussed the derivation of the distribution of linear combinations of the components of a Dirichlet random vector $\boldsymbol{Y}$.

If we now let these parameters have *general* values, so that Z_j is distributed as a standard beta variable with parameters a_j, b_j, the corresponding Y's have a *generalized Dirichlet distribution* described by Connor and Mosimann (1969). Further discussion is given in Section 8.5. The joint density function is (replacing Y's by X's)

$$
p_{\boldsymbol{X}}(\boldsymbol{x}) = \left(\prod_{j=1}^{m} B(a_j, b_j) \right)^{-1} \left(1 - \sum_{j=1}^{m} x_j \right)^{b_m - 1}
$$
$$
\times \prod_{j=1}^{m} \left[x_j^{a_j - 1} \left(1 - \sum_{i=1}^{j-1} x_i \right)^{b_{j-1} + (a_j + b)} \right] \left(0 \leq x_j; \sum_{j=1}^{m} x_j \leq 1 \right).
$$

$$(49.12)$$

We have

$$
E[X_j] = \frac{a_j}{a_j + b_j} \prod_{i=1}^{j-1} \frac{b_i + 1}{a_i + b_i},
$$

$$(49.13)$$

$$
\mathrm{var}(X_j) = E[X_j] \left\{ \frac{a_j + 1}{a_j + b_j + 1} \prod_{i=1}^{j-1} \frac{b_i + 1}{a_i + b_i + 1} - E[X_j] \right\},
$$

$$(49.14)$$

$$
\mathrm{cov}(X_j, X_k) = E[X_k] \left\{ \frac{a_j}{a_j + b_j + 1} \prod_{i=1}^{j-1} \frac{b_i + 1}{a_i + b_i + 1} - E[X_j] \right\},
$$

$$(j > k) \qquad (49.15)$$

If $b_{j-1} = a_j + b_j$ $(j = 1, 2, \ldots, m)$, we obtain a Dirichlet distribution. Note that in general the marginal distributions corresponding to (49.12) (except that of X_1) are *not* beta distributions.

A different kind of generalization can be obtained [Craiu and Craiu (1969)] by supposing $X_0, X_1, \ldots, X_m$, at the beginning of this section, to have a *generalized gamma distribution* (rather than a χ^2 distribution) with density function

$$
f(x_i; a_i, \alpha_i, \beta_i) = \frac{\alpha_i}{(a_i)^{\beta_i/\alpha_i} \Gamma(\beta_i/\alpha_i)} \, x_i^{\beta_i - 1} \exp\left(-\frac{x_i^{\alpha_i}}{a_i^{\alpha_i}} \right),
$$
$$
x_i > 0, \; \alpha_i > 0, \; \beta_i > 0.
$$

Ma (1996) has discussed *multivariate rescaled Dirichlet distribution* [denoted by $\boldsymbol{X} \sim MRD_k(a, \theta_1, \theta_2, \ldots, \theta_k)$] with continuous survival function

$$S(\boldsymbol{x}) = \begin{cases} \left(1 - \sum_{i=1}^{k} \theta_i x_i\right)^a, & 0 \leq \sum_{i=1}^{k} \theta_i x_i \leq 1, \\ 0, & \text{otherwise.} \end{cases}$$

Here, $a, \theta_1, \ldots, \theta_k$ are positive constants. This distribution (along with the multivariate Lomax distribution with the survival function [Nayak (1987)]

$$S(\boldsymbol{x}) = \left(1 + \sum_{i=1}^{k} \theta_i x_i\right)^{-a}$$

where $a, \theta_1, \ldots, \theta_k$ are as above) enjoys a strong property involving residual life distribution.

Define $S(\boldsymbol{y}; \boldsymbol{x}) = \Pr[\boldsymbol{X} > \boldsymbol{x} + \boldsymbol{y} | \boldsymbol{X} > \boldsymbol{x}] = \frac{S(\boldsymbol{x}+\boldsymbol{y})}{S(\boldsymbol{x})}$ – the residual life distribution; then

$$S(\theta(\boldsymbol{x})\boldsymbol{y}; \boldsymbol{x}) = S(\boldsymbol{y}),$$

where, for the MRD_k, $\theta(\boldsymbol{x}) = \left(1 - \sum_{i=1}^{k} \theta_i x_i\right)$. Thus, the residual life at age $\boldsymbol{x}$ follows an accelerated life model in terms of the original distribution.

It follows from this property that if a random variable on $\mathbb{R}_+^k$ has a continuous survival function $S(\boldsymbol{x})$, whose partial derivatives with respect to x_i $(1, \ldots, k)$ exist and $S(\boldsymbol{0}) = 1$, it has a multivariate rescaled Dirichlet distribution iff it is IHR (or, equivalently, $0 < \theta(\boldsymbol{x}) < 1$ except for $\boldsymbol{x} = \boldsymbol{0}$).

Let $\boldsymbol{X}_1$ be a $(k \times k)$ random stochastic matrix such that the rows of $\boldsymbol{X}_1$ are independent with Dirichlet distributions. The rows of the $(k \times k)$ matrix $\boldsymbol{A}$ are the parameters of these Dirichlet distributions. Suppose the sums of the rows and columns of $\boldsymbol{A}$ provide the same vector $\boldsymbol{r} = (r_1, \ldots, r_k)$. If $(\boldsymbol{X}_n)_{n=1}^{\infty}$ are independent and identically distributed, Chamayou and Letac (1994) have proved that $\lim_{n \to \infty} (\boldsymbol{X}_n \cdots \boldsymbol{X}_1)$ almost surely has identical rows that are Dirichlet distributed with parameter $\boldsymbol{r}$.

2   INVERTED DIRICHLET DISTRIBUTION

If $X_0, X_1, \ldots, X_k$ are independent random variables, with X_j distributed as χ^2 with v_j degrees of freedom $(j = 0, 1, \ldots, k)$, then the joint distribution of

$$Y_j = \frac{X_j}{X_0} \qquad (j = 1, \ldots, k)$$

has the density function

$$p_{Y_1,\ldots,Y_k}(y_1,\ldots,y_k)$$
$$= \frac{\Gamma\left(\frac{1}{2}v\right)}{\prod_{j=0}^{k}\Gamma\left(\frac{1}{2}v_j\right)}\frac{\prod_{j=1}^{k}y_j^{(v_j/2)-1}}{\left(1+\sum_{j=1}^{k}y_j\right)^{(v/2)}}$$
$$(0 \le y_j;\ j = 1,\ldots,m) \tag{49.16}$$

with $v = \sum_{j=0}^{k}v_j$.

Since (49.16) is a probability density function, we have

$$\int_0^\infty \cdots \int_0^\infty \left(1 + \sum_{j=1}^{k}y_j\right)^{-(v/2)} \prod_{j=1}^{k} y_j^{(v_j/2)-1} dy_1 \cdots dy_k$$
$$= \frac{\prod_{j=0}^{k}\Gamma\left(\frac{1}{2}v_j\right)}{\Gamma\left(\frac{1}{2}v\right)}.$$

Hence, provided that $\sum_{j=1}^{k}r_j < \frac{1}{2}v$, we obtain

$$\mu'_{r_1,r_2,\ldots,r_k} = E\left[\prod_{j=1}^{k}Y_j^{r_j}\right]$$
$$= \frac{\Gamma\left(\frac{1}{2}v-r\right)}{\Gamma\left(\frac{1}{2}v\right)}\prod_{j=1}^{k}\frac{\Gamma\left(\frac{1}{2}v_j+r_j\right)}{\Gamma\left(\frac{1}{2}v_j\right)}$$
$$= \prod_{j=1}^{k}\left(\frac{1}{2}v_j\right)^{[r_j]} \Big/ \left(\frac{1}{2}v-1\right)^{(r)}, \tag{49.17}$$

where $r_1, r_2, \ldots, r_k$ are positive integers, while $r = \sum_{j=1}^{k}r_j$, $M^{[N]} = M(M+1)\cdots(M+N-1)$, and $M^{(N)} = M(M-1)\cdots(M-N+1)$. Formula (49.17) can also be obtained by noting that

$$E\left[\prod_{j=1}^{k}Y_j^{r_j}\right] = E\left[X_0^{-r}\prod_{j=1}^{k}X_j^{r_j}\right] = E[X_0^{-r}]\prod_{j=1}^{k}E[X_j^{r_j}].$$

If $\frac{1}{2}v_j$ in (49.16) is replaced by θ_j $(j = 0,\ldots,k)$, we obtain the *standard inverted Dirichlet distribution* [Tiao and Guttman (1965)]

$$p(y_1,\ldots,y_k) = \frac{\Gamma\left(\sum_{j=0}^{k}\theta_j\right)}{\prod_{j=0}^{k}\Gamma(\theta_j)}\frac{\prod_{j=1}^{k}y_j^{\theta_j-1}}{\left(1+\sum_{j=1}^{k}y_j\right)^{\sum_{j=0}^{k}\theta_j}} \quad (0 < y_j).$$

$$\tag{49.18}$$

Here $\theta_j > 0$ for all j, but the θ_j's need not be integers or integers plus $\frac{1}{2}$ (as perhaps implied by the derivation from χ^2.) The distribution (49.18) can be obtained by supposing X_j to have a standard gamma distribution with parameter θ_j. This distribution is called the *multivariate inverted beta distribution*. It is also called *Type II Dirichlet distribution*.

Note that $Y_j/Y_{j'} = X_j/X_{j'}$ (for $1 \le j \ne j' \le k$) so the distribution of this ratio is that of the ratio of two independent gamma variables. Also, since the conditional distribution of X_0, given that $X_j/X_0 = y_j$, is that of $(1 + y_j)^{-1} \times$ (standard gamma variable with parameter $(\theta_0 + \theta_j)$), the conditional distribution of $Y_{j'}$, given that $Y_j = y_j$, is that of $(1 + y_j) \times$ (ratio of two independent standard gamma variables with parameters $\theta_{j'}, \theta_0 + \theta_j$). Hence,

$$E[Y_{j'}|Y_j = y_j] = (1 + y_j)\theta_{j'}(\theta_0 + \theta_j - 1)^{-1}(\theta_0 + \theta_j > 1) \qquad (49.19)$$

and

$$\begin{aligned}
&\mathrm{var}(Y_{j'}|Y_j = y_j) \\
&= (1 - y_j)^2\theta_{j'}(\theta_0 + \theta_j + \theta_{j'} - 1)(\theta_0 + \theta_j - 1)^{-2}(\theta_0 + \theta_j - 2)^{-1}.
\end{aligned} \qquad (49.20)$$

Also,

$$\left.\begin{aligned}
\mathrm{cov}(Y_j, Y_{j'}) &= \theta_j\theta_{j'}(\theta_0 - 1)^{-2}(\theta_0 - 2)^{-1} \\
\mathrm{corr}(Y_j, Y_{j'}) &= [\theta_j\theta_{j'}(\theta_0 + \theta_j - 1)^{-1}(\theta_0 + \theta_{j'} - 1)^{-1}]^{1/2}
\end{aligned}\right\}$$
$$(\theta_0 > 2). \qquad (49.21)$$

From the structure of the inverted Dirichlet distribution, we see that the conditional distribution of $Y_1, \ldots, Y_m$ given $X_0 = x_0$ is simply the product of independent $x_0^{-1}\chi^2_{v_j}$ densities, so that

$$p_{Y_1,\ldots,Y_m|X_0}(y_1, \ldots, y_m|x_0)$$
$$= \prod_{j=1}^{m}\left[\left(\frac{1}{2}x_0\right)^{(v_j/2)}\left\{\Gamma\left(\frac{1}{2}v_j\right)\right\}^{-1}y_j^{(v_j/2)-1}e^{-(1/2)x_0y_j}\right],$$
$$\left(x_0 > 0;\ y_j > 0;\ \sum_{j=1}^{m}y_j < 1\right), \qquad (49.22)$$

and

$$p_{X_0,Y_1,\ldots,Y_m}(x_0, y_1, \ldots, y_m)$$

$$= \left[\prod_{j=0}^{m}\left\{\Gamma\left(\frac{1}{2}\,v_j\right)\right\}^{-1}\right] x_0^{(1/2)\sum_{j=0}^{m} v_j-1} \left[\prod_{j=1}^{m} y_j^{(v_j/2)-1}\right]$$

$$\times \exp\left[-\frac{1}{2}\,x_0\left(1+\sum_{j=1}^{m} y_j\right)\right],$$

$$\left(x_0 > 0;\ y_j > 0;\ \sum_{j=1}^{m} y_j < 1\right).$$

This is termed by Roux (1971) a *Dirichlet-gamma distribution*. Roux (1971) has discussed some multivariate exponential-type properties of the distribution (49.16) and (49.22).

Sometimes [see, for example, Yassaee (1974)] a slightly different notation is used. A random vector $(X_1, X_2, \ldots, X_k)^T$ with the p.d.f.

$$p(x_1, x_2, \ldots, x_k) = \left[\Gamma\left(\sum_{j=1}^{k+1} v_j\right) \bigg/ \prod_{j=1}^{k+1} \Gamma(v_j)\right] \prod_{i=1}^{k} x_i^{v_i-1}$$

$$\times \left(1 + \sum_{i=1}^{k} x_i\right)^{-\sum_{j=1}^{k+1} v_j} \qquad (49.23)$$

for $x_i > 0$ $(i = 1, \ldots, k)$ is called a random vector with k-variate *inverted Dirichlet distribution* $D'(v_1, v_2, \ldots, v_k, v_{k+1})$.

In this definition, given $k + 1$ independent variables $X_1, X_2, \ldots, X_{k+1}$ having gamma distributions with same scale but different shape parameters $v_1, v_2, \ldots, v_{k+1}$, the joint distribution of

$$(Y_1, \ldots, Y_k)^T,$$

where $Y_i = X_i/X_{k+1}$, $i = 1, 2, \ldots, k$, is, of course, the inverted Dirichlet distribution $D'(v_1, v_2, \ldots, v_k, v_{k+1})$.

The marginal distribution of any set

$$(X_{1_i}, X_{2_i}, \ldots, X_{s_i})^T$$

is an s-variate inverted Dirichlet distribution $(s \leq k)$, $D'(v_{1_i}, \ldots, v_{s_i}, v_{k+1})$, $i = 1, 2, \ldots, \frac{k!}{s!(k-s)!}$, but the conditional distribution of $(X_1, \ldots, X_s)^T$ given $X_{s+1}, \ldots, X_k$ is not a multivariate inverted Dirichlet distribution; see Yassaee (1974).

An important property of use in calculation of Dirichlet probabilities is that if $(X_1, \ldots, X_k)^T$ has a k-variate *inverted* Dirichlet distribution

$D'(v_1, v_2, \ldots, v_k, v_{k+1})$, then the vector $(Y_1, \ldots, Y_{k-1})^T$ has a $(k-1)$-variate Dirichlet distribution, where

$$Y_i = X_i / \sum_{j=1}^{k} X_j, \qquad i = 1, 2, \ldots, k-1.$$

Finally, if $\boldsymbol{X} = (X_1, X_2, \ldots, X_k)^T$ has a k-variate inverted Dirichlet distribution $D'(1, 1, \ldots, 1, 1)$, then

$$\boldsymbol{Y} = (Y_1, Y_2, \ldots, Y_k)^T$$

where $Y_i = -\ln X_i$, $i = 1, 2, \ldots, k$, has a *k-variate logistic distribution* with the joint density function

$$p(y_1, y_2, \ldots, y_k) = k! \exp\left\{ -\sum_{i=1}^{k} y_i \right\} \left[1 + \sum_{i=1}^{k} \exp(-y_i) \right]^{-(k+1)},$$

$$-\infty < y_i < \infty;$$

see Chapter 52 for more details.

3 CHARACTERISTIC FUNCTIONS

Let

$$p(x_1, x_2, \ldots, x_k) = K \left(\prod_{j=1}^{k} x_j^{v_j - 1} \right) \left(1 - \sum_{j=1}^{k} x_j \right)^{v_{k+1} - 1}$$

in the simplex S_k, where $K = \Gamma\left(\sum_{s=1}^{k+1} v_s \right) \big/ \prod_{s=1}^{k+1} \Gamma(v_s)$, and v_s ($s = 1, 2, \ldots, k+1$) are arbitrary positive real numbers, be the probability density function of a Dirichlet random variable.

The characteristic function is

$$\phi(t_1, t_2, \ldots, t_k)$$
$$= K' \sum_{n_j=0}^{\infty} \cdots \sum \frac{\left(\sum_{j=1}^{k} v_j, \sum_{i=1}^{k-1} n_i \right)}{\left(\sum_{j=1}^{k+1} v_j, \sum_{i=1}^{k-1} n_i \right)}$$
$$\times F_B^* \left(\sum_{j=1}^{k} v_j + \sum_{i=1}^{k-1} n_i; \sum_{j=1}^{k+1} v_j + \sum_{i=1}^{k-1} n_i; it_k \right)$$
$$\times \left[\prod_{j=1}^{k-1} \frac{\{i(t_j - t_k)\}^{n_j}}{n_j!} \right] \left(\prod_{j=1}^{k-1} y_j^{n_j + v_j - 1} \right) \left(1 - \sum_{i=1}^{k-1} y_i \right)^{v_k - 1} \prod_{i=1}^{k-1} dy_i,$$

where

$$K' = \frac{\Gamma\left(\sum_{s=1}^k v_s\right)}{\prod_{s=1}^k \Gamma(v_s)}$$

and

$$F_B^*\left(v_1, v_2, \ldots, v_k; \sum_{s=1}^{k+1} v_s; it_1, it_2, \ldots, it_k\right)$$

$$= \sum_{n_j=0}^\infty \cdots \sum \frac{\prod_{j=1}^k (v_j, n_j)}{\left(\sum_{s=1}^{k+1} v_s, \sum_{j=1}^{k-1} n_j\right) \prod_{j=1}^k n_j!} \prod_{j=1}^k (it_j)^{n_j}$$

with $(v_j, n_j) = \frac{\Gamma(v_j+n_j)}{\Gamma(v_j)} = v_j^{[n_j]}$ and, in particular, $(1, n_j) = 1^{[n_j]} = n_j!$.
Furthermore,

$$F_B^*\left(v_1, v_2, \ldots, v_k; \sum_{s=1}^{k+1} v_s; it_1, \ldots, it_k\right)$$

$$= \sum_{n_j=0}^\infty \cdots \sum \frac{\prod_{j=1}^{k-1}(v_j, n_j)}{\left(\sum_{j=1}^{k+1} v_j, \sum_{i=1}^{k-1} n_i\right) \prod_{j=1}^{k-1} n_j!}$$

$$\times \prod_{j=1}^{k-1} [i(t_j - t_k)]^{n_j} F_B^*\left(\sum_{j=1}^k v_j + \sum_{i=1}^{k-1} n_i; \sum_{j=1}^{k+1} v_j + \sum_{i=1}^{k-1} n_i; it_k\right).$$

$$(49.24)$$

For inverted Dirichlet random variables with joint density function

$$p(x_1, x_2, \ldots, x_k)$$

$$= \begin{cases} K\left(\prod_{j=1}^k x_j^{v_j-1}\right)\left(1 + \sum_{j=1}^k x_j\right)^{-\sum_{s=1}^{k+1} v_s}, & \text{for } 0 < x_j < \infty, \\ 0, & \text{otherwise,} \end{cases}$$

the characteristic function is

$$\phi(t_1, t_2, \ldots, t_k) = \frac{2\pi}{\Gamma(v_{k+1})}$$

$$\times \sum_{m=0}^\infty \sum_{n_r}^\infty \cdots \sum \frac{\exp\left\{i\pi\left(-v_{k+1} + \sum_{r=1}^{k-1} n_r - \frac{1}{2}\right)\right\}}{\left(\sum_{s=1}^{k+1} v_s, m\right)}$$

$$\times \frac{\left\{\prod_{r=1}^{k-1}(v_r, n_r)\right\}\left(\sum_{j=1}^k v_j + \sum_{r=1}^{k-1} n_r, m\right)}{\prod_{r=1}^{k-1}(1, n_r)(1, m)}$$

$$\times \prod_{r=1}^{k-1}\{i(t_r - t_k)\}^{n_k}(1 - it_k)^m.$$

$(v_1, \ldots, v_{k+1}$ may be nonintegers.)

Since the limiting distribution of the inverted Dirichlet distribution $D'(v_1, v_2, \ldots, v_k, v_{k+1})$ as $v_{k+1} \to \infty$ is the product of k independent single parameter gamma distributions with parameters v_i's, $i = 1, 2, \ldots, k$, we have

$$\lim_{v_{k+1} \to \infty} \phi(t_1, t_2, \ldots, t_k) = \prod_{j=1}^{k} (1 - it_j)^{-v_j}.$$

Also,

$$\lim_{v_{k+1} \to \infty} F_B^*(v_1, v_2, \ldots, v_k; \sum_{s=1}^{k+1} v_s; it_1, it_2, \ldots, it_k) = \prod_{j=1}^{k} (1 - it_j)^{-v_j}$$

since likewise the limiting distribution of the Dirichlet distribution $D(v_1 \ldots, v_k, v_{k+1})$ when $v_{k+1} \to \infty$ is the product of k independent single parameter gamma distributions with parameters v_i's, $i = 1, \ldots, k$ [Yassaee (1978)].

In the bivariate case, the characteristic function is [Lee (1971)]

$$
\begin{aligned}
& E[\exp(it_1 X_1 + it_2 X_2)] \\
&= \frac{\Gamma(v_1 + v_2 + v_3)}{\Gamma(v_1)\Gamma(v_2)\Gamma(v_3)} \int_{x_1 \geq 0, x_2 \geq 0, x_1 + x_2 \leq 1} \int e^{it_1 x_1 + it_2 x_2} \\
&\quad \times x_1^{v_1 - 1} x_2^{v_2 - 1} (1 - x_1 - x_2)^{v_3 - 1} dx_1 dx_2 \\
&= \Phi_2[v_1, v_2; v_1 + v_2 + v_3; it_1, it_2], \qquad\qquad (49.25)
\end{aligned}
$$

where the Humbert series Φ_2 is

$$\Phi_2[\beta, \beta'; \gamma; x, y] = \sum_{m=0}^{\infty} \sum_{n=0}^{\infty} \frac{(\beta)_m (\beta')_n}{(\gamma)_{m+n} m! n!} x^m y^n; \qquad\qquad (49.26)$$

see Humbert (1920–1921). The double series in (49.26) is convergent for all values of x and y independently of the parameters, except when γ equals a negative integer or zero.

4 EVALUATION OF PROBABILITY INTEGRALS OF DIRICHLET DISTRIBUTIONS

Let $\boldsymbol{X} = (X_1, X_2, \ldots, X_k)^T$ have a k-variate Dirichlet distribution $D(v_1, \ldots, v_k, v_{k+1})$ with the probability density function

$$p(x_1, x_2, \ldots, x_k) = \frac{\Gamma\left(\sum_{j=1}^{k+1} v_j\right)}{\prod_{j=1}^{k+1} \Gamma(v_j)} \left(\prod_{i=1}^{k} x_i^{v_i - 1}\right) \left(1 - \sum_{i=1}^{k} x_i\right)^{v_{k+1} - 1}$$

in the simplex $S_k = \{x_i : x_i \geq 0, \sum_{i=1}^{k} x_i \leq 1, i = 1, 2, \ldots, k\}$. Let R be the region $(X_i \leq a_i, i = 1, 2, \ldots, k)$. To compute $I = \Pr[\boldsymbol{X} \in R] = \int_0^{a_1} \cdots \int_0^{a_k} f(x_1, \ldots, x_k) \prod_{i=1}^{k} dx$, we define

$$b_i = \frac{a_i}{1 - \sum_{j=1}^{k} a_j},$$

and then use the identity [Yassaee (1979)]

$$\Pr[\boldsymbol{X} \in R] = \Pr[\boldsymbol{Y} \in R'],$$

where $\boldsymbol{Y} = (Y_1, \ldots, Y_k)^T$ is a k-variate inverted Dirichlet random variable with density function

$$p(y_1, y_2, \ldots, y_k) = \frac{\Gamma\left(\sum_{i=1}^{k+1} v_i\right)}{\prod_{i=1}^{k+1} \Gamma(v_i)} \left(\prod_{i=1}^{k} y_i^{v_i - 1}\right) \left(1 + \sum_{i=1}^{k} y_i\right)^{-\sum_{j=1}^{k+1} v_j}$$

for $y_i > 0$, and the region R' is defined as $(0 \leq Y_i \leq b_i, i = 1, 2, \ldots, k)$ and $b_i = \frac{a_i}{1 - \sum_{j=1}^{k} a_j}$, and finally use a program for the probability integral of *inverted* Dirichlet distribution given by Yassaee (1976).

Using the inequality

$$1 + \sum_{i=1}^{k} x_i \leq \prod_{i=1}^{k} (1 + x_i), \qquad 0 < x_i < 1,$$

it can be shown that the probability integral of the inverted Dirichlet distribution is greater than or equal to

$$\frac{\left[\Gamma\left(\sum_{j=2}^{k+1} v_j\right)\right]^k}{\left[\Gamma\left(\sum_{j=1}^{k+1} v_j\right)\right]^{k+1} \Gamma(v_{k+1})} \prod_{i=1}^{k} I_{b_i}\left(v_i, \sum_{\substack{j=1 \\ j \neq i}}^{k+1} v_j\right),$$

where $I_{b_i}(v_i, v)$ is the value of incomplete beta function ratio of the second kind given by

$$I_{b_i}(v_i, v) = \frac{1}{B(v_i, v)} \int_0^{b_i} \frac{t^{v_i - 1}}{(1 + t)^{v + v_i}} \, dt.$$

For the evaluation of

$$\Pr[X_1 \leq a_1, X_2 \leq a_2, \ldots, X_k \leq a_k] = P(a_1, a_2, \ldots, a_k; v_1, v_2, \ldots, v_{k+1}),$$

where

$$P(a_1, a_2, \ldots, a_k; v_1, v_2, \ldots, v_k; v_{k+1})$$
$$= \int_0^{a_1} \cdots \int_0^{a_k} f(t_1, t_2, \ldots, t_k) \prod_{i=1}^{k} dt_i, \qquad (49.27)$$

with

$$f(x_1, x_2, \ldots, x_k)$$
$$= \begin{cases} K \left(\prod_{i=1}^{k} x_i^{v_i-1} \right) \left(1 + \sum_{i=1}^{k} x_i \right)^{-\sum_{j=1}^{k+1} v_j}, & 0 < x_i < \infty \\ & (i = 1, \ldots, k) \\ 0, & \text{otherwise,} \end{cases}$$

and

$$K = \frac{\Gamma \left(\sum_{j=1}^{k+1} v_j \right)}{\prod_{j=1}^{k} \Gamma(v_j)}, \qquad (49.28)$$

Yassaee (1976) proposed a generalized Gaussian procedure to evaluate the integral in (49.27). Denoting

$$y_i = \frac{2x_i}{a_i} - 1, \qquad i = 1, 2, \ldots, k,$$

we write (49.27) in the form

$$P(a_1, a_2, \ldots, a_k; v_1, v_2, \ldots, v_k, v_{k+1})$$
$$= 2^{v_k+1} K \prod_{i=1}^{k} a_i^{v_i} \int_{-1}^{1} \cdots \int_{-1}^{1} \left\{ \prod_{i=1}^{k} (1 + y_i)^{v_i-1} \right\}$$
$$\times \left\{ 2 + \sum_{i=1}^{k} a_i + \sum_{i=1}^{k} a_i y_i \right\}^{-\sum_{i=1}^{k+1} v_i} \prod_{i=1}^{k} dy_i.$$

Then using the expansion, for a given n,

$$\int_{-1}^{1} x_1^{\alpha_1} dx_1 = \sum_{i=1}^{n} R_i u_i^{\alpha_1}, \qquad \alpha_1 = 1, 2, \ldots,$$

where the coefficients R_i and u_i are given in standard textbooks on numerical analysis, Yassaee extended this procedure to

$$I = \int_{-1}^{1} \cdots \int_{-1}^{1} \left[\prod_{i=1}^{k} x_i^{\alpha_i} \right] \prod_{i=1}^{k} dx_i = \prod_{i=1}^{k} \int_{-1}^{1} x_i^{\alpha_i} dx_i$$
$$= \sum_{i_1=1}^{n} \cdots \sum_{i_k=1}^{n} R_{i_1} \ldots R_{i_k} u_{i_1}^{\alpha_1} \cdots u_{i_k}^{\alpha_k}$$

and, using m points for n dimensional hypercube, selected m^n points on the hypercube.

Yassaee (1979) used four methods to calculate the probability integral of inverted Dirichlet distribution in the trivariate case. These methods are:

1. exact
2. Gaussian quadrature
3. asymptotic expansion
4. Taylor expansion.

Gaussian quadrature yields remarkably accurate results with the least amount of computing time. Evaluation of incomplete Dirichlet integrals as proposed by Sobel, Uppuluri, and Frankowski (1977, 1985) is discussed in Chapter 35 (Section 4) of Johnson, Kotz, and Balakrishnan (1997).

Lee (1971) has provided a diagonal expansion for the bivariate Dirichlet distribution, developing a bilinear summation formula in shifted (orthonormal) Jacobi polynomials.

5   CHARACTERIZATIONS

Suppose X_1 and X_2 are nonnegative variables such that $X_1 + X_2 \leq 1$ (i.e., random proportions), and suppose X_1 and X_2 are continuous. Then, X_i is said to be *neutral* if X_i and $X_j/(1 - X_i)$ (for $i \neq j$) are independent – that is, the size of X_i does not affect the proportion of X_j in $1 - X_i$, or X_j is (in a sense) "independent" of X_i except for the bound on $X_1 + X_2$. It is easy to verify that if $(X_1, X_2)^T$ jointly follow a Dirichlet distribution, then both X_1 and X_2 are neutral, and the converse is also true. Apart from Dirichlet distributions, there is a scarcity of suitable, easily handled multivariate distributions defined for continuous random *proportions* and, therefore, a scarcity of distributions for proportions that are *not* neutral.

Fabius (1973b) provided the following two characterizations of the Dirichlet distribution. These involve the concepts of $(CM)_i$-neutrality (or i-neutrality) and $(DR)_i$-neutrality (or complete neutrality), which are, in essence, independence concepts.

The first means that the fractions $\frac{X_j}{1-X_i}$ with $j \neq i$ are independent of X_i, while the second means that $\frac{X_i}{1-\sum_{j \neq i} X_j}$ is independent of the X_j with $j \neq i$. These properties arise naturally in statistical problems in biology, chemistry, and geology. Fabius (1973b), among others, defined $(CM)_i$-neutrality as follows: $\boldsymbol{X}$ is $(CM)_i$-neutral for a given $i \in \{1, \ldots, k\}$ iff, for any integers $r_j \geq 0$, $j \neq i$, there is a constant c, such that

$$E \left[\prod_{j \neq i} X_j^{r_j} | X_i \right] = c(1 - X_i)^{\sum_{j \neq i} r_j} \quad \text{a.s.} \, ,$$

they defined $(DR)_i$ (complete)-neutrality as follows: $\boldsymbol{X}$ is $(DR)_i$-neutral for a given $i \in \{1, \ldots, k\}$ iff, for any integer $r \geq 0$, there is a constant c such that

$$E[X_i^r | X_j, j \neq i] = c \left(1 - \sum_{j \neq i} X_j \right)^r \quad \text{a.s.}$$

The equivalence of these two neutralities under certain regularity conditions involving continuous densities was proved earlier by Darroch and Ratcliff (1971) using Connor and Mosimann's (1969) definitions.

Fabius's (1973a,b) characterization asserts that $(CM)_i$-neutrality *for all i* is equivalent to $(DR)_i$ neutrality *for all i*, and both are equivalent to the fact that $\boldsymbol{X} = (X_1, \ldots, X_k)^T$ has a Dirichlet distribution or is a limit of Dirichlet distributions. A crucial step in Fabius's (1973a,b) proof is to show that all mixed moments can be expressed in terms of marginal moments or, more precisely, for any i any mixed moment of $X_1, \ldots, X_i$ can be expressed in terms of marginal moments of these same random variables (which is, of course, a property of the Dirichlet distribution); see Section 1.

For readers' convenience, we summarize these concepts as originally introduced by Connor and Mosimann (1969) and their relation to the Dirichlet distribution. Let Q be the distribution of a random vector $\boldsymbol{X} = (X_1, \ldots, X_r)^T$ with $X_j \geq 0$ for all j and $\sum X_j = 1$.

Complete Neutrality

Intuitive Definition

The distribution Q and the random vector $\boldsymbol{X}$ are *completely neutral* if for every i, $X_j / (1 - \sum_{k=1}^{i} X_k)$ with $i < j \leq r$ are independent of $(X_1, \ldots, X_i)$. (Complete neutrality involves the order of components of a vector – it can be introduced or destroyed by a permutation of components.) Any Dirichlet distribution is completely neutral and remains so under *all* permutations of the components [see, for example, Wilks (1962)]. Connor and Mosimann (1969) quote an unpublished result of W. H. Kruskal, according to which, under certain regularity conditions, the Dirichlet distributions are the *only* distributions with this property.

i-Neutrality

Intuitive Definition

The distribution Q and the random vector $\boldsymbol{X}$ are i-*neutral* for a given i if $X_j/(1 - X_i)$ $(j \neq i)$ are independent of X_i. (Random vector $\boldsymbol{X}$ describes how a given quantity is divided into r fractions by means of some random mechanism.)

Formal Definition

The distribution A and the random vector $\boldsymbol{X}$ are i-*neutral* for a given i if

$$E\left[\prod_{j \neq i} X_j^{k_j} | X_i\right] = c(1 - X_i)^{\sum^*} \quad \text{a.s.},$$

where

$$\textstyle\sum^* = \sum_{j \neq i} k_j$$

for all nonnegative integers k_j, $j \neq i$, where c is a constant depending on k_j. (In particular, $E[X_j | X_i] = c_{ij}(1 - X_i)$ a.s.) This definition is identical to Fabius' (1973b) definition of $(CM)_i$-neutrality.

Characterization of i-Neutrality via Multinomial Distributions

If for any integer n, the random vector

$$\boldsymbol{Z}_n = (Z_{n1}, \ldots, Z_{nr})^T$$

has a multinomial distribution with parameters n and $\boldsymbol{p} = (p_1, \ldots, p_r)$, then Q is i-neutral iff for each n the posterior distribution of p_i based on prior distribution Q and $\boldsymbol{Z}_n$ depends on $\boldsymbol{Z}_n$ only through Z_{ni}. In this characterization, the distribution Q of $\boldsymbol{X}$ is viewed as a prior distribution for the unknown probability vector of a multinomial distribution. This property is widely used in Bayesian applications of Dirichlet distributions; see, for example, Lange (1995) mentioned below.

Fabius's (1973b) Characterization of the Dirichlet Distribution

Let $r \geq 3$ and $EX_i > 0$ for all i. Q and X are i-neutral for all i iff Q is a Dirichlet distribution or a limit of Dirichlet distributions. This characterization is a precise statement of Kruskal's result mentioned above.

Rao and Sinha (1988) characterized Dirichlet distributions within the class of Liouville-type distributions by the linearity of a certain regression (see the next chapter). More precisely, assume that $(X_1, \ldots, X_n)^T$ is distributed over a simplex $\mathcal{S}_n$, with continuous density function $\phi(x_1, \ldots, x_n)$. Assume that

(i) $\phi(x_1, \ldots, x_n) > 0$ on $\mathcal{S}_n = \{(x_1, \ldots, x_n) : x_i > 0, 1 \leq i \leq n; \sum x_i < 1\}$.

(ii) $\phi(x_1, \ldots, x_n)$ is of the product form

$$\phi(x_1, \ldots, x_n) = f_{n+1}\left(1 - \sum_{i=1}^{n} x_i\right) \prod_{i=1}^{n} f_i(x_i). \qquad (49.29)$$

(iii) At least one f_i in (49.29) is a homogeneous (or power) function.

(iv) For all $i = 1, \ldots, n$, the regression $E[X_i | \sum_{j \neq i} X_j = t]$ is a linear function of t.

Then $(X_1, \ldots, X_n)^T$ has a Dirichlet distribution. This result is an extension of an earlier result by Gupta and Richards (1987) in which the structure of the form

$$\phi(x_1, \ldots, x_n) = f\left(\sum_{i=1}^{n} x_i\right) \prod_{i=1}^{n} x_i^{a_i - 1}, \qquad a_i > 0$$

was assumed.

Gupta and Richards (1990) have shown that Rao and Sinha's result remains valid when hypothesis (iv) is replaced by (iv)′:

(iv)′ For all $i \neq j$, the regression $E[X_i^{k_i} | \sum_{j \neq i} X_j = t]$ is a polynomial of degree k_i in t for some sequence $k_1, \ldots, k_n$ of positive integers.

The same authors have also shown that within the class of random variables $X_1, \ldots, X_n$ having the joint density of form

$$\phi(x_1, \ldots, x_n) = g_n\left(\sum_{j=1}^{n} x_j\right) \prod_{i=1}^{n} f_i(x_i)$$

ranging over the orthant

$$R_+^n = \{(x_1, \ldots, x_n) : x_i > 0, \ i = 1, \ldots, n\},$$

and the variables $Y_1, \ldots, Y_n$ defined by the transformation

$$(X_1, \ldots, X_n) = \left(Y_1, \ldots, Y_{n-1}, 1 - \sum_{i=1}^{n-1} Y_i\right) Y_n,$$

$(Y_1, \ldots, Y_{n-1})^T$ and Y_n are mutually independent if and only if $f_1, \ldots, f_n$ are homogeneous. In that case, $(Y_1, \ldots, Y_{n-1})^T$ has a Dirichlet distribution.

An earlier result of James (1975) provided the following characterization of *bivariate* Dirichlet distribution:

> The conditional distributions of X_1 given X_2 and of X_2 given X_1 are both beta, and (at least) one of the distributions of X_1 or X_2 is beta if and only if X_1 and X_2 are jointly Dirichlet.

6   ESTIMATION

As noted on several occasions in this chapter, a random vector $X = (X_1, \ldots, X_{k-1})^T$ follows a Dirichlet distribution with parameter vector $\alpha = \{\alpha_1, \ldots, \alpha_k\}$ if the joint density function is

$$
\begin{aligned}
& p_{X_1, \ldots, X_k}(x_1, \ldots, x_k) \\
& = \frac{\Gamma\left(\sum_i \alpha_i\right)}{\prod_{j=1}^k \Gamma(\alpha_j)} \prod_{j=1}^{k-1} x_j^{\alpha_j - 1} \left(1 - \sum_{j=1}^{k-1} x_j\right)^{\alpha_k - 1}, \qquad x \in R^k,
\end{aligned}
$$

where $R^k = [x_j, \ j = 1, \ldots, k-1; \ x_j > 0, \ \sum_{j=1}^{k-1} x_j \le 1]$. This distribution is a multivariate extension of the two-parameter beta distribution.

The population moments are

$$
E[X_i] = \frac{\alpha_i}{\alpha}, \qquad E[X_i^2] = \frac{\alpha_i(\alpha_i + 1)}{\alpha(\alpha + 1)}, \tag{49.30}
$$

where $\alpha = \sum_{i=1}^k \alpha_i$. Since there are $k-1$ first order moments and $k-1$ second-order moments, there are a total of $\binom{2k-2}{k}$ possible combinations of equations to solve for the k parameters. Fielitz and Myers (1975) recommend to choose the $(k-1)$ first-order equations and the first second-order equation for solving these equations.

Denoting the sample moments

$$
M'_{1j} = \frac{1}{n} \sum_{i=1}^n X_{ij}, \qquad j = 1, \ldots, k-1
$$

(where X_{ij} is the ith observation on the jth component) and

$$
M'_{21} = \frac{1}{n} \sum_{i=1}^n X_{i1}^2
$$

and solving for α_i with sample analogs M'_{1j} and M'_{21} of (49.30), we obtain

$$\hat{\alpha}_i = \frac{(M'_{11} - M'_{21})M'_{1i}}{M'_{21} - (M'_{11})^2}, \qquad i = 1, \ldots, k-1,$$

and

$$\hat{\alpha}_k = \frac{(M'_{11} - M'_{21})(1 - \sum_{i=1}^{k-1} M'_{1i})}{M'_{21} - (M'_{11})^2},$$

which can serve as starting values in an iterative solution of the maximum likelihood equations.

The log-likelihood function can be written as

$$\log L = n \left\{ \log \Gamma \left(\sum_{j=1}^{k} \alpha_j \right) - \sum_{j=1}^{k} \log \Gamma(\alpha_j) \right\} + n \sum_{j=1}^{k} (\alpha_j - 1) \log G_j,$$

where

$$G_j = \left[\prod_{i=1}^{n} X_{ij} \right]^{1/n}, \ j = 1, 2, \ldots, k-1, \ \text{and} \ \ G_k = \left[\prod_{i=1}^{n} \left(\sum_{j=1}^{k-1} X_{ij} \right) \right]^{1/n}$$

$$(49.31)$$

are geometric means of the observed values of the variables $X_1, \ldots, X_{k-1}$ and $1 - \sum_{j=1}^{k-1} X_j$. Taking the derivatives of the log-likelihood function, the likelihood equations become

$$\frac{\partial \log L}{\partial \alpha_j} = n \Psi \left(\sum_{m=1}^{k} \alpha_m \right) - n\Psi(\alpha_j) + n \log G_j, \qquad j = 1, \ldots, k,$$

where $\Psi(\cdot)$ is the digamma function [see Eq. (1.37) in Chapter 1 of Johnson, Kotz, and Kemp (1992)]. The second partial and mixed partial derivatives are given by

$$\frac{\partial^2 \log L}{\partial \alpha_j^2} = n \Psi' \left(\sum_{m=1}^{k} \alpha_m \right) - n\Psi'(\alpha_j), \qquad j = 1, \ldots, k,$$

$$\frac{\partial^2 \log L}{\partial \alpha_i \partial \alpha_j} = n \Psi' \left(\sum_{m=1}^{k} \alpha_m \right).$$

The information matrix $(\boldsymbol{I})$ is

$$\boldsymbol{I} = \{I_{ij}\} = -E \left[\frac{\partial^2 \log L}{\partial \alpha_i \partial \alpha_j} \right],$$

where

$$I_{ij} = -n\Psi'\left(\sum_{m=1}^{k} \alpha_m\right), \quad i \neq j$$

and

$$I_{ii} = n\Psi'(\alpha_i) - n\Psi'\left(\sum_{m=1}^{k} \alpha_m\right).$$

Denote

$$D = \text{diag}[n\Psi'(\alpha_1), \ldots, n\Psi'(\alpha_k)],$$
$$G = -n\Psi'\left(\sum_{m=1}^{k} \alpha_m\right),$$

where $\Psi'(\cdot)$ is the trigamma function [see Eq. (1.38) in Chapter 1 of Johnson, Kotz, and Kemp (1992)]. Then

$$\boldsymbol{I} = D + G\boldsymbol{1}\boldsymbol{1}'$$

and

$$\boldsymbol{V} = \boldsymbol{I}^{-1} = D^* + \beta \boldsymbol{a}^* \boldsymbol{a}^{*'}, \tag{49.32}$$

where

$$D^* = \text{diag}\left[\frac{1}{n\Psi'(\alpha_1)}, \ldots, \frac{1}{n\Psi'(\alpha_k)}\right],$$
$$\boldsymbol{a}^{*'} = \left[\frac{1}{\Psi'(\alpha_1)}, \ldots, \frac{1}{\Psi'(\alpha_k)}\right],$$
$$\beta = n\Psi'\left(\sum_{j=1}^{k} \alpha_j\right)\left[1 - \Psi'\left(\sum_{j=1}^{k} \alpha_j\right)\sum_{j=1}^{k} \frac{1}{\Psi'(\alpha_j)}\right]^{-1},$$

and $\Psi'(\cdot)$ is the trigamma function.

To maximize numerically the likelihood function (49.31), Narayanan (1991a) used a Newton–Raphson procedure. For the initial estimates, the moment estimates above ($\hat{\alpha}_i$'s) were used. Ronning (1989) suggested setting all $\alpha_j = \min\{X_{ij}\}$, $i = 1, \ldots, n$, for the initial estimates (preventing the α_j's from becoming negative in the course of iterations.)

Equation (49.32) for calculating $\boldsymbol{V}$ (the variance–covariance matrix) does not require inversion of the information matrix $\boldsymbol{I}$ at each stage of the Newton–Raphson algorithm.

Algorithms for calculating the digamma function $\Psi(\cdot)$ and the trigamma function $\Psi'(\cdot)$ are now widely available; see, for example, Bernardo (1976) and Schneider (1978).

Fisher's scoring method [see, for example, Rao (1952)] can be used for iterations:

$$
\begin{bmatrix} \hat{\alpha}_1 \\ \vdots \\ \hat{\alpha}_k \end{bmatrix}_{(i)}
$$

$$
= \begin{bmatrix} \hat{\alpha}_1 \\ \vdots \\ \hat{\alpha}_k \end{bmatrix}_{(i-1)} + \begin{bmatrix} \text{var}(\hat{\alpha}_1) & \cdots & \text{cov}(\hat{\alpha}_1, \hat{\alpha}_k) \\ \vdots & \ddots & \vdots \\ \text{cov}(\hat{\alpha}_k, \hat{\alpha}_1) & \cdots & \text{var}(\hat{\alpha}_k) \end{bmatrix}_{(i-1)} \begin{bmatrix} g_1(\hat{\boldsymbol{\alpha}}) \\ \vdots \\ g_k(\hat{\boldsymbol{\alpha}}) \end{bmatrix}_{(i-1)},
$$

where $\hat{\boldsymbol{\alpha}}_{[0]} = [\hat{\alpha}_{1(0)}, \ldots, \hat{\alpha}_{k(0)}]^T$ are the initial estimates.

A test of convergence can be carried out by means of the test statistic

$$
S = [g_1(\hat{\boldsymbol{\alpha}}), \ldots, g_k(\hat{\boldsymbol{\alpha}})] \begin{bmatrix} \text{var}(\hat{\alpha}_1) & \cdots & \text{cov}(\hat{\alpha}_1, \hat{\alpha}_k) \\ \vdots & \ddots & \vdots \\ \text{cov}(\hat{\alpha}_k, \hat{\alpha}_1) & \cdots & \text{var}(\hat{\alpha}_k) \end{bmatrix} \begin{bmatrix} g_1(\hat{\boldsymbol{\alpha}}) \\ \vdots \\ g_k(\hat{\boldsymbol{\alpha}}) \end{bmatrix}.
$$

which is distributed asymptotically as a χ^2 random variable with k degrees of freedom [see Serfling (1980)]. This is a large sample test, but seems to work well in samples of moderate sizes as well.

Narayanan (1992) illustrated the above-described procedure by means of two numerical examples: One deals with brand choice for regular ground coffee (5 brands), and the other is an elaboration of Mosimann's (1962) classical data of frequency of occurrence of four types of grains falling under different levels of core.

A Fortran program for Narayanan's algorithm is presented in Narayanan (1991b).

7 APPLICATIONS

This distribution has wide-ranging applications:

1. Approximating multinomial distribution using a joint Dirichlet density function; see Johnson (1960). This is discussed in some detail in Johnson, Kotz, and Balakrishnan (1997, Chapter 35, Section 5).

2. Spurious correlations or correlations among proportions studied, by Mosimann (1962) among others, in relation to various types of pollen, grains and types of vegetation in general (see also a more recent work by Narayanan (1992) for an illuminating numerical example mentioned above).

3. Modeling the activity times in a PERT (Program Evaluation and Review Technique) network. A PERT network involves a collection of activities and each activity is often modelled as a random variable following a beta distribution [see, for example, Monhor (1987)].

4. Modeling the heterogeneous buyer behavior in the context of multi-store purchasing in a city. Applying the model to a spatially disaggregate consumer-panel survey conducted in Cardiff, Wrigley, and Dunn (1984) displayed that the Dirichlet model provides a good fit and hence is useful for studies of urban consumer behavior.

5. Inverted Dirichlet distribution is required for calculation of confidence regions for variance ratios of random models for balanced data [see, for example, Sahai and Anderson (1973)]. Specifically, let the model

$$Y = \alpha \boldsymbol{\mu} + \sum \boldsymbol{X}_i \boldsymbol{\beta}_i + \boldsymbol{\varepsilon},$$

where α is a scalar, $\boldsymbol{\mu}$ is an $n \times 1$ vector of ones, $\boldsymbol{X}_i$ is an $n \times m_i$ matrix, $\boldsymbol{\beta}_i$ is an $m_i \times 1$ vector of independent variables from $N(0, \sigma_i^2)$, and $\boldsymbol{\varepsilon}$ is an $n \times 1$ vector of independent variables with common $N(0, \sigma_0^2)$, and $\boldsymbol{\beta}_i$, $i = 1, 2, \ldots, k$ and $\boldsymbol{\varepsilon}$ mutually independent, be given. Let S_i^2 and v_i be the mean squares and corresponding degrees of freedom associated with random effects $\boldsymbol{\beta}_i$, $i = 1, 2, \ldots, k$ and let S_0^2 and v_0 be the mean square and degrees of freedom associated with the residual variance.

To find a confidence region for variance ratios $\rho_i = \frac{\sigma_i^2}{\sigma_0^2}$, $i = 1, \ldots, r$, one needs to compute

$$P \left[\frac{S_0^2 \sigma_i^2}{S_i^2 \sigma_0^2} \leq F(\alpha_i, v_0, v_i); \ i = 1, 2, \ldots, k \right]$$

$$= \int_{a_1}^{\infty} \cdots \int_{a_k}^{\infty} p(x_1, x_2, \ldots, x_k) \prod dx_i, \qquad (49.33)$$

where $F(\alpha_i, v_0, v_i)$ is the upper $100\alpha_i$ percentage point of the F-distribution with (v_0, v_i) degrees of freedom, α_i's satisfy $1 - \alpha = \prod_{i=1}^{k}(1 - \alpha_i)$, $a_i = v_i / \{v_0 F(\alpha_i, v_0, v_i)\}$, and $p(x_1, x_2, \ldots, x_k)$ is the density of an inverted Dirichlet distribution, as in (49.18).

A similar application is presented in Hurlburt and Spiegel (1976) in connection with the model

$$Y = \mu + \alpha_i + \beta_j + \varepsilon_{ij},$$

for testing $\beta_j = 0$ on the condition that $\alpha_i = 0$, and the model

$$Y = \mu + \alpha_i + \beta_j + \gamma_k + \varepsilon_{ijk},$$

for testing $\gamma_k = 0$ on condition that $\alpha_i = 0$ and $\beta_j = 0$.

6. Dirichlet distribution appears prominently in the distributional aspects of the slippage tests for the χ^2 distribution. Specifically, given a random sample of size n, $x_1, \ldots, x_n$, we test

$$H_0: \text{ all } x_i \sim \sigma^2 \chi_m^2$$

(where m is known and σ is unknown) versus the alternative of slippage to the right:

$$
\begin{aligned}
H_1 : x_1 &\sim \lambda^2 \sigma^2 \chi_m^2 \\
x_j &\sim \sigma^2 \chi_m^2, \ j \neq 1
\end{aligned}
$$

where $\lambda > 1$ is some unknown constant. Cochran's (1941) test statistic

$$S = \max_j \frac{x_j}{\sum_{k=1}^{n} x_k}$$

has a distribution closely related to a Dirichlet distribution [see Hawkins (1972)].

Indeed, denoting $z = \sum_{k=1}^{n} x_k$ and $s_i = x_i / z$ $(i = 1, 2, \ldots, n)$, the joint density function of $(s_1, \ldots, s_n)^T$ is

$$p(s_1, \ldots, s_n) = \frac{\Gamma\left(\frac{mn}{2}\right)}{\left\{\Gamma\left(\frac{m}{2}\right)\right\}^n} \prod_{i=1}^{n} s_i^{\frac{m}{2} - 1}, \quad s_i \geq 0, \ \sum_{i=1}^{n} s_i = 1,$$

which is independent of z.

7. One of the most prominent applications of Dirichlet distribution is in a model of buying behavior popularized by Goodhardt, Ehrenberg, and Chatfield; see, for example, Chatfield and Goodhardt (1975) and Goodhardt, Ehrenberg, and Chatfield (1984). The Dirichlet model specifies probabilistically how many purchases each customer makes in a time period and which brand is bought on each occasion. It combines both purchase incidence and brand-choice aspects of buyer behavior into one model.

Suppose one particular consumer has a particular probability p_x of choosing Brand X each time she makes a purchase in the product field. Then the probability that she chooses some brand other than X is $1 - p_x$.

To explain how the population of consumers buy Brand X, we need to know how many consumers have probabilities close to the particular values p_x and $1 - p_x$ of buying X and all other brands, respectively, and how many have quite different probabilities. This is described for each brand by a beta-distribution that gives the proportion of the populations with probability of choosing Brand X close to p_x as

$$\text{proportion} = C' p_x^{\alpha_x - 1} (1 - p_x)^{S - \alpha_x - 1},$$

where S is a parameter of the product field, α_x is S times the market share of Brand X, and C' is a constant independent of the particular value of p_x.

If now, instead of considering just Brand X and all other brands together, we examine the buyers of each of the separate brands X, Y, Z, W, and so on, in the market, we have to concern ourselves with the probabilities of choosing each of these brands, which for a particular consumer may be a set of values p_x, p_y, p_z, p_w, and so on. Again we need to know the proportion of consumers who have similar probabilities. This proportion is given by the Dirichlet distribution as

$$\text{proportion} = C \, p_x^{\alpha_x - 1} \, p_y^{\alpha_y - 1} \, p_z^{\alpha_z - 1} \, p_w^{\alpha_w - 1} \, \cdots \, ,$$

where the α's are again S times the market shares of the brands, and C is a constant multiplier that does not depend on the p's.

The brand-choice part of the Dirichlet model can therefore be fully specified by using only the brands' shares within the market together with the single product-field parameter S. The latter is a measure of the overall amount of switching or multibrand buying in the product field. No characteristics of the individual brands other than their market shares are required.

A special feature of the model is that it is not materially affected by the way any particular brand is defined, or even by the definition of the market or product-class as a whole.

Although in many practical applications the assumptions leading to the Dirichlet model are not quite true, empirically the model works pretty well [Goodhardt, Ehrenberg, and Chatfield (1984)]. While discussing that paper, Kemp and Kemp have suggested alternative formulations that would lead to the same multivariate (Dirichlet) distribution in a single time period.

8. Sobel and Uppuluri (1974) utilized a Dirichlet distribution for the distribution of *sparse* and *crowded* cells, closely related to occupancy models. A multinomial distribution with k cells is given with b cells ($1 \leq b \leq k$)

having common cell probability p $(0 < p \le 1/b)$; these are called *blue cells*. (Dual concepts of sparseness and crowdedness are introduced for these b blue cells, based on a fixed number n of observations.) A Dirichlet distribution is used to evaluate the cumulative distribution functions, the moments, the joint probability law, the joint moments of the number S of *sparse* blue cells and the number C of *crowded* blue cells. Let $\min(j, n) \ge v$ (v an integer) denote the event that the minimum frequency (based on n observations) in a specified set of j blue cells is at least v; then

$$
\begin{aligned}
\Pr[\min(j, n) \ge v | p] &= I_p^{(j)}(v, n) \\
&= \frac{\Gamma(n+1)}{(\Gamma(v))^j \Gamma(n+1-jv)} \\
&\quad \times \int_0^p \cdots \int_0^p \left(1 - \sum_{\alpha=1}^j x_\alpha\right)^{n-jv} \prod_{\alpha=1}^j x_\alpha^{v-1} \, dx_\alpha,
\end{aligned}
$$

(49.34)

where $0 \le p \le 1/b \le 1/j$, since $j \le b$. (For $j = 1$, $I_p^{(1)}(v, n) = I_p(v, n - v + 1)$ an incomplete Beta function ratio.)

As a generalization of (49.34), Sobel and Uppuluri (1974) defined the *I*-function

$$
\begin{aligned}
I_p^{(\alpha+\beta)} &\left((t)_\alpha, (v)_\beta, n\right) \\
&= \frac{\Gamma(n+1)}{\Gamma^\alpha(t)\Gamma^\beta(v)\Gamma(n+1-\alpha t - \beta v)} \\
&\quad \times \int_0^p \cdots \int_0^p \left(1 - \sum_{i=1}^{\alpha+\beta} x_i\right)^{n-\alpha t - \beta v} \prod_{i=1}^\alpha x_i^{t-1} dx_i \prod_{j=1}^\beta x_{\alpha+j}^{v-1} \, dx_{\alpha+j},
\end{aligned}
$$

where $(t)_\alpha$ and $(v)_\beta$ stand for $t, \ldots, t$ repeated α times and $v, \ldots, v$ repeated β times, respectively. This represents the probability that a specified set of α blue cells each have frequency $\ge t$ and another disjoint specified set of β blue cells each have frequency $\ge v$, when there are n observations in all, and all the blue cells have common probability p, with $p \le 1/(\alpha + \beta)$ [see also Olkin and Sobel (1965)].

9. Lange (1995) applied the Dirichlet distribution to forensic match probabilities. Utilizing the fact that the Dirichlet distribution is a conjugate prior for Bayesian analysis involving multinomial properties, he recommended to use it in computing match probabilities, which would take into account the presence of genetic heterogeneity in the population. The Dirichlet distribution is also relevant to the related problem of allele frequency estimation.

8   GENERALIZATIONS

8.1   Generalized Dirichlet Distribution as a Prior for Multinomial Parameters

A generalization of the Dirichlet distribution, given by Lochner (1975), is as follows:

Let $F(x)$ be the cumulative distribution function for some population of life times. For real numbers $t_1 < t_2 < \cdots < t_k$, where $F(t_1) > 0$ and $F(t_k) < 1$, let $p_1 = F(t_1)$ and $p_i = F(t_i) - F(t_{i-1})$ for $i = 2, 3, \ldots, k$. Lochner's generalized Dirichlet distribution of $\boldsymbol{p} = (p_1, \ldots, p_k)^T$ has a density function $f(\boldsymbol{p}) \propto \prod_{i=1}^{k} p_i^{\alpha_i - 1} (1 - p_1 - \cdots - p_i)^{\gamma_i}$. Specifically, the proportionality coefficient is $B(\alpha_i, \beta_i)^{-1}$ and

$$
\gamma_i = \begin{cases} \beta_i - \alpha_{i+1} - \beta_{i+1}, & i = 1, 2, \ldots, k-1, \\ \beta_k - 1, & i = k. \end{cases}
$$

The initial motivation stemmed from Bayesian life testing.

Many authors have assigned a Dirichlet prior distribution to the parameter vector of a multinomial distribution [e.g., Novick and Grizzle (1965), Altham (1969), and Lochner and Basu (1972)]. In application, the probability vector

$$
\boldsymbol{p} = (p_1, \ldots, p_k)^T
$$

has a multinomial distribution in the following situation. Suppose we wish to make an inference concerning $F(x)$ based on a random sample from a parent population. Let p_i be as above. The posterior density of $\boldsymbol{p}$ allows us to make inference about $\boldsymbol{p}$ and hence about $F(x)$. Given a random sample from a population with cumulative distribution function $F(x)$, let y_i denote the number of sample observations having values between t_{i-1} and t_i, hence $\boldsymbol{y} = (y_1, \ldots, y_{k+1})^T$ is a multinomial random variable ($y_1 =$ number of sample observations less than t_1 and $y_{k+1} =$ number of sample observations greater than t_k). [Lindley (1971) has objected to using a Dirichlet prior, since it does not take the relative position of intervals into consideration (the relationship between p_i and p_j is not a function of how close i and j are to each other).]

The density

$$
f(\boldsymbol{p}) \propto \prod_{i=1}^{k} p_i^{\alpha_i - 1} (1 - p_1 - \cdots - p_i)^{\gamma_i}
$$

over $\mathcal{P} = \{\boldsymbol{p} : 0 < p_i, \; i = 1, 2, \ldots, k, \text{ and } p_1 + \cdots + p_k \leq 1\}$ has properties that seem to overcome these objections. Indeed, rather than obtaining a prior for $\boldsymbol{p}$ directly, we define a prior on $\Delta_i = p_i - p_{i-1}$. Lochner (1975) has shown that the appropriate density on Δ_i which takes the distances between p_i and p_{i-1} into account is related to the beta density of the first type; specifically, if

$$B_i = \left\{1 - \sum_{j=1}^{i-1}(i-j)\Delta_j\right\}^{-1}$$

and

$$A_i = (\Delta_1 + \cdots + \Delta_{i-1})B_i$$

and

$$w_i = A_i + B_i\Delta_i, \qquad \text{for } i = 2, \ldots, k,$$
$$w_1 = \Delta_1,$$

then it is reasonable to assign the density $f(w_i) = \{B(\alpha_i, \beta_i)\}^{-1}w_i^{\alpha_i-1}$ $(1 - w_i)^{\beta_i-1}$ for $0 < w_i < 1$ $(i = 1, \ldots, k)$. Now

$$f(w_1, \ldots, w_k) = \prod_{i=1}^{k}\{B(\alpha_i, \beta_i)\}^{-1}w_i^{\alpha_i-1}(1 - w_i)^{\beta_i-1}$$

implies that

$$f(\boldsymbol{p}) = \prod_{i=1}^{k}\{B(\alpha_i, \beta_i)\}^{-1}p_i^{\alpha_i-1}(1 - p_1 - \cdots - p_i)^{\gamma_i},$$

where $\gamma_i = \beta_i - \alpha_{i+1} - \beta_{i+1}$ for $i = 1, 2, \ldots, k-1$ and $\gamma_k = \beta_k - 1$. This is a *generalized* Dirichlet density. If we set $\gamma_1 = \gamma_2 = \cdots = \gamma_{k-1} = 0$, we obtain a Dirichlet density. The moments of p_i $(i = 1, 2, \ldots, k)$ are

$$E[p_i^m] = \left[\prod_{j=1}^{i-1}E[(1 - w_j)^m]\right]E[w_i^m]$$

$$= \left\{\prod_{j=1}^{i-1}\frac{(\beta_j)_m}{(\alpha_j + \beta_j)_m}\right\}\frac{(\alpha_i)_m}{(\alpha_i + \beta_i)_m}, \qquad (49.35)$$

where, as above, $(a)_m = a(a+1)\cdots(a+m-1)$. Hence,

$$E[p_i] = \left(\prod_{j=1}^{i-1}\frac{\beta_j}{\alpha_j + \beta_j}\right)\left(\frac{\alpha_i}{\alpha_i + \beta_i}\right) \qquad (49.36)$$

and

$$
\begin{aligned}
\operatorname{var}(p_i) = & \left(\prod_{j=1}^{i-1} \frac{\beta_j}{\alpha_j + \beta_j}\right)\left(\frac{\alpha_i}{\alpha_i + \beta_i}\right)\left\{\left(\prod_{j=1}^{i-1} \frac{\beta_j + 1}{\alpha_j + \beta_j + 1}\right)\right. \\
& \times \left(\frac{\alpha_i + 1}{\alpha_i + \beta_i + 1}\right) - \left.\left(\prod_{j=1}^{i-1} \frac{\beta_j}{\alpha_j + \beta_j}\right)\left(\frac{\alpha_i}{\alpha_i + \beta_i}\right)\right\}.
\end{aligned}
$$

$$(49.37)$$

Lochner (1975) suggested that, to reduce the number of parameters involved, it might be assumed that

$$E[p_i] = 1/(k+1)$$

holds for $i = 1, 2, \ldots, k$. In that case,

(i) $\operatorname{cov}(p_r, p_s) = \frac{1}{k+1}\left\{\left(\prod_{j=1}^{r-1} \frac{\beta_j + 1}{\alpha_j + \beta_j + 1}\right) \frac{\alpha_r}{\alpha_r + \beta_r + 1} - \frac{1}{k+1}\right\}$

(ii) $\operatorname{cov}(p_r, p_s) \gtrless \operatorname{cov}(p_{r+1}, p_s)$ according as $a_r \gtrless a_{r+1}$.

Note that $\operatorname{cov}(p_r, p_s)$ is independent of s; but for the ordinary Dirichlet, $\operatorname{cov}(p_r, p_s)$ is independent of both r and s.

The most interesting properties of this generalized Dirichlet distribution concern the posterior means. As before, let $\boldsymbol{y} = (y_1, \ldots, y_{k+1})^T$, where y_i is the number of sample observations from the parent population that failed in the time interval (t_{i-1}, t_i) and $t_{k+1} = \infty$. Then $E[p_i|\boldsymbol{y}]$ is determined once time t_i has elapsed and does not get affected by what happens after time t_i.

8.2   Antelman's Generalization

The restrictive nature of the Dirichlet distribution in relation to interrelated Bernoulli processes concerned Antelman (1972), who developed the trivariate generalization with the density kernel

$$
\left[\prod_{i=1}^{3} x_i^{a_i}\right](1 - x_1 - x_2 - x_3)^b (x_1 + x_2)^{c_1}(1 - x_1 - x_2)^{c_2}
$$
$$
\times (x_1 + x_3)^{c_3}(1 - x_1 - x_3)^{c_4}.
$$

Unfortunately, except in some special cases, the constant of integration for this distribution is intractable for practical purposes, but Antelman provided some simpler approximations.

8.3   Johnson and Kotz's Generalization

In their attempt to generalize the univariate symmetric Tukey's λ distribution [see Eq. (12.75) in Chapter 12 of Johnson, Kotz, and Balakrishnan (1994)] to the multivariate case, Johnson and Kotz (1973) have proposed the joint distribution of

$$Y_i = \lambda_i^{-1}\left[T_i^{\lambda_i} - (1 - T_i)^{\lambda_i}\right], \quad i = 1, \ldots, n,$$

where $T_1, \ldots, T_n$ are jointly distributed as Dirichlet. The distribution is found to be almost intractable, but some properties can be obtained from the conditional distributions, utilizing the fact that the T's have beta conditional distributions. Their form of analysis may thus be extended to cover all distributions for $T_1, \ldots, T_n$ which have beta conditional distributions (although it is also desirable that each T_i has a beta distribution).

8.4   Delta-Dirichlet Distributions

Lewy (1996) extended Dirichlet distributions in the following manner. Let $H_1, H_2, \ldots, H_m$ be a set of stochastic variables satisfying $0 < H_j < 1$, $j = 1, \ldots, m$ and $\sum_{j=1}^m H_j = 1$, such that $\Pr[H_j = 1 | \bigcup_{i=1}^m (H_i = 1)] = d'_j$, and

$$\Pr[I_1 h_1 < H_1 < I_1(h_1 + dh_1), \ldots, I_m h_m < H_m < I_m(h_m + dh_m)]$$
$$= d(I_1, \ldots, I_m) \frac{\Gamma(\sum I_j p_j)}{\Gamma(I_1 p_1) \times \cdots \times \Gamma(I_m p_m)} \, h_1^{I_1(p_1 - 1)}$$
$$\times \cdots \times h_m^{I_m(p_m - 1)} (dh_1)^{I_1} \cdots (dh_m)^{I_m} \qquad \text{if } H_j < 1 \text{ for all } j,$$

$$\tag{49.38}$$

where

$$I_j = \begin{cases} 0 & \text{if } H_j = 0, \\ 1 & \text{if } 0 < H_j \leq 1, \end{cases}$$

and d'_j and $d(I_1, \ldots, I_m)$ are parameters for which

$$\sum_j d'_j + \sum_{(I_1, \ldots, I_m) \in [0,1]^m} d(I_1, \ldots, I_m) = 1.$$

The event $(I_1 = 0, \ldots, I_{j-1} = 0, I_j = 1, I_{j+1} = 0, \ldots, I_m = 0)$ is not defined, implying that the parameter $d(0, \ldots, 0, I_j = 1, 0, \ldots, 0)$ is not defined. As a matter of convenience, such parameters $d(0, \ldots, 0, 1, 0, \ldots, 0)$, characterized by $m - 1$ zeros and a single one, are defined to be equal to d'_j:

$$d(0, \ldots, 0, 1, 0, \ldots, 0) = d'_j.$$

According to the latter definition, the parameters, d, including both types of events $(I_1, I_2, \ldots, I_m)$ and $(0, \ldots, 0, 1, 0, \ldots, 0)$, will be used in the following to describe the singularities at both 0 and 1. Lewy (1996) called distribution (49.38) a *delta-Dirichlet distribution*. It is implicitly assumed here that the events that one or more of the individual variables H_j are equal to zero are independent of the other variables when these are positive and strictly less than 1.

Motivation for this distribution was from the Danish industrial fishery when H_j, representing proportions of species in a sample, may be singular at 0 and 1; that is, the events that some H_j are 0 or that one variable is 1 have positive probability. The development of delta-Dirichlet distributions originated in sampling problems related to the estimation of the species composition of the biomass within the Danish industrial fishery and with evaluation of the accuracy of the estimates. The species composition of biomass is needed in order to monitor catch quota limitations by fish species. This is a different objective than estimating species composition by numbers, which is done for tracking cohorts and estimating mortality.

The parameters d'_j denote the probabilities that a sample consists of one species only while $d(I_1, \ldots, I_m) = \Pr(I_1, \ldots, I_m)$ denotes the probability of a set $(I_1, I_2, \ldots, I_m)$, given that at least two species are included in the sample. H_j and p_j are restricted as in the Dirichlet distribution:

$$f(h_1, \ldots, h_{m-1}|p_1, \ldots, p_m)$$
$$= \frac{\Gamma(p.)}{\Gamma(p_1) \cdots \Gamma(p_m)} \prod_{i=1}^{m-1} h_i^{p_i-1}(1 - h_1 - \cdots - h_{m-1})^{p_m-1}, \quad (49.39)$$

$0 < h_j < 1$, $\sum_{j=1}^{m} h_j = 1$, $p_j > 0$ and $p. = \sum_{j=1}^{m} p_j$.

Delta-Dirichlet distributions may be considered as varying mixtures of multinomial distribution and Dirichlet distribution, as these distributions correspond to the delta-Dirichlet distribution with $d(1, 0, \ldots, 0) + \cdots + d(0, 0, \ldots, 1) = 1$ and $d(1, 1, \ldots, 1) = 1$, respectively. Let

$d0_j$ denote the parameter $d(1, \ldots, 1, I_j = 0, 1, \ldots, 1)$,
$d0_{jk}$ denote the parameter $d(1, \ldots, 1, I_j = 0, 1, \ldots, 1, I_k = 0, 1, \ldots, 1)$,
$d0_{jkl}$ denote the parameter $d(1, \ldots, 1, I_j = 0, 1, \ldots, 1, I_k = 0, 1, \ldots, 1, I_l = 0, 1, \ldots, 1)$,
$d1_j$ denote the parameter $d(0, \ldots, 0, I_j = 1, 0, \ldots, 0) \ (= d'_j)$,
$d1_{ji}$ denote the parameter $d(0, \ldots, 0, I_j = 1, 0, \ldots, 0, I_i = 1, 0, \ldots, 0)$,
$d1_{jik}$ denote the parameter $d(0, \ldots, 0, I_j = 1, 0, \ldots, 0, I_i = 1, 0, \ldots, 0, I_k = 1, 0, \ldots, 0)$,
$A1_j$ denote the set $\{i|1 \le i \le m \wedge i \ne j\}$,
$A2_j$ denote $\{i, k|1 \le i, k \le m \wedge i \ne j \wedge k \ne j \wedge i \ne k\}$,

$A3_j$ denote $\{i, k, l | 1 \leq i, k, l \leq m \wedge i \neq j \wedge k \neq j \wedge l \neq j \wedge i \neq k \neq l\}$,
$B1_{jk}$ denote the set $\{i | 1 \leq i \leq m \wedge i \neq (j \wedge k)\}$,
$B2_{jk}$ denote $\{i, l | 1 \leq i, l \leq m \wedge i \neq (j \wedge k) \wedge l \neq (j \wedge k)\}$,
$B3_{jk}$ denote $\{i, l, h | 1 \leq i, l, h \leq m \wedge i \neq (j \wedge k) \wedge l \neq (j \wedge k) \wedge h \neq (j \wedge k)\}$.

In general, $d0_{i,k,l,\ldots,z}$ denotes the probability that all species except for the species $i, k, l, \ldots,$ and z are included in the sample. Correspondingly, $d1_{i,k,l,\ldots,z}$ denotes the probability that the species $i, k, l, \ldots,$ and z are the only species included in the sample. The number of parameters $d(I_1, I_2, \ldots, I_m)$ is $2^m - 1$ ($d(0, 0, \ldots, 0)$ is not defined). The marginal distributions of one and two variables have rather complicated expressions. The distribution of the singular component is

$$
\begin{aligned}
\Pr[H_j = 1] &= d1_j \\
\Pr[H_j = 0] &= \sum_{\substack{(I_1, \ldots, I_{j-1}, I_{j+1}, \ldots, I_m) \in (0,1)^{m-1} \\ I_j = 0}} d(I_1, \ldots, I_m).
\end{aligned}
$$

With this notation, it is tedious but straightforward to show that

$$
\begin{aligned}
\Pr[h_j &< H_j < h_j + dh_j] \\
&= d(1, \ldots, 1) f(h_j | p_j, p_. - p_j) + \sum_{i \in A1_j} d0_i f(h_j | p_j, p_. - p_j - p_i) \\
&\quad + \sum_{i,k \in A2_j} d0_{ik} f(h_j | p_j, p_. - p_j - p_i - p_k) \\
&\quad + \cdots + \sum_{i \in A1_j} d1_{ji} f(h_j | p_j, p_i), \qquad (0 < h_j < 1),
\end{aligned}
$$

where f denotes the density of the Dirichlet distribution $f(h_1, \ldots, h_{m-1} | p_1, \ldots, p_m)$. Moreover,

$$
\begin{aligned}
E[H_j] &= d1_j + p_j \left[\frac{d(1, \ldots, 1)}{p_.} + \sum_{i \in A1_j} \frac{d0_i}{p_. - p_i} \right. \\
&\quad + \sum_{i,k \in A2_j} \frac{d0_{ik}}{p_. - p_i - p_k} + \sum_{i,k,l \in A3_j} \frac{d0_{ikl}}{p_. - p_i - p_k - p_l} \\
&\quad \left. + \cdots + \sum_{i,k \in A2_j} \frac{d1_{jik}}{p_j + p_i + p_k} + \sum_{i \in A1_j} \frac{d1_{ji}}{p_j + p_i} \right], \qquad (49.40)
\end{aligned}
$$

and analogous but more complicated expressions are straightforwardly derived for $E[H_j^2]$ and $E[H_j H_k]$. See Lewy (1996) for details.

The maximum likelihood estimator for $d(I_1, \ldots, I_m)$ is

$$
\hat{d}(I_1, \ldots, I_m) = \frac{n(I_1, \ldots, I_m)}{n},
$$

where n is the number of stochastically independent observations, $h_i = (h_{i,1}, \ldots, h_{i,m})$, $i = 1, 2, \ldots, n$, identically delta-Dirichlet distributed. For each observation, there corresponds an event defined by $(I_1, \ldots, I_m)$ or $(0, \ldots, 0, 1, 0, \ldots, 0)$ and n_j denotes the number of observations for which species j is included in the n samples and $n(I_1, \ldots, I_m)$ is the number of events $(I_1, \ldots, I_m)$ or $(0, \ldots, 0, 1, 0, \ldots, 0)$. Analogous to the ordinary Dirichlet distribution, the likelihood function of the delta-Dirichlet distribution is also concave; see Ronning (1989) and Lewy (1996).

Although convenient formulas for asymptotic variances and covariances of the maximum likelihood estimator $\hat{p}$ are available for the case of Dirichlet distributions [see, for example, Narayanan (1991a,b)], the corresponding expression for delta-Dirichlet distributions are given by Lewy (1996) for the case $m = 2$ only, corresponding to delta-beta distributions. These are:

$$V(\hat{p}_1) = \frac{\sum_{n=0}^{\infty} \frac{1}{(p_2+n)^2} - \sum_{n=0}^{\infty} \frac{1}{(p.+n)^2}}{nD},$$

$$\text{cov}(\hat{p}_1, \hat{p}_2) = \frac{\sum_{n=0}^{\infty} \frac{1}{(p.+n)^2}}{nD},$$

where

$$D = \left(\sum_{n=0}^{\infty} \frac{1}{(p_1+n)^2} - \sum_{n=0}^{\infty} \frac{1}{(p.+n)^2} \right)$$
$$\times \left(\sum_{n=0}^{\infty} \frac{1}{(p_2+n)^2} - \sum_{n=0}^{\infty} \frac{1}{(p.+n)^2} \right) - \left(\sum_{n=0}^{\infty} \frac{1}{(p.+n)^2} \right)^2.$$

Let $n(I_1, \ldots, I_m)$ denote the number of events $(I_1, \ldots, I_m)$ or $(0, \ldots, 0, 1, 0, \ldots, 0)$ and let n_j denote the number of observations for which species j is included in the n samples. Then

$$n_j = \sum_{I_1, \ldots, I_{j-1}, I_{j+1}, \ldots, I_m} n(I_1, \ldots, I_{j-1}, I_j = 1, I_{j+1}, \ldots, I_m), \quad j = 1, \ldots, m.$$

For a given n the set of $((n(I_1, \ldots, I_m))|n)$ is multinomially distributed. For a given n the likelihood function L is

$$L = \binom{n}{(n(I_1, \ldots, I_m))} \prod_{(I_1, \ldots, I_m)} d(I_1, \ldots, I_m)^{n(I_1, \ldots, I_m)}$$
$$\times \frac{\prod_{(I_1, \ldots, I_m)} \Gamma^{n(I_1, \ldots, I_m)} \left(\sum I_j p_j \right)}{\Gamma^{n_1}(p_1) \cdots \Gamma^{n_m}(p_m)}$$

$$\times \left(\prod_{I_{i1}=1}^{i} h_{i1} \right)^{p_1-1} \cdots \left(\prod_{I_{im}=1}^{i} h_{im} \right)^{p_m-1}. \qquad (49.41)$$

The likelihood function shows that $(n(I_1,\ldots,I_m), \prod_{I_{ij}=1}^{i} h_{ij}, j = 1,\ldots,m)$ are sufficient statistics for $(d(I_1,\ldots,I_m), p_j, j = 1,\ldots,m)$.

Since (49.41) is concave in $(p_1,\ldots,p_m)$, a local maximum of (49.41) is the unique global maximum.

Simulation results obtained by Lewy (1996) (for $n = 5, 10, 20, \ldots, 200$ with four groups) show that the MLEs of the parameters are biased and the bias can be approximated by

$$E[\hat{p}_j] \doteq \left(1 + \frac{k}{n} \right) p_j, \qquad j = 1,\ldots,4, \qquad (49.42)$$

where the constant k may vary from 2 to 20, depending on the parameter $d(I_1, I_2, I_3, I_4)$. Formula probably holds also for the cases when the number of groups is greater than 4. For the ordinary Dirichlet distribution, (49.42) is valid with $k = 3$.

Detailed investigations by Lewy (1996) indicate that: The superiority of MLE compared to estimation based on empirical moments for Dirichlet distributions in the small sample case tends to disappear in the case of a delta-Dirichlet distribution when $d(1,1,1,1)$ converges to zero; as a general rule, the efficiency of the MLE relative to the empirical moment estimator is close to 1 when $d(1,1,1,1)$ is greater than 0.5, and when the parameters p_j are less than 5 the MLE may be substantially more efficient than the empirical moment estimator.

8.5 Connor and Mosimann's Generalization

As mentioned earlier, Connor and Mosimann noted that Dirichlet distributions are restricted to being "neutral," and they generalized these distributions as follows.

They defined the vector $(P_1,\ldots,P_m)^T$ of proportions, where $\sum_{i=1}^{m} P_i < 1$, to be "completely neutral" if the ratios $P_1, P_2/(1 - P_1),\ldots, P_m/(1 - P_1 - \cdots - P_{m-1})$ are independent. If, further, it is assumed that each of these ratios is marginally beta with parameters (a_i, b_i) respectively, then the proportions $(P_1,\ldots,P_m)^T$ have the joint density function

$$\left[\prod_{i=1}^{m} B(a_i, b_i) \right]^{-1} \left[1 - \sum_{i=1}^{m} P_i \right]^{b_m-1}$$

$$\times \prod_{i=1}^{m} \left[p_i^{a_i-1} \left(1 - \sum_{j=1}^{i-1} p_j \right)^{b_i-1-(a_i+b_i)} \right], \qquad (49.43)$$

where $B(\cdot, \cdot)$ is the beta function, $a_i, b_i > 0$, and $\sum_{j=1}^{i} p_j$ is taken to be zero if $i < 1$. As already indicated, Connor and Mosimann call (49.43) the *generalized Dirichlet density function*.

An alternate parametrization leads to the joint density

$$\prod_{i=1}^{m} \frac{1}{B(\alpha_1, \beta_i)} \, p_i^{\alpha_i-1} (1 - p_1 - \cdots - p_i)^{\gamma_i}$$

for $p_i \geq 0$ and $\sum_{i=1}^{m} p_i = 1$, $\gamma_i = \beta_i - \alpha_{i+1} - \beta_{i+1}$ ($i = 1, \ldots, m-1$) for $\gamma_m = \beta_m - 1$. In this distribution, P_1 is always negatively correlated with other random variables; however, P_j and P_ℓ can be positively correlated for $j, \ell > 1$. If there exists some $\ell > j$ such that P_j and P_ℓ are positively (negatively) correlated, then P_j will be positively (negatively) correlated with P_k for all $k > j$. Since the generalized Dirichlet distribution has a more general covariance structure than the Dirichlet distribution, this makes the generalized Dirichlet distribution to be more practical and useful.

Wong (1998) studied the generalized Dirichlet distribution and showed that

$$E[P_1^{r_1} \cdots P_m^{r_m}] = \prod_{j=1}^{m} \left\{ \frac{\Gamma(\alpha_j + \beta_j)\Gamma(\alpha_j + \gamma_j)\Gamma(\beta_j + \delta_j)}{\Gamma(\alpha_j)\Gamma(\beta_j)\Gamma(\alpha_j + \beta_j + \gamma_j + \delta_j)} \right\},$$

where $\delta_j = r_{j+1} + \cdots + r_m$ for $j = 1, \ldots, m-1$, and $\delta_m = 0$. It has also been shown that the joint distribution of $(X_1, \ldots, X_\ell)^T$, for any $\ell < m$, is an ℓ-variate generalized Dirichlet distribution. Wong (1998) has also shown that the property that Dirichlet distribution is conjugate to multinomial sampling is naturally carried over to generalized Dirichlet distribution as well. Wong has noted that the order of variables in a generalized Dirichlet random vector is important.

The completely neutral property implies conditional independence among random variables. When the distribution of a random vector is a Dirichlet distribution, every permutation of the variables in the random vector is completely neutral. Thus, the conditions for a random vector to have a Dirichlet distribution are restrictive (although the Dirichlet distribution is easy to construct and has some good computational properties). The conditional independence in a generalized Dirichlet random vector is weaker. This suggests that the generalized Dirichlet distribution is a more suitable prior for realistic situations, but the construction and the computation for such a prior are more complex.

If $b_{i-1} = a_i + b_i$ for all $i = 2, \ldots, m$ in (49.53), this generalized Dirichlet becomes the standard Dirichlet distribution with parameters $(a_1, a_2, \ldots, a_m; b_m)$; namely, with density function

$$\frac{\Gamma(a_1 + a_2 + \cdots + a_m + b_m)}{\Gamma(a_1) \cdots \Gamma(a_m)\Gamma(b_m)} \left(1 - \sum_{i=1}^{m} p_i\right)^{b_m - 1}$$

$$\times \prod_{i=1}^{m} p_i^{a_i - 1}. \tag{49.44}$$

Note that if $(P_1, \ldots, P_m)^T$ follows a standard Dirichlet distribution, then complete neutrality holds for all permutations of these proportions and each of the ratios in the neutrality definition has a beta distribution. James (1972) proved a result that substantially restricts the Connor-Mosimann generalization. He has shown that if $(P_1, \ldots, P_m)^T$ has a generalized Dirichlet density and each of the variables $U_i = P_i/(1 - \sum_{j \neq i} P_j)$, $i = 1, \ldots, m - 1$, have marginal beta distributions, then $b_i = a_{i+1} + b_{i+1}$ for all $i = 1, \ldots, m-1$ and $(P_1, \ldots, P_m)^T$ has the regular Dirichlet density. Note that in case $m = 2$ this implies that if $(P_1, P_2)^T$ follows a generalized Dirichlet distribution [i.e., P_1 and $\frac{P_2}{1-P_1}$ are independent Beta variables and if $P_1/(1 - P_2)$ is marginally Beta], then $(P_1, P_2)^T$ is a Dirichlet random variable and hence P_2 and $\frac{P_1}{1-P_2}$ are also independent.

James (1972, 1975) provided the following generalization of Dirichlet distributions in the spirit of Connor and Mosimann's (1969) generalization. Let $X_1, \ldots, X_m$, $n \geq 2$, be nonnegative random variables with $\sum X_i \leq 1$. The conditional distribution of each X_i given $X_j = x_j$, $j \neq i$, $i = 1, \ldots, n$, is beta on $(0, 1 - \sum_{j \neq i} x_j)$ if and only if $X_1, \ldots, X_m$ have a joint density function of the form

$$p(x_1, \ldots, x_m)$$
$$= \mu \left[\prod_{i=1}^{m} x_i^{\alpha_i - 1}\right] (1 - x_1 - \cdots - x_m)^{\gamma - 1}$$
$$\times \exp[\phi(x_1, \ldots, x_m)], \tag{49.45}$$

where

$$\phi(x_1, \ldots, x_m) = \sum_{i<j} a_{ij} \log x_i \log x_j$$

$$+ \sum_{i<j<k} b_{ijk} \log x_i \log x_j \log x_k + \cdots + c \prod_{i=1}^{m} \log x_i.$$

The marginal densities of (49.45), apart from the $(m - 1)$-dimensional marginals, are not, in general, available in simple form.

If, however, we assume that the conditional distributions are Dirichlet, then the joint distribution reduces to a Dirichlet distribution. More precisely, James (1975) has established the following result: Suppose $m \geq 3$. Then the conditional distribution of $\{X_j; j \neq i\}$ given X_i is Dirichlet for $i = 1, 2, 3$ if and only if $X_1, \ldots, X_m$ are jointly Dirichlet. [James (1975) has also emphasized the strong affinity between neutrality and the Beta distribution and has warned against unrestricted use of the Beta distribution in cases when neutrality may not be applicable.]

Let $(T, \leq)$ be a tree and let $j \in t'_\phi$ be any nonterminal node of T. For every $j \in t'_\phi$, let the set of mode labels $X_j = (X_k : k \in s(j))$ denote a vector random variable having a $\rho(j)$-variate Dirichlet Type 1 distribution with parametric vector $\alpha_j = (\alpha_k > 0 : k \in s(j))$ given by the density function

$$p(x_j) = \frac{1}{B(\alpha_j)} \prod_{k \in s(j)} x_k^{\alpha_k - 1}, \qquad j \in t'_\phi$$

at any point x_j in the simplex $S_{\rho(j)} = \{(x_k : k \in s(j)) : x_k > 0$ and $\sum_{k \in s(j)} x_k = 1\}$ in $\mathbb{R}_{\rho(j)}$, where

$$B(\alpha_j) = \frac{1}{\Gamma(\alpha_{s(j)})} \prod_{k \in s(j)} \Gamma(\alpha_k)$$

and $\alpha_{s(j)} = \sum_{k \in s(j)} \alpha_k$.

If the random vectors X_j $(j \in t'_\phi)$ are independent, then the joint density function of the random vector $X = \{X_j : j \in t'_\phi\}$ is of the form

$$p(x) = \prod_{j \in t'_\phi} p(x_j) = \frac{1}{\prod_{j \in t'_\phi} B(\alpha_j)} \prod_{j \in t'_\phi} \left(\prod_{k \in s(j)} x_k^{\alpha_k - 1} \right).$$

$$(49.46)$$

Without loss of any generality, let $s(j) = \{\rho(0), \ldots, j\rho(j)\}$ for all $j \in t'_\phi$. Let $Y = (Y_i : i \in t)$ denote the vector of terminal random variables of the tree T, where Y_i $(i \in t)$ are defined by the set of transformation equations

$$T_\phi : Y_i = \prod_{j \leq i} X_j, \qquad i \in t.$$

An application of the transformation T_ϕ to (49.46) yields the density function of the random vector $Y = (Y_i : i \in t)$ as

$$p(y) = \frac{1}{\prod_{j \in t'_\phi} B(\alpha_j)} \prod_{i \in t} y_i^{\alpha_i - 1} \prod_{j \in t'} y_{t(j)}^{\beta_j} \qquad (49.47)$$

at any point in the simplex $S_{\text{card}(t)} = \{(y_i : i \in t) : y_i > 0 \text{ and } \sum_{i \in t} y_i = 1\}$ in $\mathbb{R}_{\text{card}(t)}$, where $\text{card}(t)$ denotes the cardinality of the set of terminal sequences, and $\beta_j = \alpha_j - \sum_{k \in s(j)} \alpha_k$ for all $j \in t'$. Dennis (1992) termed the distribution in (49.47) over the tree T a *hyper-Dirichlet Type 1* distribution. This distribution is not only a generalization of the simple Dirichlet Type 1 distribution, but is also a generalization of the generalized Dirichlet distribution described by Connor and Mosimann (1969).

BIBLIOGRAPHY

(Some bibliographical items not mentioned in the text are included here for completeness.)

Altham, P. M. E. (1969). Exact Bayesian analysis of a 2×2 contingency table, and Fisher's "Exact" significance test, *Journal of the Royal Statistical Society, Series B*, **31**, 261–269.

Antelman, G. R. (1972). Interrelated Bernoulli processes, *Journal of the American Statistical Association*, **67**, 831–841.

Bernardo, J. M. (1976). Algorithm AS 103: PSI (Digamma) function, *Applied Statistics*, **25**, 315–317.

Chamayou, J.-F., and Letac, G. (1994). A transient random walk on stochastic matrices with Dirichlet distributions, *Annals of Probability*, **22**, 424–430.

Chatfield, C. R., and Goodhardt, G. J. (1975). Results concerning brand choice, *Journal of Marketing Research*, **12**, 110–113.

Cochran, W. G. (1941). The distribution of the largest of a set of estimated variances as a fraction of their total, *Annals of Eugenics*, **11**, 47–52.

Connor, R. J., and Mosimann, J. E. (1969). Concepts of independence for proportions with a generalization of the Dirichlet distribution, *Journal of the American Statistical Association*, **64**, 194–206.

Craiu, M. and Craiu, V. (1969). Repartitia Dirichlet generalizatá, *Analele Universitatii Bucuresti, Mathematicá-Mecanicá*, **18**, 9–11.

Cramér, H. (1946). *Mathematical Methods of Statistics*, Princeton, New Jersey: Princeton University Press.

Darroch, J. N., and Ratcliff, D. (1971). A characterization of the Dirichlet distribution, *Journal of the American Statistical Association*, **66**, 641–643.

Dennis, S. Y. (1992). On the hyper-Dirichlet type 1 distribution, Unpublished manuscript.

Dirichlet, P. G. L. (1839). Sur une nouvelle méthode pour la determination des intégrales multiples, *Liouville Journal des Mathematiques, Ser. I*, **4**, 164–168.

Fabius, J. (1973a). Two characterizations of the Dirichlet distribution, *Annals of Statistics*, **1**, 583–587.

Fabius, J. (1973b). Neutrality and Dirichlet distributions, *Transactions of the Sixth Prague Conference on Information Theory*, pp. 175–181, Czechoslovakia: Academia, Prague.

Fielitz, B., and Myers, B. L. (1975). Estimation of parameters in the beta distribution, *Decision Sciences*, **6**, 1–13.

Goodhardt, G. J., Ehrenberg, A. S. C., and Chatfield, C. (1984). The Dirichlet: A comprehensive model of buying behaviour, *Journal of the Royal Statistical Society, Series A*, **147**, 621-643; Discussion, 643–655.

Gupta, R. D., and Richards, D. St. P. (1987). Multivariate Liouville distributions, *Journal of Multivariate Analysis*, **23**, 233–256.

Gupta, R. D., and Richards, D. St. P. (1990). The Dirichlet distributions and polynomial regression, *Journal of Multivariate Analysis*, **32**, 95–102.

Hawkins, D. M. (1972). Analysis of a slippage test for the chisquared distribution, *South African Statistical Journal*, **6**, 11–17.

Humbert, P. (1920–1921). The confluent hypergeometric function of two variables, *Proceedings of the Royal Society of Edinburgh, Section A*, **41**, 73–96.

Hurlburt, R. T., and Spiegel, D. K. (1976). Dependence of F-ratios sharing a common denominator mean square, *The American Statistician*, **30**, 74–78.

James, I. R. (1972). Products of independent beta variables with application to Connor and Mosimann's generalized Dirichlet distribution, *Journal of the American Statistical Association*, **67**, 910-912.

James, I. R. (1975). Multivariate distributions which have beta conditional distributions, *Journal of the American Statistical Association*, **70**, 681-684.

Johnson, N. L. (1960). An approximation to the multinomial distribution; some properties and applications, *Biometrika*, **47**, 93–103.

Johnson, N. L., and Kotz, S. (1973). Extended and multivariate Tukey lambda distributions, *Biometrika*, **60**, 655–661.

Johnson, N. L., Kotz, S., and Balakrishnan, N. (1994). *Continuous Univariate Distributions*, Vol. 1, Second edition, New York: John Wiley & Sons.

Johnson, N. L., Kotz, S., and Balakrishnan, N. (1995). *Continuous Univariate Distributions*, Vol. 2, Second edition, New York: John Wiley & Sons.

Johnson, N. L., Kotz, S., and Balakrishnan, N. (1997). *Discrete Multivariate Distributions*, New York: John Wiley & Sons.

Johnson, N. L., Kotz, S., and Kemp, A. W. (1992). *Univariate Discrete Distributions*, second edition, New York: John Wiley & Sons.

Lange, K. (1995). Application of the Dirichlet distribution to forensic match probabilities, *Genetica*, **96**, 107–117.

Lee, P. A. (1971). A diagonal expansion for the 2-variate Dirichlet probability density function, *SIAM Journal of Applied Mathematics*, **21**, 155–165.

Lewy, P. (1996). A generalized Dirichlet distribution accounting for singularities of the variables, *Biometrics*, **52**, 1394–1409.

Lindley, D. V. (1971). *Bayesian Statistics, A Review*, Philadelphia: Society for Industrial and Applied Mathematics.

Lochner, R. H. A. (1975). A generalized Dirichlet distribution in Bayesian life testing, *Journal of the Royal Statistical Society, Series B*, **37**, 103–113.

Lochner, R. H. A., and Basu, A. P. (1972). Bayesian analysis of the two-sample problem with incomplete data, *Journal of the American Statistical Association*, **67**, 432–438.

Ma, C. (1996). Multivariate survival functions characterized by constant product of mean remaining lives and hazard rates, *Metrika*, **44**, 71–83.

Mathai, A. M., and Moschopoulos, P. G. (1997). A multivariate inverted beta model, *Statistica*, **57**, 189–197.

Monhor, D. (1987). An approach to PERT: Application of Dirichlet distribution, *Optimization*, **18**, 113–118.

Mosimann, J. E. (1962). On the compound multinomial distribution, the multivariate β-distribution and correlations among proportions, *Biometrika*, **49**, 65–82.

Narayanan, A. (1991a). Algorithm AS 266: Maximum likelihood estimation of the parameters of the Dirichlet distribution, *Applied Statistics*, **40**, 365–374.

Narayanan, A. (1991b). Small sample properties of parameter estimation in the Dirichlet distribution, *Communications in Statistics—Simulation and Computation*, **20**, 647–666.

Narayanan, A. (1992). A note on parameter estimation in the multivariate beta distribution, *Computer Mathematics and Applications*, **24**, 11–17.

Nayak, T. K. (1987). Multivariate Lomax distribution: Properties and usefulness in reliability theory, *Journal of Applied Probability*, **24**, 170–171.

Novick, M. R., and Grizzle, J. E. (1965). A Bayesian approach to the analysis of data from clinical trials, *Journal of the American Statistical Association*, **60**, 81–96.

Olkin, I., and Sobel, M. (1965). Integral expressions for tail probabilities of the multinomial and negative multinomial distribution, *Biometrika*, **52**, 167–179.

Provost, S. B., and Cheong, Y.-H. (2000). On the distribution of linear combinations of the components of a Dirichlet random vector, *Canadian Journal of Statistics* (to appear).

Rao, B. V., and Sinha, B. K. (1988). A characterization of Dirichlet distributions, *Journal of Multivariate Analysis*, **25**, 25–30.

Rao, C. R. (1952). *Advanced Statistical Methods in Biometric Research*, New York: John Wiley & Sons.

Ronning, G. (1989). Maximum likelihood estimation of Dirichlet distributions, *Communications in Statistics—Simulation and Computation*, **32**, 215–221.

Roux, J. J. J. (1971). A characterization of multivariate distributions, *South African Statistical Journal*, **5**, 27–36.

Sahai, H., and Anderson, R. L. (1973). Confidence regions for variance ratios of random models for balanced data, *Journal of the American Statistical Association*, **68**, 951–952.

Schneider, B. E. (1978). Algorithm AS 121: Trigamma function, *Applied Statistics*, **27**, 97–98.

Serfling, R. J. (1980). *Approximation Theorems of Mathematical Statistics*, New York: John Wiley & Sons.

Sobel, M., and Uppuluri, V. R. R. (1974). Sparse and crowded cells and Dirichlet distribution, *Annals of Statistics*, **2**, 977–987.

Sobel, M., Uppuluri, V. R. R., and Frankowski, K. (1977). Dirichlet distribution – Type 1, *Selected Tables in Mathematical Statistics* – **4**, Providence, Rhode Island: American Mathematical Society.

Sobel, M., Uppuluri, V. R. R., and Frankowski, K. (1985). Dirichlet integrals of Type 2 and their applications, *Selected Tables in Mathematical Statistics* – **9**, Providence, Rhode Island: American Mathematical Society.

Tiao, G. C., and Afonja, B. (1969). Some approximations and uses of the Dirichlet distributions, Preliminary report, *Annals of Mathematical Statistics*, **40**, 1514 (Abstract).

Tiao, G. C., and Guttman, I. J. (1965). The inverted Dirichlet distribution with applications, *Journal of the American Statistical Association*, **60**, 793–805; Correction, **60**, 1251–1252.

Wilks, S. S. (1962). *Mathematical Statistics*, New York: John Wiley & Sons.

Wong, T.-T. (1998). Generalized Dirichlet distribution in Bayesian analysis, *Applied Mathematics and Computation*, **97**, 165–181.

Wrigley, N., and Dunn, R. (1984). Stochastic panel-data models of urban shopping behavior: 2. Multistore purchasing patterns and the Dirichlet model, *Environment and Planning A*, **16**, 759–778.

Yassaee, H. (1974). Inverted Dirichlet distribution and multivariate logistic distributions, *Canadian Journal of Statistics*, **2**, 99–105.

Yassaee, H. (1976). Probability integral of inverted Dirichlet distribution and its applications, *Compstat*, **76**, pp. 64–71, Vienna: Physica-Verlag.

Yassaee, H. (1978). Characteristic functions of Dirichlet distributions, *Transactions of the 8th Prague Conference on Information Theory, Statistical Decision Functions and Random Processes*, Volume B, pp. 383–392, Prague, Czechoslovakia: Academia.

Yassaee, H. (1979). On probability integral of Dirichlet distributions and their applications, Preprint, Arya-Mehr University of Technology, Tehran, Iran.

CHAPTER 50

Multivariate Liouville Distributions

1 INTRODUCTION

Liouville distributions seem to be one of those classes of distributions which have attracted great attention from researchers during the past twenty years, motivated mainly by theoretical (rather than practical) considerations of generalizing results based on multivariate normal distributions. As mentioned already in Chapter 45, multivariate normal distributions, in addition to having many nice theoretical properties, served as a cornerstone in data analysis with several components. Multivariate Liouville distributions are motivated by Joseph Liouville's (1809–1882) integral, which is a generalization of the Dirichlet integral; see, for example, Fang, Kotz, and Ng (1989) for details.

Liouville distributions arise in a variety of probabilistic and statistical contexts. These include multivariate majorization [see Marshall and Olkin (1979, p. 308) and Diaconis and Perlman (1987)], generalizations of Dirichlet and inverted Dirichlet distributions (see Chapter 50), total positivity and correlation inequalities [see Gupta and Richards (1987, 1991)], and statistical reliability theory [see Gupta and Richards (1992)].

There are at least two approaches to define multivariate Liouville distributions. The first one is based on the Liouville multiple integral defined over the positive orthant $\mathbb{R}_+^k = \{(x_1, \ldots, x_k)^T : x_i \geq 0, \ i = 1, \ldots, k\}$, which is an extension of the Dirichlet integral. Marshall and Olkin (1979) were the first to use this integral in order to define what they called the *Liouville–Dirichlet distribution*. The second approach, taken by Fang, Kotz, and Ng (1989), presents the multivariate Liouville distribution as a uniform base with a constraint multiplied by a positive generating variate.

Marshall and Olkin's (1979) treatise is perhaps the first place where Liouville distributions were discussed briefly in their relation to Schur-convex functions. Subsequently, Sivazlian (1981) presented results on marginal distributions and transformation properties of Liouville distributions. Anderson and Fang (1982, 1987) discussed Liouville distributions arising from the distributions of quadratic forms. The first comprehensive discussion of these distributions was provided by Fang, Kotz, and Ng (1989, Chapter 6) wherein some new results due to Ng were also included. A series of six papers by Gupta and Richards (1987–1997), along with one co-authored with Misiewicz (1996), constitutes a rich source of information on Liouville distributions and their properties, matrix extensions, other generalizations, and their applications in statistical reliability theory. Finally, Ma and Yue (1995) and Ma, Yue, and Balakrishnan (1996) presented generalizations of multivariate Liouville distributions in the framework of ℓ_p-norm symmetric distributions; ℓ_p-norm symmetric distributions were introduced by Fang and Fang (1988, 1989) [see also Yue and Ma (1995)] and have been discussed in great detail by Fang, Kotz, and Ng (1989).

2 DEFINITIONS

An absolutely continuous random vector $\boldsymbol{X} = (X_1, \ldots, X_k)^T$ has a *multivariate Liouville distribution* if its joint density function is proportional to [Gupta and Richards (1987)]

$$f\left(\sum_{i=1}^{k} x_i\right) \prod_{i=1}^{k} x_i^{a_i - 1}, \tag{50.1}$$

where the variables range over the orthant

$$\mathbb{R}_+^k = \{(x_1, \ldots, x_k) : x_i > 0, \ i = 1, 2, \ldots, k\},$$

a_i's $(i = 1, \ldots, k)$ are positive numbers, and the function f is positive, continuous and appropriately integrable. Gupta and Richards (1987) have used the notation $L_k[f(\cdot); a_1, \ldots, a_k]$ to denote this distibution.

If the support is noncompact, $\boldsymbol{X}$ is said to have a *Liouville distribution of the first kind*. If, on the other hand, the support is compact $[(0, 1)^k$, without loss of generality], then the variables range over the simplex

$$\mathcal{S}_k = \left\{(x_1, \ldots, x_k) : x_i > 0 \text{ for } i = 1, \ldots, k, \ \sum_{i=1}^{k} x_i < 1\right\}$$

and in this case $\boldsymbol{X}$ is said to have a *Liouville distribution of the second kind*. In order to distinguish the two forms, the notations $L_k^{(1)}[f(\cdot); a_1, \ldots, a_k]$ and $L_k^{(2)}[f(\cdot); a_1, \ldots, a_k]$ are used to denote them.

An alternate definition, as given by Fang, Kotz, and Ng (1989), is as follows. Recall that the Dirichlet distribution is a distribution on the hyperplane

$$B_k = \left\{ (y_1, \ldots, y_k) : \sum_{i=1}^k y_i = 1 \right\} \in R_+^k$$

or a distribution inside the simplex

$$A_{k-1} = \left\{ (y_1, \ldots, y_{k-1}) : \sum_{i=1}^{k-1} y_i \le 1 \right\} \in R_+^{k-1},$$

with the density function

$$p(y_1, \ldots, y_k) = \frac{\Gamma\left(\sum_{i=1}^k a_i\right)}{\prod_{i=1}^k \Gamma(a_i)} \prod_{i=1}^{k-1} y_i^{a_i-1} \left(1 - \sum_{i=1}^{k-1} y_i\right)^{a_k-1},$$

$$y_i \ge 0, \ \sum_{i=1}^{k-1} y_i \le 1, \ y_k = 1 - \sum_{i=1}^{k-1} y_i, \ a_i > 0. \quad (50.2)$$

Now, $\boldsymbol{X} \in \mathbb{R}_+^k$ has a *Liouville distribution* if $\boldsymbol{X} \stackrel{d}{=} R\boldsymbol{Y}$, where $R = \sum_{i=1}^k X_i$ has an univariate Liouville distribution $L_1[f(\cdot); a]$ and $Y = (Y_1, \ldots, Y_k)^T$ is independent of R and possesses the above Dirichlet distribution. Here, $f(\cdot)$ is said to be the *generating density*, R is the *generating variable*, $\boldsymbol{Y}$ is the *Dirichlet base*, and $\boldsymbol{a} = (a_1, \ldots, a_k)^T$ is the *Dirichlet parameter*.

3 PROPERTIES

Using the stochastic representation $\boldsymbol{X} \stackrel{d}{=} R\boldsymbol{Y}$, it can be shown that several properties of the Dirichlet distributions remain valid for the Liouville distributions of the second kind; see Gupta and Richards (1999). These properties include the amalgamation, subcompositional and partition properties; see, for example, Sections 2.5–2.7 of Aitchison (1986). In order to see the subcompositional properties of the Liouville distributions of the second kind, for instance, let us assume that $\boldsymbol{X} = (X_1, \ldots, X_k)^T \stackrel{d}{=} L_k^{(2)}[f(\cdot); a_1, \ldots, a_k]$. For $\ell < k$, let $\{X_{i_1}, \ldots, X_{i_\ell}\}$ be a subset of $\{X_1, \ldots, X_k\}$, and let us define

$$Y_{i_j} = \frac{X_{i_j}}{X_{i_1} + \cdots + X_{i_\ell}} \quad \text{for } j = 1, 2, \ldots, \ell.$$

Since $(X_{i_1}, \ldots, X_{i_\ell})^T$ has a marginal Liouville distribution [Gupta and Richards (1987)], a stochastic representation of the form $\boldsymbol{X} \stackrel{d}{=} R\boldsymbol{Y}$ also holds for $(X_{i_1}, \ldots, X_{i_\ell})^T$. We then deduce that $(Y_{i_1}, \ldots, Y_{i_\ell})^T$ has a singular Dirichlet distribution with parameters $(a_{i_1}, \ldots, a_{i_\ell})^T$. Therefore, in the terminology of Aitchison (1986, Section 2.5), we have shown that every subcomposition of a Liouville-distributed vector is a Dirichlet decomposition.

One of the properties of multivariate Liouville distributions is that if $\boldsymbol{X} \stackrel{d}{=} L_k[f(\cdot); a_1, \ldots, a_k]$, $\phi : \mathbb{R}^+ \rightarrow \mathbb{R}^+$ is a Borel function, and $\boldsymbol{V} = (V_1, \ldots, V_k)^T$ where $V_i = X_i/\phi(X_1, \ldots, X_k)$ for $i = 1, 2, \ldots, k$, then $\boldsymbol{Y} = \left(\frac{V_1}{\sum_{i=1}^{k} V_i}, \ldots, \frac{V_k}{\sum_{i=1}^{k} V_i} \right)^T = \left(\frac{X_1}{\sum_{i=1}^{k} X_i}, \ldots, \frac{X_k}{\sum_{i=1}^{k} X_i} \right)^T$ will have a singular Dirichlet distribution. If V_i's are as defined above based on independent random variables $X_1, \ldots, X_k$ having Liouville distributions, Wesolowski (1993) has investigated the issue whether the only possible distribution of the X_i's is gamma distribution. Wesolowski (1993) has shown that this is so for independent positive random variables $X_1, \ldots, X_k$ (for $k \geq 3$), a Borel function $\phi : \mathbb{R}^+ \rightarrow \mathbb{R}^+$, and the vector $\boldsymbol{V}$ as defined earlier. His solution is through the result that if $U_1, U_2,$ and U_3 are independent gamma random variables with a common scale parameter, then the random vector $\left(\frac{U_1}{U_3}, \frac{U_2}{U_3} \right)$ has a bivariate beta distribution of the second kind (see Chapter 49) and hence the result of Kotlarski (1967) on characterization of the gamma distribution can be used here; also see Chapter 17 of Johnson, Kotz, and Balakrishnan (1994). (The case $k = 2$ has been shown not to be valid; Wesolowski (1993) has given a counterexample with $\phi(x) = x$.) In this connection, it needs to be mentioned that Ng (1989), in the monograph of Fang, Kotz and Ng (1989), has shown that if $\boldsymbol{X}$ has a Liouville distribution and $\Pr[\boldsymbol{X} = \boldsymbol{0}] = 0$, then the condition that $X_1, \ldots, X_k$ are independently distributed is equivalent to the condition that $X_1, \ldots, X_k$ are distributed as gamma with a common scale parameter.

From the alternative definition of the multivariate Liouville distribution given by Fang, Kotz, and Ng (1989), it turns out that when $L_k[f(\cdot); a_1, \ldots, a_k]$ has a generating density, then the density function of the Liouville distribution is

$$\frac{\Gamma(a)}{\prod_{i=1}^{k} \Gamma(a_i)} \frac{\prod_{i=1}^{k} x_i^{a_i-1}}{\left(\sum_{i=1}^{k} x_i\right)^{a-1}} f\left(\sum_{i=1}^{k} x_i\right), \quad a = \sum_{i=1}^{k} a_i \qquad (50.3)$$

defined over the simplex $\left\{ (x_1, \ldots, x_k) : x_i \geq 0, \ 0 \leq \sum_{i=1}^{k} x_i \leq c \right\}$ if and only if $f(\cdot)$ is defined on $(0, c)$. We will use these two forms interchangeably from now on. Now, for the sake of convenience, let us introduce the density generator as

$$g(t) = \frac{\Gamma(a)}{t^{a-1}} f(t), \qquad t > 0. \tag{50.4}$$

Using this generator, we may rewrite the density function of the multivariate Liouville distribution in (50.3) as

$$\prod_{i=1}^{k} \left(\frac{x_i^{a_i-1}}{\Gamma(a_i)} \right) g \left(\sum_{i=1}^{k} x_i \right). \tag{50.5}$$

[Compare this form with (50.1).] Observe that the generator $g(\cdot)$ in (50.4) satisfies the condition

$$\int_0^\infty \frac{t^{a-1}}{\Gamma(a)} g(t)\, dt = 1, \qquad a = \sum_{i=1}^{k} a_i. \tag{50.6}$$

Let us now consider some special cases:

1. From (50.1), upon setting $f(t) = t^{a-1} e^{-bt}$ for $t > 0$, $a > 0$, $b > 0$, we obtain the multivariate Liouville distribution of the first kind with joint density proportional to

$$\left(\sum_{i=1}^{k} x_i \right)^{a-1} \prod_{i=1}^{k} \left\{ e^{-bx_i}\, x_i^{a_i-1} \right\}. \tag{50.7}$$

   This form corresponds to *distribution of correlated gamma variables*; see, for example, Marshall and Olkin (1979).

2. From (50.1), upon setting $f(t) = (1 - t)^{a_{k+1}-1}$ for $0 < t < 1$, we obtain the multivariate Liouville distribution of the first kind with joint density proportional to

$$\left(\prod_{i=1}^{k} x_i^{a_i-1} \right) \left(1 - \sum_{i=1}^{k} x_i \right)^{a_{k+1}-1}, \tag{50.8}$$

   which is simply the *Dirichlet distribution*.

3. From (50.1), upon setting $f(t) = (1 + t)^{-\sum_{i=1}^{k+1} a_i}$ for $t > 0$ and $a_{k+1} > 0$, we obtain the multivariate Liouville distribution of the first kind with joint density proportional to

$$\frac{\prod_{i=1}^{k} x_i^{a_i-1}}{\left(1 + \sum_{i=1}^{k} x_i \right)^{\sum_{i=1}^{k+1} a_i}}, \qquad x_i > 0, \tag{50.9}$$

   which is the *inverted Dirichlet distribution*.

Gupta and Richards (1987) have shown that if $\boldsymbol{Y} = (Y_1, \ldots, Y_k)^T \overset{d}{=} L_k^{(2)}[g(\cdot); a_1, \ldots, a_k]$, then random vector $\boldsymbol{X} = (X_1, \ldots, X_k)^T \overset{d}{=} L_k^{(1)}[f(\cdot); a_1, \ldots, a_k]$, where $X_i = \frac{Y_i}{1 - \sum_{j=1}^{k} Y_j}$ for $i = 1, 2, \ldots, k$, and $f(t) = (1 + t)^{-\left(\sum_{i=1}^{k} a_i + 1\right)} g\left(\frac{t}{1+t}\right)$ for $t > 0$; the correspondence between $f(\cdot)$ and $g(\cdot)$ is one-to-one.

Furthermore, if $\boldsymbol{X} = (X_1, \ldots, X_k)^T \overset{d}{=} L_k[f(\cdot); a_1, \ldots, a_k]$, then:

(i) $(X_1, \ldots, X_k)^T \overset{d}{=} (Y_1, \ldots, Y_{k-1}, 1 - \sum_{i=1}^{k-1} Y_i)^T Y_k$, where $(Y_1, \ldots, Y_{k-1})^T$ and Y_k are statistically independent, and $(Y_1, \ldots, Y_{k-1})^T$ has a Dirichlet distribution with density function

$$\frac{\Gamma\left(\sum_{i=1}^{k} a_i\right)}{\prod_{i=1}^{k} \Gamma(a_i)} \left(\prod_{i=1}^{k-1} y_i^{a_i - 1}\right) \left(1 - \sum_{i=1}^{k-1} y_i\right)^{a_k - 1},$$

$$y_i \geq 0, \ \sum_{i=1}^{k-1} y_i \leq 1, \ a_i > 0. \tag{50.10}$$

(ii) $(X_1, \ldots, X_k)^T \overset{d}{=} \left(\prod_{i=1}^{k-1} Y_i, (1 - Y_1) \prod_{i=2}^{k-1} Y_i, \ldots, 1 - Y_{k-1}\right) Y_k$, where $Y_1, \ldots, Y_k$ are mutually independent random variables with Y_i (for $i = 1, 2, \ldots, k - 1$) having a beta distribution with density function

$$\frac{\Gamma\left(\sum_{j=1}^{i+1} a_j\right)}{\Gamma\left(\sum_{j=1}^{i} a_j\right) \Gamma(a_{i+1})} \, y_i^{\sum_{j=1}^{i} a_j - 1} (1 - y_i)^{a_{i+1} - 1}, \quad 0 < y_i < 1.$$

$$\tag{50.11}$$

Here, $Y_k \overset{d}{=} \sum_{i=1}^{k} X_i \overset{d}{=} L_1\left[f(\cdot); \sum_{i=1}^{k} a_i\right]$. Hence, the joint distribution of $(X_1, \ldots, X_k)^T$ is uniquely determined by the distribution of Y_k. The result also implies that Y_k can have a gamma distribution only if $(X_1, \ldots, X_k)^T$ has joint density proportional to

$$\left(\sum_{i=1}^{k} x_i\right)^{\sum_{i=1}^{k} a_i - 1} \prod_{i=1}^{k} \left\{e^{-bx_i} x_i^{a_i - 1}\right\}.$$

Furthermore, if $(X_1, \ldots, X_k)^T \overset{d}{=} L_k[f(\cdot); a_1, \ldots, a_k]$, then $Z_j = \sum_{i=1}^{j} X_i / \sum_{i=1}^{k} X_i$ (for $j = 1, 2, \ldots, k - 1$) have a $(k - 1)$-variate beta distribution with density function (see Chapter 49)

$$\frac{\Gamma\left(\sum_{i=1}^{k} a_i\right)}{\Gamma\left(\sum_{i=1}^{j} a_i\right) \Gamma\left(\sum_{i=j+1}^{k} a_i\right)} \, z_j^{\sum_{i=1}^{j} a_i - 1} (1 - z_j)^{\sum_{i=j+1}^{k} a_i - 1}, \ 0 < z_j < 1.$$

$$\tag{50.12}$$

In order to discuss the marginal distributions of multivariate Liouville distributions, we need the following concept of *Weyl fractional integral*. If a continuous function $f : R_+^{m \times m} \to R$ satisfies the condition

$$\int |\mathbf{T}|^{a-p} f(\mathbf{T}) \, d\mathbf{T} < \infty, \tag{50.13}$$

where $p = \frac{m+1}{2}$, $a_i > p - 1$ (for $i = 1, \ldots, k$), $a = \sum_{i=1}^{k} a_i$ and $\alpha > p - 1$, then the *Weyl fractional integral of order* α of $f(\cdot)$ is

$$W^\alpha f(\mathbf{T}) = \frac{1}{\Gamma_m(\alpha)} \int_{\mathbf{S} > \mathbf{T}} |\mathbf{S} - \mathbf{T}|^{\alpha - p} f(\mathbf{S}) \, d\mathbf{S}, \tag{50.14}$$

where $\Gamma_m(\cdot)$ is the multidimensional gamma function, and "$\mathbf{S} > \mathbf{T}$" means that $\mathbf{S} - \mathbf{T}$ is positive definite. Two main properties of the Weyl fractional integral are as follows:

(i) If a continuous function $f(\cdot)$ satisfies (50.13), then there is a one-to-one correspondence between $f(\cdot)$ and its "fractional integral" $W^\alpha f(\cdot)$;

and

(ii) W^α satisfies the semigroup property $W^{\alpha+\beta} = W^\alpha W^\beta$ for $\alpha > p - 1$ and $\beta > p - 1$.

Now if $\mathbf{X} = (X_1, \ldots, X_k)^T \stackrel{d}{=} L_k[f(\cdot); a_1, \ldots, a_k]$, then $(X_1, \ldots, X_\ell)^T \stackrel{d}{=} L_\ell[f_\ell(\cdot); a_1, \ldots, a_\ell]$ for $\ell < k$, where $a = \sum_{i=\ell+1}^{k} a_i$ and $f_\ell(t) = W^a f(t)$ is the Weyl fractional integral of order a of $f(\cdot)$. In the extreme case of $\ell = 1$, the distribution of $(X_1, \ldots, X_k)^T$ is uniquely determined by the distribution of X_1. In the class $L_k^{(2)}$, at most one univariate marginal can be uniformly distributed on (0,1). [Compare this with corresponding result for Dirichlet distributions.] A simpler version of the above-stated property of marginal distributions can be given by the Liouville distribution with the density function

$$c_k \, \theta^{-a} f \left(\frac{1}{\theta} \sum_{i=1}^{k} x_i \right) \prod_{i=1}^{k} x_i^{a_i - 1}, \tag{50.15}$$

with $\theta > 0$, $a_i > 1$ ($i = 1, \ldots, k$), $a = \sum_{i=1}^{k} a_i$, and the variables $x_1, \ldots, x_k$ range over the orthant $R_+^k = \{(x_1, \ldots, x_k) : x_i > 0 \text{ for } i = 1, 2, \ldots, k\}$; $f(\cdot)$ is continuous, positive on R_+^k such that, for $\alpha > 0$, $\int_0^\infty t^{\alpha-1} f(t) \, dt < \infty$. The Weyl fractional integral of order α ($\alpha > 0$) of $f(\cdot)$ is

$$f_\alpha(t) = \frac{1}{\Gamma(\alpha)} \int_0^\infty s^{\alpha-1} f(t + s) \, ds, \qquad t > 0, \tag{50.16}$$

with $f_0(t) \equiv f(t)$. Also,

$$f_{\alpha+\beta}(t) = \frac{1}{\Gamma(\beta)} \int_0^\infty s^{\beta-1} f_\alpha(t+s) \, ds, \qquad t > 0, \qquad (50.17)$$

which is the semigroup property $f_{\alpha+\beta}(t) = (f_\alpha)_\beta$. In this case, c_k in (50.15) is given by $c_k^{-1} = \{\prod_{i=1}^k f(a_i)\} f_a(0)$. Using this definition of the Weyl fractional integral, if $(X_1, \ldots, X_k)^T \stackrel{d}{=} L_k[f(\cdot), \theta; \alpha, \ldots, \alpha]$, then $(X_1, \ldots, X_i)^T \stackrel{d}{=} L_i[f_{(n-i)}\alpha(\cdot), \theta; \alpha, \ldots, \alpha]$. In the special case when $(X_1, \ldots, X_k)^T \stackrel{d}{=} L_k[f(\cdot), \theta; 1, \ldots, 1]$ and $t \geq 0$,

$$\Pr\left[\bigcap_{i=1}^k (X_i \leq t)\right] = \sum_{j=0}^k (-1)^j \binom{k}{j} \frac{f_k(jt/\theta)}{f_k(\theta)}. \qquad (50.18)$$

4   MOMENTS AND COVARIANCE STRUCTURE

Let $\boldsymbol{X}$ have a multivariate Liouville distribution with joint density function (50.1), and let $a = \sum_{i=1}^k a_i$. Then, from the stochastic representation that $\boldsymbol{X} \stackrel{d}{=} R\boldsymbol{Y}$, where $\boldsymbol{Y}$ has a Dirichlet distribution with density as in (50.2), the moments and the covariance structure of the multivariate Liouville distributions can be derived easily; see Gupta and Richards (1999). Since the marginal distribution of Y_i is beta (for $i = 1, 2, \ldots, k$), we readily obtain (since Y_i and R are independent).

$$E[X_i] = E[RY_i] = E[Y_i]E[R] = \frac{a_i}{a} E[R] \qquad (50.19)$$

and

$$E[X_i^2] = E[R^2 Y_i^2] = E[Y_i^2]E[R^2] = \frac{a_i(a_i+1)}{a(a+1)} E[R^2], \qquad (50.20)$$

so the variance of X_i is given by

$$\begin{aligned}
\operatorname{var}(X_i) &= E[X_i^2] - \{E[X_i]\}^2 \\
&= \frac{a_i}{a^2(a+1)} \left\{ a(a_i+1)E[R^2] - a_i(a+1)(E[R])^2 \right\} \\
&= \frac{a_i}{a^2(a+1)} \left\{ a(a_i+1)\operatorname{var}(R) + (a - a_i)(E[R])^2 \right\}
\end{aligned}$$

$$(50.21)$$

for $i = 1, 2, \ldots, k$. Similarly, for $1 \leq i < j \leq k$, we find

$$
\begin{aligned}
\operatorname{cov}(X_i, X_j) &= E[R^2 Y_i Y_j] - E[RY_i]E[RY_j] \\
&= E[R^2]E[Y_i Y_j] - (E[R])^2 E[Y_i]E[Y_j] \\
&= \frac{a_i a_j}{a^2(a+1)} \left\{ a\, E[R^2] - (a+1)(E[R])^2 \right\} \\
&= \frac{a_i a_j}{a^2(a+1)} \left\{ a\, \operatorname{var}(R) - (E[R])^2 \right\}.
\end{aligned}
\tag{50.22}
$$

Note that for each $i < j$, the covariances all have the same sign. Furthermore, the covariance is negative if and only if the coefficient of variation of R, namely, $CV(R) = \frac{\sqrt{\operatorname{var}(R)}}{E[R]}$, satisfies the condition

$$
CV(R) < \frac{1}{\sqrt{a}}.
\tag{50.23}
$$

Gupta and Richards (1999) have presented a sufficient condition for the inequality in (50.23) to hold. Suppose X has a multivariate Liouville $(L_k[f(\cdot); a_1, \ldots, a_k])$ distribution, where $f(\cdot)$ is differentiable, strictly log-concave, and

$$
\lim_{u \to 0^+} u^a f(u) = \lim_{u \to \infty} u^{a+1} f(u) = 0.
\tag{50.24}
$$

Then, the inequality in (50.23) holds.

For example, if we choose $f(t) = (1-t)^{a_{k+1}-1}$ for $0 < t < 1$ and $a_{k+1} > 1$, we have X following a Dirichlet distribution with joint density function

$$
\frac{\Gamma(a_1 + \cdots + a_{k+1})}{\Gamma(a_1) \cdots \Gamma(a_{k+1})} \prod_{i=1}^{k} x_i^{a_i - 1} \left(1 - \sum_{i=1}^{k} x_i \right)^{a_{k+1}-1},
$$

$$
x_i \geq 0, \ \sum_{i=1}^{k} x_i \leq 1, \ a_i > 0.
$$

In this case, it can be easily verified that the above sufficient condition holds; consequently, we have $\operatorname{cov}(X_i, X_j) < 0$ for all $1 \leq i < j \leq k$, a well-known result for Dirichlet distributions.

We also note from (50.22) that $CV(R) = \frac{1}{\sqrt{a}}$ iff $\operatorname{cov}(X_i, X_j) = 0$ for some, and hence for all, $1 \leq i < j \leq k$. We may now choose $f(t) = t^\alpha (1-t)^\beta$, $0 < t < 1$, where α and β are chosen such that $CV(R) = \frac{1}{\sqrt{a}}$. We see that the X_i's are all pairwise uncorrelated, but no two of them are mutually independent.

If we next choose $f(t) = e^{-t}t^{\alpha}$, $t > 0$, it can be shown that the condition $CV(R) = \frac{1}{\sqrt{a}}$ (or pairwise uncorrelatedness of the X_i's) implies $\alpha = 0$ which is indeed equivalent to the mutual independence of the X_i's.

Gupta and Richards (1999) have also established that if $\boldsymbol{X}$ has the multivariate Liouville distribution in (50.1), the covariance between any two distinct log ratios $\log(X_i/X_j)$ and $\log(X_{i'}/X_{j'})$ is nonnegative. They have also discussed the covariance structure of the random vector $(W_1, \ldots, W_k)^T$, where $W_i = (X_i/\alpha_i)^{\beta_i}$ is a power-scale transformation of X_i and the constants α_i and β_i (for $i = 1, \ldots, k$) are all positive.

5 CHARACTERIZATIONS

When $\boldsymbol{X} = (X_1, \ldots, X_k)^T$ has a multivariate Liouville distribution, many characterization results are available. The following concept of *complete neutrality* is necessary to discuss these results.

A random vector $(X_1, X_2, \ldots, X_k)^T$ taking on values in a simplex $\mathcal{S}_k$ is said to be *completely neutral* if there exist nonnegative mutually independent random variables $Y_1, \ldots, Y_k$ such that

$$(X_1, \ldots, X_k)^T \stackrel{d}{=} \left(Y_1, Y_2(1 - Y_1), \ldots, Y_k \prod_{i=1}^{k-1}(1 - Y_i) \right)^T. \quad (50.25)$$

Then, if $\boldsymbol{X} = (X_1, \ldots, X_k)^T \stackrel{d}{=} L_k^{(2)}[f(\cdot); a_1, \ldots, a_k]$, the random vector $\boldsymbol{X}$ is completely neutral if and only if $\boldsymbol{X}$ has the Dirichlet distribution with joint density function

$$\frac{\Gamma\left(\sum_{i=1}^{k+1} a_i\right)}{\prod_{i=1}^{k+1} \Gamma(a_i)} \left\{ \prod_{i=1}^{k} x_i^{a_i-1} \right\} \left(1 - \sum_{i=1}^{k} x_i \right)^{a_{k+1}-1},$$

$$x_i \geq 0, \ 0 \leq \sum_{i=1}^{k} x_i \leq 1, \ a_i > 0$$

for some $a_{k+1} > 0$.

Similarly, suppose that $\boldsymbol{X} = (X_1, \ldots, X_k)^T \stackrel{d}{=} L_k^{(1)}[f(\cdot); a_1, \ldots, a_k]$. Then there exist nonnegative mutually independent random variables $Y_1, \ldots, Y_k$ such that

$$\boldsymbol{X} = (X_1, \ldots, X_k)^T \stackrel{d}{=} \left(Y_1, Y_2(1 + Y_1), \ldots, Y_k \prod_{i=1}^{k}(1 + Y_i) \right)^T \quad (50.26)$$

if and only if $\boldsymbol{X}$ has an inverted Dirichlet distribution with joint density function being proportional to

$$\frac{\prod_{i=1}^{k} x_i^{a_i-1}}{\left(1 + \sum_{i=1}^{k} x_i\right)^{\sum_{i=1}^{k+1} a_i}}, \qquad x_i > 0,$$

for some $a_{k+1} > 0$.

Gupta and Richards (1992) have derived conditions on the functions $f(\cdot)$ and $g(\cdot)$ and the parameters a_i and b_i (for $i = 1, 2, \ldots, k$) so that, given $\boldsymbol{X} = (X_1, \ldots, X_k)^T \overset{d}{=} L_k[f(\cdot); a_1, \ldots, a_k]$ and $\boldsymbol{Y} = (Y_1, \ldots, Y_k)^T \overset{d}{=} L_k[g(\cdot); b_1, \ldots, b_k]$, we have

$$(X_1, \ldots, X_k)^T \overset{\text{st}}{\geq} (Y_1, \ldots, Y_k)^T, \qquad (50.27)$$

that is,

$$E[\psi(X_1, \ldots, X_k)] \geq E[\psi(Y_1, \ldots, Y_k)] \qquad (50.28)$$

for any function $\psi : \mathbb{R}_+^k \to \mathbb{R}$ such that ψ is monotone increasing in each component. Their result is that if $a_i \geq b_i$ for $i = 1, 2, \ldots, k$ with $b_k \geq 1$ and that the function $\frac{f(x+t)}{g(x)}$ is monotone increasing in $x > 0$ for any $t \geq 0$, then $(X_1, \ldots, X_k)^T \overset{\text{st}}{\geq} (Y_1, \ldots, Y_k)^T$. If, however, $a_i = b_i$ $(i = 1, \ldots, k)$, then it is required that $\frac{f}{g}$ is monotone increasing on $\mathbb{R}_+$. Let us choose again $f(t) = t^a\, e^{-\alpha t}$ (for $t > 0$, $\alpha > 0$ and $a \geq 0$) and $g(t) = e^{-\beta t}$ (for $t > 0$ and $\beta > 0$). Then, the variables $X_1, \ldots, X_k$ are correlated if $a > 0$ while the variables $Y_1, \ldots, Y_k$ are independent gamma variables. Now if $a_i \geq b_i$ (for $i = 1, 2, \ldots, k$) with $b_k \geq 1$ and also $\beta \geq \alpha$, we have $(X_1, \ldots, X_k)^T \overset{\text{st}}{\geq} (Y_1, \ldots, Y_k)$.

The concept of *positively dependent by mixture* was first proposed by Shaked (1977); see Section 12 of Chapter 44. In the present case, it means that $(X_1, \ldots, X_k)^T$ is stochastically equal to a mixture of independent and identically distributed random variables. Gupta and Richards (1992) have then established that if the multivariate Liouville vector $(X_1, \ldots, X_k)^T$ is positively dependent by mixture and that the mixing is complete, then $(X_1, \ldots, X_k)^T$ is a mixture of independent and identically distributed gamma variables.

Gupta and Richards (1995), in a later paper, observed that a far-reaching generalization of Liouville distributions can be obtained by defining a Liouville density function to be of the form

$$f\left(\sum_{i=1}^{k} x_i\right) \prod_{i=1}^{k} \phi_{a_i}(x_i), \qquad (50.29)$$

where the monomial $x_i^{a_i-1}$ has been replaced by the density ϕ_a $(a > 0)$ satisfying the convolution property

$$\phi_{a_1} \star \phi_{a_2} = \phi_{a_1 + a_2} \tag{50.30}$$

for all $a_1, a_2, > 0$. In fact, Gupta and Richards (1995) have worked with set $\mathcal{F}$ of densities ϕ_a (or probability measures μ_a), where the indices a belong to an abstract Abelian semigroup I, instead of the positive real line $\mathbb{R}_+$, and the elements of $\mathcal{F}$ satisfy the convolution property in (50.30). This construction naturally produces a larger class of distributions because of more flexibility we have in the choice of the densities ϕ_a (or measures μ_a), and the Abelian semigroup I.

It should be noted that this is related to the work of Barndorff-Nielsen and Jørgensen (1991) wherein new classes of parametric models on the unit simplex have been introduced by conditioning independent generalized inverse Gaussian random variables on their sum. These models can, in fact, be derived from the above-described model of Gupta and Richards (1995) by choosing the densities ϕ_a from a convolution family of generalized inverse Gaussian densities and conditioning on their sum.

One of the methods used extensively in the construction of multivariate distributions is the method of *compounding* (or *mixing*); see, for example, Chapter 34 of Johnson, Kotz, and Balakrishnan (1997). Consider a random vector $\boldsymbol{Y} = (Y_1, \ldots, Y_k)^T$ with a given distribution and a random vector $\boldsymbol{Z} = (Z_1, \ldots, Z_k)^T$ such that the conditional distribution of $\boldsymbol{Z}$ given $\boldsymbol{Y} = \boldsymbol{y}$ is specified. Then the method of compounding simply amounts to determining the unconditional distribution of $\boldsymbol{Z}$. Many of the examples given in Chapter 34 of Johnson, Kotz, and Balakrishnan (1997) deal with the situation where the marginal distribution of $\boldsymbol{Y}$ as well as the conditional distribution of $\boldsymbol{Z}$ given $\boldsymbol{Y} = \boldsymbol{y}$ are special cases of multivariate Liouville distributions. More generally, let us suppose that the conditional distribution of $\boldsymbol{Z}$ given $\boldsymbol{Y} = \boldsymbol{y}$ is $L_k[f(\cdot); y_1, \ldots, y_k]$ where each $\mu_{\boldsymbol{y}_i}$ is a Poisson distribution with mean $\boldsymbol{y}_i$, and the distribution of $\boldsymbol{Y}$ is $L_k[g(\cdot); a_1, \ldots, a_k]$ where the corresponding μ_{a_i} has a gamma distribution with shape parameter a_i. By adopting the fractional calculus techniques in order to reduce multiple integrals to a single integral, the unconditional distribution of $\boldsymbol{Z}$ is found to be

$$\tilde{f}\left(\sum_{i=1}^{k} z_i\right) \prod_{i=1}^{k} \frac{\Gamma(a_i + z_i)}{z_i! \, \Gamma(a_i)} , \tag{50.31}$$

where

$$\tilde{f}(t) = \frac{f(t)}{\Gamma\left(t + \sum_{i=1}^{k} a_i\right)} \int_0^\infty e^{-2y} \, y^{t + \sum_{i=1}^{k} a_i - 1} g(y) \, dy. \tag{50.32}$$

6 ESTIMATION AND APPLICATIONS

The forms of $L_k[f(\cdot), \theta; 1, \ldots, 1]$ and $L_k[f(\cdot), \theta; a_1, \ldots, a_k]$ have been exploited successfully by Gupta and Richards (1991) in order to derive results concerning the uniformly minimum variance unbiased estimators (UMVUEs) of some reliability functions when the observed data have been assumed to have arisen from a multivariate Liouville distribution.

If $\boldsymbol{X} = (X_1, \ldots, X_k)^T \stackrel{d}{=} L_k[f(\cdot), \theta; a_1, \ldots, a_k]$, then $U = \sum_{i=1}^{k} X_i$ is a sufficient statistic for the parameter θ. Also, if $f(\cdot)$ is complete and c is a specified constant, then the UMVUE of the reliability function $R_1(\theta, c) = \Pr[X_1 > c]$ is

$$\hat{R}_1 = \begin{cases} 1 & \text{if } c \leq 0, \\ 1 - I\left(\frac{c}{u}; a_1, a - a_1\right) & \text{if } 0 < c < u, \\ 0 & \text{if } u \leq c, \end{cases} \tag{50.33}$$

where

$$I(t; \alpha, \beta) = \frac{1}{B(\alpha, \beta)} \int_0^t x^{\alpha-1}(1-x)^{\beta-1} dx, \quad 0 < t < 1 \tag{50.34}$$

is the incomplete beta function ratio, $a = \sum_{i=1}^{k} a_i$ and $u = \sum_{i=1}^{k} x_i$.

Suppose we have $\boldsymbol{X} = (X_1, \ldots, X_k)^T \stackrel{d}{=} L_k[f(\cdot), \theta_1; a_1, \ldots, a_k]$ and $\boldsymbol{Y} = (Y_1, \ldots, Y_{k'})^T \stackrel{d}{=} L_{k'}[g(\cdot), \theta_2; b_1, \ldots, b_{k'}]$ with $f(\cdot)$ and $g(\cdot)$ being complete. If $a - a_1 = \sum_{i=2}^{k} a_i$ and $b - b_1 = \sum_{i=2}^{k'} b_i$ are both integers, then the UMVUE of $R_2 = \Pr[X_1 > Y_1]$ is given by

$$\hat{R}_2 = \begin{cases} \frac{\Gamma(a)}{\Gamma(a_1)B(b_1, b-b_1)} \sum_{i=0}^{b-b_1-1} (-1)^i \binom{b-b_1-1}{i} \\ \quad \times \frac{\Gamma(a_1+b_1+i)}{\Gamma(a+b_1+i)(b_1+i)} \left(\frac{u}{v}\right)^{b_1+i} & \text{if } u < v, \\ 1 - \frac{\Gamma(b)}{\Gamma(b_1)B(a_1, a-a_1)} \sum_{i=0}^{a-a_1-1} (-1)^i \binom{a-a_1-1}{i} \\ \quad \times \frac{\Gamma(a_1+b_1+i)}{\Gamma(a_1+b+i)(a_1+i)} \left(\frac{v}{u}\right)^{a_1+i} & \text{if } u \geq v, \end{cases} \tag{50.35}$$

where $u = \sum_{i=1}^{k} x_i$ and $v = \sum_{i=1}^{k'} y_i$. If either $a - a_1$ or $b - b_1$ is not an integer, then the UMVUE of $R_2 = \Pr[X_1 > Y_1]$ is

$$\hat{R}_2 = \begin{cases} \frac{\Gamma(a)\Gamma(b)\Gamma(a_1+b_1)}{\Gamma(a_1)\Gamma(b_1+1)\Gamma(a+b_1)\Gamma(b-b_1+1)} \left(\frac{u}{v}\right)^{b_1} \\ \quad \times {}_3F_2\left[\begin{matrix} -b+b_1, a_1+b_1, b_1 \\ a+b_1, b_1+1 \end{matrix} \middle| \frac{u}{v}\right] & \text{if } u < v, \\ 1 - \frac{\Gamma(a)\Gamma(b)\Gamma(a_1+b_1)}{\Gamma(a_1+1)\Gamma(b_1)\Gamma(a_1+b)\Gamma(a-a_1+1)} \left(\frac{v}{u}\right)^{a_1} \\ \quad \times {}_3F_2\left[\begin{matrix} -a+a_1, a_1+b_1, a_1 \\ a_1+b, a_1+1 \end{matrix} \middle| \frac{v}{u}\right] & \text{if } u \geq v, \end{cases} \tag{50.36}$$

where $_3F_2$ is a generalized hypergeometric function; see Chapter 1 (Section A8) of Johnson, Kotz, and Kemp (1992). When $a - a_1$ and $b - b_1$ are both integers, $_3F_2$ reduces to finite sums, of course.

When $\boldsymbol{X} = (X_1, \ldots, X_k)^T \stackrel{d}{=} L_k[f(\cdot), \theta_1; 1, \ldots, 1]$ and $X_{(1)}$ denotes $\min(X_1, \ldots, X_k)$, we have $kX_{(1)} \stackrel{d}{=} L_1[f_{k-1}(\cdot), \theta_1; 1]$ and, hence, $kX_{(1)} \stackrel{d}{=} X_1$. Note that this property is analogous to a property of the exponential distribution; see, for example, Chapter 19 of Johnson, Kotz, and Balakrishnan (1994).

If $\boldsymbol{X} = (X_1, \ldots, X_k)^T \stackrel{d}{=} L_k[f(\cdot), \theta; a_1, \ldots, a_k]$, then the log-likelihood function is

$$\log L(\theta) = c_k' - a \log \theta + \log f\left(\frac{u}{\theta}\right), \qquad (50.37)$$

where $u = \sum_{i=1}^k x_i$, $a = \sum_{i=1}^k a_i$, and c_k' is just a constant. From (50.37), we note that the maximum likelihood estimator (MLE) of θ exists if and only if the function $h(t) = t^a f(t)$ (for $t > 0$) has a unique positive maximum. Also, if $f(\cdot)$ is twice differentiable, then the MLE of θ (say, $\hat{\theta}$) satisfies the equation

$$a\hat{\theta} f\left(\frac{u}{\hat{\theta}}\right) + u f'\left(\frac{u}{\hat{\theta}}\right) = 0. \qquad (50.38)$$

For the special case when $f(t) = e^{-t} t^\alpha$ ($t > 0$ and $a + \alpha > 0$), in which case $X_1, \ldots, X_k$ are correlated gamma variables (see Section 3), we have the MLE of θ to be

$$\hat{\theta} = u/(a + \alpha) \qquad (50.39)$$

which is also an unbiased estimator of θ; see Gupta and Richards (1991).

When $\boldsymbol{X} = (X_1, \ldots, X_k)^T \stackrel{d}{=} L_k[f(\cdot); 1, \ldots, 1]$, the following three results have been shown to be equivalent:

(i) $\Pr[X_1 \geq t_1, \ldots, X_k \geq t_k] \geq \prod_{i=1}^k \Pr[X_i \geq t_i]$ for $t_i \geq 0$, $i = 1, \ldots, k$;

(ii) the function $h(t) = -\log\left\{\frac{f_k(t)}{f_k(0)}\right\}$ (for $t \geq 0$) is subadditive, that is, $h(t_1 + t_2) \leq h(t_1) + h(t_2)$ for $t_1, t_2 > 0$;

and

(iii) $X_{(1)} = \min(X_1, \ldots, X_k)$ is new-worse-than-used, that is,

$$\Pr[X_{(1)} \geq t_1 + t_2] \geq \Pr[X_{(1)} \geq t_1] \Pr[X_{(1)} \geq t_2] \qquad \text{for } t_1, t_2 > 0.$$

In addition, the following three results have also been shown to be equivalent:

(a) X_k is stochastically increasing in $X_1, \ldots, X_{k-1}$;

(b) $X_1, \ldots, X_k$ are conditionally increasing in sequence;

and

(c) $(X_1, \ldots, X_{k-1})^T$ has multivariate total positivity of order 2 [for a definition, see Johnson, Kotz, and Balakrishnan (1997, p. 276)].

Aitchison (1986) has presented a lucid discussion on the theory of compositional data analysis. Compositional data are observations of a random vector $(X_1, \ldots, X_k, X_{k+1})^T$, where $(X_1, \ldots, X_k)^T$ is supported on the simplex $\mathcal{S}_k$ and $X_{k+1} = 1 - \sum_{i=1}^{k} X_i$ is the "fill-up value." While discussing possible parametric families of distributions appropriate for modeling compositional data, Aitchison (1986) has mentioned that it is necessary that the off-diagonal entries of the correlation matrix of the assumed parametric family exhibit arbitrary sign patterns. He then pointed out that the Dirichlet distributions are not suitable for modeling compositional data, because all the off-diagonal entries in its correlation matrix are all negative and also due to its invariance property under subcompositions. Since these are exactly the same properties satisfied by the multivariate Liouville distributions as previously established in Section 4, any Liouville distribution or any power-scale transformation of it (in which the powers are positive) also will not be suitable for modeling compositional data, as mentioned by Gupta and Richards (1999). In this regard, note that the support of the power-scale transformed variables $(U_1, \ldots, U_k)$ is the region

$$\left\{ u_i > 0 \ (i = 1, \ldots, k) : \sum_{i=1}^{k} \alpha_i u_i^{1/\beta_i} < 1 \right\}, \qquad (50.40)$$

which is different from the simplex support set $(\mathcal{S}_k)$ for compositional data. For this reason, Rayens and Srinivasan (1994) proposed the usage of *generalized Liouville distributions* having joint density functions proportional to

$$f\left(\sum_{i=1}^{k} \left(\frac{u_i}{q_i} \right)^{\beta_i} \right) \prod_{i=1}^{k} u_i^{a_i - 1}, \qquad (u_1, \ldots, u_k)^T \in \mathcal{S}_k, \qquad (50.41)$$

where $f : \mathbb{R}_+ \rightarrow \mathbb{R}_+$. However, as pointed out by Gupta and Richards (1999), the classical Liouville densities on the simplex $\mathcal{S}_k$ cannot be transformed into the densities given in (50.41) by means of a power-scale transformation.

7 SIGN-SYMMETRIC DIRICHLET AND LIOUVILLE DISTRIBUTIONS

Let Z_i $(i = 1, 2, \ldots, k)$ be mutually independent real-valued random variables with probability density function

$$\frac{\alpha_i}{2\Gamma\left(\frac{\beta_i}{\alpha_i}\right)} \, |z_i|^{\beta_i - 1} \, e^{-|z|^{\alpha_i}}, \qquad z_i \in \mathbb{R}, \; \alpha_i > 0, \; \beta_i > 0. \qquad (50.42)$$

Let us define

$$U_i = \frac{Z_i}{\left(\sum_{j=1}^{k} |Z_j|^{\alpha_j}\right)^{1/\alpha_i}} \qquad \text{for } i = 1, 2, \ldots, k. \qquad (50.43)$$

Then, the distribution of the random vector $\boldsymbol{U} = (U_1, \ldots, U_k)^T$ is called the *sign-symmetric Dirichlet distribution* and is denoted by $\mathrm{SD}(\alpha_1, \ldots, \alpha_k; \beta_1, \ldots, \beta_k)$.

If we now consider the random vector

$$\boldsymbol{X} = (X_1, \ldots, X_k)^T \overset{d}{=} \left(U_1 \Theta^{1/\alpha_1}, \ldots, U_k \Theta^{1/\alpha_k}\right)^T, \qquad (50.44)$$

where Θ is a positive random variable distributed independently of $\boldsymbol{U}$, the distribution of $\boldsymbol{X}$ is called the *sign-symmetric Liouville distribution* and is denoted by $\mathrm{SL}(\alpha_1, \ldots, \alpha_k; \beta_1, \ldots, \beta_k; \Theta)$. For the special case when $\alpha_i = \alpha$ and $\beta_i = 1$ for $i = 1, 2, \ldots, k$, the distribution becomes ℓ_α-*isotropic Liouville distribution*. The general sign-symmetric Liouville distributions were introduced by Gupta, Misiewicz, and Richards (1996).

The distribution of the random vector $\boldsymbol{X}$ is said to be sign-symmetric if it has the same distribution as $(r_1 X_1, \ldots, r_k X_k)^T$, where $r_1, \ldots, r_k$ is the so-called Rademacher sequence of independent random variables with $\Pr[r_i = 1] = \Pr[r_i = -1] = \frac{1}{2}$ and are distributed independently of $\boldsymbol{X}$. Here, the terms *sign-symmetric Dirichlet distribution* and *sign-symmetric Liouville distribution* are used in order to underline that these distributions differ from the Dirichlet and Liouville distributions not only by their sign-symmetry but also by the shape of their supports.

Specifically, a random vector $\boldsymbol{U}$ is said to have a *sign-symmetric Dirichlet distribution* with parameters $(\alpha_1, \ldots, \alpha_k)$ and $(\beta_1, \ldots, \beta_k)$, namely, $\boldsymbol{U} \overset{d}{=} \mathrm{SD}(\alpha_1, \ldots, \alpha_k; \beta_1, \ldots, \beta_k)$, if

(i) U_k is a symmetric random variable,

(ii) $\sum_{i=1}^{k} |U_i|^{\alpha_i} = 1$ almost surely,

and

(iii) the joint density function of $(U_1, \ldots, U_{k-1})^T$ is

$$\frac{\Gamma(p_k)}{\Gamma\left(\frac{\beta_k}{\alpha_k}\right)} \left\{ \prod_{i=1}^{k-1} \frac{\alpha_i}{2\Gamma\left(\frac{\beta_i}{\alpha_i}\right)} |u_i|^{\beta_i-1} \right\} \left(1 - \sum_{i=1}^{k-1} |u_i|^{\alpha_i} \right)_+^{\frac{\beta_k}{\alpha_k}-1}, \quad (50.45)$$

where $p_i = \sum_{j=1}^{i} (\beta_j/\alpha_j)$ and $(a)_+ = a$ or 0 according as $a > 0$ or $a \le 0$, respectively.

Suppose $\boldsymbol{U} \stackrel{d}{=} \mathrm{SD}(\alpha_1, \ldots, \alpha_k; \beta_1, \ldots, \beta_k)$. Then, the marginal density function of $(U_1, \ldots, U_\ell)^T$ (for $1 \le \ell < k$) is

$$\frac{\Gamma(p_k)}{\Gamma(p_k - p_\ell)} \left\{ \prod_{i=1}^{\ell} \frac{\alpha_i}{2\Gamma\left(\frac{\beta_i}{\alpha_i}\right)} |u_i|^{\beta_i-1} \right\} \left(1 - \sum_{i=1}^{\ell} |u_i|^{\alpha_i} \right)_+^{p_k-p_\ell-1}. \quad (50.46)$$

Moreover, for $1 < \ell < k$, the conditional density function $(U_1, \ldots, U_{\ell-1})^T$ given $(U_{\ell+1} = u_{\ell+1}, \ldots, U_k = u_k)$ is

$$\frac{\Gamma(p_\ell)}{\Gamma\left(\frac{\beta_\ell}{\alpha_\ell}\right)} \left\{ \prod_{i=1}^{\ell-1} \frac{\alpha_i}{2\Gamma\left(\frac{\beta_i}{\alpha_i}\right)} |u_i|^{\beta_i-1} \right\} \left(1 - \sum_{\substack{i=1 \\ i \ne \ell}}^{k} |u_i|^{\alpha_i} \right)_+^{\frac{\beta_\ell}{\alpha_\ell}-1}$$

$$\times \left(1 - \sum_{i=\ell+1}^{k} |u_i|^{\alpha_i} \right)_+^{1-p_\ell}. \quad (50.47)$$

The moments of U_i can also be determined easily. For example, for $h_1, \ldots, h_k > 0$ we have

$$E \left[\prod_{i=1}^{k} |U_i|^{h_i} \right] = \frac{\Gamma(p_k)}{\Gamma\left(p_k + \sum_{i=1}^{k} \frac{h_i}{\alpha_i}\right)} \prod_{i=1}^{k} \frac{\Gamma\left(\frac{\beta_i+h_i}{\alpha_i}\right)}{\Gamma\left(\frac{\beta_i}{\alpha_i}\right)}; \quad (50.48)$$

moreover,

$$E[U_i] = 0, \quad \mathrm{var}(U_i) = \frac{\Gamma(p_k)}{\Gamma\left(p_k + \frac{2}{\alpha_i}\right)} \frac{\Gamma\left(\frac{\beta_i+2}{\alpha_i}\right)}{\Gamma\left(\frac{\beta_i}{\alpha_i}\right)}, \quad i = 1, \ldots, k,$$

and

$$\mathrm{cov}(U_i, U_j) = 0 \quad \text{for } i \ne j.$$

Note that if Z_i $(i = 1, \ldots, k)$ are independent random variables with density function as in (50.42) and U_i are as defined as in (50.43), it may be directly seen that U_k is a symmetric random variable, $\sum_{i=1}^{k} |U_i|^{\alpha_i} = 1$ almost surely, and the joint density function of $(U_1, \ldots, U_{k-1})^T$ determined from the joint density function of $(Z_1, \ldots, Z_k)$ is as given in Eq. (50.45). Thence, as stated above, $\boldsymbol{U} = (U_1, \ldots, U_k)^T \stackrel{d}{=} \text{SD}(\alpha_1, \ldots, \alpha_k; \beta_1, \ldots, b_k)$.

Now let us suppose that $\boldsymbol{X} = (X_1, \ldots, X_k)^T \stackrel{d}{=} \text{SL}(\alpha_1, \ldots, \alpha_k; \beta_1, \ldots, \beta_k; \Theta)$ and that Θ is absolutely continuous with probability density function $g(\cdot)$. Then, the joint density function of $\boldsymbol{X}$ is given by

$$\Gamma(p_k) \left\{ \prod_{i=1}^{k} \frac{\alpha_i}{2\Gamma\left(\frac{\beta_i}{\alpha_i}\right)} |x_i|^{\beta_i - 1} \right\} \left(\sum_{i=1}^{k} |x_i|^{\alpha_i} \right)^{1 - p_k} g\left(\sum_{i=1}^{k} |x_i|^{\alpha_i} \right), \quad (50.49)$$

where $x_1, \ldots, x_k \in \mathbb{R}$. If $\alpha_i = \alpha$ and $\beta_i = 1$ (for $i = 1, 2, \ldots, k$), then $\boldsymbol{X}$ has ℓ_α-isotropic distribution with density function as constant on the sphere $\{\boldsymbol{x} \in \mathbb{R}^k : \sum_{i=1}^{k} |x_i|^\alpha = c\}$. Furthermore, the random variables $X_1, \ldots, X_k$ are mutually independent if and only if X_i's have densities of the form

$$\frac{\alpha_i \, b^{\beta_i/\alpha_i}}{2\Gamma\left(\frac{\beta_i}{\alpha_i}\right)} \, |x_i|^{\beta_i - 1} \, e^{-b|x_i|^{\alpha_i}} \quad (50.50)$$

for $x_i \in \mathbb{R}$ $(i = 1, 2, \ldots, k)$. In that case, Θ has gamma density function

$$\frac{b^{p_k}}{\Gamma(p_k)} \, r^{p_k - 1} \, e^{-br} \quad \text{for } r > 0. \quad (50.51)$$

Suppose $\boldsymbol{X} \stackrel{d}{=} \text{SL}(\alpha_1, \ldots, \alpha_k; \beta_1, \ldots, \beta_k; \Theta)$. Then, the marginal density function of $(X_1, \ldots, X_\ell)^T$ (for $1 \leq \ell < k$) $\stackrel{d}{=} \text{SL}(\alpha_1, \ldots, \alpha_\ell; \beta_1, \ldots, \beta_\ell; \Theta_\ell)$ where $\Theta_\ell = \left(1 - \sum_{i=\ell+1}^{k} |U_i|^{\alpha_i}\right) \Theta$. Moreover, Θ_ℓ is absolutely continuous (with respect to the Lebesgue measure) and the density function g_ℓ of Θ_ℓ is

$$g_\ell(r) = \frac{\Gamma(p_k)}{\Gamma(p_\ell)\Gamma(p_k - p_\ell)} \, r^{p_\ell - 1} \int_0^\infty (t - r)_+^{p_k - p_\ell - 1} \, t^{1 - p_k} \lambda(dt), \ r > 0,$$

$$(50.52)$$

where $\lambda(\cdot)$ is the distribution function of Θ. The representation $\boldsymbol{X} \stackrel{d}{=} \text{SL}(\alpha_1, \ldots, \alpha_k; \beta_1, \ldots, \beta_k; \Theta)$ of a sign-symmetric Liouville random vector is unique if $n = 2$ and Θ has a continuous density or if $n \geq 3$. The parameters α_i and β_i $(i = 1, 2, \ldots, k)$ are uniquely determined by the distribution of $\boldsymbol{X}$, and Θ is uniquely determined by the constraint $\Theta = \sum_{i=1}^{k} |X_i|^{\alpha_i}$ almost surely.

The conditional distributions of sign-symmetric Liouville random vectors are once again sign-symmetric Liouville as Gupta, Misiewicz, and Richards (1996) have shown. Let $\boldsymbol{X} \stackrel{d}{=} SL(\alpha_1, \ldots, \alpha_k; \beta_1, \ldots, b_k; \Theta)$, where Θ is absolutely continuous with density function $g(\cdot)$. Then, for $1 \leq \ell < k$, the conditional distribution of $(X_1, \ldots, X_\ell)^T$ given $(X_{\ell+1} = x_{\ell+1}, \ldots, x_k = x_k)$ is $SL(\alpha_1, \ldots, \alpha_\ell; \beta_1, \ldots, \beta_\ell; \Theta_c)$, where Θ_c is the conditional random variable $\sum_{i=1}^{\ell} |X_i|^{\alpha_i}$, given $\sum_{i=\ell+1}^{k} |X_i|^{\alpha_i} = \sum_{i=\ell+1}^{k} |x_i|^{\alpha_i}$, whose density is given by

$$x^{p_\ell - 1} \, r^{p_k - p_\ell - 1} (x + r)^{1 - p_k} g(x + r) / g_{k-\ell}(r) \qquad \left(\text{with } r = \sum_{i=\ell+1}^{k} |x_i|^{\alpha_i} \right). \tag{50.53}$$

The moments of the sign-symmetric Liouville distribution can be determined easily from the moments of the sign-symmetric Dirichlet distribution presented earlier through the relationship in (50.44). In particular, $E[\Theta^{2h+1}] < \infty$ implies that $E[X_i^{2h+1}] = 0$ (for $i = 1, \ldots, k$) for any integer h.

8   MULTIVARIATE p-ORDER LIOUVILLE DISTRIBUTIONS

An absolutely continuous random vector $\boldsymbol{X} = (X_1, \ldots, X_k)^T$ with nonnegative components has a *multivariate p-order Liouville distribution* with order $p > 0$, parameter $\theta > 0$ and density generator $f(\cdot)$, denoted by $L_{k,p}[f(\cdot), \theta; a_1, \ldots, a_k]$ if the joint density function of $\boldsymbol{X}$ is of the form

$$\frac{c}{\theta^a} \left\{ \prod_{i=1}^{k} x_i^{a_i - 1} \right\} f\left(\frac{\|\boldsymbol{x}\|}{\theta} \right), \tag{50.54}$$

where $a_i > 0$ (for $i = 1, 2, \ldots, k$), $a = \sum_{i=1}^{k} a_i$, c is the normalizing constant, $\|\boldsymbol{x}\| = \left(\sum_{i=1}^{k} x_i^p \right)^{1/p}$ is the ℓ_p-norm of $\boldsymbol{x}$, and $f(\cdot)$ is a nonnegative measurable function on $\mathbb{R}_+$ such that $0 < \int_0^\infty t^{a-1} f(t) \, dt < \infty$. This definition, generalizing the multivariate Liouville distribution (case when $p = 1$), is due to Ma and Yue (1995). It also generalizes the multivariate ℓ_p-norm symmetric distribution (case when $a_i = p$ for $i = 1, \ldots, k$); see Yue and Ma (1995).

The components of $\boldsymbol{X}$ can be viewed as an univariate dependent sample of random lifetimes of k nonrenewable components or a coherent system or of proportional hazards model when the joint density of X is given by (50.54). Ma, Yue and Balakrishnan (1996), in addition to discussing the basic properties and the dependence structure of multivariate p-order Liouville distributions, also discussed the multivariate order statistics induced by ordering the ℓ_p-norm.

Ma and Yue (1995) have considered estimation of the parameter θ. Let $X \stackrel{d}{=} L_{k,p}[f(\cdot), \theta; a_1, \ldots, a_k]$, $r = \|x\|$, and $m(t) = t^a f(t)$. If $f(t)$ is continuous on $\mathbb{R}_+$ and the maximum of $m(t)$ is attained at a finite positive point t_m, then $\frac{r}{t_m}$ is the MLE of θ. r is also a sufficient statistic for θ. If $f(t)$ is decreasing for sufficiently large t, $m(t)$ has a finite positive maximum point, and $m(t)$ is integrable on $\mathbb{R}_+$, then there exists an unbiased estimator of θ.

As a special case, let us consider the multivariate Lomax distribution of Nayak (1987) which has the joint density function

$$\frac{c}{\theta^a} \left\{ \prod_{i=1}^{k} x_i^{a_i-1} \right\} \left(1 + \frac{1}{\theta} \sum_{i=1}^{k} x_i \right)^{-(a+\ell)}, \qquad (50.55)$$

where $\ell > 0$. In this case, $m(t) = t^a (1+t)^{-(a+\ell)}$ attains its maximum at $t_m = a/\ell$ and the MLE of θ is $\hat{\theta} = \frac{\ell}{a} \sum_{i=1}^{k} x_i$, which is also unbiased. Note that $f(t) = (1+t)^{-(a+\ell)}$ is complete and that $\hat{\theta}$ is UMVUE of θ.

BIBLIOGRAPHY

(Some bibliographical items not mentioned in the text are included here for completeness.)

Aitchison, J. (1986). *The Statistical Analysis of Compositional Data*, London: Chapman and Hall.

Anderson, T. W., and Fang, K. T. (1982). Distributions of quadratic forms and Cochran's theorem for elliptically contoured distributions and their applications, Technical Report No. 53, Department of Statistics, Stanford University, California.

Anderson, T. W., and Fang, K. T. (1987). Cochran's theorem for elliptically contoured distributions, *Sankhyā, Series A*, **49**, 305–315.

Barndorff-Nielsen, O. E., and Jørgensen, B. (1991). Some parametric models on the simplex, *Journal of Multivariate Analysis*, **39**, 106–116.

Diaconis, P., and Perlman, M. D. (1987). Bounds for tail probabilities of weighted sums of independent gamma random variables, in *Topics in Statistical Dependence*, IMS Lecture Notes—**16**, pp. 147–166, Hayward, California: Institute of Mathematical Statistics.

Fang, K. T., and Fang, B. Q. (1988). Some families of multivariate symmetric distributions related to exponential distribution, *Journal of Multivariate Analysis*, **24**, 109–122.

Fang, K. T., and Fang, B. Q. (1989). A characterization of multivariate ℓ_1-norm symmetric distribution, *Statistics & Probability Letters*, **7**, 297–299.

Fang, K. T., Kotz, S., and Ng, K. W. (1989). *Symmetric Multivariate and Related Distributions*, London: Chapman and Hall.

Gupta, R. D., Misiewicz, J. K., and Richards, D. St. P. (1996). Infinite sequences with sign-symmetric Liouville distributions, *Probability and Mathematical Statistics*, **16**, 29–44.

Gupta, R. D., and Richards, D. St. P. (1987). Multivariate Liouville distributions, *Journal of Multivariate Analysis*, **23**, 233–256.

Gupta, R. D., and Richards, D. St. P. (1990). The Dirichlet distributions and polynomial regression, *Journal of Multivariate Analysis*, **32**, 95–102.

Gupta, R. D., and Richards, D. St. P. (1991). Multivariate Liouville distributions, II, *Probability and Mathematical Statistics*, **12**, 291–309.

Gupta, R. D., and Richards, D. St. P. (1992). Multivariate Liouville distributions, III, *Journal of Multivariate Analysis*, **43**, 29–57.

Gupta, R. D., and Richards, D. St. P. (1995). Multivariate Liouville distributions, IV, *Journal of Multivariate Analysis*, **54**, 1–17.

Gupta, R. D., and Richards, D. St. P. (1997). Multivariate Liouville distributions, V, in *Advances in the Theory and Practice of Statistics: A Volume in Honor of Samuel Kotz* (N. L. Johnson and N. Balakrishnan, eds.), pp. 377–396, New York: John Wiley & Sons.

Gupta, R. D., and Richards, D. St. P. (1999). The covariance structure of the multivariate Liouville distributions, *Journal of Statistical Planning and Inference* (to appear).

Johnson, N. L., Kotz, S., and Balakrishnan, N. (1994). *Continuous Univariate Distributions*, Vol. 1, second edition, New York: John Wiley & Sons.

Johnson, N. L., Kotz, S., and Balakrishnan, N. (1995). *Continuous Univariate Distributions*, Vol. 2, second edition, New York: John Wiley & Sons.

Johnson, N. L., Kotz, S., and Balakrishnan, N. (1997). *Discrete Multivariate Distributions*, New York: John Wiley & Sons.

Johnson, N. L., Kotz, S., and Kemp, A. W. (1992). *Discrete Univariate Distributions*, second edition, New York: John Wiley & Sons.

Kotlarski, I. I. (1967). On characterizing the gamma and the normal distribution, *Pacific Journal of Mathematics*, **20**, 69–76.

Ma, C., and Yue, X. (1995). Multivariate p-order Liouville distributions: Parameter estimation and hypothesis testing, *Chinese Journal of Applied Probability and Statistics*, **11**, 425–431.

Ma, C., Yue, X., and Balakrishnan, N. (1996). Multivariate p-order Liouville distributions: Definitions, properties and multivariate order statistics induced by ordering ℓ_p-norm, Technical Report, McMaster University, Hamilton, Ontario, Canada.

Marshall, A. W., and Olkin, I. (1979). *Inequalities: Theory of Majorization and Its Applications*, New York: Academic Press.

Nayak, T. (1987). Multivariate Lomax distribution: Properties and usefulness in reliability theory, *Journal of Applied Probability*, **24**, 170–177.

Ng, K. W. (1989). In *Symmetric Multivariate and Related Distributions* by K.-T. Fang, S. Kotz, and K. W. Ng, London: Chapman and Hall.

Rayens, W. S., and Srinivasan, C. (1994). Dependence properties of generalized Liouville distributions on the simplex, *Journal of the American Statistical Association*, **89**, 1465–1470.

Shaked, M. (1977). A concept of positive dependence for exchangeable random variables, *Annals of Statistics*, **5**, 505–515.

Sivazlian, B. D. (1981). On a multivariate extension of the gamma and beta distributions, *SIAM Journal of Applied Mathematics*, **41**, 205–209.

Wesolowski, J. (1993). Some remarks on the multivariate Liouville distribution, *Statistics*, **24**, 167–170.

Yue, X., and Ma, C. (1995). Multivariate ℓ_p-norm symmetric distributions, *Statistics & Probability Letters*, **24**, 281-283.

CHAPTER 51

Multivariate Logistic Distributions

1 INTRODUCTION

The univariate logistic distribution has been studied rather extensively and, in fact, many of its developments through the years were motivated by considering the logistic distribution as an alternative to the normal distribution; for details, see the handbook of Balakrishnan (1992). However, work on multivariate logistic distributions has been rather skimpy compared to the voluminous work that has been carried out on bivariate and multivariate normal distributions. A casual glance over this chapter and Chapters 45 and 46 will likely provide ample testimony to this statement.

As a matter of fact, the first attempt to define bivariate logistic distributions was made only in 1961 by Gumbel (1961). In his pioneering paper, Gumbel (1961) actually proposed three bivariate logistic distributions in the following forms [with $(x, y) \in \mathbb{R}^2$]:

$$F_{X,Y}(x, y) = \frac{1}{1 + e^{-x} + e^{-y}}, \tag{51.1}$$

$$F_{X,Y}(x, y) = \exp\left[-\left\{\left(\log(1 + e^{-x})\right)^{1/\alpha} + \left(\log(1 + e^{-y})\right)^{1/\alpha}\right\}^{\alpha}\right], \tag{51.2}$$

and

$$F_{X,Y}(x,y)$$

$$= \frac{1}{(1+e^{-x})(1+e^{-y})} \left\{ 1 + \frac{\alpha\, e^{-x-y}}{(1+e^{-x})(1+e^{-y})} \right\}, \quad -1 < \alpha < 1.$$

$$(51.3)$$

Of course, location and scale parameters can easily be introduced in all these models. It is quite clear that these models have been constructed by exploiting some specific property or form of the univariate logistic distribution and that the model is very specific with regard to the nature of the dependence between the random variables X and Y. While the bivariate logistic distribution in (51.1) is the simplest and the most extensively studied one, the distribution in (51.3) is in the familiar Farlie–Gumbel–Morgenstern form (see Section 12 of Chapter 44 for details).

Twelve years after Gumbel's (1961) work on bivariate logistic distributions, a natural multivariate extension of the bivariate logistic distribution in (51.1) was given by Malik and Abraham (1973). Though quite restrictive in its correlation structure, the work of Malik and Abraham (1973) seemed to have generated renewed interest on the construction, properties and applications of bivariate and multivariate logistic distributions. A lucid review of multivariate logistic distributions has been prepared by Arnold (1992) [Chapter 11 in the handbook of Balakrishnan (1992)], and naturally it served as a basis for much of the discussion in this chapter.

2 GUMBEL–MALIK–ABRAHAM DISTRIBUTION

Gumbel (1961) considered the bivariate logistic distribution with joint distribution function as

$$F_{X,Y}(x,y) = \frac{1}{1 + e^{-(x-\mu_1)/\sigma_1} + e^{-(y-\mu_2)/\sigma_2}}, \quad (x,y) \in \mathbb{R}^2, \qquad (51.4)$$

and corresponding joint density function as

$$p_{X,Y}(x,y) = \frac{2\, e^{-(x-\mu_1)/\sigma_1} e^{-(y-\mu_2)/\sigma_2}}{\sigma_1 \sigma_2 \{1 + e^{-(x-\mu_1)/\sigma_1} + e^{-(y-\mu_2)/\sigma_2}\}^3}, \quad (x,y) \in \mathbb{R}^2. \ (51.5)$$

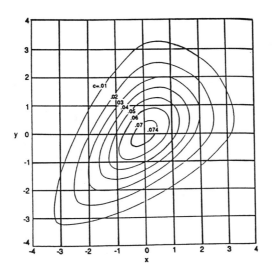

FIGURE 51.1

Contours of Bivariate Logistic Distribution with $p_{X,Y}(x, y) = c$.

The standard case of $\mu_1 = \mu_2 = 0$ and $\sigma_1 = \sigma_2 = 1$, of course, corresponds to the distribution in (51.1).

From (51.4), by letting x or y go to ∞, we readily obtain the marginal distribution functions of X and Y as

$$F_X(x) = \frac{1}{1 + e^{-(x-\mu_1)/\sigma_1}} \quad \text{and} \quad F_Y(y) = \frac{1}{1 + e^{-(y-\mu_2)/\sigma_2}} \qquad (51.6)$$

which are the univariate logistic distributions with mean μ_i and variance $\pi^2 \sigma_i^2 / 3$, $(i = 1, 2)$, respectively; see Balakrishnan (1992).

Contours of the bivariate density function (51.5) for the standard case of $\mu_1 = \mu_2 = 0$ and $\sigma_1 = \sigma_2 = 1$, taken from Gumbel (1961), are presented in Figure 51.1. Positive correlation of this distribution is clearly evident from the arrowhead-shaped contours.

Using (51.5) and (51.6), we obtain the conditional density function of $X|Y$ as

$$p_{X|Y}(x|y) = \frac{2\, e^{-(x-\mu_1)/\sigma_1} \left(1 + e^{-(y-\mu_2)/\sigma_2}\right)^2}{\sigma_1 \left\{1 + e^{-(x-\mu_1)/\sigma_1} + e^{-(y-\mu_2)/\sigma_2}\right\}^3} \qquad (51.7)$$

and the conditional density function of $Y|X$ as

$$p_{Y|X}(y|x) = \frac{2\, e^{-(y-\mu_2)/\sigma_2} \left(1 + e^{-(x-\mu_1)/\sigma_1}\right)^2}{\sigma_2 \left\{1 + e^{-(x-\mu_1)/\sigma_1} + e^{-(y-\mu_2)/\sigma_2}\right\}^3} \cdot \qquad (51.8)$$

From (51.7), we derive the conditional moment generating function of X given $Y = y$ as

$$
\begin{aligned}
E\left[e^{t_1 X} | Y = y\right] \\
&= 2\left(1 + e^{-(y-\mu_2)/\sigma_2}\right)^2 \int_{-\infty}^{\infty} \frac{e^{t_1 x}\, e^{-(x-\mu_1)/\sigma_1}}{\sigma_1 \left\{1 + e^{-(x-\mu_1)/\sigma_1} + e^{-(y-\mu_2)/\sigma_2}\right\}^3}\, dx \\
&= 2\left(1 + e^{-(y-\mu_2)/\sigma_2}\right)^2 e^{t_1 \mu_1} \\
&\quad \times \int_0^{1/(1+e^{-(y-\mu_2)/\sigma_2})} v\left\{\frac{1}{v} - \left(1 + e^{-(y-\mu_2)/\sigma_2}\right)\right\}^{-t_1 \sigma_1}\, dv \\
&\qquad \left(\text{with } v = \frac{1}{1 + e^{-(x-\mu_1)/\sigma_1} + e^{-(y-\mu_2)/\sigma_2}}\right) \\
&= \frac{2\, e^{t_1 \mu_1}}{\left(1 + e^{-(y-\mu_2)/\sigma_2}\right)^{t_1 \sigma_1}} \int_0^1 w^{t_1 \sigma_1 + 1}(1 - w)^{-t_1 \sigma_1}\, dw \\
&\qquad \left(\text{with } w = v\left(1 + e^{-(y-\mu_2)/\sigma_2}\right)\right) \\
&= \left(1 + e^{-(y-\mu_2)/\sigma_2}\right)^{-t_1 \sigma_1} e^{t_1 \mu_1} \Gamma(2 + t_1 \sigma_1) \Gamma(1 - t_1 \sigma_1). \qquad (51.9)
\end{aligned}
$$

From (51.9), we readily obtain the conditional mean and the conditional second moment of X given $Y = y$ as

$$
\begin{aligned}
E[X | Y = y] &= \mu_1 + \sigma_1 \Gamma'(2) - \sigma_1 \Gamma'(1) - \sigma_1 \ln\left(1 + e^{-(y-\mu_2)/\sigma_2}\right) \\
&= \mu_1 + \sigma_1 - \sigma_1 \ln\left(1 + e^{-(y-\mu_2)/\sigma_2}\right) \qquad (51.10)
\end{aligned}
$$

and

$$
\begin{aligned}
E[X^2 | Y = y] &= \mu_1^2 + \sigma_1^2 \Gamma''(2) + \sigma_1^2 \Gamma''(1) + \sigma_1^2 \left\{\ln\left(1 + e^{-(y-\mu_2)/\sigma_2}\right)\right\}^2 \\
&= \mu_1^2 + 2\sigma_1^2 \Gamma''(1) + 2\sigma_1^2 \Gamma'(1) + \sigma_1^2 \left\{\ln\left(1 + e^{-(y-\mu_2)/\sigma_2}\right)\right\}^2. \\
&\qquad\qquad\qquad (51.11)
\end{aligned}
$$

For the standard case, for example, (51.10) reduces to

$$
E[X | Y = y] = 1 - \ln(1 + e^{-y}). \qquad (51.12)
$$

Analogous expressions can be similarly presented for the conditional moments of Y given $X = x$. The resulting curvilinear regression curves, taken from Gumbel (1961), are presented in Figure 51.2, wherein they are compared with the corresponding regression lines for the bivariate normal distribution with correlation coefficient equal to that of this bivariate logistic distribution.

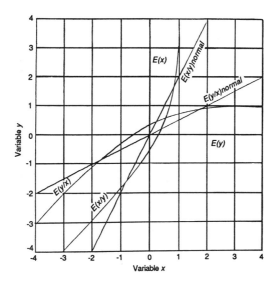

FIGURE 51.2

Regression Curves for the Bivariate Logistic and Bivariate Normal
Distributions.

Let U, V, and W be independent and identically distributed extreme
value random variables with density function (see Chapter 22)

$$p_U(u) = e^{-u}\, e^{-e^{-u}}, \qquad -\infty < u < \infty, \tag{51.13}$$

and moment-generating function

$$M(t) = \Gamma(1 - t) \qquad \text{for } |t| < 1.$$

Let us now consider the transformation

$$X = V - U \quad \text{and} \quad Y = W - U. \tag{51.14}$$

Then, by change of variables, we obtain the joint density function of X
and Y as

$$
\begin{aligned}
p_{X,Y}(x, y) &= e^{-x-y} \int_{-\infty}^{\infty} e^{-3u}\, e^{-(1+e^{-x}+e^{-y})}\, du \\
&= \frac{2\, e^{-x} e^{-y}}{(1 + e^{-x} + e^{-y})^3}, \qquad x, y \in \mathbb{R}^2,
\end{aligned}
$$

which is the standard bivariate logistic density function corresponding to
(51.5).

From the representation in (51.14), we readily get

$$\operatorname{cov}(X,Y) = \operatorname{cov}(V - U, W - U) = \operatorname{var}(U) = \frac{\pi^2}{6},$$

which, together with the fact that $\operatorname{var}(X) = \operatorname{var}(Y) = \pi^2/3$, yields the correlation coefficient between X and Y as

$$\operatorname{corr}(X,Y) = \frac{1}{2}. \qquad (51.15)$$

This clearly reflects the restrictive nature of this bivariate logistic distribution.

For the standard bivariate logistic distribution with joint cumulative distribution function

$$F_{X,Y}(x,y) = \frac{1}{1 + e^{-x} + e^{-y}}$$

and joint density function

$$p_{X,Y}(x,y) = \frac{2e^{-x-y}}{(1 + e^{-x} + e^{-y})^3},$$

we have the corresponding copula

$$C(u,v) = \frac{2uv}{(u + v - uv)^3}.$$

Now let us consider Mardia's generalized bivariate Pareto distribution with joint survival function (see Chapter 52)

$$\bar{F}_{X,Y}(x,y) = \left(\frac{\alpha_1 \alpha_2}{\alpha_1 x + \alpha_2 y - \alpha_1 \alpha_2}\right)^a \quad \text{for } x > \alpha_1 > 0, \ y > \alpha_2 > 0, \ a > 0.$$

The corresponding copula, for the case $a = 1$, is obtained to be

$$C(u,v) = \frac{2(1-u)(1-v)}{\{(1-u) + (1-v) - (1-u)(1-v)\}^3}.$$

Rotating this surface about $\left(\frac{1}{2}, \frac{1}{2}\right)$ by π radians (or $180°$), we obtain the surface given above for the bivariate logistic distribution. As a matter of fact, the dependence imposed between X and Y in the bivariate logistic distribution is equivalent to the dependence imposed between $\frac{\alpha_1 X}{X - \alpha_1}$ and $\frac{\alpha_2 Y}{Y - \alpha_2}$ when $(X,Y)^T$ have the above-given Mardia's bivariate Pareto distribution. This bivariate Pareto distribution can be transformed to the bivariate logistic distribution upon transforming the marginal variables X to $1 + e^{-aX}$.

Generalization of this bivariate distribution to the k-dimensional case is fairly straightforward. From (51.4), we may say that $\boldsymbol{X} = (X_1, \ldots, X_k)^T$ has a k-variate logistic distribution if its joint distribution function is of the form

$$F_{\boldsymbol{X}}(\boldsymbol{x}) = \frac{1}{1 + \sum_{i=1}^{k} e^{-(x_i - \mu_i)/\sigma_i}}, \qquad -\infty < x_i < \infty, \qquad (51.16)$$

and its joint density function is of the form

$$F_{\boldsymbol{X}}(\boldsymbol{x}) = \frac{k! \, e^{-\sum_{i=1}^{k}(x_i - \mu_i)/\sigma_i}}{\left(\prod_{i=1}^{k} \sigma_i\right)\left\{1 + \sum_{i=1}^{k} e^{-(x_i - \mu_i)/\sigma_i}\right\}^{k+1}}, \qquad -\infty < x_i < \infty. \tag{51.17}$$

The marginal distribution of X_i in this case is univariate logistic with mean μ_i and variance $\pi^2 \sigma_i^2 / 3$ for $i = 1, \ldots, k$. For the multivariate logistic distribution in (51.17), Ahmed and Gokhale (1989) have derived a formula for the entropy.

In the standard case, of course, the above forms reduce to

$$F_{\boldsymbol{X}}(\boldsymbol{x}) = \frac{1}{1 + \sum_{i=1}^{k} e^{-x_i}}, \qquad -\infty < x_i < \infty, \qquad (51.18)$$

and

$$p_{\boldsymbol{X}}(\boldsymbol{x}) = \frac{k! e^{-\sum_{i=1}^{k} x_i}}{\left(1 + \sum_{i=1}^{k} e^{-x_i}\right)^{k+1}}, \qquad -\infty < x_i < \infty. \tag{51.19}$$

From (51.19), we obtain the joint moment-generating function of the standard k-variate logistic distribution as

$$\begin{aligned}
M_{\boldsymbol{X}}(\boldsymbol{t}) &= \int_{-\infty}^{\infty} \cdots \int_{-\infty}^{\infty} \frac{k! \, e^{-\sum_{i=1}^{k}(1 - t_i)x_i}}{\left\{1 + \sum_{i=1}^{k} e^{-x_i}\right\}^{k+1}} \, dx_1 \cdots dx_k \\
&= \Gamma\left(1 + \sum_{i=1}^{k} t_i\right) \prod_{i=1}^{k} \Gamma(1 - t_i) \qquad \text{for } |t_i| < 1, \ i = 1, \ldots, k.
\end{aligned} \tag{51.20}$$

This expression, in fact, suggests the following representation of this standard k-variate logistic distribution, which is an extension of (51.14). With

$U_0, U_1, \ldots, U_k$ being independent and identically distributed extreme value random variables with common density function (51.13), the random variables

$$X_i = U_i - U_0, \qquad i = 1, 2, \ldots, k, \tag{51.21}$$

are jointly distributed as the standard k-variate logistic distribution in (51.18). It is then clear that

$$\mathrm{var}(X_i) \;=\; \pi^2/3 \qquad \text{for all } i,$$

$$\mathrm{cov}(X_i, X_j) \;=\; \mathrm{cov}(U_i - U_0, U_j - U_0) = \mathrm{var}(U_0) = \frac{\pi^2}{6}, \qquad i \ne j,$$

and, hence,

$$\mathrm{corr}(X_i, X_j) = \frac{\mathrm{cov}(X_i, X_j)}{\sqrt{\mathrm{var}(X_i)\mathrm{var}(X_j)}} = \frac{\pi^2/6}{\pi^2/3} = \frac{1}{2}.$$

From (51.19), we obtain the conditional joint density function of $(X_{\ell+1}, \ldots, X_k)$, given $(X_1 = x_1, \ldots, X_\ell = x_\ell)$, as

$$
\begin{aligned}
&p(x_{\ell+1}, \ldots, x_k | x_1, \ldots, x_\ell) \\[4pt]
&= \; \frac{k!}{\ell!} \, \frac{\left(1 + \sum_{i=1}^{\ell} e^{-x_i}\right)^{\ell+1}}{\left(1 + \sum_{i=1}^{k} e^{-x_i}\right)^{k+1}} \exp\left(-\sum_{i=\ell+1}^{k} x_i\right) \\[6pt]
&= \; k^{(k-\ell)} \left(1 + \sum_{i=1}^{k} e^{-x_i}\right)^{-(k-\ell)} \left\{ 1 + \frac{\sum_{i=\ell+1}^{k} e^{-x_i}}{1 + \sum_{i=1}^{\ell} e^{-x_i}} \right\}^{-(\ell+1)} \\[6pt]
&\qquad\qquad\qquad \times \exp\left(-\sum_{i=\ell+1}^{k} x_i\right).
\end{aligned}
$$

Denoting $Y_i = X_i + \log\left(1 + \sum_{i=1}^{\ell} e^{-x_i}\right)$ for $i = \ell+1, \ldots, k$, we obtain the conditional joint density function of $(Y_{\ell+1}, \ldots, Y_k)$, given $(X_1 = x_1, \ldots, X_\ell = x_\ell)$, as

$$
\begin{aligned}
&p(y_{\ell+1}, \ldots, y_k | x_1, \ldots, x_\ell) \\[4pt]
&= \; k^{(k-\ell)} \left(1 + \sum_{i=\ell+1}^{k} e^{-y_i}\right)^{-(k+1)} \exp\left(-\sum_{i=\ell+1}^{k} y_i\right).
\end{aligned}
$$

This simply reveals that the conditional joint distribution of $(X_{\ell+1}, \ldots, X_k)$, given $(X_1, \ldots, X_\ell)$, is always of the same shape and scale, and varies only in regard to location parameter; in other words, the array variation is homoscedastic. Also, since

$$E[Y_1|X_2,\ldots,X_k] = \psi(1) - \psi(k),$$

we obtain

$$E[X_1|X_2,\ldots,X_k] = \psi(1) - \psi(k) - \log\left(1 + \sum_{i=2}^{k} e^{-x_i}\right),$$

where $\psi(\cdot)$ is the digamma function.

Moore (1969) showed how estimators of improved accuracy can be obtained by using grouped (contingency table) multivariate data. He considered the bivariate case with $\sigma_1 = \sigma_2 = 1$ and discussed the estimation of μ_1 and μ_2 by applying a general method of estimation using data from two-way contingency tables with boundaries defined by assuming σ to be known (and equal to 1). Suppose that the data are divided into a 4×4 table with boundaries at the $(1+e)^{-1}$, $\frac{1}{2}$, and $(1+e^{-1})^{-1}$ sample quantiles and that the proportions in each cell (ij) are observed. Then, estimators for the parameters μ_1 and μ_2 can be constructed based on the observed proportions and the sample quantiles. These estimators turn out to be more efficient than the corresponding marginal sample means or medians.

3   FRAILTY AND ARCHIMEDEAN DISTRIBUTIONS

Frailty models are of great interest in the analysis of survival data; see, for example, Oakes (1989). Although, in this context, these models are applied to random variables whose support is in the positive real line, the models may be extended to random variables whose support is in the entire real line, even though the practical motivation for such models may be lacking.

Let X denote a standard univariate logistic random variable. Then, a frailty representation of X corresponding to a specified distribution P defined on $(0, \infty)$ is of the form

$$\frac{1}{1+e^{-x}} = \Pr[X \le x] = \int_0^\infty \{\Pr[U \le x]\}^\theta \, dP(\theta)$$
$$= \int_0^\infty e^{-\theta\{-\log \Pr[U \le x]\}} \, dP(\theta). \quad (51.22)$$

Since the right-hand side of (51.22) is in the form of the Laplace transform of the distribution $P(\theta)$, it is clear that the distribution of U is uniquely determined once the distribution $P(\theta)$ is specified. In fact, we have

$$\Pr[U \le x] = e^{-L_P^{-1}\left(\frac{1}{1+e^{-x}}\right)}, \tag{51.23}$$

where $L_P(t) = \int_0^\infty e^{-\theta t} dP(\theta)$ is the Laplace transform of $P(\theta)$.

For a specified distribution P on $(0, \infty)$, let us consider a standard k-variate logistic distribution as one with joint cumulative distribution function as

$$F_{\boldsymbol{X}}(\boldsymbol{x}) = \int_0^\infty \left\{ \prod_{i=1}^k \Pr[U \le x_i] \right\}^\theta dP(\theta), \tag{51.24}$$

where the distribution of U is related to P as given in (51.23). We can then write

$$\begin{aligned}
F_{\boldsymbol{X}}(\boldsymbol{x}) &= \int_0^\infty e^{-\theta \sum_{i=1}^k L_P^{-1}\left(\frac{1}{1+e^{-x_i}}\right)} dP(\theta) \\
&= L_P\left(\sum_{i=1}^k L_P^{-1}\left(\frac{1}{1+e^{-x_i}}\right) \right),
\end{aligned} \tag{51.25}$$

where, as before, $L_P(t)$ is the Laplace transform of the distribution $P(\theta)$. Clearly, the resulting multivariate logistic distribution will have all its marginal distributions as univariate logistic, since (51.22) holds when the distribution U and $P(\theta)$ are related as in (51.23).

As an example, let us now choose $P(\theta)$ to be Gamma$(\alpha, 1)$ distribution with probability density function (see Chapter 17)

$$dP(\theta) = \frac{1}{\Gamma(\alpha)} e^{-\theta} \theta^{\alpha-1} \, d\theta, \qquad \theta > 0, \ \alpha > 0$$

whose Laplace transform is given by

$$L_P(t) = \int_0^\infty e^{-\theta t} \frac{1}{\Gamma(\alpha)} e^{-\theta} \theta^{\alpha-1} \, d\theta = \frac{1}{(1+t)^\alpha}$$

and as a result

$$L_P^{-1}(u) = u^{-1/\alpha} - 1.$$

Using these expressions in (51.25), we obtain the corresponding multivariate logistic distribution function as

$$F_{\boldsymbol{X}}(\boldsymbol{x}) = \frac{1}{\left\{ 1 + \sum_{i=1}^k (1+e^{-x_i})^{1/\alpha} - k \right\}^\alpha} \qquad \text{with } \alpha > 0. \tag{51.26}$$

For the case when $\alpha = 1$, the multivariate logistic distribution in (51.26) simply reduces to the Gumbel–Malik–Abraham distribution in (51.18). Therefore, the multivariate logistic distribution in (51.26) may be regarded as a one-parameter extension of the Gumbel–Malik–Abraham distribution in (51.18).

Next, let us consider the distribution $P(\theta)$ for which the Laplace transform is $L_P(t) = e^{-t^\alpha}$ for $\alpha \leq 1$. In this case, we have $L_P^{-1}(u) = (-\log u)^{1/\alpha}$ and the corresponding multivariate logistic distribution function is obtained from (51.25) as

$$F_{\boldsymbol{X}}(\boldsymbol{x}) = \exp\left[-\left\{\sum_{i=1}^{k}\left(\log(1 + e^{-x_i})\right)^{1/\alpha}\right\}^\alpha\right], \qquad \alpha \leq 1. \qquad (51.27)$$

Observe that this is the k-dimensional version of the bivariate logistic distribution in (51.2) introduced by Gumbel (1961).

It should be mentioned that $L_P(\cdot)$ in (51.25) need not be a Laplace transform. It is sufficient if it is a nonnegative decreasing function with $L_P(0) = 1$ and has a nonnegative second derivative. In such a general form, the distribution in (51.25) becomes an *Archimedean distribution* as termed by Genest and MacKay (1986); see also Chapter 4 of Nelsen (1998).

4   FARLIE–GUMBEL–MORGENSTERN DISTRIBUTIONS

As seen earlier in Section 12 of Chapter 44, the Farlie–Gumbel–Morgenstern family of one-parameter k-variate logistic distributions has its joint cumulative distribution function as

$$F_{\boldsymbol{X}}(\boldsymbol{x}) = \left\{\prod_{i=1}^{k}\frac{1}{1 + e^{-x_i}}\right\}\left\{1 + \alpha\prod_{i=1}^{k}\frac{e^{-x_i}}{1 + e^{-x_i}}\right\}, \qquad -1 < \alpha < 1.$$

$$(51.28)$$

For the distribution in (51.28), it may be easily shown that

$$\text{corr}(X_i, X_j) = \frac{3\alpha}{\pi^2}$$

which in absolute value is less than 0.304. Such a restricted range for the correlation coefficient has limited the use of the multivariate logistic distribution in (51.28). Another limitation of this distribution is the lack of a stochastic model that can possibly account for data with such a distribution. Note that Gumbel's third bivariate logistic distribution in (51.3) belongs to the family (51.28). In the bivariate case, Smith and Moffatt (1999) discussed the properties of the maximum likelihood estimator of the correlation. By considering the situations when both variables are fully observed, when both variables are censored at zero, and when both variables are observed only in sign, they examined how the Fisher information corresponding to the correlation coefficient will be reduced due to censoring.

The Farlie generalization of the multivariate logistic distribution in (51.28) will allow the terms $e^{-x_i}/(1 + e^{-x_i})$ to be replaced by appropriate functions of the form $\phi_i(x_i)$ subject only to the condition that the resulting expression is a proper k-dimensional distribution function.

Further generalizations of the Farlie–Gumbel–Morgenstern distributions can be presented. For example, let S denote the class of all possible k-dimensional vectors of 0's and 1's with at least two of its elements equal to 1; thus, S contains $2^k - k - 1$ vectors with a generic element denoted by $s = (s_1, \ldots, s_k)$. Then, a general k-dimensional logistic distribution can be defined as one with its joint distribution function as

$$F_{\boldsymbol{X}}(\boldsymbol{x}) = \left\{ \prod_{i=1}^{k} \frac{1}{1 + e^{-x_i}} \right\} \left\{ 1 + \sum_{\boldsymbol{s} \in S} \alpha_{\boldsymbol{s}} \prod_{i=1}^{k} \left(\frac{e^{-x_i}}{1 + e^{-x_i}} \right)^{s_i} \right\}. \quad (51.29)$$

All the marginal distributions of the k-variate logistic distribution in (51.29) are univariate standard logistic distributions. Once again, a more general family of multivariate logistic distributions can be obtained by replacing the terms $e^{-x_i}/(1 + e^{-x_i})$ by appropriate functions subject only to the condition that the resulting expression is a proper k-dimensional distribution function.

5 DIFFERENCES OF EXTREME VALUE VARIABLES

It is well known that if U and V are independent and identically distributed as extreme value [see Section 16 of Chapter 22 of Johnson, Kotz, and Balakrishnan (1995)] with density function as in (51.13), then $U - V$ has a standard univariate logistic distribution. This immediately suggests some ways of constructing multivariate logistic distributions.

If $\boldsymbol{U}$ and $\boldsymbol{V}$ are independent k-dimensional random vectors with all their marginal distributions as extreme value, then the vector $\boldsymbol{X} = \boldsymbol{U} - \boldsymbol{V}$

has a multivariate logistic distribution. Specifically, if we choose $U_1, \ldots, U_k$ to be independent extreme value random variables, and $V_1 = V_2 = \cdots = V_k$ possesses an extreme value distribution, then $\boldsymbol{X}$ has the k-variate Gumbel–Malik–Abraham's version of logistic distribution in (51.18). If we switch the above setting and let $U_1 = U_2 = \cdots = U_k$ to be distributed as extreme value and $V_1, V_2, \ldots, V_k$ to be independent extreme value random variables, then $\boldsymbol{X}$ has its joint survival function to be

$$\Pr[\boldsymbol{X} \geq \boldsymbol{x}] = \frac{1}{1 + \sum_{i=1}^{k} e^{x_i}} \, . \tag{51.30}$$

Observe that this is the distribution of a random vector whose negative has the Gumbel–Malik–Abraham distribution in (51.18). This distribution has been discussed by Lindley and Singpurwalla (1986) in the context of reliability.

Next, let us assume that the random vector $\boldsymbol{U}$ has Gumbel's (1958) multivariate extreme value distribution with joint distribution function

$$\Pr[\boldsymbol{U} \leq \boldsymbol{u}] = e^{-\left(\sum_{i=1}^{k} e^{-\alpha u_i}\right)^{1/\alpha}} \qquad \text{for } \alpha \geq 1. \tag{51.31}$$

If we now choose $V_1 = V_2 = \cdots = V_k$ to be distributed as extreme value independently of $\boldsymbol{U}$, then $\boldsymbol{X} = \boldsymbol{U} - \boldsymbol{V}$ has a one-parameter k-variate logistic distribution with joint distribution function

$$
\begin{aligned}
F_{\boldsymbol{X}}(\boldsymbol{x}) &= \int_{-\infty}^{\infty} \Pr(U_1 \leq x_1 + v, \ldots, U_k \leq x_k + v) f_V(v) \, dv \\
&= \int_{-\infty}^{\infty} e^{-\left(\sum_{i=1}^{k} e^{-\alpha(x_i+v)}\right)^{1/\alpha}} e^{-v} \, e^{-e^{-v}} \, dv \\
&= \int_{-\infty}^{\infty} e^{-v} e^{-e^{-v}\left\{1+\left(\sum_{i=1}^{k} e^{-\alpha x_i}\right)^{1/\alpha}\right\}} \, dv \\
&= \frac{1}{1 + \left(\sum_{i=1}^{k} e^{-\alpha x_i}\right)^{1/\alpha}} \, .
\end{aligned}
\tag{51.32}
$$

This multivariate logistic distribution, discussed in a random utility context by Strauss (1979), clearly includes the Gumbel–Malik–Abraham's multivariate logistic distribution in (51.18) as a special case when $\alpha = 1$. For the multivariate extreme value distribution in (51.31), Tiago de Oliveira (1961) showed that $\text{corr}(U_i, U_j) = 1 - \frac{1}{\alpha^2}$. Using this result, it can be readily shown for the multivariate logistic distribution in (51.32) that $\text{corr}(X_i, X_j) = 1 - \frac{1}{2\alpha^2}$, which reveals that a strong correlation (at least 0.5) is always present.

Instead, if we assume that $\boldsymbol{U}$ and $\boldsymbol{V}$ are independently distributed as multivariate extreme value [as in (51.31)] with parameter α and α', respectively, then the k-dimensional random vector $\boldsymbol{X} = \boldsymbol{U} - \boldsymbol{V}$ will possess a more general form of multivariate logistic distribution. In this case, upon utilizing Tiago de Oliveira's (1961) result once again, it can be shown that $\text{corr}(X_i, X_j) = 1 - \frac{1}{2\alpha^2} - \frac{1}{2\alpha'^2}$, which allows any nonnegative correlation. Consequently, this general multivariate logistic distribution is more flexible in terms of its correlation structure; however, the distribution function cannot be written in this case in an explicit form unfortunately.

Somewhat similar but more involved constructions of multivariate logistic distributions are possible. For example, George and Mudholkar (1981) have shown that a standard univariate logistic random variable X is such that

$$X \stackrel{d}{=} \sum_{i=1}^{\infty} \frac{W_i}{i} - \sum_{i=1}^{\infty} \frac{W_i'}{i} , \qquad (51.33)$$

where W_i's and W_i''s are independent and identically distributed as standard exponential. The representation in (51.33) can be exploited to construct multivariate logistic distributions that will allow the whole range of -1 to $+1$ for correlation coefficients; see, for example, Arnold (1992).

6   MIXTURE FORMS

In order to construct a multivariate distribution with all its marginal distributions as standard logistic, we may consider the scale-mixture form

$$X_i = U V_i, \qquad i = 1, 2, \ldots, k, \qquad (51.34)$$

where U is a nonnegative random variable independent of $\boldsymbol{V}$, $V_1, \ldots, V_k$ are independent and identically distributed random variables, and X_i's are univariate standard logistic random variables. The scale-mixture model in (51.34) will be completely specified if either the distribution of U or the common distribution of V_i's is given. Naturally, there are some restrictions on the possible choices for the distributions of U and V_i's. A convenient choice for the distribution of U is Uniform(0,1), of course! In this case, since $E[U^{2r}] = \frac{1}{2r+1}$ and $E[X^{2r}] = 2(2r)! \sum_{n=1}^{\infty} \frac{(-1)^{n-1}}{n^{2r}}$, we need to find a symmetric distribution for V_i's such that its even moments are given by

$$E[V^{2r}] = 2(2r + 1)! \sum_{n=1}^{\infty} \frac{(-1)^{n-1}}{n^{2r}} . \qquad (51.35)$$

A suitable density for V is given by

$$p_V(v) = \frac{1}{2} \sum_{n=1}^{\infty} (-1)^{n-1} n^2 |v| e^{-n|v|}, \qquad -\infty < v < \infty,$$

which can be alternatively written as

$$p_V(v) = \frac{1}{4}\left\{\frac{|v|}{2}\tanh\left(\frac{|v|}{2}\right)\text{sech}^2\left(\frac{|v|}{2}\right)\right\}, \quad -\infty < v < \infty. \quad (51.36)$$

The corresponding k-variate density function of $\boldsymbol{X}$ is given by [see Arnold (1992)]

$$p_{\boldsymbol{X}}(\boldsymbol{x}) = \frac{1}{4^k}\int_0^1\prod_{i=1}^k\left\{\frac{|v_i|}{2u}\tanh\left(\frac{|v_i|}{2u}\right)\text{sech}^2\left(\frac{|v_i|}{2u}\right)\right\}\frac{1}{u^k}\,du, \quad (51.37)$$

which will naturally have standard logistic marginal distributions.

The scale-mixture form in (51.34) has a major drawback in that $\text{corr}(X_i, X_j) = 0$ no matter what our choice for the distribution of U is. This is easily observed from the fact that, due to the symmetry of the standard logistic distribution, the common distribution of the V_i's is necessarily symmetric about zero.

A more general form of the scale-mixture model in (51.34) is possible if we consider

$$X_i = U_i V_i, \qquad i = 1, 2, \ldots, k, \quad (51.38)$$

where $\boldsymbol{U}$ is a nonnegative random vector, $\boldsymbol{V}$ is a random vector independent of $\boldsymbol{U}$, and X_i's are univariate standard logistic random variables. For example, we could choose $\boldsymbol{U}$ to be distributed as multivariate uniform or Dirichlet and examine the resulting multivariate logistic distribution. Once again, this multivariate distribution will possess all its marginal distributions to be standard logistic as we have indeed prescribed.

Instead of such scale-mixture forms, we can also consider additive-mixture models of the form

$$X_i = U_i + V_i, \qquad i = 1, 2, \ldots, k,$$

where X_i's are standard logistic random variables and $\boldsymbol{U}$ and $\boldsymbol{V}$ are independent k-dimensional random vectors. In Section 5, we have already discussed some models of this form based on extreme value distributions.

7   GEOMETRIC MINIMA AND MAXIMA

Let $X_1, X_2, \ldots, X_N$ be independent and identically distributed standard logistic random variables and N itself be distributed (independently of the X_i's) as geometric with probability mass function

$$\Pr(N = n) = p(1 - p)^{n-1}, \qquad n = 1, 2, \ldots. \quad (51.39)$$

Let us now define

$$X^* = \min_{1 \leq i \leq N} X_i - \log p. \qquad (51.40)$$

Then, we can obtain the survival function of X^* as

$$
\begin{aligned}
\Pr(X^* \geq x) &= \Pr\left(\min_{1 \leq i \leq N} X_i \geq x + \log p\right) \\
&= \sum_{n=1}^{\infty} p(1-p)^{n-1} \Pr\left(\min_{1 \leq i \leq n} X_i \geq x + \log p\right) \\
&= \sum_{n=1}^{\infty} p(1-p)^{n-1} \left(\frac{e^{-x}}{p + e^{-x}}\right)^n \\
&= \frac{e^{-x}}{1 + e^{-x}}
\end{aligned}
$$

which simply reveals that X^* again has a standard logistic distribution. In other words, the logistic distribution is closed under the operation of geometric minimization.

Similarly, if we define

$$X^{**} = \max_{1 \leq i \leq N} X_i + \log p,$$

we obtain its cumulative distribution function as

$$
\begin{aligned}
\Pr(X^{**} \leq x) &= \Pr\left(\max_{1 \leq i \leq N} X_i \leq x - \log p\right) \\
&= \sum_{n=1}^{\infty} p(1-p)^{n-1} \Pr\left(\max_{1 \leq i \leq n} X_i \leq x - \log p\right) \\
&= \sum_{n=1}^{\infty} p(1-p)^{n-1} \left(\frac{1}{1 + p e^{-x}}\right)^n \\
&= \frac{1}{1 + e^{-x}}.
\end{aligned}
$$

This reveals that X^{**} is distributed again as standard logistic, which means that the logistic distribution is closed under the operation of geometric maximization.

Naturally, this property can be exploited to construct multivariate logistic distributions through geometric minima and maxima. For example, let us consider a sequence of independent trials taking on values $0, 1, \ldots, k$ with corresponding probabilities $p_0, p_1, \ldots, p_k$, respectively. Let $\boldsymbol{N} = (N_1, \ldots, N_k)$ be a random vector where N_i denotes the number of times i appeared before the first occurrence of 0. The probability generating function of N is

$$G_N(t) = E\left[\prod_{i=1}^{k} t_i^{N_i}\right] = \frac{p_0}{1 - \sum_{i=1}^{k} p_i t_i}. \qquad (51.41)$$

From this, we note that $N_i + 1$ (for $i = 1, \ldots, k$) has a geometric distribution in (51.39) with p_i in place of p. Let $Y_i^{(j)}$, $j = 1, 2, \ldots, k$, $i = 1, 2, \ldots$, be k independent sequences of independent standard logistic random variables. Let the k-dimensional random vector $X = (X_1, \ldots, X_k)$ be defined as

$$X_j = \min_{1 \le i \le N_j + 1} Y_i^{(j)} \qquad \text{for } j = 1, \ldots, k. \qquad (51.42)$$

As seen earlier, the marginal distributions of X_i's will be logistic. The joint survival function of X is

$$\begin{aligned}
\Pr(X \ge x) &= E[\Pr(X \ge x \mid N] \\
&= \sum_n \Pr(N = n) \prod_{j=1}^{k} \frac{1}{(1 + e^{x_j})^{n_j + 1}} \\
&= \left\{\prod_{j=1}^{k} \frac{1}{1 + e^{x_j}}\right\} E\left[\prod_{j=1}^{k} \frac{1}{(1 + e^{x_j})^{N_j}}\right] \\
&= \frac{p_0}{\left(1 - \sum_{j=1}^{k} \frac{p_j}{1 + e^{x_j}}\right) \prod_{j=1}^{k}(1 + e^{x_j})} \qquad (51.43)
\end{aligned}$$

upon using (51.41). The above joint survival function can be reparameterized and written as [see Arnold (1992)]

$$\begin{aligned}
\Pr(X \ge x) = \Big\{ 1 + \sum_{j=1}^{k} e^{x_j} + \sum_{j_1 \ne j_2} \sum c_{j_1 j_2} \, e^{x_{j_1} + x_{j_2}} + \cdots \\
+ c_{12\ldots k} \, e^{x_1 + x_2 + \cdots + x_k} \Big\}^{-1}. \qquad (51.44)
\end{aligned}$$

Arnold (1990) has discussed the conditions that need to be placed on the coefficients c's for (51.44) to be a valid joint survival function. In the bivariate case, (51.44) gives rise to the model

$$\Pr(X_1 \ge x_1, X_2 \ge x_2) = \frac{1}{1 + e^{x_1} + e^{x_2} + \theta \, e^{x_1 + x_2}} \qquad \text{for } 0 \le \theta \le 2. \qquad (51.45)$$

Instead of geometric minimization, if we perform geometric maximization and proceed similarly, we obtain the bivariate model

$$\Pr(X_1 \le x_1, X_2 \le x_2) = \frac{1}{1 + e^{-x_1} + e^{-x_2} + \theta\, e^{-x_1 - x_2}} \qquad \text{for } 0 \le \theta \le 2.$$
(51.46)

The Gumbel–Malik–Abraham bivariate distribution in (51.18) (with $k = 2$) is seen to be a special case of the geometric maximization model in (51.46) when $\theta = 0$. It is of interest to mention here that the bivariate model in (51.46) was derived by Ali, Mikhail, and Haq (1978) as the solution to the equation

$$\frac{\partial^2 H(x_1, x_2)}{\partial H(x_1, \infty)\partial H(\infty, x_2)} = \theta,$$

where

$$H(x_1, x_2) = \frac{1}{\Pr[X_1 \le x_1, X_2 \le x_2]} - 1.$$

The hierarchy of bivariate geometric distributions detailed by Arnold (1975) can be utilized with geometric minima and maxima in order to develop even more complicated bivariate logistic distributions.

8   A GENERAL FLEXIBLE MODEL

In (51.32) (with $k = 2$), we derived a bivariate logistic distribution with joint distribution function

$$F_{X,Y}(x, y) = \frac{1}{1 + (e^{-\alpha x} + e^{-\alpha y})^{1/\alpha}}$$
(51.47)

as a difference of two independent extreme value vectors. Through the geometric maximization, we derived in (51.46) a bivariate logistic distribution with joint distribution function

$$F_{X,Y}(x, y) = \frac{1}{1 + e^{-x} + e^{-y} + \theta e^{-x-y}}.$$
(51.48)

These two one-parameter bivariate logistic distributions can be combined to form a general two-parameter flexible model as

$$F_{X,Y}(x, y) = \frac{1}{1 + (e^{-\alpha x} + e^{-\alpha y} + \theta e^{-\alpha x - \alpha y})^{1/\alpha}},$$
(51.49)

where θ and α are chosen so that (51.49) represents a valid bivariate distribution function. This simply means that θ and α should be such that

$$p_{X,Y}(x,y) = \frac{\partial^2}{\partial x \partial y} F(x,y) \geq 0 \qquad \text{for all } x, y,$$

which is equivalent to the condition that

$$\frac{\partial^2}{\partial x \partial y} \left\{ \frac{1}{1 + (x + y + \theta xy)^{1/\alpha}} \right\} \geq 0 \qquad \text{for all } x > 0, \ y > 0.$$

From the above condition, it can be readily shown that α must be at least 1 and for a given choice of α we must have

$$0 \leq \theta \leq \left\{ (\alpha^2 - 1)^{1/\alpha} + (1 + \alpha)(\alpha^2 - 1)^{(1-\alpha)/\alpha} \right\}^{\alpha}. \qquad (51.50)$$

The upper limit in (51.50) is at least 2.

9 CONDITIONALLY SPECIFIED LOGISTIC MODEL

Suppose we seek a bivariate density function $p(x, y)$ such that the conditional density of X, given $Y = y$, is logistic for every y and that the conditional density of Y, given $X = x$, is logistic for every x. In the general scenario, we may let the conditional density of X, given $Y = y$, to be logistic in which both location and scale parameters depend on y; and similarly we may let the conditional density of Y, given $X = x$, to be logistic in which both location and scale parameters depend on x. The determination of such a bivariate distribution is still an open problem.

However, if we let only the location parameter of the logistic distribution depend on the value of the conditioning variable, then we are seeking a bivariate distribution such that

$$\Pr[X \geq x \mid Y = y] = \frac{1}{1 + e^{x - a(y)}} \quad \text{for all } y \text{ and some function } a(y)$$

$$(51.51)$$

and

$$\Pr[Y \geq y \mid X = x] = \frac{1}{1 + e^{y - b(x)}} \quad \text{for all } x \text{ and some function } b(x).$$

$$(51.52)$$

Upon making the change of variable $U = e^X$ and $V = e^Y$, (51.51) and (51.52) simply imply that U and V should have Pareto conditionals, as discussed already by Arnold (1987). From Arnold's (1987) work, it then follows that the joint density function [for which (51.51) and (51.52) hold] is given by

$$p_{X,Y}(x,y) = \left(\frac{1-\phi}{-\log\phi}\right) \frac{e^{(x-\mu_1)+(y-\mu_2)}}{\left(1 + e^{x-\mu_1} + e^{y-\mu_2} + \phi\, e^{(x-\mu_1)+(y-\mu_2)}\right)^2}$$

$$\text{for } \phi > 0. \quad (51.53)$$

From (51.53), we readily observe that this bivariate distribution will possess independent logistic marginal distributions when $\phi = 1$.

10   SOME OTHER GENERALIZATIONS

Satterthwaite and Hutchinson (1978) have considered a generalization of Gumbel's bivariate logistic distribution in (51.1) of the following form:

$$F_{X,Y}(x,y) = \frac{1}{(1 + e^{-x} + e^{-y})^\gamma} \qquad \text{where } \gamma > 0. \quad (51.54)$$

Though this bivariate distribution has a more flexible correlation structure than Gumbel's bivariate distribution in (51.1), its marginal distributions are not logistic. In fact, the marginal distributions of (51.24) are Type I generalized logistic distributions, as termed by Balakrishnan and Leung (1988) and Zelterman and Balakrishnan (1992). The k-variate version of the distribution in (51.54) can be written as

$$F_{\mathbf{X}}(\mathbf{x}) = \frac{1}{\left(1 + \sum_{i=1}^{k} e^{-x_i}\right)^\gamma} \qquad \text{where } \gamma > 0, \quad (51.55)$$

which will naturally have all its marginals to be Type I generalized logistic distributions. Clearly, this includes the Gumbel–Malik–Abraham k-variate logistic distribution as a special case when $\gamma = 1$.

Cook and Johnson (1986) considered a bivariate family of distributions of the form

$$F_{X,Y}(x,y) = \frac{1+\beta}{(1 + e^{-x} + e^{-y})^\alpha} + \frac{\beta}{(1 + 2e^{-x} + 2e^{-y})^\alpha}$$

$$- \frac{\beta}{(1 + 2e^{-x} + e^{-y})^\alpha} - \frac{\beta}{(1 + e^{-x} + 2e^{-y})^\alpha},$$

$$\alpha > 0, \ -1 \le \beta \le 1. \quad (51.56)$$

This bivariate logistic distribution has logistic marginals only when $\alpha = 1$. Symanowski and Koehler (1989) proposed a variation on the above distribution of the form

$$
\begin{aligned}
F_{X,Y}(x,y) \;=\; & \frac{1+\beta}{\{(1+e^{-x})^{1/\alpha} + (1+e^{-y})^{1/\alpha} - 1\}^{\alpha}} \\
& + \frac{\beta}{\{2(1+e^{-x})^{1/\alpha} + 2(1+e^{-y})^{1/\alpha} - 3\}^{\alpha}} \\
& - \frac{\beta}{\{2(1+e^{-x})^{1/\alpha} + (1+e^{-y})^{1/\alpha} - 2\}^{\alpha}} \\
& - \frac{\beta}{\{(1+e^{-x})^{1/\alpha} + 2(1+e^{-y})^{1/\alpha} - 2\}^{\alpha}}\,, \\
& \hspace{4cm} \alpha > 0,\ -1 \le \beta \le 1. \quad (51.57)
\end{aligned}
$$

This bivariate logistic distribution always has logistic marginals. The correlation coefficient ρ tends to 1 as α tends to 0 (regardless of β) and tends to a minimum value of $-\frac{3}{\pi^2}$ for $\beta = -1$ as $\alpha \to \infty$. Independence between X and Y is approached as $\alpha \to \infty$ when $\beta = 0$. Koehler and Symanowski (1992) have mentioned that the distribution can be conveniently reparameterized through a single unrestricted dependency parameter λ in which case β and α may be expressed as[1]

$$
\beta(\lambda) = \frac{1 - e^{-\lambda}}{1 + e^{-\lambda}} \quad \text{and} \quad \alpha(\lambda) = 40\left(\frac{e^{-2.5\lambda} - 1}{e^{2.5\lambda} + 1} + 1\right),
$$

where $-\infty < \lambda < \infty$. In this case, the correlation ρ may be approximated by

$$
\rho(\lambda) = \frac{-c_1 + (c_1 + a_1\lambda + a_2\lambda^2)e^{a_3\lambda}}{1 + (c_1 + a_1\lambda + a_2\lambda^2)e^{a_3\lambda}},
$$

where $c_1 = \frac{3}{\pi^2}$, $a_1 = -0.2131$, $a_2 = 0.0930$ and $a_3 = 1.3739$. This expression of correlation also reveals that $\lim_{\lambda \to \infty} \rho(\lambda) = 1$ and $\lim_{\lambda \to -\infty} \rho(\lambda) = -\frac{3}{\pi^2}$, exactly the same range for ρ mentioned earlier. In Figure 51.3, contours of constant density for various values of λ are presented for the case when the marginal distributions have means 0 and variances 1. Figure 51.3(a) corresponds to a correlation of 0.9 (large value of λ) and a peaked density surface; Figure 51.3(b) corresponds to a correlation of 0.5 (for $\lambda = 1.5326$) and is interestingly very similar to the contour plot of Gumbel's bivariate logistic distribution presented in Figure 51.1; Figure 51.3(c) corresponds to $\lambda = 0$ and the case

---

[1]Since $e^{-\lambda} = \frac{1-\beta}{1+\beta}$, we have $\alpha = 40\left[\frac{\left(\frac{1-\beta}{1+\beta}\right)^{2.5} - 1}{\left(\frac{1-\beta}{1+\beta}\right)^{2.5} + 1} + 1\right]$ and so it appears that only a

subclass of (51.57) is being considered.

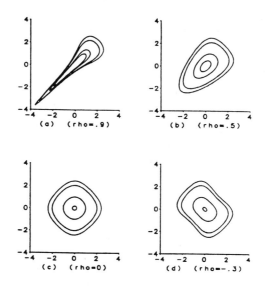

FIGURE 51.3

Contours of Constant Density Corresponding to (51.57) for Selected
Values of λ. [With permission from Koehler and Symanowski (1992).]

when X and Y are nearly independent; and (d) corresponds to a correlation of -0.3 (large negative value of λ) and less peaked density surface.
Koehler and Symanowski (1992) have also discussed applications of this
bivariate distribution in the analysis of two-way contingency tables with
ordered categories.

   Marshall and Olkin (1988) have given a multivariate logistic distribution generated from mixtures of convolution and product families. Specifically, they have considered multivariate distributions of the form

$$F(\boldsymbol{x}) = \int \prod_{i=1}^{k} P_i(x_i|\theta) \; dG(\theta),$$

where $G(\cdot)$ and each $P_i(\cdot)$ is an univariate distribution function. Clearly,
the multivariate distribution $F(\boldsymbol{x})$ has univariate marginals of the same
form. Now, if F_i's are taken to be iterated exponential extreme value
distributions for minima with survival function $\bar{F}_i(x_i|\theta) = e^{-\theta\, e^{x_i}}$, $-\infty <
x_i < \infty$ and $\theta > 0$ (for $i = 1, 2, \ldots, k$), and if $G(\theta)$ is a gamma distribution
with shape parameter α and scale parameter λ, then the distribution $F(\boldsymbol{x})$
given above has joint survival function

$$\bar{F}(\boldsymbol{x}) = \Pr[X_1 > x_1, \ldots, X_k > x_k] = \frac{\lambda^\alpha}{\left(\lambda + \sum_{i=1}^{k} e^{x_i}\right)^\alpha}, \qquad \lambda, \alpha > 0.$$

Note that this is the generalized logistic distribution seen in the beginning of this section.

Volodin (1999) presented exact formulae for the probability density function of the spherically symmetric distribution with logistic marginals. For odd values of k (the dimension), this spherically symmetric logistic density can be expressed in terms of elementary functions; but, for even values of k, the density can be expressed only as an infinite series of functions. To be specific, let $\boldsymbol{x} = (x_1, \ldots, x_k)^T$ and $s = \boldsymbol{x}^T \boldsymbol{V}^{-1} \boldsymbol{x}$, where $\boldsymbol{V}$ is not the covariance matrix but is proportional to it. For the case when $k = 2\ell + 1$ ($\ell = 1, 2, \ldots$), the joint density is

$$p(x_1, \ldots, x_k) = \frac{(-1)^\ell}{\pi^\ell |\boldsymbol{V}|^{1/2}} \frac{d^\ell}{ds^\ell} \left(\frac{1}{2\{1 + \cosh \sqrt{s}\}} \right).$$

For example, when $k = 3, 5$, the spherically symmetric logistic density can be written explicitly as

$$p(x_1, x_2, x_3) = \frac{\sinh \sqrt{s}}{4\pi |\boldsymbol{V}|^{1/2} \sqrt{s} \{1 + \cosh \sqrt{s}\}^2}$$

and

$$p(x_1, \ldots, x_5) = \frac{\sinh \sqrt{s} - \sqrt{s}(2 - \cosh \sqrt{s})}{8\pi^2 |\boldsymbol{V}|^{1/2} s \sqrt{s} \{1 + \cosh \sqrt{s}\}^2},$$

respectively. Arnold and Robertson (1989b) and Arnold (1996) had earlier derived the above density for the case $k = 3$. For the case when $k = 2\ell$ ($\ell = 1, 2, \ldots$), the joint density can be expressed as

$$p(x_1, \ldots, x_{2\ell}) = \frac{(2\ell - 1)!!}{2^{\ell-1} \pi^{\ell-1} |\boldsymbol{V}|^{1/2}} \sum_{j=1}^{\infty} \frac{2\ell \pi^2 (2j-1)^2 - s}{\{\pi^2 (2j-1)^2 + s\}^{(2\ell+3)/2}}.$$

Final mention should be made regarding the fact that we need to work on stationary logistic processes that implicitly involve various multivariate logistic distributions; see, for example, Arnold and Robertson (1989a) and Arnold (1989).

BIBLIOGRAPHY

(Some bibliographical items not mentioned in the text are included here for completeness.)

Ahmed, N. A., and Gokhale, D. V. (1989). Entropy expressions and their estimators for multivariate distributions, *IEEE Transactions on Information Theory*, **35**, 688–692.

Ali, M. M., Mikhail, N. N., and Haq, M. S. (1978). A class of bivariate distributions including the bivariate logistic, *Journal of Multivariate Analysis*, **8**, 405–412.

Arnold, B. C. (1975). Multivariate exponential distributions based on hierarchical successive damage, *Journal of Applied Probability*, **12**, 142–147.

Arnold, B. C. (1987). Bivariate distributions with Pareto conditionals, *Statistics & Probability Letters*, **5**, 263–266.

Arnold, B. C. (1989). A logistic process constructed using geometric minimization, *Statistics & Probability Letters*, **7**, 253–257.

Arnold, B. C. (1990). A flexible family of multivariate Pareto distributions, *Journal of Statistical Planning and Inference*, **24**, 249–258.

Arnold, B. C. (1992). Multivariate logistic distributions, Chapter 11 in *Handbook of the Logistic Distribution* (N. Balakrishnan, ed.), pp. 237–261, New York: Marcel Dekker.

Arnold, B. C. (1996). Distributions with logistic marginals and/or conditionals, in *Distributions with Fixed Marginals and Related Topics* (L. Rüschendorf, B. Schweitzer, and M. D. Taylor, eds.), pp. 15–32, IMS Lecture Notes Monograph Series, Vol. 28, Hayward, California: Institute of Mathematical Statistics.

Arnold, B. C., and Robertson, C. A. (1989a). Autoregressive logistic processes, *Journal of Applied Probability*, **26**, 524–531.

Arnold, B. C., and Robertson, C. A. (1989b). Elliptically contoured distributions with logistic marginals, Technical Report No. 180, Department of Statistics, University of California, Riverside, California.

Balakrishnan, N. (ed.) (1992). *Handbook of the Logistic Distribution*, New York: Marcel Dekker.

Balakrishnan, N., and Leung, M. Y. (1988). Order statistics from the Type I generalized logistic distribution, *Communications in Statistics—Simulation and Computation*, **17**, 25–50.

Cook, R. D., and Johnson, M. E. (1986). Generalized Burr-Pareto-logistic distributions with applications to a uranium exploration data set, *Technometrics*, **28**, 123–131.

Genest, C., and MacKay, J. (1986). The joy of copulas: Bivariate distributions with uniform marginals, *The American Statistician*, **40**, 280–285.

George, E. O., and Mudholkar, G. S. (1981). Some relationships between the logistic and the exponential distributions, in *Statistical Distributions in Scientific Work*, Vol. 4 (C. Taillie, G. P. Patil, and B. Baldessari, eds.), pp. 401–409, Dordrecht: D. Reidel.

Gumbel, E. J. (1958). *Statistics of Extremes*, New York: Columbia University Press.

Gumbel, E. J. (1961). Bivariate logistic distributions, *Journal of the American Statistical Association*, **56**, 335–349.

Koehler, K. J., and Symanowski, J. T. (1992). Applications to ordered categorical data, Section 18.1 in *Handbook of the Logistic Distribution* (N. Balakrishnan, ed.), pp. 495–512, New York: Marcel Dekker.

Lindley, D. V., and Singpurwalla, N. D. (1986). Multivariate distributions for the life lengths of components of a system sharing a common environment, *Journal of Applied Probability*, **23**, 418–431.

Malik, H. J., and Abraham, B. (1973). Multivariate logistic distributions, *Annals of Statistics*, **1**, 588–590.

Marshall, A. W., and Olkin, I. (1988). Multivariate distributions generated from mixtures of convolution and product families, in *Topics in Statistical Dependence* (H. W. Block, A. R. Sampson, and T. H. Savits, eds.), IMS Lecture Notes/Monograph Series, Hayward, California.

Moore, D. S. (1969). Asymptotically nearly efficient estimators of multivariate location parameters, *Annals of Mathematical Statistics*, **40**, 1809–1823.

Nelsen, R. B. (1998). *An Introduction to Copulas*, Lecture Notes in Statistics—**139**, New York: Springer-Verlag.

Oakes, D. (1989). Bivariate survival models induced by frailties, *Journal of the American Statistical Association*, **84**, 487–493.

Satterthwaite, S. P., and Hutchinson, T. P. (1978). A generalization of Gumbel's bivariate logistic distribution, *Metrika*, **25**, 163–170.

Smith, M. D., and Moffatt, P. G. (1999). Fisher's information on the correlation coefficient in bivariate logistic models, *The Australian & New Zealand Journal of Statistics*, **41**, 315–330.

Strauss, D. J. (1979). Some results on random utility, *Journal of Mathematical Psychology*, **20**, 35–51.

Symanowski, J. T., and Koehler, K. J. (1989). A bivariate logistic distribution with applications to categorical responses, Technical Report No. 89-29, Department of Statistics, Iowa State University, Ames, Iowa.

Tiago de Oliveira, J. (1961). La répresentation des distributions extrêmales bivariées, *Bulletin of the International Statistical Institute*, **33**, 477–480.

Volodin, N. A. (1999). Spherically symmetric logistic distribution, *Journal of Multivariate Analysis*, **70**, 202–206.

Zelterman, D., and Balakrishnan, N. (1992). Univariate generalized distributions, Chapter 9 in *Handbook of the Logistic Distribution* (N. Balakrishnan, ed.), pp. 209–221, New York: Marcel Dekker.

CHAPTER 52

Multivariate Pareto Distributions

1 INTRODUCTION

As in the case of univariate Pareto distributions [see, for example, Chapter 20 of Johnson, Kotz, and Balakrishnan (1994)], mathematical simplicity and tractability have generated a lot of interest in the theory and applications of multivariate Pareto distributions. An early detailed and lucid discussion on multivariate Pareto distributions can be found in the monograph by Arnold (1983). Numerous papers dealing with bivariate and multivariate Pareto distributions have subsequently appeared in the literature many of which, of course, are due to Arnold and his associates. In addition to presenting a concise discussion on all these developments in this chapter, we have included a comprehensive bibliography that could assist interested readers in pursuing further studies. We recommend that the reader browse through Chapter 20 of Johnson, Kotz, and Balakrishnan (1994) before reading this chapter.

2 FORMS OF UNIVARIATE PARETO DISTRIBUTIONS

For ease of reference, we summarize the various forms of univariate Pareto distributions and describe some of their distinctive properties that will be utilized extensively throughout this chapter; also see Section 3 in Chapter 20 of Johnson, Kotz, and Balakrishnan (1994, pp. 574–576), where slightly different notations are used.

A random variable Z is said to have a *standard Pareto distribution* if

$$\Pr[Z \geq z] = \frac{1}{1+z} \quad \text{for } z > 0. \tag{52.1}$$

A random variable X is said to have a *Pareto(I) distribution* if

$$\Pr[X \geq x] = \left(\frac{x}{\sigma}\right)^{-\alpha} \quad \text{for } x \geq \sigma, \ \sigma > 0; \tag{52.2}$$

the distribution is denoted by $P(I)(\sigma, \alpha)$; here, α is referred to as the *Pareto index of inequality*.

A random variable X is said to have a *Pareto(II) distribution* if

$$\Pr[X \geq x] = \left\{1 + \frac{x - \mu}{\sigma}\right\}^{-\alpha}$$
$$\text{for } x \geq \mu, \ \sigma > 0, \ \alpha > 0, \ \mu \in \mathbb{R}; \tag{52.3}$$

the distribution is denoted by $P(II)(\mu, \sigma, \alpha)$.

A random variable X is said to have a *Pareto(III) distribution* if

$$\Pr[X \geq x] = \left\{1 + \left(\frac{x - \mu}{\sigma}\right)^{1/\gamma}\right\}^{-1}$$
$$\text{for } x \geq \mu, \ \sigma > 0, \ \gamma > 0, \ \mu \in \mathbb{R}; \tag{52.4}$$

the notation $P(III)(\mu, \sigma, \gamma)$ is used for this distribution; here, $\gamma > 0$ is the *Gini index of inequality*. This distribution is also called the *log-logistic distribution*; see, for example, Eq. (23.88) in Chapter 23 of Johnson, Kotz, and Balakrishnan (1995). Note that if Z is distributed as standard Pareto in (52.1), then $X = \mu + \sigma Z^\gamma \stackrel{d}{=} P(III)(\mu, \sigma, \gamma)$. Also, it can be verified that geometric minima of independent Pareto(III) random variables are also distributed as Pareto(III).

A random variable X is said to have a *Pareto(IV) distribution* if

$$\Pr[X \geq x] = \left\{1 + \left(\frac{x - \mu}{\sigma}\right)^{1/\gamma}\right\}^{-\alpha}$$
$$\text{for } x \geq \mu, \ \sigma > 0, \ \gamma > 0, \ \alpha > 0, \ \mu \in \mathbb{R}; \tag{52.5}$$

the distribution is denoted by $P(IV)(\mu, \sigma, \gamma, \alpha)$. This distribution is often referred to as the *generalized Pareto distribution* and is, in fact, a member of the Burr family of distributions; see, for example, Chapter 12 of Johnson, Kotz, and Balakrishnan (1994). This distribution plays an important role in the theory and applications of extreme value distributions; see Chapter 22 of Johnson, Kotz, and Balakrishnan (1995).

The following relationships between these four families of Pareto distributions may be easily noted:

$$P(I)(\sigma, \alpha) \equiv P(IV)(\sigma, \sigma, 1, \alpha), \qquad (52.6)$$

$$P(II)(\mu, \sigma, \alpha) \equiv P(IV)(\mu, \sigma, 1, \alpha), \qquad (52.7)$$

$$P(III)(\mu, \sigma, \gamma) \equiv P(IV)(\mu, \sigma, \gamma, 1). \qquad (52.8)$$

Finally, the *Feller–Pareto family* of distributions may be constructed as follows. Let Y be a beta random variable with probability density function [see Chapter 25 of Johnson, Kotz, and Balakrishnan (1995)]

$$p_Y(y) = \frac{1}{B(\gamma_1, \gamma_2)} y^{\gamma_1 - 1}(1 - y)^{\gamma_2 - 1}, \quad 0 < y < 1, \ \gamma_1 > 0, \ \gamma_2 > 0.$$

Then, $W = \mu + \sigma(\frac{1}{Y} - 1)^\gamma$ has a *Feller–Pareto distribution* and is denoted by $FP(\mu, \sigma, \gamma, \gamma_1, \gamma_2)$. If $\gamma_2 = 1$, then

$$\Pr[W \geq w] = \left\{ 1 + \left(\frac{w - \mu}{\sigma} \right)^{1/\gamma} \right\}^{-\gamma_1} \qquad (52.9)$$

so that W has a $P(IV)(\mu, \sigma, \gamma, \gamma_1)$ distribution. If $\gamma_1 = 1$, then

$$\Pr[W \geq w] = 1 - \left\{ 1 - \left(\frac{w - \mu}{\sigma} \right)^{-1/\gamma} \right\}^{-\gamma_2}. \qquad (52.10)$$

If $\gamma_1 = \gamma_2 = 1$, then W has a $P(III)(\mu, \sigma, \gamma)$ distribution. Occasionally, the variable $U = \frac{1}{Y} - 1$ is referred to as a Feller–Pareto variable.

3   BIVARIATE PARETO DISTRIBUTIONS

3.1   Bivariate Pareto of the First Kind

The bivariate distribution with joint density function [Mardia (1962)]

$$p_{X_1, X_2}(x_1, x_2) = (a + 1)a(\theta_1\theta_2)^{a+1}(\theta_2 x_1 + \theta_1 x_2 - \theta_1\theta_2)^{-(a+2)},$$
$$x_1 \geq \theta_1 > 0, \ x_2 \geq \theta_2 > 0, \ a > 0, \qquad (52.11)$$

may be called a *bivariate Pareto distribution of the first kind*, since the marginal distributions have density functions

$$p_{X_i}(x_i) = a\theta_i^a x_i^{-(a+1)}, \quad x_i \geq \theta_i > 0, \ i = 1, 2 \qquad (52.12)$$

– that is, $X_i \stackrel{d}{=} P(I)(\frac{1}{\theta_i}, a)$. Note that the marginal distributions share a common value of the shape parameter a. From these marginal Pareto distributions, we readily have [see Eqs. (20.11a) and (20.11b) in Chapter 20 of Johnson, Kotz, and Balakrishnan (1994)]

$$E[X_i] = \frac{a}{a-1} \, \theta_i \qquad \text{for } a > 1, \; i = 1, 2, \tag{52.13}$$

and

$$\text{var}(X_i) = \frac{a}{(a-1)^2(a-2)} \, \theta_i^2 \qquad \text{for } a > 2, \; i = 1, 2. \tag{52.14}$$

The conditional density function of X_2, given $X_1 = x_1$, is

$$p_{X_2|X_1}(x_2|x_1) = (a+1)\theta_1(\theta_2 x_1)^{a+1}(\theta_1 x_2 + \theta_2 x_1 - \theta_1\theta_2)^{-(a+2)},$$
$$x_2 \geq \theta_2 > 0, \; \theta_1 > 0, \; a > 0. \tag{52.15}$$

From (52.15), we may note that the conditional distribution of $\theta_1 X_2 + \theta_2(x_1 - \theta_1)$, given $X_1 = x_1$, is a $P(I)(\frac{1}{\theta_2 x_1}, a+1)$ distribution. From (52.15), we also find

$$E[X_2 \mid (X_1 = x_1)] = \theta_2 \left(1 + \frac{x_1}{a\theta_1}\right) \tag{52.16}$$

and

$$\text{var}(X_2 \mid (X_1 = x_1)) = \left(\frac{\theta_2}{\theta_1}\right)^2 \frac{(a+1)}{a^2(a-1)} \, x_1^2. \tag{52.17}$$

Using (52.16), we find, for $a > 2$,

$$\text{cov}(X_1, X_2) = \theta_2 E[X_1] + \frac{\theta_2}{a\theta_1} E[X_1^2] - \frac{a^2}{(a-1)^2} \theta_1\theta_2$$
$$= \frac{\theta_1\theta_2}{(a-1)^2(a-2)} \tag{52.18}$$

and, consequently,

$$\text{corr}(X_1, X_2) = \frac{1}{a}. \tag{52.19}$$

This shows that X_1 and X_2 are positively correlated.

Consider a bivariate logistic distribution with joint cumulative distribution function (see Chapter 51)

$$F(x_1, x_2) = 1/(1 + e^{-x_1} + e^{-x_2})$$

and joint density function

$$p(x_1, x_2) = \frac{2e^{-x_1-x_2}}{(1 + e^{-x_1} + e^{-x_2})^3} , \qquad x_1, x_2 \in \mathbb{R},$$

with corresponding copula

$$q(u, v) = \frac{2uv}{(u + v - uv)^3} .$$

On the other hand, consider the bivariate Pareto distribution of the first kind in (52.11) for the special case when $a = 1$ with the corresponding copula given by

$$q^*(u, v) = \frac{2(1 - u)(1 - v)}{\{(1 - u) + (1 - v) - (1 - u)(1 - v)\}^3} .$$

If we rotate the surface of $q^*(u, v)$ about $(\frac{1}{2}, \frac{1}{2})$ by $180°$, we will obtain the surface of the copula $q(u, v)$ corresponding to the bivariate logistic distribution. Thus, the dependence imposed between X_1 and X_2 in the above bivariate logistic distribution is equivalent to the dependence between $\frac{\theta_1 X_1}{X_1 - \theta_1}$ and $\frac{\theta_2 X_2}{X_2 - \theta_2}$ in the bivariate Pareto distribution of the first kind in (52.11).

Given observed values of n independent pairs of random variables $(x_{1i}, x_{2i})^T$, each having the joint density function in (52.11), the maximum likelihood estimators of θ_1, θ_2 and a are

$$\left. \begin{array}{l} \hat{\theta}_1 = \min(x_{11}, x_{12}, \ldots, x_{1n}), \\ \hat{\theta}_2 = \min(x_{21}, x_{22}, \ldots, x_{2n}), \\ \hat{a} = \left(\frac{1}{S} - \frac{1}{2}\right) + \sqrt{\frac{1}{S^2} + \frac{1}{4}} , \end{array} \right\} \tag{52.20}$$

where

$$S = \frac{1}{n} \sum_{i=1}^{n} \log \left(\frac{x_{1i}}{\hat{\theta}_1} + \frac{x_{2i}}{\hat{\theta}_2} - 1 \right). \tag{52.21}$$

Observe that, for $i = 1, 2$, $\hat{\theta}_i$ has a Pareto(I) distribution with shape parameter na, and consequently

$$E[\hat{\theta}_i] = \frac{na}{na - 1} \theta_i \quad \text{for } na > 1 \tag{52.22}$$

and

$$\text{var}(\hat{\theta}_i) = \frac{na}{(na - 1)^2(na - 2)} \theta_i^2 \quad \text{for } na > 2. \tag{52.23}$$

Furthermore, the joint distribution of $\hat{\theta}_1$ and $\hat{\theta}_2$ is

$$\Pr[(\hat{\theta}_1 \geq c_1) \cap (\hat{\theta}_2 \geq c_2)] = (\theta_1\theta_2)^{na}(\theta_2 c_1 + \theta_1 c_2 - c_1 c_2)^{-na} \quad (52.24)$$

which is of the same form as the bivariate Pareto(I) distribution corresponding to (52.11) with a replaced by na. Hence, the correlation coefficient between $\hat{\theta}_1$ and $\hat{\theta}_2$ [see (52.19)] is $1/(na)$. See Mardia (1962) for more details.

Krishnan (1985) has used the bivariate Pareto distribution in (52.11) to model jointly the crude birth rate and the crude death/infant mortality rate, thus revealing its usefulness in demographic studies.

If a positive random variable X has finite expected value, then it has a Pareto distribution if and only if

$$E[X \mid X > x] = h + gx \text{ with } g > 1,$$

as established by Revankar, Hartley, and Pagano (1974). [The motivation for this characterization was in connection with modeling income distribution. If a constant proportion of excess of income over tax-exemption limit is underreported (for tax purposes), then average amount of underreporting is a linear function of income if and only if the income distribution is Paretian.]

Arnold, Castillo, and Sarabia (1992) have shown that if

$$X_1|(X_2 = x_2) \overset{d}{=} P(I)(\sigma_1(x_2), \alpha + 1)$$

and

$$X_2|(X_1 = x_1) \overset{d}{=} P(I)(\sigma_2(x_1), \alpha + 1)$$

and the regression functions $E[X_1|(X_2 = x_2)]$ and $E[X_2|(X_1 = x_1)]$ are linear, then the joint density function of $(X_1, X_2)^T$ must be the bivariate Pareto distribution of the first kind of the form

$$p_{X_1,X_2}(x_1, x_2) = \frac{(\alpha+1)\alpha}{\sigma_1\sigma_2}\left(1 + \frac{x_1}{\sigma_1} + \frac{x_2}{\sigma_2}\right)^{-(\alpha+2)}, \quad x_1 > 0, \ x_2 > 0. \tag{52.25}$$

In fact, it is sufficient to assume Pareto conditionals and one nonconstant linear regression function. Analogously, if the conditional distributions are Pareto and the marginals are Pareto, then the joint distribution is as given in (52.25).

Wesolowski (1994) has extended the results of Arnold, Castillo, and Sarabia (1992) by stipulating that

$$p_{X_2|X_1}(x_2|x_1) = \frac{\alpha(a + bx_1)^\alpha}{(a + bx_1 + x_2)^{\alpha+1}},$$
$$x_2 \geq 0, \ x_1 \geq 0, \ a \geq 0, \ b > 0, \ \alpha > 1. \quad (52.26)$$

Since Pareto(II) distribution is given by (52.3), the above conditional specification of Wesolowski (1994) is simply that

$$X_2|(X_1 = x_1) \overset{d}{=} P(II)(0, a + bx_1, \alpha).$$

He has shown *inter alia* that for an arbitrary bivariate random vector $(X_1, X_2)^T$ satisfying (52.26), the distribution is uniquely determined by $E[X_1|(X_2 = x_2)]$. Specifically, a bivariate random vector $(X_1, X_2)^T$, with the conditional distribution of X_1 given $X_2 = x_2$ as in (52.26) and with a linear regression function $E[X_1|(X_2 = x_2)] = \frac{a+x_2}{\alpha-1}$, possesses a bivariate Pareto distribution of the first kind of the form

$$p_{X_1,X_2}(x_1, x_2) = \frac{C}{(a + bx_1 + y)^{\alpha+1}}. \quad (52.27)$$

For Mardia's bivariate Pareto distribution of the first kind in (52.11), Malik and Trudel (1985) have shown that the density function of the product $U = X_1 X_2$ is

$$p_U(u) = (a+1)a(\theta_1\theta_2)^{a+1} \int_{\theta_2}^{\infty} \frac{v^{a+1}}{(\theta_2 u + \theta_1 v - \theta_1\theta_2 v)^{a+2}} \, dv,$$
$$u \geq \theta_1\theta_2 > 0, \ a > 0. \quad (52.28)$$

The integral in (52.28) can be evaluated for a fixed a with the aid of integration formulas given, for example, in Gradshteyn and Ryzhik (1965).

The density function of the quotient $V = X_1/X_2$ takes on a simpler form derived by Malik and Trudel (1985). It is

$$p_V(v) = \frac{\theta_1^{a+1}\theta_2}{(\theta_2 v)^a (\theta_2 v + \theta_1)} \left(\frac{a}{\theta_2 v} + \frac{1}{\theta_2 v + \theta_1} \right),$$
$$v > 0, \ \theta_1 > 0, \ \theta_2 > 0, \ a > 0. \quad (52.29)$$

Lindley and Singpurwalla (1986) considered the distribution of life lengths measured in a laboratory environment as independent exponentials and proved that when they work in a different environment that may be harsher, same or gentler than the original, the resulting survival distribution of life lengths is bivariate Pareto with joint survival function

$$\bar{F}(x_1, x_2) = \Pr[X_1 > x_1, X_2 > x_2] = (1 + a_1 x_1 + a_2 x_2)^{-b}.$$

3.2 Mardia's Bivariate Pareto of the Second Kind

Mardia (1962) started with $(Y_1, Y_2)^T$ having a specific bivariate exponential distribution with joint density function

$$p_{Y_1,Y_2}(y_1, y_2) = \frac{1}{1 - \rho^2} I_0 \left(\frac{2\rho\sqrt{y_1 y_2}}{1 - \rho^2} \right) e^{-(y_1+y_2)/(1-\rho^2)},$$

$$y_1 > 0, \ y_2 > 0, \qquad (52.30)$$

(the so-called *Wicksell–Kibble-type* bivariate distribution), where $I_0(\cdot)$ is a modified Bessel function of the first kind of order 0; see Eq. (1.103) in Chpater 1 of Johnson, Kotz, and Kemp (1992). Setting $X_1 = \theta_1 e^{Y_1/a_1}$ and $X_2 = \theta_2 e^{Y_2/a_2}$, we readily obtain from (52.30) the joint density function of X_1 and X_2 to be

$$p_{X_1,X_2}(x_1, x_2)$$

$$= \frac{a_1 a_2}{(1 - \rho^2)x_1 x_2} \left\{ \left(\frac{\theta_1}{x_1} \right)^{a_1} \left(\frac{\theta_2}{x_2} \right)^{a_2} \right\}^{1/(1-\rho^2)}$$

$$\times I_0 \left[\frac{2\rho\sqrt{a_1 a_2 \log\left(\frac{x_1}{\theta_1}\right) \log\left(\frac{x_2}{\theta_2}\right)}}{1 - \rho^2} \right], x_1 \geq \theta_1, \qquad x_2 \geq \theta_2.$$

$$(52.31)$$

Recall that if $p_X(x) = a\theta^a x^{-(a+1)}$, then $\log(X/\theta)$ has a standard exponential distribution; see Chapter 20 of Johnson, Kotz, and Balakrishnan (1994). This is referred to as *Mardia's bivariate Pareto distribution of the second kind*.

In this case, we have

$$E[X_2|(X_1 = x_1)] = \frac{a_2 \theta_2}{a_2 - 1 + \rho^2} \left(\frac{x_1}{\theta_1} \right)^{a_1 \rho^2/(a_1 - 1 + \rho^2)}, \qquad (52.32)$$

$$\mathrm{var}(X_2|(X_1 = x_1))$$

$$= a_2 \theta_2^2 \left\{ \frac{\left(\frac{x_1}{\theta_1} \right)^{2a_1\rho^2/(a_1-2+2\rho^2)}}{a_1 - 2 + 2\rho^2} - \frac{\left(\frac{x_1}{\theta_1} \right)^{2a_1\rho^2/(a_1-1+\rho^2)}}{a_1 - 1 + \rho^2} \right\},$$

$$(52.33)$$

and

$$\mathrm{corr}(X_1, X_2) = \frac{\rho^2\sqrt{a_1 a_2 (a_1 - 2)(a_2 - 2)}}{(a_1 - 1)(a_2 - 1) - \rho^2} \quad \text{for } a_1 > 2, \quad a_2 > 2, \ \rho^2 < 1.$$

$$(52.34)$$

As in the case of Mardia's Type I family, the correlation between X_1 and X_2 is positive here as well.

The maximum likelihood estimators of θ_1, θ_2, a_1, and a_2 are exactly the same as those for the corresponding univariate Pareto distributions; see Chapter 20 of Johnson, Kotz, and Balakrishnan (1994). Mardia (1962) has shown that the maximum likelihood estimator of ρ^2 satisfies the equation

$$\left| (\hat{\rho}^2)^{1/2} \right| = \frac{1}{n} \sum_{j=1}^{n} g_j \left\{ \frac{I_1(2\hat{\rho} g_j/(1-\hat{\rho}^2))}{I_0(2\hat{\rho} g_j/(1-\hat{\rho}^2))} \right\}, \tag{52.35}$$

where

$$g_j = \sqrt{\frac{\log\left(\frac{X_{1j}}{\theta_1}\right) \log\left(\frac{X_{2j}}{\theta_2}\right)}{\log\left(\frac{\hat{G}_1}{\theta_1}\right) \log\left(\frac{\hat{G}_2}{\theta_2}\right)}}, \tag{52.36}$$

$$\hat{G}_j = \prod_{i=1}^{n} x_{ji}^{1/n} \qquad \text{for } j = 1, 2, \tag{52.37}$$

and $I_1(\cdot)$ is the modified Bessel function of the first kind of order 1. A consistent estimator of ρ^2, which is easier to compute than the maximum likelihood estimator in (52.35), is the sample product moment correlation between $\log x_{1j}$ and $\log x_{2j}$. This estimator has approximate variance

$$\frac{1}{n}(1-\rho^4)(2\rho^4 + 6\rho^2 + 1). \tag{52.38}$$

3.3 Bivariate Pareto of the Fourth Kind

Arnold (1983) has presented three basic methods of generating a bivariate Pareto distribution of the fourth kind.

Mixture of Weibull and Gamma

It is quite straightforward to verify that given $Z = z$, if X has a Weibull survival function [see Chapter 21 of Johnson, Kotz, and Balakrishnan (1994)]

$$\exp\left\{ -z \left(\frac{x-\mu}{\sigma}\right)^{1/\gamma} \right\}$$

and Z has a standard gamma distribution with shape parameter α, then the unconditional distribution of X is $P(IV)(\mu, \sigma, \gamma, \alpha)$. This property can be utilized to generate bivariate Pareto distributions of the fourth kind.

For example, let $(U_1, U_2)^T$ have Marshall and Olkin's (1967a,b) bivariate exponential distribution with joint survival function [see Chapter 47]

$$\Pr[U_1 > u_1, U_2 > u_2] = e^{-u_1 - u_2 - \lambda \max(u_1, u_2)}, \quad u_1 > 0, \ u_2 > 0, \ \lambda > 0.$$

Let $X_i = \mu_i + \sigma_i \left(\frac{U_i}{Z}\right)^{\gamma_i}$ for $i = 1, 2$, where Z has a standard gamma distribution (independently of U_1 and U_2) with shape parameter α. Then, X_i is distributed as $P(IV)(\mu_i, \sigma_i, \gamma_i, \alpha)$ and

$$\begin{aligned}
\bar{F}_{X_1, X_2}(x_1, x_2) &= \Pr[X_1 > x_1, X_2 > x_2] \\
&= \Bigg[1 + \left(\frac{x_1 - \mu_1}{\sigma_1}\right)^{1/\gamma_1} + \left(\frac{x_2 - \mu_2}{\sigma_2}\right)^{1/\gamma_2} \\
&\quad + \lambda \max \left\{ \left(\frac{x_1 - \mu_1}{\sigma_1}\right)^{1/\gamma_1}, \left(\frac{x_2 - \mu_2}{\sigma_2}\right)^{1/\gamma_2} \right\} \Bigg]^{-\alpha}, \\
&\qquad x_1 \geq \mu_1, \ x_2 \geq \mu_2.
\end{aligned} \tag{52.39}$$

Transformation of Exponential

If U has a standard exponential distribution, then $X = \mu + \sigma(e^{U/\alpha} - 1)^\gamma$ has a $P(IV)(\mu, \sigma, \gamma, \alpha)$ distribution. Hence, if we start with a bivariate random vector $(U_1, U_2)^T$ having standard exponential marginals, and then define X_1 and X_2 by the above transformation, then $(X_1, X_2)^T$ will have a bivariate Pareto distribution of the fourth kind. Obviously, both marginal distributions will be $P(IV)$. There are indeed many choices of bivariate distributions for $(U_1, U_2)^T$ with standard exponential marginals; Mardia (1962) was the first to utilize this construction, using Wicksell-Kibble [Wicksell (1933) and Kibble (1941)]-type bivariate exponential distributions discussed earlier in Section 3.2.

Trivariate Reduction

Given three independent random variables U_i $(i = 1, 2, 3)$ distributed as $P(IV)(\mu, \sigma, \gamma, \alpha_i)$, respectively, the random vector $(X_1, X_2)^T = (\min(U_1, U_3), \min(U_2, U_3))^T$ has a bivariate Pareto distribution of the fourth kind with joint survival function

$$\begin{aligned}
&\bar{F}_{X_1, X_2}(x_1, x_2) \\
&= \Pr[X_1 > x_1, X_2 > x_2] \\
&= \left\{ 1 + \left(\frac{\max(x_1, x_2) - \mu}{\sigma}\right)^{1/\gamma} \right\}^{-\alpha_3} \left\{ 1 + \left(\frac{x_1 - \mu}{\sigma}\right)^{1/\gamma} \right\}^{-\alpha_1}
\end{aligned}$$

$$\times \left\{ 1 + \left(\frac{x_2 - \mu}{\sigma} \right)^{1/\gamma} \right\}^{-\alpha_2}, \qquad x_1 \geq \mu, \; x_2 \geq \mu. \tag{52.40}$$

Clearly, both marginal distributions are $P(IV)$. Moreover, $\min(X_1, X_2)$ is distributed as $P(IV)(\mu, \sigma, \gamma, \alpha_1 + \alpha_2 + \alpha_3)$. This family of bivariate distributions, however, has the undesirable restrictive property that the two marginal distributions must share common values for μ, σ, and γ. This is due to the fact that $\min(X_1, X_2)$ has a Pareto distribution only if the marginal distributions have common values for μ, σ and γ.

3.4   Conditionally Specified Bivariate Pareto

Let the conditional densities $p(x_1 | x_2)$ and $p(x_2 | x_1)$ be members of Pareto$(II)(0, \sigma, \alpha)$ family of distributions with density function

$$p(x) = \frac{\alpha}{\sigma} \left(1 + \frac{x}{\sigma} \right)^{-\alpha - 1} \qquad \text{for } x > 0, \; \sigma > 0, \; \alpha > 0.$$

Then, the joint density function of $(X_1, X_2)^T$ is necessarily of the form [Arnold (1987, 1989) and Arnold, Castillo, and Sarabia (1992)]

$$p(x_1, x_2) \propto (\lambda_0 + \lambda_1 x_1 + \lambda_2 x_2 + \lambda_3 x_1 x_2)^{-\alpha - 1} \qquad \text{for } x_1 > 0, \; x_2 > 0. \tag{52.41}$$

An expression for the normalizing constant is given by Arnold (1987) and Arnold, Castillo, and Sarabia (1992). In order for $p(x_1, x_2)$ in (52.41) to be a valid bivariate density function, we must have $\lambda_0 > 0$, $\lambda_1 > 0$ and $\lambda_2 > 0$ while $\lambda_3 \geq 0$ (with $\lambda_3 > 0$ for $0 < \alpha \leq 1$). The case $\lambda_3 = 0$ leads to Mardia's bivariate Pareto distribution of the second kind with $\alpha > 1$. Clearly, (52.41) represents a general bivariate Pareto family which comprises all bivariate densities for which both conditional densities are Pareto$(II)(0, \sigma, \alpha)$.

From the joint density function in (52.41), we have

$$\Pr[X_1 > x_1 | X_2 = x_2] = \left(1 + \frac{\lambda_1 + \lambda_3 x_2}{\lambda_0 + \lambda_2 x_2} x_1 \right)^{-\alpha} \tag{52.42}$$

and the sign of $\frac{d}{dx_2} \Pr[X_1 > x_1 | X_2 = x_2]$ depends on $\lambda_0 \lambda_3 - \lambda_1 \lambda_2$; that is, X_1 is either stochastically increasing or decreasing in X_2. Bivariate distributions with either $\lambda_0 = 0$ or $\lambda_3 = 0$ have always positive correlation coefficient.

From (52.41), we also find the marginal density functions of X_1 and X_2 to be

$$p_{X_1}(x_1) \propto \frac{1}{(\lambda_2 + \lambda_3 x_1)(\lambda_0 + \lambda_1 x_1)^\alpha} \quad \text{and}$$

$$(52.43)$$

$$p_{X_2}(x_2) \propto \frac{1}{(\lambda_1 + \lambda_3 x_2)(\lambda_0 + \lambda_2 x_2)^\alpha} \, .$$

For inferential purposes, we may reparameterize (52.41) to the form

$$p(x_1, x_2) \propto (1 + \lambda_1 x_1 + \lambda_2 x_2 + \lambda_{12} x_1 x_2)^{-\alpha - 1} \qquad (52.44)$$

(assuming $\lambda_0 = 0$ without loss of any generality). Now setting $\phi = \frac{\lambda_{12}}{\lambda_1 \lambda_2}$, we have the following exact expression in the case when $\alpha = 1$:

$$p(x_1, x_2) = \frac{\lambda_1 \lambda_2 (1 - \phi)}{- \log \phi} (1 + \lambda_1 x_1 + \lambda_2 x_2 + \phi \lambda_1 \lambda_2 x_1 x_2)^{-2},$$
$$x_1 > 0, \ x_2 > 0, \ \lambda_1 > 0, \ \lambda_2 > 0 \text{ and } 0 < \phi < 1.$$

$$(52.45)$$

Based on a random sample $(x_{1i}, x_{2i})^T$, $i = 1, 2, \ldots, n$, from the bivariate distribution in (52.45), the log-likelihood function is

$$\log L(\lambda_1, \lambda_2, \phi) = n \log \lambda_1 + n \log \lambda_2 + n \log(1 - \phi) - n \log(- \log \phi)$$
$$- 2 \sum_{i=1}^{n} \log(1 + \lambda_1 x_{1i} + \lambda_2 x_{2i} + \phi \lambda_1 \lambda_2 x_{1i} x_{2i}).$$

For a fixed value of ϕ, this log-likelihood function will be maximized by λ_1 and λ_2 satisfying

$$\frac{n}{\lambda_1} = 2 \sum_{i=1}^{n} \frac{x_{1i} + \phi \lambda_2 x_{1i} x_{2i}}{1 + \lambda_1 x_{1i} + \lambda_2 x_{2i} + \phi \lambda_1 \lambda_2 x_{1i} x_{2i}} \qquad (52.46)$$

and

$$\frac{n}{\lambda_2} = 2 \sum_{i=1}^{n} \frac{x_{2i} + \phi \lambda_1 x_{1i} x_{2i}}{1 + \lambda_1 x_{1i} + \lambda_2 x_{2i} + \phi \lambda_1 \lambda_2 x_{1i} x_{2i}} \, . \qquad (52.47)$$

Equations (52.46) and (52.47) can be solved iteratively, and a simple search procedure can be used to find the optimal value of ϕ.

If we make the transformations $Y_1 = \log X_1$ and $Y_2 = \log X_2$ in (52.45), then $(Y_1, Y_2)^T$ will have a bivariate logistic distribution with both its conditional distributions of logistic form with unit scale parameters. If, however, we consider the transformations $Y_1 = \frac{1}{X_1}$ and $Y_2 = \frac{1}{X_2}$, then the bivariate density function of $(Y_1, Y_2)^T$ is

$$p_{Y_1,Y_2}(y_1, y_2) \propto \left\{ 1 + \frac{\lambda_2}{\lambda_{12}} y_1 + \frac{\lambda_1}{\lambda_{12}} y_2 + \frac{1}{\lambda_{12}} y_1 y_2 \right\}^{-2}. \qquad (52.48)$$

Thus, if $\lambda_1 = \lambda_2 = 1$ in (52.45), then

$$\left(\frac{1}{\lambda_{12} X_1}, \frac{1}{\lambda_{12} X_2} \right)^T \stackrel{d}{=} (X_1, X_2)^T.$$

Arnold, Castillo, and Sarabia (1992) have further generalized the conditionally specified bivariate Pareto distribution of the second kind in (52.41) by specifying that both $p(x_1|x_2)$ and $p(x_2|x_1)$ be beta density functions of the second kind of the form [see Eq. (25.79) in Chapter 25 of Johnson, Kotz, and Balakrishnan (1995, p. 248)]

$$p(x) = \frac{\sigma^b}{B(a,b)} \cdot \frac{x^{a-1}}{(\sigma + x)^{a+b}}, \qquad x > 0, \ \sigma > 0, \ a > 0, \ b > 0$$

leading to the bivariate density function

$$p(x_1, x_2) \propto \frac{x_1^{a-1} x_2^{a-1}}{(\lambda_0 + \lambda_1 x_1 + \lambda_2 x_2 + \lambda_3 x_1 x_2)^{a+b}}, \qquad x_1, x_2 > 0,$$
$$a, b > 0, \ \lambda_0, \lambda_1, \lambda_2 > 0, \ \lambda_3 \geq 0. \qquad (52.49)$$

Observe that the bivariate density function in (52.41) is a special case of (52.49) with $a = 1$.

Next, extending (52.41), Arnold, Castillo, and Sarabia (1993) identified the bivariate distributions for which $X_1|(X_2 = x_2)$ is Pareto(IV), with parameters $(\sigma(x_2), \delta(x_2), \alpha(x_2))$ and $X_2|(X_1 = x_1)$ is also Pareto(IV), with parameters $(\tau(x_1), \gamma(x_1), \beta(x_1))$. Denoting the corresponding marginal density functions of X_1 and X_2 by $p_{X_1}(x_1)$ and $p_{X_2}(x_2)$, respectively, the following functional equation is then valid:

$$a_1(x_1) x_2^{\gamma(x_1)} \left\{ 1 + b_1(x_1) x_2^{\gamma(x_1)} \right\}^{c_1(x_1)}$$
$$= a_2(x_2) x_1^{\delta(x_2)} \left\{ 1 + b_2(x_2) x_1^{\delta(x_2)} \right\}^{c_2(x_2)}, \qquad x_1 > 0, \ x_2 > 0,$$

$$(52.50)$$

where

$$
\begin{aligned}
a_1(x_1) &= x_1 p_{X_1}(x_1)\beta(x_1)\gamma(x_1)/\{\tau(x_1)\}^{\gamma(x_1)}, \\
b_1(x_1) &= \{\tau(x_1)\}^{-\gamma(x_1)}, \\
c_1(x_1) &= -\{\beta(x_1)+1\}, \\
a_2(x_2) &= x_2 p_{X_2}(x_2)\alpha(x_2)\delta(x_2)/\{\sigma(x_2)\}^{\delta(x_2)}, \\
b_2(x_2) &= \{\sigma(x_2)\}^{-\delta(x_2)}, \\
c_2(x_2) &= -\{\alpha(x_2)+1\};
\end{aligned}
$$

here, $a_1, b_1, \gamma, a_2, b_2$, and δ are all positive, and c_1 and c_2 are both less than -1.

Two cases are of special interest:

(i) $\gamma(x_1)$ and $\delta(x_2)$ are constant functions leads to

Model I:

$$
\begin{aligned}
p_{X_1,X_2}&(x_1 x_2) \\
&= x_1^{\delta-1} x_2^{\gamma-1} \left\{\lambda_1 + \lambda_2 x_1^{\delta} + \lambda_3 x_2^{\gamma} + \lambda_4 x_1^{\delta} s_2^{\gamma}\right\}^{\lambda_5} \\
&\quad \text{for } x_1, x_2 > 0, \ \lambda_1 > 0, \ \lambda_2 > 0, \ \lambda_3 > 0, \ \lambda_4 \geq 0, \lambda_5 < -1.
\end{aligned}
$$
(52.51)

Model II:

$$
\begin{aligned}
p_{X_1,X_2}&(x_1 x_2) \\
&= x_1^{\delta-1} x_2^{\gamma-1} \exp\{\theta_1 + \theta_2 \log(\theta_5 + x_1^{\delta}) \\
&\quad + \theta_3 \log(\theta_6 + x_2^{\gamma}) + \theta_4 \log(\theta_5 + x_1^{\delta}) \log(\theta_6 + x_2^{\gamma})\} \\
&\quad \text{for } x_1, x_2 > 0, \ \theta_5, \theta_6 > 0, \ \theta_2, \theta_3 < -1, \ \theta_4 \leq 0.
\end{aligned}
$$
(52.52)

The marginal density functions of Model I in (52.51) are

$$
p_{X_1}(x_1) = \frac{1}{\delta(-1-\lambda_5)} \, x_1^{\delta-1}(\lambda_3 + \lambda_4 x_1^{\delta})^{-1}(\lambda_1 + \lambda_2 x_1^{\delta})^{\lambda_5+1}, \quad x_1 > 0,
$$

and

$$
p_{X_2}(x_2) = \frac{1}{\lambda(-1-\lambda_5)} \, x_2^{\delta-1}(\lambda_2 + \lambda_4 x_2^{\gamma})^{-1}(\lambda_1 + \lambda_3 x_2^{\gamma})^{\lambda_5+1}, \quad x_2 > 0.
$$

These are Pareto(IV) distributions only when $\lambda_4 = 0$. Arnold, Castillo, and Sarabia (1993) have provided three-dimensional plots and contour plots for the joint density function $p_{X_1,X_2}(x_1, x_2)$ in (52.51). The bivariate density function is unbounded as $x_1, x_2 \to 0$ for $\gamma < 1$ and $\delta < 1$, and the density is unimodal if $\gamma \geq 1$ and $\delta \geq 1$.

The marginal density functions of Model II in (52.52) are

$$p_{X_1}(x_1)$$

$$= \frac{-x_1^{\delta-1}}{1 + \theta_3 + \theta_4 \log(\theta_5 + x_1^\delta)} \exp\{\theta_1 + (1+\theta_3)\log\theta_6$$

$$+ (\theta_2 + \theta_4 \log\theta_6)\log(\theta_5 + x_1^\delta)\}, \quad x_1 > \left\{e^{-(1+\theta_3)/\theta_4} - \theta_5\right\}^{1/\delta},$$

and

$$p_{X_2}(x_2)$$

$$= \frac{-x_2^{\gamma-1}}{1 + \theta_2 + \theta_4 \log(\theta_6 + x_2^\gamma)} \exp\{\theta_1 + (1+\theta_2)\log\theta_5$$

$$+ (\theta_3 + \theta_4 \log\theta_5)\log(\theta_6 + x_2^\delta)\}, \quad x_2 > \left\{e^{-(1+\theta_2)/\theta_4} - \theta_6\right\}^{1/\gamma}.$$

Arnold, Castillo, and Sarabia (1993) have also provided three-dimensional plots and contour plots for the joint density function $p_{X_1,X_2}(x_1, x_2)$ in (52.52). This model yields a nonpositive correlation.

(ii) $\beta(x_1)$ *and* $\alpha(x_2)$ *are constant functions* leads to

$$p_{X_1,X_2}(x_1, x_2)$$

$$= \frac{\exp\{\theta_0 + \theta_2 \log x_1 + \theta_3 \log x_2 + \theta_4 \log x_1 \log x_2\}}{x_1 x_2 [1 + \exp\{\theta_1 + \theta_2 \log x_1 + \theta_3 \log x_2 + \theta_4 \log x_1 \log x_2\}]^{\alpha+1}}$$
$$\text{for } x_1 > 0, \ x_2 > 0, \ \alpha > 0. \tag{52.53}$$

Compare this with bivariate logistic distribution discussed in Chapter 51.

Arnold, Castillo, and Sarabia (1993) have noted that generalization to the case where it is postulated that

$$X_1 | (X_2 = x_2) \overset{d}{=} GP^*(\mu, \sigma(x_2), \delta(x_2), \alpha(x_2))$$

and

$$X_2 | (X_1 = x_1) \overset{d}{=} GP^*(\nu, \tau(x_1), \gamma(x_1), \beta(x_1)),$$

where $GP^*(\mu, \sigma, \delta, \alpha)$ is a distribution with survival function $\{1 + (\frac{x-\mu}{\sigma})^\delta\}^{-\alpha}$ (for $x \geq \mu$), leads to translated versions of Models I and II. The solution to the problem when μ and ν are functions of x_2 and x_1, respectively, has not yet been obtained.

Arnold (1995) observed that the specification

$$\Pr[X_1 > x_1 | X_2 > x_2] = \left\{1 + \left(\frac{x_1}{\sigma_1(x_2)}\right)^{c_1(x_2)}\right\}^{-k_1(x_2)}, \qquad x_1 > 0,$$

and

$$\Pr[X_2 > x_2 | X_1 > x_1] = \left\{1 + \left(\frac{x_2}{\sigma_2(x_1)}\right)^{c_2(x_1)}\right\}^{-k_2(x_1)}, \qquad x_2 > 0,$$

for some positive functions $c_1(x_2)$, $c_2(x_1)$, $k_1(x_2)$, $k_2(x_1)$, $\sigma_1(x_2)$ and $\sigma_2(x_1)$, leads to the following equation for the marginal survival functions:

$$\bar{F}_{X_2}(x_2)\left\{1 + b_1(x_2)x_1^{c_1(x_2)}\right\}^{-k_1(x_2)}$$
$$= \bar{F}_{X_1}(x_1)\left\{1 + b_2(x_1)x_2^{c_2(x_1)}\right\}^{-k_2(x_1)}, \qquad (52.54)$$

where $b_1(x_2) = \{\sigma_1(x_2)\}^{-c_1(x_2)}$ and $b_2(x_1) = \{\sigma_2(x_1)\}^{-c_2(x_1)}$.

In the case when $c_1(x_2) \equiv c_1$ and $c_2(x_1) \equiv c_2$, which was investigated in detail by Arnold (1995), there are two sets of solutions for the joint survival function, as follows:

1.

$$\bar{F}(x_1, x_2) = \left\{1 + \left(\frac{x_1}{\sigma_1}\right)^{c_1} + \left(\frac{x_2}{\sigma_2}\right)^{c_2} + \theta\left(\frac{x_1}{\sigma_1}\right)^{c_1}\left(\frac{x_2}{\sigma_2}\right)^{c_2}\right\}^{-k},$$
$$x_1, x_2 > 0, \ c_1, c_2, \sigma_1, \sigma_2, k > 0, \ \text{and} \ 0 \leq \theta \leq 2.$$
$$(52.55)$$

The condition $0 \leq \theta \leq 2$ is needed in order to ensure positivity of the density.

2.

$$\bar{F}(x_1, x_2) = \exp\left[-\theta_1 \log\left\{1 + \left(\frac{x_1}{\sigma_1}\right)^{c_1}\right\} - \theta_2 \log\left\{1 + \left(\frac{x_2}{\sigma_2}\right)^{c_2}\right\}\right.$$
$$\left. - \theta_3 \log\left\{1 + \left(\frac{x_1}{\sigma_1}\right)^{c_1}\right\}\log\left\{1 + \left(\frac{x_2}{\sigma_2}\right)^{c_2}\right\}\right],$$
$$x_1, x_2 > 0, \ \theta_1, \theta_2, \sigma_1, \sigma_2 > 0, \ c_1, c_2 > 0, \ \theta_3 \geq 0.$$
$$(52.56)$$

Imposing conditions $k_1(x_2) \equiv k_1$ and $k_2(x_1) \equiv k_2$ in the conditional specification yields $k_1 = k_2 = k$, and then, setting $c_1(x_2) \equiv c_1$ and $c_2(x_1) \equiv c_2$ results in the bivariate survival function in (52.55).

Finally, we mention the work of Arnold, Sarabia, and Castillo (1995) in which all bivariate distributions, whose conditional distributions have the so-called *Pickands–de Haan's generalized Pareto distribution* with survival function

$$F(x; \alpha, k) = \left(1 - \frac{kx}{\alpha}\right)^{1/k}, \qquad 0 < x < \frac{\alpha}{\max(0, k)}, \quad k \in \mathbb{R}, \ \alpha > 0,$$

(52.57)

are identified. Note that the exponential distribution is obtained as a limiting case of (52.57) when $k \to 0$. Specifically, by stipulating that $X_1|(X_2 = x_2)$ is distributed as (52.57) with parameters $k_1(x_2)$ and $\alpha_1(x_2)$ for each x_2 and that $X_2|(X_1 = x_1)$ is distributed as (52.57) with parameters $k_2(x_1)$ and $\alpha_2(x_1)$ for each x_1, where $\alpha_1(x_2) > 0 \ \forall \ x_2$ and $\alpha_2(x_1) > 0 \ \forall \ x_1$, Arnold, Sarabia and Castillo (1995) have shown that the joint density function of $(X_1, X_2)^T$ will have one of the following forms:

Model I: If $k_1(x_2) = k_2(x_1 \equiv k$, then

$$p(x_1, x_2) = \frac{\theta_0 k^2}{\alpha_1 \alpha_2} \left(1 - \frac{kx_1}{\alpha_1} - \frac{kx_2}{\alpha_2} + \frac{\delta k^2 x_1 x_2}{\alpha_1 \alpha_2}\right)^{\frac{1}{k} - 1},$$

$$0 < x_1 < \frac{\alpha_1}{\max(0, k)}, \ 0 < x_2 < \frac{\alpha_2}{\max(0, k)},$$

$$1 - \frac{kx_1}{\alpha_1} - \frac{kx_2}{\alpha_2} + \frac{\delta k^2 x_1 x_2}{\alpha_1 \alpha_2} > 0. \qquad (52.58)$$

Here, θ_0 is a normalizing constant. The choice $\delta = 1$ corresponds to independence of X_1 and X_2, and $k \to 0$ corresponds to the bivariate exponential distribution of Arnold and Strauss (1988) with exponential conditionals. Restrictions on δ depend on whether k is positive or negative.

Model II:

$$p(x_1, x_2) = \frac{\alpha_0 k_1 k_2}{\alpha_1 \alpha_2} \left(1 - \frac{k_1 x_1}{\alpha_1}\right)^{\frac{1}{k_1} - 1} \left(1 - \frac{k_2 x_2}{\alpha_2}\right)^{\frac{1}{k_2} - 1}$$

$$\times \exp\left\{\xi \log\left(1 - \frac{k_1 x_1}{\alpha_1}\right) \log\left(1 - \frac{k_2 x_2}{\alpha_2}\right)\right\},$$

$$x_1, x_2 > 0, \ 1 - \frac{k_1 x_1}{\alpha_1} > 0, \ 1 - \frac{k_2 x_2}{\alpha_2} > 0,$$

$$(52.59)$$

where $\alpha_1, \alpha_2 > 0$, $k_1, k_2 \in \mathbb{R}$. The parameter $\xi \leq 0$ governs dependence. If $\xi = 0$, we have the case of X_1 and X_2 being independent; it is only in this case that k_1 and k_2 can have opposite signs. α_0 is a normalizing constant.

For the joint density function in (52.58), corresponding to Model I, if the parameter k is positive, then $(X_1^*, X_2^*)^T \equiv \left(\frac{kX_1}{\alpha_1}, \frac{kX_2}{\alpha_2}\right)^T$ has the joint density function

$$p_{X_1^*, X_2^*}(x_1, x_2) = \theta_0 (1 - x_1 - x_2 + \delta x_1 x_2)^{\frac{1}{k}-1},$$

$$0 < x_1 < 1, \ 0 < x_2 < 1, \ 1 - x_1 - x_2 + \delta x_1 x_2 > 0.$$

$$(52.60)$$

From (52.60), we observe that X_1^* and $\frac{X_2^*(1-\delta X_1^*)}{1-X_1^*}$ are statistically independent, and the latter has a $\text{Beta}(1, \frac{1}{k})$ distribution, which facilitates the derivation of the following expressions, when $\delta \neq 0$:

$$E[X_j] = \frac{\alpha_j}{k\delta}\left(1 - \frac{k^2 \theta_0}{k+1}\right),$$

$$E[X_j^2] = \left(\frac{\alpha_j}{k\delta}\right)^2 \left\{1 - \left(1 + \frac{\delta k}{1+2k}\right)\frac{k^2 \theta_0}{k+1}\right\}, \quad (j = 1, 2),$$

and

$$E[X_1 X_2] = \left(\frac{\alpha_1}{k\delta}\right)\left(\frac{\alpha_2}{k\delta}\right)\left\{\frac{2k+1}{k+1}\left(1 - \frac{k^2\theta_0}{k+1}\right) - \frac{\delta k}{k+1}\right\},$$

where $\theta_0 = \sum_{j=0}^{\infty} k\delta^j B(j+1, \frac{1}{k}+1)$ if $|\delta| \leq 1$.

For the case when $k > 0$ and $\delta = 0$, we have $\theta_0 = \frac{k+1}{k^2}$ and

$$E[X_j] = \frac{\alpha_j}{1+2k},$$

$$E[X_j^2] = \frac{2\alpha_j^2}{(1+2k)(1+3k)} \quad (j = 1, 2),$$

and

$$E[X_1 X_2] = \frac{\alpha_1 \alpha_2}{(1+2k)(1+3k)}$$

so that $\text{corr}(X_1, X_2) = -k/(k+1)$.

3.5 Muliere and Scarsini's Bivariate Pareto

The assumption of independence in Lindley and Singpurwalla's (1986) model is somewhat restrictive because in many systems the component lifelengths have a well-defined dependence structure. This has led to consideration of well-known bivariate exponential distributions as the initial model which, when placed in a varying environment, produces the corresponding bivariate Pareto distributions.

Muliere and Scarsini (1987) proposed a bivariate Pareto distribution with joint survival function

$$
\begin{aligned}
\bar{F}_{X_1, X_2}(x_1, x_2) &= \Pr[X_1 > x_1, X_2 > x_2] \\
&= \left(\frac{x_1}{\beta}\right)^{-\lambda_1} \left(\frac{x_2}{\beta}\right)^{-\lambda_2} \left\{\max\left(\frac{x_1}{\beta}, \frac{x_2}{\beta}\right)\right\}^{-\lambda_0},
\end{aligned}
$$
$$
\beta \leq \min(x_1, x_2) < \infty. \tag{52.61}
$$

Note the similarity of this distribution with Marshall and Olkin's (1967a,b) bivariate exponential distribution. Jeevanand and Padamadan (1996) established the following characterization of this bivariate Pareto distribution. Let $Z = \min(X_1, X_2)$, $\delta_1 = I(X_1 < X_2)$ and $\delta_2 = I(X_1 > X_2)$, where $I(A)$ denotes the indicator function of event A. Let $\lambda = \sum_{i=0}^{2} \lambda_i$. Then, the bivariate random vector $(X_1, X_2)^T$ has the Muliere–Scarsini bivariate Pareto distribution in (52.61) iff there exist independent random variables $U_0, U_1,$ and U_2 each distributed as $P(I)(\beta, \lambda_i)$ $(i = 0, 1, 2)$, respectively, such that $X_i = \min(U_0, U_i)$ for $i = 1, 2$. Moreover, Z has a $P(I)(\beta, \lambda)$ distribution, $(\delta_1, \delta_2)^T$ has a multinomial distribution with parameters $\left(1; \frac{\lambda_1}{\lambda}, \frac{\lambda_2}{\lambda}\right)$ [see Chapter 35 of Johnson, Kotz, and Balakrishnan (1997)], and the variables Z and $(\delta_1, \delta_2)^T$ are statistically independent. The proof is based on the observation that the event $(X_1 > x_1, X_2 > x_2)$ is equivalent to the event $(U_1 > x_1, U_2 > x_2, U_0 > \max(x_1, x_2))$.

When the parameter β is known, then Jeevanand and Padamadan (1996) have suggested a conjugate prior for the λ_i's as one with density function

$$
f(\lambda_0, \lambda_1, \lambda_2) = c\lambda_0^{\alpha_0 - 1} \lambda_1^{\alpha_1 - 1} \lambda_2^{\alpha_2 - 1} e^{-\lambda u_0}, \quad u_0 > 0, \ \lambda_i > 0 \ (i = 0, 1, 2). \tag{52.62}
$$

When u_0 and the α_i's $(i = 0, 1, 2,)$ all tend to 0, we obtain Bayes estimates under quadratic loss corresponding to the noninformative Jeffreys's (1961) prior, which are

$$
\hat{\lambda}_j = \frac{n_j}{t - n \log \beta}, \qquad j = 0, 1, 2, \tag{52.63}
$$

where z_i are independent observations on the random variable Z, d_{1i} and d_{2i} $(i = 1, 2, \ldots, n)$ are observations on δ_1 and δ_2, respectively, $d_{0i} = 1 - d_{1i} - d_{2i}$, $t = \sum_{i=1}^{n} \log z_i$, $n_j = \sum_{i=1}^{n} d_{ji}$ $(j = 0, 1, 2)$, and $n = n_0 + n_1 + n_2$. These are also the maximum likelihood estimators of λ_i $(i = 0, 1, 2)$.

When the parameter β is unknown, the estimators are quite complicated; they have been discussed by Jeevanand and Padamadan (1996). It turns out that the biases of the estimators are substantially less when β is known than when β is unknown.

Jeevanand (1997) has shown in the case of (52.61) that

$$R = \Pr[X_2 < X_1] = \lambda_2/\lambda$$

and has presented an expression for the Bayes estimator of R.

Padamadan and Nair (1994) have characterized the bivariate Pareto distribution in (52.61) by the property that the marginal distributions are $P(I)$ with survival functions

$$\Pr[X_1 > x_1] = \left(\frac{x_1}{\beta}\right)^{-\theta_1}, \quad x_1 \geq \beta > 0, \ \theta_1 > 0 \qquad (52.64)$$

and

$$\Pr[X_2 > x_2] = \left(\frac{x_2}{\beta}\right)^{-\theta_2}, \quad x_2 \geq \beta > 0, \ \theta_2 > 0, \qquad (52.65)$$

and a "lack of memory property"

$$\Pr[X_1 > x_1 t\beta, X_2 > x_2 t\beta \mid X_1 > t\beta, X_2 > t\beta]$$
$$= \Pr[X_1 > x_1\beta, X_2 > x_2\beta] \qquad (52.66)$$

holds for all $x_1, x_2, t \geq 1$. Here, $\lambda_1 = \delta - \theta_2$, $\lambda_2 = \delta - \theta_1$ and $\lambda_{12} = \theta_1 + \theta_2 - \delta$ for some $\delta > 0$. As Padamadan and Nair (1994) have mentioned, it is essential that the lower limits for the supports of X_1 and X_2 be the same (namely, β).

3.6   Bilateral Bivariate Pareto

For a two-parameter uniform distribution with probability density function [see Chapter 26 of Johnson, Kotz and Balakrishnan (1995)]

$$p(x|\alpha, \beta) = \frac{1}{\beta - \alpha} I(\alpha < x < \beta), \qquad (52.67)$$

a family of conjugate priors is the so-called *bilateral bivariate Pareto distribution* with density function [see Section 9.7 of DeGroot (1970) and Lee (1989, pp. 246–247)]

$$p(x_1, x_2 \mid \xi, \eta, \gamma)$$
$$= \begin{cases} (\gamma+1)\gamma(\xi-\eta)^\gamma(x_2-x_1)^{-\gamma-2}, & x_1 < \eta < \xi < x_2, \ \gamma > 1, \\ 0, & \text{otherwise.} \end{cases}$$
$$(52.68)$$

In terms of indicator functions, we may write (52.68) as

$$p(x_1, x_2 \mid \xi, \eta, \gamma) \propto (x_2 - x_1)^{-\gamma-2} I_{(-\infty,\eta)}(x_1) I_{(\xi,\infty)}(x_2). \qquad (52.69)$$

If indeed the prior on (α, β), according to (52.69), is

$$p(\alpha, \beta) \propto (\beta - \alpha)^{-\gamma-2} I_{(-\infty,\eta)}(\alpha) I_{(\xi,\infty)}(\beta),$$

then, upon noting that the likelihood function is

$$\ell(\alpha, \beta \mid \boldsymbol{x}) = (\beta - \alpha)^{-n} I_{(-\infty,m)}(\alpha) I_{(M,\infty)}(\beta),$$

where $m = \min(x_1, \ldots, x_n)$ and $M = \max(x_1, \ldots, x_n)$, we have the posterior distribution:

$$p(\alpha, \beta \mid \boldsymbol{x}) \propto (\beta - \alpha)^{-(\gamma+n)-2} I_{(-\infty,\eta)}(\alpha) I_{(-\infty,m)}(\alpha)$$
$$\times I_{(\xi,\infty)}(\beta) I_{(M,\infty)}(\beta). \qquad (52.70)$$

From the bilateral bivariate Pareto distribution in (52.68), it can be shown that

$$E[X_1] = \frac{\gamma\xi - \eta}{\gamma - 1}, \ E[X_2] = \frac{\gamma\eta - \xi}{\gamma - 1} \qquad \text{for } \gamma > 1,$$

$$\operatorname{var}(X_1) = \operatorname{var}(X_2) = \frac{\gamma(\xi - \eta)^2}{(\gamma - 1)^2(\gamma - 2)} \qquad \text{for } \gamma > 2,$$

and

$$\operatorname{corr}(X_1, X_2) = -\frac{1}{\gamma} \qquad \text{for } \gamma > 2.$$

The marginal distribution of X_2 is $P(III)$ with cumulative distribution function

$$F_{X_2}(x_2) = \left\{ 1 - \left(\frac{\xi - \eta}{x_2 - \eta} \right)^\gamma \right\} I_{(\xi,\infty)}(x_2). \qquad (52.71)$$

In particular, median$(X_2) = \eta + 2^{1/\gamma}(\xi - \eta)$. Similarly, the marginal distribution function of X_1 is given by

$$F_{X_1}(x_1) = \left(\frac{\xi - \eta}{\eta - x_1}\right)^{\gamma} I_{(-\infty,\eta)}(x_1) \tag{52.72}$$

with median$(X_1) = \xi - 2^{-1/\gamma}(\xi - \eta)$.

3.7   Bivariate Semi-Pareto

A random vector $(X_1, X_2)^T$ is said to have a *bivariate semi-Pareto distribution* with parameters α_1, α_2, and p if its survival function is of the form

$$\bar{F}_{X_1,X_2}(x_1, x_2) = \Pr[X_1 > x_1, X_2 > x_2] = \frac{1}{1 + \psi(x_1, x_2)}, \tag{52.73}$$

where $\psi(x_1, x_2)$ satisfies the functional equation

$$\psi(x_1, x_2) = \frac{1}{p}\, \psi(p^{1/\alpha_1}x_1, p^{1/\alpha_2}x_2),$$
$$0 < p < 1, \ \alpha_i > 0 \ (i = 1, 2), \ x_1 > 0, \ x_2 > 0. \tag{52.74}$$

The solution of this functional equation is

$$\psi(x_1, x_2) = x_1^{\alpha_1} h_1(x_1) + x_2^{\alpha_2} h_2(x_2),$$

where $h_1(\cdot)$ and $h_2(\cdot)$ are periodic functions in $\log x_1$ and $\log x_2$, respectively, with periods $\frac{2\pi\alpha_i}{-\log p}$ for $i = 1, 2$ [see, for example, Kagan, Linnik, and Rao (1973, p. 163)]; in the particular case when $h_1(\cdot) \equiv h_2(\cdot) \equiv 1$, we obtain from (52.73)

$$\bar{F}_{X_1,X_2}(x_1, x_2) = \frac{1}{1 + x_1^{\alpha_1} + x_2^{\alpha_2}}, \qquad x_i > 0, \ \alpha_i > 0 \ (i = 1, 2), \tag{52.75}$$

which is a special case of the bivariate Pareto distribution in (52.39).

Balakrishna and Jayakumar (1997) have established the following characterization result via geometric minimization. Let $\{(X_{1i}, X_{2i})^T, \ i = 1, 2, \ldots, \}$ be a sequence of independent and identically distributed bivariate random vectors with joint survival function

$$\bar{F}(x_1, x_2) = \Pr[X_1 > x_1, X_2 > x_2] = \frac{1}{1 + \psi(x_1, x_2)};$$

let N be a random variable with parameter p and with geometric distribution

$$\Pr[N = n] = pq^{n-1}, \qquad n = 1, 2, \ldots, \quad 0 < p < 1, \quad p = 1 - q,$$

independently of $(X_{1i}, X_{2i})^T$. Finally, let

$$L_{N1} = \min_{1 \le i \le N} X_{1i} \quad \text{and} \quad L_{N2} = \min_{1 \le i \le N} X_{2i}.$$

Then, Balakrishna and Jayakumar (1997) have shown that the vectors $(p^{-1/\alpha_1} L_{N1}, p^{-1/\alpha_2} L_{N2})^T$ and $(X_{1i}, X_{2i})^T$ are identically distributed iff $(X_{1i}, X_{2i})^T$ have a bivariate semi-Pareto distribution as given in (52.73). Compare this with Arnold's (1975) construction of multivariate exponential distributions using geometric compounding.

(Note that *any* survival function $\bar{F}(x_1, x_2)$ can be represented in the form $\frac{1}{1+\psi(x_1,x_2)}$, where $\psi(x_1, x_2)$ is a monotonically increasing function in both x_1 and x_2 while $\lim_{x_1 \to 0} \lim_{x_2 \to 0} \psi(x_1, x_2) = 0$ and $\lim_{x_1 \to \infty} \lim_{x_2 \to \infty} \psi(x_1, x_2) = \infty$.)

A stronger characterization result is obtained if it is additionally assumed that $\bar{F}(x_1, x_2)$ is of the form $\frac{1}{1+\psi(x_1,x_2)}$, where $\psi(x_1, x_2)$ is such that

$$\lim_{x_1 \to 0+} \lim_{x_2 \to 0+} \frac{\psi(x_1, x_2)}{x_1^{\alpha_1} + x_2^{\alpha_2}} = 1.$$

Then the identical distribution of $(p^{-1/\alpha_1} L_{N1}, p^{-1/\alpha_2} L_{N2})^T$ and $(X_{1i}, X_{2i})^T$ implies that $(X_{1i}, X_{2i})^T$ has a simplified bivariate Pareto distribution with joint survival function as in (52.75).

4 MULTIVARIATE PARETO DISTRIBUTIONS

4.1 Multivariate Pareto of the First Kind

Mardia's (1962) *multivariate Pareto distribution of the first kind* has joint density function

$$
\begin{aligned}
p_{\mathbf{X}}(\mathbf{x}) &= p_{X_1,\ldots,X_k}(x_1, \ldots, x_k) \\
&= a(a+1)\cdots(a+k-1)\left(\prod_{i=1}^{k}\theta_i\right)^{-1}\left(\sum_{i=1}^{k}\frac{x_i}{\theta_i} - k + 1\right)^{-(a+k)},
\end{aligned}
$$

$$x_i > \theta_i > 0, \ a > 0, \tag{52.76}$$

and is denoted by $\boldsymbol{X} \overset{d}{=} MP^{(k)}(I)(\boldsymbol{\theta}, a)$. Any subset of $\boldsymbol{X} = (X_1, \ldots, X_k)^T$ has a joint density of the same form as (52.76). The conditional density function of $(X_{s+1}, \ldots, X_k)^T$, given $X_1 = x_1, X_2 = x_2, \ldots, X_s = x_s$, is also of the same form as (52.76) with a replaced by $(a + s)$, θ_j by $\theta_j(\sum_{i=1}^{s} \frac{x_i}{\theta_i} - s + 1)$, and k by $(k - s)$.

Arnold (1983) observed that the joint survival function of the above $MP^{(k)}(I)(\boldsymbol{\theta}, a)$ is

$$\Pr[\boldsymbol{X} \geq \boldsymbol{x}] = \Pr[X_1 \geq x_1, \ldots, X_k \geq x_k]$$
$$= \left(\sum_{i=1}^{k} \frac{x_i}{\theta_i} - k + 1 \right)^{-a}, \quad x_i > \theta_i > 0, \ a > 0$$

(52.77)

and obscures the dual role played by $\boldsymbol{\theta}$ as both location and scale parameters, and therefore he recommended the representation

$$\Pr[\boldsymbol{X} \geq \boldsymbol{x}] = \left(1 + \sum_{i=1}^{k} \frac{x_i - \theta_i}{\theta_i} \right)^{-a}, \quad x_i > \theta_i > 0, \ a > 0. \quad (52.78)$$

It can be easily verified that coordinatewise minima of random samples of $MP^{(k)}(I)(\boldsymbol{\theta}, a)$ are themselves $MP(I)$.

Arnold and Pourahmadi (1988) and Wesolowski and Ahsanullah (1995) discussed some characterization results for the $MP^{(k)}(I)(\boldsymbol{1}, a)$ distribution (also called the *standard* multivariate Pareto distribution of the first kind). Arnold and Pourahmadi (1988) showed that if $\boldsymbol{X} = (X_1, \ldots, X_k)^T$ is a random vector satisfying the equidistribution condition

$$(X_1, \ldots, X_{k-1})^T \overset{d}{=} (X_2, \ldots, X_k)^T$$

and is such that the conditional distribution of X_k, given $(X_1, \ldots, X_{k-1})^T$, is Pareto(II) with parameters $a + k - 1$ and $\sum_{i=1}^{k-1} X_i + 1$ $(a > 0)$, then $\boldsymbol{X}$ follows $MP^{(k)}(\boldsymbol{1}, a)$.

Wesolowski and Ahsanullah (1995) modified this result slightly and established the characterization that if $\boldsymbol{X} = (X_1, \ldots, X_k)^T$ is a random vector satisfying

$$(X_0, X_1, \ldots, X_\ell)^T \overset{d}{=} (X_0, X_1, \ldots, X_{\ell-1}, X_{\ell+1})^T, \quad \ell = 1, \ldots, k - 1$$

(with $X_0 = 0$ a.s.) and is such that the conditional distribution of X_k, given $(X_1, \ldots, X_{k-1})^T$, is $P(II)$ with parameters $a + k - 1$ and $\sum_{i=1}^{k-1} X_i + 1$ $(a > 0)$, then $\boldsymbol{X}$ is distributed as $MP^{(k)}(I)(\boldsymbol{1}, a)$. Wesolowski and

Ahsanullah's (1995) proof utilizes the solution of a homogeneous Fredholm integral equation of the second kind, as shown below. If for all $x > 0$,

$$p(x) = \int_0^\infty \frac{(\beta + 1)(\gamma + y)^{\beta+1}}{(\gamma + x + y)^{\beta+2}} \, p(y) \, dy, \qquad \beta > 0, \ \gamma > 0,$$

where $p(\cdot)$ is a probability density function, then

$$p(x) = \frac{\beta \gamma^\beta}{(\gamma + x)^{\beta+1}}, \qquad x > 0.$$

Jupp and Mardia (1982) [also see Ruiz, Marin, and Zoroa (1993)] provided the following characterization of the $MP^{(k)}(I)(\boldsymbol{\theta}, a)$ distribution. They showed that $\boldsymbol{X} \overset{d}{=} MP^{(k)}(I)(\boldsymbol{\theta}, a)$ (with $a > 1$) if and only if

$$E[\boldsymbol{X} \mid (\boldsymbol{X} > \boldsymbol{x})] = (\Psi_1(\boldsymbol{x}), \ldots, \Psi_k(\boldsymbol{x}))^T,$$

where

$$\Psi_i(\boldsymbol{x}) = \frac{x_i}{\theta_i} + \frac{1}{a-1} \left(\sum_{i=1}^k \frac{x_i}{\theta_i} - k + 1 \right), \qquad x_i > \theta_i > 0, \ i = 1, \ldots, k.$$

Jupp and Mardia (1982) have also given an economic interpretation of this result.

Discussions on inferential methods for various forms of multivariate Pareto distributions have been somewhat limited. The pioneering work of Mardia (1962) on multivariate Pareto distribution of the first kind was followed by Arnold's (1983) discussion on estimation procedures for multivariate Pareto distributions of the second, third, and fourth kinds and also on the Bayesian estimation for multivariate Pareto distribution of the second kind. The contributions of Targhetta (1979), Tajvidi (1996), and Hanagal (1996) are also noteworthy in this direction. Tajvidi (1996) and Hanagal (1996) have specifically discussed the maximum likelihood estimation of parameters of some particular multivariate Pareto forms. Targhetta's (1979) method is based on a variant of the integral transform. If $X_1, X_2, \ldots, X_n$ are independent and identically distributed as $F_X(x|\theta)$, then for each θ we have

$$-\sum_{i=1}^n \log F_X(X_i|\theta) \overset{d}{=} \text{Gamma}(n, 1).$$

Then, the set

$$\left\{ \theta \, \middle| -\sum_{i=1}^n \log F_X(X_i|\theta) \in (C_1, C_2) \right\}$$

where C_1 and C_2 are $\frac{\alpha}{2}$ and $1 - \frac{\alpha}{2}$ percentiles of the Gamma$(n, 1)$ distribution, respectively, will provide a $100(1 - \alpha)\%$ confidence interval. Considering now a sample from k-dimensional distribution of $\boldsymbol{X} = (X_1, \ldots, X_k)^T$ with properly defined conditional distributions, let

$$U_{ij}(\boldsymbol{\theta}) = F_{X_j | X_1, \ldots, X_{j-1}} \left(X_j^{(i)} | X_1^{(i)}, \ldots, X_{j-1}^{(i)} | \boldsymbol{\theta} \right),$$
$$i = 1, \ldots, n, \ \ j = 2, \ldots, k$$

and

$$U_{i1}(\boldsymbol{\theta}) = F_{X_1}(X_1^{(i)} | \boldsymbol{\theta}).$$

Then,

$$Z_n(\boldsymbol{\theta}) = - \sum_{i=1}^{n} \sum_{j=1}^{k} \log U_{ij}(\boldsymbol{\theta})$$

is distributed as Gamma$(kn, 1)$, and this can be used to construct a confidence region for $\boldsymbol{\theta}$. Arnold has simplified this expression by utilizing the survival function instead of the distribution function.

4.2   Multivariate Pareto of the Second Kind

Mardia (1962) defined *multivariate Pareto distribution of the second kind* in the following way. We first note that if

$$p_X(x) = a\theta^a x^{-a-1}, \qquad x > \theta > 0, \ a > 0,$$

then $\log(X/\theta)$ has a standard exponential distribution; see Chapter 19 of Johnson, Kotz, and Balakrishnan (1994). If $X_1, X_2, \ldots, X_k$ each have Pareto distributions with densities

$$p_{X_i}(x_i) = a_i \theta_i^{a_i} x_i^{-a_i - 1}, \qquad x_i > \theta_i > 0, \ a_i > 0,$$

then the variables

$$Y_i = a_i \log(X_i/\theta_i), \qquad i = 1, 2, \ldots, k$$

will have some form of standard multivariate exponential distribution. To each such form of joint distribution will correspond a multivariate Pareto distribution for the X_i's. For the general multivariate case, Mardia (1962) in fact assumed that $(Y_1, \ldots, Y_k)^T$ have the multivariate exponential distribution of Krishnamoorthy and Parthasarathy (1951) (see Chapter 47).

Note that the a_i's need not be equal for the multivariate Pareto distribution of the second kind. It is worthwhile to determine the multivariate exponential distribution corresponding to the random variables

$Y_i = a \log(X_i/\theta_i)$ when $\boldsymbol{X} = (X_1, \ldots, X_k)^T$ follows a multivariate Pareto distribution of the first kind. From (52.76), we readily find that

$$p_{Y_1,\ldots,Y_k}(y_1, \ldots, y_k)$$
$$= \frac{a(a+1)\cdots(a+k-1)}{a^k} \left(\sum_{i=1}^{k} e^{y_i/a} - k + 1 \right)^{-(a+k)} e^{\sum_{i=1}^{k} y_i/a},$$
$$y_i > 0, \ a > 0, \ i = 1, 2, \ldots, k. \tag{52.79}$$

From the form of the joint survival function in (52.78), Arnold (1983) considered a natural generalization (by introducing a location parameter $\boldsymbol{\mu}$)

$$\Pr[\boldsymbol{X} \geq \boldsymbol{x}] = \left(1 + \sum_{i=1}^{k} \frac{x_i - \mu_i}{\theta_i} \right)^{-a}, \qquad x_i \geq \mu_i, \ \theta_i > 0, \ a > 0,$$
$$\tag{52.80}$$

which is denoted by $\boldsymbol{X} \stackrel{d}{=} MP^{(k)}(II)(\boldsymbol{\mu}, \boldsymbol{\theta}, a)$. This distribution has marginal expected values

$$E[X_i] = \mu_i + \frac{\theta_i}{a-1}$$

while the variances and covariances are exactly the same as those of $MP^{(k)}(I)(\boldsymbol{\theta}, a)$. The regression remains linear with

$$E[X_i|(X_j = x_j)] = \mu_i + \frac{\theta_i}{a} \left(1 + \frac{x_j - \mu_j}{\theta_j} \right)$$

and

$$\text{var}\,(X_i|(X_j = x_j)) = \theta_i^2 \frac{(a+1)}{a^2(a-1)} \left(1 + \frac{x_j - \mu_j}{\theta_j} \right)^2.$$

Ahmed and Gokhale (1989) have derived an expression for the entropy of the distribution in (52.80).

For the $MP^{(k)}(II)(\boldsymbol{\mu}, \boldsymbol{\theta}, a)$ distribution in (52.80), we have the representation

$$X_i = \mu_i + \theta_i(W_i/Z), \qquad i = 1, 2, \ldots, k,$$

where W_i's are independent and identically distributed as standard exponential, and Z is distributed as Gamma$(a, 1)$, independently of W_i's. For the special case when $\boldsymbol{\mu} = \boldsymbol{0}$ in (52.80), we obtain the $MP^{(k)}(II)(\boldsymbol{0}, \boldsymbol{\theta}, a)$ distribution with joint survival function

$$\Pr[\boldsymbol{X} > \boldsymbol{x}] = \left(1 + \sum_{i=1}^{k} \frac{x_i}{\theta_i} \right)^{-a}, \qquad x_i > 0, \ \theta_i > 0, \ a > 0; \tag{52.81}$$

see also Lindley and Singpurwalla (1986) and Nayak (1987). It is easy to show in this case that X_i/X_j is distributed as $P(II)(0, \frac{\theta_i}{\theta_j}, 1)$ and that X_i/X_j and X_j are independent, a property reminiscent of the gamma distribution; see Chapter 17 of Johnson, Kotz, and Balakrishnan (1994). Evidently, X_i is distributed as $P(II)(0, \theta_i, a)$.

Arnold, Castillo, and Sarabia (1994b) discussed multivariate Pareto distribuiton, $MP^{(k)}(II)(0, \boldsymbol{\theta}, a)$, with joint survival function of the form (52.81). They showed that, if $\boldsymbol{X} = (\boldsymbol{X}^*, \boldsymbol{X}^{**})^T$, where $\boldsymbol{X}^*$ is k_1-dimensional and $\boldsymbol{X}^{**}$ is $(k-k_1)$-dimensional, then $\boldsymbol{X}^* \overset{d}{=} MP^{(k_1)}(0, \boldsymbol{\theta}^*, a)$, where $\boldsymbol{\theta}^*$ denotes the first k_1 coordinates of $\boldsymbol{\theta}$. Thus, $MP^{(k)}(II)(0, \boldsymbol{\theta}, a)$ has $MP(II)$-type marginals. Also, $\boldsymbol{X}^*|(\boldsymbol{X}^{**} = \boldsymbol{x}^{**})$ is distributed as

$$MP^{(k_1)}(II)(0, c(\boldsymbol{x}^{**})\boldsymbol{\theta}, a + k - k_1),$$

where $c(\boldsymbol{x}^{**}) = \left(1 + \sum_{i=k_1+1}^{k} x_i/\theta_i\right)$. They characterized this distribution based on the properties of bivariate conditional distributions. Specifically, they assumed that, for every i and j, the conditional distribution of $(X_i, X_j)^T$, given $\boldsymbol{X}_{(i,j)} = \boldsymbol{x}_{(i,j)}$, is bivariate Pareto $MP^{(2)}(II)$ distribution for each $\boldsymbol{x}_{(i,j)}$. Here, $\boldsymbol{X}_{(i,j)}$ denotes the vector $\boldsymbol{X}$ with the i-th and j-th coordinates deleted and an analogous notation is used for $\boldsymbol{x}$. Under the above assumption, the joint density corresponds to $MP^{(k)}(II)(0, \boldsymbol{\theta}, a)$. This result is similar to the characterization of the multivariate normal distribution due to Arnold, Castillo, and Sarabia (1994a).

Restricting further to the case when $\boldsymbol{\theta} = \boldsymbol{1}$ (the standard form of the distribution) with joint survival function

$$\Pr[\boldsymbol{X} > \boldsymbol{x}] = \left(1 + \sum_{i=1}^{k} x_i\right)^{-a}, \qquad x_i > 0, \; a > 0 \qquad (52.82)$$

and joint density function

$$p_{\boldsymbol{X}}(\boldsymbol{x}) = a(a+1)\cdots(a+k-1)\left(1 + \sum_{i=1}^{k} x_i\right)^{-(a+k)}, \qquad x_i > 0, \qquad (52.83)$$

one finds that $Y = \sum_{i=1}^{k} X_i \overset{d}{=} FP(0, 1, 1, a, k)$. In fact, $Y' = \sum_{i=1}^{k'} X_i \overset{d}{=} FP(0, 1, 1, a, k)$ for any $k' \leq k$, and $W_j = X_j/(1 + \sum_{i=1}^{j-1} X_i)$ are independently distributed as $P(II)(0, 1, a + j - 1)$. Thus, $MP^{(k)}(II)$ can be constructed directly from independent univariate $P(II)$ random variables. Specifically, if $W_j \overset{d}{=} P(II)(0, 1, a+j-1)$ for $j = 1, 2, \ldots, k$ independently, then by defining $X_1 = W_1$ and $X_j = W_j \prod_{i=1}^{j-1}(1 + W_i)$ for $2 \leq j \leq k$, we arrive at $\boldsymbol{X} = (X_1, \ldots, X_k)^T \stackrel{d}{=} MP^{(k)}(II)(\boldsymbol{0}, 1, a)$. Of course, location parameter $\boldsymbol{\mu}$ and scale parameter $\boldsymbol{\theta}$ can then be introduced by linear transformation. Arnold (1983) has referred to this result as an "enticing siren call" for constructing plausible stochastic models involving multivariate Pareto distributions.

Order statistics $X_{1:k} \le X_{2:k} \le \cdots \le X_{k:k}$ constructed from the vector $\boldsymbol{X} = (X_1, \ldots, X_k)^T \stackrel{d}{=} MP^{(k)}(II)(\boldsymbol{0}, \boldsymbol{\theta}, a)$ and the corresponding scaled spacings $S_i = (k - i + 1)(X_{i:k} - X_{i-1:k})$ possess some elegant properties reminiscent of those for an exponential distribution; see Chapter 19 of Johnson, Kotz, and Balakrishnan (1994). If $\boldsymbol{X} \stackrel{d}{=} MP^{(k)}(II)(\boldsymbol{0}, \boldsymbol{\theta}, a)$, then

$$X_{1:k} \stackrel{d}{=} P(II)\left(0, \left(\sum_{i=1}^{k} 1/\theta_i\right)^{-1}, a\right).$$

For higher-order statistics, simple expressions are obtained only if we assume scale homogeneity—that is, all θ_i's to be equal. So, let us assume $\boldsymbol{X} \stackrel{d}{=} MP^{(k)}(II)(\boldsymbol{0}, 1, a)$, and let

$$S_i = (k - i + 1)(X_{i:k} - X_{i-1:k}) \quad \text{with } X_{0:k} \equiv 0$$

denote the scaled spacings. Then, $\boldsymbol{S} = (S_1, \ldots, S_k)^T \stackrel{d}{=} MP^{(k)}(II)(\boldsymbol{0}, 1, a)$, that is, $\boldsymbol{S} \stackrel{d}{=} \boldsymbol{X}$. The conclusion also extends incidentally to the case when $\boldsymbol{X}$ is a scale mixture of independent exponential random variables; see Arnold (1983).

4.3 Multivariate Pareto of the Third Kind

From (52.80), Arnold (1983) suggested a further extension that results in *multivariate Pareto distribution of the third kind* with joint survival function

$$\Pr[\boldsymbol{X} > \boldsymbol{x}] = \left\{1 + \sum_{i=1}^{k} \left(\frac{x_i - \mu_i}{\theta_i}\right)^{1/\gamma_i}\right\}^{-1}, \quad x_i > \mu_i, \qquad i = 1, \ldots, k,$$

$$(52.84)$$

which is denoted by $MP^{(k)}(III)(\boldsymbol{\mu}, \boldsymbol{\theta}, \boldsymbol{\gamma})$. This distribution has Pareto(III) marginals, but is not closed with respect to conditional distributions.

Assuming that the basic survival function is of the standard Pareto form

$$\Pr[Z > z] = (1 + z)^{-1}, \qquad z > 0,$$

a simple method of defining a multivariate distribution with Pareto marginals is to postulate that the multivariate odds-ratio function

$$\phi_{\mathbf{Z}}(\mathbf{z}) = \frac{1 - \Pr[\mathbf{Z} > \mathbf{z}]}{\Pr[\mathbf{Z} > \mathbf{z}]} \qquad \text{for } \mathbf{z} > 0$$

is the sum of marginal odds-ratios given by

$$\phi_{Z_i}(z_i) = \frac{1 - \Pr[Z_i > z_i]}{\Pr[Z_i > z_i]} \ ,$$

that is, $\phi_{\mathbf{Z}}(\mathbf{z}) = \sum_{i=1}^{k} \phi_{Z_i}(z_i)$, where $\mathbf{Z} = (Z_1, \ldots, Z_k)^T$. For the standard Pareto case, this yields [see Arnold (1996)]

$$\Pr[\mathbf{Z} > \mathbf{z}] = \left(1 + \sum_{i=1}^{k} z_i\right)^{-1}, \qquad z_i > 0,$$

which is a standard multivariate Pareto distribution of the first kind. If we now define $X_i = \mu_i + \theta_i Z_i^{\gamma_i}$ for $i = 1, \ldots, k$, we arrive at the joint survival function in (52.84) corresponding to the $MP^{(k)}(III)(\boldsymbol{\mu}, \boldsymbol{\theta}, \boldsymbol{\gamma})$ distribution.

4.4   Multivariate Pareto of the Fourth Kind

From (52.84), a simple extension leads to *multivariate Pareto distribution of the fourth kind* with joint survival function

$$\Pr[\mathbf{X} > \mathbf{x}] = \left\{1 + \sum_{i=1}^{k} \left(\frac{x_i - \mu_i}{\theta_i}\right)^{1/\gamma_i}\right\}^{-a}, \quad x_i > \mu_i, \qquad i = 1, \ldots, k,$$

$$(52.85)$$

which is denoted by $MP^{(k)}(IV)(\boldsymbol{\mu}, \boldsymbol{\theta}, \boldsymbol{\gamma}, a)$. This distribution does possess the conditional distribution closure property, with a replaced by $a + k - k_1$ (k_1 being the number of components in $\mathbf{X}_1$ where $\mathbf{X} = (\mathbf{X}_1, \mathbf{X}_2)^T$) and the components of $\boldsymbol{\theta}$ are replaced by $\tau_i = \theta_i \left\{1 + \sum_{j=k_1+1}^{k} \left(\frac{x_j - \mu_j}{\theta_j}\right)^{1/\gamma_j}\right\}^{\gamma_i}$. For the case when $\boldsymbol{\mu} = \mathbf{0}$ and $\boldsymbol{\theta} = \mathbf{1}$, we arrive at Takahasi's (1965) multivariate Burr distribution. Note that $MP^{(k)}(IV)$ distribution, under the restriction of homogeneous γ_i's (that is, $\gamma_1 = \cdots = \gamma_k = \gamma$), can be obtained from $MP^{(k)}(II)$ distribution by applying the same power transformation to all the coordinates.

Many structural properties of $MP^{(k)}(II)$ distributions presented earlier were extended by Yeh (1994) to $MP^{(k)}(IV)$ distributions. First of all, if $\mathbf{X}$ has $MP^{(k)}(IV)$ distribution, then $\mathbf{X}$ has the representation

$$X_i \overset{d}{=} \mu_i + \theta_i (Y_i/Z)^{\gamma_i}, \qquad i = 1, \ldots, k, \qquad (52.86)$$

where Y_i's $(i = 1, \ldots, k)$ are independent and identically distributed as standard exponential and Z is distributed as $\text{Gamma}(a, 1)$, independently of Y_i's. Also,

$$X_i \mid (Z = z) \overset{d}{=} \text{Weibull}\left(\frac{\theta_i}{z}, \frac{1}{\gamma_i}\right), \quad i = 1, \ldots, k.$$

Recall that the random variable $W = \mu + \theta\left(\frac{1}{V} - 1\right)^\gamma$, where V is distributed as $\text{Beta}(\lambda_1, \lambda_2)$, has Feller–Pareto (FP) distribution with probability density function

$$p_W(w) = \frac{1}{B(\lambda_1, \lambda_2)\gamma\theta}\left(\frac{w - \mu}{\theta}\right)^{\frac{1}{\gamma}(\lambda_2+1)-2}\left\{1 + \left(\frac{w - \mu}{\theta}\right)^{\frac{1}{\gamma}}\right\}^{-(\lambda_1+\lambda_2)},$$
$$w \geq \mu,$$

and is denoted by $FP(\mu, \theta, \gamma, \lambda_1, \lambda_2)$. Evidently, $P(IV)(\mu, \theta, \gamma, a) = FP(\mu, \theta, \gamma, a, 1)$, and $FP(0, 1, 1, \lambda_1, \lambda_2)$ (the so-called "scaled F-distribution") has probability density function

$$p_W(w) = \frac{1}{B(\lambda_1, \lambda_2)} w^{\lambda_2-1}(1 + w)^{-(\lambda_1+\lambda_2)}, \quad w > 0.$$

If $\boldsymbol{X}$ is distributed as $MP^{(k)}(IV)(\boldsymbol{0}, \boldsymbol{1}, \gamma, a)$, then Yeh (1994) observed that $\sum_{i=1}^{k} X_i^{1/\gamma}$ is distributed as $FP(0, 1, 1, a, k)$, $\sum_{i=1}^{\ell} X_i^{1/\gamma}$ (for $1 \leq \ell \leq k$) is distributed as $FP(0, 1, 1, a, \ell)$, and $X_j^{1/\gamma}/\{1+\sum_{i=1}^{j-1} X_i^{1/\gamma}\}$ (for $1 \leq j \leq k$) is independently distributed as $FP(0, 1, 1, a + j - 1, 1)$. Conversely, if Y_j's $(j = 1, \ldots, k)$ are independently distributed as $FP(0, 1, 1, a + j - 1, 1)$, then for any fixed $\gamma > 0$,

$$X_1 = Y_1^\gamma \quad \text{and} \quad X_j = Y_j^\gamma(1 + Y_{j-1})^\gamma \cdots (1 + Y_1)^\gamma \quad \text{for } j = 2, \ldots, k$$

form a k-dimensional vector that is distributed as $MP^{(k)}(IV)(\boldsymbol{0}, \boldsymbol{1}, \gamma, a)$. Yeh (1994) also noted that the smallest order statistic $X_{1:k}$ from $\boldsymbol{X} \overset{d}{=} MP^{(k)}(IV)(\boldsymbol{0}, \boldsymbol{\theta}, \gamma, a)$ is distributed as

$$P(IV)\left(0, \left(\frac{1}{\sum_{i=1}^{k}(1/\theta_i)^{1/\gamma}}\right)^\gamma, \gamma, a\right);$$

the distribution of $X_{i:k}$ (for $2 \leq i \leq k$) assumes a simple form provided that $\theta_1 = \theta_2 = \cdots = \theta_k$. Further, if $\boldsymbol{X} \overset{d}{=} MP^{(k)}(IV)(\boldsymbol{0}, \boldsymbol{1}, \gamma, a)$, then the vector $(S_1, \ldots, S_k)^T$ of the associated scaled spacings, where $S_i = (k - i + 1)\left(X_{i:k}^{1/\gamma} - X_{i-1:k}^{1/\gamma}\right)$, is distributed as $MP^{(k)}(II)(\boldsymbol{0}, \boldsymbol{1}, a)$.

Let $\boldsymbol{X}^{(1)}, \ldots, \boldsymbol{X}^{(n)}$ be independent and identically distributed as $MP^{(k)}(IV)(\mu, \theta, \boldsymbol{\gamma}, a)$. If $\boldsymbol{Y} = \min_j \boldsymbol{X}^{(j)}$ (the coordinatewise minimum), then $\boldsymbol{Y} \stackrel{d}{=} MP^{(k)}(IV)(\mu, \theta, \boldsymbol{\gamma}, na)$. Furthermore, rescaled geometric minima of standard Pareto random variables are again distributed as standard Pareto. This property has been exploited by Arnold (1983, 1990) for constructing multivariate Pareto distributions in various ways. Let $Z_i^{(j)}$ ($j = 1, 2, \ldots, k$; $i = 1, 2, \ldots$) be k independent sequences of independent standard Pareto random variables, and let $\boldsymbol{N} = (N_1, \ldots, N_k)^T$ be a k-variate geometric random variable [denoted by $G^{(k)}(\boldsymbol{p})$] with joint probability generating function

$$E\left[\prod_{i=1}^{k} t_i^{N_i}\right] = \frac{p_0}{1 - \sum_{i=1}^{k} p_i t_i}, \qquad \sum_{i=0}^{k} p_i = 1.$$

Define a k-dimensional random vector $\boldsymbol{U} = (U_1, \ldots, U_k)^T$ by

$$U_j = \min_{i \leq N_j + 1} Z_i^{(j)}, \qquad j = 1, \ldots, k,$$

where $\boldsymbol{N}$ is distributed as $G^{(k)}(\boldsymbol{p})$; see Chapter 36 of Johnson, Kotz and Balakrishnan (1997). Reparameterizing $\delta_i = p_i/p_0$, $i = 1, \ldots, k$, Arnold (1990) has shown that

$$
\begin{aligned}
\Pr[\boldsymbol{U} \geq \boldsymbol{u}] = \Bigg\{ & 1 + \sum_{i=1}^{k}(1 + \delta_i)u_i + \sum\sum_{i_1 \neq i_2}(1 + \delta_{i_1} + \delta_{i_2})u_{i_1}u_{i_2} \\
& + \sum\sum\sum_{i_1 \neq i_2 \neq i_3}(1 + \delta_{i_1} + \delta_{i_2} + \delta_{i_3})u_{i_1}u_{i_2}u_{i_3} \\
& + \cdots + \left(1 + \sum_{i=1}^{k}\delta_i\right)u_1 u_2 \cdots u_k \Bigg\}^{-1}.
\end{aligned}
\tag{52.87}
$$

Finally, upon defining $V_i = (1 + \delta_i)U_i$ for $i = 1, \ldots, k$, we obtain

$$
\begin{aligned}
\Pr[\boldsymbol{V} \geq \boldsymbol{v}] = \Bigg\{ & 1 + \sum_{i=1}^{k}v_i + \sum\sum_{i_1 \neq i_2}\eta_{i_1 i_2}v_{i_1}v_{i_2} + \sum\sum\sum_{i_1 \neq i_2 \neq i_3}\eta_{i_1 i_2 i_3}v_{i_1}v_{i_2}v_{i_3} \\
& + \cdots + \eta_{1 \cdots k}v_1 v_2 \cdots v_k \Bigg\}^{-1},
\end{aligned}
\tag{52.88}
$$

where

$$\eta_{i_1 i_2 \cdots i_\ell} = \left(1 + \sum_{j=1}^{\ell}\delta_{i_j}\right) \Bigg/ \prod_{j=1}^{\ell}\left(1 + \delta_{i_j}\right).$$

The parameters η's satisfy

$$0 \leq \eta_{i_1 i_2} \leq 1 \quad \forall \, i_1, i_2,$$
$$0 \leq \eta_{i_1 i_2 i_3}^2 \leq \eta_{i_1 i_2} \eta_{i_1 i_3} \eta_{i_2 i_3} \quad \forall \, i_1, i_2, i_3,$$

and so on. In fact, there are only k functionally independent η's even though (52.88) continues to be a valid k-variate survival function for a wider range of η's. The transformation $W_j = -\log V_j$ (for $j = 1, \ldots, k$) yields a multivariate logistic distribution that includes the Malik–Abraham (1973) form of the multivariate logistic distribution as a special case when $\boldsymbol{\eta} = \mathbf{0}$; see Chapter 49 for more details.

Denoting by $\bar{F}_{\boldsymbol{\eta}}(\boldsymbol{v})$ the survival function in (52.88), Arnold (1990) defined a flexible family of generalized multivariate Pareto (IV) distributions as the one with joint survival function of the form

$$\left\{ \bar{F}_{\boldsymbol{\eta}} \left(\left(\frac{v_1 - \mu_1}{\theta_1} \right)^{1/\gamma_1}, \left(\frac{v_2 - \mu_2}{\theta_2} \right)^{1/\gamma_2}, \ldots, \left(\frac{v_k - \mu_k}{\theta_k} \right)^{1/\gamma_k} \right) \right\}^a, \quad v_i \geq \mu_i,$$
(52.89)

where $\boldsymbol{\mu} \in \mathbb{R}^k$, $\boldsymbol{\theta} \in \mathbb{R}_+^k$, $\boldsymbol{\gamma} \in \mathbb{R}_+^k$, and $a > 0$. In order for (52.89) to represent a k-variate survival function, modified constraints are required on the parameters η's.

For the case $k = 2$, Durling (1975) called the distribution in (52.88) a *bivariate Burr distribution*. Particular bivariate cases have also been discussed by Arnold (1983, pp. 260–263). The independent case is obtained when all η's are equal to 1. If we set all η's equal to 0, we obtain the multivariate Burr family discussed by Takahasi (1965). The classical Mardia's (1962) multivariate Pareto distribution is obtained if we set $\boldsymbol{\mu} = \mathbf{0}$, $\boldsymbol{\gamma} = \mathbf{1}$ and $\boldsymbol{\eta} = \mathbf{0}$. It is easily seen that a subvector of dimension ℓ ($\ell < k$) of $\boldsymbol{V} = (V_1, \ldots, V_k)^T$, where $\boldsymbol{V}$ is distributed as (52.88), has a joint survival function once again of the form (52.88); but the conditional distribution remains generalized Pareto only when $\boldsymbol{\eta} = \mathbf{0}$ or $\boldsymbol{\eta} = \mathbf{1}$.

In the bivariate case of (52.89), when $\boldsymbol{\gamma}$ is known and assumed to be **1** without loss of generality, we have

$$\Pr[X_1 > x_1, X_2 > x_2]$$
$$= \left\{ 1 + \frac{x_1 - \mu_1}{\theta_1} + \frac{x_2 - \mu_2}{\theta_2} + \eta \left(\frac{x_1 - \mu_1}{\theta_1} \right) \left(\frac{x_2 - \mu_2}{\theta_2} \right) \right\}^{-a},$$
$$x_1 > \mu_1, x_2 > \mu_2,$$
(52.90)

where $\mu_i \in \mathbb{R}$, $\theta_i \in \mathbb{R}_+$ ($i = 1, 2$), $a > 0$, and $0 \leq \eta \leq a+1$. Evidently, the marginal minima (i.e., $\min_i X_{1i}$ and $\min_i X_{2i}$) are appropriate estimators of μ_1 and μ_2. When $\mu_1 = \mu_2 = 0$, the density corresponding to (52.90) is

$$
\begin{aligned}
&p_{X_1,X_2}(x_1, x_2) \\
&= \frac{(a^2 + a - a\eta) + a^2\eta\left(\frac{x_1}{\theta_1} + \frac{x_2}{\theta_2} + \eta\frac{x_1 x_2}{\theta_1 \theta_2}\right)}{\theta_1 \theta_2 \left(1 + \frac{x_1}{\theta_1} + \frac{x_2}{\theta_2} + \eta\frac{x_1 x_2}{\theta_1 \theta_2}\right)^{a+2}},
\end{aligned}
$$
$$
x_1 > 0, \quad x_2 > 0. \qquad (52.91)
$$

Arnold and Ganeshalingam (1987), in an unpublished technical report, presented the four likelihood equations for estimating the four parameters θ_1, θ_2, a, and η in (52.91). These equations need to be solved iteratively, but the iterative method is time-consuming and also may not converge. Therefore, Arnold (1990) proposed a simpler hybrid method in which marginal information is first used to estimate θ_1, θ_2, and a. Specifically, we choose $\tilde{\theta}_1, \tilde{\theta}_2$, and $\tilde{a}$ to maximize the product of the marginal likelihoods, $\prod_{i=1}^n p_{X_1}(x_{1i}; \theta_1, a) \prod_{i=1}^n p_{X_2}(x_{2i}; \theta_2, a)$. The corresponding marginal likelihood equations are

$$
\frac{1}{\tilde{a}} = \frac{1}{2n}\left\{\sum_{i=1}^n \log\left(1 + \frac{x_{1i}}{\tilde{\theta}_1}\right) + \sum_{i=1}^n \log\left(1 + \frac{x_{2i}}{\tilde{\theta}_2}\right)\right\},
$$

$$
\tilde{\theta}_1 = \frac{\tilde{a}+1}{n}\sum_{i=1}^n \frac{x_{1i}\tilde{\theta}_1}{\tilde{\theta}_1 + x_{1i}}, \qquad \tilde{\theta}_2 = \frac{\tilde{a}+1}{n}\sum_{i=1}^n \frac{x_{2i}\tilde{\theta}_2}{\tilde{\theta}_2 + x_{2i}}. \qquad (52.92)
$$

This system of equations possesses a rapid convergence.

If θ_1 and θ_2 are known, upon defining $Z = \min\left(\frac{X_1}{\theta_1}, \frac{X_2}{\theta_2}\right)$, we have

$$
\Pr[Z > z] = (1 + 2z + \eta z^2)^{-a}, \qquad z > 0
$$

and

$$
p_Z(z) = \frac{2a(1 + \eta z)}{(1 + 2z + \eta z^2)^{a+1}}, \qquad z > 0.
$$

In this case, for known a, the log-likelihood function is

$$
\log L(\eta) = c + \sum_{i=1}^n \log(1 + \eta z_i) - (a+1)\sum_{i=1}^n \log(1 + 2z_i + \eta z_i^2).
$$
$$
(52.93)
$$

Since $\eta \in [0, a+1]$, it is straightforward to identify the maximizing value of η. The hybrid procedure suggested by Arnold (1990) thus consists of the following two steps:

(i) Using marginal likelihoods, determine the estimates $\tilde{\theta}_1$, $\tilde{\theta}_2$, and $\tilde{a}$ satisfying (52.92).

(ii) Compute $z_i = \min\left(\frac{x_{1i}}{\theta_1}, \frac{x_{2i}}{\theta_2}\right)$ and then use the log-likelihood, $\log L(\eta)$ in (52.93), with $a = \tilde{a}$, to determine the MLE of η.

The bivariate case in (52.88) corresponds to the joint survival function

$$\Pr[V_1 \geq v_1, V_2 \geq v_2] = (1 + v_1 + v_2 + \eta v_1 v_2)^{-1}, \quad v_1 > 0, \; v_2 > 0 \tag{52.94}$$

and the joint density function

$$p(v_1, v_2) = \eta(1 + v_1 + v_2 + \eta v_1 v_2)^{-2} + 2(1 - \eta)(1 + v_1 + v_2 + \eta v_1 v_2)^{-3},$$
$$v_1 > 0, \; v_2 > 0. \tag{52.95}$$

Arnold (1990) has presented plots of the density in (52.95) when $\eta = 0, 0.5, 1.0, 1.5, 1.75, 2.0$. When $\eta < 1 \, (> 1)$, the distribution is positive (negative) quadrant-dependent. If $\eta \leq 1.5$, the mode is at the origin; if, however, $\eta \in (1.5, 2]$, the density has multiple modes whose locus is on the curve

$$x + y + \eta xy = 2 - \frac{3}{\eta}.$$

There are applications of multivariate Pareto distributions in the area of reliability [see, for example, Hutchinson (1979), Lindley and Singpurwalla (1986), Nayak (1987), and Sankaran and Unnikrishnan Nair (1993)] as well as in the study of multivariate income distributions. Arnold (1983, 1990) has also mentioned that multivariate geometric minima of Pareto(III) variables might provide suitable models for incomes accruing to related individuals.

4.5   Conditionally Specified Multivariate Pareto

Proceeding along the lines of Section 3.4, one can derive multivariate Pareto distributions by specifying the conditional distributions. For example, Arnold, Castillo, and Sarabia (1993) have proposed the following multivariate extensions of bivariate Models I and II in (52.51) and (52.52), respectively:

Model I: The k-dimensional version of the bivariate density in (52.51) is of the form

$$p_{\boldsymbol{X}}(\boldsymbol{x}) = \left\{\prod_{i=1}^{k} x_i^{\delta_i - 1}\right\}\left\{\sum_{\boldsymbol{s} \in \xi_k} \lambda_{\boldsymbol{s}} \prod_{i=1}^{k} x_i^{s_i \delta_i}\right\}^{-(a+1)},$$

$$x_i > 0 \ (i = 1, \ldots, k), \qquad (52.96)$$

where ξ_k is the set of all vectors of 0's and 1's of dimension k.

Model II: The k-dimensional version of the bivariate density in (52.52) is of the form

$$p_{\boldsymbol{X}}(\boldsymbol{x}) = \left\{ \prod_{i=1}^{k} x_i^{\delta_i - 1} \right\} \exp \left\{ \sum_{\boldsymbol{s} \in \xi_k} \lambda_{\boldsymbol{s}} \prod_{i=1}^{k} \log(\theta_i + x_i^{\delta_i}) \right\},$$
$$x_i > 0 \ (i = 1, \ldots, k), \qquad (52.97)$$

where ξ_k is once again the set of all vectors of 0's and 1's of dimension k.

Similar multivariate extensions can be proposed for other forms of bivariate Pareto distributions as well. For example, the k-dimensional version of the bivariate density in (52.58), is of the form

$$p_{\boldsymbol{X}}(\boldsymbol{x}) = \left\{ \sum_{\boldsymbol{s} \in \xi_k} \lambda_{\boldsymbol{s}} \prod_{i=1}^{k} (\alpha x_i)^{s_i} \right\}^{\frac{1}{\alpha} - 1}, \qquad \boldsymbol{x} \in D, \qquad (52.98)$$

where ξ_k is the set of all vectors of 0's and 1's of dimension k, and D is the set of x_i's for which the expression in (52.98) remains positive.

4.6   Marshall–Olkin Type Multivariate Pareto

The joint survival function of a Marshall–Olkin-type multivariate Pareto distribution is given by

$$\begin{aligned} \bar{F}_{\boldsymbol{X}}(\boldsymbol{x}) &= \Pr[X_1 > x_1, \ldots, X_k > x_k] \\ &= \left(\frac{x_1}{\theta}\right)^{-\lambda_1} \cdots \left(\frac{x_k}{\theta}\right)^{-\lambda_k} \left\{ \frac{\max(x_1, \ldots, x_k)}{\theta} \right\}^{-\lambda_0}, \end{aligned}$$
$$(52.99)$$

where $\lambda_0, \ldots, \lambda_k > 0$ and $\theta > 0$. It is obtained from Marshall and Olkin's multivariate exponential distribution with joint survival function (see Chapter 48)

$$\begin{aligned} \bar{F}_{\boldsymbol{Y}}(\boldsymbol{y}) &= \Pr[Y_1 > y_1, \ldots, Y_k > y_k] \\ &= \exp\{-\lambda_1 y_1 - \cdots - \lambda_k y_k - \lambda_0 \max(y_1, \ldots, y_k)\} \end{aligned}$$
$$(52.100)$$

by the transformation $X_i = \theta\, e^{Y_i}$ for $i = 1, \ldots, k$. The multivariate Pareto distribution in (52.99) is not absolutely continuous with respect to the Lebesgue measure on $\mathbb{R}_+^k$. The marginal distribution of X_i (for $i = 1, 2, \ldots, k$) is given by

$$\bar{F}_{X_i}(x_i) = \Pr[X_i > x_i] = \left(\frac{x_i}{\theta}\right)^{-(\lambda_i + \lambda_0)}. \tag{52.101}$$

X_i's are independent if $\lambda_0 = 0$, and they are identically distributed if $\lambda_1 = \cdots = \lambda_k$. Moreover,

$$\Pr[X_1 = \cdots = X_k] = \lambda_0/\lambda$$

and

$$\Pr[\min(X_1, \ldots, X_k) > x] = \left(\frac{x}{\theta}\right)^{-\lambda}, \tag{52.102}$$

where $\lambda = \sum_{i=0}^k \lambda_i$.

Hanagal (1996) has termed the distribution in (52.99) (Marshall–Olkin-type) *multivariate Pareto distribution of Type 1*, disregarding the nomenclature established by Arnold (1983). When $\theta = 1$, we arrive at

$$\bar{F}_{\boldsymbol{X}}(\boldsymbol{x}) = \prod_{i=1}^k x_i^{-\lambda_i}\{\max(x_1, \ldots, x_k)\}^{-\lambda_0} \tag{52.103}$$

which has been termed the (Marshall–Olkin-type) *multivariate Pareto distribution of Type 2* by Hanagal (1996). The Type 2 family in (52.103) satisfy the "dullness property" that

$$\Pr[\boldsymbol{X} > t\boldsymbol{s} \mid \boldsymbol{X} \geq \boldsymbol{t}] = \Pr[\boldsymbol{X} > \boldsymbol{s}] \text{ for all } \boldsymbol{s} \geq \boldsymbol{1}, \tag{52.104}$$

where $\boldsymbol{s} = (s_1, \ldots, s_k)^T$, $\boldsymbol{t} = (t, \ldots, t)$ and $\boldsymbol{1} = (1, \ldots, 1)$. Note that this is a parallel property to the "loss of memory property" of Marshall and Olkin's multivariate exponential distribution. Hanagal (1996) has also extended the characterization result of Padamadan and Nair (1994) [based on the condition (52.66)] to the Marshall–Olkin-type multivariate Pareto distribution of Type 1 in (52.99). The same author has also discussed the maximum likelihood estimation of the parameters of Type 2 distribution in (52.103). The analysis parallels the maximum likelihood estimation of the parameters of Marshall and Olkin's multivariate exponential distribution; see Chapter 48. The likelihood equations are not easy to solve. Consistent estimators $(u_0, u_1, \ldots, u_k)$ of the parameters $(\lambda_0, \lambda_1, \ldots, \lambda_k)$ are used as initial estimates in the Newton–Raphson procedure for determining the maximum likelihood estimates. These consistent estimators are

$$u_i = r_i / \sum_{j=1}^{n} \min(x_{1j}, \ldots, x_{kj}), \qquad i = 0, 1, \ldots, k, \qquad (52.105)$$

where $(x_{1j}, \ldots, x_{kj})^T$ $(j = 1, \ldots, n)$ are random observations from the distribution in (52.103), r_i $(i = 1, \ldots, k)$ is the number of observations with $x_{ij} < \min_{\ell \neq i} x_{\ell j}$, and r_0 is the number of observations with $x_{1j} = \cdots = x_{kj}$. The distribution of $(r_1, \ldots, r_k)^T$ is Multinomial$(n; \frac{\lambda_1}{\lambda}, \ldots, \frac{\lambda_k}{\lambda})$, and we have $\sum_{i=0}^{k} r_i = n$. The estimators are $u_i \xrightarrow{p} \lambda_i$ for $i = 0, 1, \ldots, k$.

4.7   Multivariate Semi-Pareto

The multivariate extension of the bivariate semi-Pareto distribution in (52.73) is rather straightforward. A random vector $\boldsymbol{X} = (X_1, \ldots, X_k)^T$ is said to have a *multivariate semi-Pareto distribution* with parameters $\alpha_1, \ldots, \alpha_k$ and p if its joint survival function is of the form [Balakrishna and Jayakumar (1997)]

$$\bar{F}_{\boldsymbol{X}}(\boldsymbol{x}) = \Pr[X_1 > x_1, \ldots, X_k > x_k] = \frac{1}{1 + \psi(x_1, \ldots, x_k)}, \qquad (52.106)$$

where $\psi(x_1, \ldots, x_k)$ satisfies the functional equation

$$\psi(x_1, \ldots, x_k) = \frac{1}{p}\, \psi\left(p^{1/\alpha_1} x_1, \ldots, p^{1/\alpha_k} x_k\right),$$
$$0 < p < 1, \ \alpha_i > 0, \ x_i > 0 \ (i = 1, \ldots, k). \qquad (52.107)$$

The solution of this functional equation is

$$\psi(x_1, \ldots, x_k) = \sum_{i=1}^{k} x_i^{\alpha_i} h_i(x_i), \qquad (52.108)$$

where $h_i(x_i)$ is a periodic function in $\log x_i$ with period $\frac{2\pi \alpha_i}{-\log p}$ (for $i = 1, \ldots, k$); see, for example, Kagan, Linnik, and Rao (1973, p. 163). In the special case when $h_i(x_i) \equiv 1$ $(i = 1, \ldots, k)$, we arrive at

$$\bar{F}_{\boldsymbol{X}}(\boldsymbol{x}) = \frac{1}{1 + \sum_{i=1}^{k} x_i^{\alpha_i}}, \qquad x_i > 0, \ \alpha_i > 0 \ (i = 1, \ldots, k), \qquad (52.109)$$

which is the joint survival function of the multivariate Pareto distribution of the third kind as in (52.84).

BIBLIOGRAPHY

(Some bibliographical items not mentioned in the text are included here for completeness.)

Ahmed, N. A., and Gokhale, D. V. (1989). Entropy expressions and their estimators for multivariate distributions, *IEEE Transactions on Information Theory*, **35**, 688–692.

Arnold, B. C. (1975). A characterization of the exponential distribution by multivariate compounding, *Sankhyā, Series A*, **37**, 164–173.

Arnold, B. C. (1983). *Pareto Distributions*, Silver Spring, Maryland: International Cooperative Publishing House.

Arnold, B. C. (1987). Bivariate distributions with Pareto conditionals, *Statistics & Probability Letters*, **5**, 263–266.

Arnold, B. C. (1989). Dependence in conditionally specified distributions, in *Topics in Statistical Dependence*, IMS Lecture Notes Monograph No. **16**, pp. 13–18, Hayward, California: Institute of Mathematical Statistics.

Arnold, B. C. (1990). A flexible family of multivariate Pareto distributions, *Journal of Statistical Planning and Inference*, **24**, 249–258.

Arnold, B. C. (1995). Conditional survival models, in *Recent Advances in Life-Testing and Reliability* (N. Balakrishnan, ed.), pp. 589–601, Boca Raton, Florida: CRC Press.

Arnold, B. C. (1996). Marginally and conditionally specified multivariate survival models: A survey, in *Statistics of Quality* (S. Ghosh, W. J. Kennedy, and W. B. Smith, eds.), pp. 233–252, New York: Marcel Dekker.

Arnold, B. C., Castillo, E., and Sarabia, J. M. (1992). *Conditionally Specified Distributions*, Lecture Notes in Statistics—**73**, New York: Springer-Verlag.

Arnold, B. C., Castillo, E., and Sarabia, J. M. (1993). Multivariate distributions with generalized Pareto conditionals, *Statistics & Probability Letters*, **17**, 361–368.

Arnold, B. C., Castillo, E., and Sarabia, J. M. (1994a). A conditional characterization of the multivariate normal distribution, *Statistics & Probability Letters*, **19**, 313–315.

Arnold, B. C., Castillo, E., and Sarabia, J. M. (1994b). Conditional characterization of the Mardia multivariate Pareto distribution, *Pakistan Journal of Statistics*, **10**, 143–145.

Arnold, B. C., and Ganeshalingam, S. (1987). Estimation for multivariate Pareto distributions: a simulation study, Technical Report No. 162, Department of Statistics, University of California, Riverside, California.

Arnold, B. C., and Pourahmadi, M. (1988). Conditional characterizations of multivariate distributions, *Metrika*, **35**, 99-108.

Arnold, B. C., Sarabia, J. M., and Castillo, E. (1995). Distributions with conditionals in the Pickands–de Haan generalized Pareto family, *IAPQR Transactions*, **20**, 27–35.

Arnold, B. C., and Strauss, D. J. (1988). Bivariate distributions with exponential conditionals, *Journal of the American Statistical Association*, **83**, 522-527.

Balakrishna, N., and Jayakumar, K. (1997). Bivariate semi-Pareto distributions and processes, *Statistical Papers*, **38**, 149–165.

DeGroot, M. H. (1970). *Optimal Statistical Decisions*, New York: McGraw-Hill.

Durling, F. C. (1975). The bivariate Burr distribution, in *Statistical Distributions in Scientific Work*, Vol. 1 (G. P. Patil, S. Kotz, and J. K. Ord, eds.), pp. 329–335, Dordrecht, The Netherlands: D. Reidel.

Gradshteyn, I. S., and Ryzhik, I. M. (1965). *Table of Integrals, Series, and Products*, fifth edition, Boston: Academic Press.

Hanagal, D. D. (1996). A multivariate Pareto distribution, *Communications in Statistics—Theory and Methods*, **25**, 1471–1488.

Hutchinson, T. P. (1979). Four applications of a bivariate Pareto distribution, *Biometrical Journal*, **21**, 553–563.

Jeevanand, E. S. (1997). Bayes estimation of $P(X_2 < X_1)$ for a bivariate Pareto distribution, *Statistician*, **46**, 93–100.

Jeevanand, E. S., and Padamadan, V. (1996). Parameter estimation for a bivariate Pareto distribution, *Statistical Papers*, **37**, 153–164.

Jeffreys, H. (1961). *Theory of Probability*, London: Oxford University Press.

Johnson, N. L., Kotz, S., and Balakrishnan, N. (1994). *Continuous Univariate Distributions*, Vol. 1, second edition, New York: John Wiley & Sons.

Johnson, N. L., Kotz, S., and Balakrishnan, N. (1995). *Continuous Univariate Distributions*, Vol. 2, second edition, New York: John Wiley & Sons.

Johnson, N. L., Kotz, S., and Balakrishnan, N. (1997). *Discrete Multivariate Distributions*, New York: John Wiley & Sons.

Johnson, N. L., Kotz, S., and Kemp, A. W. (1992). *Univariate Discrete Distributions*, Second edition, New York: John Wiley & Sons.

Jupp, P. E., and Mardia, K. V. (1982). A characterization of the multivariate Pareto distribution, *Annals of Statistics*, **10**, 1021–1024.

Kagan, A. M., Linnik, Yu. V., and Rao, C. R. (1973). *Characterization Problems in Mathematical Statistics*, New York: John Wiley & Sons.

Kibble, W. F. (1941). A two-variable gamma type distribution, *Sankhyā*, **5**, 137–150.

Krishnamoorthy, A. S., and Parthasarathy, M. (1951). A multivariate gamma-type distribution, *Annals of Mathematical Statistics*, **22**, 549–557. Correction, **31**, 229.

Krishnan, P. (1985). Some demographic applications of the univariate and the bivariate forms of the Pareto distribution, *Proceedings of the 45th Session of the International Statistical Institute*, **1**, 267–268.

Lee, P. M. (1989). *Bayesian Statistics: An Introduction*, London: Arnold.

Lindley, D. V., and Singpurwalla, N. D. (1986). Multivariate distributions for the life lengths of components of a system sharing a common environment, *Journal of Applied Probability*, **23**, 418–431.

Malik, H. J., and Abraham, B. (1973). Multivariate logistic distributions, *Annals of Statistics*, **1**, 588–590.

Malik, H. J., and Trudel, R. (1985). Distribution of the product and the quotient from bivariate t, F and Pareto distributions, *Communications in Statistics—Theory and Methods*, **14**, 2951–2602.

Mardia, K. V. (1962). Multivariate Pareto distributions, *Annals of Mathematical Statistics*, **33**, 1008-1015.

Mardia, K. V. (1964). Some results on the order statistics of the multivariate normal and Pareto type I populations, *Annals of Mathematical Statistics*, **35**, 1815-1818.

Marshall, A. W., and Olkin, I. (1967a). A generalized bivariate exponential distribution, *Journal of Applied Probability*, **4**, 291-302.

Marshall, A. W., and Olkin, I. (1967b). A multivariate exponential distribution, *Journal of the American Statistical Association*, **62**, 30-44.

Muliere, P., and Scarsini, M. (1987). Characterization of a Marshall-Olkin type class of distributions, *Annals of the Institute of Statistical Mathematics*, **39**, 429–441.

Nayak, T. K. (1987). Multivariate Lomax distribution: Properties and usefulness in reliability theory, *Journal of Applied Probability*, **24**, 170–171.

Padamadan, V., and Nair, K. R. M. (1994). Characterization of a bivariate Pareto distribution, *Journal of the Indian Statistical Association*, **20**, 15–20.

Revankar, N. S., Hartley, M. J., and Pagano, M. (1974). A characterization of the Pareto distribution, *Annals of Statistics*, **2**, 599–601.

Roy, D. (1989). A characterization of Gumbel's bivariate exponential and Lindley and Singpurwalla's bivariate Lomax distribution, *Journal of Applied Probability*, **26**, 886–891; Correction, **27**, 736.

Roy, D., and Gupta, R. P. (1996). Bivariate extension of Lomax and finite range distributions through characterization approach, *Journal of Multivariate Analysis*, **59**, 22–33.

Ruiz, J. M., Marin, J., and Zoroa, P. (1993). A characterization of continuous multivariate distributions by conditional expectations, *Journal of Statistical Planning and Inference*, **37**, 13–21.

Sankaran, P. G., and Unnikrishnan Nair, N. (1993). A bivariate Pareto model and its applications to reliability, *Naval Research Logistics*, **40**, 1013–1020.

Sankaran, P. G., and Unnikrishnan Nair, N. (1996). On a bivariate finite range distribution, *Journal of Indian Statistical Association*, **34**, 119–124.

Tajvidi, N. (1996). Characterisation and some statistical aspects of univariate and multivariate generalised Pareto distributions, Report, Department of Mathematics, Chalmers University of Technology, Göteborg, Sweden.

Takahasi, K. (1965). Note on the multivariate Burr's distribution, *Annals of the Institute of Statistical Mathematics*, **17**, 257–260.

Targhetta, M. L. (1979). Confidence intervals for a one parameter family in a mixture of distributions, *Biometrika*, **65**, 687–688.

Wesolowski, J. (1994). Bivariate distributions via the Pareto conditional distribution and a regression function, Preprint, Mathematical Institute, Warsaw University of Technology, Warsaw, Poland.

Wesolowski, J., and Ahsanullah, M. (1995). Conditional specification of multi-variate Pareto and Student distributions, *Communications in Statistics—Theory and Methods*, **24**, 1023–1031.

Wicksell, S. D. (1933). On correlation surfaces of Type III, *Biometrika*, **25**, 121–133.

Yeh, H.-C. (1994). Some properties of the homogeneous multivariate Pareto(IV) distribution, *Journal of Multivariate Analysis*, **51**, 46–52.

CHAPTER 53

Bivariate and Multivariate Extreme Value Distributions

During the past two decades, a considerable amount of work has been carried out on bivariate and multivariate extreme value distributions; see, for example, Smith (1990) and Galambos (1987). Recently, excellent review articles, as well as some research monographs, have been prepared on this topic of research. For example, the book by Joe (1996) pays special attention to various forms of bivariate and multivariate extreme value distributions, their properties, inferential issues, and more importantly their applications to practical problems. Due to the vastness of the literature as well as the current nature of the above reference, the coverage of this chapter should be regarded as a basic introduction into the theoretical and practical aspects of bivariate and multivariate extreme value distributions, and not as a complete account.

A general theoretical treatment of weak asymptotic convergence of multivariate extreme values was provided by Deheuvels (1978). For the bivariate case, some results were available earlier and were due to Finkelshteyn (1953), Geffroy (1958, 1959), Sibuya (1969), and Tiago de Oliveira (1958, 1975). The book by Galambos (1978, 1987) provides detailed description of these results.

621

1 GENERAL BIVARIATE EXTREME VALUE DISTRIBUTIONS

1.1 Definition

Let $(X_i, Y_i)^T$ be independent pairs of random variables, each having the same continuous joint cumulative distribution function $F(x, y)$. We then consider the joint distribution of $X_{\max} = \max(X_1, \ldots, X_n)$, $Y_{\max} = (Y_1, \ldots, Y_n)$. Since $X_1, \ldots, X_n$ are independent and identically distributed continuous random variables, it will usually be possible [as described earlier in Chapter 22 of Johnson, Kotz, and Balakrishnan (1995)] to find linear transformations

$$X_{(n)} = a_n X_{\max} + b_n \qquad (a_n > 0)$$

such that $X_{(n)}$ has a limiting distribution (as $n \to \infty$) that is one of the three types of extreme value distributions. There will also, of course, be a transformation

$$Y_{(n)} = c_n Y_{\max} + d_n \qquad (c_n > 0)$$

with similar properties.

The limiting joint distribution of $X_{(n)}$ and $Y_{(n)}$ as $n \to \infty$ is a *bivariate extreme value distribution*.

Consider a random sample $(X_1, Y_1)^T, \ldots, (X_n, Y_n)^T$ from a bivariate extreme value distribution $L(x, y)$ with fixed marginals $F(x)$ and $G(y)$. It has been known since the work of Pickands (1981) that the copula associated with L can be expressed in the form

$$
\begin{aligned}
C(u, v) &= \Pr\{F(X) \le u, \; G(Y) \le v\} \\
&= \exp[\log(uv) A\{\log(u)/\log(uv)\}]
\end{aligned}
\qquad (53.1)
$$

for all $0 \le u, v \le 1$ in terms of a convex function A defined on [0,1] in such a way that $\max(t, 1 - t) \le A(t) \le 1$ for all $0 \le t \le 1$. See the discussion at the end of this section concerning the meaning of the function A.

1.2 Properties

The joint cumulative distribution function of $X_{\max}$ and $Y_{\max}$ is $\{F(x, y)\}^n$. Denoting the bivariate extreme value cumulative distribution function by $F_{(\infty)}(x, y)$, we have

$$F_{(\infty)}(x, y) = \lim_{n \to \infty} \{F(a_n x + b_n, c_n y + d_n)\}^n. \qquad (53.2)$$

This equation is commonly referred to as the "postulate of stability." It is a natural extension of the univariate Fréchet–Fisher–Tippett equation [see Chapter 22 of Johnson, Kotz, and Balakrishnan (1995)]. Clearly, if X_i and Y_i are mutually independent, so will be $X_{\max}$ and $Y_{\max}$, and $X_{(n)}$ and $Y_{(n)}$, and the limiting distribution will also be that of two independent random variables. The converse, however, is not necessarily true. Geffroy (1958, 1959) has shown that the condition

$$\lim_{x,y\to\infty} \frac{1 - F_X(x) - F_Y(y) + F_{X,Y}(x,y)}{1 - F_{X,Y}(x,y)} = 0 \qquad (53.3)$$

is sufficient for the asymptotic independence of $X_{\max}$ and $Y_{\max}$, even though $F_{X,Y}(x,y) \neq F_X(x)F_Y(y)$. Condition (53.3) is satisfied, for example, by the following:

(a) The bivariate normal distribution with $|\rho| \neq 1$ [Sibuya (1960)].
(b) Bivariate distributions of type

$$F_{X,Y}(x,y) = F_X(x)F_Y(y)\{1 + \alpha[1 - F_X(x)][1 - F_Y(y)]\}$$

[Gumbel (1958)] and generalizations due to Farlie (1960).

(c) Bivariate exponential distributions of type

$$F_{X,Y}(x,y) \;=\; 1 - e^{-x} - e^{-y} - e^{-x-y-\theta xy}$$
$$(x, y \geq 0; \; 0 \leq \theta \leq 1)$$

[see Gumbel (1960) and Chapter 47].

(d) The bivariate logistic distribution [Gumbel (1961b)]

$$F_{X,Y}(x,y) = (1 + e^{-x} + e^{-y})^{-1}$$

and, indeed, any joint distribution satisfying the relation

$$\frac{1}{F_{X,Y}(x,y)} = \frac{1}{F_X(x)} + \frac{1}{F_Y(y)} - 1. \qquad (53.4)$$

The marginal distribution of a bivariate extreme value distribution may be any one of the three extreme value types. There are thus $\binom{3}{2} + 3 = 6$ possible pairs of types. The two marginal types do not completely specify the joint distribution (just as, for example, marginal normality does not ensure bivariate normality), but they do imply quite severe restrictions on it, since the cumulative distribution function must also satisfy the stability postulate (53.2).

Geffroy (1958, 1959), Gumbel (1958), Sibuya (1960), and Tiago de Oliveira (1958, 1961, 1962/1963) have obtained a number of more or less equivalent results on the forms of bivariate extreme value distributions. Their results may be summarized as follows.

If $F_{(\infty)}(x, y)$ is the joint cumulative distribution function of a bivariate extreme value distribution, with $F_{1(\infty)}(x)$ and $F_{2(\infty)}(y)$ the corresponding univariate functions, then

$$-\log F_{(\infty)}(x, y) = -\log F_{1(\infty)}(x) - \log F_{2(\infty)}(y)$$
$$+g\left(\frac{\log F_{2(\infty)}(y)}{\log F_{1(\infty)}(x)}\right)(-\log F_{1(\infty)}(x)), \quad (53.5)$$

where $g(t)$ is a continuous convex function with $\max(-t, -1) \leq g(t) \leq 0$. Note that it follows from (53.5) that

$$F_{(\infty)}(x, y) \geq F_{1(\infty)}(x) F_{2(\infty)}(y) \qquad (53.6)$$

for any bivariate extreme value distribution.

It is of interest to note that if $F_{(\infty)}(x, y)$ and $G_{(\infty)}(x, y)$ are two bivariate extreme value distributions, so is their weighted geometric mean

$$[F_{(\infty)}(x, y)]^{\beta}[G_{(\infty)}(x, y)]^{1-\beta} \qquad (0 \leq \beta \leq 1);$$

see Gumbel and Goldstein (1964).

Asymptotic distribution of minimum values are equivalent to (minimum) extreme value distributions, as can be seen by changing the signs of all variables. They, therefore, do not require a separate treatment.

2   SPECIAL BIVARIATE EXTREME VALUE DISTRIBUTIONS

Gumbel (1958, 1965) has described two general forms for bivariate extreme value distributions in terms of the marginal (univariate extreme value) distributions:

1. Type A

$$F_{(\infty)}(x, y) = F_{1(\infty)}(x) F_{2(\infty)}(y)$$
$$\times \exp\left[-\theta\left\{\frac{1}{\log F_{1(\infty)}(x)} + \frac{1}{\log F_{2(\infty)}(y)}\right\}^{-1}\right].$$
$$(53.7)$$

2. Type B

$$F_{(\infty)}(x, y) = \exp[-\{(-\log F_{1(\infty)}(x))^m + (-\log F_{2(\infty)}(y))^m\}^{1/m}].$$
(53.8)

In (53.7), θ is a parameter $(0 \le \theta < 1)$, and in (53.8), m is a parameter $(m \ge 1)$. If $\theta = 0$ in (53.7), X and Y are independent; if $m = 1$ in (53.8), X and Y are independent.

We now suppose that each marginal distribution is of the standard Type I extreme value form, so that

$$F_{1(\infty)}(x) = \exp(-e^{-x}), \quad F_{2(\infty)}(y) = \exp(-e^{-y}).$$

Each of these distributions has expected value $\gamma \ (= 0.577\cdots)$ and variance $\pi^2/6$. Since Types II and III can be obtained from Type I by simple transformations, much of our analysis will be relevant to bivariate extreme value distributions with marginal distributions of these other types.

Tiago de Oliveira (1958, 1961) showed that a bivariate distribution with standard Type I extreme value marginal distributions can be defined by a cumulative distribution function of the form

$$F_{X_1, X_2}(x_1, x_2) = \exp\{-(e^{-x_1} + e^{-x_2})g(x_2 - x_1)\}.$$
(53.9)

If a density function exists, the function $g(\cdot)$ must satisfy the conditions

$$\lim_{t \to \pm\infty} g(t) = 1,$$
(53.10)

$$\frac{d}{dt}\{(1 + e^{-t})g(t)\} \le 0,$$
(53.11)

$$\frac{d}{dt}\{(1 + e^{t})g(t)\} \ge 0,$$
(53.12)

$$(1 + e^{-t})g''(t) + (1 - e^{-t})g'(t) \ge 0.$$
(53.13)

Type A is obtained by taking

$$g(t) = 1 - \frac{1}{4}\theta \operatorname{sech}^2 \frac{1}{2}t.$$
(53.14)

Type B is obtained by taking

$$g(t) = (e^{mt} + 1)^{1/m}(e^t + 1)^{-1}.$$
(53.15)

A third type [see Tiago de Oliveira (1970)], which we shall call Type C, is obtained by taking

$$g(t) = (e^t + 1)^{-1}\{1 - \phi + \max(e^t, \phi)\} \qquad (0 < \phi < 1).$$
(53.16)

2.1   Type A Distributions

For these distributions (also known as *mixed model*),

$$F_{X,Y}(x,y) = \exp[-e^{-x} - e^{-y} + \theta(e^x + e^y)^{-1}] \qquad (0 \le \theta \le 1)$$

$$\text{(53.17)}$$

and the joint density function is

$$
\begin{aligned}
p_{X,Y}(x,y) &= e^{-(x+y)}[1 - \theta(e^{2x} + e^{2y})(e^x + e^y)^{-2} + 2\theta\, e^{2(x+y)}(e^x + e^y)^{-3} \\
&\quad + \theta^2\, e^{2(x+y)}(e^x + e^y)^{-4}]\, \exp[-e^{-x} - e^{-y} + \theta(e^x + e^y)^{-1}].
\end{aligned}
$$

$$\text{(53.18)}$$

$F_{X,Y}(x,y)$ is an increasing function of θ. The median of the common distribution of X and Y is

$$\mu = -\log(\log 2) = 0.36651. \qquad \text{(53.19)}$$

We note that, since $\exp(-e^{-\mu}) = \frac{1}{2}$,

$$
\begin{aligned}
F_{X,Y}(\mu,\mu) &= \exp\left(-2e^{-\mu} + \frac{1}{2}\,\theta\, e^{-\mu}\right) \\
&= \left(\frac{1}{4}\right)^{1-\theta/4}
\end{aligned}
$$

$$\text{(53.20)}$$

(while $F_X(\mu)F_Y(\mu) = \frac{1}{4}$). Also

$$F_{X,Y}(0,0) = (e^{-2})^{1-\theta/4}. \qquad \text{(53.21)}$$

The value $\tilde{\mu}$ is such that $F_{X,Y}(\tilde{\mu}, \tilde{\mu}) = \frac{1}{4}$ satisfies the equation

$$\left(2 - \frac{1}{2}\theta\right) e^{-\tilde{\mu}} = 2\log 2$$

and so

$$\tilde{\mu} = \log\left(1 - \frac{1}{4}\theta\right) - \log(\log 2) \doteq \log\left(1 - \frac{1}{4}\theta\right) + 0.3665. \qquad \text{(53.22)}$$

(Since $0 \le \theta \le 1$; $0.3665 - \log(\frac{4}{3}) = 0.0787 \le \tilde{\mu} \le 0.3665$.) The mode of the common distribution of X and Y is at zero. The mode of the joint distribution is at

$$x = y = \log\left[\frac{(2-\theta)(4-\theta)}{2\theta}\left\{\sqrt{\frac{1}{2} + \frac{2}{(2-\theta)^2}} - 1\right\}\right]. \qquad \text{(53.23)}$$

Some numerical values are given in Table 53.1.

The shape of the density function in (53.18) is not easily assessed from its analytical expression, nor is it easy to obtain expressions for product moments. However, Gumbel and Mustafi (1967) have shown that a useful idea of the way in which the density varies with θ can be obtained by studying the behavior of $p_{X,Y}(x, x)$ (i.e., on the diagonal line $y = x$) as θ varies. Figure 53.1, taken from Gumbel and Mustafi (1967), shows values of $p_{X,Y}(x, x)$ for $\theta = 0$ and $\theta = 1$. For intermediate values of θ, $p_{X,Y}(x, x)$ lies between the values shown. As θ increases, the joint distribution tends to concentrate along the diagonal $x = y$.

Although the population product moment correlation is difficult to evaluate, the so-called *grade correlation* [Konijn (1959)] defined by

$$\tilde{\rho} = 12 \int_{-\infty}^{\infty} \int_{-\infty}^{\infty} F_{X,Y}(x, y) \, dF_X(x)dF_Y(y) - 3$$

TABLE 53.1
Modes of the Bivariate Extreme Value Distributions

Type A Distribution			Type B Distribution		
Parameter θ	Mode at $x = y =$	Grade Correlation	Parameter m	Mode at $x = y =$	Correlation
0	0.0000	0.0000	1.0	0.0000	0.0000
0.1	−0.0125	0.0509	1.5	−0.1150	0.5556
0.2	−0.0255	0.1031	2.0	−0.1346	0.7500
0.3	−0.0385	0.1571	2.5	−0.1282	0.8400
0.4	−0.0514	0.2127	3.0	−0.1155	0.8889
0.5	−0.0649	0.2702	3.5	−0.1026	0.9184
0.6	−0.0790	0.3296	4.0	−0.0912	0.9375
0.7	−0.0926	0.3909	4.5	−0.0815	0.9506
0.8	−0.1071	0.4542	5.0	−0.0733	0.9600
0.9	−0.1219	0.5198	6.0	−0.0606	0.9722
1.0	−0.1362	0.5894	7.0	−0.0514	0.9796
			8.0	−0.0444	0.9844
			9.0	−0.0391	0.9877
			10.0	−0.0348	0.9900
			∞	0.0000	1.0000

is much simpler. In fact,

$$\tilde{\rho} = 3\left(2 - \frac{1}{4}\theta\right)^{-1}$$
$$\times \left[1 + 2\left(2\theta - \frac{1}{4}\theta^2\right)^{-1} \tan^{-1}\left\{\left(2\theta - \frac{1}{4}\theta^2\right)^{1/2}\left(2 - \frac{\theta}{2}\right)^{-1}\right\}\right] - 3.$$

$$(53.24)$$

There is a misprint in the formula given by Gumbel and Mustafi (1967). Table 53.1 shows some values of $\tilde{\rho}$ for a few values of θ. Tawn (1988) extended this model by adding one more parameter ϕ to provide further flexibility.

2.2   Type B Distributions

For these distributions (also known as *logistic models*) we have

$$F_{X,Y}(x,y) = \exp[-(e^{-mx} + e^{-my})^{1/m}] \qquad (m \geq 1) \qquad (53.25)$$

and the joint density function is

$$p_{X,Y}(x,y) = e^{-m(x+y)}(e^{-mx} + e^{-my})^{-2+1/m}$$
$$\times \{m - 1 + (e^{-mx} + e^{-my})^{1/m}\} \exp[-(e^{-mx} + e^{-my})^{1/m}].$$

$$(53.26)$$

Since $\lim_{m\to\infty}(e^{-mx} + e^{-my})^{1/m} = \max(e^{-x}, e^{-y})$, we obtain

$$\lim_{m\to\infty} F_{X,Y}(x,y) = \min[\exp(-e^{-x}), \exp(-e^{-y})]$$
$$= \min(F_X(x), F_Y(y)). \qquad (53.27)$$

With the univariate median value μ defined in (53.14) and (53.15), we find, for Type B distributions,

$$F_{X,Y}(\mu,\mu) = \left(\frac{1}{2}\right)^{2^{1/m}} \qquad (53.28)$$

and also

$$F_{X,Y}(0,0) = (e^{-1})^{2^{1/m}} \qquad (53.29)$$

[compare with (53.20) and (53.21)].

The value $\tilde{\mu}$ such that $F_{X,Y}(\tilde{\mu}, \tilde{\mu}) = \frac{1}{4}$ satisfies the equation

$$\exp[-2^{1/m} e^{-\tilde{\mu}}] = \frac{1}{4}$$

and so

$$\tilde{\mu} = -\log_e(\log_e 2) - \frac{m-1}{m} \log_e 2. \qquad (53.30)$$

(Since $m \geq 1$; $0.3665 - \log_e 2 = -0.3266 \leq \tilde{\mu} \leq 0.3665$.)
    The mode of the joint distribution is at

$$x = y = (1 + m^{-1}) \log_e 2 - \log_e \left[\sqrt{(m-1)^2 + 4} - m + 3 \right]. \qquad (53.31)$$

Some numerical values are given in Table 53.1.
    Tiago de Oliveira (1961) has shown that m and the population product
moment correlation, ρ, are related by the simple formula

$$m = (1 - \rho)^{-1/2}; \qquad \rho = 1 - m^{-2}. \qquad (53.32)$$

Values of ρ for $m = 1.0(0.5)5.0(1)10, \infty$ are given in Table 53.1.
    Figure 53.1, taken from Gumbel and Mustafi (1967), shows $p_{X,Y}(x, x)$
(i.e., values on the diagonal $x = y$) for $m = 1, 2, \infty$. Note that, for $m = 1$,
the graph is the same as that for a Type A distribution with $\theta = 0$ shown
in Figure 53.1. Both correspond to independence, with

$$p_{X,Y}(x, y) = p_X(x) p_Y(y) = e^{-(x+y)} \exp(-e^{-x} - e^{-y}).$$

It is interesting to note that $(X - Y)$ has a logistic distribution, with

$$\Pr[X - Y \leq t] = (1 + e^{-mt})^{-1}. \qquad (53.33)$$

[We have already noted this for the mutual independence case $(m = 1)$
in Chapter 22 of Johnson, Kotz, and Balakrishnan (1995).] This property
does not hold for Type A distributions.
    The logistic model was used by Hougaard (1986) to analyze tumor
data. Tawn (1988) extended this model by adding an extra parameter.

2.3 Type C Distributions

For these distributions (also known as *biextremal model*),

$$F_{X,Y}(x, y) = \exp[-\max\{e^{-x} + (1 - \phi)e^{-y}, e^{-y}\}] \quad (0 < \phi < 1).$$

$$(53.34)$$

The distribution (53.34) can be generated as the joint distribution of X and

$$Y = \max(X + \log \phi, Z + \log(1 - \phi)),$$

where X, Z are mutually independent and each has a standard Type I extreme value distribution.

A notable feature of this distribution is that it has a singular component on the line $Y = X + \log \phi$, since

$$
\begin{aligned}
\Pr[Y = X + \log \phi] &= \Pr[X + \log \phi \geq Z + \log(1 - \phi)] \\
&= \Pr[Z - X \leq \log\{\phi/(1 - \phi)\}] = \phi;
\end{aligned}
$$

$$(53.35)$$

see Eq. (53.33) with $m = 1$.

The correlation between X and Y is

$$-6\pi^{-2} \int_0^\phi (1 - t)^{-1} \log t \, dt \tag{53.36}$$

and the grade correlation is

$$3\phi/(2 + \phi). \tag{53.37}$$

The median of each marginal distribution is, of course, $\mu = -\log \log 2$. Note that

$$F_{X,Y}(\mu, \mu) = \frac{1}{4}(2^\phi) \tag{53.38}$$

[compare with (53.20) and (53.28)] and

$$F_{X,Y}(0, 0) = (e^{-2})^{1-\phi/2} \tag{53.39}$$

[compare with (53.21) and (53.29)].

The value $\tilde{\mu}$ such that $F_{X,X}(\tilde{\mu}, \tilde{\mu}) = \frac{1}{4}$ satisfies the equation

$$\exp[-(2 - \phi)e^{-\tilde{\mu}}] = \frac{1}{4}$$

and so

$$\tilde{\mu} = -\log\left(\frac{\log 2}{1 - \frac{1}{2}\phi}\right). \tag{53.40}$$

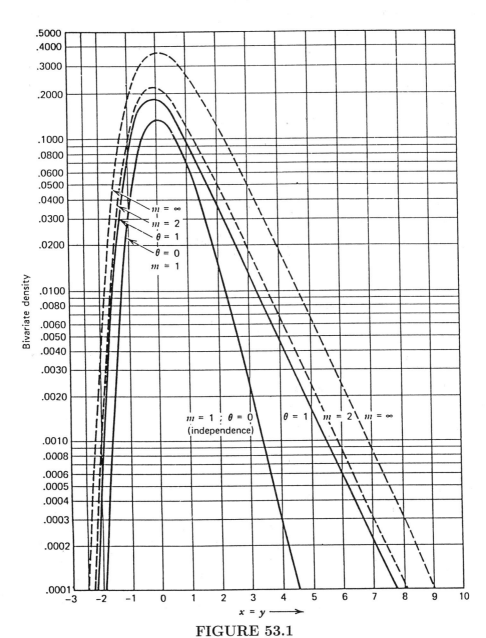

FIGURE 53.1

Bivariate Density Along the Diagonal $(x = y)$—Type A ———.

Bivariate Density Along the Diagonal $(x = y)$—Type B - - - -.

2.4   Normal-Like Bivariate Extreme Value Distributions

A new type of bivariate extreme value distribution was proposed by Hüsler and Reiss (1989) and was also extended to the multivariate case by the same authors. The derivation was recently simplified by Joe (1994).

A MEV distribution with margins transformed to a survival function with exponential survival functions as univariate margins is a *min-stable exponential distribution*. Let G be a min-stable m-dimensional exponential distribution. From extreme value theory, $A = -\log G$ satisfies

$$A(tz_1, \ldots, tz_m) = tA(z_1, \ldots, z_m) \qquad \forall\, t > 0. \tag{53.41}$$

Let G_S be a marginal survival function of G. Then

$$G_S(z_S) = \exp\{-A_S(z_S)\},$$

where A_S is obtained from A by setting $z_j = 0$ for $j \notin S$.

Hüsler and Reiss (1989) have taken

$$
\begin{aligned}
&A(z_1, z_2, \lambda) \\
&\quad = z_1 \Phi\left(\lambda + (2\lambda)^{-1} \log \frac{z_1}{z_2}\right) + z_2 \Phi\left(\lambda + (2\lambda)^{-1} \log \frac{z_2}{z_1}\right), \; \lambda \geq 0,
\end{aligned}
\tag{53.42}
$$

where Φ is the standard normal c.d.f. The multivariate extension of Hüsler and Reiss (1989) is closed under marginals and has (53.42) with parameters $\lambda_{ij} = \lambda_{ji}$ for the (i, j)th bivariate marginal thus providing dependence structure that is the same as for the multivariate normal distribution (except that there is no negative dependence). Oakes and Manatunga (1992) have studied bivariate extreme value distribution with marginal distributions parametrized as two-parameter Weibull distributions with possibly different indices. The survival function is given by

$$
\begin{aligned}
\bar{F}(t_1, t_2) &= \Pr[T_1 > t_1, T_2 > t_2] \\
&= \exp[-\{(\eta_1^{\kappa_1} t_1^{\kappa_1})^\phi + (\eta_2^{\kappa_2} t_2^{\kappa_2})^\phi\}^\alpha].
\end{aligned}
\tag{53.43}
$$

The parameter $\alpha = 1/\phi$ represents the degree of dependence in the two-component random variable. The case $\alpha = 0$ and $\alpha = 1$ correspond to maximal positive dependence (the upper Frechét bound) and independence, respectively.

The marginal survival functions are

$$F_i(t_i) = \Pr[T_i > t_i] = \exp(-\eta_i^{\kappa_i} t_i^{\kappa_i}), \quad i = 1, 2$$

with scale parameters η_i and "index" parameters κ_i $(i = 1, 2)$.

As shown by Lee (1979), the transformation

$$Z = \{(\eta_1^{\kappa_1} T_1^{\kappa_1})^\phi + (\eta_2^{\kappa_2} T_2^{\kappa_2})^\phi\}^\alpha$$

and

$$U = \frac{(\eta_1^{\kappa_1} T_1^{\kappa_1})^\phi}{(\eta_1^{\kappa_1} T_1^{\kappa_1})^\phi + (\eta_2^{\kappa_2} T_2^{\kappa_2})^\phi}$$

results in a pair of independent variables, U being uniform over $(0,1)$ and Z having mixed gamma distribution with the density

$$f_Z(z) = (1 - \alpha + \alpha z)e^{-z}, \qquad z > 0.$$

This provides a convenient method for generating random variables with the distribution (53.43).

The joint density of T_1 and T_2 is

$$f(t_1, t_2) = \phi \kappa_1 \kappa_2 \eta_1^{\kappa_1 \phi} \eta_2^{\kappa_2 \phi} t_1^{\kappa_1 \phi - 1} t_2^{\kappa_2 \phi - 1} s^{\alpha - 2}(1 - \alpha + \alpha z)e^{-z},$$

where

$$s = (\eta_1^{\kappa_1} t_1^{\kappa_1})^\phi + (\eta_2^{\kappa_2} t_2^{\kappa_2})^\phi, \qquad z = s^\alpha.$$

It is assumed that the parameters $\kappa_1, \kappa_2, \eta_1$, and η_2 are not functionally related to α.

Using Lee's transformation, Oakes and Manatunga (1992) have derived an explicit formula for the elements of the Fisher information matrix. The elements not involving the index parameters κ_1 and κ_2 are expressed as elementary functions and the exponential integral

$$E_1(x) = \int_x^\infty u^{-1} e^{-u} \, du$$

tabulated in Abramovitz and Stegun (1965, p. 228) and in Thompson (1997, Table 5.11), while the remaining elements of the Fisher information matrix involving α are expressed as incomplete digamma and trigamma integrals.

Oakes and Manatunga (1992) calculated numerically the asymptotic variance (as $n \to \infty$) of the maximum likelihood estimator $\hat{\alpha}$ of α. Calculations reveal that estimator of the scale parameters η_1 and η_2 are almost exactly orthogonal to that of the dependence parameter α. Numerical results suggest that considerable variation may be expected in samples of small to moderate size.

An extensive study of bivariate extreme value distributions was carried out by Tawn (1988). He dealt with the following models. Let $(X, Y)^T$ be a random vector and let

$$G(x, y) = \Pr[X > x, Y > y].$$

$(X, Y)^T$ follows an extreme value distribution (with unit exponential marginals) if $\Pr[X > x] = e^{-x}$, $\Pr[Y > y] = e^{-y}$ and

$$G^n(x, y) = G(nx, ny) \qquad (x > 0, \ y > 0)$$

for all n.

Pickands (1981) has shown that $(X, Y)^T$ has this distribution if and only if

$$G(x, y) = \exp\left\{ -(x + y)A\left(\frac{y}{x + y}\right) \right\} \qquad (x > 0, y > 0), \qquad (53.44)$$

where the Pickands dependence function $A(\cdot)$ is

$$A(w) = \int_0^1 \max\{(1 - w)q, w(1 - q)\} \, dH(q).$$

$A(\cdot)$ is not to be confused with the *dependence* function introduced, for example, by Oakes and Manatunga. Here, H is a positive finite measure on [0,1] satisfying

$$1 = \int_0^1 q \, dH(q) = \int_0^1 (1 - q) \, dH(q).$$

The case $A(w) = 1$ $(0 \le w \le 1)$ corresponds to $(X, Y)^T$ being an independent vector. The measure $H(\cdot)$ puts one at each of the endpoints 0 and 1. If $A(w) = \max(w, 1 - w)$, then $Pr[X = Y] = 1$. If the univariate distributions are unspecified, then the most general form for $G(x, y)$ is

$$G(x, y) = \exp\left(\int_0^1 \log[\min\{G_1^q(x), G_2^{1-q}(y)\}] \right) dH(q),$$

where G_1 and G_2 are univariate extreme value distributions given by the (univariate) generalized extreme value distribution:

$$G(z; \mu, \sigma, \kappa) = \exp\left[-\left\{ 1 - \kappa\left(\frac{z - \mu}{\sigma}\right) \right\}^{1/\kappa} \right],$$

where $\sigma > 0$ and the range of z is determined by $1 - \kappa(z - \mu)/\sigma > 0$.

Particular cases of (53.44) are:
(a) The *mixed model* with the survival function

$$G_{X,Y}(x,y) = \exp\left\{-(x+y) + \frac{\theta xy}{x+y}\right\}$$

corresponding to $A(w) = \theta w^2 - \theta w + 1$ $(0 \le \theta \le 1)$. For $\theta = 0$, we have independence but cannot achieve complete dependence. The variables X and Y are exchangeable and the correlation between X and Y is

$$\rho = \left(1 - \frac{1}{4}\theta\right)^{-3/2} \frac{1}{\sqrt{\theta}} \left\{\sin^{-1}\left(\frac{1}{2\sqrt{\theta}}\right) - \frac{1}{2}\sqrt{\theta}\left(1 - \frac{1}{4}\theta\right)^{\frac{1}{2}}\left(1 - \frac{1}{2}\theta\right)\right\}.$$

(b) The *logistic model* with survival function [studied originally by Hougaard (1986)]

$$G(x,y) = \exp\left\{-(x^r + y^r)^{1/r}\right\}$$

corresponding to $A(w) = \{(1-w)^r + w^r\}^{1/r}$ for $r \ge 1$. Here, independence corresponds to $r = 1$ and complete dependence to $r = \infty$. The variables X and Y are exchangeable and

$$\rho = r^{-1}\{\Gamma(2/r)\}^{-1}\{\Gamma(1/r)\}^2 - 1.$$

Quite often, the parameterization $\nu = 1/r$ $(0 \le \nu \le 1)$ is used in this case.

(c) *Asymmetric non-exchangeable mixed model* with the dependence function

$$A(w) = \phi w^3 + \theta w^2 - (\theta+\phi)w + 1 \ (\theta \ge 0, \ \theta+\phi \le 1, \ \theta+2\phi \le 1, \ \theta+3\phi \ge 0)$$

and the corresponding joint survival function as

$$G(x,y) = \exp[-(x+y) + xy\{x(\theta+\phi) + y(2\phi+\theta)\}(x+y)^{-2}].$$

Independence in this case corresponds to the situation when $\theta = \phi = 0$.

(d) *Asymmetric nonexchangeable logistic model* with the dependence function

$$A(w) = \{\theta^r(1-w)^r + \phi^r w^r\}^{1/r} + (\theta - \phi)w + 1 - \theta \quad (0 \le \theta, \ \phi \le 1, \ r \ge 1)$$

or

$$A(1-w) = (1-\phi)(1-w) + (1-\theta)w + \{\phi^r(1-w)^r + \theta^r w^r\}^{1/r}$$

and the corresponding joint survival function

$$G(x, y) = \exp\{-(1 - \theta)x - (1 - \phi)y - (x^r\theta^r + y^r\phi^r)^{1/r}\}.$$

When $\theta = \phi = 1$, we have the logistic model, but this model includes other models. If $\theta = \phi$, we get a mixture of logistic and independence and the variables are exchangeable. If $r \to +\infty$, we have

$$A_\infty(w) = \max\{1 - \phi w, 1 - \theta(1 - w)\}, \qquad (53.45)$$

a new nondifferentiable model with $\Pr[Y\phi = X\theta] = \phi\theta/(\theta + \phi - \theta\phi)$.

In (53.45), when $\theta = 1$ and $\phi = \alpha$ we have the biextremal (α) model, whereas when $\theta = \alpha$ and $\phi = 1$ we have the dual of the biextremal (α) model, which corresponds to X and Y being exchanged. If $\theta = \phi = \alpha$, we have the Gumbel model. Complete dependence corresponds to $\theta = \phi = 1$ and $r = +\infty$, whereas independence corresponds to $\theta = 0$ or $\phi = 0$ or $r = 1$.

The asymmetric mixed model is a crude model, but has the advantage of having a single parameter clearly identified with nonexchangeability. On the other hand, the asymmetric logistic model is flexible and simply expressible, but it has identifiability problems.

For the logistic model with $G(x, y) = \exp\{-(x^r + y^r)^{1/r}\}$ and the density

$$g(x, y; r) = (xy)^{r-1}(x^r + y^r)^{-2+1/r}\{(x^r + y^r)^{1/r} + r - 1\}\exp\{-(x^r + y^r)^{1/r}\},$$

the behavior of the maximum likelihood estimator of $r, \bar{r}_n$, was studied by Tawn (1988). At $r = 1$, $\bar{r}_n$ has a nonregular behavior. Tawn (1988) obtained the asymptotic distribution of

$$U_n(1) = \frac{dL_n(1)}{dr} = \frac{d\sum_{i=1}^n \log g(x_i, y_i; r)}{ar}.$$

The score statistic

$$\left(\frac{1}{2} n \log n\right)^{-1/2} U_n(1)$$

converges in distribution to a standard normal variable, and the asymptotic behavior of $\bar{r}_n$ for $r = 1$ is as follows:

$$(\bar{r}_n - 1)\left(\frac{1}{2} n \log n\right)^{1/2}$$

converges in distribution to a nonnegative random variable S such that $\Pr[S \leq s] = h(s)\Phi(s)$, where $h(\cdot)$ is the Heaviside function and $\Phi(\cdot)$ is the standard normal cumulative distribution function. Similar results are cited for the other models. For asymmetric mixed models, the score vector converges to a bivariate normal distribution while the maximum likelihood estimator converges to a truncated normal distribution.

By observing that the general form of a bivariate extreme value distribution is

$$F(x,y) = \exp\{-V(x,y)\}, \qquad x \geq 0, \ y \geq 0,$$

where

$$V(x,y) = \int_{[0,1]} \max\left(\frac{w}{x}, \ \frac{1-w}{y}\right) H_*(dw),$$

and H_* is a finite nonnegative measure on $[0,1]$ with a total mass of 2 and unit means, Nadarajah (1999a) has proposed a new flexible model by choosing

$$H_*(\{0\}) = \gamma_0, \quad H_*(\{1\}) = \gamma_1, \quad \text{and} \quad H_*(\{\theta\}) = \gamma_\theta$$

for $\theta \in (0,1)$ and

$$h(w) = \frac{\partial H_*([0,w])}{dw} = \begin{cases} \alpha w^r & \text{if } 0 < w < \theta, \\ \beta(1-w)^s & \text{if } \theta < w < 1. \end{cases}$$

Thus, the bivariate extreme value model here is described polynomially in the interiors $(0,\theta)$ and $(\theta,1)$ and has atoms of mass at the endpoints $w = 0,1$ as well as at the interior point θ. To ensure nonnegativity of $h(\cdot)$, we need to take $\alpha \geq 0$ and $\beta \geq 0$, and for the continuity at θ, we need to impose $\alpha\theta^r = \beta(1-\theta)^s$. Nadarajah (1999a) has used this model to analyze the data on experimental releases of ionized air from a fixed source discussed earlier by Mole and Jones (1994).

Nadarajah (1999c) has presented a general method of simulating bivariate extreme value distributions and has also discussed its generalization to the multivariate case.

Additional results on bivariate extreme value distributions are scattered in the second half of this chapter, devoted to multivariate extreme value distributions.

2.5   Estimation

For distributions of each type (A, B, and C), four further parameters (making five in all) can be introduced as location and scale parameters for the two variables. Their values can be calculated from the separate marginal distributions of the variables, using the methods described in Chapter 22 of Johnson, Kotz, and Balakrishnan (1995).

In order to estimate the "association parameter" (θ or m, as the case may be) from a random sample of size n, we can use the observed frequencies in the 2×2 table formed by dichotomizing each variable at its sample median. Using formulas (53.20) and (53.28), we obtain estimators θ^*, m^* of θ, m, respectively, by equating the observed proportions, $\hat{p}$, of the n pairs of values, for which both X and Y exceed their sample medians, to

$$\left(\frac{1}{4}\right)^{1-\theta^*/4} \qquad \text{for Type A,}$$

$$\left(\frac{1}{2}\right)^{2^{1/m^*}} \qquad \text{for Type B,}$$

leading to estimators

$$\theta^* = 4 + \frac{2\log\hat{p}}{\log 2} = 4 + 2.8854 \log\hat{p}, \tag{53.46}$$

$$\begin{aligned} m^* &= \left\{\log\left(-\frac{\log\hat{p}}{\log 2}\right)\right\}^{-1} \log 2 \\ &= \{0.5288 + 1.4427\log(-\log\hat{p})\}^{-1}. \end{aligned} \tag{53.47}$$

The variance of θ^* is actually infinite, since $\Pr[\hat{p} = 0] > 0$, but if $(\hat{p} = 0)$ is excluded from the distribution of $\hat{p}$, then

$$\text{var}(\theta^*) \doteq \frac{4}{n(\log 2)^2} \left[\frac{1}{2}\{F_{(\infty)}(\mu,\mu)\}^{-1} - 1\right], \tag{53.48}$$

where n is the sample size. This may be estimated as

$$8.3255 \left(\frac{1}{2}\hat{p}^{-1} - 1\right) n^{-1}.$$

The variance of m^{*-1} is also infinite, but excluding $(\hat{p} = 0)$ we have

$$\text{var}(m^{*-1}) \doteq \frac{1}{n(\log 2)^2} \left[\frac{1}{2}\{F_\infty(\mu,\mu)\}^{-1} - 1\right] \frac{1}{[-\log F_\infty(\mu,\mu)]^2}. \tag{53.49}$$

This may be estimated as

$$2.0814 \left(\frac{1}{2}\hat{p}^{-1} - 1\right)(-\log\hat{p})^{-2}n^{-1}.$$

Some values for (53.48) and (53.49), taken from Gumbel and Mustafi (1967), are given in Table 53.2.

TABLE 53.2

Variances of Estimators of the Parameters

	Approximate Value			Approximate Value
	(A)	(B)		
$F_{X,Y}(\mu, \mu)$	$n\mathrm{var}(\theta^*)$	$n\mathrm{var}(m^{*-1})$	$F_{X,Y}(\mu, \mu)$	$n\mathrm{var}(m^{*-1})$
0.25	8.32548	1.08304	0.36	0.77548
0.26	7.68506	1.05879	0.37	0.73976
0.27	7.09208	1.03423	0.38	0.70207
0.28	6.54145	1.00919	0.39	0.66213
0.29	6.02880	0.98359	0.40	0.61975
0.30	5.55032	0.95722	0.41	0.57473
0.31	5.10271	0.93004	0.42	0.52682
0.32	4.68308	0.90175	0.43	0.47570
0.33	4.28888	0.87234	0.44	0.42108
0.34	3.91787	0.84158	0.45	0.36270
0.35	3.56806	0.80934	0.46	0.30015
			0.47	0.23305
			0.48	0.16099
			0.49	0.08347
			0.50	0.00000

From (53.20) (noting that $0 \leq \theta \leq 1$) it can be seen that for Type A, $F_{X,Y}(\mu, \mu)$ lies between $\frac{1}{4}$ and $(\frac{1}{4})^{0.75} = 0.35355$.

For Type B, $F_{X,Y}(\mu, \mu)$ lies between $\frac{1}{4}$ and $\frac{1}{2}$. If $\hat{p}$ lies outside these limits, formulas (53.48) and (53.49) cannot be used. Of course, observations of this kind might well be regarded as evidence that a Type A, or a Type B, extreme value distribution is not appropriate.

An alternative estimator of θ, for Type A distributions, is obtained by solving (53.24) for θ, with $\bar{\rho}$ replaced by the sample grade correlation. For Type B distributions, m is easily estimated by replacing ρ in (53.32) by the sample (product moment) correlation coefficient.

Posner *et al.* (1969) have proposed a method of estimation of θ or m based on the distribution of $|X_1 - X_2|$. They show that, in the general case (53.9),

$$P[\alpha < X_1 - X_2 < \beta] = h(\beta) - h(\alpha)$$

with

$$h(t) = (1 + e^{-t})^{-1} + g'(t)/g(t).$$

In particular, for example

$$\Pr[|X_1 - X_2| < \delta] = \frac{e^\delta - 1}{e^\delta + 1} + 2\frac{g'(\delta)}{g(\delta)} = P(\delta). \tag{53.50}$$

The variance of the estimator so obtained depends markedly on δ. For large sample size n and Type A distributions,

$$n \, \text{var}(\hat{\theta}) \doteq P(\delta)\{1 - P(\delta)\}g(\delta). \tag{53.51}$$

Tiago de Oliveira (1970) has described several methods of estimating ϕ for Type C distributions. An estimator that he has noted to be particularly useful is

$$\tilde{\phi} = \min\left(\exp\left\{\min_j(Y_j - X_j)\right\}, 1\right). \tag{53.52}$$

The cumulative distribution function of $\tilde{\phi}$ is

$$\Pr[\tilde{\phi} \le t] = \begin{cases} 0 & \text{for } t < \phi, \\ 1 - [(1-\phi)^{-1}t + 1]^{-n} & \text{for } \phi \le t < 1, \\ 1 & \text{for } t \ge 1. \end{cases} \tag{53.53}$$

Note that

$$\Pr[\tilde{\phi} = \phi] = 1 - (1-\phi)^n$$

and

$$\Pr[\tilde{\phi} = 1] = (1-\phi)^n(2-\phi)^{-n}.$$

The estimator $\tilde{\phi}$ necessarily has a positive bias but

$$\lim_{n\to\infty} \Pr[\tilde{\phi} = \phi] = 1.$$

The expected value of $\tilde{\phi}$ is

$$\phi + \frac{1}{n-1}(1-\phi)^n\{1 - (2-\phi)^{-(n-1)}\}. \tag{53.54}$$

3 MULTIVARIATE EXTREME VALUE DISTRIBUTIONS

Multivariate extreme value distributions arise in connection with extremes from several dependent populations. The classical definition is expressed in terms of asymptotic joint distribution of normalized componentwise maxima. Let $(Y_{i1}, \ldots, Y_{ik})^T$ $(i = 1, \ldots, n)$ denote independent and identically distributed random vectors. For $j = 1, \ldots, k$, we define $M_{nj} = \max(Y_{1j}, \ldots, Y_{nj})$. Suppose there exist $a_n = (a_{n1}, \ldots, a_{nk})$ with each $a_{nj} > 0$, and $b_n = (b_{n1}, \ldots, b_{nk})$, such that

$$\lim_{n\to\infty} \Pr\left(\frac{M_{ni} - b_{ni}}{a_{ni}} < x_i, \ i = 1, \ldots, k\right) = G(x_1, \ldots, x_k),$$

where G is a nondegenerate k-variate distribution function. Then, G is called a *multivariate extreme value distribution*. From univariate extreme value theory, the univariate marginals of G must be generalized extreme value distributions

$$H(x; \mu, \sigma, \xi) = \exp\left\{-\left(1 - \xi \frac{x - \mu}{\sigma}\right)_+^{1/xi}\right\},$$

where $y_+ = \max(0, y)$. This includes all three types of the extreme value distribution: $\xi = 0$, $\xi > 0$ and $\xi < 0$ corresponding, respectively, to the Gumbel, Fréchet, and Weibull distributions; see Chapter 22 of Johnson, Kotz, and Balakrishnan (1995).

de Haan and Resnick (1977) and Pickands (1981) have shown that any k-variate extreme value distribution $G(x_1, \ldots, x_k)$ depends on an arbitrary positive measure over a $(k - 1)$-dimensional simplex. A number of parametric models have been given by Coles and Tawn (1991). One is the logistic model. This has joint distribution function

$$G(x_1, \ldots, x_k) = \exp\left[-\left\{\sum_{j=1}^{k}\left(1 - \xi_j \frac{x_j - \mu_j}{\sigma_j}\right)^{1/(\alpha\xi_j)}\right\}^\alpha\right], \quad (53.55)$$

where $0 \le \alpha \le 1$ measures the dependence between the variates, and the limits $\alpha \to 1$, $\alpha \to 0$ correspond, respectively, to independence and complete dependence.

For notational convenience, we set

$$y_i = \left(1 - \xi_i \frac{x_i - \mu_i}{\sigma_i}\right)^{1/\xi_i} \quad (i = 1, \ldots, k), \quad z = \left(\sum_{i=1}^{k} y_i^{1/\alpha}\right)^\alpha.$$

This transformation was suggested by Lee (1979).

The k-variate extreme value distribution has joint density function

$$g(x_1, \ldots, x_k) = \frac{\partial^k G}{\partial x_1 \ldots \partial x_k} = \left(\prod_{i=1}^{k} \frac{y_i^{1/\alpha - \xi_i}}{\sigma_i}\right) z^{1-k/\alpha} Q_k(z, \alpha) e^{-z},$$

where

$$Q_k(z, \alpha) = \left(\frac{k - 1}{\alpha} - 1 + z\right) Q_{k-1}(z, \alpha) - z \frac{\partial Q_{k-1}(z, \alpha)}{\partial z}, \quad Q_1(z, \alpha) = 1.$$

In general, $Q_k(z, \alpha)$ is a $(k - 1)$-order polynomial in z. Shi (1995a) noted that if we let $T_i = \left(\frac{Y_i}{Z}\right)^{1/\alpha}$, the joint density function of $(Z, T_1, \ldots, T_{k-1})^T$

becomes

$$p(z, t_1, \ldots, t_{k-1}) = \frac{\alpha^{k-1}}{(k-1)!} \, Q_k(z, \alpha) e^{-z} (k-1)!,$$

$$z > 0, \ 0 < t_i < 1 \text{ for } i = 1, \ldots, k-1, \ 0 < \sum_{i=1}^{k-1} t_i < 1.$$

Thus, $(T_1, \ldots, T_{k-1})^T$ has a multivariate beta distribution $\beta_k(1, 1, \ldots, 1)$ (see Chapter 49) and Z has a mixed gamma distribution with density function

$$p(z) = \frac{\alpha^{k-1}}{(k-1)!} \, Q_k(z, \alpha) e^{-z}, \quad z > 0,$$

and Z and $(T_1, \ldots, T - k - 1)^T$ are mutually independent.

Tawn (1990) has dealt with the multivariate case of the logistic (see Chapter 51) and related models. Let $G(y_1, y_2, \ldots, y_k) = \Pr[Y_1 > y_1, \ldots, Y_k > y_k]$. The condition

$$\bar{G}^s(y_1, \ldots, y_k) = \bar{G}(s y_1, \ldots, s y_k)$$

implies that

$$\bar{G}(y_1, \ldots, y_k) = \exp\{-t B(w_1, \ldots, w_{k-1})\}$$

for $y_i \geq 0$, where $w_i = y_i/t$, $t = \sum_{i=1}^{k} y_i$, and

$$B(w_1, \ldots, w_{k-1}) = \int_{S_k} \max_{1 \leq i \leq k} (w_i q_i) \, dH(q_1, \ldots, q_{k-1}), \qquad (53.56)$$

with H being an arbitrary positive finite measure over the unit simplex,

$$S_k = \{q \in \mathbf{R}^{k-1} : q_1 + \cdots + q_{k-1} \leq 1, \ q_i \geq 0, \ i = 1, \ldots, k-1\},$$

satisfying

$$1 = \int_{S_k} q_i \, dH(q_1, \ldots, q_{k-1}) \qquad (i = 1, \ldots, k). \qquad (53.57)$$

Here, $j_k = 1 - (j_1 + \cdots + j_{k-1})$ for $j \equiv q$ and $j \equiv w$. Following Pickands (1981), B will be called the *dependence function*. It is a convex function satisfying

$$\max(w_1, \ldots, w_k) \leq B(w_1, \ldots, w_{k-1}) \leq 1. \qquad (53.58)$$

The properties of B are very similar to those of the bivariate dependence function. However, a key difference is that for $k = 2$, (53.58) and convexity are necessary and sufficient conditions for (53.56), whereas for $k \geq 3$ they are only *necessary* conditions.

For the logistic (or *Gumbel*) model, we have

$$B(w_1, \ldots, w_{k-1}) = \left(\sum_{i=1}^{k} w_i^r \right)^{1/r}, \qquad r \geq 1.$$

The corresponding survival function has a density, along with single parameter that governs association among the exchangeable variables.

Shi (1995a) has derived explicit algebraic expressions for the Fisher information matrix of the multivariate extreme value distribution with generalized extreme value marginals and logistic dependence structure; see also Shi (1995b). Moment estimation of the parameters has been discussed by Shi (1995c), who has also derived the asymptotic variance-covariance matrix of these estimators and compared the relative efficiencies of these moment estimators with those of the maximum likelihood estimators and the stepwise estimators. The author has specifically shown that when there is strong dependence among the components, the generalized variance of the moment estimators is much smaller than that of the stepwise estimators, particularly when the dimension is large. As an alternative to full maximum likelihood method based on the joint distribution, Shi, Smith and Coles (1992) have considered a "marginal estimation" method in which the marginal and dependence parameters are estimated separately. This method is simpler to use, but is somewhat inefficient. The authors have examined the relative efficiency of this estimation method through asymptotic results as well as in finite-sample case through simulations.

In discrete choice literature, McFadden (1978) [see also Dagsvik (1988)] suggested two models for the dependence function B:

$$B(w_1, \ldots, w_{k-1}) = \sum_{m=1}^{M} a_m \left(\sum_{i \in C_m} w_i^{r_m} \right)^{1/r_m}, \tag{53.59}$$

$$B(w_1, \ldots, w_{k-1}) = \sum_{m=1}^{M} a_m \left\{ \sum_{q \in D_m} \left(\sum_{i \in C_q} w_i^{t_q} \right)^{r_m/t_q} \right\}^{1/r_m}. \tag{53.60}$$

In each case, $\bigcup C_m = \{1, \ldots, k\}$, $r_m \geq 1$, $a_m \geq 0$; and for (53.60), $t_q \geq r_m$ for $q \in D_m$, where D_m is an arbitrary subset of $\{1, \ldots, k\}$. Necessary additional constraints on the a_m were not given, but these are apparent from

(53.57). Although not a requirement of the model, the C_m are typically taken as disjoint. Neither model was physically motivated, and so only limited interpretation could be given to the parameters.

Tawn (1990) has provided some physical motivation of these models, in terms of occurrence of storms of differing severity, and the following generalization. Let C be an index variable over the set S, the class of all nonempty subsets of $\{1, \ldots, k\}$. Let $V_{i,C}^{(j)}$ be the size of the j-th realization at site i, of an extreme spatial storm of the type which occurs only at the collection C of sites. Here, $V_{i,C}^{(j)}$ $(j = 1, \ldots, N_C)$ are assumed to be conditionally independent, given N_C, where the random variable N_C is taken to have a Poisson distribution with mean τ_C. Also α_C denotes the unrecorded covariate information variable, which has a positive stable distribution and characteristic exponent $0 < 1/r_C \le 1$. The α_C are assumed to be mutually independent.

Denote

$$V_{i,C} = \max(V_{i,C}^{(1)}, \ldots, V_{i,C}^{(N_C)})$$

for $N_C > 0$. We take

$$\Pr[V_{i,C}^{(j)} < v \mid V_{i,C}^{(j)} > u_i] = 1 - \{1 - p_i(v - u_i)/\sigma_i\}^{1/p_i},$$

where $v > u_i$, $1 - p_i(v - u_i)/\sigma_i > 0$, $\sigma_i > 0$ and $p_i \in \mathbb{R}$ (a generalized Pareto distribution) is taken for the univariate exceedances.

Then the $Z_1, \ldots, Z_k$, where for $i = 1, \ldots, k$

$$Z_i = \max_{C \in S_{(i)}} (V_{i,C}),$$

and $S_{(i)}$ is the subclass of S containing all nonempty subsets that include $V_{i,C}$, $i \in C$, are dependent generalized extreme value random variables. Tawn (1990) has shown, using a conditioning argument, that their joint distribution function G is

$$G_{\mathbf{Z}}(z_1, \ldots, z_k) = \exp\left(-\sum_{C \in S} \tau_C \left[\sum_{i \in C} \{1 - p_i(z_i - u_i)/\sigma_i\}^{r_C/p_i}\right]^{1/r_C}\right).$$

Letting $Y_i = \sum \tau_C \{1 - p_i(Z_i - u_i)/\sigma_i\}^{1/p_i}$, where the summation is over $C \in S_{(i)}$, the marginal distribution of Y_i is unit exponential for $i = 1, \ldots, k$. Also, $Y_1, \ldots, Y_k$ have joint survival function

$$\bar{G}_Y(y_1, \ldots, y_k) = \exp\left[-\sum_{C \in S} \left\{\sum_{i \in C} (\theta_{i,C} \, y_i)^{r_C}\right\}^{1/r_C}\right], \qquad (53.61)$$

where $r_C \geq 1$ and $\theta_{i,C} = \tau_C / \sum \tau_C$, the summation being over $C \in S_{(i)}$. With $\theta_{i,C} = 0$ if $i \notin C$, then for $i = 1, \ldots, k$, we obtain $0 \leq \theta_{i,C} \leq 1$ and $\sum_{C \in S} \theta_{i,C} = 1$. This is a multivariate extreme value distribution with unit exponential marginals and the associate dependence function

$$B(w_1, \ldots, w_{k-1}) = \sum_{C \in S} \left\{ \sum_{i \in C} (\theta_{i,C} \ w_i)^{r_C} \right\}^{1/r_C},$$

which has $2^{k-1}(k+2) - (2k+1)$ parameters.

When $r_C \to \infty$ for all $C \in S$, we get the particular case corresponding to McFadden and logistic models. By letting only certain $r_C \to \infty$, we obtain cases where only some variables have singular components in their dependence structure. For $k = 2$, the model becomes the bivariate asymmetric logistic model discussed earlier. Further extensions using two-stage conditioning are briefly mentioned by Tawn (1990).

In view of its substantial practical importance, we present here a real-world trivariate data example provided by Tawn (1990).

The data are 40 years of trivariate sea-level annual maxima at Kings Lynn, Southend, and Sheerness, three sites on the southeast coast of England. For each site, the parameters of the marginal generalized extreme value distribution were estimated and the observations transformed to unit exponential. For all possible pairs of sites, the logistic model was found to be the best-fitting bivariate parametric model. The estimated dependence parameter values for the logistic models are 1.33 (0.17) for Sheerness with Kings Lynn, 1.66 (0.23) for Kings Lynn with Southend, and 2.52 (0.31) for Southend with Sheerness; the figures in parentheses here are the standard errors of the estimates.

Because of their locations, we expect stronger dependence between Southend and Sheerness than between the other pairs. The greater dependence between Kings Lynn and Southend than between Kings Lynn and Sheerness is not significant and does not really tie in with physical knowledge of the North Sea dynamics. Essentially, the only types of storm are spatial ones that affect either all three sites or just Sheerness and Southend. There are also local storms that affect only Kings Lynn. Suitable subfamilies of models that correspond to this are

$$B(w_1, w_2) = (1 - \theta_3) w_3 + \left[\sum_{i=1}^{2} \{(1 - \theta_i) w_i\}^r \right]^{1/r} + \left\{ \sum_{i=1}^{3} (\theta_i w_i)^s \right\}^{1/s},$$

$$(53.62)$$

$$B(w_1, w_2) = \phi\{(w_1^{rs} + w_2^{rs})^{1/r} + w_3^s\}^{1/s} + (1 - \phi)\{(w_1^t + w_2^t)^{1/t} + w_3\},$$

$$(53.63)$$

where $r, s, t \geq 1$ and $0 \leq \theta_i, \phi \leq 1$, for $i = 1, 2, 3$. Subscripts 1, 2 and 3 correspond to the sites Southend, Sheerness and Kings Lynn, respectively. The subfamily considered by Smith, Tawn, and Yuen (1990) is (53.62) with $r = 1$. Results obtained from estimating model (53.62), model (53.63), and *restricted subfamilies* are contained in Table 53.3.

TABLE 53.3
Trivariate Estimation Results

Model	Constraints	Likelihood Ratio Test Statistic	Parameter Estimates
(a) independence	–	−119.23	–
(b) (53.62)	$\theta_1 = \theta_2 = \theta_3 = 1$	−95.93	$s = 1.59$
(c) (53.62)	$\theta_1 = \theta_2 = 1$	−88.85	$(s, \theta_3) = (2.48, 0.25)$
(d) (53.62)	$\theta_1 = \theta_2 = \theta$	−86.15	$(s, r, \theta, \theta_3) = (7.44, 2.21, 0, .23, 0.55)$
(e) (53.63)	$\phi = 1$	−89.26	$(s, r) = (1.59, 1.27)$
(f) (53.63)	–	−86.09	$(s, r, t, \phi) = (1.69, 1.25, 7.44, 0.74)$

Models (a)–(d) are a nested series of subfamilies of (53.62) and hence can be sequentially tested by likelihood ratio tests. Testing Gumbel's logistic model (b), where all the variables are exchangeable, against independence (a) gives a highly significant value for the log-likelihood ratio statistic when compared with the suitable squared stable distribution. The nonexchangeable model (c) is highly significant against model (b), when compared with a one-half chi-squared with one degree of freedom. Model (c) was presented by Smith, Tawn, and Yuen (1990). This corresponds to extremes at Southend and Sheerness arising only from storms that affect all three sites, whereas extremes at Kings Lynn can also occur due to storms that affect only Kings Lynn. Physically this is not realistic, as storm surges are often generated in the southern North Sea, leading to storms that occur only at Southend and Sheerness among the three sites. To account for this, model (d) was proposed. A more detailed analysis of the data has been given by Tawn (1990). Coles and Tawn (1991, 1994) have similarly analyzed two more environmental data sets from U.K.

Joe's (1990) "negative" *asymmetric logistic* multivariate extreme value model has the distribution function

$$G(\boldsymbol{x}) = \exp\left[-\sum_{j=1}^{k} \frac{1}{x_i} + \sum_{c \in C : |c| \geq 2} (-1)^{|c|} \left\{ \sum_{i \in C} \left(\frac{\theta_{i,c}}{x_i} \right)^{r_c} \right\}^{1/r_c} \right]$$

with parameter constraints given by $r_c \leq 0$ for all $c \in C$, $\theta_{i,c} = 0$ if $i \notin c$, $\theta_{i,c} \geq 0$ for all $c \in C$, $\sum_{c \in C}(-1)^{|c|}\theta_{i,c} \leq 1$. Here, C is the set of all non-empty subsets of $\{1, \ldots, k\}$; also see Coles and Tawn (1990).

Compare with Tawn's (1990) asymmetric logistic model

$$G(\boldsymbol{x}) = \exp\left[-\sum_{c \in C}\left\{\sum_{i \in C}\left(\frac{\theta_{i,c}}{x_i}\right)^{r_c}\right\}^{1/r_c}\right],$$

where the condition on the $\theta_{i,c}$ is $\sum_{c \in C}\theta_{i,c} = 1$. The main achievement of Tawn's work is the development of techniques that simplify the difficult problem of generating *parametric measures* satisfying

$$\int_{S_k} w_j \, dH(\boldsymbol{w}) = 1, \qquad j = 1, \ldots, k \tag{53.64}$$

where S_k is a $(k-1)$-dimensional unit simplex

$$S_k = \{(w_1, \ldots, w_k) : \sum w_j = 1, \; w_j \geq 0, \; j = 1, \ldots, k\}$$

and H is an arbitrary finite positive measure. Since the only constraints on H are given by (53.64), no finite parameterization exists for this measure. On the other hand, relation (53.64) is required so that the marginals have the correct form, the unit Fréchet distribution, with cumulative distribution function

$$\exp(-1/x), \qquad x > 0,$$

without loss of generality.

As pointed out by Coles and Tawn (1994), there is no reason a priori why one particular model for the dependence measure should fit better than any other. A natural procedure, therefore, is to work with several parametric families, each of which has a flexible dependence structure determined by the parameter configuration.

Nadarajah, Anderson, and Tawn (1998) have discussed multivariate extreme value models and associated inferential methods for data with vector observations whose components are subject to an order restriction. This method, which extends the multivariate threshold methodology of Coles and Tawn, has also been applied by the authors to analyze extreme rainfalls of different durations.

Gomes and Alperin (1986) have defined a multivariate generalized extreme value model as follows: Assume a dependent sample $\boldsymbol{X} = (X_1, X_2, \ldots, X_{m+1})^T$, where $Z_j = (X_j - \lambda)/\delta$, $1 \leq j \leq m+1$, $\lambda \in \mathbb{R}^1$, and $\delta \geq 0$ being unknown location and scale parameters, respectively, have a joint probability density function given by

$$h_\theta(z_1, \ldots, z_{m+1})$$
$$= g_\theta(z_{m+1})\prod_{j=1}^{m}\{g_\theta(z_j)/G_\theta(z_j)\}, \quad z_1 > \cdots > z_{m+1}, \; \theta \in \mathbb{R};$$

$$\tag{53.65}$$

here,

$$G_\theta(z) = \begin{cases} \exp\{-(1-\theta z)^{1/\theta}, & 1-\theta z > 0, \ z \in \mathbb{R} \text{ if } \theta \neq 0, \\ \exp\{-\exp(-z)\}, & z \in \mathbb{R} \text{ if } \theta = 0 \end{cases} \tag{53.66}$$

is the GEV (θ) or *von Mises–Jenkinson* (generalized form of limiting distribution of maxima values, suitably normalized, in a classical framework) distribution function, and $g_\theta(z) = \partial G_\theta(z)/\partial z$; see Chapter 22 of Johnson, Kotz, and Balakrishnan (1995).

The classical definition of multivariate extreme value distributions [see, for example, Takahashi (1987) and Marshall and Olkin (1983)] is as follows:

For $\boldsymbol{a}, \boldsymbol{b}, \boldsymbol{x} \in \mathbb{R}^k$, write $\boldsymbol{a}\boldsymbol{x} + \boldsymbol{b}$ to denote the vector

$$(a_1 x_1 + b_1, \ldots, a_k x_k + b_k).$$

(Basic arithmetical operations are always meant componentwise.) Let $\boldsymbol{X}^{(1)}, \boldsymbol{X}^{(2)}, \ldots$ be a sequence of independent k-dimensional random vectors with common distribution function F, and let

$$Z_j^{(n)} = \max_{1 \le i \le n} X_j^{(i)}, \qquad j = 1, \ldots, k.$$

If there exist $\boldsymbol{a}^{(n)} > 0$, $\boldsymbol{b}^{(n)} \in \mathbb{R}^k$, $n = 1, 2, \ldots$ ($\boldsymbol{a}^{(n)} > 0$ means $a_j^{(n)} > 0$, $j = 1, \ldots, k$) such that $(\boldsymbol{Z}^{(n)} - \boldsymbol{b}^{(n)})/\boldsymbol{a}^{(n)}$ converges in distribution to a random vector $\boldsymbol{U}$ with nondegenerate distribution function H (i.e., all univariate marginals of H are nondegenerate), then F is said to be in the *domain of attraction* of H with the notation $F \in \mathbf{D}(H)$ and H is said to be *a multivariate extreme value distribution*. The convergence in distribution is equivalent to the condition

$$\lim_{n\to\infty} F^n(\boldsymbol{a}^{(n)}\boldsymbol{x} + \boldsymbol{b}^{(n)}) = H(\boldsymbol{x}) \tag{53.67}$$

for all $\boldsymbol{x}$, because multivariate extreme value distributions are continuous but not always absolutely continuous; see Theorem 5.2.2 of Galambos (1978).

If $(\boldsymbol{Z}^{(n)} - \boldsymbol{b}^{(n)})/\boldsymbol{a}^{(n)}$ converges in distribution to $\boldsymbol{U}$, then the jth component of $(\boldsymbol{Z}^{(n)} - \boldsymbol{b}^{(n)})/\boldsymbol{a}^{(n)}$ converges to the jth component of $\boldsymbol{U}$ and thus the normalizing constants $\{a_j^{(n)}\}$, $\{b_j^{(n)}\}$ can be determined from univariate considerations, $j = 1, \ldots, k$.

Marshall and Olkin (1983) [see also Theorem 5.3.1 of Galambos (1978)] have shown that Eq. (53.67) is equivalent to

$$\lim_{n\to\infty} n\{1 - F(\boldsymbol{a}^{(n)}\boldsymbol{x} + \boldsymbol{b}^{(n)})\} = -\log H(\boldsymbol{x})$$

for all x such that $0 < H(x) < 1$.

Recall that [Johnson, Kotz, and Balakrishnan (1995); Chapter 22] *univariate* extreme value distributions can only be one of the following types:

$$\Phi_\alpha(x) = \exp(-x^{-\alpha}), \quad x > 0 \ (\alpha > 0),$$
$$\Psi_\alpha(x) = \exp(-(-x)^\alpha), \quad x \leq 0 \ (\alpha > 0),$$
$$\Lambda(x) = \exp(-e^{-x}), \quad -\infty < x < \infty.$$

Takahashi (1987) has established the following result: Let H be a nondegenerate k-dimensional distribution function. Then a necessary and sufficient condition that H is an (multivariate) extreme value distribution is that for all $s > 0$ there exist vectors $\boldsymbol{A}^{(s)} > 0$ and $\boldsymbol{B}^{(s)}$ such that

$$H^s(\boldsymbol{A}^{(s)}\boldsymbol{x} + \boldsymbol{B}^{(s)}) = H(\boldsymbol{x}) \tag{53.68}$$

for all $\boldsymbol{x} \in \mathbb{R}^k$.

This result implies that if H is a (multivariate) extreme value distribution, then so is H^t for any $t > 0$. Moreover, the structure of the vectors $\boldsymbol{A}^{(s)} > 0$ and $\boldsymbol{B}^{(s)}$ corresponding to the three types of extreme value distributions is as follows, wherein H_i denotes the ith marginal of H:

(i) $H_i = \Phi_{\alpha_i}$, $i = 1, \ldots, k$, then $\boldsymbol{A}^{(s)} = (s^{1/\alpha_1}, \ldots, s^{1/\alpha_k})$ and $\boldsymbol{B}^{(s)} = 0$;

(ii) $H_i = \Psi_{\alpha_i}$, $i = 1, \ldots, k$, then $\boldsymbol{A}^{(s)} = (s^{-1/\alpha_1}, \ldots, s^{-1/\alpha_k})$ and $\boldsymbol{B}^{(s)} = 0$;

(iii) $H_i = \Lambda$, $i = 1, \ldots, k$, then $\boldsymbol{A}^{(s)} = 1 = (1, \ldots, 1)$ and $\boldsymbol{B}^{(s)} = (\log s, \ldots, \log s)$,

where $\alpha_i > 0$, $i = 1, \ldots, k$. Takahashi (1987) has pointed out that

$$H(x_1, x_2, \ldots, x_k) = \exp\{-\exp[-\min(x_1, x_2, \ldots, x_k)]\}$$

is an extreme value distribution with $H_i = \Lambda$, $i = 1, \ldots, k$, and moreover

$$H^s(x_1 + \log s, x_2 + \log s, \ldots, x_k + \log s) = H(x_1, \ldots, x_k)$$

for any $s > 0$, while the Farlie–Gumbel–Morgenstern (FGM) distribution constructed from $\Lambda(x) = \exp(-e^{-x})$ is not, since

$$H(x_1, x_2) = \Lambda(x_1)\Lambda(x_2)\left[1 + \frac{1}{2}(1 - \Lambda(x_1))(1 - \Lambda(x_2))\right]$$

does not satisfy

$$H^s(x_1 + \log s, x_2 + \log s) = H(x_1, x_2) \quad \text{for } s \neq 1.$$

In fact, Marshall and Olkin (1983) have observed that for the FGM distribution

$$F(x_1, x_2) = F_1(x_1)F_2(x_2)[1 + \alpha\{1 - F_1(x_1)\}\{1 - F_2(x_2)\}],$$

where $F_i(x)$ are univariate extreme value distributions, $F(x_1, x_2)$ is a bivariate extreme value distribution only if $\alpha = 0$. Joe (1990) has discussed some connections between families of min-stable multivariate exponential and multivariate extreme value distributions. Some of these multivariate extreme value models have been presented already, based on the work of Tawn (1988).

Tiago de Oliveira (1962/1963) noted that multivariate extreme value distributions satisfy the inequality

$$H(z_1, z_2, \ldots, z_k) \geq H_1(z_1)H_2(z_2) \cdots H_k(z_k),$$

namely, the components of a random vector with the c.d.f. H are positively correlated; this property for the case $k = 2$ is known as *positive quadrant dependence*; see Lehmann (1966).

Marshall and Olkin (1983) have shown that if $\boldsymbol{X} = (X_1, \ldots, X_k)^T$ have multivariate extreme value distribution, then they are *associated* in the sense of Esary, Proschan and Walkap (1967) [see Chapter 33 of Johnson, Kotz, and Balakrishnan (1995)], namely,

$$\text{cov}(\theta(\boldsymbol{X}), \psi(\boldsymbol{X})) \geq 0$$

for any pair θ and ψ of nondecreasing functions defined on $\mathbb{R}^k$. They also noted that the study of independence in multivariate extreme value distributions is substantially simplified by the fact, proved by Berman (1961/1962), that *pairwise* independent random variables $X_1, \ldots, X_k$ having a multivariate extreme value distribution are *mutually* independent.

Galambos (1985) provided the following bounds. Denote by $H_{ij}(z_i, z_j)$ the bivariate marginal of $H(z_1, \ldots, z_k)$ corresponding to the ith and jth components, and let

$$r_{ij}(z_i, z_j) = \frac{H_{ij}(z_i, z_j)}{H_i(z_i)H_j(z_j)}.$$

Then,

$$H(z_1, \ldots, z_k) \leq \prod_{j=1}^{k} H_j(z_j) \prod_{1 \leq i < j \leq k} r_{ij}(z_i, z_j);$$

this follows from Bonferroni-type inequalities applied to

$$F^n(a_n^{(1)} + b_n^{(1)} z_1, \ldots, a_n^{(k)} + b_n^{(k)} z_k).$$

Moreover, set k_0 equal to the integer part of the expression

$$\frac{2 \log \prod_{1 \leq i < j \leq k} r_{ij}(z_i, z_j)}{k}. \tag{53.69}$$

Takahashi (1987) has pointed out that for an extreme value distribution with $H_i = \Lambda$, $i = 1, \ldots, k$, a necessary and sufficient condition that

$$H(\boldsymbol{x}) = \Lambda(x_1) \cdots \Lambda(x_k) \tag{53.70}$$

for any $\boldsymbol{x} = (x_1, \ldots, x_k) \in \mathbb{R}^k$ is that

$$H(0, \ldots, 0) = \Lambda(0)^k. \tag{53.71}$$

Similarly, in the case $H_i = \Phi_{\alpha_i}$, $\alpha_i > 0$, $i = 1, \ldots, k$, a necessary and sufficient condition for

$$H(\boldsymbol{x}) = \prod_{i=1}^{k} \Phi_{\alpha_i}(x_i)$$

for any $\boldsymbol{x} \in \mathbb{R}^k$ is that

$$H(\mathbf{1}) = \prod_{i=1}^{k} \Phi_{\alpha_i}(1),$$

and in the case $H_i = \Psi_{\alpha_i}$, $\alpha_i > 0$, $i = 1, \ldots, k$, an analogous condition is

$$H(-\mathbf{1}) = \prod_{i=1}^{k} \Psi_{\alpha_i}(-1).$$

In an excellent survey paper, Joe (1994) has advocated a general approach to deriving multivariate exponential distributions based on a family of copulas

$$G(y_1, \ldots, y_k) = C(G_1(y_1), \ldots, G_k(y_k))$$

taking univariate marginals to be exponential with mean 1. Let the starting multivariate distribution be denoted by F, and let $\bar{F}$ denote its survival function. For a subset S of $\{1, \ldots, k\}$ with cardinality at least 2, let F_S and $\bar{F}_S$ denote the marginal distribution and the survival function based on the subset of random variables with indices in S. The limiting MEV distribution is

$$\lim_{n \to \infty} F^n(x_1 + \log n, \ldots, x_k + \log n), \tag{53.72}$$

provided that the limit exists; the linear transform $x + \log n$ comes from univariate extreme value theory. The univariate margins of the limiting distribution are the extreme value distributions, $H(x_j) = \exp\{-e^{-x_j}\}$.

Nadarajah (1999b) has used a similar idea and considered the copula function $C : [0,1]^3 \to [0,1]$ such that

$$C^k(x^{1/k}, y^{1/k}, z^{1/k}) = C(x, y, z) \qquad \text{for } (x, y, z) \in [0,1]^3, \; k > 0,$$

using which a trivariate extreme value distribution can be reaidly formed as

$$F(x_1, x_2, x_3) = C\left(F_A(x_1, x_2), F_B(x_1, x_3), F_C(x_2, x_3)\right), \; x_1, x_2, x_3 \geq 0,$$

where $F_A, F_B,$ and F_C are bivariate extreme value distributions.

In attempting to generalize the model of Tiago de Oliveira (1980, 1982), Dener and Sungur (1991) considered a general multivariate extreme value model of the form

$$Y_i = \max(X_i - \mu_i, U), \qquad i = 1, 2, \ldots, n,$$

where $\mu_i \in \mathbb{R}$ and $X_1, X_2, \ldots, X_n$ are independent and identically distributed Gumbel random variables and U is an independent (of X_i's) Gumbel random variable. These authors then discussed some basic properties of this model and studied the form of the dependence function and its relation to the correlation coefficients.

Finally, we conclude this chapter by referring to Chapter 6 of Joe (1996) in which special emphasis is assigned to various form of multivariate extreme value distributions, their properties, inferential issues, and to their applications. Some interesting data sets (pertaining to environmental problems) are analyzed by the author, using multivariate extreme value distributions.

BIBLIOGRAPHY

(Some bibliographical items not mentioned in the text are included here for completeness.)

Abramowitz, M., and Stegun, I. A. (eds.) (1965). *Handbook of Mathematical Functions with Formulas, Graphs, and Mathematical Tables*, New York: Dover Publications.

Berman, S. M. (1961/1962). Convergence to bivariate limiting extreme value distributions, *Annals of the Institute of Statistical Mathematics*, **13**, 217–233.

Capéraà, P., Fougères, A.-L., and Genest, C. (1996). A nonparametric estimation procedure for bivariate extreme value copulas, Preprint, Département de mathématiques et de statistique, Université Laval, Québec.

Coles, S. G., and Tawn, J. A. (1991). Modelling extreme multivariate events, *Journal of the Royal Statistical Society, Series B*, **52**, 377–392.

Coles, S. G., and Tawn, J. A. (1994). Statistical methods for multivariate extremes: an application to structural design, *Applied Statistics*, **43**, 1–48.

Dagsvik, J. (1988). Markov chains generated by maximizing components of multidimensional extremal processes, *Stochastic Processes and Applications*, **28**, 31–45.

de Haan, L., and de Ronde, J. (1998). Sea and wind: Multivariate extremes at work, *Extremes*, **1**, 7–45.

de Haan, L., and Resnick, S. I. (1977). Limit theory for multivariate sample extremes, *Zeit. Wahr. verw. Geb.*, **40**, 317–337.

Deheuvels, P. (1978). Caracterisation complete des lois extremes multivariees et de la convergence des types extremes, *Publications of the Institute of Statistics, Université de Paris*, **23**, 1–36.

Dener, A., and Sungur, E. (1991). Some new models for multivariate extremes, in *The Frontiers of Statistical Computation, Simulation, and Modeling* (P. R. Nelson, E. J. Dudewicz, A. Öztürk, and E. C. van der Meulen, eds.), pp. 185–196, Columbus, Ohio: American Sciences Press.

Esary, J. D., Proschan, F., and Walkap, D. W. (1967). Association of random variables with applications, *Annals of Mathematical Statistics*, **38**, 1466–1474.

Farlie, D. J. G. (1960). The performance of some correlation coefficients for a general bivariate distribution, *Biometrika*, **47**, 307–323.

Finkelshteyn, A. (1953). On the limiting distribution of the extreme terms of a variational series of a two dimensional random quantity, *Soviet Doklady Akad. Nauk SSSR*, **91**, 209–211.

Galambos, J. (1978). *The Asymptotic Theory of Extreme Order Statistics*, New York: John Wiley & Sons.

Galambos, J. (1985). A new bound on multivariate extreme value distributions, *Ann. Univ. Scient. Budapest, Rolando Eötvös, Nomin Sect. Math.*, **27**, 37–40.

Galambos, J. (1987). *The Asymptotic Theory of Extreme Order Statistics*, second edition, Malabar, Florida: Kreiger.

Geffroy, J. (1958, 1959). Contribution à la théorie des valeurs extrêmes, *Publications de l'Institut de Statistique de l'Université de Paris*, **7**, 37–121; **8**, 123–184.

Gomes, M. I., and Alperin, M. T. (1986). Inference in a multivariate generalized extreme value model—asymptotic properties of two test statistics, *Scandinavian Journal of Statistics*, **13**, 291–300.

Gumbel, E. J. (1958). Distributions à plusieurs variables dont les marges sont données (with remarks by M. Fréchet), *Computes Rendus de l'Académie des Sciences, Paris*, **246**, 2717–2720.

Gumbel, E. J. (1960). Bivariate exponential distributions, *Journal of the American Statistical Association*, **55**, 698–707.

Gumbel, E. J. (1961a). Multivariate exponential distributions, *Bulletin of the International Statistical Institute*, **39**, 469–475.

Gumbel, E. J. (1961b). Bivariate logistic distributions, *Journal of the American Statistical Association*, **56**, 335–349.

Gumbel, E. J. (1965). *Two Systems of Bivariate Extremal Distributions*, 35th Session of the International Statistical Institute, Beograd, No. 69.

Gumbel, E. J., and Goldstein, N. (1964). Analysis of empirical bivariate extremal distributions, *Journal of the American Statistical Association*, **59**, 794–816.

Gumbel, E. J., and Mustafi, C. K. (1967). Some analytical properties of bivariate extremal distributions, *Journal of the American Statistical Association*, **62**, 569–588.

Harris, R. (1966). *Reliability Applications of a Bivariate Exponential Distribution*, University of California Operations Research Center, ORC 66-36.

Hougaard, P. (1986). A class of multivariate failure time distributions, *Biometrika*, **73**, 671–678.

Hüsler, J., and Reiss, R.-D. (1989). Maxima of normal random vectors: between independence and complete dependence, *Statistics & Probability Letters*, **7**, 282–286.

Joe, H. (1990). Families of min-stable multivariate exponential and multivariate extreme value distributions, *Statistics & Probability Letters*, **9**, 75–81.

Joe, H. (1994). Multivariate extreme value distributions with applications to environmental data, *Canadian Journal of Statistics*, **22**, 47–64.

Joe, H. (1996). *Multivariate Models and Dependence Concepts*, London: Chapman & Hall.

Johnson, N. L., Kotz, S., and Balakrishnan, N. (1995). *Continuous Univariate Distributions*, Vol. 2, second edition, New York: John Wiley & Sons.

Konijn, H. S. (1957). On a class of two-dimensional random variables and distribution functions, *Sankhyā*, **18**, 167–172.

Konijn, H. S. (1959). Positive and negative dependence of two random variables, *Sankhyā*, **21**, 269–280.

Lee, L. (1979). Multivariate distributions having Weibull properties, *Journal of Multivariate Analysis*, **9**, 267–277.

Lehmann, E. L. (1966). On concepts of dependence, *Annals of Mathematical Statistics*, **37**, 1137–1153.

Marshall, A. W., and Olkin, I. (1983). Domains of attraction of multivariate extreme value distributions, *Annals of Probability*, **11**, 168–177.

McFadden, D. (1978). Modelling the choice of residential location, in *Spatial Interaction Theory and Planning Models* (A. Karlqvist, L. Lundquist, F. Snickers, and J. Weibull, eds.), pp. 75–96, Amsterdam: North-Holland.

Mole, N., and Jones, C. D. (1994). Concentration fluctuation data from dispersion experiments carried out in stable and unstable conditions, *Boundary-Layer Meteorology*, **67**, 41–47.

Nadarajah, S. (1999a). A new model for bivariate extreme value distributions, Preprint.

Nadarajah, S. (1999b). Multivariate extreme value models based on bivariate structures, Preprint.

Nadarajah, S. (1999c). Simulation of multivariate extreme values, Preprint.

Nadarajah, S., Anderson, C. W., and Tawn, J. A. (1998). Ordered multivariate extremes, *Journal of the Royal Statistical Society, Series B*, **60**, 473–496.

Oakes, D. (1989). Bivariate survival models induced by frailties, *Journal of the American Statistical Association*, **84**, 487–493.

Oakes, D., and Manatunga, A. K. (1992). Fisher information for a bivariate extreme value distribution, *Biometrika*, **79**, 827–832.

Pickands, J. (1981). Multivariate extreme value distributions, in *Proceedings of the 43rd Session of the International Statistical Institute, Buenos Aires*, pp. 859–878, Amsterdam: International Statistical Institute.

Posner, E. C., Rodemich, E. R., Ashlock, J. C., and Lurie, S. (1969). Application of an estimator of high efficiency in bivariate extreme value theory, *Journal of the American Statistical Association*, **64**, 1403–1414.

Shi, D. (1995a). Fisher information for a multivariate extreme value distribution, *Biometrika*, **82**, 644–649.

Shi, D. (1995b). Multivariate extreme value distribution and its Fisher information matrix, *Acta Mathematicae Applicatae Sinica*, **11**, 421–428.

Shi, D. (1995c). Moment estimation for multivariate extreme value distribution, *Applied Mathematics–JCU*, **10B**, 61–68.

Shi, D., Smith, R. L., and Coles, S. G. (1992). Joint versus marginal estimation for bivariate extremes, Technical Report No. 2074, Department of Statistics, University of North Carolina, Chapel Hill, NC.

Sibuya, M. (1960). Bivariate extreme statistics, I, *Annals of the Institute of Statistical Mathematics*, **19**, 195–210.

Smith, R. L. (1990). Extreme value theory, in *Handbook of Applicable Mathematics*, **7**, pp. 437–472, New York: John Wiley & Sons.

Smith, R. L., Tawn, J. A., and Yuen, H. K. (1990). Statistics of multivariate extremes, *International Statistical Review*, **58**, 47–58.

Takahashi, R. (1987). Some properties of multivariate extreme value distributions and multivariate tail equivalence, *Annals of the Institute of Statistical Mathematics*, **39**, 637–649.

Tawn, J. A. (1988). Bivariate extreme value theory: Models and estimation, *Biometrika*, **75**, 397–415.

Tawn, J. A. (1990). Modelling multivariate extreme value distributions, *Biometrika*, **77**, 245–253.

Thompson, W. J. (1997). *Atlas for Computing Mathematical Functions*, New York: John Wiley & Sons.

Tiago de Oliveira, J. (1958). Extremal distributions, *Revista da Faculdade de Ciencias, Lisboa, Series A*, **7**, 215–227.

Tiago de Oliveira, J. (1961). La représentation des distributions extrêmales bivariées, *Bulletin of the International Statistical Institute*, **33**, 477–480.

Tiago de Oliveira, J. (1962/1963). Structure theory of bivariate extremes; extensions, *Estudos de Matemática, Estaística e Econometria*, **7**, 165–195.

Tiago de Oliveira, J. (1964). L'indépendance dans les distributions extrêmales bivariées, *Publications de l'Institut de Statistique de l'Université de Paris*, **13**, 137–141.

Tiago de Oliveira, J. (1968). Extremal processes: Definitions and properties, *Publications de l'Institut de Statistique de l'Université de Paris*, **18**, 25–36.

Tiago de Oliveira, J. (1970). Biextremal distributions: Statistical decision, *Trabajos de Estadística*, **21**, 107–117.

Tiago de Oliveira, J. (1975). Bivariate extremes: extensions, *Bulletin of the International Statistical Institute, Proceedings of the 40th Session*, Warsaw, **46**, Book 2.

Tiago de Oliveira, J. (1980). Bivariate extreme; foundations and statistics, in *Multivariate Analysis-V* (P. R. Krishnaiah, ed.), pp. 349–366, Amsterdam: North-Holland.

Tiago de Oliveira, J. (1982). Decision and modelling for extremes, in *Some Recent Advances in Statistics*, pp. 101-110, London: Academic Press.

CHAPTER 54

Natural Exponential Families[1]

1 INTRODUCTION

This chapter presents some results on multivariate distributions belonging to exponential families. Different definitions of exponential families have been given in the literature—some involving few assumptions and some involving more assumptions on the statistical model; see Section 6 of Chapter 44 for some preliminary details. In particular, Carl Morris (1982), in a pioneering paper, made an important distinction between general and natural exponential families. Here, we will concentrate on the latter, as most of the basic statistical properties of exponential models can be derived from the *natural exponential families* (NEF) that can be associated with them. For an extensive discussion on NEF, we refer the readers to the *Lecture Notes* of Letac (1992); the books of Barndorff-Nielsen (1978) and Brown (1986) will similarly provide elaborate discussions on *general exponential families* (GEF).

Throughout this chapter, we will denote the differentiation of any function f defined on $\mathbb{R}^k$ with values in $\mathbb{R}$ by $f' : \boldsymbol{x} \to \left(\frac{\partial f}{\partial x_1}, \cdots, \frac{\partial f}{\partial x_k} \right)$. Similarly, we will denote a function $\boldsymbol{f}$ with values in $\mathbb{R}^d$ ($d > 1$) by $\boldsymbol{f}(\boldsymbol{x}) = (f_1(\boldsymbol{x}), \ldots, f_d(\boldsymbol{x}))$ and its derivative by $\boldsymbol{f}' = \left(\left(\frac{\partial f_i}{\partial x_j} \right) \right)_{i,j}$, $i = 1, \ldots, d$, $j = 1, \ldots, k$. The Greek letters $\mu, \nu, \ldots$ will represent measures on $\mathbb{R}^k$ or on any measurable set with their mass being not necessarily finite, and $d\boldsymbol{x}$ will denote the Lebesgue measure. Finally, P will be used to denote a

[1]This chapter has been prepared by Dr. Muriel Casalis of the Laboratory of Probability and Statistics, Université Paul Sabatier, Toulouse, France, and edited by the authors of this volume. Dr. Casalis expresses her thanks to Professor Gérard Letac and Samuel Kotz for their valuable suggestions and comments.

probability measure and m for its mean (note that μ is already used to denote a measure).

2 MULTIVARIATE NATURAL EXPONENTIAL FAMILIES

In Chapter 44, we described a family of distributions [proposed originally by Bildikar and Patil (1968)] as a class of multivariate exponential type distributions with joint density function of the form

$$f_{\boldsymbol{\theta}}(\boldsymbol{x}) = h(\boldsymbol{x}) \, e^{\boldsymbol{x}^T\boldsymbol{\theta} - q(\boldsymbol{\theta})} \tag{54.1}$$

with respect to a positive measure ν on $\mathbb{R}^k$. Here, $\boldsymbol{\theta} = (\theta_1, \ldots, \theta_k)^T$ represents a vector parameter and ν can be a Lebesgue measure, a counting measure, a combination of the two, or a transformation of such a measure by some mapping. This includes the discrete multivariate exponential families described in Chapter 34 by Johnson, Kotz, and Balakrishnan (1997). Note that the univariate natural exponential family is included as the case when $k = 1$. From (54.1), we also observe that the function $h(\boldsymbol{x})$ does not play any role in statistical inference relating to the vector parameter $\boldsymbol{\theta}$. For the purpose of simplification, let us write $h(\boldsymbol{x})d\nu(\boldsymbol{x})$ as $d\mu(\boldsymbol{x})$. For a rigorous definition of NEF, let us use $D(\mu)$ to denote the whole set of $\boldsymbol{\theta}$ for which the Laplace transform

$$L_\mu(\boldsymbol{\theta}) = \int e^{\boldsymbol{x}^T\boldsymbol{\theta}} d\mu(\boldsymbol{x}) \tag{54.2}$$

is finite, and use $\Theta(\mu)$ to denote the interior of the set $D(\mu)$.

Now, we impose the following two fundamental conditions on the measure μ:

(i) μ is not concentrated on an affine hyperplane of $\mathbb{R}^k$, i.e., on a set $\{\boldsymbol{x} = (x_1, \ldots, x_k)^T : \sum_{i=1}^k a_i x_i + b = 0\}$ for $(a_1, \ldots, a_k, b) \in \mathbb{R}^{k+1}$;

(ii) The open set $\Theta(\mu)$ of parameters is not empty. This condition is needed to avoid, for example, the case of the Cauchy distribution with density $\frac{dx}{\pi(1+x^2)}$ for which $\int e^{\theta x} \frac{dx}{\pi(1+x^2)} < \infty$ only for $\theta = 0$; see Chapter 16 of Johnson, Kotz, and Balakrishnan (1994).

Under these two conditions, the family of distributions $P_{\boldsymbol{\theta},\mu}$ with density function

$$e^{\boldsymbol{x}^T\boldsymbol{\theta} - q(\boldsymbol{\theta})} \tag{54.3}$$

with respect to μ, where $\boldsymbol{\theta}$ lies in $\boldsymbol{\Theta}(\mu)$, is called the *natural exponential family* (NEF) generated by μ; see Morris (1982). We will denote this family by $F(\mu)$. Occasionally, this family is also called a *standard exponential family* or a *linear exponential family*, with the latter emphasizing that $\boldsymbol{x}$ and $\boldsymbol{\theta}$ appear in (54.3) in a bilinear form, but not the above two conditions on the measure μ. The two conditions, however, specify the exact domain of the parameters and also that the underlying spaces for the observation $\boldsymbol{X}$ and for the parameters are the smallest ones having the same dimension. Barndorff-Nielsen (1978) and Brown (1986), therefore, describe this case as "minimal representation" for the exponential model. Efron (1978) and Morris (1982) were the first to identify this concept as the most appropriate one for the development of theory. The vector parameter $\boldsymbol{\theta}$ in (54.3) is referred to as the *canonical parameter*. Though some distributions in the literature involve a different parameter $\boldsymbol{\alpha}$ (say), if the transformation $\boldsymbol{\alpha} \to \boldsymbol{\theta}$ is a diffeomorphism, meaning that $\boldsymbol{\alpha} \to \boldsymbol{\theta}$ is one-to-one and differentiable and so also is the inverse function $\boldsymbol{\theta} \to \boldsymbol{\alpha}$, this only results in a superficial generalization from a theoretical point of view. If $\boldsymbol{\alpha}$ is a d-dimensional vector with $d < k$, however, the family becomes more complicated and corresponds to the so-called *curved exponential family* discussed, for example, by Barndorff-Nielsen and Cox (1994).

Let us now consider some specific examples of the NEF.

1. *Multivariate normal family* $N(\boldsymbol{m}, \boldsymbol{\Sigma})$ with fixed covariance matrix $\boldsymbol{\Sigma}$ and mean vector $\boldsymbol{m} \in \mathbb{R}^k$.

The density function with respect to the Lebesgue measure on $\mathbb{R}^k$ is (see Chapter 45)

$$\frac{1}{(2\pi)^{k/2} \, |\boldsymbol{\Sigma}|^{1/2}} \, \exp\left\{ -\frac{1}{2} (\boldsymbol{x} - \boldsymbol{m})^T \boldsymbol{\Sigma}^{-1} (\boldsymbol{x} - \boldsymbol{m}) \right\}. \tag{54.4}$$

The set $\{ N(\boldsymbol{m}, \boldsymbol{\Sigma}) : \boldsymbol{m} \in \mathbb{R}^k \}$ is a NEF with

$$
\begin{aligned}
d\mu(\boldsymbol{x}) &= \frac{1}{(2\pi)^{k/2} \, |\boldsymbol{\Sigma}|^{1/2}} \, \exp\left\{ -\frac{1}{2} \boldsymbol{x}^T \boldsymbol{\Sigma}^{-1} \boldsymbol{x} \right\} d\boldsymbol{x}, \\
\boldsymbol{\theta}^T &= \boldsymbol{\Sigma}^{-1} \boldsymbol{m},
\end{aligned}
$$

and

$$q(\boldsymbol{\theta}) = \frac{1}{2} \, \boldsymbol{m}^T \boldsymbol{\Sigma}^{-1} \boldsymbol{m} = \frac{1}{2} \boldsymbol{\theta}^T \boldsymbol{\Sigma} \boldsymbol{\theta}.$$

2. *Discrete multivariate power series distributions family.*

The joint probability mass function is of the form

$$\frac{1}{A(\boldsymbol{\beta})} \, a(\boldsymbol{n}) \prod_{i=1}^{k} \beta_i^{n_i}$$

with $\boldsymbol{\beta} \in \mathbb{R}^k$ and $\boldsymbol{n} = (n_1, \ldots, n_k) \in \mathbb{N}^k$; see Chapter 38 of Johnson, Kotz, and Balakrishnan (1997). They form a NEF with

$$\mu = \sum_{\boldsymbol{n} \in \mathbb{N}^k} a(\boldsymbol{n}) \delta_{\boldsymbol{n}},$$

$$\boldsymbol{\theta} = (\log \beta_1, \ldots, \log \beta_k),$$

and

$$q(\boldsymbol{\theta}) = \log A(\boldsymbol{\beta}) = \log A\left(e^{\theta_1}, \ldots, e^{\theta_k}\right)$$

with suitable conditions on the real numbers $a(\boldsymbol{n})$ to satisfy the two conditions stated earlier; $\delta_{\boldsymbol{n}}$ is the indicator function.

3. *Wishart distributions.*

Recall first that the family of gamma distributions, with fixed shape parameter $\alpha > 0$ and scale parameter σ describing $(0, \infty)$, includes distributions on $\mathbb{R}$ with density function

$$\frac{1}{\Gamma(\alpha)\sigma^\alpha} \, e^{-x/\sigma} \, x^{\alpha-1} I_{(0,\infty)}(x),$$

where $I_{(0,\infty)}(x)$ denotes an indicator function taking on the value 1 if $x \in (0, \infty)$ and 0 otherwise; see Chapter 17 of Johnson, Kotz, and Balakrishnan (1994). In this case,

$$d\mu(x) = \frac{1}{\Gamma(\alpha)} \, x^{\alpha-1} I_{(0,\infty)}(x) \, dx,$$

$$\theta = -1/\sigma,$$

and

$$q(\theta) = \alpha \log \sigma = -\alpha \log(-\theta).$$

The multivariate version of this gamma family is, in fact, the family of Wishart distributions on the space $\boldsymbol{V}$ of real symmetric $k \times k$ matrices $\boldsymbol{x} = ((x_{ij}))$. Note that the dimension of $\boldsymbol{V}$ is $N = k(k+1)/2$, and a matrix $\boldsymbol{x}$ in this case may be considered as a vector $(x_{11}, x_{12}, \ldots, x_{1k}, x_{22}, \ldots, x_{2k}, \ldots,$

x_{kk}). If V_+ denotes the set of positive definite matrices of V, the Wishart distribution $W_k(n, \Sigma)$ (for $n \geq k$) has the following density function with respect to the Lebesgue measure $dx = \prod_{1 \leq i \leq j \leq k} dx_{ij}$:

$$f_n(x; \Sigma) = \frac{1}{C_k(n)|\Sigma|^{n/2}} \exp\left\{-\frac{1}{2} \text{trace}(\Sigma^{-1}x)\right\} |x|^{(n-1-k)/2} I_{V_+}(x),$$

(54.5)

where $C_k(n)$ is a normalizing constant.

Let us fix $n \geq k$. Then, we get a NEF $F_{n/2}$ when Σ describes the set V_+ (the domain of parameters) with

$$d\mu(x) = \frac{1}{C_k(n)} |x|^{(n-1-k)/2} I_{V_+}(x) \, dx,$$

$$\theta = -\frac{1}{2} \Sigma^{-1} \text{ so that } -\frac{1}{2} \text{trace}(\Sigma^{-1}x) = \sum_{i \leq j} \theta_{ij} x_{ij},$$

and

$$q(\theta) = \frac{n}{2} \log |\Sigma| = -\frac{n}{2} \log |-2\theta|;$$

see, for example, Letac (1989b) and Casalis (1990) for a detailed discussion on this family.

4. *Bivariate Poisson distributions.*

The class of bivariate Poisson distributions introduced by Holgate (1964) as the joint distribution of $X_1 = Y_1 + Y_{12}$ and $X_2 = Y_2 + Y_{12}$, with Y_1, Y_2, and Y_{12} being independently distributed as Poisson, does not form a NEF; see Chapter 37 of Johnson, Kotz, and Balakrishnan (1997) for more details on these distributions. While the distribution of the random vector $Y = (Y_1, Y_2, Y_{12})$ is a NEF on $\mathbb{R}^3$, this is not the case with the distribution of its projection $X = (X_1, X_2)$; see Casalis (1997). The problem of the projection of a NEF has been discussed in detail by Bar-Lev *et al.* (1994); see also Section 7.2.

Some general observations on the definition of NEF in (54.3) may be made as follows.

(i) Due to the structure of the density function in (54.3), the generating measure μ is not unique. As an example, on $\mathbb{R}$, both $I_{(0,\infty)}(x) \, dx$ and $e^{-x} I_{(0,\infty)}(x) \, dx$ generate the NEF of exponential distributions given by

$$F = \left\{\frac{1}{\sigma} e^{-x/\sigma} I_{(0,\infty)}(x) \, dx; \ \sigma > 0\right\}.$$

As a matter of fact, in more general terms, $F(\mu) = F(\mu')$ holds iff there exists $(\boldsymbol{a}, b)$ in $\mathbb{R}^k \times \mathbb{R}$ such that

$$d\mu'(\boldsymbol{x}) = e^{\boldsymbol{x}^T \boldsymbol{a} + b} \, d\mu(\boldsymbol{x}). \tag{54.6}$$

Observe that when (54.6) holds, we have

$$\Theta(\mu') = \Theta(\mu) - a = \{\boldsymbol{\theta} : \boldsymbol{\theta} + a \in \Theta(\mu)\}.$$

Therefore, a change of generating measure induces a translation on the set of parameters. In view of (54.3) and (54.6), we note that any distribution $P_{\boldsymbol{\theta}_0}$ in a NEF F (with $\boldsymbol{\theta}_0$ fixed) generates the family as well.

(ii) The regularity property of the Laplace transform $\boldsymbol{\theta} \to L_\mu(\boldsymbol{\theta})$ defined on $\Theta(\mu)$ by (54.2), and hence that of the function $q : \boldsymbol{\theta} \to \log L_\mu(\boldsymbol{\theta})$ appearing in the density in (54.3) (called the log-Laplace transform), will enable us to get the cumulants of the random variable $\boldsymbol{X}$ distributed as $P_{\boldsymbol{\theta},\mu}$. In fact, the moment-generating function of $\boldsymbol{X}$ is

$$\begin{aligned} E[e^{\boldsymbol{t}^T \boldsymbol{X}}] &= e^{-q(\boldsymbol{\theta})} \int e^{\boldsymbol{x}^T (\boldsymbol{\theta} + \boldsymbol{t})} \, d\mu(\boldsymbol{x}) \\ &= L_\mu(\boldsymbol{\theta} + \boldsymbol{t}) / L_\mu(\boldsymbol{\theta}) \\ &= e^{q(\boldsymbol{\theta} + \boldsymbol{t}) - q(\boldsymbol{\theta})}, \end{aligned}$$

and so the cumulant generating function is

$$K_{\boldsymbol{X}}(\boldsymbol{t}) = q(\boldsymbol{\theta} + \boldsymbol{t}) - q(\boldsymbol{\theta}).$$

By successive differentiations with respect to $t_1, \ldots, t_k$, we obtain the following recurrence relation between the cumulants

$$\kappa_{r_1, \ldots, r_{i-1}, r_i+1, r_{i+1}, \ldots, r_k} = \frac{\partial}{\partial \theta_i} \, \kappa_{r_1, \ldots, r_k};$$

see Eq. (44.41) in Chapter 44. In particular, we obtain

$$\frac{\partial}{\partial \theta_i} \, q(\boldsymbol{\theta}) = \int x_i \, dP_{\boldsymbol{\theta},\mu}(\boldsymbol{x}) = m_i, \qquad i = 1, \ldots, k, \tag{54.7}$$

and

$$\frac{\partial^2}{\partial \theta_i \partial \theta_j} \, q(\boldsymbol{\theta}) = \int (x_i - m_i)(x_j - m_j) \, dP_{\boldsymbol{\theta},\mu}(\boldsymbol{x}) = \text{cov}(X_i, X_j),$$

$$i, j = 1, \ldots, k. \tag{54.8}$$

The first assumption (of the two stated earlier) on the generating measure implies that the variance-covariance matrix

$$\text{Var}(\boldsymbol{X}) = E[(\boldsymbol{X} - \boldsymbol{m})(\boldsymbol{X} - \boldsymbol{m})^T]$$

is positive definite.

(iii) The distributions $P_{\boldsymbol{\theta},\mu}$ corresponding to $\boldsymbol{\theta}$ being on the boundary of $D(\mu)$ have been deliberately excluded from $F(\mu)$ in order to assure the existence of all the moments of $\boldsymbol{X}$. The whole family $\bar{F}(\mu) = \{P_{\boldsymbol{\theta},\mu} : \boldsymbol{\theta} \in D(\mu)\}$ is often called the *full natural exponential family*; see, for example, Barndorff-Nielsen (1978). When $D(\mu)$ is open, i.e., $D(\mu) = \Theta(\mu)$, the family is said to be *regular*.

3 MULTIVARIATE GENERAL EXPONENTIAL FAMILIES

The *general exponential families* GEF are families of distributions $Q_{\boldsymbol{\theta}}$ defined on a measurable space Ω with densities

$$p(\omega; \boldsymbol{\theta}) = e^{\boldsymbol{\theta}^T \boldsymbol{t}(\omega) - q(\boldsymbol{\theta})} \tag{54.9}$$

with respect to some positive measure ν on Ω. Here, $\boldsymbol{\theta}$ and $\boldsymbol{t} = \boldsymbol{t}(\omega)$ are vectors of dimension k, and $\boldsymbol{t}$ is a measurable map from Ω into $\mathbb{R}^k$. We also require that the image of ν by the map $\boldsymbol{t}$—namely, $\mu = \boldsymbol{t}(\nu)$, satisfies the two basic assumptions stated in the last section. In that case, it is possible to introduce the NEF $F(\mu)$ and to work directly with it. Note that all information about the GEF with regard to estimation of the parameter $\boldsymbol{\theta}$ can be obtained from $F(\mu)$, with statistic $\boldsymbol{t}$ being sufficient for $\boldsymbol{\theta}$; see Barndorff-Nielsen (1978) and Brown (1986). Such a family is called an *associated NEF* of the GEF under consideration. Martin-Löf refers to μ as the *structure measure*; see Sundberg (1974).

Note, however, that this construction is not unique; a change of the canonical parameter results in a change of μ and $F(\mu)$. But, it can be shown that the only feasible changes are affine transformations on $\boldsymbol{\theta}$ and on $\boldsymbol{t}(\omega)$. This means that the associated NEFs of a GEF are all linked by an affine mapping $\boldsymbol{x} \to \boldsymbol{A}\boldsymbol{x} + \boldsymbol{b}$, where $\boldsymbol{A}$ is a nonsingular $k \times k$ matrix and $\boldsymbol{b}$ is a vector in $\mathbb{R}^k$; see, for example, Lemma 8.1 of Barndorff-Nielsen (1978).

The following example, taken from Letac (1992), illustrates the difference between NEF and GEF. Consider the family of normal distributions on $\mathbb{R}$. Assume that the mean m is unknown while the variance σ^2 is known. We then have a NEF

$$F = \{N(m, \sigma^2) : m \in \mathbb{R}\}$$

generated by the measure $d\mu(x) = \frac{1}{\sqrt{2\pi}\,\sigma} e^{-x^2/2\sigma^2} dx$ with canonical parameter $\theta = m/\sigma^2$; see (54.4) with $k = 1$. Next, let us assume that both mean and variance are unknown, so that the parameter vector becomes of dimension 2. We then obtain a GEF with Ω as $\mathbb{R}$, ν as $\frac{1}{\sqrt{2\pi}} I_\Omega(\omega)\, d\omega$, and $p(\omega; \boldsymbol{\theta}) = \frac{1}{\sigma} \exp\left\{-\frac{\omega^2}{2\sigma^2} + \frac{m\omega}{\sigma^2} - \frac{m^2}{2\sigma^2}\right\}$ with possibly $\theta_1 = \frac{m}{\sigma^2}$, $\theta_2 = \frac{1}{\sigma^2}$, and $\boldsymbol{t}(\omega) = (\omega, -\omega^2/2)$. Finally, defining μ as the image of $\frac{1}{\sqrt{2\pi}}\, d\omega$ by the mapping $\boldsymbol{t}$, we obtain a measure concentrated on the hyperbola $y = -x^2/2$ of $\mathbb{R}^2$. Note that though μ is singular, it is not concentrated on a straight line. Thus, $F(\mu)$ is a NEF on $\mathbb{R}^2$ and is given by

$$F(\mu) = \left\{ dP_{\boldsymbol{\theta}}(x, y) = e^{\theta_1 x + \theta_2 y - q(\theta_1, \theta_2)} : \theta_1 \in \mathbb{R},\ \theta_2 > 0 \right\}, \quad (54.10)$$

where

$$q(\theta_1, \theta_2) = \frac{m^2}{2\sigma^2} + \frac{1}{2} \log \sigma^2 = \frac{\theta_1^2}{2\theta_2} - \frac{1}{2} \log \theta_2.$$

Another example of GEF is given by beta distributions of the first type with density function [see chapter 25 of Johnson, Kotz, and Balakrishnan (1995)]

$$d\beta_{p,q}(\omega) = \frac{1}{B(p, q)}\, \omega^{p-1}(1 - \omega)^{q-1} I_{(0,1)}(\omega)\, d\omega,$$

where p and q are positive, and $B(p, q)$ denotes the complete beta function. Here,

$$\Omega = (0, 1), \qquad d\nu(\omega) = \frac{1}{\omega(1 - \omega)} I_{(0,1)}(\omega)\, d\omega,$$

$$\theta_1 = p, \qquad\qquad \theta_2 = q,$$

and

$$\boldsymbol{t}(\omega) = (\log \omega, \log(1 - \omega)).$$

The corresponding NEF will consist of the probability measures

$$dP_{\boldsymbol{\theta}}(x, y) = e^{\theta_1 x + \theta_2 y - q(\theta_1, \theta_2)},$$

where

$$q(\theta_1, \theta_2) = \log B(\theta_1, \theta_2) = \log B(p, q) \quad \text{for } (\theta_1, \theta_2) \in (0, \infty)^2.$$

As we already mentioned in the case of NEF, it is possible here, too, to generalize the definition a little bit more by introducing a different parameterization. In practice, however, the situation may demand more complicated models than this. For this reason, some authors have introduced *partly exponential models* for which the model may be written in the form

$$e^{\boldsymbol{\theta}^T \boldsymbol{x} - q\boldsymbol{\lambda}(\boldsymbol{\theta})} \mu_{\boldsymbol{\lambda}}(d\boldsymbol{x}),$$

where $\mu_{\boldsymbol{\lambda}}$ is a measure depending on a parameter vector $\boldsymbol{\lambda}$; see, for example, Zhao, Prentice, and Self (1992). In modeling situations, one can then impose desirable properties through the measure $\mu_{\boldsymbol{\lambda}}$. Such models include amongst them the NEFs (with $\boldsymbol{\lambda}$ known), the generalized linear models, and also the exponential dispersion models discussed by Jørgensen (1987, 1997); see also Section 54.6. Another way to fit a general model to data is to construct *special exponential families* using computers, without worrying about mathematical tractability. Efron and Tibshirani (1996) have carried out different inferential procedures on such families, including maximum likelihood estimation and density estimation.

4   PARAMETERIZATION BY THE MEAN AND STEEPNESS

We shall now introduce a new parametrization of NEF which is more intrinsic and is independent of the choice of the generating measure μ. As will be seen in the following section, this new parametrization of NEF will enable us to express the covariance matrix of a distribution belonging to the family in terms of the mean of the distribution and also to characterize the model in its entirety.

Let $F = F(\mu)$ be a NEF. Recall that q is the log-Laplace transform of μ. We can then show that, under the conditions stated earlier in Section 54.2, the first derivative of q, namely, $q' = \left(\frac{\partial q}{\partial \theta_1}, \cdots, \frac{\partial q}{\partial \theta_k}\right)$, is an infinitely differentiable diffeomorphism from $\Theta(\mu)$ onto its image set $q'(\Theta(\mu))$. From (54.7), we readily observe that $q'(\boldsymbol{\theta})$ is the mean vector of the distribution $P_{\boldsymbol{\theta},\mu}$ so that $q'(\Theta(\mu))$ is simply the *domain of the means* of the family. Clearly, this domain does not depend on μ since the underlying measure has no relevance to the mean, viewed as a parameter, unlike the canonical parameter. We shall denote it by M_F and use $P_{\boldsymbol{m},F}$ in place of $P_{\boldsymbol{\theta},\mu}$ under this parameterization.

Since M_F is the image of $\Theta(\mu)$ by the regular mapping q, M_F has attractive topological properties. By definition $\Theta(\mu)$ is open and may also be verified to be convex; in fact, it is without holes and connected (two points are always linked by a line which, of course, is a straight line in the case of a convex set). Likewise, M_F is also open, without holes and connected; however, M_F is not necessarily convex. Efron (1978) gave an example and another example presented by Del Castillo (1994) uses the singly truncated normal distribution [see Chapter 13 of Johnson, Kotz, and Balakrishnan (1994)]. Let Y, a normal $N(m, \sigma^2)$ variable, be restrained to be nonnegative. Then the distribution of X belongs to the GEF

$$\left\{ dP_{\boldsymbol{\theta}}(x) = \frac{1}{C(m, \sigma^2)} \, e^{-(x-m)^2/2\sigma^2} I_{(0,\infty)}(x) \, dx : m \in \mathbb{R}, \ \sigma > 0 \right\},$$

where $\theta_1 = m/\sigma^2$ and $\theta_2 = 1/\sigma^2$ as in (54.10), and $C(m, \sigma^2)$ is a normalizing constant. If μ is the image measure of $I_{(0,\infty)}(x) \, dx$ by the mapping $t : x \to (x, -x^2/2)$, Del Castillo (1994) considered the associated NEF

$$F(\mu) = \left\{ dP_{\boldsymbol{\theta}}(x, y) = e^{\theta_1 x + \theta_2 y - q(\theta_1, \theta_2)} d\mu(x, y) : \theta_1 \in \mathbb{R}, \ \theta_2 > 0 \right\}$$

which satisfies

$$M_{F(\mu)} = \left\{ (x, y) \in \mathbb{R}^2 : x > 0, \ -x^2 < y < -x^2/2 \right\}.$$

Observe that the support S of the measure $P_{\boldsymbol{\theta}}$ is the parabola $y = -x^2/2$ so that the convex hull of S (which is the smallest closed convex set containing S) is the area $\bar{C}$ under the parabola given by $\bar{C} = \{(x, y) : x \geq 0, \ y \leq -x^2/2\}$. Here, M_F is strictly included in the interior C of $\bar{C}$. The interior C here is not to be confused with the normalizing constant $C(m, \sigma^2)$.

More generally, since M_F is the domain of the means of $P_{\boldsymbol{\theta}}$, it is always included in C. However, $M_F = C$ is not always true. We will, therefore, say that a NEF is *steep* if $M_F = C$. Barndorff-Nielsen (1978) has given an equivalent formal definition as follows:

F is steep iff for any $\boldsymbol{\theta}_0 = (\theta_{01}, \ldots, \theta_{0k})$ on the boundary $D(\mu) \backslash \Theta(\mu)$ and any $\boldsymbol{\theta} \in \Theta(\mu)$, we have

$$\lim_{\lambda \searrow 0} \sum_{i=1}^{k} (\theta_{0i} - \theta_i) \frac{\partial q}{\partial \theta_i} \left((1 - \lambda) \boldsymbol{\theta}_0 + \lambda \boldsymbol{\theta} \right) = +\infty. \qquad (54.11)$$

In other words, the mapping q has an infinite slope when $\boldsymbol{\theta}$ moves closer to $\boldsymbol{\theta}_0$ on a straight line.

It should be noted that when $D(\mu)$ is open—that is, when $F(\mu)$ is *regular*, the condition (54.11) is void and is automatically satisfied, and therefore a regular NEF is always steep.

The steepness assumption provides a mixed parametrization of the model in terms of $(\boldsymbol{m}^{(1)}, \boldsymbol{\theta}^{(2)})$ when $\boldsymbol{\theta} = (\boldsymbol{\theta}^{(1)}, \boldsymbol{\theta}^{(2)})$ and $\boldsymbol{m} = (\boldsymbol{m}^{(1)}, \boldsymbol{m}^{(2)})$ with $\boldsymbol{\theta}^{(1)}, \boldsymbol{m}^{(1)} \in \mathbb{R}^d$ and $\boldsymbol{\theta}^{(2)}, \boldsymbol{m}^{(2)} \in \mathbb{R}^{k-d}$. In that case, $\boldsymbol{m}^{(1)}$ and $\boldsymbol{\theta}^{(2)}$ are in variation-independent; that is, they lie in the product of subsets $M_1 \times \boldsymbol{\Theta}_2(\mu)$, where M_1 is the projection of M_F by $\boldsymbol{x} \rightarrow \boldsymbol{x}^{(1)}$ and $\boldsymbol{\Theta}_2(\mu)$ is the projection of $\boldsymbol{\Theta}(\mu)$ by $\boldsymbol{\theta} \rightarrow \boldsymbol{\theta}^{(2)}$; see Lemma 3.1 of Barndorff-Nielsen and Blaesild (1983). Such mixed parameterizations are of particular interest in connection with the notion of *cuts* arising in problems of inferential separation; see Section 7 for details.

5 VARIANCE FUNCTION

The *variance function*, which is an important concept for NEF, is the mapping

$$
\begin{aligned}
V_F : \quad & M_F \rightarrow \mathcal{S}_+ \\
& \boldsymbol{m} \rightarrow V_F(\boldsymbol{m}) = \left(\left(\int (x_i - m_i)(x_j - m_j) \, dP_{\boldsymbol{m},F}(\boldsymbol{x}) \right) \right)_{i,j},
\end{aligned}
$$

defined on M_F with values in the subset $\mathcal{S}_+$ of positive definite symmetric real $k \times k$ matrices; V_F associates to each $\boldsymbol{m}$ the covariance matrix of the distribution $P_{\boldsymbol{m},F}$. This mapping can be a constant as is the case with the multivariate normal NEF. Statistically, these distributions are intrinsically different from the others where V_F depends on $\boldsymbol{m}$—for example, linearly in the case of Poisson distributions and quadratically in the case of binomial and negative binomial distributions; see Section 8 for details.

From (54.8), we obtain

$$
V_F(\boldsymbol{m}) = q''(\boldsymbol{\theta}) = \left(\left(\frac{\partial^2 q(\boldsymbol{\theta})}{\partial \theta_i \partial \theta_j} \right) \right)_{i,j}, \tag{54.12}
$$

where $\boldsymbol{m} = q'(\boldsymbol{\theta})$. Denoting the inverse function of q' by $\boldsymbol{\psi}_\mu$, (54.12) becomes

$$
V_F(\boldsymbol{m}) = q'' \left(\boldsymbol{\psi}_\mu(\boldsymbol{m}) \right).
$$

An alternate way to compute $V_F(\boldsymbol{m})$ is to differentiate the equation $q'(\boldsymbol{\psi}_\mu(\boldsymbol{m})) = \boldsymbol{m}$ and obtain the relation

$$
V_F(\boldsymbol{m}) = (\boldsymbol{\psi}'_\mu(\boldsymbol{m}))^{-1}. \tag{54.13}
$$

(54.13) simply shows that $V_F(\boldsymbol{m})$ is the inverse matrix of

$$\boldsymbol{\psi}'_\mu(\boldsymbol{m}) = \left(\left(\frac{\partial \psi_i}{\partial m_j}\right)\right)_{i,j} = \left(\left(\frac{\partial \theta_i}{\partial m_j}\right)\right)_{i,j}, \quad i,j = 1,\ldots,k$$

if $\boldsymbol{\theta} = \boldsymbol{\psi}_\mu(\boldsymbol{m}) = (\psi_1(\boldsymbol{m}),\ldots,\psi_k(\boldsymbol{m}))^T$.

The usefulness of the variance function V_F lies in the fact that it fully characterizes the family F. If F_1 and F_2 are two NEFs such that V_{F_1} and V_{F_2} coincide on a nonempty open set, then $F_1 = F_2$. This result follows immediately upon solving the system of ordinary differential equations in (54.12); see Morris (1982) and Letac (1992) for a detailed proof. This is also the essence of Theorem 2.1 of Jani and Singh (1996) who have dealt with the general framework of GEF. These authors have actually introduced the variance function of an associated NEF to characterize the density functions of the GEF through moments, but without mentioning this terminology.

Casalis and Letac (1996) have used the variance function V_F to characterize Wishart distributions. This is similar to the following result for gamma distributions: If U and V are independent nonnegative random variables such that $U + V$ is almost surely positive, then $U + V$ and $U/(U + V)$ are independent iff U and V are distributed as gamma with the same shape parameter; see Chapter 17 of Johnson, Kotz, and Balakrishnan (1994). The variance function V_F also provides a tool for some theoretical characterizations of NEFs; see Sections 7.2, 7.3 and 8.3.

It should also be mentioned that for certain common distributions in $\mathbb{R}$ and $\mathbb{R}^k$, the variance function V_F is quite simple. This prompted Morris (1982) to classify all NEFs on $\mathbb{R}$ with V_F as a polynomial in m of degree at most 2. This classification led to various extensions as will be seen later in Section 8.

6   AFFINE TRANSFORMATIONS AND CONVOLUTION: THE EXPONENTIAL DISPERSION MODEL

It has been mentioned earlier that the only transformations that conserve the structure of NEF are affine transformations $\varphi : \boldsymbol{x} \to \boldsymbol{A}\boldsymbol{x} + \boldsymbol{b}$ of $\mathbb{R}^k$, where $\boldsymbol{A}$ is a nondegenerate matrix and $\boldsymbol{b}$ is a vector of $\mathbb{R}^k$. Such an affine transformation transforms the NEF $F = F(\mu)$ into the NEF

$\varphi(F) = F(\varphi(\mu))$, where $\varphi(\mu)$ denotes the image of μ by φ. Some simple algebraic calculations yield the following relations:

$$
\begin{array}{rl}
\text{(i)} & \Theta(\varphi(\mu)) = \boldsymbol{A}^T(\Theta(\mu)) = \left\{ \boldsymbol{A}^T\boldsymbol{\theta} : \boldsymbol{\theta} \in \Theta(\mu) \right\}, \\[4pt]
\text{(ii)} & q_{\varphi(\mu)}(\boldsymbol{\theta}) = q_\mu(\boldsymbol{A}^T\boldsymbol{\theta}) + \boldsymbol{b}^T\boldsymbol{\theta}, \quad \boldsymbol{\theta} \in \Theta(\varphi(\mu)), \\[4pt]
\text{(iii)} & \varphi(P_{\boldsymbol{\theta},\mu}) = P_{\boldsymbol{A}^T\boldsymbol{\theta},\varphi(\mu)}, \\[4pt]
\text{(iv)} & M_{\varphi(F)} = \varphi(M_F) = \{\varphi(\boldsymbol{m}) : \boldsymbol{m} \in M_F\}, \text{ and} \\[4pt]
& V_{\varphi(F)}(\tilde{\boldsymbol{m}}) = \boldsymbol{A}V_F(\varphi^{-1}(\tilde{\boldsymbol{m}}))\boldsymbol{A}^T, \quad \tilde{\boldsymbol{m}} \in \varphi(M_F). \quad (54.14)
\end{array}
$$

A NEF is said to be *invariant* under a group of transformations if $\varphi(F) = F$ for any element φ of the group. In that case, F is called an *exponential transformation model*. Barndorff-Nielsen, Blaesild, and Eriksen (1989) discussed such NEFs, and various examples have been considered by Lloyd (1988), Letac (1989b), and Casalis (1990). For example, the family of Fisher–von Mises distributions generated by the uniform measure on the sphere of $\mathbb{R}^k$ is invariant under the group of orthogonal transformations $O(\mathbb{R}^k)$. Another example is the family of normal distributions $\{N(\boldsymbol{m}, \boldsymbol{\Sigma}) : \boldsymbol{m} \in \mathbb{R}^k\}$ which is the only NEF invariant under the group of translations; see Chapter 45.

Convolution is another important operation to consider on NEF. Specifically, let $F(\mu)$ be a NEF and $\boldsymbol{X}_1, \boldsymbol{X}_2, \ldots, \boldsymbol{X}_n$ be n independent random variables distributed as $P_{\boldsymbol{\theta},\mu}$. Then, the distribution of $\boldsymbol{S}_n = \sum_{i=1}^n \boldsymbol{X}_i$ is the convolution $(P_{\boldsymbol{\theta},\mu})^{*n}$ and is still exponential of the form $P_{\boldsymbol{\theta},(\mu)^{*n}}$, where $(\mu)^{*n}$ is the image of the product $\prod_{i=1}^n d\mu(\boldsymbol{x}_i)$ by $(\boldsymbol{x}_1, \ldots, \boldsymbol{x}_n) \to \sum_{i=1}^n \boldsymbol{x}_i = s(\boldsymbol{x}_1, \ldots, \boldsymbol{x}_n)$. In fact, $(\boldsymbol{X}_1, \ldots, \boldsymbol{X}_n)$ has the following density with respect to $\prod_{i=1}^n d\mu(\boldsymbol{x}_i)$:

$$
\exp\left\{ \sum_{i=1}^n \boldsymbol{x}_i^T\boldsymbol{\theta} - nq_\mu(\boldsymbol{\theta}) \right\} = \exp\left\{ (s(\boldsymbol{x}_1, \ldots, \boldsymbol{x}_n))^T\boldsymbol{\theta} - nq_\mu(\boldsymbol{\theta}) \right\}. \quad (54.15)
$$

If μ satisfies the conditions stated earlier in Section 54.2, so does $(\mu)^{*n}$. Furthermore, (54.15) shows that

$$
q_{(\mu)^{*n}}(\boldsymbol{\theta}) = nq_\mu(\boldsymbol{\theta}). \quad (54.16)
$$

The NEF $F(\mu^{*n})$ is called the nth *power* or the nth *convolution* of $F(\mu)$.

This construction can be generalized in the following manner. Let $\Lambda(\mu)$ be the set of all nonnegative real numbers λ for which there exists a suitable measure μ_λ such that

$$
L_{\mu_\lambda}(\boldsymbol{\theta}) = (L_\mu(\boldsymbol{\theta}))^\lambda, \quad \boldsymbol{\theta} \in \Theta(\mu), \quad (54.17)
$$

or, equivalent to (54.16), such that

$$q_{\mu_\lambda}(\boldsymbol{\theta}) = \lambda q_\mu(\boldsymbol{\theta}).$$

Then, the NEF $F_\lambda = F(\mu_\lambda)$ is called the λth *power of F* and is characterized by the following relations due to (54.12) and (54.17):

$$
\begin{aligned}
M_{F_\lambda} &= \lambda M_F, \\
V_{F_\lambda}(\boldsymbol{m}) &= \lambda V_F\left(\frac{\boldsymbol{m}}{\lambda}\right), \qquad \boldsymbol{m} \in M_{F_\lambda}.
\end{aligned}
$$

The parameter λ here is sometimes referred to as the Jørgensen parameter and $\Lambda_F = \Lambda(\mu)$ is referred to as the *Jørgensen set* of F since Jørgensen (1987) emphasised their importance in the study of exponential dispersion models. In that framework, however, λ is called the index or the precision parameter and $\sigma^2 = 1/\lambda$ is called the dispersion parameter. Jørgensen defined an exponential dispersion model as the multivariate generalization of the error distribution of Nelder and Wedderburn's (1972) generalized linear models (GLIMs). To be specific, let $F = F(\mu)$ be a NEF and let $\boldsymbol{X}$ be a random variable in $\mathbb{R}^k$ distributed as $P_{\boldsymbol{\theta},\mu_\lambda}$. Let Q_λ denote the image of μ_λ by $\boldsymbol{x} \to \boldsymbol{x}/\lambda$, and let $\boldsymbol{Y} = \boldsymbol{X}/\lambda$. Then, $\boldsymbol{Y}$ has the distribution

$$dQ_{\boldsymbol{\theta},\lambda}(\boldsymbol{y}) = e^{\lambda(\boldsymbol{\theta}^T \boldsymbol{y} - q_\mu(\boldsymbol{\theta}))} \, dQ_\lambda(\boldsymbol{y}).$$

If λ is an integer n, $Q_{\boldsymbol{\theta},\lambda}$ is in fact the distribution of the empirical mean $\bar{\boldsymbol{X}} = \frac{1}{n}\sum_{i=1}^n \boldsymbol{X}_i$ where $\boldsymbol{X}_1, \ldots, \boldsymbol{X}_n$ are independent random variables distributed as $P_{\boldsymbol{\theta},\mu}$. The model $G = \{Q_{\boldsymbol{\theta},\lambda} : \boldsymbol{\theta} \in \Theta(\mu), \ \lambda \in \Lambda\}$ is called the *exponential dispersion model* and is the union of exponential families G_λ which are characterized by the following means domain and variance function:

$$M_{G_\lambda} = M_F \qquad \text{and} \qquad V_{G_\lambda}(\boldsymbol{m}) = \frac{1}{\lambda} V_F(\boldsymbol{m}).$$

Note that here the means of $P_{\boldsymbol{\theta},\mu}$ and $Q_{\boldsymbol{\theta},\lambda}$ are the same.

An important class of real exponential dispersion models corresponds to power variance functions $V_F(m) = m^p$ for $p \in \mathbb{R}\backslash(0,1)$; this class contains, for example, the Poisson, gamma, normal, and inverse Gaussian families, and different NEF generated by extreme stable distributions; see Jørgensen (1987). Jørgensen developed two types of asymptotics and derived large-sample results (as the sample size $n \to \infty$) and "small-dispersion" results (as the index parameter $\lambda \to \infty$); see Jørgensen (1997) for details.

From (54.17), we note that $\Lambda_{F_\lambda} = \frac{1}{\lambda} \Lambda_F$. Though Λ_F is often equal to $(0,\infty)$ or proportional to $\mathbf{N}\backslash\{0\}$ for real NEFs, it may be more complex for multivariate NEFs. For example, in the case of families of Wishart distributions introduced earlier in (54.5), Λ_F is proportional to $\Lambda = \{\frac{1}{2}, 1, \ldots,$ $\frac{d-1}{2}\} \cup \left(\frac{d-1}{2}, \infty\right)$; in fact, $\Lambda_{F_{n/2}} = \frac{2}{n}\Lambda$. Casalis and Letac (1994) have proposed a method of computing Λ_F for multivariate NEFs.

7 SOME STATISTICAL RESULTS FOR NEFs

In this section, we will present briefly some well-known statistical results for NEFs and then describe the concept of *cuts* which is closely related to the *S-sufficiency* and the *S-ancillarity* of a statistic for a model. Finally, we discuss the conjugate prior families of NEFs in the Bayesian framework.

7.1 Sufficiency, MLE, and UMVUE

It is well known that the sum of n i.i.d. exponential random variables is a sufficient statistic for the model; see, for example, Chapter 19 of Johnson, Kotz, and Balakrishnan (1994). This property also holds for the whole model $\bigcup_{\lambda \in \Lambda_F} F_\lambda$, referred to as the *exponential additive model*. Specifically, let $F = F(\mu)$ be a NEF, $\boldsymbol{\theta} \in \Theta(\mu)$, and $\lambda_1, \lambda_2, \ldots, \lambda_n \in \Lambda_F$. If $\boldsymbol{X}_1, \boldsymbol{X}_2, \ldots, \boldsymbol{X}_n$ are independent random variables distributed as $P_{\boldsymbol{\theta}, \mu_{\lambda_1}}, P_{\boldsymbol{\theta}, \mu_{\lambda_2}}, \ldots, P_{\boldsymbol{\theta}, \mu_{\lambda_n}}$, respectively, then the sum $\boldsymbol{S}_n = \sum_{i=1}^n \boldsymbol{X}_i$ is distributed as $P_{\boldsymbol{\theta}, \mu_\lambda}$ with $\lambda = \sum_{i=1}^n \lambda_i$; also, the joint distribution of $(\boldsymbol{X}_1, \boldsymbol{X}_2, \ldots, \boldsymbol{X}_n)$, conditioned on $\boldsymbol{S}_n$, does not depend on $\boldsymbol{\theta}$. The MLE of $\boldsymbol{\theta}$ is $\psi_\mu(\boldsymbol{S}_n/\lambda)$ as long as $\boldsymbol{S}_n/\lambda$ belongs to M_F, where ψ_μ (as before) denotes the inverse function of q. Under the regularity conditions satisfied by the NEF, the maximum of the likelihood function corresponds to the maximum of the concave function $\boldsymbol{\theta} \to (\sum_{i=1}^n \boldsymbol{x}_i)^T \boldsymbol{\theta} - \lambda q(\boldsymbol{\theta})$ so that $\hat{\boldsymbol{\theta}}$ is the root of $\sum_{i=1}^n \boldsymbol{x}_i = \lambda q'(\boldsymbol{\theta})$.

A similar result holds for GEF as well. If $\boldsymbol{Y}_1, \boldsymbol{Y}_2, \ldots, \boldsymbol{Y}_n$ are n i.i.d. random variables distributed as $Q_{\boldsymbol{\theta}}$ as given in (54.9), then $\sum_{i=1}^n t(\boldsymbol{Y}_i)$ is sufficient for $\boldsymbol{\theta}$ and $\hat{\boldsymbol{\theta}} = \psi_\mu\left(\frac{1}{n}\sum_{i=1}^n t(\boldsymbol{Y}_i)\right)$ as long as $\frac{1}{n}\sum_{i=1}^n t(\boldsymbol{Y}_i) \in M_F$. It should be noted that, for large n, this condition is satisfied in view of the law of large numbers.

It is also well known that the sum $\boldsymbol{S}_n$ of n i.i.d. random variables distributed as $P_{\boldsymbol{\theta}, \mu}$ is a complete statistic for the model, so that the UMVUE of a real parameter $f(\boldsymbol{\theta})$ is easily determined to be $c(\boldsymbol{S}_n)$, where c is a real function such that for all $\boldsymbol{\theta} \in \Theta(\mu)$

$$\int c(\boldsymbol{x})dP_{\boldsymbol{\theta},\mu^{*n}}(\boldsymbol{x}) = f(\boldsymbol{\theta}) \quad \text{and} \quad \int (c(\boldsymbol{x}))^2\, dP_{\boldsymbol{\theta},\mu^{*n}}(\boldsymbol{x}) < \infty.$$

(54.18)

Let us consider an example to illustrate this result. Let F be a NEF concentrated on $\mathbf{N}$ with generating measure $d\mu(x) = \Sigma_k\, \mu(k)\delta_k(x)$. Let G denote the generating function of μ given by $G(z) = \Sigma\mu(i)z^i$ so that the Laplace transform of μ is given by $L_\mu(\theta) = G(e^\theta)$ on an interval $(-\infty, a)$. Similarly, corresponding to the convolution measure $\mu_n = \mu^{*n}$, let the generating function be $G^n(z) = \Sigma\mu_n(i)z^i$ and the Laplace transform be $L_{\mu_n}(\theta) = L_\mu^n(\theta) = \Sigma\mu_n(i)e^{i\theta}$. With $f(\theta)$ being the parameter to estimate, let us write $f(\theta) = \Sigma g_i e^{i\theta} = g(e^\theta)$ where $g(z) = \Sigma g_i z^i$. If the UMVUE $c(S_n)$ of $f(\theta)$ exists, then from (54.18) we have

$$f(\theta)\,(L_\mu(\theta))^n = (gG^n)(e^\theta) = \int c(x)e^{\theta x}\, d\mu_n(x).$$

Expanding gG^n in a power series as $(gG^n)(z) = \Sigma a_n(i)z^i$, we obtain for any θ in an open interval of $\mathbb{R}$

$$\sum_i a_n(i)\, e^{i\theta} = \sum_i \mu_n(i)c(i)e^{i\theta}$$

so that

$$c(i) = a_n(i)/\mu_n(i)$$

which yields $c(S_n)$. For example, the UMVUE of $p = 1-e^\theta$ for the negative binomial distribution with parameter λ is $c(S_n) = \left(\lambda - \frac{1}{n}\right) \big/ \left(\lambda - \frac{1}{n} + \frac{S_n}{n}\right)$.

Kokonendji and Seshadri (1996) have determined the UMVUE of the generalized variance function (which is the determinant of the variance of a NEF on $\mathbb{R}^k$) based on $k+1$ observations and have given its exact expression for the simple quadratic NEF discussed in Section 8 and also for the Wishart family. In passing, we mention that the saddlepoint method yields accurate approximations for the density function of the sufficient statistic and of the MLE. Daniels (1980) has characterized the situations (on $\mathbb{R}$) where this approximation is exact, which turns out to be the case only for normal, gamma and inverse Gaussian distributions. Barndorff-Nielsen and Cox (1979) have applied the saddlepoint method for conditional inference for the problem on $\mathbb{R}^k$. Fraser, Reid, and Wong (1991) have described a numerical procedure to obtain such approximations for a real parameter and applied it for a single component of a multivariate parameter using the conditional density of the sufficient statistic given the other components. Interested readers may refer to Jensen (1995) for details.

7.2 Cuts in NEFs

A statistical analysis of a multivariate model is often split into separate parts corresponding to a partition of $\boldsymbol{\theta}$. When the partition is of the form $\boldsymbol{\theta} = (\boldsymbol{\theta}^{(1)}, \boldsymbol{\theta}^{(2)}) \in \mathbb{R}^d \times \mathbb{R}^{k-d}$, it is natural to consider the marginal and conditional models. Let $F = F(\mu)$ be a NEF on $\mathbb{R}^k$ and let $\boldsymbol{X} = (\boldsymbol{X}^{(1)}, \boldsymbol{X}^{(2)})$ be a random variable with distribution in F, with $\boldsymbol{X}^{(1)} \in \mathbb{R}^d$ and $\boldsymbol{X}^{(2)} \in \mathbb{R}^{k-d}$. Let us write the generating measure μ as

$$d\mu(\boldsymbol{x}^{(1)}, \boldsymbol{x}^{(2)}) = d\pi(\boldsymbol{x}^{(1)}) K(\boldsymbol{x}^{(1)}, d\boldsymbol{x}^{(2)}).$$

Note that this is always possible when μ is a probability measure with π being the marginal distribution of $\boldsymbol{X}^{(1)}$ under μ and $K(\boldsymbol{x}^{(1)}, \cdot)$ being the conditional distribution of $\boldsymbol{X}^{(2)}$, given $\boldsymbol{X}^{(1)} = \boldsymbol{x}^{(1)}$. On the other hand, if μ is not a probability measure, we have

$$d\mu(\boldsymbol{x}^{(1)}, \boldsymbol{x}^{(2)}) = e^{-\boldsymbol{\theta}_0^{(1)T} \boldsymbol{x}^{(1)} - \boldsymbol{\theta}_0^{(2)T} \boldsymbol{x}^{(2)} + q(\boldsymbol{\theta}_0)} \, dP_{\boldsymbol{\theta}_0, \mu}(\boldsymbol{x}^{(1)}, \boldsymbol{x}^{(2)})$$

for any $\boldsymbol{\theta}_0 \in \Theta(\mu)$, in which case the measures π and $K(\boldsymbol{x}^{(1)}, \cdot)$ are defined up to a function of $\boldsymbol{x}^{(1)}$. Then, the conditional distribution of $\boldsymbol{X}^{(2)}$, given $\boldsymbol{X}^{(1)}$, will still have an exponential density function of the form

$$\frac{e^{\boldsymbol{\theta}^{(2)T} \boldsymbol{x}^{(2)}}}{\int e^{\boldsymbol{\theta}^{(2)T} \boldsymbol{x}^{(2)}} K(\boldsymbol{x}^{(1)}, d\boldsymbol{x}^{(2)})}$$

with respect to $K(\boldsymbol{x}^{(1)}, \cdot)$, which only depends on the part $\boldsymbol{\theta}^{(2)}$ of $\boldsymbol{\theta}$. Note, however, that $K(\boldsymbol{x}^{(1)}, d\boldsymbol{x}^{(2)})$ can be concentrated on an affine hyperplane of $\mathbb{R}^{k-d}$ so that the conditional model is not exactly a NEF. For example, consider the NEF associated with the GEF of normal distributions with mean and variance unknown. Clearly, if $\boldsymbol{X}^{(1)} = X$ and $\boldsymbol{X}^{(2)} = -X^2/2$, the conditional distribution of $\boldsymbol{X}^{(2)}$ given $\boldsymbol{X}^{(1)} = x$ is simply the Dirac mass $\delta_{-x^2/2}$. But, when K satisfies the basic conditions stated earlier in Section 54.2, the conditional model is a NEF parameterized by $\boldsymbol{\theta}^{(2)}$.

Furthermore, it is easy to see in general that the marginal model composed by the laws of $\boldsymbol{X}^{(1)}$ as

$$e^{\boldsymbol{\theta}^{(1)T} \boldsymbol{x}^{(1)}} \left\{ \int e^{\boldsymbol{\theta}^{(2)T} \boldsymbol{x}^{(2)}} K(\boldsymbol{x}^{(1)}, d\boldsymbol{x}^{(2)}) \right\} e^{-q(\boldsymbol{\theta})} \, d\pi(x^{(1)})$$

$$= e^{\boldsymbol{\theta}^{(1)T} \boldsymbol{x}^{(1)} + q_1(\boldsymbol{x}^{(1)}, \boldsymbol{\theta}^{(2)}) - q(\boldsymbol{\theta})} \, d\pi(\boldsymbol{x}^{(1)})$$

is not a NEF anymore. An interesting example can be seen in Barndorff-Nielsen (1978). In the last example of normal distributions with mean and variance unknown, the marginal distribution of $\boldsymbol{X}^{(1)} = X$ belongs to a GEF and not a NEF. Note that the marginal distribution of $\boldsymbol{X}^{(1)}$ in general is the image of F by the projection $\boldsymbol{x} \to \boldsymbol{x}^{(1)}$.

When the above given marginal model is indeed a NEF, $\boldsymbol{X}^{(1)}$ is said to be a *cut*. More generally, a statistic μ is said to be a cut if there exists a parameterization $(\varphi^{(1)}, \varphi^{(2)})$ of the model such that the density function $f_{\boldsymbol{\theta}}(\boldsymbol{x})$ is the product of the marginal density of u depending on $\varphi^{(1)}$ and the conditional density given u depending only on $\varphi^{(2)}$. In this case, u is said to be *S-sufficient* for $\varphi^{(1)}$ and *S-ancillary* for $\varphi^{(2)}$. Barndorff-Nielsen (1978) noted that any cut in a NEF is necessarily of the form $\boldsymbol{X}^{(1)}$ up to an affine transformation [or $t^{(1)}$ for a GEF if t is a sufficient statistic of the model in (54.9)]. Note that the parameterization $(\varphi^{(1)}, \varphi^{(2)})$ here corresponds to the mixed parameterization $(\boldsymbol{m}^{(1)}, \boldsymbol{\theta}^{(2)})$ introduced in Section 4.

It should be mentioned that Barndorff-Nielsen and Koudou (1995) have completely characterized the cuts of NEFs in eight different equivalent statements. They have also presented a natural method of constructing new NEFs with cuts. Bar-Lev *et al.* (1994) have proved that $\boldsymbol{X}^{(1)}$ is a cut for the NEF iff the marginal variance function (i.e., the principal $d \times d$ minor of V_F) is a function only of $\boldsymbol{m}^{(1)}$ and does not depend on $\boldsymbol{m}^{(2)}$.

7.3   Bayesian Theory

Developments on conjugate prior families have amply demonstrated the important role that variance functions play in this topic. Raiffa and Schlaifer (1961) explained conjugate priors as a family of priors closed under sampling. Specifically, let $\mathcal{P}$ be a statistical model parametrized by Θ. Let $\boldsymbol{X}_1, \ldots, \boldsymbol{X}_n, \ldots$ be i.i.d. random variables with distribution $P_{\boldsymbol{\theta}} \in \mathcal{P}$. Suppose $\boldsymbol{\theta}$ is random with distribution π in a set of prior measures Π on Θ. Then, Π is said to be *conjugate* to $\mathcal{P}$ if the posterior distribution of $\boldsymbol{\theta}$, given $\boldsymbol{X}_1, \ldots, \boldsymbol{X}_n$, still belongs to Π for any n. When $\mathcal{P}$ is an exponential family F, Diaconis and Ylvisaker (1979) considered a conjugate prior family for F as the set Π of probability measures

$$d\pi_{p,\boldsymbol{x}_0}(\boldsymbol{\theta}) = K_{p,\boldsymbol{x}_0} \, e^{\boldsymbol{x}_0^T \boldsymbol{\theta} - pq(\boldsymbol{\theta})} I_{\Theta(\mu)}(\boldsymbol{\theta}) \, d\boldsymbol{\theta}, \qquad (54.19)$$

where $p > 0$, $\boldsymbol{x}_0 \in pM_F$, and $K_{p,\boldsymbol{x}_0}$ is a suitable normalizing constant. Now, let $\boldsymbol{X}_1, \ldots, \boldsymbol{X}_n$ be n i.i.d. random variables distributed as $P_{\boldsymbol{\theta},\mu}$ with sum $\boldsymbol{S}_n$, and let $\boldsymbol{\theta}$ be distributed as $\pi_{p,\boldsymbol{x}_0}$. Then, the posterior distribution of $\boldsymbol{\theta}$, given $\boldsymbol{S}_n$, can be shown easily to be $\pi_{p+n,\boldsymbol{x}_0+\boldsymbol{S}_n}$. Arnold, Castillo, and Sarabia (1993) have described a most general structure for a conjugate exponential family with d parameters $\boldsymbol{b} = (b_1, \ldots, b_d)$ as

$$d\pi_{\boldsymbol{b}}(\boldsymbol{\theta}) = r(\boldsymbol{\theta})e^{\sum_{i=1}^{k} b_i\theta_i + b_{k+1}q(\boldsymbol{\theta}) + \sum_{i=k+2}^{d} b_i s_i(\boldsymbol{\theta}) + \lambda_0(\boldsymbol{b})} \, d\boldsymbol{\theta},$$

where $s_{k+2}(\boldsymbol{\theta}), \ldots, s_d(\boldsymbol{\theta})$ are arbitrary functions, $\lambda_0(\boldsymbol{b})$ is a normalizing constant, and $d \geq k+1$.

Diaconis and Ylvisaker's family Π defined in (54.19) is characterized through the property that the posterior expectation of the mean parameter is linear. Consonni and Veronese (1992) subsequently introduced a similar family Π^* on the domain of means, mimicking the form of Diaconis and Ylvisaker's prior distributions for the mean parameter in (54.19), as

$$d\pi^*_{p,\boldsymbol{x}_0}(\boldsymbol{m}) = K^*_{p,\boldsymbol{x}_0} \, e^{\boldsymbol{x}_0^T \boldsymbol{\psi}_\mu(\boldsymbol{m}) - pq(\boldsymbol{\psi}_\mu(\boldsymbol{m}))} I_{M_F}(\boldsymbol{m}) \, d\boldsymbol{m}. \qquad (54.20)$$

Such distributions are referred to as *standard conjugate prior distributions* for the parameter under consideration. Consonni and Veronese (1992) then compared two sets of priors on M_F, namely, Π^* given by (54.20) and the set $\tilde{\Pi}$ induced by Π on M_F (through the mapping q') given by

$$d\tilde{\pi}_{p,\boldsymbol{x}_0}(\boldsymbol{m}) = K_{p,\boldsymbol{x}_0} \, e^{\boldsymbol{x}_0^T \boldsymbol{\psi}_\mu(\boldsymbol{m}) - pq(\boldsymbol{\psi}_\mu(\boldsymbol{m}))} |V_F(\boldsymbol{m})|^{-1} I_{M_F}(\boldsymbol{m}) \, d\boldsymbol{m}.$$

They then showed that the two sets Π^* and $\tilde{\Pi}$ coincide in the univariate case iff the NEF F has a quadratic variance function. This result has been generalized by Gutiérrez-Peña and Smith (1995) for an arbitrary parameterization and for the multivariate case. These authors have established, in particular, that the necessary and sufficient condition on the variance function in order for Π^* and $\tilde{\Pi}$ to be identical is given by

$$|V_F(\boldsymbol{m})|^{-1} V_F(\boldsymbol{m}) \left(\frac{\partial |V_F(\boldsymbol{m})|}{\partial m_i} \right)_i = a\boldsymbol{m} + \boldsymbol{b}, \qquad (54.21)$$

where $a \in \mathbb{R}$ and $\boldsymbol{b} \in \mathbb{R}^k$.

Casalis (1996) obtained, independently, two alternative equivalent conditions as

$$\text{(i)} \qquad \sum_{j=1}^{k} \left(\frac{\partial V_F(\boldsymbol{m})}{\partial m_j} \right)_{ij} = am_i + b_i, \qquad i = 1, \ldots, k \qquad (54.22)$$

and

$$\text{(ii)} \qquad |V_F(\boldsymbol{m})| = e^{\boldsymbol{b}^T \boldsymbol{\psi}_\mu(\boldsymbol{m}) + aq(\boldsymbol{m}) + c}. \qquad (54.23)$$

Wishart distribution, as well as the simple quadratic NEF discussed in the following section, satisfy (54.21) or equivalently (54.22) and (54.23). These conditions, however, are not characteristic of the quadratic NEF, as can be seen from the family composed of direct products of binomial and negative binomial distributions with variance function as $V_F(\boldsymbol{m}) = \text{Diag}\left(m_1 - \frac{m_1^2}{N}, \ m_2 + \frac{m_2^2}{\lambda}\right)$, $N \in \mathbb{N}$, and $\lambda > 0$. As yet, the class of NEF satisfying (54.21) has not been determined in its entirety.

8 NEFs WITH QUADRATIC VARIANCE FUNCTION

8.1 Morris Class

Morris (1982) observed that only six well-known families of real-valued distributions (and only those up to affine transformations and powers) have a quadratic variance function—that is, have $V_F(\boldsymbol{m})$ to be a polynomial in $\boldsymbol{m}$ of degree at most 2. They are as follows:

Normal

$$M_F = \mathbb{R}, \qquad V_F(m) = 1,$$

$$d\mu(x) = \frac{1}{\sqrt{2\pi}} \, e^{-x^2/2} \, dx,$$

$$\Theta(\mu) = \mathbb{R}, \qquad q_\mu(\theta) = \theta^2/2,$$

$$dP_{m,F}(x) = \frac{1}{\sqrt{2\pi}} \, e^{-(x-m)^2/2} \, dx,$$

$$\Lambda = (0, \infty),$$

$$F_\lambda = \{N(m, \lambda) : \ m \in \mathbb{R}\};$$

see Chapter 13 of Johnson, Kotz, and Balakrishnan (1994).

Poisson

$$M_F = (0, \infty), \qquad V_F(m) = m,$$

$$\mu = \sum_{n=0}^{\infty} \delta_n/n!,$$

$$\Theta(\mu) = (0, \infty), \qquad q_\mu(\theta) = e^\theta,$$

$$dP_{m,F}(x) = \sum_{n=0}^{\infty} \frac{e^{-m} m^n}{n!} \delta_n(x),$$

$$\Lambda = (0, \infty),$$

$$F_\lambda = F, \qquad \mu_\lambda = \sum_{n=0}^{\infty} \frac{\lambda^n}{n!} \delta_n;$$

see Chapter 4 of Johnson, Kotz, and Kemp (1992).

Binomial

$$M_F = (0, 1), \qquad V_F(m) = m - m^2,$$

$$\mu = \delta_0 + \delta_1,$$

$$\Theta(\mu) = \mathbb{R}, \qquad q_\mu(\theta) = 1 + e^\theta,$$

$$dP_{m,F}(x) = m\delta_0(x) + (1 - m)\delta_1(x),$$

$$\Lambda = \mathbb{N},$$

$$F_n = \{\mathrm{Bin}(n, p) : \ 0 < p < 1\};$$

see Chapter 3 of Johnson, Kotz, and Kemp (1992).

Negative binomial

$$M_F = (0, \infty), \qquad V_F(m) = m + m^2,$$

$$\mu = \Sigma \delta_n,$$

$$\Theta(\mu) = (-\infty, 0), \qquad q_\mu(\theta) = -\log(1 - e^\theta),$$

$$dP_{m,F}(x) = \frac{1}{1+m} \sum_{n=0}^{\infty} \left(\frac{m}{1+m}\right)^n \delta_n(x),$$

$$\Lambda = (0, \infty),$$

$$dP_{m,F_\lambda} = \sum_{n=0}^{\infty} \frac{\Gamma(n+\lambda)}{\Gamma(\lambda)n!} \left(\frac{m}{m+\lambda}\right)^n \left(\frac{\lambda}{m+\lambda}\right)^\lambda \delta_n;$$

see Chapter 5 of Johnson, Kotz, and Kemp (1992).

Gamma

$$M_F = (0, \infty), \qquad V_F(m) = m^2,$$

$$d\mu(x) = I_{(0,\infty)}(x)\, dx,$$

$$\Theta(\mu) = (-\infty, 0), \qquad q_\mu(\theta) = -\log(-\theta),$$

$$dP_{m,F}(x) = \frac{1}{m}\, e^{-x/m} I_{(0,\infty)}(x)\, dx,$$

$$\Lambda = (0, \infty),$$

$$dP_{m,F_\lambda}(x) = \frac{e^{-x/m} x^{\lambda-1}}{\Gamma(\lambda) m^\lambda}\, I_{(0,\infty)}(x)\, dx;$$

see Chapter 17 of Johnson, Kotz, and Balakrishnan (1994).

Hyperbolic cosine

$$M_F = \mathbb{R}, \qquad V_F(m) = 1 + m^2,$$

$$d\mu(x) = \frac{1}{2\cosh(\pi x/2)}\, dx,$$

$$\Theta(\mu) = \left(-\frac{\pi}{2}, \frac{\pi}{2}\right), \qquad q_\mu(\theta) = -\log(\cos\theta),$$

$$dP_{m,F}(x) = \cos\theta\, e^{\theta x} \mu(dx) \text{ with } m = \tan\theta,$$

$$\Lambda = (0, \infty),$$

$$d\mu_\lambda(x) = \frac{2^{\lambda-1}}{\Gamma(\lambda)} \prod_{j=0}^{\infty} \left\{ 1 + \frac{x^2}{(\lambda+2j)^2} \right\}^{-1} I_{\mathbb{R}}(x) \ dx.$$

Several extensions of this classification have been developed in the literature. On the real line, for example, the following need to be mentioned.

(i) *Mora class* consisting of cubic variance functions [see Letac and Mora (1990)] and including the well-known inverse Gaussian distributions [see Chapter 15 of Johnson, Kotz, and Balakrishnan (1994)]

$$dIG_{m,\lambda^2}(x) = \frac{\lambda}{\sqrt{2\pi}x^{3/2}} \ e^{-\frac{\lambda^2}{2m^2x}(x-m)^2} I_{(0,\infty)}(x)dx,$$

where $m > 0$ and $\lambda > 0$. For fixed λ, we obtain a NEF with variance function $V_{F_\lambda}(m) = m^3/\lambda^2$. Included in this class are also the Ressel–Kendall, strict arcsine, and large arcsine distributions.

(ii) *Babel class* of variance functions of the form $V_F(m) = P\Delta + Q\sqrt{\Delta}$, where P, Q, and Δ are polynomials in m of degree at most 1, 2, and 2, respectively. The term "Babel" has been formed from the names Bar-Lev, Bshouty and Enis (1991), and Letac [Bar-Lev *et al.* (1994)], who were the first authors to consider this class [Letac (1992)]; see also Jørgensen (1987) and Kokonendji (1993). Babel was actually introduced by Letac (1987) in his discussion of Jørgensen's (1987) paper; see the recent review of Babel variance functions prepared by Jørgensen and Letac (1999).

(iii) *Class of power exponential families* with $V_F(m) = cm^\gamma$. This class has been studied by a number of authors including Tweedie (1984). Incidentally, Jørgensen (1987) has cited it as an example of exponential dispersion model.

8.2 Multivariate Case

It is natural to consider the extension of the Morris class to $\mathbb{R}^k$. The first such extension is obtained by replacing the polynomial $V_F(m) = am^2 + bm + c$ by a matrix function whose entries are quadratic polynomials in the coefficients $m_1, m_2, \ldots, m_k$ of $\boldsymbol{m}$. Such a polynomial is of the form

$$\sum_{r=1}^{k} \sum_{s=1}^{k} a^{rs} m_r m_s + \sum_{r=1}^{k} b^r m_r + c,$$

where a^{rs}, b^r, and c are arbitrary real numbers. Thus, this multivariate extension of Morris class is composed of NEF with variance function

$$V_F(\boldsymbol{m}) = \left(\left(\sum_{r=1}^{k}\sum_{s=1}^{k} a_{ij}^{rs} m_r m_s + \sum_{r=1}^{k} b_{ij}^r m_r + c_{ij}\right)\right)_{i,j},$$

or equivalently

$$V_F(\boldsymbol{m}) = \sum_{r=1}^{k}\sum_{s=1}^{k} \boldsymbol{A}^{rs} m_r m_s + \sum_{r=1}^{k} \boldsymbol{B}^r + \boldsymbol{C}, \qquad (54.24)$$

where $\boldsymbol{A}^{rs} = ((a_{ij}^{rs}))_{i,j}$, $\boldsymbol{B}^r = ((b_{ij}^r))_{i,j}$, and $\boldsymbol{C} = ((c_{ij}))_{i,j}$ are real symmetric $k \times k$ matrices.

Of course, trivial examples of multivariate quadratic NEFs can be obtained by direct products of real quadratic NEFs. A NEF F is said to be the *direct product* of two NEFs F_1 and F_2 on $\mathbb{R}^d$ and $\mathbb{R}^{k-d}$, respectively, if F is composed of the laws of two independent random variables $\boldsymbol{X}_1$ and $\boldsymbol{X}_2$ with distributions belonging to F_1 and F_2, respectively. A NEF that is an affine transformation of a direct product is said to be *reducible*; it is said to be *irreducible* otherwise.

Until now, the classification of quadratic irreducible NEFs is not complete in its generality because there are some technical difficulties in this general case. On $\mathbb{R}$, one way to recover a generating measure from the variance function is to start from (54.13) given by

$$\psi_\mu'(m) = (V_F(m))^{-1} = \frac{1}{V_F(m)}, \qquad (54.25)$$

which, when integrated with respect to m, yields

$$\theta = \psi_\mu(m) = \int_{m_0}^{m} \frac{dt}{V_F(t)}.$$

Upon inverting the above mapping ψ_μ, we get $m = q_\mu'(\theta)$ which, when integrated once, gives rise to $q_\mu(\theta)$. The principal difficulty is now in recognizing when the function $q_\mu(\theta)$ is in fact the log-Laplace transform of a suitable distribution. Of course, tables of Laplace transforms [see, for example, Hladik (1969)] will be very useful for this purpose. Kokonendji (1993) has adopted a geometrical approach. In the case of $\mathbb{R}^k$, however, problems arise even in the first stage of integration from $\boldsymbol{\psi}_\mu'$ to $\boldsymbol{\psi}_\mu$. To overcome this difficulty, we observe that (54.13) and (54.25) yield symmetry conditions after differentiation. In other words, if $\boldsymbol{\psi}_\mu(\boldsymbol{m}) = (\psi_1(\boldsymbol{m}), \dots, \psi_k(\boldsymbol{m}))$, we translate the obvious relations $\frac{\partial^2 \psi_\ell}{\partial m_i \partial m_j} = \frac{\partial^2 \psi_\ell}{\partial m_j \partial m_i}$ (for $i, j, \ell = 1, \ldots, k$) to V_F through (54.24). For example, in the case of $\mathbb{R}^2$, by writing $\boldsymbol{m} = (x, y)$ and

$$V_F(\boldsymbol{m}) = \begin{pmatrix} A(x,y) & F(x,y) \\ F(x,y) & B(x,y) \end{pmatrix},$$

we obtain the following differential system for A, B, and F:

$$A \frac{\partial B}{\partial x} - B \frac{\partial F}{\partial y} + F \left(\frac{\partial B}{\partial y} - \frac{\partial F}{\partial x} \right) = 0,$$

$$-A \frac{\partial F}{\partial y} + B \frac{\partial A}{\partial x} + F \left(\frac{\partial A}{\partial x} - \frac{\partial F}{\partial y} \right) = 0. \qquad (54.26)$$

Thus, V_F satisfies the following properties:

(i) $V_F(\boldsymbol{m})$ is a symmetric matrix,

(ii) $V_F(\boldsymbol{m})$ satisfies the symmetry conditions [similar to (54.26)],

and

(iii) $V_F(\boldsymbol{m})$ is positive definite on an open set of $\mathbb{R}^k$.

Affine transformations have also been used to simplify the relations [of the type (54.26)] as much as possible through (54.14). This has led to the determination of two subclasses of multivariate quadratic NEFs which we present now, but the symmetry conditions have not yet been solved completely for the general case.

Homogeneous Quadratic Variance Functions

In this case, the matrices $\boldsymbol{B}^r$, $r = 1, 2, \ldots, k$, and $\boldsymbol{C}$ in (54.24) are zero so that

$$V_F(\boldsymbol{m}) = \sum_{r=1}^{k} \sum_{s=1}^{k} \boldsymbol{A}^{rs} m_r m_s. \qquad (54.27)$$

On the real line, of course, this corresponds to the variance function $V_F(m) = m^2/\lambda$, which is that of the NEF of gamma distributions with shape parameter λ.

On $\mathbb{R}^k$, we obtain all NEF of Wishart distributions that can be defined

(i) on the revolution cone of $\mathbb{R}^k$, namely, $\{(x_1, \ldots, x_k) : x_1 > 0, \ x_1^2 - \cdots -x_k^2 > 0\}$,

(ii) on the set of real positive definite symmetric $k \times k$ matrices,

and

(iii) on the set of positive definite Hermitian matrices with coefficients in the complex plane, in the quaternion field (four-dimensional hypercomplex numbers of the form $x_0 1 + x_1 i + x_2 j + x_3 k$, $x_t \in \mathbb{R}^1$) and also in the so-called Cayley algebra (eight-dimensional hypergeometric numbers).

Such domains are called *symmetric cones*. Note that we have already briefly discussed the Wishart family of distributions on the space of real symmetric matrices in Section 2.1.

The variance function on the set V_+ of positive definite symmetric $k \times k$ matrices can be applied to a matrix $\boldsymbol{x}$ of V, which will yield

$$V_{F_\lambda}(\boldsymbol{m})\boldsymbol{x} = \frac{1}{\lambda}\,\boldsymbol{m}\boldsymbol{x}\boldsymbol{m}.$$

Here, λ belongs to $\left\{\frac{1}{2}, 1, \ldots, \frac{k-1}{2}\right\} \cup \left(\frac{k-1}{2}, \infty\right)$ and $\boldsymbol{m} \in V_+$. Writing the matrix $\boldsymbol{m} = ((m_{ij}))$ as a vector $\boldsymbol{m} = (m_{11}, m_{12}, \ldots, m_{1k}, m_{22}, \ldots, m_{2k}, \ldots, m_{kk})$ of dimension $k(k+1)/2$ and $\boldsymbol{x}$ in the same way, we obtain the matrix form of the covariance. On $\mathbb{R}^2$, for example, this readily gives

$$V_F(\boldsymbol{m}) = \frac{2}{\lambda} \begin{pmatrix} m_{11}^2 & m_{11}m_{12} & m_{12}^2 \\ m_{11}m_{12} & \frac{1}{2}(m_{22}^2 + m_{11}m_{22}) & m_{12}m_{22} \\ m_{12}^2 & m_{12}m_{22} & m_{22}^2 \end{pmatrix};$$

see Letac (1989b).

Wishart distributions are the natural generalizations on $\mathbb{R}^k$ of the gamma family as NEF, while the multivariate gamma distributions of Holgate type discussed in Chapter 48 go beyond this framework. Wishart distributions have been discussed extensively in the context of characterizations, in particular, by invariance property [see Letac (1989b) and Casalis (1990)] and by properties of linear and inverse-linear regression [see Letac and Massam (1997)]. Wishart distributions have also been utilized to define generalized inverse Gaussian distributions on a symmetric cone [see Bernadac (1995)] as well as to define Dirichlet distributions [see Casalis and Letac (1996) and Massam (1994)].

Simple Quadratic Variance Functions

These are of the special form

$$V_F(\boldsymbol{m}) = a\boldsymbol{m}^T\boldsymbol{m} + \sum_{r=1}^{k} \boldsymbol{B}^r m_r + \boldsymbol{C}, \tag{54.28}$$

where a is a real number; in this case, the matrices $\boldsymbol{A}^{rs}$ in fact are $\boldsymbol{A}^{rs} = ((a_{ij}^{rs})) = (a\delta_i^r\delta_j^s)$. Casalis (1996) has shown that there are only $2k + 4$ simple quadratic NEFs up to affine transformations and powers. In this instance, we will say that two NEFs F_1 and F_2 are of the *same type* if F_2 is an affine transformation of a power of F_1 and, as a result, a type will be described entirely by one of its representations.

The $2k + 4$ types of quadratic NEFs are as follows.

(a) $(k + 1)$ *Poisson–Gaussian types* $(PG)_d$, $d = 0, \ldots, k$

They are composed of NEFs with an affine variance function $V_F(\boldsymbol{m}) = \sum_{r=1}^{k} \boldsymbol{B}^r m_r + \boldsymbol{C}$, wherein the real number a in (54.28) has been taken to be zero. Recall here that on $\mathbb{R}$, the variance function $V_F(m) = bm + c$ yielded the normal NEF (when $b = 0$) and the Poisson NEF (when $b = 1$ and $c = 0$) with their affine transformations. It is quite disappointing in the case of $\mathbb{R}^k$, however, since we just obtain the types corresponding to direct products of d univariate Poisson NEFs and $k - d$ normal NEFs on $\mathbb{R}$ $(d = 0, 1, \ldots, k)$. The cases $d = 0$ and $d = k$ correspond simply to k normal NEFs and k Poisson NEFs, respectively.

This NEF has, therefore,

$$M_F = (0, \infty)^d \times \mathbb{R}^{k-d},$$

and

$$V_F(\boldsymbol{m}) = \text{diag}(m_1, \ldots, m_d, 1, \ldots, 1), \quad \boldsymbol{m} \in M_F.$$

Consequently, all the NEF of Poisson–Gaussian types are reducible.

The remaining $k + 3$ NEFs correspond to the case $a \neq 0$ in (54.28) and they are irreducible.

(b) $(k + 1)$ *Negative Multinomial-Gamma Types* $(NM - Ga)_d$, $d = 0, \ldots, k$

The distributions in these types are combinations of negative multinomial, gamma, and normal distributions, and naturally three different subclasses arise:

1. **Negative multinomial type** $(NM - Ga)_k$: This consists of the NEFs of negative multinomial distributions which are defined by their Laplace transform

$$\left(S + 1 - \sum_{i=1}^{k} m_{0i} e^{\theta_i}\right)^{-p}$$

for fixed $m_{0i} > 0$ $(i = 1, \ldots, k)$, $p > 0$ and $S = \sum_{i=1}^{k} m_{0i}$; see Chapter 36 of Johnson, Kotz, and Balakrishnan (1997) with small changes in notations (the Laplace transform instead of the probability generating function). From this formula, with fixed $p > 0$, we obtain the probability mass function (with parameter $\boldsymbol{m}$) defined on $\mathbb{N}^k$ as

$$P_{\boldsymbol{m},p}(\boldsymbol{n}) = \frac{\Gamma(p + \sum_{i=1}^{k} n_i)}{\Gamma(p) \prod_{i=1}^{k} n_i!} (S+1)^{-p} \prod_{i=1}^{k} \left(\frac{m_i}{S+1}\right)^{n_i}.$$

Thus, for fixed p, negative multinomial distributions form a NEF with generating measure on $\mathbb{N}^k$ as

$$\nu_p^{(k)} = \sum_{i=1}^{k} \frac{\Gamma(p + \sum_{i=1}^{k} n_i)}{\Gamma(p) \prod_{i=1}^{k} n_i!} \delta_{\boldsymbol{n}}. \tag{54.29}$$

Note that the parameter p here is the Jørgensen parameter. The variance function of $F_p = F(\nu_p)$ is given on $M_{F_p} = \{\boldsymbol{m} : m_i > 0, \ i = 1, \ldots, k\}$ as

$$V_{F(\nu_p)}(\boldsymbol{m}) = \frac{1}{p} \boldsymbol{m}\boldsymbol{m}^T + \text{diag}(m_1, \ldots, m_k).$$

2. $(NM - Ga)_{k-1}$ **NEF:** Let us now consider the combination of a $(k-1)$-dimensional negative multinomial family with a gamma family on $\mathbb{R}$, constructed as follows.

   Let $\nu_p^{(k-1)}$ be the measure on $\mathbb{R}^{k-1}$ as defined in (54.29) with k replaced by $k-1$ and with $p > 0$. Let γ_s be the measure on $\mathbb{R}$ generating the gamma NEF (with shape parameter s) given by

$$d\gamma_s(x) = \frac{1}{\Gamma(s)} x^{s-1} I_{(0,\infty)}(x) \, dx. \tag{54.30}$$

For fixed $p > 0$, let us introduce

$$d\mu_p^{(k-1)}(x_1, \ldots, x_k) = d\nu_p^{(k-1)}(x_1, \ldots, x_{k-1}) d\gamma_{\sum_{i=1}^{k-1} x_i + p}(x_k)$$

with its Laplace transform

$$L_{\mu_p^{(k-1)}}(\boldsymbol{\theta}) = \left(-\theta_k - \sum_{i=1}^{k-1} e^{\theta_i}\right)^{-p}$$

on $\Theta\left(\mu_p^{(k-1)}\right) = \left\{\boldsymbol{\theta} \in \mathbb{R}^k : \sum_{i=1}^{k-1} e^{\theta_i} + \theta_k < 0\right\}$. The variance function of the NEF $F_p = F(\mu_p^{(k-1)})$ is defined on $M_{F_p} = (0, \infty)^k$ as

$$V_{F(\mu_p^{(k-1)})}(\boldsymbol{m}) = \frac{1}{p}\,\boldsymbol{mm}^T + \text{diag}(m_1, \ldots, m_{k-1}, 0).$$

The above construction simply means that the distributions forming this NEF are the distributions of the random variable $(X_1, \ldots, X_k)$, where $(X_1, \ldots, X_{k-1})$ has a negative multinomial distribution with parameters $p, m_1, \ldots, m_{k-1}$, and X_k conditionally on $(X_1, \ldots, X_{k-1})$ has a gamma distribution with shape parameter $\sum_{i=1}^{k-1} X_i + p$ and mean m_k.

3. $(NM-Ga)_d$, $d = 0, \ldots, k-2$, **NEF:** Let d be fixed in $\{0, 1, \ldots, k-2\}$ and $p > 0$, and let $\nu_p^{(d)}$ and γ_s be measures on $\mathbb{R}^d$ and $\mathbb{R}$ as defined in (54.29) and (54.30), respectively. Moreover, let $\lambda_p^{(k-d-1)}$ be the multivariate normal distribution $N(\boldsymbol{0}, p\boldsymbol{I}_{k-d-1})$ on $\mathbb{R}^{k-d-1}$, where $\boldsymbol{I}_{k-d-1}$ denotes the identity matrix of order $k - d - 1$. Then, let us consider for $d \geq 1$

$$\begin{aligned} d\mu_p^{(d)}(x_1, \ldots, x_k) &= d\nu_p^{(d)}(x_1, \ldots, x_d)\,d\gamma_{\sum_{i=1}^d x_i + p}(x_{d+1}) \\ &\quad \cdot d\lambda_{x_{d+1}}^{(k-d-1)}(x_{d+2}, \ldots, x_k) \end{aligned}$$

and for $d = 0$

$$d\mu_p^{(0)}(x_1, \ldots, x_k) = d\gamma_p(x_1)\,d\lambda_{x_1}^{(k-1)}(x_2, \ldots, x_k).$$

We have in this case

$$\Theta_{\mu_p^{(d)}} = \left\{\boldsymbol{\theta} \in \mathbb{R}^k : \theta_{d+1} + \frac{1}{2}\sum_{i=d+2}^k \theta_i^2 + \sum_{i=1}^d e^{\theta_i} < 0\right\}$$

and

$$L_{\mu_p^{(d)}}(\boldsymbol{\theta}) = \left(-\theta_{d+1} - \frac{1}{2}\sum_{i=d+2}^k \theta_i^2 - \sum_{i=1}^d e^{\theta_i}\right)^{-p}.$$

This NEF $F_p = F\left(\mu_p^{(d)}\right)$ is thus characterized by

$$M_{F_p} = (0, \infty)^{d+1} \times \mathbb{R}^{k-d-1}$$

and

$$V_{F_p}(\boldsymbol{m}) = \frac{1}{p}\,\boldsymbol{mm}^T + \text{diag}(m_1, \ldots, m_d, 0, m_{d+1}, \ldots, m_{d+1}).$$

Here again, the NEF consists of distributions of random variables $(X_1, \ldots, X_k)$, where $(X_1, \ldots, X_d)$ has a negative multinomial distribution with parameters $p, m_1, \ldots, m_d$, X_{d+1} conditional on $(X_1, \ldots, X_d)$ is distributed as gamma with shape parameter $\sum_{i=1}^{d} X_i + p$ and mean m_{d+1}, and $(X_{d+2}, \ldots, X_k)$ conditional on $(X_1, \ldots, X_{d+1})$ is distributed as $k-d-1$-dimensional Gaussian $N(\mathbf{0}, X_{d+1}\boldsymbol{I}_{k-d-1})$. Note that this Gaussian distribution depends only on X_{d+1} and not on $(X_1, \ldots, X_d)$.

For $d = 0$, observe that the negative multinomial component disappears. On $\mathbb{R}^2$, the three families of negative binomial, gamma, and normal are never combined together. The family $F(\mu_p^{(1)})$ has been mentioned by Bar-Lev *et al.* (1994) as NEF whose marginals are in two different Morris families. Such an NEF is called *diagonal quadratic*; see Section 54.8.4.

(c) *Multinomial Type M*

Let $\boldsymbol{e}_1, \ldots, \boldsymbol{e}_k$ denote the vectors of the canonical basis of $\mathbb{R}^k$ and $\boldsymbol{e}_0$ the null vector. Consider the multinomial distributions with probability mass function

$$P\boldsymbol{m}, F_p(n_1, \ldots, n_k) = \binom{p}{n_0, n_1, \ldots, n_k} \left(1 - \frac{\sum_{i=1}^{k} m_i}{p}\right)^{n_0} \prod_{i=1}^{k} \left(\frac{m_i}{p}\right)^{n_i},$$

where $n_0, n_1, \ldots, n_k$ are positive integers, $\sum_{i=0}^{k} n_i = p$, and $\binom{p}{n_0, n_1, \ldots, n_k}$ $= \frac{p!}{n_0! n_1! \cdots n_k!}$; see Chapter 35 of Johnson, Kotz, and Balakrishnan (1997). When $\boldsymbol{m}$ varies in $M_{F_p} = \{\boldsymbol{m} \in \mathbb{R}^k : m_i > 0, \sum_{i=1}^{k} m_i < 1\}$, we obtain the NEF F_p generated by the p-th convolution of $\mu = \sum_{i=0}^{k} \delta_{\boldsymbol{e}_i}$ with its Laplace transform as $(1 + \sum_{i=1}^{k} e^{\theta_i})^p$. The variance function of this NEF is given by

$$V_{F_p}(\boldsymbol{m}) = -\frac{1}{p} \boldsymbol{m}\boldsymbol{m}^T + \mathrm{diag}(m_1, \ldots, m_k).$$

(d) *Hyperbolic Type H*

As in the case of $(NM - Ga)_{k-1}$ type, this last type is a combination of a negative multinomial family on $\mathbb{R}^{k-1}$ and the hyperbolic cosine family on $\mathbb{R}$. Let $\nu_p^{(k-1)}$ be the measure on $\mathbb{R}^{k-1}$ as defined in (54.29) with $p > 0$, and let α_p be the measure defined by its Laplace transform on $(-\frac{\pi}{2}, \frac{\pi}{2})$ as $L_{\alpha_p}(\theta) = (\cos \theta)^{-p}$. Let us now consider

$$d\mu_p(x_1, \ldots, x_k) = d\nu_p^{(k-1)}(x_1, \ldots, x_{k-1}) d\alpha_{\sum_{i=1}^{k-1} x_i + p}(x_k).$$

We then have

$$\Theta(\mu_p) = \left\{ \boldsymbol{\theta} \in \mathbb{R}^k : \sum_{i=1}^{k-1} e^{\theta_i} < \cos \theta_k \right\},$$

$$L_\mu(\boldsymbol{\theta}) = \left(\cos \theta_k - \sum_{i=1}^{k-1} e^{\theta_i} \right)^{-p},$$

and

$$M_{F_p} = (0, \infty)^{k-1} \times \mathbb{R},$$

and the variance function of this NEF is given by

$$V_{F_p}(\boldsymbol{m}) = \frac{1}{p}\, \boldsymbol{mm}^T + \mathrm{diag}\left(m_1, \ldots, m_{k-1}, \sum_{i=1}^{k-1} m_i + p \right).$$

Once again, this family consists of distributions of random variables $(X_1, \ldots, X_k)$, where $(X_1, \ldots, X_{k-1})$ has a negative multinomial distribution and X_k conditioned on $(X_1, \ldots, X_{k-1})$ has a hyperbolic cosine distribution with parameter $\sum_{i=1}^{k-1} X_i + p$.

In concluding this subsection, it is worth pointing out the following interesting structural property of quadratic NEFs. If $(X_1, \ldots, X_k)$ belongs to a simple quadratic NEF, then the marginal distribution of X_1 belongs to the Morris class, and for any $d = 2, \ldots, k$, the conditional distribution of X_d, conditioned on $(X_1, \ldots, X_{d-1})$, also belongs to the Morris class with Jørgensen parameter, depending on an affine transformation of $(X_1, \ldots, X_{d-1})$. However, such combinations do not always result in simple quadratic NEFs.

8.3   Characterizations

As seen already, simple quadratic class does not contain any new distribution and produces only combinations of conditional distributions of univariate Morris class. However, it appears in a natural way when generalizing various characterization results known for the univariate quadratic class of distributions. Different characterizations of Morris class of distributions are available in the literature, with some of them based on orthogonal polynomials; see, for example, Meixner (1934), Shanbhag (1979), and Feinsilver (1991). Meixner (1934), using different terminology, established that if $F = F(\mu)$ is a NEF on $\mathbb{R}$ where μ is a probability measure with mean 0 (note that such a measure μ may exist after a translation), then F is quadratic if the sequence of μ-orthogonal polynomials Q_n, where Q_n

is of degree n and monic (meaning that the coefficient of x^n is 1), has an exponential generating function of the form

$$\sum_{n=0}^{\infty} \frac{z^n}{n!} \, Q_n(x) = e^{a(z)x+b(z)}.$$

Feinsilver's (1991) characterization specifies one particular sequence of μ-orthogonal polynomials: If $f_\mu(x, m)$ is the density function of the probability measure $P_{m,F}$ with respect to μ and if $P_n(x) = \frac{\partial^n}{\partial m^n} \, f_\mu(x, m)|_{m=0}$, then P_n is a monic polynomial of degree n and the sequence P_n is μ-orthogonal iff F is quadratic. Pommerêt (1997) has extended these two characterization results to $\mathbb{R}^k$ through polynomials obtained by differentiating the density function $f_\mu(\boldsymbol{x}, \boldsymbol{m})$ with respect to the mean in specific directions. Note that the structure of the simple quadratic NEF can be seen from the polynomials themselves. Indeed, each simple quadratic NEF is a combination of quadratic NEFs on $\mathbb{R}$, and its polynomials are also combinations of the associated polynomials on $\mathbb{R}$. For example, the polynomials of the $(NM - Ga)_0$ NEF on $\mathbb{R}^2$ (which is a combination of gamma and normal NEFs) are combinations of Laguerre and Hermite polynomials; similarly, the polynomials of the $(NM - Ga)_1$ NEF (which is a combination of negative binomial and gamma NEFs) are combinations of Meixner and Laguerre polynomials.

Shanbhag (1979), in a similar vein, used Bhattacharrya matrices (whose coefficients also involve derivatives of densities) and proved that these matrices are diagonal only in the case of real quadratic NEFs. Pommerêt (1997) has established the multivariate extension of this result, namely, that the Bhattacharrya matrices are diagonal only for simple quadratic NEFs. He has also introduced a weaker condition, called the *pseudodiagonality*, for quadratic NEFs.

Pommerêt (1997) derived the variance of the UMVUE of any real function of the parameter in the case of quadratic NEFs. As mentioned earlier, Kokonendji and Seshadri (1996) have derived explicitly the generalized variance function $|V_F(\boldsymbol{m})|$. The UMVUE of the variance itself is simply $\frac{n}{n+a}V_F(\bar{\boldsymbol{X}}_n)$, when $\bar{\boldsymbol{X}}_n$ is the sample mean of $\boldsymbol{X}_1, \ldots, \boldsymbol{X}_n$, each distributed as $P(\boldsymbol{m}, F)$. This result is not true for general quadratic NEFs on $\mathbb{R}^k$ [see Casalis (1992)] and, in fact, remains as a conjecture even in the one-dimensional case that this characterizes the simple quadratic class [see Letac (1992)].

8.4 Extensions

As in the univariate case, it will be natural to consider cubic variance functions on $\mathbb{R}^k$. With this purpose in mind, Hassairi (1994) observed that all cubic variance functions on $\mathbb{R}$ can be obtained from quadratic ones by a specific action on the linear group of $\mathbb{R}^2$ defined by the following transformation T_g: If h_g is the homography $h_g(x) = \frac{\gamma + \delta x}{\alpha + \beta x}$, then if F_1 is a NEF, one can define another NEF F_2 through its variance function

$$V_{F_2}(m) = T_g(V_{F_1})(m) = \frac{(\alpha + \beta m)^3}{(\alpha \delta - \beta \gamma)^2} \, V_{F_1} \left(\frac{\gamma + \delta m}{\alpha + \beta m} \right);$$

when F_1 is quadratic, F_2 is cubic. Use of such transformations on $\mathbb{R}^k$ enabled Hassairi (1994) to obtain a subclass of cubic variance functions on $\mathbb{R}^k$.

Another extension of the Morris class is through all diagonal variance functions, namely, V_F whose diagonal is of the form $(a_1(m_1), \ldots, a_k(m_k))$ for any $\boldsymbol{m} = (m_1, \ldots, m_k) \in M_F$. Note here that the ith term in the diagonal is a function of m_i only. This concept is linked to the cuts discussed earlier in Section 54.7. Thus, in this case, all the marginal distributions of the corresponding NEF belong to NEFs as well. But, this assumption is rather restrictive since Bar-Lev *et al.* (1994) have derived only six types of diagonal NEFs. It also turns out that all the corresponding marginal distributions belong to the Morris class.

BIBLIOGRAPHY

(Some bibliographical items not mentioned in the text are included here for completeness.)

Arnold, B. C., Castillo, E., and Sarabia, J. M. (1993). Conjugate exponential family priors for exponential family likelihoods, *Statistics*, **25**, 71–77.

Bar-Lev, S. K., Bshouty, D., and Enis, P. (1991). Variance functions with meromorphic means, *Annals of Probability*, **19**, 1349–1366.

Bar-Lev, S. K., Bshouty, D., Enis, P., Letac, G., Li-Lu, I., and Richards, D. (1994). The diagonal multivariate natural exponential families and their classification, *Journal of Theoretical Probability*, **7**, 883–929.

Barndorff-Nielsen, O. E. (1978). *Information and Exponential Families in Statistical Theory*, Chichester, England: John Wiley & Sons.

Barndorff-Nielsen, O. E., and Blaesild, P. (1983). Exponential models with affine dual foliations, *Annals of Statistics*, **11**, 753–769.

Barndorff-Nielsen, O. E., Blaesild, P., and Eriksen, P. S. (1989). *Decomposition and Invariance of Measures, and Statistical Transformation Models*, Lecture Notes in Statistics—**58**, Heidelberg: Springer-Verlag.

Barndorff-Nielsen, O. E., and Cox, D. R. (1979). Edgeworth and saddle-point approximations with statistical applications, *Journal of the Royal Statistical Society, Series B*, **41**, 279–312.

Barndorff-Nielsen, O. E., and Cox, D. R. (1994). *Inference and Asymptotics*, London: Chapman and Hall.

Barndorff-Nielsen, O. E., and Koudou, A. E. (1995). Cuts in natural exponential families, *Theory of Probability and Its Applications*, **40**, 361–372.

Bernadac, E. (1995). Random continued fractions and inverse Gaussian distribution on a symmetric cone, *journal of Theoretical Probability*, **8**, 221–260.

Bildikar, S., and Patil, G. P. (1968). Multivariate exponential-type distributions, *Annals of Mathematical Statistics*, **39**, 1316–1326.

Brown, L. D. (1986). *Fundamentals of Statistical Exponential Families*, Lecture Notes – Monographs Series No. **9**, Hayward, California: Institute of Mathematical Statistics.

Casalis, M. (1990). Familles exponentielles naturelles invariantes par un sous-groupe affine, Thèse de l'Université Paul Sabatier, Toulouse, France.

Casalis, M. (1992). Un estimateur de la variance pour une famille exponentielle naturelle à fonction variance quadratique, *Comptes Rendus, Academy of Sciences, Paris*, **314**, Série I, 143–146.

Casalis, M. (1996). The $2d + 4$ simple quadratic natural exponential families on $\mathbb{R}^d$, *Annals of Statistics*, **24**, 1828–1854.

Casalis, M. (1997). Private communication.

Casalis, M., and Letac, G. (1994). Characterization of the Jørgensen set in generalized linear models, *Test*, **3**, 145–162.

Casalis, M., and Letac, G. (1996). The Lukacs–Olkin–Rubin characterization of Wishart distributions on symmetric cones, *Annals of Statistics*, **24**, 763–786.

Consonni, C., and Veronese, P. (1992). Conjugate priors for exponential families on $\mathbb{R}^d$, *Journal of the American Statistical Association*, **87**, 1123–1127.

Daniels, H. E. (1980). Exact saddlepoint approximations, *Biometrika*, **67**, 59–63.

Del Castillo, J. (1994). The singly truncated normal distribution: a non-steep exponential family, *Annals of the Institute of Statistical Mathematics*, **46**, 57–66.

Diaconis, P., and Ylvisaker, D. (1979). Conjugate priors for exponential families, *Annals of Statistics*, **7**, 269–281.

Efron, B. (1978). The geometry of exponential families, *Annals of Statistics*, **6**, 362–376.

Efron, B., and Tibshirani, R. (1996). Using specially designed exponential families for density estimation, *Annals of Statistics*, **24**, 2431–2461.

Feinsilver, P. (1991). Some classes of orthogonal polynomials associated with martingales, *Proceedings of the American Mathematical Society*, **98**, 298–302.

Fraser, D. A. S., Reid, N., and Wong, A. (1991). Exponential linear models: A two-pass procedure for saddlepoint approximation, Technical Report, University of Toronto, Toronto, Canada.

Gutiérrez-Peña, E., and Smith, A. F. M. (1995). Conjugate parameterizations for natural exponential families, *Journal of the American Statistical Association*, **90**, 1347–1356.

Hassairi, A. (1994). Classification des familles exponentielles naturelles dans $\mathbb{R}^d$ de variance cubique du type Mora, Thèse de l'Université Paul Sabatier, Toulouse, France.

Hladik, J. (1969). *La transformation de Laplace à plusieurs variables*, Paris: Masson & Cie.

Holgate, P. (1964). Estimation for the bivariate Poisson distribution, *Biometrika*, **51**, 241–245.

Jani, P. N., and Singh, A. K. (1996). On the characterization of multiparameter exponential family and of some irregular families of distributions, *Calcutta Statistical Association Bulletin*, **46**, 181–182.

Jensen, J. L. (1995). *Saddlepoint Approximations*, Oxford, England: Oxford University Press.

Johnson, N. L., Kotz, S., and Balakrishnan, N. (1994). *Continuous Univariate Distributions*, Vol. 1, second edition, New York: John Wiley & Sons.

Johnson, N. L., Kotz, S., and Balakrishnan, N. (1995). *Continuous Univariate Distributions*, Vol. 2, second edition, New York: John Wiley & Sons.

Johnson, N. L., Kotz, S., and Balakrishnan, N. (1997). *Discrete Multivariate Distributions*, New York: John Wiley & Sons.

Johnson, N. L., Kotz, S., and Kemp, A. W. (1992). *Univariate Discrete Distributions*, second edition, New York: John Wiley & Sons.

Jørgensen, B. (1987). Exponential dispersion models, *Journal of the Royal Statistical Society, Series B*, **49**, 127–162.

Jørgensen, B. (1997). *The Theory of Dispersion Models*, New York: Chapman and Hall.

Jørgensen, B. and Letac, G. (1999). Babel variance functions, Preprint.

Kokonendji, C. C. (1993). Familles exponentielles naturelles réelles de fonction variance en $R\Delta + Q\sqrt{\Delta}$, Thèse de l'Université Paul Sabatier, Toulouse, France.

Kokonendji, C. C. (1994). Exponential families with variance functions in $\sqrt{\Delta}P(\sqrt{\Delta})$: Seshadri's class, *Test*, **3**, 123–172.

Kokonendji, C. C. (1995). Sur les familles exponentielles naturelles de grand-Babel, *Annales de la Faculté des Sciences de Toulouse*, IV **4**, 763–799.

Kokonendji, C. C., and Seshadri, V. (1996). On the determinant of the second derivative of a Laplace transform, *Annals of Statistics*, **24**, 1813–1827.

Letac, G. (1986). La réciprocité des familles exponentielles sur **R**, *Comptes Rendus, Academy of Sciences, Paris*, **303**, Série I, 61–64.

Letac, G. (1987). Discussion of B. Jørgensen, "Exponential dispersion models", *Journal of the Royal Statistical Society, Series B*, **49**, 154.

Letac, G. (1989a). Le problème de la classification des familles exponentielles naturelles dans $\mathbb{R}^d$ ayant une fonction variance quadratique, in *Probability Measures on Groups*, Lecture Notes in Mathematics—**1306**, pp. 194–215, Berlin: Springer-Verlag.

Letac, G. (1989b). A characterization of the Wishart exponential families by an invariance property, *Journal of Theoretical Probability*, **2**, 71–86.

Letac, G. (1992). *Lectures on Natural Exponential Families and their Variance Functions*, Monografias de Matemática No. **50**, Instituto de Matemática Pura e Aplicada, Rio de Janeiro, Brazil.

Letac, G., and Massam, H. (1998). Quadratic and inverse regressions for Wishart distributions, *Annals of Statistics*, **26**, 573–595.

Letac, G., and Mora, M. (1990). Natural real exponential families with cubic variance functions, *Annals of Statistics*, **18**, 1–37.

Lloyd, C. J. (1988). Bivariate normal transformation models, *Scandinavian Journal of Statistics*, **15**, 177–185.

Massam, H. (1994). An exact decomposition theorem and a unified view of some related distributions for a class of exponential transformation models on symmetric cones, *Annals of Statistics*, **22**, 369–394.

Meixner, J. (1934). Orthogonal Polynomsysteme mit einer besonderen Gestalt der erzeugenden Function, *Journal of the London Mathematical Society*, **9**, 6–13.

Mora, M. (1986). Classification des fonctions variance cubiques des familles exponentielles sur **R**, *Comptes Rendus, Academy of Sciences, Paris*, **302**, Série I, 587–590.

Morris, C. N. (1982). Natural exponential families with quadratic variance function, *Annals of Statistics*, **10**, 65–80.

Nelder, J. A., and Wedderburn, R. W. M. (1972). Generalized linear models, *Journal of the Royal Statistical Society, Series A*, **135**, 370–384.

Pommerêt, D. (1996). Orthogonal polynomials and natural exponential families, *Test*, **5**, 77–111.

Pommerêt, D. (1997). Multidimensional Bhattacharrya matrices and natural exponential families, *Journal of Multivariate Analysis*, **63**, 105–118.

Raiffa, H., and Schlaifer, R. (1961). *Applied Statistical Decision Theory*, Boston: Harvard University Press.

Shanbhag, D. N. (1979). Diagonality of the Bhattacharrya matrix as a characterization, *Theory of Probability and Its Applications*, **24**, 430–433.

Sundberg, R. (1974). Maximum likelihood theory for incomplete data from an exponential family, *Scandinavian Journal of Statistics*, **1**, 49–58.

Tweedie, M. C. K. (1984). An index which distinguishes between some important exponential families, in *Statistics: Applications and New Directions* (J. K. Ghosh and J. Roy, eds.), Proceedings of the Indian Statistical Institute Golden Jubilee International Conference, pp. 579–604, Calcutta, India: Indian Statistical Institute.

Zhao, L. P., Prentice, R., and Self, S. G. (1992). Multivariate mean parameter estimation by using a partly exponential model, *Journal of the Royal Statistical Society, Series B*, **54**, 805–811.

Author Index

The purpose of this index is to provide readers with quick and easy access to the contributions (pertinent to this volume) of any individual author, and not to highlight any particular author's contribution.

Abazliev, A. K. J., 3, 4, 85
Abbe, E. N., 122, 223
Abdul-Hamid, H., 19, 85
Abraham, B., 552, 575, 609, 617
Abrahamson, I. G., 148, 223
Abramowitz, M., 348, 633, 652
Achcar, J. A., 362, 416
Adelfang, S. I., 443, 444, 483
Adler, R., 98
Adrian, R., 252, 333
Afifi, A. A., 79, 80, 96, 183, 223, 298, 333
Afonja, B., 487, 527
Aggarwala, R., 388, 417
Ahmed, N. A., 557, 573, 603, 615
Ahmed, S. E., 191, 192, 223
Ahsanullah, M., 157, 159, 223, 278-280, 297, 333, 600, 601, 619
Aitchison, J., 197, 223, 531, 532, 543, 548
Aitken, A. C., 204, 224
Aitkin, M. A., 312, 333
Akesson, O. A., 326, 334
Albers, W., 276, 334
Alhumound, J. M., 384, 427
Ali, A. M., 191, 224
Ali, M. M., 568, 574
Al-Mousawi, J. S., 195, 241
Al-Mutairi, D. K., 381, 416
Alperin, M. T., 647, 654
Al-Saadi, S. D., 373, 400, 416, 417
Altham, P. M. E., 512, 523
Amos, D. E., 270, 334
Ananda, M. M. A., 193, 244
Anderson, C. W., 647, 655
Anderson, D. E., 126, 224

Anderson, D. N., 33, 72, 73, 85
Anderson, M. R., 153, 224
Anderson, R. L., 508, 526
Anderson, T. W., 113, 114, 183, 224, 231, 253, 292, 300, 334, 530, 548
Andrews, F. C., 205, 227
Anis, A. A., 141, 224
Antelman, G. R., 514, 523
Aoshima, M., 195, 235
Apostolakis, G. E., 367, 417
Ardanuy, R., 79, 85
Arnold, B. C., 71-73, 85, 157-159, 216, 224, 225, 292, 316, 317, 334, 365, 370, 371, 392, 395, 417, 552, 564, 565, 567, 568, 570, 573, 574, 577, 582, 583, 585, 587, 589-593, 599-601, 603-606, 608-611, 613, 615, 616, 676, 691
Asha, G., 387, 417
Ashford, J. R., 270, 346
Ashlock, J. C., 639, 656
Assaf, D., 62, 85, 403, 417
Avérous, J., 77, 84-86
Awad, A. M., 366, 417
Azzalini, A., 86, 215-217, 225, 260, 316, 317, 331, 334
Azzam, M. M., 366, 417

Bacon, R. H., 136, 139, 140, 225
Badahdah, S. O., 191, 192, 223
Baggs, G. E., 358, 365, 403, 417, 426
Bain, L. J., 366, 418
Bairamov, I. G., 32, 86, 121, 225
Balakrishna, N., 598, 599, 614, 616

697

Balakrishnan, N., 33, 38, 40, 43, 81, 92, 93, 106, 110, 115, 119-121, 151, 155, 156, 160, 161, 196, 200-202, 206, 211, 212, 215, 219, 220, 225, 236, 260, 274, 279, 281, 286, 287, 294, 306, 310, 316, 328, 330, 334, 337, 340, 347, 349, 363, 365, 372, 374, 375, 388, 403, 407, 417, 418, 423, 424, 426, 428, 431, 433, 434, 436, 440, 443, 459, 470, 475, 480, 485, 487-489, 500, 507, 515, 525, 530, 532, 540, 542, 543, 548-551, 553, 562, 570, 574-580, 584, 585, 589, 595, 596, 602, 604, 605, 608, 615-617, 622, 623, 629, 638, 641, 648-650, 655, 660, 662, 663, 666, 668, 670, 673, 678, 680, 681, 686, 688, 694

Balanda, K. P., 84, 86
Balasubramanian, K., 368, 418
Baldessari, B., 94, 429, 575
Bansal, N., 279, 333
Banys, M. I., 86
Bapat, R. B., 477, 478
Barakat, R., 212, 225
Baranchik, A. J., 162, 190, 225
Bargmann, R. E., 183, 248, 298, 347
Baringhaus, L., 78-80, 86
Bar-Lev, S. K., 663, 676, 681, 688, 691
Barlow, R. E., 387, 418, 419
Barndorff-Nielsen, O. E., 540, 548, 659, 661, 665, 668, 669, 671, 674-676, 691, 692
Barnes, J. W., 432, 481
Barnett, V., 352, 418
Basford, K. E., 221, 240
Basu, A. P., 283, 334, 349, 362, 363, 386, 391, 396-398, 404, 405, 417-419, 423, 424, 426, 428, 429, 512, 525
Basu, D., 151, 152, 225
Beattie, A. W., 207, 226
Beaver, R. J., 216, 224, 316, 317, 334
Becker, A., 175, 176, 226
Becker, P. J., 446, 478
Beg, M. I., 368, 418
Begum A. A., 412, 418

Bélisle, C. J. P., 38, 86
Bemis, B. M., 366, 418
Bens, G. B., 472, 478
Berge, P. O., 64, 86
Berger, J., 163-166, 226
Berland, R., 445, 446, 479
Berman, S. M., 650, 652
Bernadac, E., 684, 692
Bernardo, J. M., 507, 523
Bernstein, R. I., 472, 482
Bhargava, R. P., 183, 195, 226, 246
Bhattacharya, P. K., 164, 226, 285, 334
Bhattacharya, S. K., 381, 418
Bhattacharyya, A., 280, 334, 379, 418
Bhattacharyya, G. K., 355, 366, 367, 415, 418, 419, 425
Bhoj, D. S., 172, 239
Bickel, P. J., 121, 226
Bikelis, A., 14, 15, 86, 87
Bildikar, S., 28, 29, 87, 153, 226, 253, 335, 660, 692
Bilodeau, M., 354, 419
Birnbaum, Z. W., 205, 226, 227, 318, 325, 335
Bischoff, W., 158, 227
Bjarnason, H., 415, 419
Bjerve, S., 332, 335
Blaesild, P., 212, 227, 669, 671, 692
Bland, R. P., 135, 137, 227
Block, H. W., 362, 364, 375, 386, 389, 396, 397, 404, 419, 575
Blumenson, L. E., 462, 472, 476, 478, 482
Bock, M. E., 172, 236
Bofinger, E., 271, 335
Bofinger, V. J., 271, 335
Boland, P. J., 364, 419
Bol'shev, L. N., 270, 345
Bordes, L., 388, 419
Borth, D. M., 274, 335
Bose, S. S., 452, 453, 478
Boswell, M. T., 154, 242
Box, G. E. P., 43, 81, 87
Boyer, J. E., 57, 100
Bratoeva, Z. N., 13, 100
Bravais, A., 253, 335
Braverman, M. S., 281, 335
Breslow, N. E., 367, 427
Brewer, D. W., 444, 445, 478

Brewster, J. F., 193, 227
Briden, J., 170, 237
Brockett, P. L., 369, 419
Broffitt, J. D., 113, 227
Brown, J. L., 127, 227
Brown, L. D., 193, 227, 659, 661, 665, 692
Browne, M. W., 87
Brucker, J., 277, 278, 335
Bryc, W., 34, 87
Bshouty, D., 663, 676, 681, 688
Buchberger, S. G., 287, 346
Burdick, G. R., 429
Burington, R. S., 272, 335
Burkhart, H. E., 22, 93
Butler, J., 137, 227
Byczkowski, T., 18, 87

Cadwell, J. H., 269, 335
Cain, M., 282, 283, 335
Cambanis, S., 52, 53, 87
Camp, B. H., 323, 335
Campbell, L. L., 160, 227
Capéraà, P., 653
Capitanio, A., 217, 225
Cardullo, M. J., 212, 242
Carter, E. M., 183, 246
Carter, G. M., 173, 227
Cartinhour, J., 207, 209, 227
Casalis, M., 659, 663, 670, 671, 673, 677, 684, 685, 690, 692
Casella, G., 163, 228
Castellan, N. J., 323, 335
Castillo, E., 71, 72, 85, 87, 158, 159, 224, 225, 228, 292, 334, 352, 353, 411, 419, 582, 583, 587, 589-591, 593, 604, 611, 615, 616, 676, 691
Chakak, A., 50, 87
Chamayou, J.-F., 491, 523
Chambers, J. M., 6, 14, 87
Chang, C.-H., 170-172, 242
Charlier, C. V. L., 326, 335
Charnley, F., 326, 336
Chatfield, C. R., 509, 510, 523, 524
Chaudhuri, G., 170-172, 242
Chaudhuri, P., 84, 88
Chen, C.-C., 376, 429
Chen, D., 367, 420

Chen, L., 174, 175, 228
Chen, S. Y., 166, 168, 228
Cheng, M. C., 145, 147, 228
Cheong, Y.-H., 490, 526
Cheriyan, K. C., 433, 435, 456, 478
Chew, V., 202, 228, 336
Chikuse, Y., 17, 88
Childs, D. R., 129, 228
Chiou, W., 176, 228
Chou, Y.-M., 315, 317, 336
Chow, Y. S., 195, 228
Chukova, S., 69, 88, 89
Church, A. E. R., 266, 337
Cirel'son, B. S., 116, 228
Clarke, R. T., 432, 478
Clayton, D. G., 413, 414, 420
Cloninger, C. R., 275, 344
Cochran, W. G., 509, 523
Cohen, A., 176, 228
Cohen, A. C., 121, 225, 228, 336
Cohen, L., 30, 88
Cole, B. F., 32, 88
Coles, S. G., 641, 643, 646, 647, 653, 656
Connor, R. J., 490, 501, 521, 523
Consonni, C., 677, 693
Conway, D., 88
Cook, R. D., 382, 420, 570, 574
Cooper, B. E., 274, 336
Cooper, P. W., 88
Cowan, R., 385, 420
Cox, D. R., 80, 88, 412, 420, 661, 674, 692
Coxeter, H. S. M., 228
Craiu, M., 490, 523
Craiu, V., 490, 523
Cramer, E., 389, 420
Cramér, H., 487, 523
Crofts, A. E., 22, 88
Crow, E. L., 28, 88
Crowder, M., 410, 420
Crowley, J., 356, 360, 425
Csörgő, M., 155, 228
Cuadras, C. M., 46, 88
Curnow, R. N., 134, 229
Currit, A., 381, 420
Czuber, E., 253, 336

Dagsvik, J., 643, 653

Dahel, S., 185, 186, 229
Dahiya, R. C., 305, 336
Daley, D. J., 274, 336
Dall'Aglio, G., 46, 88, 94
Dalla Valle, A., 86, 215, 216, 225, 260, 331, 334
Damaraju, C. V., 384, 427
Daniels, H. E., 674, 693
Darroch, J. N., 489, 501, 523
Das, S. C., 134, 136, 229
DaSilva, A. G., 191, 229
David, F. N., 140, 148, 149, 229, 433, 445, 478
David, H. A., 121, 136, 229, 283, 285, 286, 336, 337, 342
Davis, A. W., 82, 89
Day, N. E., 30, 89, 220, 221, 229, 326, 337
Deàk, I., 139, 229
Deddens, J. A., 287, 346
Deemer, W. L., 319, 347
D'Este, G. M., 441, 479
DeGroot, M. H., 162, 163, 229, 597, 616
de Haan, L., 641, 653
Deheuvels, P., 621, 653
Del Castillo, J., 668, 693
DeLury, D. B., 297, 337
Dener, A., 652, 653
Dennis, S. Y., 523, 524
de Ronde, J., 653
De Silva, B. M., 17, 89
Des Raj, S., 314, 337
Dey, D. K., 194, 195, 229, 381, 411, 420, 423, 427
Diaconis, P., 529, 549, 676, 693
Dickey, J. M., 180, 181, 230
Dickson, I. D. H., 253, 337
DiDonato, A. R., 139, 230
Dimitrov, B., 69, 88, 89
Dirichlet, P. G. L., 487, 524
D'jachenko, Z. N., 431, 436, 439, 479
Doksum, K., 332, 335
Doktorov, B. Z., 188, 230
Donnelly, T. G., 139, 230, 274, 337
Downton, F., 365, 371, 372, 420
Dowson, D. C., 110, 230
Drezner, Z., 148, 149, 151, 230, 272, 273, 276, 337
Dudewicz, E. J., 653

Dunn, O. J., 113, 230
Dunn, R., 527
Dunnage, J. E. A., 16, 17, 89
Dunnett, C. W., 124, 125, 134, 139, 229, 230, 253, 266, 337
Dunsmore, I. R., 197, 223
Durling, F. C., 609, 616
Dussauchoy, A., 445, 446, 479
Dykstra, R. L., 113, 243

Eagleson, G. K., 11, 89, 431, 479
Eaton, M. L., 114, 152, 157, 230, 231
Ebrahimi, N., 352, 362, 420
Edgeworth, F. Y., 6, 89
Efron, B., 168, 170, 173, 231, 661, 667, 668, 693
Ehrenberg, A. S. C., 509, 510, 524
Eichenauer-Hermann, J., 174, 175, 228
Eidemiller, R. L., 315, 341
Elandt-Johnson, R. C., 58, 89
Elashoff, R. M., 183, 223, 298, 333
Elderton, E. M., 266, 337
Elderton, W. P., 8, 9, 89
Elfessi, A., 168, 179, 180, 242
Elston, R. C., 275, 341
Enis, P., 663, 676, 681, 688, 691
Eriksen, P. S., 671, 692
Ernst, M. D., 215, 231
Esary, J. D., 396, 420, 650, 653
Escoufier, Y., 122, 231
Evans, M., 129, 130, 231
Everitt, P. F., 266, 337

Fabius, J., 500-502, 524
Fairweather, W. R., 155, 231
Fang, B. Q., 530, 549
Fang, K. T., 44, 94, 114, 115, 156, 231, 529-532, 548-550
Farewell, V. T., 367, 427
Farlie, D. J. G., 52, 89, 623, 653
Feinsilver, P., 689, 690, 693
Feldheim, M. E., 17, 89
Feldman, R. E., 98
Fernandez, C., 20, 89
Fieger, W., 158, 227
Fielitz, B., 504, 524
Fieller, E. C., 219, 231, 266, 272, 327, 337, 338
Filus, J. K., 217, 218, 231

Filus, L., 217, 218, 231
Findeisen, P., 160, 231
Finkelshteyn, A., 621, 653
Finney, D. J., 232, 452, 453, 478
Fisher, R. A., 338
Fisk, P. R., 153, 232
Fix, E., 433, 445, 478
Fleishman, A. I., 36, 37, 90
Flournoy, N., 367, 427
Fomin, Ye. A., 472, 478
Foster, K., 78, 96
Fougères, A.-L., 653
Frankowski, F., 500, 527
Fraser, D. A. S., 200, 232, 278, 338, 674, 693
Fréchet, M., 44, 47, 90, 151, 232, 654
Freeman, P. R., 98
Freund, R. J., 350, 355, 361, 421, 446, 447, 479
Friday, D. S., 369, 370, 421
Fuchs, C., 202, 204, 232
Fujisawa, H., 183, 185, 232
Fussell, J. B., 419, 429

Gail, M., 367, 421
Galambos, J., 66, 71, 87, 90, 158, 228, 286, 336, 387, 411, 419, 421, 621, 648, 650, 653, 654
Galton, F., 4, 90, 253, 338
Ganeshalingam, S., 610, 616
Gastwirth, J. L., 79, 90
Gauss, C. F., 338
Gaver, D. P., 437, 456, 479
Gayen, A. K., 38, 90
Geary, R. C., 152, 232, 328, 338
Gebizlioglu, O. L., 121, 225
Geffroy, J., 621, 623, 624, 654
Gehrlein, W. V., 232
Geisser, S., 188, 232
Gelfand, A. E., 194, 195, 229
Gelman, A., 280, 338
Genest, C., 413, 421, 561, 574, 653
Genz, A., 144, 232
George, E. O., 564, 575
Ghirtis, G. C., 435, 479
Ghosh, J. K., 283, 334, 696
Ghosh, M., 179, 194, 245
Ghosh, P., 19, 90
Ghosh, S., 615

Ghurye, S. G., 151, 196, 232, 292, 334, 383, 421
Gill, P. S., 306, 347
Giri, N., 169, 170, 185, 186, 229, 243
Gleser, L. J., 190, 233
Godwin, H. J., 65, 90
Gokhale, D. V., 557, 573, 603, 615
Goldstein, N., 624, 654
Gomes, M. I., 647, 654
Good, I. J., 25, 90
Goodhardt, G. J., 509, 510, 523, 524
Goodman, I., 90
Goodman, N. R., 223, 233
Gould, S. J., 220, 233
Gradshteyn, I. S., 583, 616
Graybill, F. A., 155, 237
Greenland, S., 326, 338
Griffith, W. S., 423
Griffiths, R. C., 12, 90, 445, 457, 464, 477, 479
Grizzle, J. E., 512, 526
Groeneveld, R. A., 216, 224, 316, 317, 334
Grundy, P. M., 338
Guldberg, S., 13, 90
Gumbel, E. J., 350, 355, 421, 551-554, 561, 563, 575, 623, 624, 627-629, 638, 654
Gunst, R. F., 439, 480
Gupta, A. K., 18, 91, 210, 233, 441, 442, 480
Gupta, P. L., 116, 118, 233, 421
Gupta, R. C., 116, 118, 233, 338, 421
Gupta, R. D., 503, 524, 529-532, 534, 536-544, 547, 549
Gupta, R. P., 618
Gupta, S. D., 113, 233
Gupta, S. S., 132, 134, 233, 285, 338
Gutiérrez-Peña, E., 677, 693
Guttman, I., 200-202, 232, 233, 492, 527

Hadi, A. S., 352, 353, 419
Haff, L. R., 173, 179, 233, 234
Hafley, W. L., 22, 100
Hageman, R. K., 139, 230
Hagen, E. W., 367, 421
Hagis, P., 451, 453, 481
Haider, A. M., 183, 234

Haldane, J. B. S., 264, 338
Hall, I. J., 202, 234
Hamdan, M. A., 314, 339, 344, 366, 417, 437
Hamedani, G. G., 157, 161, 234, 278, 279, 333, 339
Han, C. P., 191, 229, 307-309, 339, 340
Hanagal, D. D., 359, 360, 366, 370, 397, 421, 422, 601, 613, 616
Hannan, J. F., 234
Haq, M. S., 568, 574
Haro-López, R. A., 411, 422
Harrell, F. E., 306, 339
Harris, B., 122, 234
Harris, R., 91
Harris, R., 654
Harris, W. A., 68, 91
Hartley, H. O., 91 (see also Hirschfeld, H. O.)
Hartley, M. J., 582, 618
Hashino, M., 361, 422
Hashorva, E., 54, 91
Hassairi, A., 691, 693
Hawkes, A. G., 375, 422
Hawkins, D. M., 509, 524
Hayakawa, Y., 379, 422
Healy, M. J. R., 338
Heinrich, G., 361, 422
Helemäe, H.-L., 47, 91
Helmert, F. R., 253, 339
Helvig, T. N., 68, 91
Henze, N., 78-81, 86, 91, 215, 234
Hickernell, F. J., 144, 234
Higgins, J. J., 366, 418
Hill, R. C., 172, 236
Hinkley, D. V., 327, 328, 339
Hirschfeld, H. O., 11, 91 (see also Hartley, H. O.)
Hladik, J., 682, 693
Hocking, R. R., 234, 246, 298, 300, 339, 345
Hoeffding, W., 309, 339, 416, 422
Holgate, P., 663, 693
Holland, P. W., 73-75, 91, 280, 339
Holmes, P. T., 361, 429
Holmquist, B., 108, 109, 234
Hong, H. S., 144, 234
Horswell, R. L., 83, 91
Hotelling, H., 234

Houdré, C., 115, 234
Hougaard, P., 409, 412, 414-416, 419, 422, 423, 629, 635, 654
Hsu, J. C., 125, 246
Huang, J. S., 52, 57, 62, 91
Hudson, H. M., 163, 235
Hughes, H. M., 339
Hughes-Oliver, J. M., 367, 420
Humbert, P., 497, 524
Hume, M. W., 333
Hurlburt, R. T., 508, 524
Hüsler, J., 53, 54, 91, 632, 654
Hutchinson, T. P., 71, 92, 369, 423, 454, 480, 570, 575, 611, 616
Hyakutake, H., 195, 235, 368, 423
Hyrenius, H., 326, 339

Ibragimov, I. A., 116, 228
Iglewicz, B., 131, 133, 135, 238
Ihm, P., 133, 134, 235
Isii, K., 65, 92
Isogai, T., 81, 83, 92
Ivshin, V. V., 199, 200, 235
Iwase, K., 28, 92
Iyengar, S., 260, 339
Iyer, P. V. K., 272, 339

Jaisingh, L. R., 423
Jaiswal, M. C., 319, 339, 340
Jakeman, E., 212, 235
James, A. T., 154, 235
James, I. R., 504, 521, 522, 524
James, W., 161, 169, 235
Jani, P. N., 428, 670, 693
Jao, J. K., 212, 249
Jarnagin, M. P., 139, 230
Jayakumar, B., 598, 599, 614, 616
Jeevanand, E. S., 595, 596, 616
Jeffreys, H., 595, 616
Jensen, D. R., 11, 92, 435, 438, 439, 462, 472-476, 478, 480
Jensen, J. L., 212, 227, 674, 693
Jensen, U., 361, 422
Jinadasa, K. G., 183-185, 235
Joarder, A. H., 9, 92
Joe, H., 116, 143, 144, 235, 621, 632, 646, 650-652, 654, 655
Jogdeo, K., 113, 236
John, S., 122, 126, 201, 202, 236

Johnson, M. E., 8, 73, 92, 214, 241, 382, 420, 570, 574
Johnson, N. L., 8, 9, 20, 22-24, 32, 43, 52, 53, 57, 81, 88, 89, 92, 93, 100, 106, 110, 113, 115, 116, 140, 151, 155, 156, 160, 161, 196, 199-202, 206, 211, 212, 215, 219, 220, 236, 260, 274, 279, 281, 294, 310, 316, 328, 330, 340, 349, 350, 363, 365, 372, 403, 406, 407, 423, 431, 433, 434, 436, 440, 443, 444, 452, 453, 455, 459, 470, 471, 473, 475, 480, 481, 485, 487-489, 500, 505-507, 515, 524, 525, 532, 540, 542, 543, 549, 550, 562, 577-580, 584, 585, 589, 595, 596, 602, 604, 605, 608, 616, 617, 622, 623, 629, 638, 641, 648-650, 655, 660, 662, 663, 666, 668, 670, 673, 678-681, 686, 688, 694
Johnson, R. A., 366, 367, 418, 419
Jones, C. D., 637, 655
Jones, M. C., 75, 93
Jones, M. Q., 472, 480
Jones, R. M., 27, 93
Jørgensen, B., 540, 548, 667, 672, 681, 694
Jørgensen, N. R., 6, 93
Joshi, S. N., 278, 343
Judge, G. G., 172, 236
Jupp, P. E., 601, 617

Kabe, D. G., 196, 236
Kadoya, M., 350, 371, 373, 375, 426
Kagan, A. M., 34, 36, 93, 153, 156, 236, 280, 340, 598, 614, 617
Kalbfleisch, J. D., 367, 427
Kale, B. K., 359, 360, 366, 397, 422
Kalinauskaité, N., 17, 93
Kallenberg, W. C. M., 276, 334
Kamat, A. R., 261-263, 340
Kamps, U., 388, 389, 420, 423
Kanazawa, M., 78, 96
Kariya, T., 215, 238, 354, 419
Karlin, S., 66, 93, 460, 481
Karlqvist, A., 655
Karunamuni, R. J., 165, 237

Kátai, I., 90
Kellogg, S. D., 432, 481
Kemp, A. W., 140, 199, 236, 350, 423, 444, 453, 455, 473, 481, 489, 505, 506, 510, 525, 542, 550, 584, 617, 679, 680, 694
Kemp, C. D., 510
Kemp, J. F., 46, 93
Kendall, M. G., 128, 145, 154, 237, 261, 296, 340, 423
Kenett, R. S., 202, 204, 232
Kennedy, W. J., 615
Kent, J. T., 170, 237, 403, 423
Khalil, Z., 69, 88, 89
Khan, A. H., 412, 418
Khan, R. A., 195, 237
Khatri, C. G., 113, 153, 223, 237, 319, 339, 340, 465, 481
Kibble, W. F., 127, 237, 424, 436, 481, 586, 617
Kingman, A., 155, 237
Kingman, J. F. C., 424
Klebanoff, A. D., 264, 344
Klein, J. P., 358, 424
Kleinbaum, D., 298, 340
Knoebel, B. R., 22, 93
Kocherlakota, K., 38, 41, 93
Kocherlakota, S., 38, 41, 93
Koehler, K. J., 50, 76, 87, 93, 571, 572, 575
Kohberger, R. C., 364, 375, 419, 427
Kokonendji, C. C., 674, 681, 682, 690, 694
Koll, P., 229
Kong, J. A., 212, 249
Konijn, H. S., 627, 655
Koopmans, L. H., 3, 94
Korwar, R. M., 305, 336
Kotlarski, I. I., 532, 550
Kotz, S., 8, 32, 33, 43, 44, 47, 52, 53, 57, 62, 81, 86, 88, 91-94, 100, 106, 110, 113, 115, 116, 140, 151, 155, 156, 160, 161, 196, 199-202, 206, 211, 212, 215, 219, 220, 231, 236, 260, 274, 279, 281, 294, 310, 316, 328, 330, 340, 349, 350, 363, 365, 372, 377, 387, 403, 406, 407, 421, 423, 424, 431, 433, 434,

436, 440, 443, 444, 453, 455, 459, 470, 473, 475, 479, 480, 481, 485, 487-489, 500, 505-507, 515, 525, 529-532, 540, 542, 543, 549, 550, 562, 577-580, 584, 585, 589, 595, 596, 602, 604, 605, 608, 616, 617, 622, 623, 629, 638, 641, 648-650, 655, 659, 660, 662, 663, 666, 668, 670, 673, 678-681, 686, 688, 694

Koudou, A. E., 676, 692
Kovacevic, M. S., 209, 247
Kowalczyk, T., 459, 461, 481
Kowalski, C. J., 113, 237
Koziol, J. A., 81, 94
Krishnaiah, P. R., 451, 453, 465, 481, 657
Krishnamoorthy, A. S., 403, 424, 454, 457, 458, 472, 481, 602, 617
Krishnamoorthy, K., 176, 177, 186, 189, 190, 202, 204, 238
Krishnan, P., 582, 617
Kruskal, W. H., 501
Kubokawa, T., 180, 238
Kudô, A., 238
Kumar, S., 381, 418
Kunte, S., 162, 163, 238
Kusunori, K., 188, 238
Kuwana, Y., 215, 238
Kwong, K. S., 133, 135, 238

Laha, R. G., 152, 238, 240, 472, 481
Lai, C. D., 71, 92, 369, 423, 454, 480
Lakshminarayan, C. K., 309, 340
Lal, D. N., 65, 94
Lal, R., 450, 483
Lamm, R. A., 230, 266, 337
Lancaster, H. O., 3, 10, 11, 94, 113, 239, 340
Landau, B. V., 110, 230
Langberg, N., 367, 396, 403, 417, 424
Langberg, N. A., 62, 85
Lange, K., 502, 511, 525
Laplace, P. S., 340
Lau, K.-S., 407, 424
Lawley, D. N., 205, 239
Le, H., 20, 95, 98
Leandro, R. A., 362, 416

Lee, A., 266, 322, 340
Lee, L., 407, 415, 424, 633, 641, 655
Lee, M.-L. T., 30-32, 60, 61, 88, 95, 381, 429, 431, 438, 481
Lee, M.-Y., 66, 90, 95
Lee, P. A., 489, 497, 500, 525
Lee, P. M., 597, 617
Lefevre, C., 381, 425
Lehmann, E. L., 310, 341, 650, 655
Lehn, J., 174, 175, 228
Leonard, T., 173, 239
Lepage, Y., 185, 186, 229
le Roux, N. J., 446-448, 450, 483
Lesaffre, E., 51, 75, 97
Leser, C. E. V., 66, 95
Letac, G., 156, 239, 491, 523, 659, 663, 665, 670, 671, 673, 676, 681, 684, 688, 690-692, 694, 695
Le Toan, T., 212, 249
Leung, M. Y., 570, 574
Leurgans, S., 356, 360, 425
Lévy, P., 17, 95
Lewin, L., 146, 239
Lewis, T., 272, 338
Lewy, P., 515-519, 525
Li, C.-S., 367, 420
Li, H. C., 155, 156, 239
Li, T. F., 172, 239
Liang, J.-J., 44, 94
Lien, D.-H. D., 26, 95
Li-Lu, I., 663, 676, 681, 688, 691
Lin, G. D., 52, 95
Lin, J.-T., 277, 341
Lin, P. E., 168, 239
Lindley, D. V., 180, 181, 230, 381, 425, 512, 525, 563, 575, 583, 595, 604, 611, 617
Ling, C., 194, 242
Linnik, Yu. V., 33, 95, 153, 154, 236, 250, 598, 614, 617
Liouville, J., 529
Lipow, M., 315, 341
Little, R. J. A., 183, 239
Liu, P.-S. L., 381, 420
Lloyd, C. J., 671, 695
Lloyd, E. H., 141, 224
Lochner, R. H. A., 512-514, 525
Lohr, S. L., 138, 139, 239
Looney, S. W., 83, 91

Lord, F. M., 183, 239
Lu, J., 355, 362, 415, 425
Lu, J.-C., 367, 420
Lukacs, E., 152, 239, 240, 472, 481
Lukatzkaya, M. L., 341
Lukomski, J., 95
Lumel'skii, Ya. P., 196, 199, 200, 235, 240
Lundquist, L., 655
Lurie, S., 639, 656
Lütkepohl, H., 80, 95

Ma, C., 116, 240, 410, 430, 491, 525, 530, 547, 548, 550
MacGillivray, H. L., 84, 86, 96
MacKay, J., 413, 421, 561, 574
Madansky, A., 296, 299, 341
Mahmoud, M. W., 25, 26, 97
Makov, U. E., 221, 248
Malice, M. P., 381, 425
Malik, H. J., 552, 575, 583, 609, 617
Malkovich, J. F., 79, 80, 96
Mallows, C. L., 25, 96, 148, 149, 229, 341
Mamatkulov, K. K., 96
Manatunga, A., 415, 425, 632, 633, 655
Manton, K. G., 367, 412, 429
Mardia, K. V., 5, 47, 71, 77, 78, 96, 170, 237, 382, 425, 579, 582, 584-586, 599, 601, 602, 609, 617
Marin, J., 601, 618
Maritz, J. S., 341
Marsaglia, G., 68, 96, 126, 136, 240, 328, 341
Marshall, A. W., 73, 96, 362, 383, 391, 396, 399, 407, 408, 420, 421, 425, 426, 529, 530, 533, 550, 572, 575, 586, 595, 617, 648, 650, 655
Martinson, N., 437
Martynov, G. V., 122, 240
Maruyama, Y., 240
Massam, H., 684, 695
Masuda, K., 198, 240
Mathai, A. M., 465-468, 481, 525
Mathew, T., 202, 204, 238
Maung, K., 97
Maurelli, V. A., 36, 101
May, D. C., 272, 335

McFadden, D., 643, 655
McFadden, J. A., 148, 240
McKay, A. T., 432, 482
McLachlan, G. J., 221, 240, 426
Mee, R. W., 272, 341
Meeker, W. Q., 216, 224, 316, 317, 334
Mehler, F. G., 127, 240
Mehrotra, K. G., 426
Meixner, J., 13, 97, 689, 695
Melnick, E. L., 113, 240, 341
Mendell, N. R., 275, 341
Meng, X.-L., 280, 338
Meste, M., 84-86
Meyer, P. L., 226
Meyer, R. M., 63, 97
Michalek, J. E., 426
Mihaïla, I. M., 12, 97
Mikhail, N. N., 568, 574
Miller, K. S., 27, 93, 223, 240, 241, 451, 462, 472, 476, 478, 482
Milton, R. C., 124, 241, 341
Minder, C. E., 274, 347
Mises, R. von, 241
Misiewicz, J. K., 530, 544, 547, 549
Miyashita, J. 255, 292, 342
Modarres, R., 18, 97
Moeschberger, M. L., 358, 424
Moffatt, P. G., 562, 575
Moffit, R., 137, 227
Mogyoródi, J., 15, 87
Mole, N., 637, 655
Molenberghs, G., 51, 75, 97
Monhor, D., 508, 525
Moore, D. S., 559, 575
Mora, M., 681, 695
Moran, P. A. P., 123, 134, 135, 140-142, 145, 241, 371, 426, 434, 436, 452, 472, 482
Morgenstern, D., 52, 97
Móri, T. F., 81, 82, 97
Morris, C., 168, 170, 173, 231, 659, 661, 670, 678, 695
Morrison, D., 341, 426
Moschopoulos, P. G., 465-468, 481, 525
Mosimann, J. E., 490, 501, 507, 521, 523, 525
Mostafa, M. D., 25, 26, 97
Mosteller, F., 302, 341
Moul, M., 266, 337

Mudholkar, G. S., 564, 575
Muirhead, R. J., 213, 241
Mukherjea, A., 255, 290, 292, 342
Mukhopadhyay, N., 195, 241
Muliere, P., 595, 618
Muller, M. E., 43, 87
Murray, G. D., 196-198, 241
Mustafi, C. K., 627-629, 638, 654
Myers, B. L., 504, 524

Nabeya, S., 261, 263, 342
Nachtsheim, C. J., 214, 241
Nadarajah, S., 637, 647, 652, 655
Nadler, J., 297, 300, 342
Nagao, H., 195, 241
Nagao, M., 350, 371, 373, 375, 426
Nagaraja, H. N., 284, 286, 337, 342, 358, 365, 403, 426
Nair, K. R. M., 352, 426, 596, 613, 618
Nair, N. U., 352, 387, 417, 426
Nakassis, A., 255, 292, 342
Narayanan, A., 506-508, 518, 526
Narumi, S., 4, 97
Nataf, A., 47, 97
Nath, G. B., 319, 342
Nayak, T. K., 382, 426, 491, 526, 548, 550, 604, 611, 618
Nelder, J. A., 672, 695
Nelsen, R. B., 561, 575
Nelsen, R. G., 57, 97
Nelson, P. R., 135, 241, 653
Neuts, M. F., 62, 97
Neyman, J., 4, 98
Ng, H. K. T., 374, 375, 418
Ng, K. W., 156, 231, 529-532, 549, 550
Nguyen, T. T., 18, 19, 91, 98, 156, 157, 242
Nicholson, C., 268, 327, 342
Nikulin, M., 388, 419
Nolan, J. P., 17-19, 85, 87, 97, 98
Novak, L. M., 212, 242
Novick, M. R., 512, 526

Oakes, D., 412, 413, 415, 425, 426, 559, 575, 632, 633, 655
Obenchain, R. L., 28
O'Cinneide, C. A., 62, 98, 401, 403, 426
O'Hagan, A., 20, 95, 98
Oja, H., 83, 84, 98

Olkin, I., 65, 73, 96, 98, 118, 152, 182, 183, 196, 224, 232, 242, 311, 342, 362, 391, 398, 399, 407, 408, 425-427, 511, 526, 529, 530, 533, 550, 572, 575, 586, 595, 617, 648, 650, 655
Olson, J. M., 275, 276, 342
O'Neill, T. J., 360, 427
Ord, J. K., 71, 98, 122, 154, 247, 479, 616
Osiewalski, J., 20, 89
Ostrovskii, I. V., 34, 98
Owen, D. B., 134, 135, 137, 227, 246, 269, 270, 272, 273, 288, 315, 317, 336, 341-343
Owens, M. E. B., 79, 90
Öztürk, A., 653

Padamadan, V., 595, 596, 613, 616, 618
Pagano, M., 582, 618
Pal, N., 168, 170-172, 179, 180, 194, 242
Pan, E., 283, 335
Pannala, M. K., 186, 238
Panorska, A. K., 19, 98
Papadoulos, A. S., 358, 427
Paranjape, S. R., 478, 482
Parikh, N. T., 314, 315, 343, 345
Parr, W. C., 57, 100
Parthasarathy, M., 403, 424, 454, 457, 458, 472, 481, 602, 617
Patil, G. P., 28-30, 87, 94, 100, 153, 154, 226, 242, 253, 278, 335, 343, 369, 370, 387, 421, 428, 429, 479, 575, 616, 660, 692
Patra, K., 411, 427
Paulauskas, V. J., 16, 17, 99
Paulson, A. S., 364, 375, 419, 427
Paulson, E., 205, 227, 343
Pearson, E. S., 272, 338
Pearson, K., 4, 99, 205, 242, 253, 261, 266, 320-324, 343
Peristiani, S., 53, 99, 137, 138, 243
Perlman, M. D., 114, 231, 529, 549
Perron, F., 169, 170, 180, 243
Peterson, A. V., 367, 427
Pickands, J., 396, 427, 622, 634, 641, 642, 656
Pierce, D. A., 113, 243
Pillai, K. C. S., 285, 338

Pitman, E. J. G., 255, 343
Plackett, R. L., 48, 49, 99, 122, 125, 243, 259, 343
Plana, G. A. A., 253, 343
Plucińska, A., 34, 87
Pólya, G., 266, 343
Pommerêt, D., 690, 695
Posner, E. C., 639, 656
Poss, S. S., 367, 429
Pourahmadi, M., 71, 85, 157-159, 225, 600, 616
Poznyakov, V. V., 125, 149, 243
Pratt, J. W., 65, 98, 182, 242, 311, 342
Predorr, A., 229
Prékopa, A., 442, 458, 459, 482
Prentice, R. L., 367, 427, 667, 696
Press, S. J., 17, 72, 85, 99, 173, 174, 180-182, 230, 243, 298, 344
Pretorius, S. J., 6, 99, 266, 337
Prince, J., 325, 343
Proschan, F., 366, 367, 370, 387, 393, 395, 396, 418, 424, 427, 650, 653
Provost, S. B., 186, 243, 490, 526
Puente, C. E., 264, 344
Pusey, P. N., 212, 235

Quenouille, M. H., 243
Quinzi, A. J., 367, 396, 424

Rafferty, J. A., 319, 347
Raftery, A. E., 377, 401-403, 426, 427
Raiffa, H., 676, 695
Raja Rao, B., 384, 427
Rajput, B., 18, 19, 87, 98
Ramabhadran, V. R., 433, 435, 456, 458, 482
Rao, B. V., 502, 526
Rao, C. R., 110, 151, 153, 156, 236, 243, 337, 407, 424, 428, 507, 526, 598, 614, 617
Rao, M. M., 481
Rastogi, S. C., 298, 344
Ratcliff, D., 489, 501, 523
Ratkowsky, D. A., 303, 304, 344
Rattihalli, R. N., 162, 163, 238, 244
Rayens, W. S., 543, 550
Read, C. B., 100
Rearden, D., 26, 95

Rees, D. H., 338
Regier, M. H., 314, 344
Reich, T., 275, 344
Reid, N., 674, 693
Reiss, R.-D., 632, 654
Resnick, S. I., 641, 653
Reuter, G. E. H., 424
Revankar, N. S., 582, 618
Rhodes, E. C., 5, 99
Rice, J., 275, 344
Richards, D. St. P., 503, 524, 529-532, 534, 536-544, 547, 549, 663, 676, 681, 688, 691
Rinott, Y., 117, 244, 460, 481
Riordan, J., 25, 96
Risser, R., 7, 99
Robbins, H., 195, 228
Robertson, C. A., 573, 574
Rödel, E., 310, 344
Rodemich, E. R., 639, 656
Rohatgi, V. K., 81, 82, 97, 189, 190, 238, 298, 344
Rolph, J. E., 173, 174, 227, 243
Romeijn, H. E., 38, 86
Ronning, G., 506, 518, 526
Rosenbaum, S., 314, 318, 344
Roux, J. J. J., 99, 175, 176, 226, 446, 478, 494, 526
Roy, D., 618
Roy, J., 696
Roy, S. N., 242
Royen, T., 440, 458, 461, 462, 464, 482, 483
Ruben, H., 134, 135, 244, 287, 344
Rubin, D. B., 183, 239
Ruiz, J. M., 601, 618
Rukhin, A. L., 193, 244
Rüschendorf, L., 47, 100, 574
Rustagi, J. S., 226
Rvaceva, E. L., 385, 428
Ryzhik, I. M., 583, 616

Sackrowitz, H., 241, 476, 482
Sagae, M., 27, 101
Sagrista, S. N., 8, 100
Sahai, H., 508, 526
Saleh, A. K. Md. E., 191, 224
Samanta, M., 428
Samorodnitsky, G., 18, 100

Sampson, A. R., 98, 156, 157, 242, 575
Sampson, P. D., 219, 244
Samuel-Cahn, E., 117, 244
Sánchez, J. M., 79, 85
Sankaran, P. G., 611, 618
Santander, L. A. M., 362, 416
Sapozhnikov, P. N., 196, 240
Sapra, S. K., 344
Sarabia, J.-M., 71, 72, 85, 158, 159, 211,
        224, 225, 244, 292, 328, 331,
        334, 345, 352, 419, 582, 583,
        587, 589-591, 593, 604, 611,
        615, 616, 676, 691
Sarkar, S. K., 368, 428
Sarkar, T. K., 367, 428
Sarmanov, O. V., 3, 13, 30, 100, 431,
        436, 437, 457, 472, 476, 483
Sato, K., 17, 100
Satterthwaite, S. P., 570, 575
Savage, R., 244
Savits, T. H., 62, 85, 403, 417, 575
Saw, J. G., 368, 428
Saxena, R. K., 468, 481
Sazonov, V. V., 15, 100
Scarsini, M., 595, 618
Schach, S., 482
Scheaffer, R. L., 428
Schervish, M. J., 124, 139, 209, 244
Schläfli, L., 127, 140, 244
Schlaifer, R., 676, 695
Schmeiser, G. B. W., 450, 483
Schmuland, B., 165, 237
Schneider, B. E., 507, 526
Schols, C. M., 253, 345
Schreuder, H. T., 22, 100
Schucany, W. R., 57, 100
Schwager, S. J., 83, 100
Schweitzer, B., 574
Scott, A., 113, 244
Scrimshaw, D. F., 373, 416
Sechtin, M. B., 212, 242
Seeger, J. P., 52, 94
Self, S. G., 667, 696
Sen, P. K., 306, 339, 369, 426, 429, 465,
        481
Seneta, E., 142, 143, 245
SenGupta, A., 360, 428
Serfling, R. J., 424, 427, 507, 526
Seshadri, V., 30, 100, 153, 155, 228,

242, 245, 387, 428, 674, 690,
        694
Sevastjanov, B. A., 472, 476, 483
Severo, N. C., 267, 348
Shah, S. M., 314, 345
Shaked, M., 47, 59, 62, 76, 85, 101, 403,
        417, 539, 550
Shanbhag, D. N., 407, 428, 689, 690,
        695
Shanmugalingam, S., 328, 345
Shanubhogue, A., 428
Shapiro, A., 87
Sheldon, D. D., 202, 234
Sheppard, W. F., 263, 266, 267, 345
Sheth, R. J., 315, 343
Shi, D., 641, 643, 656
Shimakura, S. E., 369, 429
Shimi, I. N., 421
Shimizu, K., 27, 28, 88, 92, 101
Shimizu, R., 152, 245
Shin, R. T., 212, 249
Shinozaki, N., 163, 245
Sibuya, M., 259, 345, 621, 623, 624, 656
Šidák, Z., 113, 245
Siegel, A. F., 117, 219, 244, 245
Simha, P. S., 272, 339
Simonelli, I., 66, 90
Singh, A. K., 670, 693
Singh, N., 245, 345
Singpurwalla, N. D., 377, 381, 419, 420,
        424, 425, 428, 563, 575, 583,
        595, 604, 611, 617
Sinha, B. K., 157, 159, 170-172, 179,
        194, 223, 242, 245, 502, 526
Siotani, M., 202, 203, 245
Sivazlian, B. D., 530, 550
Six, F. B., 136, 229
Skitovič, V. P., 156, 245
Sklar, A., 50, 101
Skovgaard, Ib. M., 6, 101
Slepian, D., 113, 246
Small, N. J. H., 80, 88
Smirnov, N. V., 270, 345
Smith, A. F. M., 98, 221, 248, 411, 422,
        677, 693
Smith, M. D., 562, 575
Smith, O. E., 443-445, 478, 483
Smith, R. L., 38, 86, 621, 643, 646, 656

Smith, W. B., 234, 246, 298, 300, 301, 339, 345, 615
Snickers, F., 655
Sobel, M., 134, 230, 500, 510, 511, 526, 527
Solow, A. R., 130-132, 143, 246
Somerville, P. N., 253, 287, 345
Soms, A. P., 122, 234
Sondhauss, U., 285, 346
Sondhi, M. M., 148, 149, 246
Song, R., 287, 346
Soong, W. C., 125, 246
Soper, H. E., 324, 346
Sowden, R. R., 270, 346
Spiegel, D. K., 508, 524
Srinivasan, C., 543, 550
Srivastava, J. N., 301
Srivastava, M. S., 80, 101, 183, 195, 246
Stadje, W., 160, 246
Stallari, E., 412, 429
Steck, G. P., 116, 122, 125, 134, 135, 246, 288, 289, 346
Steel, M. F. J., 20, 89
Steel, S. J., 446-448, 450, 483
Stegun, I. A., 348, 633, 652
Stein, C., 161, 163, 169, 179, 193, 235, 246, 247
Steinberg, L., 451, 453, 481
Stephens, D. A., 411, 429
Stephens, R., 290, 292, 342
Steyn, H. S., 8, 37, 101
Stoyanov, J., 279, 346
Strauss, D. J., 73, 85, 370, 371, 417, 563, 575, 593, 616
Strawderman, W. E., 172, 193, 247
Streit, F., 278, 338
Stuart, A., 122, 134, 154, 237, 247, 261, 296, 340
Studden, W. J., 66, 93
Styan, G. P. H., 189, 247
Subramanian, S., 338
Sudakov, V. N., 116, 228
Sullo, P., 366, 370, 393, 395, 396, 427
Sun, H.-J., 128, 129, 247
Sun, K., 391, 404, 405, 418
Sundberg, R., 665, 695
Sungur, E. A., 209, 247, 255, 259, 293, 346, 652, 653
Suzuki, M., 28, 92

Swartz, T., 129, 130, 231
Symanowski, J. T., 76, 93, 571, 572, 575
Szabłowski, P. J., 281, 346
Szántai, T., 434, 442, 458, 459, 482, 483
Székely, G. J., 81, 82, 97

Tadikamalla, P. R., 37, 101
Taillie, C., 94, 429, 575
Tajvidi, N., 601, 618
Takada, Y., 195, 235
Takahashi, R., 648, 649, 651, 656
Takahasi, K., 606, 609, 618
Tallis, G. M., 182, 205, 207, 210, 247, 248, 319, 346
Tan, W. Y., 306, 347, 431, 483
Taqqu, M. S., 18, 98, 100
Targhetta, M. L., 601, 618
Tate, R. F., 234, 324, 325, 344, 346
Tawn, J. A., 628, 629, 634, 636, 641, 642, 644-647, 650, 653, 655, 656
Taylor, M. D., 574
Teicher, H., 160, 248
Teichroew, D., 287, 346
Tenenbein, A., 73, 92, 113, 240, 341
Terza, J. V., 277, 346
Theilen, B., 80, 95
Thomas, D. H., 154, 248
Thompson, W. A., Jr., 407, 424
Thompson, W. J., 633, 656
Tiago de Oliveira, J., 563, 564, 576, 621, 624, 625, 629, 640, 650, ·652, 656, 657
Tiao, G. C., 487, 492, 527
Tibshirani, R., 667, 693
Tihansky, D. P., 271, 272, 346
Tiit, E.-M., 46, 47, 91, 94, 101
Tiku, M. L., 306, 347
Titterington, D. M., 221, 248
Todhunter, J., 110, 248
Tolley, H. D., 367, 429
Tong, Y. L., 47, 101, 114, 122, 248, 260, 339, 398, 399, 427
Tosch, T. J., 361, 429
Tough, R. J. A., 212, 235
Tracy, D. S., 183-185, 210, 233, 235
Trawinski, I., 183, 248, 298, 347
Traynard, C. E., 7, 99
Trenkler, G., 482

Troutt, M. D., 44, 94, 101
Trudel, R., 583, 617
Tsai, H. L., 168, 239
Tsai, T. W.-Y., 356, 360, 425
Tsokos, C. P., 421
Tubbs, J. D., 443-445, 478, 483
Tweedie, M. C. K., 681, 696
Tyrcha, J., 459, 461, 481

Ulrich, G., 376, 429
Unnikrishnan Nair, N., 611, 618
Uppuluri, V. R. R., 500, 510, 511, 527
Urzúa, C. M., 214, 248

Vale, C. D., 36, 101
van der Meulen, E. C., 653
van der Vaart, H. R., 122, 248
van Uven, M. J., 7, 102
Vaupel, J. W., 412, 429
Vere-Jones, D., 436, 452, 472, 482, 483
Veronese, P., 677, 693
Vesely, W. E., 367, 429
Viana, M., 118, 242
Vijverberg, W. P. M., 145, 248
Vitale, R. A., 118, 248
Voinov, V., 388, 419
Volodin, N. A., 573, 576
Votaw, D. F., 319, 347

Wada, C. Y., 369, 429
Wald, A., 202, 248
Waldman, D. M., 210, 249
Walkap, D. W., 650, 653
Walker, S. G., 411, 429
Wang, L., 192, 249
Wang, Y. J., 73-75, 91, 102, 160, 249, 280, 339
Wani, J. K., 30, 102
Warmuth, W., 45, 46, 102
Watson, G. S., 81, 87
Watterson, G. A., 249, 285, 347
Webster, J. T., 136, 249, 439, 480
Wedderburn, R. W. M., 672, 695
Weibull, J., 655
Weier, D. R., 361, 398, 429
Weiler, H., 318, 347
Weinman, D. G., 388, 429
Weiss, L., 102
Weiss, M. C., 249

Weissfeld, L. A., 275, 276, 342
Welland, U., 277, 346
Welsh, G. S., 347
Wesolowski, J., 34, 36, 102, 159, 223, 278-280, 333, 340, 532, 550, 583, 600, 618, 619
Wesolowsky, G. O., 272, 273, 276, 337
Whitmore, G. A., 32, 88, 381, 429
Wicksell, S. D., 6, 102, 326, 335, 436, 483, 586, 619
Wiesen, J. M., 270, 343
Wilks, S. S., 300, 347, 501, 527
Williams, J. M., 347
Wold, H., 272, 347
Wolfe, J. H., 221, 249
Wolfowitz, J., 102
Wong, A., 674, 693
Wong, C. F., 441, 442, 480
Wong, T.-T., 520, 527
Wooding, R. A., 222, 249
Woodroofe, M., 195, 249
Wrigge, S., 278, 347
Wrigley, N., 527
Wu, C., 387, 429

Xu, J.-L., 114, 115, 231

Yang, G. L., 430
Yassaee, H., 494, 497-500, 527
Yatchev, A., 219, 249
Yeh, H.-C., 606, 607, 619
Ylvisaker, D., 676, 693
Yoneda, K., 249
Young, A. W., 261, 343
Young, D. H., 373, 400, 416, 417
Young, J. C., 223, 249, 274, 347
Young, S. S. Y., 318, 347
Youngren, M. A., 377, 428
Yue, X., 410, 430, 530, 547, 548, 550
Yueh, S. H., 212, 249
Yuen, H. K., 646, 656
Yule, G. U., 4, 102, 322, 347

Zaatar, R. L., 301
Zaharov, V. K., 472, 476, 483
Zahedi, H., 352, 420
Zaslavsky, A. M., 32, 88
Zebker, H. A., 212, 249
Zelen, M., 267, 348

Zelterman, D., 570, 576
Zemroch, P. J., 78, 96
Zeng, W.-B., 18, 19, 91, 103
Zhang, H., 279, 333
Zhao, L. P., 667, 696

Zidek, J. V., 193, 227
Zinger, A. A., 154, 250
Zolotarev, V. M., 16, 103
Zoroa, P., 601, 618

Subject Index

Absolute continuity, 379, 385, 387, 648
Absolute moments, 260
Algorithms, 200
    for computing skewness and kurto-
        sis, 78
    "hit and run", 38
    MULNOR (Schervish), 124, 139, 209
    MVNPRD (Dunnett), 124, 139
    Narayana, 507
Almost-lack of memory (ALM), 69
Ancillarity, S-, 673, 676
Antithetic variates, 139
Applications
    adulteration in citrus juice, 204
    allele frequency estimation, 571
    artillery fire control, 252
    binary store display data, 32
    communication theory, 223
    dam runoffs, 465
    deer antler measurements, 220
    engineering, 223
    enumeration of labelled trees, 25
    forensic matching, 511
    hydrology, 287, 432
    industrial fishery, 516
    market research, 508
    microelectronics industry, 204
    nuclear reactor safety, 367
    psychology, 325
    quality control, 204
    rainmaking experiments, 436
    reliability, 367
    stand structure of tree lengths, 22
    stochastic routing, 466
    temperature measurements, 645
    tumor data analysis, 629
Approximations, 139, 143, 145, 200, 202,
        203, 252, 316, 322, 328, 445,
        471, 487, 489
    Bikelis, 15

Dirichlet probabilities, 497
Edgeworth, 14
Geary, 328
Mendell, 32
Mendell-Elston, 276
modal values, 445
multivariate (and bivariate) normal
    densities and probabilities, 75,
    131, 259, 264, 270
multivariate stable densities, 18
trivariate normal probabilities, 287
Archimedean copulas, 413
Archimedean distributions, 559, 561
Array, 4, 7, 22
Association parameter, 638

Bayesian analysis, 20, 162, 358, 362,
    510, 673, 676
Bayesian estimators, 172, 595
Bernoulli process, 531
Biextremal model, 629
Bimodality, 22, 280, 328, 330
Biserial correlation, 324
Bonferroni intervals, 472, 650
Boole's formula, 140
Bounds, Fréchet, 44

Canonical parameter, 661, 663
Cell probabilities, 73, 476, 510
    prior distribution, 511
    sparse/crowded, 512
Censored data, 305
Censoring, 379, 385
Centered normal conditionals, 328
Central limit theorems, 465
    multivariate, 10, 15
Characteristic coefficients, 3
Characteristic exponent, 17, 644
Characteristic function, 9, 15, 17, 33,
    34, 39, 372, 386, 436, 438, 456,

461, 472, 495
Characterization, 70, 151, 175, 253, 277,
    292, 352, 386, 387, 400, 405,
    489, 500, 582, 595, 598-600,
    604, 613, 670, 687, 689
  "maximum likelihood", 160
  of stable densities, 19
Clisy function (Clitic), 2
Clock back to zero, 385
Coefficient of variation, 26
Competing risks, 367
Complex distribution, 251
Compound, 448, 540
Compounding, geometric, 33, 599
Computation model, 447
Concentration, 114
Concomitants of order statistics, 285,
    412
Concordance coefficient, 415
Conditional distributions, specified, 70
Confidence interval, 195
Confidence region, 186, 188, 508
Contingency coefficient, 11
Contingency tables, 559, 572
Continuous bivariate distributions
    beta, 30
    Burr, 609
    Cauchy, 72, 74, 75
    chi, 453
    chi-square, 438, 451
    circular normal, 255, 265, 268
    delta-beta, 518
    double exponential, 376
    double gamma, 435
    Edgeworth series, 40
    exponential (BED), 349, 595
        absolutely continuous (ACBVE)
           (Block and Basu), 387
        Arnold and Strauss, 370
        Block and Basu, 380
        Cowan, 385
        extension (BEE), 356
        Freund, 355
        Friday and Patil, 369, 379
        Ghurye's extended, 383
        Gompertz, 369
        Gumbel, 350, 623
        Hayakawa, 379
        Lindley and Singpurwalla, 380

        Moran and Downton, 350, 371
        Proschan and Sullo, 370
        truncated, 70
        -type, 253
    extreme value, 76, 622
        normal-like, 632
    Farlie-Gumbel-Morgenstern (FGM),
        353, 623, 649
    gamma, 431
        Becker and Roux, 446
        Cheriyan and Ramabhadran, 432
        Dussauchoy and Berland, 445
        FGM-type, 441
        Jensen, 438
        Kibble and Moran, 436
        Le Roux and Steel, 446
        McKay, 432
        Prékopa and Szántai, 442
        Royen, 440
        Sarmanov, 437
        Schneider and Lal, 450
        Smith-Adelfang-Tubbs, 443
    logistic, 551
        Gumbel, 551
        Gumbel-Malik-Abraham, 552, 561,
          563, 568
        Satterthwaite and Hutchinson, 570
        Symanowski and Koehler, 571
    loglogistic, 596
    lognormal, 27
    Makeham, 369
    mixtures, 326
    normal, 25, 74, 115, 116, 161, 251,
        623
        ratios, 327
    normed, 307
    Pareto, 577
        I-IV, 578
        bilateral, 596
        conditionally specified, 587
        Feller, 579, 607
        generalized, 578
        Mardia, 579, 582, 584, 587
        Muliere and Scarsini, 595
        semi-, 598
    Pearson, 9, 432
    Sarmanov, 30
    skew-normal, 30
    t, 20

Weibull, 362, 407
Continuous multivariate distributions
  beta, 487, 532, 534
  Burr, 382, 574, 578, 600
  Cauchy, 219
  chi-square, 471
   noncentral, 475
  Dirichlet, 130, 470, 485, 533, 537,
    684
   generalized, 490
   inverted, 491, 533
   rescaled, 491
   Type II, 492
  Dirichlet-gamma, 494
  ellipsoidal, 78, 118
  exponential, 349, 651
   Block and Basu, 397, 404
   Freund, 388
   Freund-Weinman, 389
   Freund-Weinman-Block, 391, 404
   Gumbel, 387
   Krishnamoorthy-Parthasarathy, 403,
    602
   Marshall-Olkin (MOMED), 391,
    404, 611
   min-stable, 632
   Olkin and Tong, 398
   Proschan and Sullo, 393
   Raftery, 401
   Wiksell-Kibble type, 584
  exponential families, 659
   associated, 665
   curved, 661
   full, 665
   linear, 661
   Morris class, 691
   multinomial type, 618
   negative multinomial-gamma, 685
   $(NM - Ga)_{k-1}$, 686
   $(NM - Ga)_d$, 687
  extreme value, 640
   classical definition, 641
  Farlie-Gumbel-Morgenstern (FGM),
    4, 30
  gamma, 431
   Cheriyan and Ramabhadran, 454
   Dussauchoy and Berland, 470
   Gaver, 456
   Kowalczyk and Tyrcha, 459

   Krishnamoorthy and Parthasarathy,
    457
   Mathai and Moschopoulos, 465
   Prékopa and Szántai, 458
   Royen, 461
  generalized hyperbolic, 212
   Laplace, 215
   uniform, 70
  Gumbel, 406
  inverse beta, 493
   Gaussian, 540
   Wishart, 180
  Kagan, 34
  K-distribution, 211
  Laplace, 33
  Linnik, 33
  Liouville, 529
   ℓ_α-isotropic, 544
   -Dirichlet, 529
   1st kind, 530
   2nd kind, 531
   p-order, 547
   sign-symmetric, 544
  logistic, 495, 551, 609, 642
  lognormal, 27, 219
  Lomax, 382, 491, 548
  nonnormal, 1, 36, 113
  normal, 1, 105, 107, 604, 661, 687
  Pareto, 599
   conditionally specified, 611
   first kind, 599
   fourth kind, 606
   generalized, 609
   Mardia, 599
   Marshall-Olkin, 612
   second kind, 601
   semi-, 614
   third kind, 605, 614
  phase-type (MPH), 62, 403
  pseudo-normal, 217
  Q-exponential, 214
  Rayleigh, 461
   biased generalized, 476
   Blumenson and Miller, 462
  Rhodes, 5
  Sarmanov, 30, 32
  semi-logarithmic, 6
  skew-normal, 216
  stable, 18

Weibull, 407
    Patra and Dey (mixture), 411
Wishart, 161, 176, 200, 457, 471,
    473, 662, 674, 684
Continuous univariate distributions
    beta, 30, 200, 434, 470, 508, 510,
        513, 519, 534, 593, 666
        of second kind, 589
    Cauchy, 279
    chi-square, 44, 360
        noncentral, 81
    Edgeworth series, 38
    exponential, 363, 381, 388, 542, 605
    extreme value, 555, 578, 622, 649
        differences, 562
        generalized, 634
    F, 186, 191, 452
        noncentral, 202
        scaled, 607
    folded normal, 326
    gamma, 5, 31, 365, 400, 414, 433,
        446, 455, 534, 560, 601, 683
    generalized Laplace, 214
    inverse Gaussian, 381
    Morris, 678, 689
    natural exponential family, 660
    Pareto, 578
    Pickands-de Haan, 593
    Poisson, 644
    S-, 20
    skew-normal, 215, 260, 316
    t, 20, 308
    uniform, 44, 352, 450, 564, 596
    Weibull, 61, 362, 632, 641
Contour, 272
    plots, 591
Convolution, 670
Correlation, 271, 354, 402, 487, 564
    biserial, 324
    canonical, 457
    coefficient, 75, 111, 351, 561
        biserial, 324
        estimators, 182
        generalized, 11
        grade, 627
        intra-class, 181, 196
        multiple, 254, 310, 469
        partial, 254
        Spearman's, 57

curve, 332
    inequalities, 529
    in truncated distributions, 312
    spurious, 507
    structure, 52, 183
        ℓ-, 137
    tetrachoric, 320
Covariance structure, 536
Cross-ratio, 413
Cumulant, 83, 107, 260, 466, 468, 664
    generating function, 14, 469
Cuts, 675

Darmois-Skitovitch theorem, 151, 156
Data analysis, 529
    compositional, 543
    exploratory, 19
Decision theory, 134
Dependence, 581
    by mixture, 58
    coefficient, 415
    function, 642, 652
        local, 73
        Pickands, 634
    positive, 310
        by mixture, 59, 539
    quadrant, 460
    structure, 402, 647
    with zero correlation, 255
Dichotomized variables, 320, 324
Dimensionally coherent, 411
Discrete choice, 643
Discrete distributions
    Bernoulli, 378
    binomial, 669
    bivariate, 372, 443
        Neyman Type A, 432
        Poisson, 432, 663
    geometric, 372, 400
    hypergeometric, 389
    multinomial, 106, 365, 488, 489, 502,
        518, 595
    multivariate power series, 662
    negative binomial, 669, 674
    negative multinomial, 689
Dispersion parameters, 672
Distributions with specified marginals,
    70
Domain of attraction, 648

means, 667

Efficiency, relative, 307
Elliptical truncation, 205
Entropy, 110, 557, 603
Equidistributional condition, 600
    contours, 271
Estimation, 161, 293, 352, 504, 637
    based on ranks, 309
    common mean, 189, 298
    correlation coefficient, 182, 295
    density functions, 196
        marginal, 643
    maximum likelihood, 22, 25, 183,
        310, 324, 601
        modified, 306
    moment, 643
    predictive, 197
    simultaneous, 196
    stepwise, 643
Estimative fit, 196
Estimators
    adaptive preliminary test, 309
    based on ranks, 310
    Bayesian, 186, 601
        empirical, 173
        generalized, 193
    Brewster-Zidek type, 194
    Camp, 323
    consistent, 371
    equivariant, 179
    inadmissible, 178
    James-Stein, 173, 180
    maximum likelihood (MLE), 190,
        294, 303, 330, 352, 359, 366,
        455, 519, 542, 562, 585, 613,
        636
        profile, 317
        unrestricted (UMLE), 190
    minimax, 164, 175, 193
    minimum variance, 395
        uniform unbiased (UMVUE), 381,
        541, 674
        unbiased (MVUE), 196, 198
    moment, 221, 319, 326, 352, 366,
        370, 373, 394, 519, 643
        -type, 366
    regression, 307
    Smith's, 301

Stein type, 193, 194
    weighting function, 309
Exchangeability, 3
Exchangeable distributions, 3, 54, 378,
    379, 399, 402, 643
Exchangeable variates, 59
Expected value vector, 16
Extreme values (FGM), 53

Factorization theorem, 292
    model, 361
    multiple causes, 367
Failure rate, bivariate, 381
    total, 405
    vector-valued, 405
    See also Hazard rate
Fit, estimative, 196
Fourfold tables, 322
    Camp's estimator, 322
    See also Contingency tables
Fractal interpolation, 264
Frailties, 412, 559
Frailty models, 412, 414
    bivariate, 414
Fréchet bounds, 44, 110, 379, 414
    marginal, 46
Frequency surface, 6
    logarithmic, 6
    semi-logarithmic, 6
    skew, 4
Futures market, 26

Gaussian conditional structure, 35
    of second order, 35
Generating density, 582
Generating measure, 663, 682
Generation of multivariate nonnormal
        distributions, 36, 38, 43
    parametric measures, 647
Geometric compounding, 599
Geometric maxima and minima, 565
Geometric mean, 505, 624
Geometric minimization, 598
Gibbs sampling, 362
Gini index of inequality, 577
Goodness-of-fit, 413
Grade correlation, 627, 630
Grouped data, 300, 320, 559

Hazard gradient, 116

Hazard rate, 58, 116, 232, 393
    proportional, 59, 543
    See also Failure rate
Heavy-tailed distribution, 20
Hedging, 26
History, 253
Homography, 691
Homoscedasticity, 112

Identifiability, 317, 369
Identity, Euler-Schläfli, 118
    Plackett, 260
    Siegel, 117
Income distribution, 532, 611
Independence, 650
Index parameter, 672
Inequality (multivariate)
    Berge, 64
    Bonferroni, 63
    Chebyshev-type, 63, 650
    correlation, 529
    Gini index of, 577
    Leser, 66
    Olkin and Pratt, 65
    Pareto index of, 578
    Pólya, 266
Infant mortality rate, 582
Inference, 370, 601
Infinite divisibility, 364, 385, 464, 477
Information
    Fisher, 415, 432
    Kullback-Liebler, 198
    matrix, 306, 397, 506, 632
    variable, 644
Integration
    generalized Gaussian, 499
    numerical, 123
    Romberg, 125
    Simpson's rule, 124, 274
    trapezoidal rule, 138
Intraclass correlation, 181, 196
Invariance
    homothetic, 84
    translational, 84
    under linear combination, 152
Isotropy, 156

Jackknife, 374

Kendall's tau, 415

Kurtic function, 2
Kurtosis, 37, 77, 304
    measure, 80

Lack of memory (LOM), 69, 383, 595
    almost- (ALOM), 69
    bivariate (BLOM), 365, 369
    extended, 385
    multivariate, 390
    weak, 406
Laplace model, Gumbel type, 33
Life tests, accelerated, 362, 410, 491
Lifetime, 367, 381
    future (residual), 357, 491
        equations, 395, 397, 610
Likelihood function, 299, 304, 317, 365,
        505, 518, 542, 588, 673
Likelihood ratio, 646
Limited memory, 399
Location-scale family, 157
Logistic models, 628
Logistic processes, stationary, 573
Log-likelihood *See* Likelihood
Loss function, 163, 180, 192
    quadratic, 174, 595

Majorization, 399, 529
Marginal distributions, 12, 73, 113, 207,
        517, 521, 588, 623, 632, 691
Marginal estimation, 643
Marginal iterative replacement, 273
Markov process, 72, 403
Markovian properties, 157
Mathematics
    Abelian semigroup, 540
    Bessel function, 212, 279
        modified, 350, 401, 444, 462, 473,
        584
    beta function, 668
    Bhattacharyya matrices, 690
    Borel measure, 160
        set, 142
    Cauchy functional equation, 407
    Cayley algebra, 662
    Choleski decomposition, 80, 138, 145
    convex function, 622
    convex hull, 668
    cyclic products, 463
    differential equations, 7

digamma function, 455, 505
digamma integral, 663
Dirichlet integral, 487
eigenvalue, 80, 136
ellipse, 253
elliptical integral, 72
Euclidean norm, 34, 121, 129, 440
exponential integral, $Ei(x)$, 351
Fourier coefficients, 309
fractional calculus, 540
functional equations, 407, 596, 614
Hermite polynomials, 127, 322, 427
Hermitian matrix, 684
Humbert series, 497
hyperboloid, 212
hypergeometric function, 261
    confluent, 329, 446
    Gaussian, 199, 389
    generalized, 10, 458, 462
image measure, 668
incomplete beta function ratio, 454,
    497, 541
incomplete gamma function, 358
indicator function, 662
integral transform, 601
Jacobi polynomials, 489
Laguerre polynomials, 434, 436, 438,
    443, 457, 690
Laplace transform, 387, 414, 415,
    464, 477, 560, 660, 665, 675
Lauricella function, 468
Lebesgue measure, 34, 44, 160, 546,
    659, 661
Legendre polynomials, 52
log-Laplace transform, 664
ℓ_p-norm, 548
matrix decomposition, 24, 48
M-matrix, 477
mapping, 668
Meixner polynomials, 690
Newton-Raphson iteration, 304, 506
orthonormal polynomials, 489
orthonormal set, 10
pseudo-inverse matrix (Penrose), 68
saddlepoint method, 674
signature matrix, 477
simplex, 641
Steiner points, 124
subadditive function, 542

tensors, 81
trigamma function, 506
trigamma integral, 633
Weyl fractional integral, 535
Mean residual lifetime, 352, 361
Mean square error (MSE), 307
Median, 628
    regression, 24, 48
Mills' ratio, 115, 325, 332
    multivariate, 114
Minimal sufficient statistic, 178
Minimax estimator, 164, 175, 193
Missing data, 183, 298, 304
Mixture, 2, 27, 28, 56, 211, 220, 326,
    352, 355, 380, 392, 397, 411,
    449, 452, 456, 475, 516, 539,
    540, 564, 572, 585, 605
    dependence by, 50
Modal regression, 26
Mode, 626
Model
    additive, 673
    buying behavior, 510
    competition, 447
    conditionally specified logistic, 569
    dispersion, 672
    exponential transformation, 671
    frailty, 412, 414, 416
    general flexible, 568
    logistic, 369, 628, 635
    log-linear, 369
    marginal, 675
    mixed, 635
    nondifferentiable, 636
    partly exponential, 667
    proportional hazards, 548
    scale-mixture, 565
    shared load, 450
    shock, 378
Moment-generating function, 261, 282,
    315, 329, 363, 377, 388, 436,
    442, 465, 469, 664
Moments, 37, 330, 536, 545, 670
    absolute, 261
    incomplete, 262
Monotone missing data pattern, 183
    sample, 183
Monte Carlo, 122, 130, 139, 144, 186,
    374

See also Simulation
Multivariate analysis of variance (MANOVA), 82
Multivariate systems of distributions
    Fréchet, 47
    Lee, 60
    Mardia, 47
    Pearson, 6
    Plackett, 47

Neutrality, 500, 519, 534
    $(CM)_i$-, 500
    complete, 501, 534
    $(DR)_i$-, 500
    i-, 502
Newton-Raphson procedure, 304, 506
Numerical integration, 123
    Simpson's rule, 124, 174
    trapezoidal rule, 138

Oblique axes, 253
Odds ratio, 50
Order statistics, 116, 281, 305, 358, 388, 607
    concomitant, 285, 412
    multivariate, 119
Ordering, 399
Orthant, 503
    probabilities, 127, 140, 145
Orthogonal transformations, 671

Parameter, 672
    dispersion, 672
    index, 672
Parameterization, 667
Parametric measures, generation, 647
Pareto index of inequality, 578
PERT (Program Evaluation and Review Technique), 508
Plackett's formula, 135
Plackett's identity, 259
Poisson process, 363, 369, 378, 385, 391, 446
Poisson weights, 475
Positively dependent by mixture, 59, 539
Posterior distribution, 31, 181, 595
Posterior risk, 174
Power, 79
Prior distribution, 31, 165, 358, 520, 595

conjugate, 677
    interclass, 173
    pseudo-conjugate, 357
Pseudo diagonality, 690

Quadrant dependent, 460, 611, 650
Quadratic form, 15, 110, 192, 271, 473
    variance function, 677, 678
Quadrature, 124
    Gaussian, 274, 500
    rank-1 lattice, 142
    (*See also* Integration)

Radial truncation, 207
Ranks used in estimation, 309
Ratios, 327
Recurrence relations, 283, 314, 664
Regression function, 4, 112, 331, 372, 375, 435, 450, 582
    curvilinear, 554
    estimator, 307
    inverse linear, 684
    linear, 603, 684
    median, 5, 22, 23, 48
    modal, 26
Relative efficiency, 642
Reliability, 218, 381, 403, 460, 530, 541, 611
Replacement theory, 468
Risks
    competing, 367
    posterior, 174
Robustness, 40

Sampling scheme, Hanagal, 370
Scale mixture, 605
Scedastic function, 2
Scoring
    Fisher, 397, 507
    Newton-Raphson, 397
Selection, general, 204
Sequential sampling, 472
Series expansion, 10, 128, 142, 271, 275, 321, 458, 497, 500
    Cornish-Fisher, 315
    Edgeworth, 465
    Gram-Charlier and Edgeworth, 6
    Humbert, 497
    Mehler, 322,

Pearson, 322
Shared load model, 450
Sheppard's correction, 182
Shot-noise process, 377
Shrinkage, 191
Siegel's identity, 117
Simulation, 144, 375, 519, 637
    of multivariate stable densities, 18
Simultaneous estimation, 196
Singular distributions, 46, 67, 364, 386,
        408, 517
    multinomial, 79
Size and shape measures, 218
Skewness, 21, 77
    measure, 80
Skitović's theorem, 156
Slippage tests, 509
Sojourn time vector, 403
Special structures, 187
Spectral measure, 17
Sphericity, 34, 156, 573
Stable distributions, 17, 414
    extreme, 672
    symmetrical, 17
Standardized variable, 13
Star-shaped region, 138
Steepness, 667, 669
Structure, 53, 474, 489, 606, 689
    bivariate Pearson, 9
    measure, 665
    special, 187
Sufficiency, S-, 673, 676
Sufficient statistic, 541
Surface
    frequency, 6
    logarithmic, 6
    semi-logarithmic, 6
Survival data, 559
Survival function, 55, 282, 358, 368, 383,
        396, 406, 415, 491, 567, 572,
        591, 598, 614, 632, 651
    marginal, 385
Symmetric cones, 684
Symmetry
    functional, 307
    radial, 154
    spherical, 34

Tail weight, 54

Taylor expansion, 500
Test
    Neyman's "smooth", 81
    multinormality, 78
    slippage, 509
    statistic, 360, 373
Tetrachoric series, 122, 141, 458
    function, 322
    table, 322
Threshold methodology, multivariate, 647
Tolerance factor, 204
Tolerance region, 204
Total positivity, 529
Transformation, 37, 48, 326, 409, 436,
        582, 588
    Lee, 632, 641
    linear, 460, 622
    orthogonal, 671
    power, 407, 606
    power-scale, 538, 543
    systems, 20
Translation (see Transformations)
Trivariate distributions, 13
    data, 646
    extreme value, 652
    factorization theorem, 292
    FGM, 58
    gamma, 440
    lognormal, 26
    normal, 252, 310, 368
Truncated distributions, 415, 668
    bivariate normal, 311
    multivariate normal, 204
Truncation
    elliptical, 205, 319
    radial, 207

Variable
    index, 644
    information, 644
Variance-covariance matrix, 2, 120
Variance function, 668
    Babel, 681
    cubic, 681, 691
    generalized, 674
    power-exponential, 681
    quadratic, 677, 678
        homogeneous, 681
        multivariate, 683

Vertical density representation, 44

Weighted geometric mean, 624
Weighting function, 309, 489
Weights, Poisson, 475

WILEY SERIES IN PROBABILITY AND STATISTICS
ESTABLISHED BY WALTER A. SHEWHART AND SAMUEL S. WILKS

Editors
*Vic Barnett, Noel A. C. Cressie, Nicholas I. Fisher,
Iain M. Johnstone, J. B. Kadane, David G. Kendall, David W. Scott,
Bernard W. Silverman, Adrian F. M. Smith, Jozef L. Teugels;
Ralph A. Bradley, Emeritus, J. Stuart Hunter, Emeritus*

Probability and Statistics Section

*ANDERSON · The Statistical Analysis of Time Series
ARNOLD, BALAKRISHNAN, and NAGARAJA · A First Course in Order Statistics
ARNOLD, BALAKRISHNAN, and NAGARAJA · Records
BACCELLI, COHEN, OLSDER, and QUADRAT · Synchronization and Linearity:
    An Algebra for Discrete Event Systems
BARNETT · Comparative Statistical Inference, *Third Edition*
BASILEVSKY · Statistical Factor Analysis and Related Methods: Theory and
    Applications
BERNARDO and SMITH · Bayesian Statistical Concepts and Theory
BILLINGSLEY · Convergence of Probability Measures, *Second Edition*
BOROVKOV · Asymptotic Methods in Queuing Theory
BOROVKOV · Ergodicity and Stability of Stochastic Processes
BRANDT, FRANKEN, and LISEK · Stationary Stochastic Models
CAINES · Linear Stochastic Systems
CAIROLI and DALANG · Sequential Stochastic Optimization
CONSTANTINE · Combinatorial Theory and Statistical Design
COOK · Regression Graphics
COVER and THOMAS · Elements of Information Theory
CSÖRGŐ and HORVÁTH · Weighted Approximations in Probability Statistics
CSÖRGŐ and HORVÁTH · Limit Theorems in Change Point Analysis
*DANIEL · Fitting Equations to Data: Computer Analysis of Multifactor Data,
    Second Edition
DETTE and STUDDEN · The Theory of Canonical Moments with Applications in
    Statistics, Probability, and Analysis
DEY and MUKERJEE · Fractional Factorial Plans
*DOOB · Stochastic Processes
DRYDEN and MARDIA · Statistical Shape Analysis
DUPUIS and ELLIS · A Weak Convergence Approach to the Theory of Large Deviations
ETHIER and KURTZ · Markov Processes: Characterization and Convergence
FELLER · An Introduction to Probability Theory and Its Applications, Volume 1,
    Third Edition, Revised; Volume II, *Second Edition*
FULLER · Introduction to Statistical Time Series, *Second Edition*
FULLER · Measurement Error Models
GHOSH, MUKHOPADHYAY, and SEN · Sequential Estimation
GIFI · Nonlinear Multivariate Analysis
GUTTORP · Statistical Inference for Branching Processes
HALL · Introduction to the Theory of Coverage Processes
HAMPEL · Robust Statistics: The Approach Based on Influence Functions
HANNAN and DEISTLER · The Statistical Theory of Linear Systems
HUBER · Robust Statistics

*Now available in a lower priced paperback edition in the Wiley Classics Library.

Probability and Statistics (Continued)

HUSKOVA, BERAN, and DUPAC · Collected Works of Jaroslav Hajek—
 with Commentary
IMAN and CONOVER · A Modern Approach to Statistics
JUREK and MASON · Operator-Limit Distributions in Probability Theory
KASS and VOS · Geometrical Foundations of Asymptotic Inference
KAUFMAN and ROUSSEEUW · Finding Groups in Data: An Introduction to Cluster
 Analysis
KELLY · Probability, Statistics, and Optimization
KENDALL, BARDEN, CARNE, and LE · Shape and Shape Theory
LINDVALL · Lectures on the Coupling Method
McFADDEN · Management of Data in Clinical Trials
MANTON, WOODBURY, and TOLLEY · Statistical Applications Using Fuzzy Sets
MORGENTHALER and TUKEY · Configural Polysampling: A Route to Practical
 Robustness
MUIRHEAD · Aspects of Multivariate Statistical Theory
OLIVER and SMITH · Influence Diagrams, Belief Nets and Decision Analysis
*PARZEN · Modern Probability Theory and Its Applications
PRESS · Bayesian Statistics: Principles, Models, and Applications
PUKELSHEIM · Optimal Experimental Design
RAO · Asymptotic Theory of Statistical Inference
RAO · Linear Statistical Inference and Its Applications, *Second Edition*
RAO and SHANBHAG · Choquet-Deny Type Functional Equations with Applications to
 Stochastic Models
ROBERTSON, WRIGHT, and DYKSTRA · Order Restricted Statistical Inference
ROGERS and WILLIAMS · Diffusions, Markov Processes, and Martingales, Volume I:
 Foundations, *Second Edition;* Volume II: Îto Calculus
RUBINSTEIN and SHAPIRO · Discrete Event Systems: Sensitivity Analysis and
 Stochastic Optimization by the Score Function Method
RUZSA and SZEKELY · Algebraic Probability Theory
SCHEFFE · The Analysis of Variance
SEBER · Linear Regression Analysis
SEBER · Multivariate Observations
SEBER and WILD · Nonlinear Regression
SERFLING · Approximation Theorems of Mathematical Statistics
SHORACK and WELLNER · Empirical Processes with Applications to Statistics
SMALL and McLEISH · Hilbert Space Methods in Probability and Statistical Inference
STAPLETON · Linear Statistical Models
STAUDTE and SHEATHER · Robust Estimation and Testing
STOYANOV · Counterexamples in Probability
TANAKA · Time Series Analysis: Nonstationary and Noninvertible Distribution Theory
THOMPSON and SEBER · Adaptive Sampling
WELSH · Aspects of Statistical Inference
WHITTAKER · Graphical Models in Applied Multivariate Statistics
YANG · The Construction Theory of Denumerable Markov Processes

Applied Probability and Statistics Section

ABRAHAM and LEDOLTER · Statistical Methods for Forecasting
AGRESTI · Analysis of Ordinal Categorical Data
AGRESTI · Categorical Data Analysis

*Now available in a lower priced paperback edition in the Wiley Classics Library.

Applied Probability and Statistics (Continued)

ANDERSON, AUQUIER, HAUCK, OAKES, VANDAELE, and WEISBERG ·
Statistical Methods for Comparative Studies

ARMITAGE and DAVID (editors) · Advances in Biometry

*ARTHANARI and DODGE · Mathematical Programming in Statistics

ASMUSSEN · Applied Probability and Queues

*BAILEY · The Elements of Stochastic Processes with Applications to the Natural
Sciences

BARNETT and LEWIS · Outliers in Statistical Data, *Third Edition*

BARTHOLOMEW, FORBES, and McLEAN · Statistical Techniques for Manpower
Planning, *Second Edition*

BASU and RIGDON · Statistical Methods for the Reliability of Repairable Systems

BATES and WATTS · Nonlinear Regression Analysis and Its Applications

BECHHOFER, SANTNER, and GOLDSMAN · Design and Analysis of Experiments for
Statistical Selection, Screening, and Multiple Comparisons

BELSLEY · Conditioning Diagnostics: Collinearity and Weak Data in Regression

BELSLEY, KUH, and WELSCH · Regression Diagnostics: Identifying Influential
Data and Sources of Collinearity

BHAT · Elements of Applied Stochastic Processes, *Second Edition*

BHATTACHARYA and WAYMIRE · Stochastic Processes with Applications

BIRKES and DODGE · Alternative Methods of Regression

BLISCHKE AND MURTHY · Reliability: Modeling, Prediction, and Optimization

BLOOMFIELD · Fourier Analysis of Time Series: An Introduction, *Second Edition*

BOLLEN · Structural Equations with Latent Variables

BOULEAU · Numerical Methods for Stochastic Processes

BOX · Bayesian Inference in Statistical Analysis

BOX and DRAPER · Empirical Model-Building and Response Surfaces

*BOX and DRAPER · Evolutionary Operation: A Statistical Method for Process
Improvement

BUCKLEW · Large Deviation Techniques in Decision, Simulation, and Estimation

BUNKE and BUNKE · Nonlinear Regression, Functional Relations and Robust
Methods: Statistical Methods of Model Building

CHATTERJEE and HADI · Sensitivity Analysis in Linear Regression

CHERNICK · Bootstrap Methods: A Practitioner's Guide

CHILÈS and DELFINER · Geostatistics: Modeling Spatial Uncertainty

CHOW and LIU · Design and Analysis of Clinical Trials: Concepts and Methodologies

CLARKE and DISNEY · Probability and Random Processes: A First Course with
Applications, *Second Edition*

*COCHRAN and COX · Experimental Designs, *Second Edition*

CONOVER · Practical Nonparametric Statistics, *Second Edition*

CORNELL · Experiments with Mixtures, Designs, Models, and the Analysis of Mixture
Data, *Second Edition*

*COX · Planning of Experiments

CRESSIE · Statistics for Spatial Data, *Revised Edition*

DANIEL · Applications of Statistics to Industrial Experimentation

DANIEL · Biostatistics: A Foundation for Analysis in the Health Sciences, *Sixth Edition*

DAVID · Order Statistics, *Second Edition*

*DEGROOT, FIENBERG, and KADANE · Statistics and the Law

DODGE · Alternative Methods of Regression

DOWDY and WEARDEN · Statistics for Research, *Second Edition*

DUNN and CLARK · Applied Statistics: Analysis of Variance and Regression, *Second
Edition*

*ELANDT-JOHNSON and JOHNSON · Survival Models and Data Analysis

EVANS, PEACOCK, and HASTINGS · Statistical Distributions, *Second Edition*

*Now available in a lower priced paperback edition in the Wiley Classics Library.

Applied Probability and Statistics (Continued)
  *FLEISS · The Design and Analysis of Clinical Experiments
  FLEISS · Statistical Methods for Rates and Proportions, *Second Edition*
  FLEMING and HARRINGTON · Counting Processes and Survival Analysis
  GALLANT · Nonlinear Statistical Models
  GLASSERMAN and YAO · Monotone Structure in Discrete-Event Systems
  GNANADESIKAN · Methods for Statistical Data Analysis of Multivariate Observations,
    Second Edition
  GOLDSTEIN and LEWIS · Assessment: Problems, Development, and Statistical Issues
  GREENWOOD and NIKULIN · A Guide to Chi-Squared Testing
  *HAHN · Statistical Models in Engineering
  HAHN and MEEKER · Statistical Intervals: A Guide for Practitioners
  HAND · Construction and Assessment of Classification Rules
  HAND · Discrimination and Classification
  HEIBERGER · Computation for the Analysis of Designed Experiments
  HEDAYAT and SINHA · Design and Inference in Finite Population Sampling
  HINKELMAN and KEMPTHORNE: · Design and Analysis of Experiments, Volume 1:
    Introduction to Experimental Design
  HOAGLIN, MOSTELLER, and TUKEY · Exploratory Approach to Analysis
    of Variance
  HOAGLIN, MOSTELLER, and TUKEY · Exploring Data Tables, Trends and Shapes
  HOAGLIN, MOSTELLER, and TUKEY · Understanding Robust and Exploratory
    Data Analysis
  HOCHBERG and TAMHANE · Multiple Comparison Procedures
  HOCKING · Methods and Applications of Linear Models: Regression and the Analysis
    of Variables
  HOGG and KLUGMAN · Loss Distributions
  HOSMER and LEMESHOW · Applied Logistic Regression
  HØYLAND and RAUSAND · System Reliability Theory: Models and Statistical Methods
  HUBERTY · Applied Discriminant Analysis
  JACKSON · A User's Guide to Principle Components
  JOHN · Statistical Methods in Engineering and Quality Assurance
  JOHNSON · Multivariate Statistical Simulation
  JOHNSON and KOTZ · Distributions in Statistics
  JOHNSON, KOTZ, and BALAKRISHNAN · Continuous Univariate Distributions,
    Volume 1, *Second Edition*
  JOHNSON, KOTZ, and BALAKRISHNAN · Continuous Univariate Distributions,
    Volume 2, *Second Edition*
  JOHNSON, KOTZ, and BALAKRISHNAN · Discrete Multivariate Distributions
  JOHNSON, KOTZ, and KEMP · Univariate Discrete Distributions, *Second Edition*
  JUREČKOVÁ and SEN · Robust Statistical Procedures: Aymptotics and Interrelations
  KADANE · Bayesian Methods and Ethics in a Clinical Trial Design
  KADANE AND SCHUM · A Probabilistic Analysis of the Sacco and Vanzetti Evidence
  KALBFLEISCH and PRENTICE · The Statistical Analysis of Failure Time Data
  KELLY · Reversability and Stochastic Networks
  KHURI, MATHEW, and SINHA · Statistical Tests for Mixed Linear Models
  KLUGMAN, PANJER, and WILLMOT · Loss Models: From Data to Decisions
  KLUGMAN, PANJER, and WILLMOT · Solutions Manual to Accompany Loss Models:
    From Data to Decisions
  KOTZ, BALAKRISHNAN, and JOHNSON · Continuous Multivariate Distributions,
    Volume 1, *Second Edition*
  KOVALENKO, KUZNETZOV, and PEGG · Mathematical Theory of Reliability of
    Time-Dependent Systems with Practical Applications
  LACHIN · Biostatistical Methods: The Assessment of Relative Risks

*Now available in a lower priced paperback edition in the Wiley Classics Library.

Applied Probability and Statistics (Continued)

LAD · Operational Subjective Statistical Methods: A Mathematical, Philosophical, and Historical Introduction

LANGE, RYAN, BILLARD, BRILLINGER, CONQUEST, and GREENHOUSE · Case Studies in Biometry

LAWLESS · Statistical Models and Methods for Lifetime Data

LEE · Statistical Methods for Survival Data Analysis, *Second Edition*

LePAGE and BILLARD · Exploring the Limits of Bootstrap

LINHART and ZUCCHINI · Model Selection

LITTLE and RUBIN · Statistical Analysis with Missing Data

LLOYD · The Statistical Analysis of Categorical Data

MAGNUS and NEUDECKER · Matrix Differential Calculus with Applications in Statistics and Econometrics, *Revised Edition*

MALLER and ZHOU · Survival Analysis with Long Term Survivors

MANN, SCHAFER, and SINGPURWALLA · Methods for Statistical Analysis of Reliability and Life Data

McLACHLAN and KRISHNAN · The EM Algorithm and Extensions

McLACHLAN · Discriminant Analysis and Statistical Pattern Recognition

McNEIL · Epidemiological Research Methods

MEEKER and ESCOBAR · Statistical Methods for Reliability Data

*MILLER · Survival Analysis, *Second Edition*

MONTGOMERY and PECK · Introduction to Linear Regression Analysis, *Second Edition*

MYERS and MONTGOMERY · Response Surface Methodology: Process and Product in Optimization Using Designed Experiments

NELSON · Accelerated Testing, Statistical Models, Test Plans, and Data Analyses

NELSON · Applied Life Data Analysis

OCHI · Applied Probability and Stochastic Processes in Engineering and Physical Sciences

OKABE, BOOTS, and SUGIHARA · Spatial Tesselations: Concepts and Applications of Voronoi Diagrams

PANKRATZ · Forecasting with Dynamic Regression Models

PANKRATZ · Forecasting with Univariate Box-Jenkins Models: Concepts and Cases

PIANTADOSI · Clinical Trials: A Methodologic Perspective

PORT · Theoretical Probability for Applications

PUTERMAN · Markov Decision Processes: Discrete Stochastic Dynamic Programming

RACHEV · Probability Metrics and the Stability of Stochastic Models

RÉNYI · A Diary on Information Theory

RIPLEY · Spatial Statistics

RIPLEY · Stochastic Simulation

ROLSKI, SCHMIDLI, SCHMIDT, and TEUGELS · Stochastic Processes for Insurance and Finance

ROUSSEEUW and LEROY · Robust Regression and Outlier Detection

RUBIN · Multiple Imputation for Nonresponse in Surveys

RUBINSTEIN · Simulation and the Monte Carlo Method

RUBINSTEIN and MELAMED · Modern Simulation and Modeling

RYAN · Statistical Methods for Quality Improvement, *Second Edition*

SCHUSS · Theory and Applications of Stochastic Differential Equations

SCOTT · Multivariate Density Estimation: Theory, Practice, and Visualization

*SEARLE · Linear Models

SEARLE · Linear Models for Unbalanced Data

SEARLE, CASELLA, and McCULLOCH · Variance Components

SENNOTT · Stochastic Dynamic Programming and the Control of Queueing Systems

STOYAN, KENDALL, and MECKE · Stochastic Geometry and Its Applications, *Second Edition*

*Now available in a lower priced paperback edition in the Wiley Classics Library.

Applied Probability and Statistics (Continued)

STOYAN and STOYAN · Fractals, Random Shapes and Point Fields: Methods of Geometrical Statistics

THOMPSON · Empirical Model Building

THOMPSON · Sampling

THOMPSON · Simulation: A Modeler's Approach

TIJMS · Stochastic Modeling and Analysis: A Computational Approach

TIJMS · Stochastic Models: An Algorithmic Approach

TITTERINGTON, SMITH, and MAKOV · Statistical Analysis of Finite Mixture Distributions

UPTON and FINGLETON · Spatial Data Analysis by Example, Volume 1: Point Pattern and Quantitative Data

UPTON and FINGLETON · Spatial Data Analysis by Example, Volume II: Categorical and Directional Data

VAN RIJCKEVORSEL and DE LEEUW · Component and Correspondence Analysis

VIDAKOVIC · Statistical Modeling by Wavelets

WEISBERG · Applied Linear Regression, *Second Edition*

WESTFALL and YOUNG · Resampling-Based Multiple Testing: Examples and Methods for p-Value Adjustment

WHITTLE · Systems in Stochastic Equilibrium

WOODING · Planning Pharmaceutical Clinical Trials: Basic Statistical Principles

WOOLSON · Statistical Methods for the Analysis of Biomedical Data

*ZELLNER · An Introduction to Bayesian Inference in Econometrics

Texts and References Section

AGRESTI · An Introduction to Categorical Data Analysis

ANDERSON · An Introduction to Multivariate Statistical Analysis, *Second Edition*

ANDERSON and LOYNES · The Teaching of Practical Statistics

ARMITAGE and COLTON · Encyclopedia of Biostatistics: Volumes 1 to 6 with Index

BARTOSZYNSKI and NIEWIADOMSKA-BUGAJ · Probability and Statistical Inference

BENDAT and PIERSOL · Random Data: Analysis and Measurement Procedures, *Third Edition*

BERRY, CHALONER, and GEWEKE · Bayesian Analysis in Statistics and Econometrics: Essays in Honor of Arnold Zellner

BHATTACHARYA and JOHNSON · Statistical Concepts and Methods

BILLINGSLEY · Probability and Measure, *Second Edition*

BOX · R. A. Fisher, the Life of a Scientist

BOX, HUNTER, and HUNTER · Statistics for Experimenters: An Introduction to Design, Data Analysis, and Model Building

BOX and LUCEÑO · Statistical Control by Monitoring and Feedback Adjustment

BROWN and HOLLANDER · Statistics: A Biomedical Introduction

CHATTERJEE and PRICE · Regression Analysis by Example, *Third Edition*

COOK and WEISBERG · Applied Regression Including Computing and Graphics

COOK and WEISBERG · An Introduction to Regression Graphics

COX · A Handbook of Introductory Statistical Methods

DILLON and GOLDSTEIN · Multivariate Analysis: Methods and Applications

*DODGE and ROMIG · Sampling Inspection Tables, *Second Edition*

DRAPER and SMITH · Applied Regression Analysis, *Third Edition*

DUDEWICZ and MISHRA · Modern Mathematical Statistics

DUNN · Basic Statistics: A Primer for the Biomedical Sciences, *Second Edition*

FISHER and VAN BELLE · Biostatistics: A Methodology for the Health Sciences

*Now available in a lower priced paperback edition in the Wiley Classics Library.

Texts and References (Continued)

FREEMAN and SMITH · Aspects of Uncertainty: A Tribute to D. V. Lindley

GROSS and HARRIS · Fundamentals of Queueing Theory, *Third Edition*

HALD · A History of Probability and Statistics and their Applications Before 1750

HALD · A History of Mathematical Statistics from 1750 to 1930

HELLER · MACSYMA for Statisticians

HOEL · Introduction to Mathematical Statistics, *Fifth Edition*

HOLLANDER and WOLFE · Nonparametric Statistical Methods, *Second Edition*

HOSMER and LEMESHOW · Applied Survival Analysis: Regression Modeling of Time to Event Data

JOHNSON and BALAKRISHNAN · Advances in the Theory and Practice of Statistics: A Volume in Honor of Samuel Kotz

JOHNSON and KOTZ (editors) · Leading Personalities in Statistical Sciences: From the Seventeenth Century to the Present

JUDGE, GRIFFITHS, HILL, LÜTKEPOHL, and LEE · The Theory and Practice of Econometrics, *Second Edition*

KHURI · Advanced Calculus with Applications in Statistics

KOTZ and JOHNSON (editors) · Encyclopedia of Statistical Sciences: Volumes 1 to 9 wtih Index

KOTZ and JOHNSON (editors) · Encyclopedia of Statistical Sciences: Supplement Volume

KOTZ, REED, and BANKS (editors) · Encyclopedia of Statistical Sciences: Update Volume 1

KOTZ, REED, and BANKS (editors) · Encyclopedia of Statistical Sciences: Update Volume 2

LAMPERTI · Probability: A Survey of the Mathematical Theory, *Second Edition*

LARSON · Introduction to Probability Theory and Statistical Inference, *Third Edition*

LE · Applied Categorical Data Analysis

LE · Applied Survival Analysis

MALLOWS · Design, Data, and Analysis by Some Friends of Cuthbert Daniel

MARDIA · The Art of Statistical Science: A Tribute to G. S. Watson

MASON, GUNST, and HESS · Statistical Design and Analysis of Experiments with Applications to Engineering and Science

MURRAY · X-STAT 2.0 Statistical Experimentation, Design Data Analysis, and Nonlinear Optimization

PURI, VILAPLANA, and WERTZ · New Perspectives in Theoretical and Applied Statistics

RENCHER · Linear Models in Statistics

RENCHER · Methods of Multivariate Analysis

RENCHER · Multivariate Statistical Inference with Applications

ROSS · Introduction to Probability and Statistics for Engineers and Scientists

ROHATGI · An Introduction to Probability Theory and Mathematical Statistics

RYAN · Modern Regression Methods

SCHOTT · Matrix Analysis for Statistics

SEARLE · Matrix Algebra Useful for Statistics

STYAN · The Collected Papers of T. W. Anderson: 1943–1985

TIERNEY · LISP-STAT: An Object-Oriented Environment for Statistical Computing and Dynamic Graphics

WONNACOTT and WONNACOTT · Econometrics, *Second Edition*

WU and HAMADA · Experiments: Planning, Analysis, and Parameter Design Optimization

*Now available in a lower priced paperback edition in the Wiley Classics Library.

WILEY SERIES IN PROBABILITY AND STATISTICS

ESTABLISHED BY WALTER A. SHEWHART AND SAMUEL S. WILKS

Editors

Robert M. Groves, Graham Kalton, J. N. K. Rao, Norbert Schwarz, Christopher Skinner

Survey Methodology Section